ASPECTS OF RADAR SIGNAL PROCESSING

The Artech House Radar Library

Radar System Analysis by David K. Barton
Introduction to Electronic Warfare by D. Curtis Schleher
Electronic Intelligence: The Analysis of Radar Signals by Richard G. Wiley
Electronic Intelligence: The Interception of Radar Signals by Richard G. Wiley
Principles of Secure Communication Systems by Don J. Torrieri
Multiple-Target Tracking with Radar Applications by Samuel S. Blackman
Solid-State Radar Transmitters by Edward D. Ostroff et al.
Logarithmic Amplification by Richard Smith Hughes
Radar Propagation at Low Altitudes by M.L. Meeks
Radar Cross Section by Eugene F. Knott, John F. Shaeffer, and Michael T. Tuley
Radar Anti-Jamming Techniques by M.V. Maksimov et al.
Introduction to Synthetic Array and Imaging Radars by S.A. Hovanessian
Radar Detection and Tracking Systems by S.A. Hovanessian
Radar System Design and Analysis by S.A. Hovanessian
Radar Signal Processing by Bernard Lewis, Frank Kretschmer, and Wesley Shelton
Radar Calculations Using the TI-59 Programmable Calculator by William A. Skillman
Radar Calculations Using Personal Computers by William A. Skillman
Techniques of Radar Reflectivity Measurement, Nicholas C. Currie, ed.
Monopulse Principles and Techniques by Samuel M. Sherman
Receiving Systems Design by Stephen J. Erst
Designing Control Systems by Olis Rubin
Advanced Mathematics for Practicing Engineers by Kurt Arbenz and Alfred Wohlhauser
Radar Reflectivity of Land and Sea by Maurice W. Long
High Resolution Radar Imaging by Dean L. Mensa
Introduction to Monopulse by Donald R. Rhodes
Probability and Information Theory, with Applications to Radar by P.M. Woodward
Radar Detection by J.V. DiFranco and W.L. Rubin
Synthetic Aperture Radar, John J. Kovaly, ed.
Infrared-to-Millimeter Wavelength Detectors, Frank R. Arams, ed.
Significant Phased Array Papers, R.C. Hansen, ed.
Handbook of Radar Measurement by David K. Barton and Harold R. Ward
Statistical Theory of Extended Radar Targets by R.V. Ostrovityanov and F.A. Basalov
Radar Technology, Eli Brookner, ed.
MTI Radar, D. Curtis Schleher, ed.

ASPECTS OF RADAR SIGNAL PROCESSING

Bernard L. Lewis
Frank F. Kretschmer, Jr.
Wesley W. Shelton

ARTECH HOUSE, INC.
685 Canton Street
Norwood, MA 02062

International Standard Book Number: 1-58053-129-6
Library of Congress Catalog Number: 86-71326

86 87 10 9 8 7 6 5 4 3 2 1

Contents

Preface

This book is intended to provide up-to-date information on important aspects of radar signal processing for use by engineering students and radar engineers already working in the field. This book is primarily an outgrowth of the research work performed by the authors and their colleagues from the early 1970s to the mid-1980s. In order to keep the book to a reasonable size, it is necessarily limited in scope and concentrates on certain aspects of radar signal processing. The main topics of this text are basic radar theory, signal enhancement techniques, clutter suppression, external interference suppression, adaptive processing, and methods for implementation of these techniques. This came to be an ambitious undertaking and it became readily apparent that we could include only a sampling of the papers dealing with these subjects. The lists of references for each of the chapters are not all-inclusive and the authors apologize for any omissions.

This book differs from previous texts by including the important references in the form of readings reprinted at the end of each chapter to which they apply. We hope that this will save time for the student or engineer and increase the probability that he or she will make use of the references. Other copyrighted papers that cannot be reproduced in this manner, however, will be referenced in the normal manner. Also, this text differs from other radar books in that it includes problems for the reader to use in developing his or her understanding of the subject.

Chapter 1
Radar Signal Processing

1.1 INTRODUCTION

The purpose of radar signal processing is to extract desired data from radar signals. The desired data usually concerns the detection of a target of interest, the location of the target in space about the radar, the time rate of change of the target's location in space and, in some cases, the identification of the target as being a particular one of a number of classes of targets.

The accuracy of the data available from a radar is limited by thermal noise introduced by the radar receiver, echoes from targets of no interest (known as *clutter*), and externally generated interference. As a consequence, radar signal processing is also used to enhance signals and to suppress clutter and externally generated interference.

Radar signal enhancement is obtained by way of transmitted signal design and filters in the receiver that are matched to the transmitted signal characteristics, which are called *matched filters*. A good example of this class of device is a radar pulse expander-compressor. Such a device divorces range resolution from transmitter pulse length and permits a radar to transmit an effective peak power greater than that which would normally cause electrical breakdown in the transmitter or antenna.

Clutter suppression is accomplished by taking advantage of differences between echoes from targets of interest and those of no interest. For example, in many cases, targets of interest have different relative velocities with respect to the radar than that of clutter. In these cases, a *moving target indicator* (MTI) or a *Doppler filter bank* (DFB) can be used to separate desired from undesired target echoes.

Externally generated interference can enter the radar antenna main lobe or its sidelobes. It can be unsynchronized with the radar *pulse repetition frequency* (PRF), synchronized with it, of low duty cycle, of high duty cycle, matched to the radar receiver, or not matched to the radar receiver. Differences between desired signals and externally generated interference can be used to suppress the interference and to augment the desired signals. Typical *interference suppressors* are known as defruiters, sidelobe blankers, coherent sidelobe cancelers, and main-lobe notchers.

In this text, emphasis will be placed on signal enhancement and the clutter and interference suppression techniques available to radar designers. Only enough radar *signal-in-noise* theory will be given to permit the reader to understand and appreciate the techniques presented herein.

1.2 BACKGROUND [1–6]

Radars transmit short time-duration pulses τ_c as electromagnetic waves and receive echoes from any electromagnetic wave reflector in the path of the radiation from the radar. The radar measures the time Δt elapsed between transmission and echo reception in order to determine the distance, or range R, separating the reflector and the radar. In this case,

$$R = \frac{c\Delta t}{2} \tag{1–1}$$

where c is the velocity of electromagnetic wave propagation (i.e., speed of light), and the 2 in the denominator takes into consideration the two-way path traveled by the pulse from radar to reflector and back again.

The nominal *range resolution* of the radar with a time duration (or length) pulse τ_c can be seen from (1–1) to be

$$\delta R = \frac{c\tau_c}{2} \tag{1–2}$$

where δR is the distance beyond which two reflectors must be separated so that their echoes can be seen as two separate pulses (i.e., one pulse ends before the second one starts). The range resolution is determined by the time duration of the pulse which equals the reciprocal of the transmitted signal's bandwidth in a matched system. In this chapter, we shall consider τ_c to represent the time duration of an uncoded wave, which is equal to the reciprocal of the bandwidth. In the next chapter, we will generalize the radar range equation for coded pulse-compression waveforms, where τ_c represents the compressed pulsewidth.

The radar pulse repetition period τ_r determines the maximum unambiguous range of the radar:

$$R_{max} = \frac{c\tau_r}{2} \tag{1–3}$$

Targets with ranges greater than R_{max} produce echoes that return to the radar after one or more pulse repetition periods and appear to have ranges between 0 and R_{max}. Typical values of τ_r and τ_c for long-range search radars are 3×10^{-3} seconds and 1×10^{-6} seconds, respectively.

Radars with pulse lengths τ_c employ receiver bandwidths on the order of $B_r = 1/\tau_c$ to preserve the greatest possible range resolution with the least possible loss in signal-to-noise ratio (SNR). Because 88% of the energy in a rectangular pulse spectrum lies within a bandwidth of $1/\tau_c$, only 12% more signal power would be made available by greatly increasing B_r. However, increasing B_r would increase the thermal noise power in the receiver by the same factor. This is a very important concept in signal processing.

An equally important concept is that the signal-to-noise ratio decreases if the radar receiver bandwidth B_r decreases below $B_t = 1/\tau_c$, where B_t is the radar's transmitted signal bandwidth. This is due to the fact that the signal-to-noise ratio is defined by the ratio of signal energy in a time-resolution cell to the noise energy in a time-resolution cell. In a receiver of bandwidth B_r, a time-resolution cell duration is $1/B_r$. The amount of signal energy of bandwidth B_t that enters a receiver having a bandwidth B_r, which is less than B_t is proportional to B_r/B_t. This energy is contained in a time-resolution cell that also contains a thermal noise energy FkT, where

$$\begin{aligned} F &= \text{Receiver noise figure} \\ &= \frac{\text{noise power out of receiver}}{\text{noise power into receiver} \times \text{receiver power gain}} \\ k &= \text{Boltzmann's constant} = 1.38 \times 10^{-23}\ \text{Joule/Kelvin} \\ T &= \text{Temperature in Kelvins} \end{aligned} \tag{1–3a}$$

Because kT is a constant that is independent of B_r and the signal energy in the time-resolution cell, which is associated with a bandwidth B_r, decreases as B_r decreases below B_t, the radar signal-to-noise ratio thus decreases as B_r decreases below B_t.

Therefore, from the preceding discussion, we can see that the ratio of the radar signal to thermal noise is maximized with $B_r = B_t$ and decreases as B_r becomes either larger or smaller than B_t.

The filter that maximizes the ratio of the peak signal power to the mean-squared noise is called a matched filter and results from a mathematical maximization procedure [1,2]. For an input waveform $s(t)$, the impulse response of the matched filter is $s(-t + \tau_0)$, where τ_0 is an arbitrary delay. In complex signal processing, which shall be discussed later, the matched filter's impulse response is $s^*(-t + \tau_0)$, where the asterisk denotes conjugation. Equivalently, the matched filter may be expressed in the frequency domain as $S^*(f)$, where $S(f)$ is the Fourier transform of the input signal $s(t)$. The conjugation in this case changes the sign of the phase spectrum.

Skolnik shows [1,2] that the ratio of the peak signal power to the mean-squared noise power at the output of a matched filter is given by $2E/N_0$, where E denotes the energy of the input signal, and N_0 denotes the one-sided noise spectral density, which is numerically equal to kT. Thus, the output peak signal to mean-squared noise ratio for a matched filter is independent of the coding or modulation of the input waveform and depends only on the signal energy and the noise spectral density.

Radar *signal detection* in thermal noise involves recognizing that a given time-resolution cell contains more energy than would be expected from thermal noise alone [1,2]. However, because the thermal noise energy varies randomly in a given resolution cell and from one time-resolution cell to another, the detection process involves statistics [1,2]. In *target detection*, a threshold is set at a given multiple of the thermal noise root-mean-square (rms) voltage, and a target detection is declared when the threshold is exceeded. The threshold voltage is chosen to provide the desired false-alarm probability.

When a target detection is declared, circuits measure the time delay between the radar's transmitted signal and the echo that was declared to be a detection. This time delay measurement determines the target range and indentifies the time-resolution cell containing the target echo for use in further signal processing on successive pulses in the same range-resolution cell.

The signal energy E_r available to a radar from a target echo is determined as described in the following paragraphs.

The radar transmits a signal with an effective peak power $P_t = E_t/\tau_c$ into a solid angle ψ, determined by the radar antenna gain

$$G = \frac{4\pi}{\psi} = \frac{4\pi A_e}{\lambda^2} \tag{1–4}$$

where A_e is the effective area of the radar antenna, λ

is the radiated wavelength, and E_t is the transmitted energy.

The peak power per unit area ϵ_t in the beam, which is at a distance R from the radar, is then

$$\epsilon_t = \frac{P_t G}{4\pi R^2} \tag{1–5}$$

due to the increase in beam diameter with range R.

The peak power incident upon a reflector having a cross sectional area A_s is then

$$P_s = \epsilon_t A_s \tag{1–6}$$

The reflector reflects P_s back to the radar with a gain G_s to produce a peak power density at the radar of

$$\epsilon_r = \frac{P_s G_s}{4\pi R^2} \tag{1–7}$$

This power density incident upon the radar antenna of effective area A_e supplies a received power P_r to the radar receiver, where

$$P_r = \epsilon_r A_e \tag{1–8}$$

From (1–4), we have

$$A_e = \frac{G\lambda^2}{4\pi} \tag{1–9}$$

and, from (1–5) through (1–9), we obtain

$$P_r = \left(\frac{P_t G}{4\pi R^2}\right)\left(\frac{A_s G_s}{4\pi R^2}\right)\left(\frac{G\lambda^2}{4\pi}\right) \tag{1–10}$$

The signal energy that is available in a radar time-resolution cell of length $\tau_c = 1/B_r$ can be found from (1–10) to be

$$E_r = P_r \tau_c = \frac{P_t G^2 A_s G_s \lambda^2 \tau_c}{(4\pi)^3 R^4} \tag{1–11}$$

Usage has defined $A_s G_s$ in (1–11) as equal to σ_s, with σ_s defined to be the equivalent-scattering-cross-section of the reflector, which is, in effect, the cross-sectional area of a sphere that could replace the reflector of interest by

$$\sigma_s = \frac{\pi D_s^2}{4} = A_s G_s \tag{1–12}$$

where D_s is the diameter of the equivalent sphere. With (1–12), equation (1–11) is usually written as

$$E_r = \frac{P_t G^2 \sigma_s \lambda^2 \tau_c}{(4\pi)^3 R^4} \tag{1–13}$$

A commonly used form of the radar-range equation can be obtained from (1–13) with E_r set equal to the thermal noise energy in a time-resolution cell of a receiver with a noise figure F, which is written as

$$E_r = FkT = \frac{P_t G^2 \sigma_s \lambda^2 \tau_c}{(4\pi)^3 R^4} \tag{1–14}$$

By solving (1–14) for R, we obtain

$$R = \left[\frac{P_t G^2 \sigma_s \lambda^2 \tau_c}{(4\pi)^3 FkT}\right]^{1/4} \tag{1–15}$$

Note that in (1–15) we have

$$\frac{P_t \tau_c}{FkT} = \frac{E_t}{E_n} \tag{1–16}$$

where E_n is the thermal noise energy in the time-resolution cell of a receiver with bandwidth B_r and noise figure F, and E_t is the transmitted energy $P_t \tau_c$.

Equation (1–15) defines the range R from which a reflector of equivalent scattering cross section σ_s will supply a radar with signal energy equal to thermal noise energy in a time-resolution cell $1/B_r$ given radar parameters P_t, G, λ, F, and B_r.

1.3 PARAMETERS DETERMINING SIGNAL LEVEL

The most important parameters determining the signal level in (1–15) are the radar peak power P_t, the radar pulse length $\tau_c = 1/B_r$, the radar antenna gain G, and the wavelength λ of the radiated radar signal.

There are limits on the maximum usable antenna gain $G = G_{\max}$ in (1–15) if a radar must scan a given volume of a space in a given time τ_s. The range component $R_{\max}$ of the volume determines the highest *pulse repetition frequency* that can be used to provide unambiguous range measurements such that $\mathrm{PRF} = c/(2R_{\max})$. The minimum scan time τ_s is determined from the solid angle ψ_s that must be scanned with at least n echoes (called *hits*) per target per scan, the radar's solid angular beamwidth $4\pi/G_{\max}$, and the $\mathrm{PRF}_{\max}$ according to the relation:

$$\tau_s = \frac{n\psi_s}{(4\pi/G_{\max})\mathrm{PRF}_{\max}} \tag{1–17}$$

From (1–17), we have

$$\tau_s = \frac{G_{\max}\psi_s n}{4\pi \text{PRF}_{\max}} \tag{1–18}$$

and, from (1–18), we hence derive

$$G_{\max} = \frac{4\pi\tau_s \text{PRF}_{\max}}{n\psi_s} \tag{1–19}$$

For example, to provide an unambiguous range of 10^5 meters, a radar's PRF must be such that the *pulse repetition interval* (PRI) times the velocity of light must equal the distance or length of the two-way path traveled by the pulse, or

$$(\text{PRI})(3 \times 10^8 \text{ m/s}) = 2 \times 10^5 \text{ m} \tag{1–20}$$

From (1–20), the PRI is 0.67×10^{-3} and the PRF is $1/\text{PRI} = 1.5 \times 10^3$ Hz. With this PRF, $n = 1$, $\psi_s = 2\pi$, and $\tau_s = 4\text{s}$, (1–19) yields $G_{\max} = 1.2 \times 10^4$. If more than one hit per scan is required, $G_{\max}$ drops correspondingly.

There are also limits on λ due to the space available for the antenna size necessary to obtain the required gain $G_{\max}$. This situation is evident from (1–4). Because most of the signal-controlling parameters are limited, we can see why signal processing to avoid these limits would be important.

In order to limit the size and scope of this book, we concentrate on certain aspects of signal processing. These are signal enhancement, clutter suppression, external interference suppression, and methods for implementation of these techniques.

To aid the reader, a short review of *coherent signal processing* and *complex number theory*, as used in digital signal processing, is presented prior to discussing the techniques of interest and their implementation.

1.4 COHERENT SIGNAL PROCESSING

Coherent signal processing requires that both the *amplitude* (magnitude) and *phase* of signals be used in the process. Therefore, in coherent systems using digital signal processing, each signal sample must specify both amplitude and phase or their equivalents. Such specification is possible by using complex numbers to represent each signal sample. Complex numbers have two orthogonal components, which may be written as $I + jQ$, where I is the real or in-phase component, Q is the imaginary or quadrature component, and j is an imaginary number defined to be $(-1)^{1/2}$.

A generalized form of a narrowband received signal $s(t)$ can be expressed as

$$s(t) = A(t) \cos [\omega_0 t + \phi(t)] \tag{1–21}$$

where ω_0 represents the IF carrier frequency, $A(t)$ and $\phi(t)$ represent narrowband (relative to ω_0) amplitude and phase modulations, respectively. The value $\phi(t)$ includes any phase modulations of the transmitted signal, Doppler effects, and constant phase shift. The I and Q components of $s(t)$ are obtained by mixing, or beating, the signal $s(t)$ with the local oscillator (LO) signal, $\cos \omega_0 t$, and with the LO signal phase shifted 90° in the other channel (Fig. 1–1). The I and Q components can be regarded as projections of a vector on two orthogonal axes at any given instant of time. It is the availability and use of the transmitter carrier frequency at the receiver which makes the process coherent.

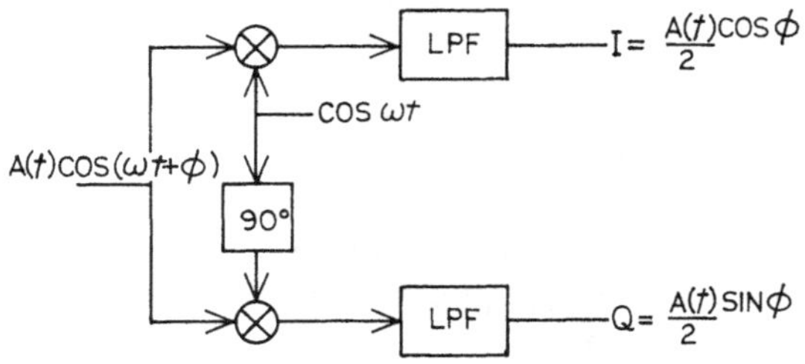

Fig. 1–1 Generation of I and Q (LPF = Low-Pass Filter, ⊗ = Multiply).

The values of I and Q in Fig. 1–1 result from the trigonometric identities that state

$$2\cos A \cos B = [\cos(A + B) + \cos(A - B)] \tag{1.22}$$

$$2\sin A \cos B = [\sin(A + B) + \sin(A - B)] \tag{1.23}$$

$$\cos(A - 90°) = \sin A \tag{1.24}$$

where $A = \omega_0 t$, $B = \omega_0 t + \phi(t)$ and the averaging performed by the low-pass filter results in the removal of the higher frequencies represented by the $(A + B)$ terms. The low-pass filters used here are designed to pass only the frequencies contained in the signal modulation $A(t)$ and $\phi(t)$. Because the circuit of Fig. 1–1 removes the signal from the carrier, it is called a detector. However, because it preserves the signal phase information, it does not destroy coherence and is called a coherent, or synchronous, detector.

Figure 1–2 demonstrates that I and Q from Fig. 1–1 and the original carrier provide enough information to reconstruct the original signal.

Figure 1–3 is a plot of an imaginary number on a coordinate system in which the horizontal is the I axis and the vertical is the Q axis.

Note that a line joining the origin and the point specified by the complex number, together with its pro-

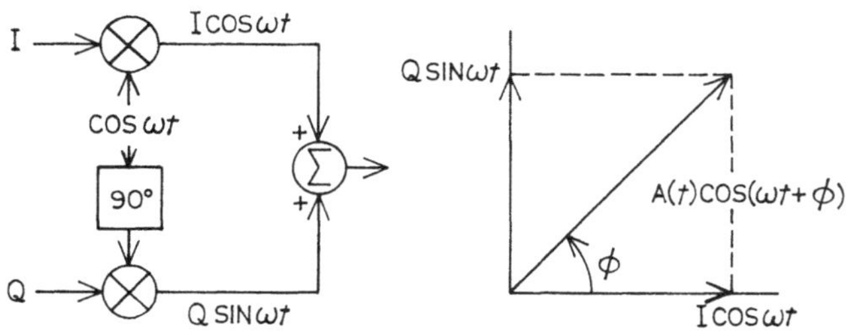

Fig. 1–2 Signal Reconstruction Using I, Q, and ωt ($\otimes$ = Multiply).

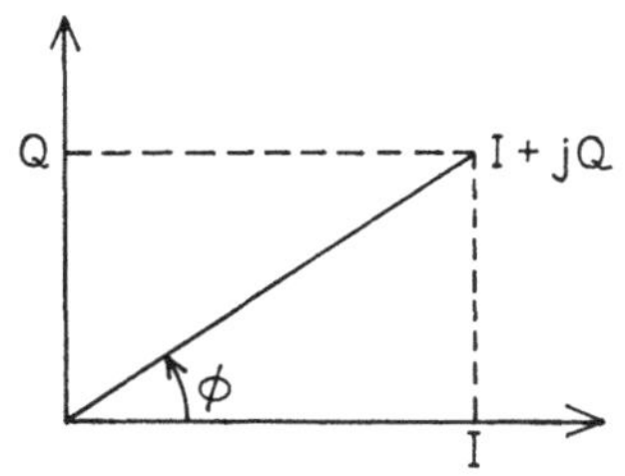

Fig. 1–3 Plot of a Complex Number (I + jQ).

jections on the I and the Q axes, form a right triangle. Therefore, the magnitude $|I + jQ|$ of a complex number equals the square root of the sum of the squares of the two components of the number, $(I^2 + Q^2)^{1/2}$. Also, from Fig. 1–3, the phase of the signal sample represented by $I + jQ$ is

$$\phi = \arctan Q/I \qquad (1\text{–}25)$$

It is often convenient to rewrite an imaginary number in exponential form as

$$Ae^{jx} = A(\cos x + j \sin x) \qquad (1\text{–}26)$$

where A is the magnitude of the number and e^{jx} defines its angle with respect to the positive I axis. The complex conjugate of a complex number is given by

$$(I + jQ)^* = I - jQ \qquad (1\text{–}27)$$

In exponential form, the complex conjugate of (1–26) is

$$Ae^{-jx} = A(\cos x - j \sin x) \qquad (1\text{–}28)$$

The complex conjugate of a complex number multiplied by the complex number yields the magnitude squared of the complex number:

$$(I + jQ)(I + jQ)^* = (I + jQ)(I - jQ) = I^2 + Q^2 \qquad (1\text{–}29)$$

and, in exponential form, we write

$$(Ae^{jx})(Ae^{jx})^* = (Ae^{jx})(Ae^{-jx}) = A^2 \qquad (1\text{–}30)$$

Before moving on to digital signal processing, we should note that conjugating a complex number changes the sign of the phase angle it represents. If this phase angle varies with time as in $e^{j\omega t}$, conjugation changes the sign of the angular frequency or time to yield $e^{-j\omega t}$.

1.5 DIGITAL SIGNAL PROCESSING

Digital signal processing is generally used in cases where it can replace analog signal processing because of the better accuracy and the ease with which accurate *time delays* can be obtained. Time delays in digital circuitry are clocked intervals between reading into and out of storage.

Digital processing can be done in *real* or *complex* data format. Real data can be obtained from the circuit shown in Fig. 1–1 if cos ωt is replaced by $\cos(\omega + B/2)t$, where B is the bandwidth of $A(t)$. With this type of circuit, the signal is not folded about the carrier. The carrier is only translated down to $B/2$. Thus, sampling and conversion to digital data must be done at a rate of at least $2B$ to preserve all the signal information. This sampling rate provides at least two samples of data per cycle of the highest frequency B in the signal. If both I and Q outputs are used, signal phase shifts can be accomplished by controlling the relative magnitudes of I and Q, and adding the resultants.

Complex data format is obtained as illustrated in Fig. 1–1. In this application, the data is folded about zero frequency by complete removal of the carrier, and the bandwidth of both I and Q are $B/2$. Thus, to preserve the information available in I or Q, each output need only be sampled and converted to digital data at a rate B in order to have two samples per cycle at the highest frequency of interest ($B/2$). The fact that I is orthogonal to Q yields the extra sample per cycle of the signal required to preserve the bandwidth B.

Because speed is costly in digital circuits, most digital processing is done with complex data. However, there are some applications where real data is used as the best way to obtain the desired result. In this text, however, only complex data processing will be considered.

In digital signal processing with complex data, the I and Q data are carried separately through the processors in time synchronization. To change the magnitude of a complex number, both the I and Q values of the number are multiplied by a real number. This requires two digital multipliers. To change the phase represented by a complex number, the number is multiplied by another complex number with unit magnitude. This requires four digital multipliers, one digital adder, and one digital subtractor to perform the operations defined by

$$[C + jD][A + jB] = [CA - DB] + j[CB + DA] \quad (1\text{–}31)$$

Of course, both the magnitude (amplitude) and phase can be changed at the same time in (1–31) by using a nonunity magnitude multiplier $(c + jd)$. A complex multiplier consists of the hardware that would implement (1–31).

PROBLEMS

Problem 1–1 Find the pulse repetition frequency of a radar if the interpulse period is 5 ms.

Problem 1–2 Find the unambiguous range of a radar if the interpulse period is 5 ms.

Problem 1–3 Find the range resolution of a radar if the pulselength is 1 μs.

Problem 1–4 Find the antenna gain of a radar if its solid angular beamwidth is 0.1 steradian.

Problem 1–5 Find the effective area of a radar antenna if the gain is 400 and the operating wavelength is 0.1 m.

Problem 1–6 Find the maximum allowable radar antenna gain to achieve a 5 s scan of 2 π steradians with an interpulse period of 5 μs and 10 hits per scan.

Problem 1–7 Determine the range at which the signal energy will equal the average thermal noise energy in a time-resolution cell if a radar's system temperature is 300 K, its peak power is 1 MW, its antenna gain is 1000, its noise figure is 4, its receiver bandwidth is 1 MHz, its operating wavelength is 0.1 m, and the target's equivalent scattering cross section is 1 m^2.

Problem 1–8 Find the phase angle represented by the complex number:
$1 + j1$

Problem 1–9 Find the phase angle represented by
(a) $(1 + 1j)^2$
(b) $(1 + 1j)^{1/2}$

Problem 1–10 Find the magnitude and angle represented by
(a) $[4 \exp(j0.5\pi)][6 \exp(j0.25\pi)]$
(b) $[4 \exp(j0.5\pi)][6 \exp(-j0.25\pi)]$

Problem 1–11 Write the following as a complex number:
(a) $6 e^{j\phi}$
(b) $4 \exp(j\phi)$

Problem 1–12 Write the following as an exponential:
(a) $\cos\theta + j \sin\theta$
(b) $\cos\theta - j \sin\theta$

Problem 1–13 What is the energy associated with a rectangular pulse of 1 μs length if its peak power is 1 MW?

Problem 1–14 What is the equivalent scattering cross section of a reflector with a power interception area of 10 m^2 and a backscattering gain of 0.1?

Problem 1–15 What is the equivalent scattering cross section of a flat, square plate of metal, 10 cm on a side, illuminated by a radar so that the power interception area is 100 cm^2 if the wavelength is 1 cm?

Problem 1–16 What would the antenna gain be if the effective area of an antenna were 1 m^2 and the wavelength employed were 1 cm?

REFERENCES

1. M. I. Skolnik, *Introduction to Radar Systems,* (2nd ed.) McGraw-Hill, New York, 1980.
2. M. I. Skolnik (ed.), *Radar Handbook,* McGraw-Hill, New York, 1970.
3. F. E. Nathanson, *Radar Design Principles,* McGraw-Hill, New York, 1969.
4. D. K. Barton, *Radar System Analysis,* Artech House, Dedham, MA, 1977.
5. R. S. Berkowitz (ed.), *Modern Radar Analysis, Evaluation, and System Design,* John Wiley and Sons, New York, 1965.
6. E. Brookner (ed.), *Radar Technology,* Artech House, Dedham, MA, 1977.

Chapter 2
Pulse Compression

2.1 BASIC CONCEPTS

Pulse compression [1–7] is employed in radar to increase the signal energy transmitted without sacrificing range resolution, nor encountering excessively high peak powers that can cause electrical breakdown. Pulse compression techniques are employed to divorce the useful signal bandwidth (range resolution) from the transmitted pulse length. The radar signal bandwidth to be transmitted is increased by modulating the signal within the transmitted pulse. This modulation may consist of amplitude, phase, or frequency changes of the signal carrier within the pulse. Target echo signals are then passed through filters matched to the transmitted signal and hence the echo energy is compressed into a pulse having a time duration τ_c, which is approximately equal to the reciprocal of the transmitted bandwidth B_t. The ratio of the transmitted to compressed pulse lengths is called the *pulse compression ratio* ρ and is given by

$$\rho = B_t\tau_t = \tau_t/\tau_c \tag{2–1}$$

where τ_t is the transmitted pulse length and $B_t = 1/\tau_c$. Values of 100 are common in radars of today and those of the future will use much higher values.

The way in which pulse compression affects the radar range equation (1–15) is illustrated by (2–2), where P_t has been replaced by $P_t\rho$, and B_r in the receiver has been replaced with $\rho/\tau_t = B_t$:

$$R = \left[\frac{P_t\rho G^2\sigma_s\lambda^2\tau_t}{(4\pi)^3 NkT\rho}\right]^{1/4} \tag{2–2}$$

Note that ρ cancels in (2–2), which leaves R proportional to $(P_t\tau_t)^{1/4}$, where the quantity in brackets is the transmitted signal energy. Note that τ_t can be increased as much as desired with constant P_t to obtain any desired range R with any desired compressed pulse length $1/(\rho/\tau_t) = 1/B_t = \tau_t/\rho$.

It should be noted that when compressing τ_t to τ_c, the compressor does not form only a single pulse in time. The pulse that is formed has precursors and followers of reduced amplitude called *range-time sidelobes*. The ratio of the main peak to sidelobe peak that is obtained in this process is dependent on the characteristics of the modulation used to spread the signal bandwidth. The compressed pulse time function is, in the absence of Doppler shift and noise, the same as the autocorrelation function of the transmitted waveform.

Time sidelobes on radar pulses are objectionable because a time sidelobe of a strong echo may hide a weaker, though more important, target return. In essence, time sidelobes compromise the radar's range resolution. Also, any clutter present in the sidelobes can leak into the range cell of interest. As a consequence, much time and effort has gone into research directed toward reducing time sidelobes.

The main peak and sidelobe responses of pulse compressors are often displayed in three dimensional plots called *ambiguity diagrams*. The three dimensions in these diagrams are power (response), normalized time, and normalized Doppler frequency. The Doppler frequency divided by the radar bandwidth axis, when multiplied by ρ, is equal to the product of the Doppler frequency times the uncompressed pulse length. This product indicates the number of cycles of Doppler frequency (or multiples of 2π phase shift) that occur over the uncompressed pulse duration. An example of an ambiguity diagram is illustrated in Fig. 2–1.

Pulse compressor responses are also plotted in two dimensions to yield more accurately readable information. These plots show power *versus* time, or sample number in units of $1/B_t$, at a single Doppler frequency, as shown in Fig. 2–2.

An intuitive understanding of pulse compressor ambiguity diagrams requires a knowledge of how they are produced. As an example, we will consider a four-element Barker code [4] implementation, which is illustrated in Fig. 2–3.

In Fig. 2–3, the three boxes marked τ_c represent time delays that form a four-tap delay line having a delay τ_c between taps. Each tap on the delay line is connected to a weighting box, marked 1 or -1, that maintains or

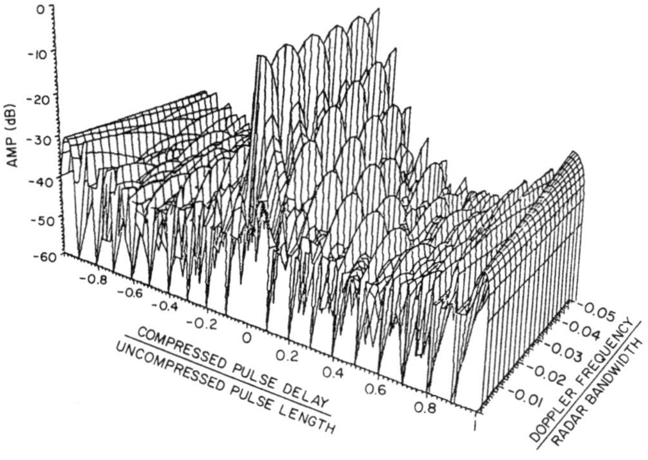

Fig. 2–1 Partial Ambiguity Diagram for a 100-Element Frank Code.

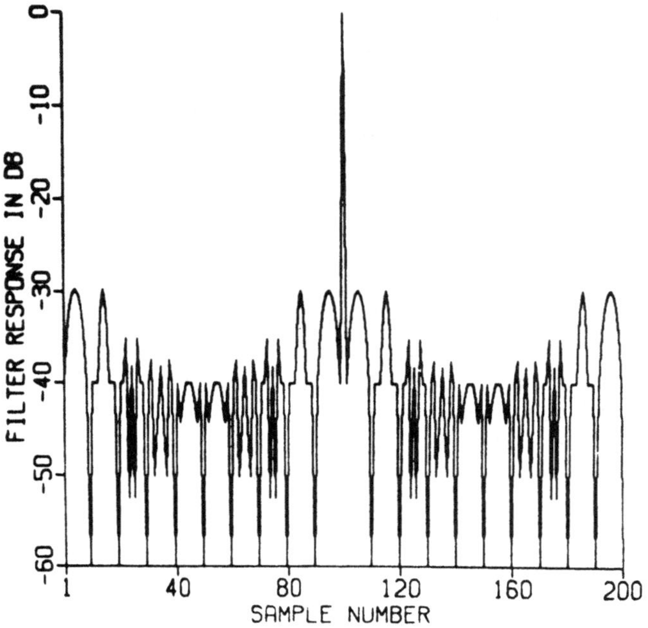

Fig. 2–2 Zero-Doppler Compressed 100-Element Frank Code.

inverts the sign of the pulse passing through the box. These boxes are said to *weight* the signals passing through them. The outputs of the weighting boxes are added together in the box marked "sum."

In operation, to form the expanded pulse code for transmission, a video pulse of time length τ_c is placed on input A as illustrated in the figure. During the first time interval τ_c, after the start of the input pulse, it enters the first delay section and exits from the sum box after passing through the weighting box 1. To aid in following the timing, the first pulse out of the A port of the sum box is marked number 1. At the end of the input pulse, the leading edge of the input pulse exits from the first delay section, passes through a box with

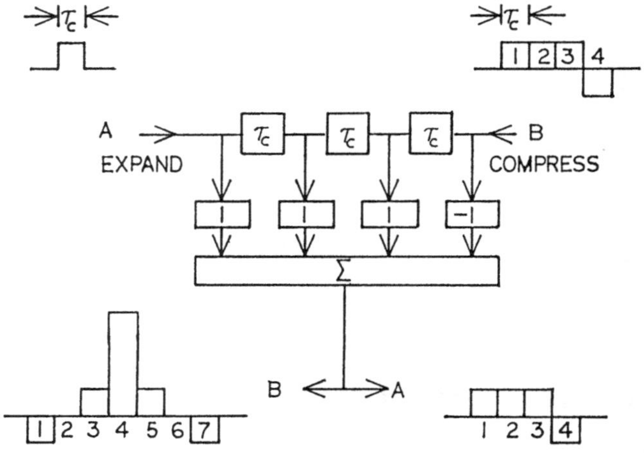

Fig. 2–3 Four-Element Barker Code Implementation.

unity weight and exits from the sum box as A output 2. This process continues until pulse #4 exits from the sum box as a negative pulse. This $4\tau_c$ long pulse is the desired expanded pulse for transmission coding. It is used for binary phase coding of a carrier.

To compress a received pulse, the pulse is sent to input B_{in} of the tapped delay line. During the first time interval τ_c, the first code element of the echo pulse exits from the sum box as a negative pulse, as illustrated on the B_{out} terminal. When the second code element arrives, the first and second code elements exit the sum box together, and have opposite polarities due to the different weights in the two weighting boxes. Because they have opposite polarities, the two code elements cancel in the summing process and a zero-amplitude element exits from the sum box as element 2 in the figure. The third code element returned to the radar produces a single-code-element high pulse out of the sum box as the result of $1 + 1 - 1$. Then, when pulse #4 returns with a polarity opposite to the first three code elements, the weighting boxes cause all inputs to sum with the same polarity and a four-code-element high pulse exits from the sum box as a pulse of length τ_c. Note that this pulse is not the last one out of the compressor. For a time interval τ_c after matching, the last three code elements of the echo pulse are still in the delay line and exit from the sum box as #5 of B_{out}. Similarly, the last two code elements received exit as pulse #6 and the last code element as pulse #7. Note that the compressed pulse is $(2\rho - 1)/\rho$ times longer than the transmitted pulse, which, in turn, is ρ times longer than the pulse of length τ_c used to form it.

If the target that had produced the above-discussed echo pulse had been moving toward or away from the radar during the time of interest, the phase of the echo would have been changing due to the time-changing range of the target. This *time-changing phase* would have changed the phase relationships in the expanded pulse echo and modified the way the various code elements combined on compression. This time-varying phase due to target motion is called *Doppler*. It is accounted for in the predicted response of a pulse compressor by computing the compressor output for all expected dopplers and plotting the resulting three-dimensional data in an ambiguity diagram as illustrated in Fig. 2–1.

2.2 TYPES OF CODES

The various types of modulation used in pulse compression are called *codes*. Some of the better known codes include:

1. Barker binary phase [1–7];
2. pseudorandom binary phase [1–7];
3. random binary phase [4];
4. step frequency modulation [1–7];
5. linear frequency modulation [1–7];
6. nonlinear frequency modulation [2, 3, 5, 6, 7];
7. step-frequency-derived polyphase (Frank and P1 codes) [8–13];
8. Butler-matrix-derived polyphase (P2 code) [9–13];
9. linear-frequency-derived polyphase (P3 and P4 codes) [10, 12–14];
10. Huffman codes [3, 15, 16];
11. Complementary codes [17–21].

The *binary phase codes* 1, 2, and 3 above exhibit what is referred to as thumbtack ambiguity diagrams. The match-point peak occurs over a very narrow range in time and in Doppler. The ambiguity diagram consists of a single narrow spike surrounded by a plateau of time and Doppler frequency sidelobes. As a consequence, these codes have little Doppler tolerance, i.e., their performance is degraded by moving targets. The *Barker codes*, however, have been called the perfect codes because the highest normalized time sidelobe at zero Doppler frequency is only one code element high. Unfortunately, the highest pulse compression ratio Barker code that has been found is 13 to 1. Other phase codes having the general properties of the Barker code have been found by using phase alphabets such as multiples of 60° [22]. The *pseudorandom* (also called maximum-length-sequence shift register codes) and the *random* binary phase codes can provide very large pulse compression ratios, but their peak sidelobe powers are only on the order of ρ below their peak [4].

The *step frequency modulation code*, which is also known as step-chirp, provides an approximation of the chirp signal, which consists of a linear frequency sweep *versus* time. The step-chirp transmitted waveform usually consists of a sequence of different tones or concatenated frequencies. This code is interesting because it was one of the first to be implemented on a radar and it is a classic example of the use of Fourier series relationships between time functions and spectra. It is well known that the spectrum of a rectangular time pulse of length t_p is a continuous spectrum of the form $[\sin(\pi f t_p)]/(\pi f t_p)$. The zeros in this spectrum occur at frequencies f, where the product (ft_p) is an integer other than zero, as illustrated by Fig. 2–4, where only the positive frequencies of the symmetrical spectrum are shown.

If the pulse is repeated periodically with a frequency ΔF, the pulse spectrum changes from a continuous to a line spectrum, where the spectral lines are separated by the repetition frequency $\Delta F t_p$, as shown in Fig. 2–4(c). When this pulse modulates a carrier frequency f_c, the spectrum becomes symmetrical about the carrier. In one form of implementation, the step frequency modulation code sequentially transmits tones corresponding to the lines in the main lobe of the spectrum between the 3 dB points so that they cannot interfere on transmission. On receive, the compressor employs filters and delays to make the various frequencies occur simultaneously so that they can reconstruct an approximation of the pulse from the spectrum of which they were taken.

In this process, each tone is allowed to last for only a time $\tau_f = 1/\Delta F$ so that the pulse formed by the tones will not periodically repeat with a frequency ΔF. Because they are limited in time, the lines become bands

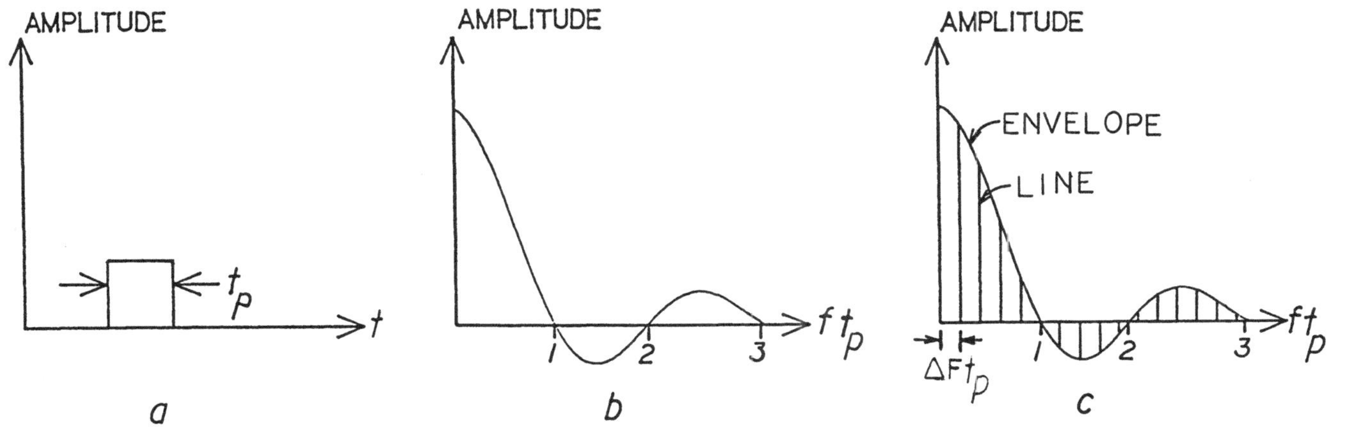

Fig. 2–4 (a) Rectangular Pulse in Time, (b) One-Sided Spectrum of Nonrepeated Rectangular Pulse in Time, (c) One-Sided Spectrum of a Rectangular Pulse in Time Repeated at a Rate ΔF.

of a bandwidth ΔF. In this form of pulse compressor, the expanded (transmitted) pulse length using N tones is

$$\tau_t = N/\Delta F = N\tau_f \tag{2–3}$$

and the transmitted bandwidth B_t is

$$B_t = N\Delta F = 1/\tau_c \tag{2–4}$$

Thus, the pulse compression ratio is

$$\rho = \tau_t/\tau_c = \tau_t B_t \tag{2–5}$$

The step frequency modulation code is not as Doppler tolerant as the linear frequency modulation code to be described next. Large grating lobes appear in the compressed pulse sidelobes at Doppler shifts that are odd multiples of $\Delta F/2$.

The linear frequency modulation code, or chirp signal, is the most Doppler tolerant code that is commonly implemented in analog form today. In this code, the frequency varies linearly with time in the transmitted expanded pulse. As in the step frequency modulation code, the pulse compression ratio of the linear frequency modulation code is equal to the product of the transmitted pulse length and the transmitted bandwidth as given by (2–5). The chirp signal has a compressed pulse, which is approximately given by $[\sin(\pi t B_t)]/(\pi t B_t)$.

Certain polyphase codes which are related to the chirp and step-chirp waveforms are described next. The Frank polyphase coded waveform may be described and generalized by considering a hypothetically sampled step-chirp waveform [11]. The *Frank code* was not originally described in this manner, but rather was given in terms of the elements of a matrix. As an example, consider a four-frequency step-chirp waveform as shown in Fig. 2–5(b), where F_i denotes the frequency tones.

In this waveform, the frequency steps are equal to the reciprocal of the tone duration $4\tau_c$, where τ_c denotes the compressed pulsewidth. Assuming that this waveform has been beat to baseband I and Q by using a synchronous oscillator having the same frequency as the first tone, the resulting phase *versus* time characteristic consists of four linear sections, as shown in Fig. 2–5(a).

The corresponding baseband frequencies are the subharmonics of the frequency $1/\tau_c$. If the baseband phases of the step-chirp waveforms are sampled every τ_c seconds and held for τ_c seconds, the phase sequence shown in Fig. 2–5(c) is obtained. This sequence of phases constitutes the Frank code phases for $N = 4$, corresponding to the four baseband frequencies of the hypothetical step-chirp waveform. The actual Frank coded waveform transmitted consists of a carrier, the phase of which is modulated according to the indicated

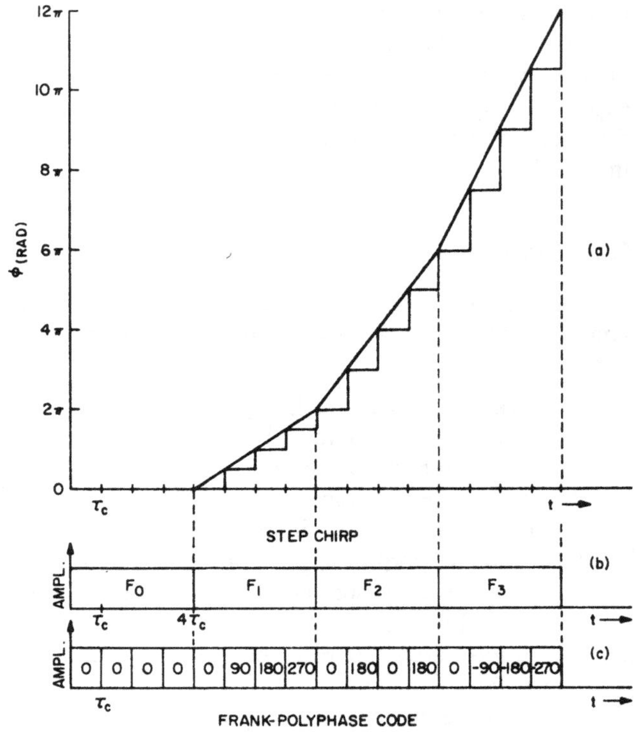

Fig. 2–5 Step-Chirp and Frank Polyphase Code Relationships.

baseband waveform sequence. For each frequency or section of the step chirp, a phase group consisting of N phase samples is obtained and the total number of code phases is N^2, which is equal to the pulse compression ratio. Note that the phase increments within the four phase groups are 0°, 90°, 180°, and 270°. However, the phases of the last group are ambiguous (>180°) and appear as −90° phase steps, or as the conjugate of the F_1 group of phases. The last phase group, because of the ambiguity, appears to complete one 360° clockwise rotation, rather than the (N–1) counterclockwise rotations of the end frequency of the step-chirp waveform.

The Frank code phases may be stated mathematically as follows. The phase of the ith code element in the jth phase group, or baseband frequency, is

$$\phi_{i\,j} = (2\pi/N)(i - 1)(j - 1) \tag{2–6}$$

where the index i ranges from 1 to N for each of the values of j ranging from 1 to N.

The *P1 code* [10, 12] is designed to be more tolerant of precompression band-limiting than the Frank code. Figure 2–6(a) depicts a Frank coded waveform, where the G_k denote the phase groups corresponding to the sampled phases of a step-chirp waveform, as previously discussed. Each group consists of N vectors beginning with a vector at a phase angle of 0°. The phase increments within the Kth group, where $0 \leq K \leq N - 1$, are

$$\Delta\phi_K = K(360°)/N \tag{2–7}$$

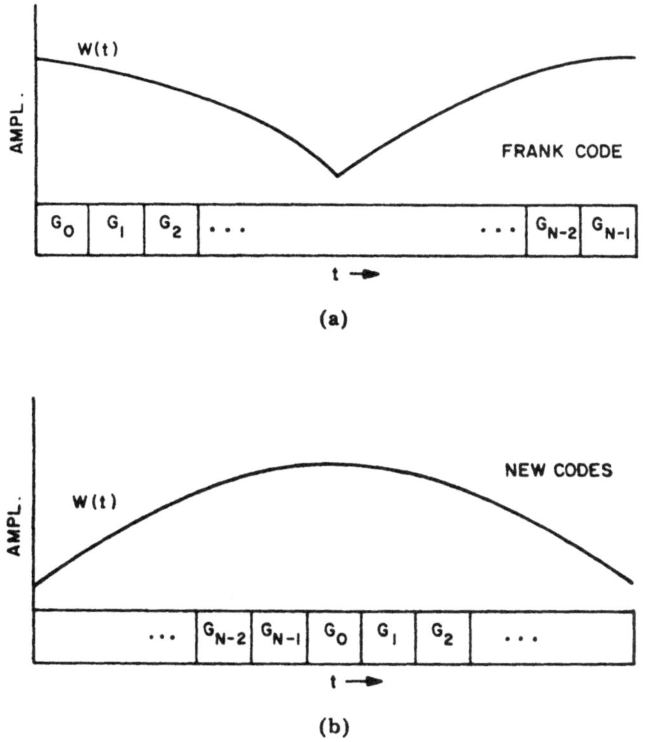

Fig. 2–6 Effect of Band-Limiting Before Pulse Compression: (a) Frank Code, (b) P1 Code.

Thus, G_0 consists of N vectors at 0°, G_1 has vectors separated by 360°/N, and so on, until at the center of the coded waveform the phase increments approach or become 180°, depending on whether N is odd or even. For increments greater than 180°, the phases are ambiguous with the result that the phasors of phase group G_{N-K} are the conjugates of the phasors of phase group G_K, so the vectors have the same increments, but rotate in opposite directions. As a result, the phase increments are small at the ends of the code and become progressively larger toward the center of the code, where the increments approach 180° from opposite directions.

If a receiver is designed so that it has an approximately rectangular bandwidth corresponding to the 3 dB bandwidth of the received waveform, then the received waveform becomes band-limited and a mismatch occurs with the compressor. This band-limiting would normally occur prior to sampling in the process of converting from analog to digital data format (A/D conversion) in order to prevent noise fold-over and aliasing. The result of any band-limiting is to average (or smooth) the vectors constituting the coded waveform. In the case of the Frank code, an effective weighting $W(t)$, such as that illustrated in Fig. 2–6(a), takes place due to the averaging process attenuating those code groups with the higher phase increments more so than the ones having the smaller phase increments. This weighting is actually the inverse of that which would reduce the pulse compressor sidelobes, thus increasing them. The desired weighting for lowering the sidelobes is shown in Fig. 2–6(b), which is obtained in the P1 code by placing the highest phase increment code groups on the ends of the transmitted signal and the lowest phase increment groups in the center.

The P1 code was derived from use of the previously described relationship between the Frank code phases and those of a sampled step-chirp waveform. The desired symmetry, having the dc or small incremental phase group at the center of the code, can be achieved by determining the phases which result from placing the hypothetical synchronous oscillator at the band center of the step-chirp waveform. For an odd number of frequencies, the synchronous oscillator frequency corresponds to one of the waveform frequencies, and the resulting phases are the same as in the Frank code, except that the phase groups are rearranged as indicated in Fig. 2–6(b).

If there is an even number of frequencies, the synchronous oscillator frequency placed at the band center does not correspond to one of the frequencies in the step-chirp signal. The phase of the ith element of the jth group is given in degrees by

$$\phi_{i\,j} = +(180°/N)[N - (2j - 1)] \\ [(j - 1)N + (i - 1)] \tag{2–8}$$

Next, we describe the *P3* and *P4 polyphase codes*, which are related to the chirp waveform. The P3 code is conceptually derived by converting a linear frequency modulation waveform to baseband by using a synchronous oscillator on one end of the frequency sweep and sampling the in-phase I and quadrature Q video at the Nyquist rate [9]. If the waveform prior to coherent detection has a pulse length τ_t and a frequency:

$$f_t = f_0 + kt \tag{2–9}$$

where k is a constant, the bandwidth of the signal will be approximately

$$B_t = k\tau_t \tag{2–10}$$

This bandwidth will support a compressed pulse length of approximately

$$\tau_c = 1/B_t \tag{2–11}$$

and the waveform will provide a pulse compression ratio:

$$\rho = \tau_t/\tau_c = B_t\tau_t \tag{2–12}$$

Assuming that the first sample of I and Q is taken at the leading edge of the waveform, the phases of successive samples taken τ_c apart are

$$\phi_i = 2\pi\int_0^{(i-1)\tau_c} [(f_0 + kt) - f_0]\,dt, \text{ radians} \tag{2–13}$$

or

$$\phi_i = \pi k(i - 1)^2\tau_c^2, \text{ radians} \tag{2–14}$$

where $i = 1, 2, \ldots, \rho$. From (2–10), $k = B_t/\tau_t$; from (2–11), $\tau_c = 1/B_t$; therefore, (2–14) can be written as

$$\phi_i = \pi(i - 1)^2/B_t\tau_t = [\pi(i - 1)^2/\rho], \text{ radians} \tag{2–15}$$

The P4 code is conceptually derived from the same waveform as the P3 code. However, in this case, the local oscillator frequency is set equal to $f_0 + k\tau_t/2$ in the I and Q detectors. With this frequency, the phases in radians of successive samples taken τ_c apart are given by

$$\phi_i = 2\pi\int_0^{(i-1)\tau_c}\left[(f_0 + kt) - \left(f_0 + \frac{k}{2}\tau_t\right)\right] dt \tag{2–16}$$

or

$$\phi_i = 2\pi\int_0^{(i-1)\tau_c} k\left(t - \frac{\tau_t}{2}\right) dt \tag{2–17}$$

Equation (2–17) can be written as

$$\begin{aligned}\phi_i &= \pi k(i - 1)^2\tau_c^2 - \pi k(i - 1)\tau_t\tau_c \\ &= [\pi(i - 1)^2/\rho] - \pi(i - 1)\end{aligned} \tag{2–18}$$

It should be noted at this point that the P3 code will share the intolerance to precompression band-limiting associated with the Frank code, whereas the P4 code will be tolerant of it as is the P1 code. Both the P3 and P4 codes, however, have the Doppler tolerance of the linear frequency modulation or chirp code.

The P3 and P4 codes can be made *palindromic* (to read the same forward and backward) by changing the upper limit on the integrals to $(i - 1/2)\tau_c$ in (2–13) and (2–16). This step is important if the codes are to be used in *moving target indicator* systems (MTI) with combined up and down chirps, which are explained in Chapter 4. The phases of the palindromic P3 code in degrees are

$$\phi_i = (45/\rho)(2i - 1)^2, \quad 1 \leq i \leq \rho, \text{ for } \rho \text{ odd} \tag{2–19}$$

and those of the palindromic P4 code in degrees are

$$\phi_i = (45/\rho)(2i - 1)^2 - 90(2i - 1), \quad 1 \leq i \leq \rho \tag{2–20}$$

The P2 code, which also has desirable features, is similar to the Butler matrix steering phases used in phased array antennas to form orthogonal beams. The P2 code is valid for N even, and each group of the code is symmetric about zero phase. The usual Butler matrix phase groups are not symmetric about zero phase, and result in higher time sidelobes in the compressed pulse. For N even, the P1 code has the same phase increments within each phase group as the P2 code, except that the starting phases are different. The ith element of the jth group of the P2 code is given in degrees by

$$\phi_i = (90/N)(N + 1 - 2i)(N + 2 - 2j) \tag{2–21}$$

where i and j are integers ranging from 1 to N as before. The requirement for N to be even in this code stems from the desire for low autocorrelation sidelobes. An odd value of N results in high sidelobes. This code has the frequency symmetry of the P1 code and also the property of being palindromic.

Huffman codes [3, 15, 16] are included for discussion in this text because they comprise an example of an amplitude-modulation compression code. They are not well suited to radars in which transmitters are operated at saturation for transmission efficiency. However, with the increased use of solid-state transmitters, it becomes more feasible to modulate the amplitude, as well as phase, of the signal from code element to code element. There is a loss of transmitted energy efficiency associated with amplitude modulation, which may not be significant in certain applications. Theoretically, for zero Doppler, Huffman codes are *idealized codes* having only a single unavoidable sidelobe at each end of the compressed signal. However, other sidelobes begin to grow with increasing Doppler [16].

Another class of codes of interest are *complementary codes* [17–21]. Complementary codes have the property that, for the zero Doppler case, the sum of the time-coincident autocorrelation functions adds up to zero, except at the match point. The sidelobes increase for Doppler-shifted signals, and thereby also limit the usefulness of complementary waveforms to low Doppler applications.

The Huffman and complementary codes are not commonly used with radar, so they will not be considered further in this text.

2.3 IMPLEMENTATION TECHNIQUES

2.3.1 Binary Phase Codes

Binary phase codes can be implemented in analog or digital format because of the availability of clocked analog (*bucket brigades or charge-coupled devices*) or digital tapped delay lines. On transmission, the desired code is normally read out of a read-only memory (ROM) and is used to modulate the carrier to be transmitted. On reception, the echo signals are I and Q detected to eliminate blind phases and to maximize the signal to thermal noise ratio. The I and Q video is then used directly or converted to digital format and sent to analog or digital matched filters, such as the one shown in Fig. 2–7, for a four-element Barker code.

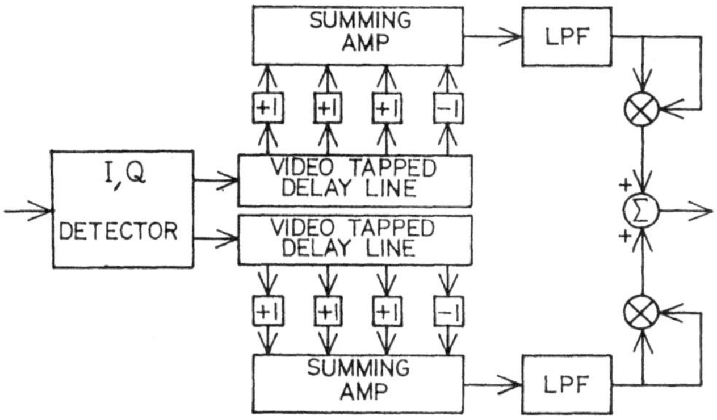

Fig. 2–7 Bipolar Video Matched Filter for Binary Phase-Coded Waveforms (LPF = Low-Pass Filter, ⊗ = Multiply).

To be the equivalent of a linear detector, the square root of the output of Fig. 2–7 can be taken to provide $(I^2 + Q^2)^{1/2}$. The device illustrated in Fig. 2–7 is called a *zero intermediate frequency* (dc − IF), or *homodyne* configuration, where bipolar video is fed into each delay line. For a digital implementation, analog-to-digital (A/D) converters are placed after the I and Q detectors, the video tapped delay lines are replaced with cascaded shift registers, the summing amplifiers become digital adders, and the low-pass filters are eliminated.

2.3.2 Step Frequency Modulation

Step frequency modulation pulse expander-compressors are implemented in analog form, as illustrated by Fig. 2–8, for the case where $\rho = 16$.

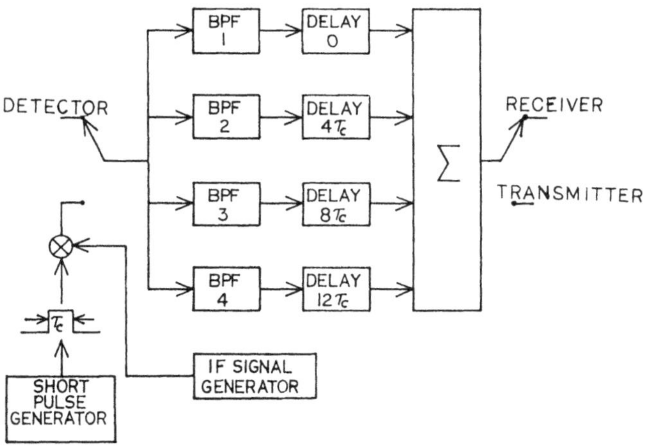

Fig. 2–8 Step Frequency Modulation Pulse Expander-Compressor; (BPF = Bandpass Filter, ⊗ = Multiply).

A short pulse of length τ_c equal to the desired compressed pulse length is used for pulse modulation of a carrier signal having a frequency which is many times the reciprocal of τ_c. The spectrum of this pulse modulated carrier is of the form of $(\sin x)/x$ with peak-to-null spacing $1/\tau_c$ as illustrated in Fig. 2–4, but the spectrum is symmetrical about the carrier frequency. This spectrum excites the bandpass filters shown in Fig. 2–8, the bandwidths of which are designed to be one-quarter of $1/\tau_c$, or one-quarter of the spectral peak-to-null spacing, which is also close to the spectral peak 3 dB bandwidth. The center frequencies of the four filters are designed to cause them to span the 3 dB bandwidth of the spectral peak. Filter 1 has the lowest center frequency, filter 2 has the next lowest, *et cetera*. Filter 2's upper 3 dB point and filter 3's lower 3 dB point are placed at the carrier frequency in this example, which uses four frequencies.

Time delays are placed at the outputs of the filters to make them contiguous in time, i.e., filter 1's output exits first, then filter 2's, then filter 3's, *et cetera*. These sequential outputs can be thought of as four spectral bands of the desired pulse transmitted sequentially. The outputs from the adder pass through a transmit-receive (TR) switch, are mixed (beat) up to a radio frequency (RF) carrier $f_0 + f_{IF}$, are power amplified and transmitted. Echoes are mixed down with a local oscillator frequency $f_0 + 2f_{IF}$ to invert the chirp and pass through a TR switch to the compressor, where the four spectral bands are made simultaneous in time so that they reform the desired short pulse.

The transmitted pulse length from the system illustrated in Fig. 2–8 will be four times the time constant of each filter $(4\tau_c)$ or $16\tau_c$. Thus, with a compressed pulse length τ_c, the pulse compression ratio is $N^2 = \rho$, where N is the number of frequencies implemented with bandpass filters.

2.3.3 Linear Frequency Modulation

Linear frequency modulation pulse expander-compressors are implemented in analog form using dispersive delay lines, as illustrated by Fig. 2–9. The principle of operation is as given in the following paragraphs.

A short pulse of length τ_c equal to the desired compressed pulse length is generated to start the radar transmission procedure. This pulse multiplies an intermediate frequency (IF) signal. The product exiting this multiplier passes through a TR switch and enters a dispersive delay line, which applies differential delay to the different spectral components f_s of the pulse modulated IF carrier. Assuming that the spectral peak is filtered to a 3 dB bandwidth $B_t = 1/\tau_c$, the pulse exiting the delay line will have a length $\tau_c + kB_t$, where k is the dispersion coefficient in seconds per cycle per second. Because kB_t is usually much greater than τ_c, the dispersive delay line expansion of the pulse length is usually taken to be kB_t.

The dispersive delay line output is a *time-varying frequency*, which changes linearly from a low to a high frequency, as illustrated by the diagram above the power amplifier in Fig. 2–9. This time-varying frequency, centered on f_{IF}, is *heterodyned* (beat) with an RF f_0 in a mixer to translate the center frequency of

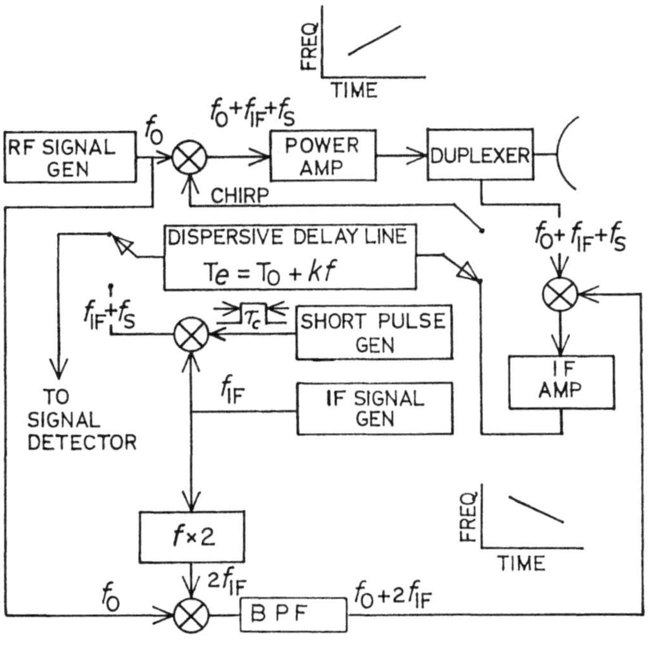

Fig. 2–9 Analog Linear Frequency Modulation Pulse Expander-Compressor (BPF = Bandpass Filter, ⊗ = Multiply).

the pulse up to $f_0 + f_{IF}$ (radio frequency). The output of the mixer enters a power amplifier and a duplexer, and is then radiated from a radar antenna. Echoes from reflectors in space re-enter the radar antenna and duplexer, and are then sent to a receiver mixer, where they are subtracted in frequency from $f_0 + 2f_{IF}$ to invert the frequency sweep or chirp direction of the echo pulses. The output of this chirp-inverting mixer is amplified and passes through the dispersive delay line, where the dispersion introduced on transmission is removed as a result of the chirp inversion. The compressed pulse exiting the dispersive delay line passes through a TR switch and is sent to the radar signal detector.

In this system, the transmitted pulse length is $\tau_t = kB_t + \tau_c$ and the compressed pulse is τ_c, therefore, the pulse compression ratio:

$$\rho = \tau_t/\tau_c \approx kB_t^2 = B_t\tau_t \tag{2–22}$$

Practical problems limit the useful pulse compression ratios that can be obtained from analog dispersive delay lines to something on the order of 10^3. These problems include attenuation of the signal in the delay line and reflections at mismatches in the circuitry.

2.3.4 Frank Polyphase Code

The Frank polyphase code elements actually consist of the conjugates of the steering weights used in a fast Fourier transform (FFT) circuit. This code can be digitally produced and compressed by using the logic illustrated in Fig. 2–10, which includes a four-point FFT [10].

Note that in this figure clocking pulses are not shown, and I and Q processing is assumed but not drawn. These simplifications were made so that the principles of operation are more obvious.

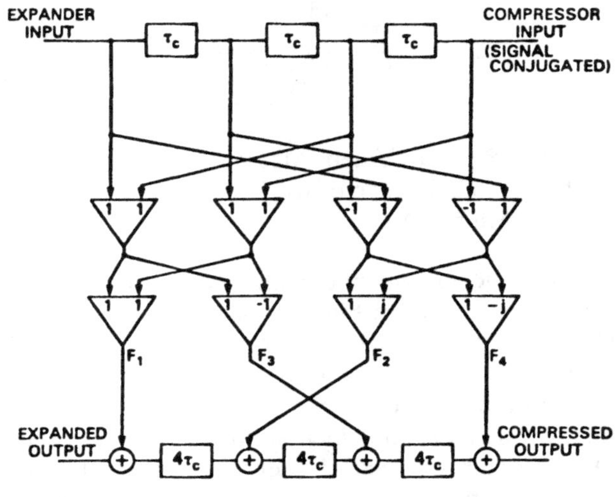

Fig. 2–10 Frank Polyphase Code Expander-Compressor for a Pulse Compression Ratio of 16.

To generate the code element phases for the expanded pulse, a unit-amplitude digital word is clocked into the expander input. This word is stored in a shift register for a clock interval τ_c and enters an input of two adders, shown as triangles after being multiplied by two + 1 operators. The outputs of these two adders each enter two more adders after being multiplied by + 1 operators. The outputs of these four adders (the FFT frequency port outputs F_1, F_2, *et cetera*) enter adders represented by a circle with a plus sign inside. On expansion, the outputs of the circle adders are assumed to exit to the left. The output of F_1 exits from the output of the circuit, while each of the outputs of F_2, F_3, and F_4 enter four cascaded shift registers that introduce delays of $4\tau_c$, as illustrated. This process continues as the input pulse shifts from tap to tap on the input tapped delay line. The first four outputs of F_1 at the expanded output are unit-amplitude, zero-phase digital words. The next four words come from the F_2 output through the $4\tau_c$ delay. These words have unit amplitude and phases of 0° (represented by an I of +1 and a Q of 0), 90° (represented by an I of 0 and a Q of +1, which denotes the multiplication of a unit-amplitude, zero-phase complex number by j), 180° (an I of −1 and a Q of 0), and 270° (an I of 0 and a Q of −1). The four words out of the F_3 port exit next, with unit amplitudes and phases of 0°, 180°, 0°, and 180°. F_4 exits last with phases of 0°, 270°, 180°, and 90° to provide $N^2 = 16$ code-element phases for transmission.

The I and Q values of these digital words are converted to analog data in a *digital-to-analog* (D/A) *converter* and are used in circuits like that of Fig. 1–2 to generate the analog transmission signal.

Frank-coded echoes from a radar are I and Q detected, conjugated (sign of Q reversed), converted to digital data, and sent to the compressor input of Fig. 2–10. Because F_1 was transmitted first, it will return first as four digital words of the same phase. When the first four code elements of an echo pulse have returned, they will exit from the F_1 port as a pulse that is four-code-element amplitudes high. At this time, it will exit from the other three ports with zero amplitude. After the next four code elements return, a word with an amplitude of four code elements will exit from F_2 and add to the four-code-element amplitude word out of the $4\tau_c$ delay from F_1. This eight-code-element amplitude word will enter the next $4\tau_c$ delay. After 12 code elements return, F_3 will peak and add to the eight-code-element output of the $4\tau_c$ delay from F_2 to produce a word that is 12 code elements in amplitude. After 16 code elements return, the 16-code-element high compressed pulse exits from the output due to the addition of four code elements from F_4 and 12 from the sums of F_1, F_2, and F_3. The desired compressed pulse can be used in digital format or converted to analog format by an A/D converter.

2.3.5 P1 Code

We have found that the intolerance of the Frank code to bandwidth limitations was due to the fact that the code groups having the largest phase changes from code element to code element were in the center of the uncompressed pulse. This caused bandwidth limitations in a radar receiver to attenuate the center of the waveform more than the ends, which adversely affected the peak to range-time sidelobe ratio obtainable in the compressed pulse.

This suggested that the Frank code groups representing the N frequencies be rearranged in time order of transmission to place the lowest phase increments from code element to code element in the center of the waveform. This triggered the invention of the P1 code, the code-element phases of which are given in degrees by

$$\phi_{i\,j} = -(180°/N)[N - (2j - 1)] \times [(j - 1)N + (i - 1)] \qquad (2\text{–}23)$$

Note that when N is odd, the P1 code is the Frank code rearranged to have conjugate symmetry about the dc term.

Implementation of the P1 code for N odd simply requires changing the taps on the output tapped delay line, which delays the successive frequency groups. Implementation for N even is illustrated in Fig. 2–11 [10].

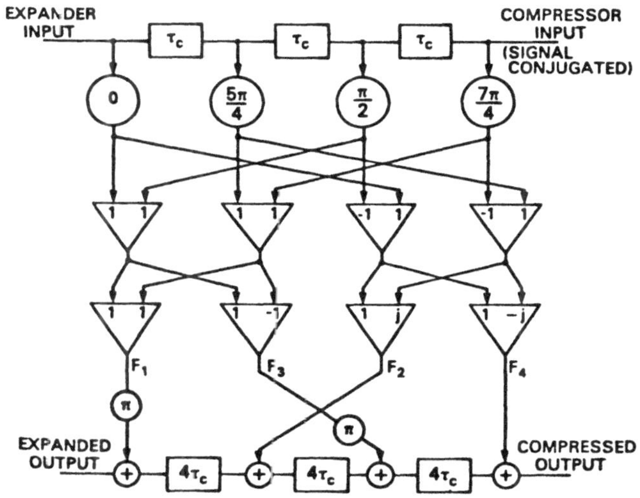

Fig. 2–11 P1 Code Expander-Compressor for a Pulse Compression Ratio of 16.

2.3.6 P3 Code

The P3 [14] code can be implemented in a pulse expander-compressor employing digital fast Fourier transform circuits similar to those discussed for the Frank and P1 codes. These expander-compressors take advantage of the fact that the P3 code only differs from the Frank code by 180° phase shifts every $\rho^{1/2}$ code elements (one frequency group) and by added phase increments that repeat every $\rho^{1/2}$ samples (every frequency group). These added phase shifts are caused by the linear frequency shift during the time that the equivalent Frank code frequency is constant. This difference between the P3 and Frank codes is illustrated in Tables 2–1 to 2–4 for $\rho = 16$. The extra phase increments can be added to the $\rho^{1/2}$ individual time samples in the FFT and their accumulated value of 180° during the time that the Frank-code frequency exists can be added to the frequency port outputs of the FFT as illustrated in Fig. 2–12. (Note that even multiples of 180° need not be added because they are equivalent to zero phase shifts.)

0	$\frac{\pi}{16}$	$\frac{4\pi}{16}$	$\frac{9\pi}{16}$
π	$\frac{25\pi}{16}$	$\frac{4\pi}{16}$	$\frac{17\pi}{16}$
0	$\frac{17\pi}{16}$	$\frac{4\pi}{16}$	$\frac{25\pi}{16}$
π	$\frac{9\pi}{16}$	$\frac{4\pi}{16}$	$\frac{\pi}{16}$

Table 2–1 P3 Code Matrix, $\rho = 16$.

0	$\frac{\pi}{16}$	$\frac{4\pi}{16}$	$\frac{9\pi}{16}$
0	$\frac{9\pi}{16}$	$\frac{20\pi}{16}$	$\frac{\pi}{16}$
0	$\frac{17\pi}{16}$	$\frac{4\pi}{16}$	$\frac{25\pi}{16}$
0	$\frac{25\pi}{16}$	$\frac{20\pi}{16}$	$\frac{17\pi}{16}$

Table 2–2 Adding π to Every Other Row of Table 2–1 and Subtracting Even Multiples of 2π.

0	0	0	0
0	$\frac{8\pi}{16}$	$\frac{16\pi}{16}$	$\frac{24\pi}{16}$
0	$\frac{16\pi}{16}$	0	$\frac{16\pi}{16}$
0	$\frac{24\pi}{16}$	$\frac{16\pi}{16}$	$\frac{8\pi}{16}$

Table 2–3 Frank Code Matrix, $\sigma = 16$.

0	$\frac{\pi}{16}$	$\frac{4\pi}{16}$	$\frac{9\pi}{16}$
0	$\frac{\pi}{16}$	$\frac{4\pi}{16}$	$\frac{9\pi}{16}$
0	$\frac{\pi}{16}$	$\frac{4\pi}{16}$	$\frac{9\pi}{16}$
0	$\frac{\pi}{16}$	$\frac{4\pi}{16}$	$\frac{9\pi}{16}$

Table 2–4 Subtracting Frank Code Phases (Table 2–3) from Phases in Table 2–2.

2.3.7 P4 Code

The P4 code [14] can be generated and compressed by using the circuit illustrated in Fig. 2–13.

2.3.8 P2 Code

The P2 [9] code implementation for $\rho = 16$ is illustrated in Fig. 2–14. This code can also be implemented by a modified FFT.

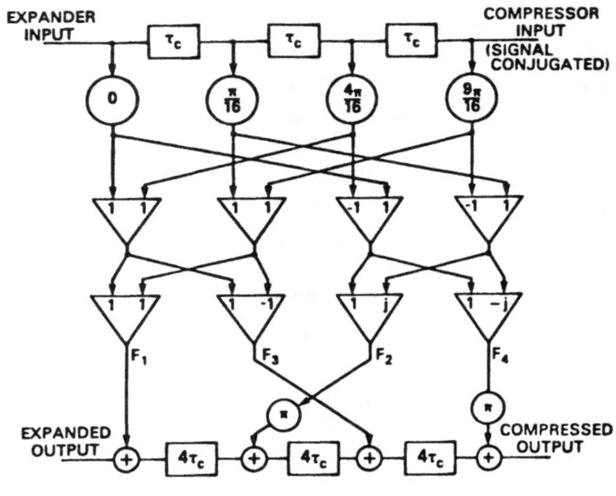

Fig. 2–12 P3 Code Expander-Compressor for a Pulse Compression Ratio of 16.

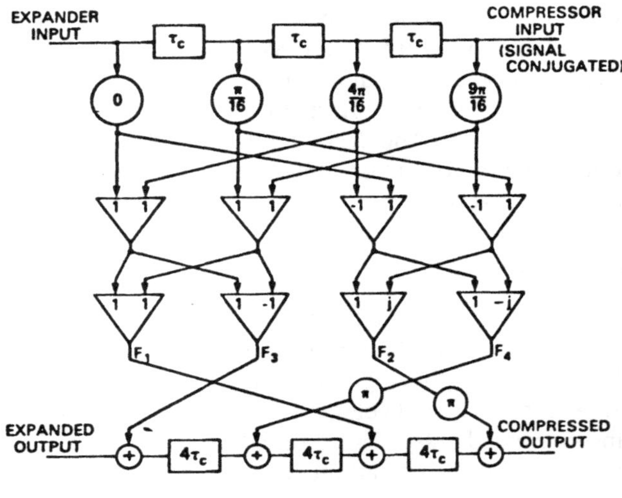

Fig. 2–13 P4 Code Expander-Compressor for a Pulse Compression Ratio of 16.

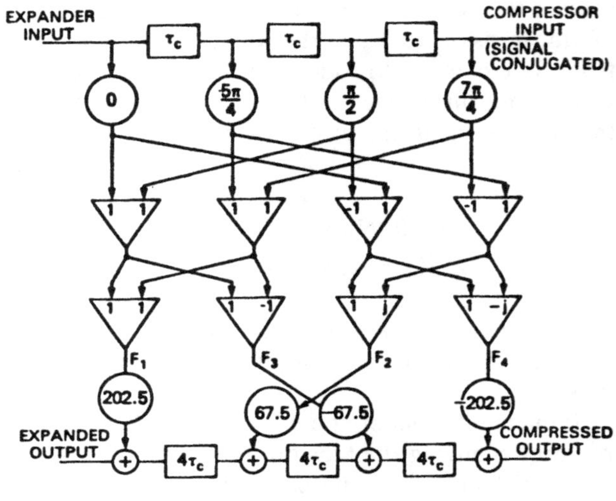

Fig. 2–14 P2 Code Expander-Compressor for a Pulse Compression Ratio of 16.

2.3.9 Alternative Implementation Techniques

For digital pulse compression waveforms having a large time-bandwidth product, other implementation techniques such as "fast convolution" [23] become useful. These methods make use of the fact that convolution in the time domain is equivalent to multiplication in the frequency domain. A limitation of time-domain processing for large time-bandwidth codes is that all multiplications and additions must be performed for each new input sample occurring every τ_c seconds.

2.4 COMPRESSED PULSE CHARACTERISTICS

2.4.1 Barker Codes

The Barker binary phase codes have maximum range-time sidelobes which are only one-code-element amplitude high. The match-point peak to maximum sidelobe power ratio for an N-element code is thus equal to N^2. Figure 2–15 is an illustration of the response of a 13-element Barker code in which a computer-drawn straight line connects the sampling points. The sample number corresponds to range cells, or time, in units of the reciprocal of the bandwidth of the waveform.

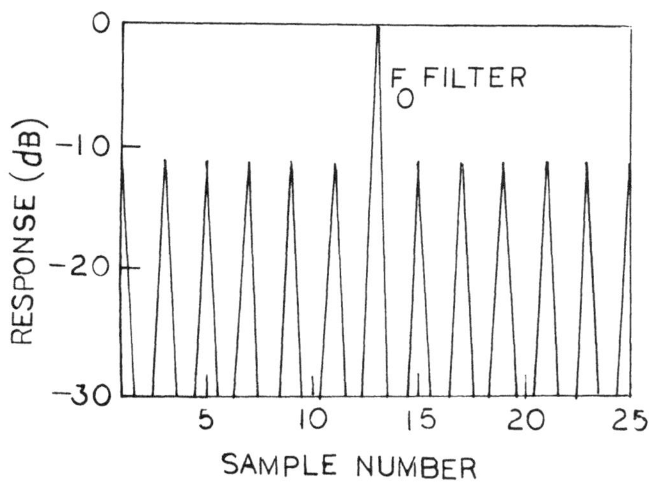

Fig. 2–15 Thirteen-Element Barker Code Autocorrelation Function.

2.4.2 Random and Pseudorandom Binary Phase Codes

The random and maximum-length sequence codes can produce compressed pulses with peak time sidelobes down in power by a factor on the order of 2ρ from the match-point peak. They can also produce mean square sidelobes down by about 10ρ from the match-point peak amplitude squared. The compressed pulse time function of a 127-element binary shift register code is illustrated in Fig. 2–16.

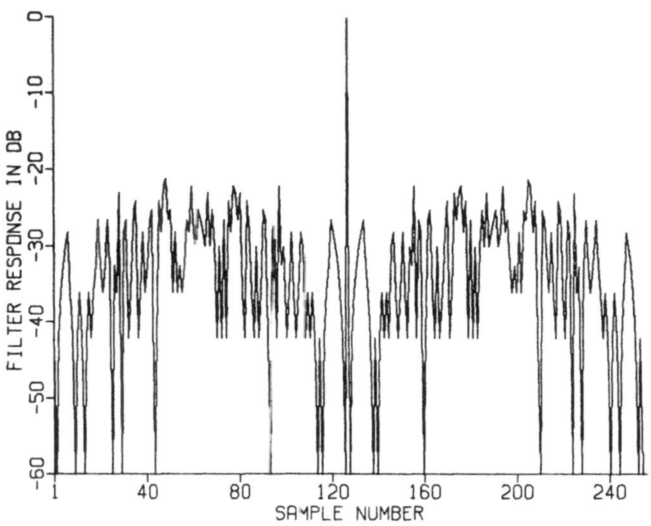

Fig. 2–16 Compressed 127-Element Shift Register Coded Waveform.

2.4.3 Step Frequency and Linear Frequency Modulation

The step frequency and linear frequency modulation waveforms compress to an approximately $(\sin x)/x$ time function in the vicinity of the main response with approximately −13 dB of peak sidelobes. These sidelobes can be reduced by amplitude weighting of the waveform prior to compression.

2.4.4 Nonlinear Frequency Modulation

The compressed pulse of the nonlinear frequency modulation waveform has much lower sidelobes than the linear frequency modulation waveform and does not normally require any weighting. The Doppler response, however, is not as good as the preceding codes.

2.4.5 Weighting Techniques

Both the step-chirp and chirp waveforms have spectra which are approximately rectangular. As the time-bandwidth product increases, the approximation becomes better. For this situation, the compressed pulse is determined by the Fourier transform of the weighting function. Much work has been performed in this area and the results are similar to the weighting used with antenna arrays. The common weightings include the Hamming, Hanning, and Chebyschev weightings [3, 4, 7, 24]. While the minimum pulsewidth is achieved with uniform weighting (corresponding to a maximum gain antenna), the sidelobes are objectionably high. The consequence of any weighting scheme is to broaden the pulsewidth and to cause a loss in the signal-to-noise ratio of approximately 1.25 dB. In digital processing using FFTs to form a Doppler filter bank, a weighting is also applied to reduce the sidelobes of the uniformly weighted filters.

2.4.6 Frank Polyphase Code

The compressed pulse of a 121-element Frank code is illustrated in Fig. 2–17.

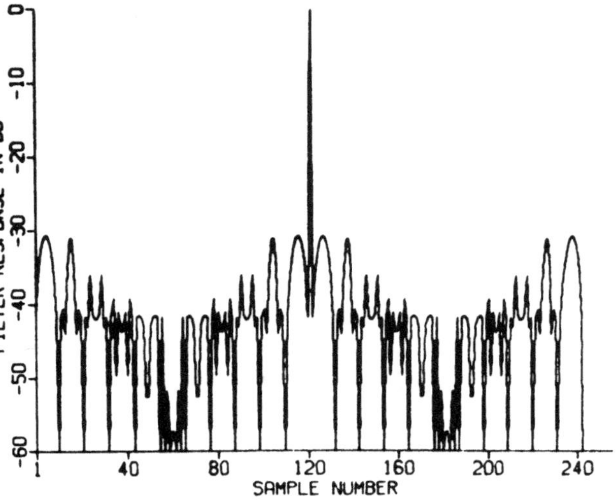

Fig. 2–17 Compressed Pulse of 121-Element Frank-Coded Waveform.

It should be noted that the highest sidelobe peaks are $\pi^2\rho$ down from the match-point peak. In the case of Fig. 2–17, this value is 1210, or 31 dB. The highest sidelobes of the Frank code are those farthest from and closest to the match-point peak. The magnitude and structure of the largest sidelobes will be explained below.

From (2–6), we find that the code group with zero-phase increments between code elements is transmitted first and returns first. On compression, this code group is first sent to the filter section, with the largest phase increments from code element to code element in time-reversed order. These largest phase increments in the case of $\rho = 100$, $N = 10$ are given by (2–6) with $j = 10$ and $i = 2$ as (36°) (9) or 324° from code element to code element. Being greater than 180°, this phase increment is ambiguous and appears as $(-1/10)$ times 360°. This ambiguity makes the phasors in the highest frequency code group rotate clockwise through 360°, from beginning to end of the code group. It also steers the returning code elements of the dc term of the Frank code as illustrated in Fig. 2–18.

When all 10 code elements of the dc term return, the vector sum of the phasors is zero, because the phasors form an approximate circle with a perimeter $N = 10$ code elements long. Because the resultant of the phasor addition is maximum when the phasors form a half circle, we can see that, in general, for N code elements this maximum will be the diameter $D = N/\pi$. Because N code groups, each having N code elements, add to form the match-point peak, the peak

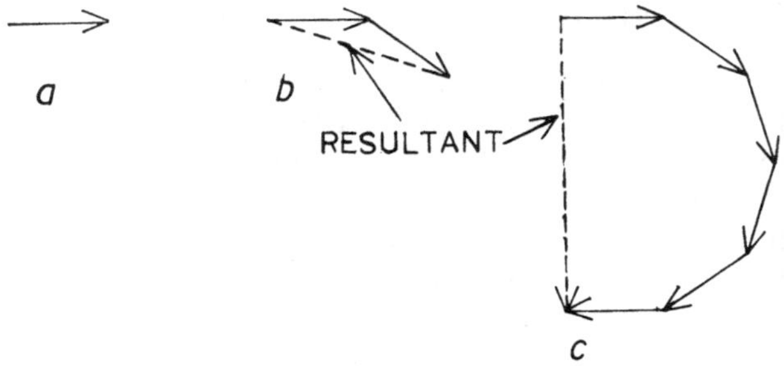

Fig. 2–18 Frank-Coded Waveform Compressor Output After the First (a) One, (b) Two, and (c) Six Code Elements of 100 Elements Return.

squared to peak sidelobe squared ratio will be

$$\frac{(\text{peak})^2}{(\text{peak sidelobes})^2} = \frac{(N^2)^2}{\left(\frac{N}{\pi}\right)^2} = \pi^2 N^2 = \pi^2 \rho \qquad (2\text{–}24)$$

For large ρ, the Frank polyphase code produces mean square sidelobes that are down on the order of $10\pi^2 N^2 = 10\pi^2\rho$ from the match-point peak squared.

The structure of the sidelobes other than the highest sidelobe is a function of whether N is even or odd in the Frank code, as can be seen by comparing Figs. 2–17 and 2–19.

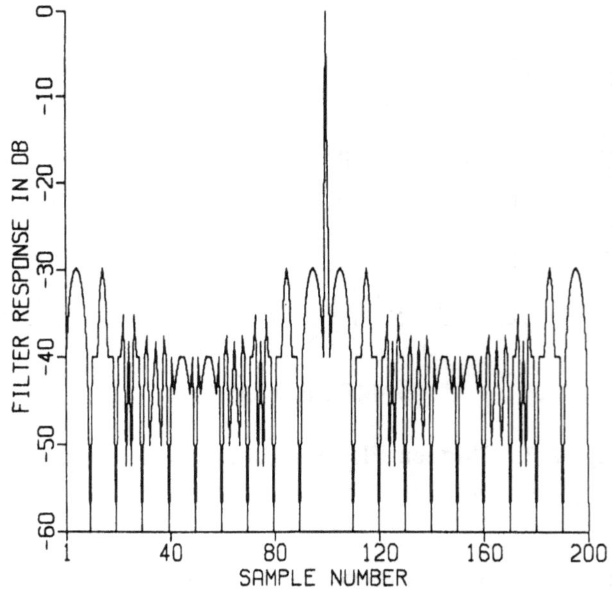

Fig. 2–19 Compressed Pulse of 100-Element Frank-Coded Waveform.

2.4.7 P1 And P2 Codes

The compressed pulses of the P1 and P2 codes are very similar to those of the Frank code with zero Doppler and no bandwidth limitations.

2.4.8 P3 And P4 Codes

The compressed pulses of the P3 and P4 codes are the same for the case of zero Doppler and no bandwidth limitations. Figure 2–20 is an example with $\rho = 100$. However, there are significant differences between the P3 or P4 code compressed pulses and that of a Frank code. These differences are evident in the sidelobes. The most significant difference is in the peak sidelobes with those of the P3 and P4 codes being on the order of 3 dB higher than the Frank code.

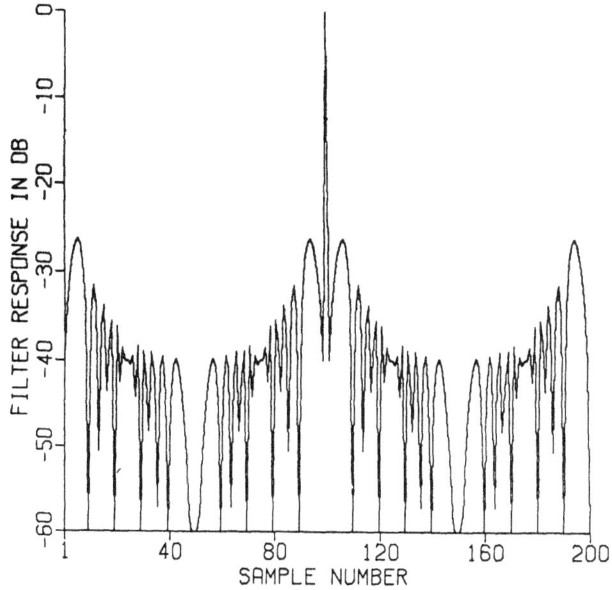

Fig. 2–20 Compressed Pulse of 100-Element P3 or P4 Coded Waveform.

Another difference is that the mean square sidelobe to match-point peak squared ratio is about 3 dB higher in the P3 and P4 codes than in the Frank code. The standard weightings can be applied to these codes to reduce the sidelobe level.

2.5 EFFECTS OF PROCESSING ERRORS [11]

Computer simulations have been performed to determine the sensitivity of the polyphase codes to phase and amplitude errors. The two types of errors considered were random errors in I and Q, and quantization errors in I and Q, which are encountered in A/D conversions.

2.5.1 Random Errors

Two types of random errors were considered as shown in Fig. 2–21. In each case, independent and uniformly distributed errors in I and Q were generated over an interval $\pm x$. For the first type shown in Fig. 2–21(a), the error ϵ was determined by letting x be a given percentage of the nominal I and Q value for each code element phasor. The resulting vector is denoted by E_R. The other type of error ϵ', shown in Fig. 2–21(b), was generated as explained above, except that x was specified as a fixed error rather than a given percentage of I or Q. In this case, the resulting vector is denoted by E_A. In determining E_R and E_A, the nominal signal amplitude is assumed to be unity.

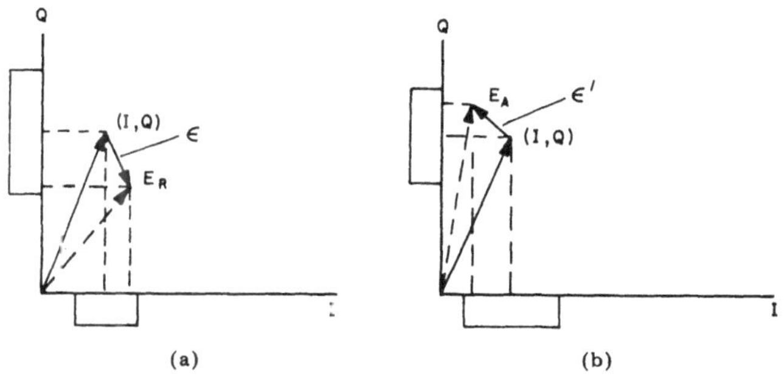

Fig. 2–21 Random Errors.

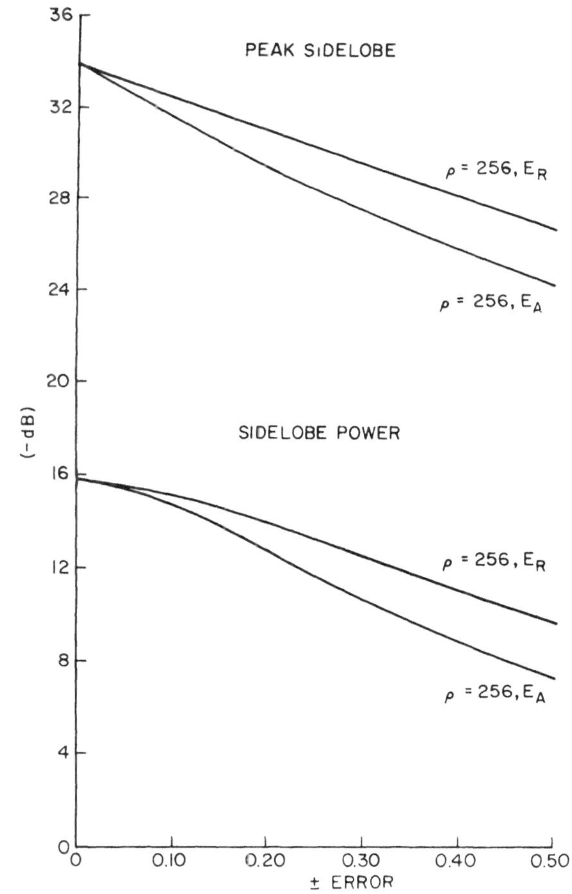

Fig. 2–22 Frank Polyphase Code Error Analysis.

Monte Carlo simulations were performed to determine the effect of the relative and absolute errors on the peak and average sidelobes of Frank codes with pulse compression ratios of 256 and 64. The results for $\rho = 256$ are shown in Fig. 2–22 with similar results obtained for $\rho = 64$. Each point, other than for zero error, was obtained by taking an average of 100 compressed pulses to compute the indicated peak sidelobe and average sidelobe levels. The errors were assumed to occur on either transmission or reception, but not both. The results of this simulation indicate that the sidelobes are not sensitive to the errors. For example, for the absolute error case, an error distribution of ± 0.10 results in an approximately 2 dB average degradation in the peak sidelobe and 1.2 dB degradation in the average sidelobe power.

2.5.2 Quantization Errors

The results of the Monte Carlo simulations previously described indicate the robustness of the polyphase codes with respect to random errors. Computer simulations were performed to quantify the effects of quantization errors. The average and peak sidelobes were determined for a symmetric A/D characteristic having the phase and amplitude specified within the accuracy of the quantization levels determined by the number of bits (including sign). Compression ratios of 144 and 36 were considered in the simulations, which did not include noise. It was assumed that the errors were due only to the A/D converters and that the phases and amplitudes of the matched filter were perfect.

The results are shown in Fig. 2–23, where each curve exhibits a knee. The knee location is seen to vary the most between the $\rho = 36$ and $\rho = 144$ peak sidelobes. The general conclusion reached from these results is that the polyphase code is relatively insensitive to the number of bits beyond a certain number. Other considerations, such as dynamic range, may dictate the use of more bits than Fig. 2–23 indicates.

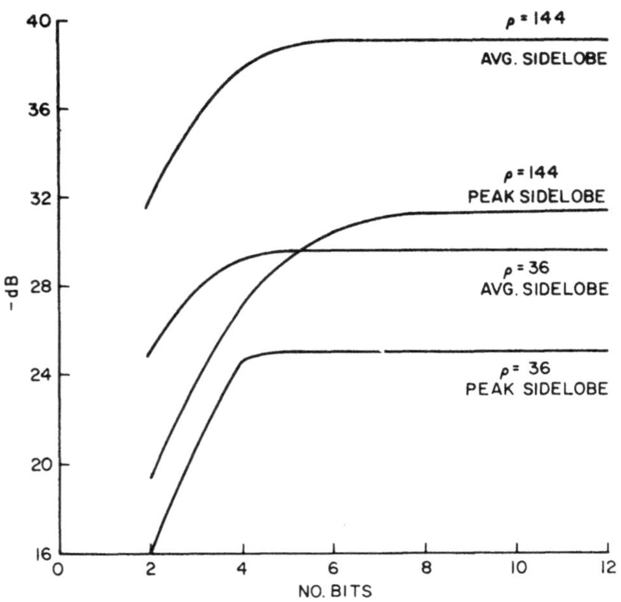

Fig. 2–23 Frank Polyphase Code A/D Quantitization Errors.

2.6 EFFECTS OF DOPPLER

2.6.1 Barker Codes

The Barker binary phase codes are so short for typical search radar bandwidths that they are not greatly affected by Doppler, unless they are lengthened by cascading codes or the radar carrier frequency is very high. However, if the Doppler produces a 360° phase shift across the uncompressed pulse, the code becomes completely mismatched to the filter that was designed to compress it. Figure 2–24 illustrates the response of a matched filter to a 13-element Barker coded signal having a Doppler shift that produces a 360° phase shift across the uncompressed pulse [25].

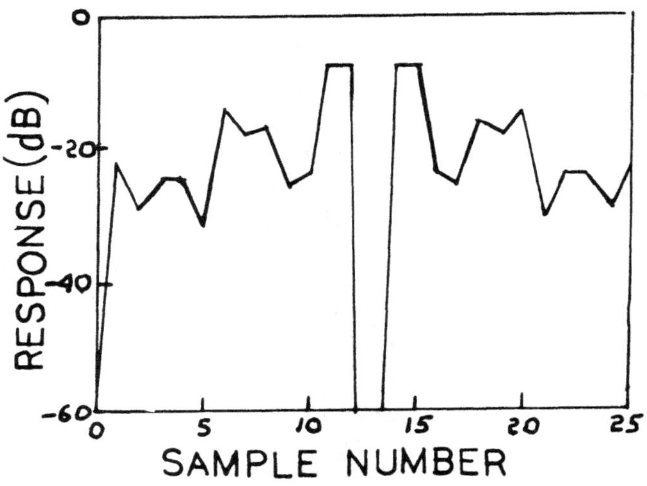

Fig. 2–24 Thirteen-Element Barker-Coded Filter Response to a Doppler Mismatched Signal.

In order to produce a 360° phase shift across an uncompressed pulse of length τ_t, the Doppler frequency shift f_d must be such that

$$2\pi f_d \tau_t = 2\pi, \text{ radians} \tag{2–25}$$

where

$$f_d = 2V/\lambda \tag{2–26}$$

where V is the target radial velocity and λ is the wavelength transmitted by the radar.

From (2–25) and (2–26), we have

$$V = \lambda/2\tau_t \tag{2–27}$$

For $\tau_t = 13 \times 10^{-6}$s and $\lambda = 0.1$ m, (2–27) yields

$$V = 3846 \text{ m/s, or about Mach 10} \tag{2–28}$$

which is faster than most expected targets.

2.6.2 Random and Pseudorandom Binary Phase Codes

The random and pseudorandom binary phase codes are very sensitive to Doppler. This is why they are said to have a thumbtack ambiguity diagram. A 180° phase shift across the uncompressed pulse reduces the match point peak by about 4 dB. A 360° phase shift across the uncompressed pulse drops the peak to zero. The latter situation is encountered at a target velocity of 393.7 m/s with a wavelength of 0.1 m and an uncompressed pulse length of 127×10^{-6} seconds as indicated by (2–27).

Figure 2–25(a) shows the compressed pulse for a 127-element maximum-length shift register code with zero Doppler, and Fig. 2–25(b) is the result of a 360° phase shift across the uncompressed pulse due to Doppler.

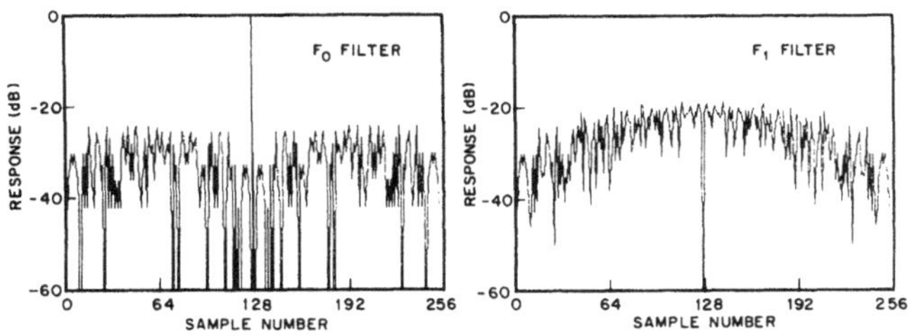

Fig. 2–25 (a) Compressed 127-Element Maximum Length Sequence with Zero Doppler; (b) Doppler-Induced 360° Phase Shift Across the Uncompressed Pulse.

2.6.3 Step Frequency Modulation Code [3–7]

The step frequency modulation code, or step-chirp, is much more tolerant of Doppler than the binary phase codes. The match-point peak of this code does not fall to zero with a 360° Doppler-induced phase shift across the uncompressed pulse. Instead, the match point moves one time (range) resolution cell in a direction determined by the sign of the Doppler with respect to the direction of the frequency modulation on transmission. If the transmitted frequency starts low and becomes higher with time, a positive Doppler shift caused by an incoming target will cause an echo to compress earlier than for a zero Doppler return. This is referred to as *range-Doppler coupling*. For the same transmission, a down, or negative, Doppler shift due to a receding target will cause the echo to compress late or become range-Doppler coupled out in range. The amplitude match-point peak also cycles in magnitude with a decreasing trend as the Doppler magnitude increases.

As Doppler increases to one-half of a frequency step, grating lobes grow in the compressed pulse and the match-point peak drops in amplitude. As the Doppler rises still higher in magnitude, the grating lobes diminish in amplitude and disappear at a Doppler shift equal to one frequency step. This process continues with an increase in Doppler magnitude and is accompanied by a continuous decrease in match-point amplitude (Fig. 2–26).

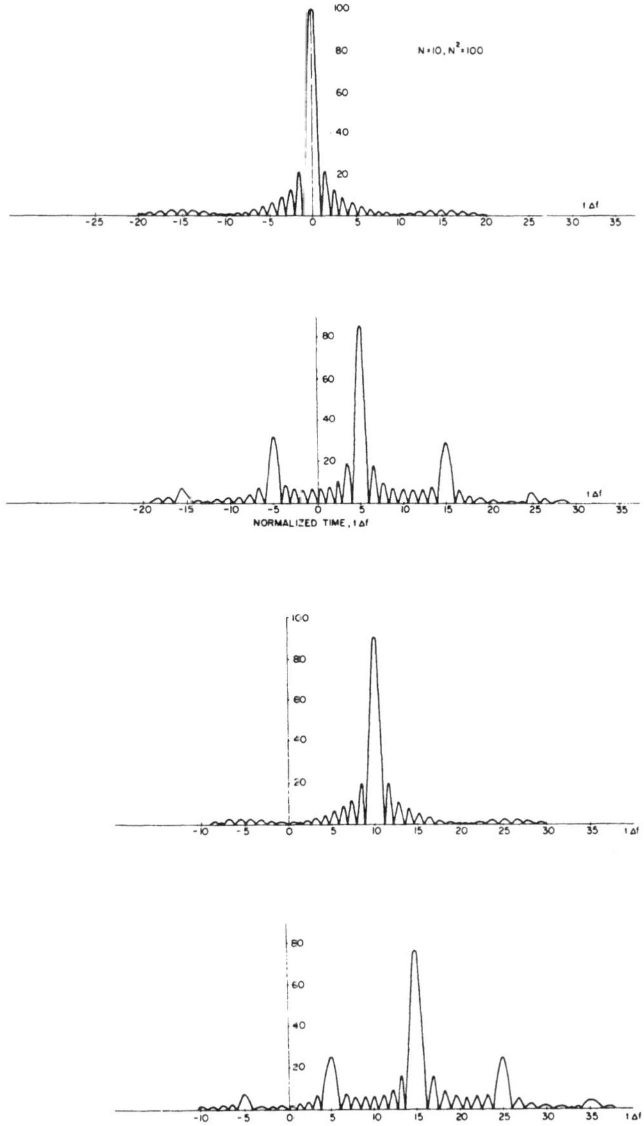

Fig. 2–26 Effect of Doppler on Compressed Pulse of Stepped-Frequency Coded Waveform. (Reprinted with Permission of Academic Press and Dr. C. E. Cook. From *Radar Signals, An Introduction to Theory and Application.*)

The continuous decrease in match-point amplitude is caused by the echo signal gradually being Doppler-shifted out of the radar receiver's passband. The cycling, however, is a grating lobe effect that varies the match-point peak between 0 and about 1.4 dB down at odd multiples of one-half a frequency step. The energy lost from the match-point peak goes into the grating lobes.

At very large Doppler frequency shifts (defined as a significant fraction of the radar bandwidth), the signal bandwidth becomes limited by the radar receiver's passband only passing a fraction of the original signal spectrum, as shown in Fig. 2–27. This bandwidth and signal energy loss causes the compressed pulse to widen in time and to drop in amplitude. The increased pulse length reduces the radar's range resolution and the reduced peak amplitude reduces the peak-to-sidelobe ratio in the compressed pulse as well as the radar's sensitivity.

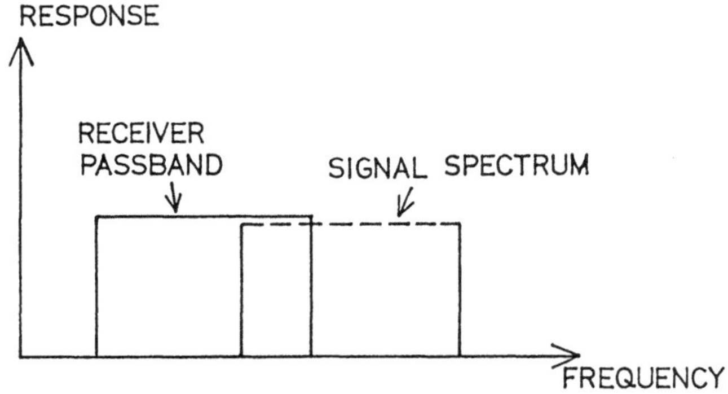

Fig. 2–27 Receiver Passband Limitation of Doppler Shifted Signal Bandwidth.

2.6.4 Linear Frequency Modulation Code

The linear frequency modulation code [1–7], or chirp waveform, is the most Doppler tolerant of any easily implemented codes. The linear code does not produce the grating lobes so evident with the step frequency modulation code, and the envelope of the match-point peak does not cycle, as it does in the case of the latter code. However, it also generates range-Doppler coupling and losses in gain as the signal spectrum shifts out of the radar passband with increasing Doppler magnitude.

2.6.5 Frank, P1, And P2 Polyphase Codes [13]

The Frank, P1, and P2 polyphase codes have the same response to Doppler as the step frequency modulation code in that grating lobes begin to appear with Doppler and maximize every odd multiple of a half frequency step. These waveforms also show the match-point peak cycling with Doppler and the loss in gain as the signal spectrum shifts out of the radar receiver's passband. In addition to these characteristics, the phase codes' match-point peak cycles by about 4 dB for every Doppler-induced 180° phase shift across the uncompressed pulse. This phenomenon was recognized only recently.

Figure 2–28 shows the compressed pulses of these codes with (a) zero Doppler and (b) Doppler-to-bandwidth (B) ratio = 0.05.

Figure 2–29 portrays the Doppler properties of the Frank, P1, and P2 codes.

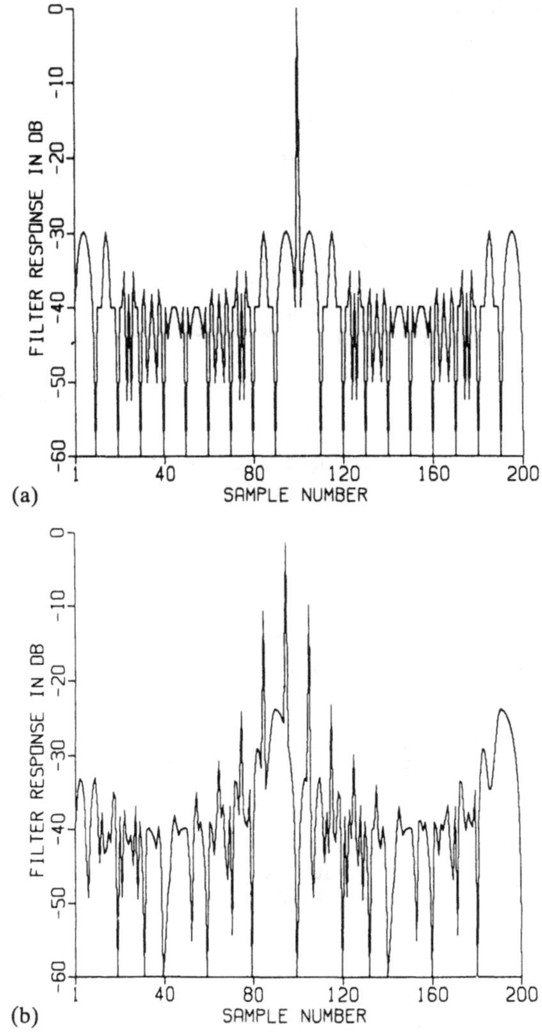

Fig. 2–28 Frank-Coded Waveform Autocorrelation Function with a Pulse Compression Ratio of 100 for (a) Zero Doppler and (b) Doppler 0.05 of the Signal Bandwidth B. *Note:* P1 and P2 Functions are Similar.

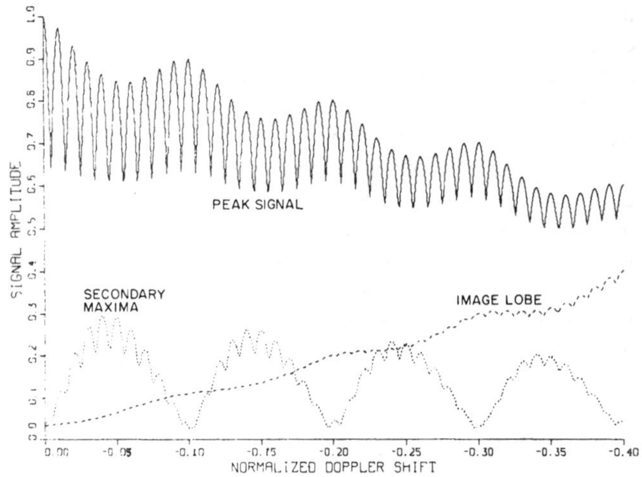

Fig. 2–29 Doppler Characteristics of 100-Element Frank, P1, or P2 Coded Waveform After Compression.

2.6.6 P3 and P4 Codes

The P3 and P4 polyphase codes [13] have the same tolerance to Doppler as the analog linear frequency modulation waveform. This is made evident by lack of grating lobes in Fig. 2–30(a,b).

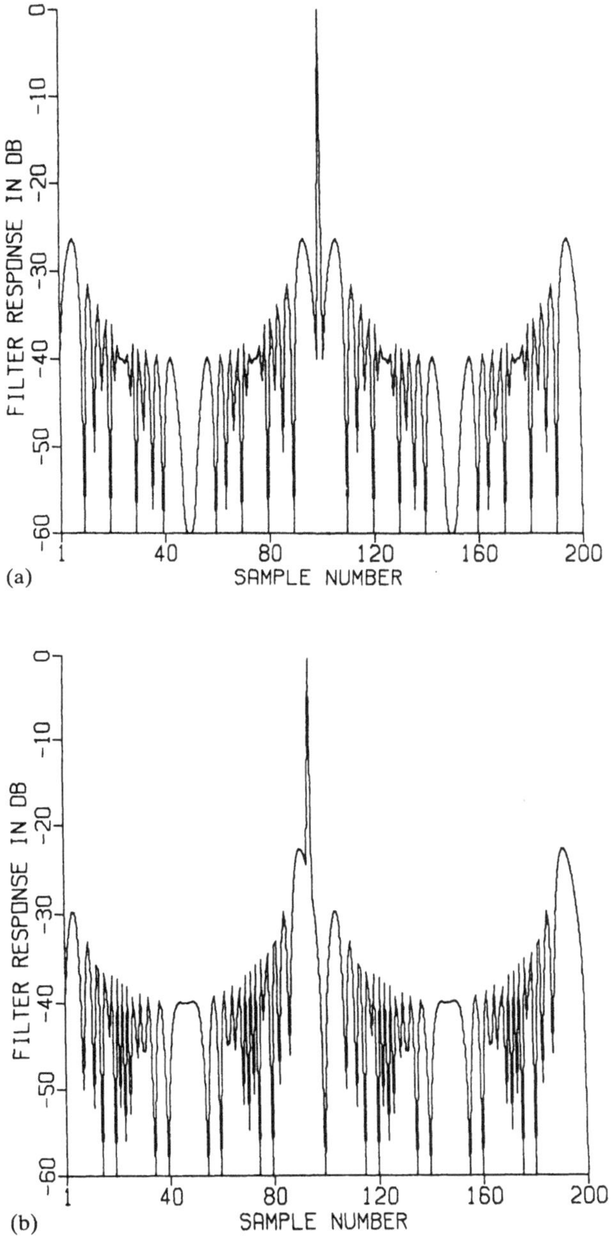

Fig. 2–30 P3 or P4 Autocorrelation Function with a Pulse Compression Ratio of 100 for (a) Zero Doppler and (b) Doppler 0.05 of the Signal Bandwidth B.

However, being discretely sampled these codes show the 4 dB cycling with Doppler shifts, as indicated in Fig. 2–31.

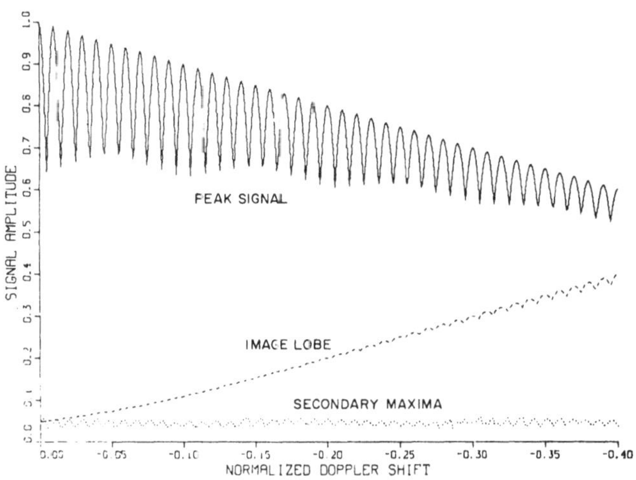

Fig. 2–31 Doppler Properties of Compressed 100-Element P3 or P4 Coded Waveforms.

2.7 EFFECT OF BANDWIDTH LIMITATIONS

2.7.1 Precompression Bandwidth Limitations

Binary Phase Codes

Bandwidth limitations in radar circuitry that is designed to maximize signal-to-noise ratio have a highly detrimental effect on the compressed pulses of binary phase coded signals. These bandwidth limitations prevent the receiver from following the fast phase shifts from 0° to 180° between code elements. The result of bandwidth limitations on these codes appears in the form of increased sidelobes and peak-gain loss.

Step Frequency Modulation and Linear Frequency Modulation Codes

Bandwidth limitations properly applied in the form of weighting as previously described can improve the autocorrelation function of these codes by reducing the time sidelobes. In the process, however, the match-point peak is widened and the peak gain is reduced [1–7].

Frank, P1, and P2 Codes [9–12]

The effect of a limited bandwidth in the IF amplifiers and the I and Q detectors preceding A/D conversion and compression of the phase codes was simulated on a digital computer. The codes to be compressed were oversampled by 5 to 1 and a sliding window average by 5 was taken to simulate the precompression bandwidth limitation. The inputs to the pulse compressor were then taken by using every fifth sample of the oversampled waveform. The particular set of inputs, therefore, is dependent on the starting time of the echo with respect to the sampling time used for the input to the compressor. To take this into consideration, all five possible time relationships were used as the inputs to the compressor and the output time functions were averaged. Figure 2–32 illustrates the resulting time functions for 100-element codes.

Note that the time sidelobes of the P1 and P2 codes drop more than the match-point peak, while the sidelobes of the Frank code are not significantly affected, but the match-point peak drops as much as in the case of the P1 and P2 codes.

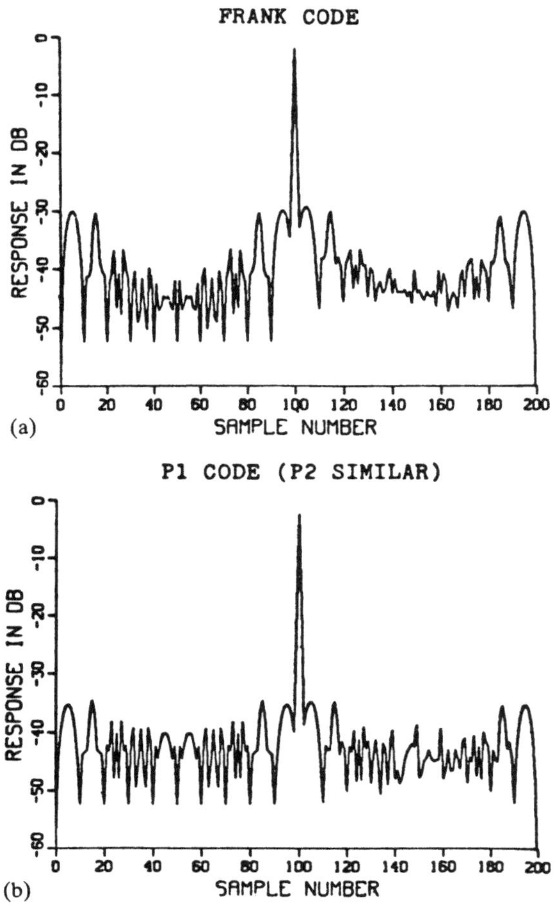

Fig. 2–32 Effect of Precompression Band Limitation on a (a) 100-Element Frank, (b) P1, or P2 Coded Waveform After Compression with Zero Doppler.

P3 and P4 Codes

Precompression band-limiting reduces the match-point peak to peak-sidelobe ratio of the P3 code as expected and improves that of the P4 code compressed pulse. This effect is illustrated in Fig. 2–33.

Precompression Band-Limiting with Doppler

When both precompression band-limiting and Doppler occur together, the bandwidth limitation does not affect the grating lobes that appear in the step-frequency-modulation-derived polyphase codes as illustrated by comparing Fig. 2–34(a) or 2–34(b) with Fig. 2–28(b).

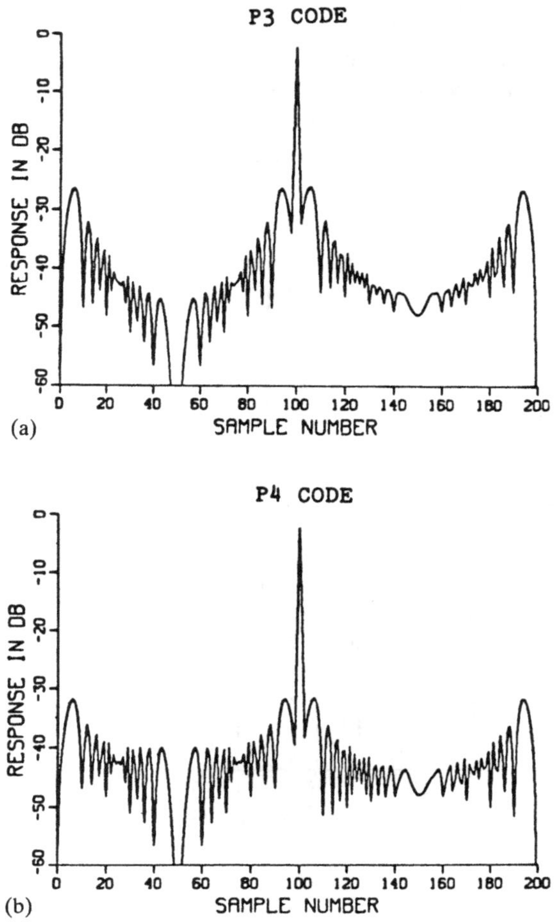

Fig. 2–33 Effect of Precompression Band Limitation on a 100-Element (a) P3 and (b) P4 Coded Waveform After Compression with Zero Doppler.

2.7.2 Postcompression Bandwidth Limitations

Binary Phase Codes

Postcompression band-limiting reduces the signal to thermal noise ratio without significantly increasing the peak to peak sidelobe ratio in compressed binary phase codes.

Step and Linear Frequency Modulation Codes.

Postcompression band-limiting reduces the signal to thermal noise ratio in compressed frequency modulation codes, but it can be used to improve the peak to peak sidelobe ratio.

Polyphase Codes

The effect of postcompression bandwidth limitations on the various compressed polyphase codes [10, 12] is illustrated in Figs. 2–35 and 2–36 for 400-element codes.

It should be noted that postcompression band-limiting via a sliding-window, two-sample addition at the output of the compressor has a significant effect on

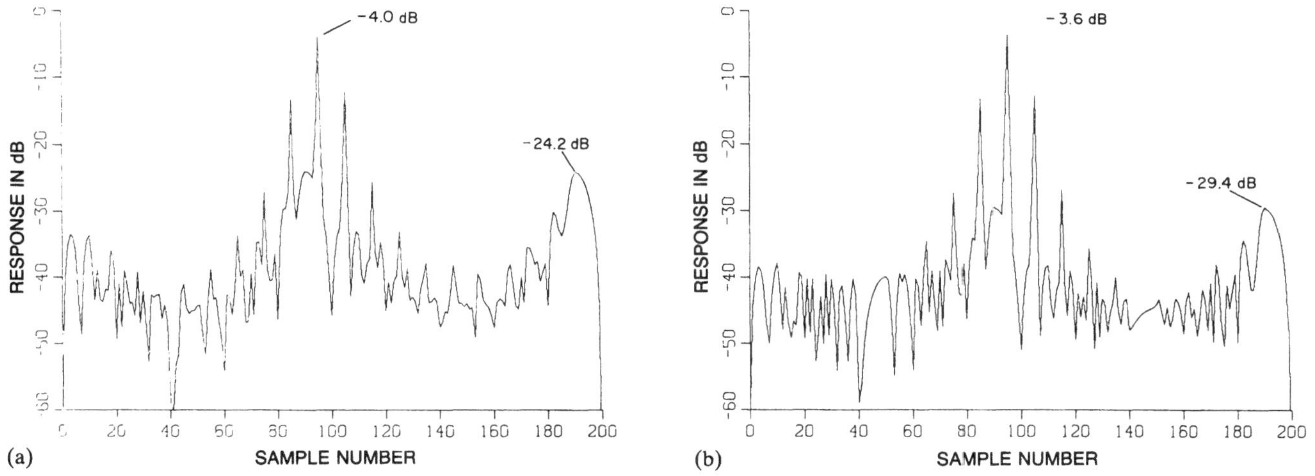

Fig. 2–34 Effect of Precompression Band Limitation on Grating Lobes Produced by Doppler on (a) Frank Code and (b) P1 or P2 Codes.

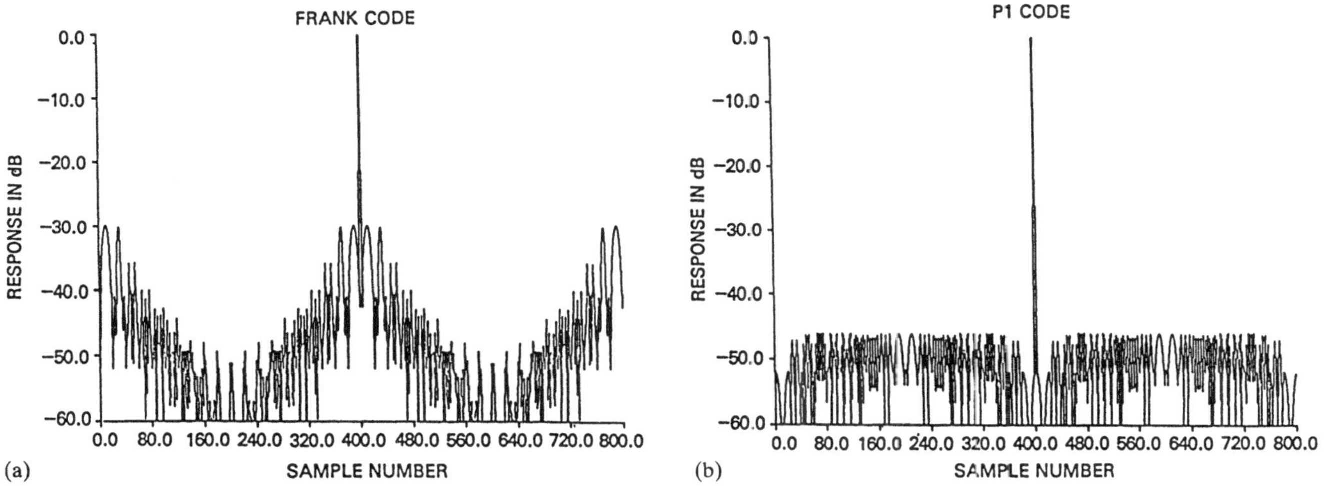

Fig. 2–35 Effect of Two-Sample Sliding-Window Adder on Compressor Output with a Pulse Compression Ratio of 400 for (a) Frank Code and (b) P1 or P2 Code.

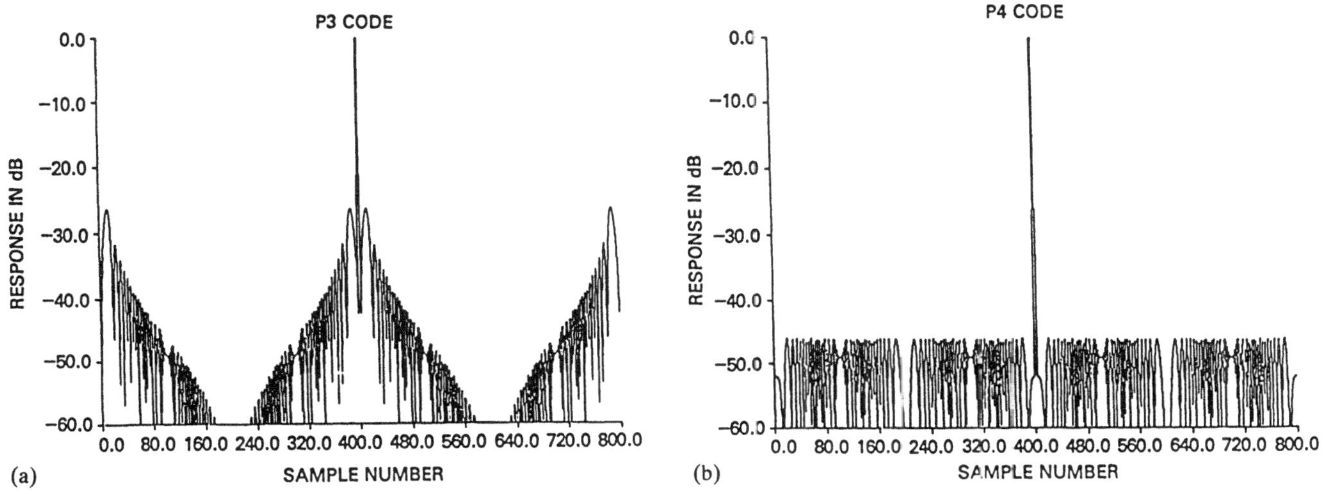

Fig. 2–36 Effect of Two-Sample Sliding-Window Adder on Compressor Output with a Pulse Compression Ratio of 400 for (a) P3 or (b) P4 Code.

the compressed pulses of the polyphase codes. In the case of the Frank and P3 codes, it increases the largest sidelobe peaks by 12 dB. In the case of the P1, P2, and P4 codes, it drastically reduces all sidelobes. However, as with any weighting method achieved by band-limiting, the signal peak is broadened and there is a loss in signal-to-noise (*S/N*) ratio. The available resolution for this weighting has been halved (peak signal is two samples wide) and there is a 3 dB loss in *S/N*, which may be partially recovered by appropriate processing. It is interesting to note that the weighting described above reduces the P1, P2, and P4 code compressed-pulse sidelobes to the Barker-code level, for an effective pulse compression ratio of 200. Ideal codes like the Barker code are defined to have maximum sidelobe peaks with an amplitude of one code element or less. One element out of 200 would yield a −46 dB ratio of the peak sidelobes to match-point peak as demonstrated in Figs. 2–35 and 2–36.

2.8 PERFORMANCE IN DISTRIBUTED CLUTTER

The performance of a pulse compression system in *distributed clutter*, such as rain or chaff, is effected by its mean square sidelobes. Distributed clutter is normally characterized by a mean equivalent scattering cross section σ_c per range and angle resolution cell. The clutter signal entering a radar through the range-time sidelobes of the pulse compressor is then given by [4] as

$$\sigma_e = \sigma_c(2\rho - 2)(\text{mean square sidelobe})/\rho^2 \qquad (2\text{–}29)$$

where σ_e is called the excess clutter, and a clutter degradation L_s is defined to be

$$L_s = 1 + (2\rho - 1)(\text{mean square sidelobe})/\rho^2 \qquad (2\text{–}30)$$

Figure 2–37 compares the mean square sidelobes of the binary and Frank polyphase codes as a function of pulse compression ratio with zero Doppler [13].

Note that the mean square sidelobe to peak squared ratio for the binary phase codes is on the order of 1/10 in Fig. 2–35 with this ratio and $\rho = 100$, L_s from (2–30) becomes 1 + 0.2 or about 1 dB. Also, these values in (2–29) result in $\sigma_e = 0.2\sigma_c$.

2.9 SHORT-RANGE BLOOMING

A phenomenon called *blooming* has been noted in radars that employ either frequency modulation or frequency-modulation-derived polyphase codes having long transmitted pulse lengths. This blooming occurs on targets at ranges short enough so that the echo arrives back at the radar before the receiver turns on after transmission [26]. When this occurs, only part of the transmitted pulse is available to the compressor and the signal bandwidth is reduced by the ratio of the received-to-transmitted pulse length. This bandwidth

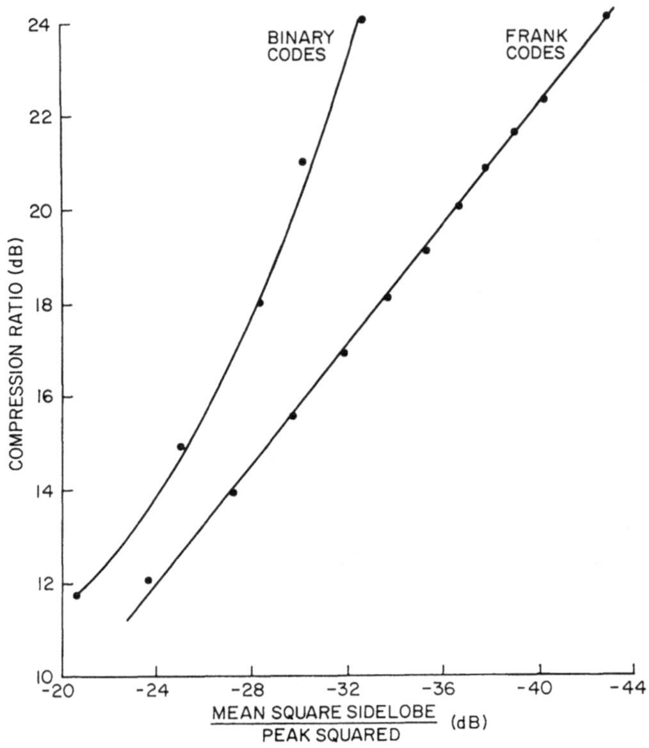

Fig. 2–37 Comparison of Mean Square Sidelobes.

reduction increases the duration of the compressed pulse peak. Also, there is a reduction in the peak of the compressed pulse, while the absolute sidelobe level does not change very much. The net result is a reduction in the relative peak to sidelobe level.

PROBLEMS

Problem 2–1 Sketch the amplitude and frequency spectra of the following:
(a) a single rectangular pulse of unity height and pulsewidth τ centered about the time origin;
(b) a single rectangular pulse of unity height and pulsewidth τ modulating the carrier waveform $A \cos 2\pi f_c t$;
(c) a pulse train consisting of pulses described in (a) with period $T = 5\tau$;
(d) a pulse train as described in (c) modulating the carrier waveform $A \cos 2\pi f_c t$.

Problem 2–2 For problem 2–1, find:
(a) the 3 dB frequency and the amplitude at that frequency.
(b) the zero crossings of the spectral amplitude.

Problem 2–3 Two reflectors along a radar line-of-sight are separated by 300 m. Find the theoretical pulsewidth required to resolve the two as separate targets.

Problem 2–4 The Doppler shift in transmitted frequency sensed by a radar receiver can be expressed as

$f = 2V_r/\lambda$, Hz

where V_r is the radial velocity of the target with respect to the radar and λ is the transmitted wavelength. If $V = 100$ m/s and θ is the angle between a line joining the target and the radar ($\theta = 0°$ means directly toward the radar), then $V_r = V \cos\theta$. Find the Doppler shift of a target for which:

(a) $\theta = 60°$
(b) $\theta = 135°$
(c) $\theta = 90°$

Problem 2–5 A radar operating at 1 GHz detects one target moving 60° relative to the line-of-sight to a radar at a speed of 400 m/s and another moving at 240° with a speed of 200 m/s. Compute the frequency difference between the two targets.

Problem 2–6 A pulse compression radar uses a transmitted pulsewidth $\tau_t = 2$ μs. Find the time duration τ_c of the compressed pulse if $\rho = 200$.

Problem 2–7 Suppose 10 contiguous pulses in a step-chirp waveform have pulsewidths of 10 μs. Calculate the pulse repetition frequency F, the transmitted bandwidth B_t, and the pulse compression ratio ρ.

Problem 2–8 Sketch the autocorrelation function of the five-element Barker code + + + − + and compute the sidelobe levels with respect to the peak lobe. Devise a tapped delay line structure to generate this code.

Problem 2–9 Find the Frank code for $N = 4$ using (2–6).

Problem 2–10 Find the P1 code for $N = 4$ using (2–8) and compare them to the Frank code in problem 2–8.

Problem 2–11 Find the P3 code using (2–15) for $N = 4$ and compare them to the results of problems 2–8 and 2–10.

Problem 2–12 Find the P2 code for $N = 4$ using (2–21) and compare the results to those of problems 2–8 through 2–10.

Problem 2–13 Find the P4 code using (2–18) for $N = 4$ and compare the results to problems 2–8 through 2–11.

Problem 2–14 Find the P4 palindromic code for $N = 4$ using (2–20) and compare the results to problem 2–12.

Problem 2–15 Compute the Frank and P1 codes for $N = 3$ and observe the conjugate symmetry of the P1 code.

Problem 2–16 (a) Using a Frank code with $\rho = 100$, graphically derive the first 10 values of its autocorrelation function.

(b) Determine the effect of a two-sample sliding-window adder on the first 10 values of the autocorrelation function of a Frank code with $\rho = 100$.

(c) What is the maximum value of the autocorrelation function of the first 10 points of a Frank code with $\rho = 100$, with and without a two-sample sliding-window adder on the output of the compressor?

(d) Repeat (b) and (c) with the two-sample sliding window adder replaced by a two-sample sliding-window subtractor.

Problem 2–17 Repeat problem 2–15 using a P1 code with $\rho = 100$.

Problem 2–18 Repeat problem 2–15 using a P3 code with $\rho = 100$.

Problem 2–19 (a) Sketch the autocorrelation function of a Frank code with $\rho = 16$.

(b) Precede the compressor in (a) with a two-sample sliding-window adder and sketch the autocorrelation function of a Frank code with $\rho = 16$ out of the compressor.

(c) repeat (b) with a two-sample sliding-window subtractor replacing the two-sample sliding-window adder.

Problem 2–20 Repeat problem 2–18 with the P1 code replacing the Frank code.

Problem 2–21 Repeat problem 2–18 with the P3 code replacing the Frank code.

Problem 2–22 Sketch the autocorrelation function of a Frank-coded waveform Doppler-shifted to produce:

(a) a π radian phase shift across the uncompressed pulse.

(b) a 2π radian phase shift across the uncompressed pulse.

Problem 2–23 Repeat problem 2–21 using a P3 code with $\rho = 16$.

REFERENCES

1. M. I. Skolnik, *Introduction to Radar Systems,* (2nd ed.) McGraw-Hill, New York, 1980.
2. E. Brookner (ed.), *Radar Technology,* Artech House, Dedham, MA, 1978.

3. C. E. Cook and M. Bernfeld, *Radar Signals: An Introduction to Theory and Application,* Academic Press, New York, 1967.
4. F. E. Nathanson, *Radar Design Principles,* McGraw-Hill, New York, 1969.
5. G. W. Dely, "Waveform Design," Ch. 3 in *Radar Handbook,* M. I. Skolnik (ed.), McGraw-Hill, New York, 1970.
6. A. W. Rihaczek, *Principles of High-Resolution Radar,* McGraw-Hill, New York, 1969.
7. D. K. Barton (ed.), "Pulse Compression," *Radars,* Vol. 3, Artech House, Dedham, MA, 1975.
8. R. L. Frank, "Polyphase Codes with Good Non-Periodic Correllation Properties," *IEEE Trans. on Information Theory,* Vol. IT-9, No. 1, Jan. 1963, pp. 43–45.
9.* B. L. Lewis and F. F. Kretschmer, Jr., "A New Class of Polyphase Pulse Compression Codes and Techniques," *IEEE Trans. on Aerospace and Electronic Systems,* Vol. AES-17, No. 3, May 1981, pp. 364–372.
10.* B. L. Lewis and F. F. Kretschmer, Jr., "New Polyphase Pulse Compression Waveforms and Implementation Techniques," IEE International Radar Conference, *Radar 82,* London, UK, Oct. 1985.
11.* F. F. Kretschmer, Jr., and B. L. Lewis, "Polyphase Pulse Compression Waveforms," NRL Report 8540, Jan. 5, 1982.
12.* B. L. Lewis, F. F. Kretschmer, Jr., and F. C. Lin, "Effects of Bandwidth Limitation on Polyphase Coded Pulse Compression Systems," NRL Report 8625, Sept. 20, 1982.
13.* F. F. Kretschmer, Jr. and B. L. Lewis, "Doppler Properties of Polyphase Coded Pulse Compression Waveforms," *IEEE Trans. on Aerospace and Electronic Systems,* Vol. AES-19, No. 4, July 1983, pp. 521–531.
14.* B. L. Lewis and F. F. Kretschmer, Jr., "Linear Frequency Modulation Derived Polyphase Pulse Compression Codes," *IEEE Trans. on Aerospace and Electronic Systems,* Vol. AES-18, No. 5, Sept. 1982, pp. 637–641.
15. D. A. Huffman, "The Generation of Impulse Equivalent Pulse Trains," *IRE Trans. on Information Theory,* Vol. IT-8, No. 5, Sept. 1962, pp. S10–S16.
16.* F. F. Kretschmer, Jr., and F. C. Lin, "Huffman-Coded Pulse Compression Waveforms," NRL Report 8894, May 23, 1985.
17. M. J. E. Golay, "Complementary Series," *IRE Trans. on Information Theory,* Vol. IT-7, April 1961, pp. 82–87.
18. C. C. Tseng and C. L. Liu, "Complementary Sets of Sequences," *IEEE Trans. on Information Theory,* Vol. IT-18, No. 5, Sept. 1972, pp. 644–651.
19. R. Sivaswamy, "Multiphase Complementary Codes," *IEEE Trans. on Information Theory,* Vol. IT-24, No. 5, Sept. 1978, pp. 546–553.
20. R. Sivaswamy, "Digital and Analog Subcomplementary Sequences for Pulse Compression," *IEEE Trans. on Aerospace and Electronic Systems,* Vol. AES-14, No. 2, March 1978, pp. 343–350.
21. G. Weathers and E. M. Holliday, "Group-Complementary Array Coding for Radar Clutter Rejection," *IEEE Trans. on Aerospace and Electronic Systems,* Vol. 19, No. 3, May 1983, pp. 369–379.
22. S. W. Golomb and R. A. Scholtz, "Generalized Barker Sequences," *IEEE Trans. on Information Theory,* Vol. IT-11, Oct. 1985, pp. 533–537.
23. H. D. Helms, "Fast Fourier Transform Method of Computing Difference Equations and Simulating Filters," *IEEE Trans. on Audio and Electroacoustics,* Vol. AU-15, No. 2, June 1967, pp. 85–90.
24. C. L. Temes, "Sidelobe Suppression in a Range-Channel Pulse-Compression Radar," *IRE Trans.,* Vol. MIL-6, 1962, pp. 162–169.
25.* F. C. Lin, B. L. Lewis, and F. F. Kretschmer, Jr., "Parameter Estimation and Target Detection in a Distributed Clutter Environment," NRL Report 8681, March 23, 1983.
26. E. R. Billam, "Eclipsing Effects With High Duty Factor Waveforms in Long Range Radar," *Record of the IEEE International Radar Conference,* 1985, pp. 6–11.

---

*Reference reprinted herein. *See* Readings, Chapter 2.

A New Class of Polyphase Pulse Compression Codes and Techniques

B.L. LEWIS, Senior Member, IEEE
F.F. KRETSCHMER, JR., Senior Member, IEEE
Naval Research Laboratory

Abstract

A new class of symmetric radar pulse compression polyphase codes is introduced which is compatible with digital signal processing. These codes share many of the useful properties of the Frank polyphase code. In contrast with the Frank code, the new codes are not subject to mainlobe to sidelobe ratio degradation caused by bandlimiting prior to sampling and digital pulse compression. It is shown that bandlimiting the new codes prior to pulse compression acts as a waveform amplitude weighting which has the effect of increasing the mainlobe to sidelobe ratios.

Manuscript received May 27, 1980; revised October 23, 1980.

Authors' address: Target Characteristics Branch, Radar Division, Naval Research Laboratory, Washington, DC 20375.

0018-9251/81/0500-0364 $00.75 ©1981 IEEE

I. Introduction

A new class of uniform amplitude polyphase codes which are amenable to digital processing is described in this paper. These codes are similar to the Frank polyphase code [1-4] in many respects. These similarities include low sidelobe levels, good Doppler tolerance for search radar applications and ease of implementation. The new codes [5] are referred to as the P1 and P2 codes in this paper. The P1 code has an autocorrelation function (ACF) magnitude which is identical to the Frank code for zero Doppler shift. While the peak sidelobes of the P2 code are the same as the Frank code for the zero Doppler case, the mean square sidelobes of the P2 code are slightly less (approximately 1 dB for a 16-element code decreasing with the pulse compression ratio).

The significant advantage of the P1 and P2 codes over the Frank code is that they are more tolerant of receiver bandlimiting prior to pulse compression. Such bandlimiting will be encountered in radars employing digital signal processing. This bandlimiting is required to avoid out-of-band noise foldover caused by sampling in conversion to digital format.

A. Relationship Between Frank Polyphase Codes and Other Pulse Compression Waveforms

In this section we compare the Frank polyphase waveform to other pulse compression waveforms; however, the comments made here also apply to the P1 and P2 codes, which are discussed in more detail subsequently. The highest peak sidelobe power level of the Frank code compressed pulse (or ACF) is approximately π^2 times the pulse compression ratio ϱ lower than the peak signal power. Thus the peak signal to peak sidelobe ratio, hereafter referred to as the peak sidelobe ratio, increases with increasing pulse compression ratios. For example, a 100-element Frank code having a pulse compression ratio of 100, has a peak sidelobe ratio of approximately 30 dB which occurs without any amplitude weighting. In contrast, the peak sidelobe ratio of a compressed chirp waveform is approximately 13 dB independent of the pulse compression ratio. Weighting is generally used in analog chirp systems to reduce the peak sidelobe ratio which results in a loss of signal to noise ratio (SNR) of approximately 1 to 2 dB. While the analog chirp peak sidelobe ratio may be reduced without weighting by use of nonlinear frequency modulated waveforms, the penalty incurred is a reduced tolerance to Doppler shifts.

Comparing the Frank polyphase code to binary phase codes, the Frank code has a peak sidelobe ratio which is approximately 10 dB better than the best pseudorandom binary or shift register codes whose peak sidelobe ratios approach ϱ [2]. Moreover the Frank polyphase code peak sidelobe ratio is superior to the "best" binary codes which have been found by search procedures [6]. Also the Frank code, or any other frequency derivable polyphase code, has better Doppler tolerance than the binary codes. An exception to this is the 4-element Frank code which is identical to the 4-element Barker code. In general, however, binary codes are known to have poor Doppler tolerance.

A search routine was used by Somaini and Ackroyd [7] which varied the Frank code phases to reduce the sidelobes and a slight improvement was obtained. For example, for the 100-element Frank code, the peak sidelobe was reduced from a magnitude of 3.24 to 1.86. However, we have found by computer simulation that, in general, the Doppler tolerance of the modified Frank code was not quite as good as that of the corresponding 100-element Frank code.

Another class of codes which are based on binary codes is known as the complementary codes [8, 9]. Under ideal conditions, these codes have lower sidelobe ratios than the Frank code. However, they have limited usefulness in radar applications. The complementary binary codes are such that when the two complementary ACFs are added together most or all of the sidelobes cancel. In practice it is necessary to separate the two complementary codes in frequency, time, or polarization to permit these ACFs to be independently measured prior to addition. This separation can cause decorrelation by radar targets or distributed clutter and can prevent cancellation of the sidelobes of the ACFs. Even for a point target a carrier frequency change will cause a differential phase shift, due to range, between the two codes and thereby prevent sidelobe elimination after combination. Also, since the complementary codes consist of binary codes, they are not tolerant of Doppler shifts caused by targets moving relative to the radar.

In view of the foregoing comments and the compatibility of polyphase codes with digital processing, it is seen that polyphase codes offer many attractive advantages.

II. Discussion

In this section we give a brief summary of the Frank code properties and the effect that bandlimiting has on the received waveform. The new polyphase codes, which are more tolerant of bandlimiting effects, are then described. Autocorrelation functions of these codes are presented in Section III.

A. Frank Polyphase Code

The Frank coded waveform consists of a constant amplitude signal whose carrier frequency is modulated by the phases of the Frank code. Each element of the Frank code is τ seconds long, which is approximately equal to the reciprocal of the waveform 3 dB bandwidth. The phases of the Frank code may be generated for transmission by multiplying the elements of the matrix

$$\begin{matrix} 0 & 0 & 0 & \cdots & 0 \\ 0 & 1 & 2 & \cdots & (N-1) \\ 0 & 2 & 4 & \cdots & 2(N-1) \\ 0 & 3 & 6 & \cdots & 3(N-1) \\ \vdots & & & & \\ 0 & (N-1) & 2(N-1) & \cdots & (N-1)^2 \end{matrix}$$

by the phase $(2\pi/N)$ and by transmitting the phases of row 1 followed by row 2 etc. Note that transmitting the phases of the rows in reverse order and properly matching does not alter the compressed pulse amplitude.

The phase of the ith code element in the jth row or code group may be expressed mathematically as

$$\phi_{i,j\ \text{Frank}} = (2\pi/N)(i-1)(j-1) \tag{1}$$

where $i = 1, 2, 3, \ldots, N$ and $j = 1, 2, 3, \ldots, N$.

In (1) the index i ranges from 1 to N for each value of j and the number of code elements formed is equal to N^2. For example, the Frank code with $N = 3$ is given by the sequence

$\phi_{1,1}$	$\phi_{2,1}$	$\phi_{3,1}$	$\phi_{1,2}$	$\phi_{2,2}$	$\phi_{3,2}$	$\phi_{1,3}$	$\phi_{2,3}$	$\phi_{3,3}$
0	0	0	0	$2\pi/3$	$4\pi/3$	0	$4\pi/3$	$8\pi/3$

The Frank code can be seen to consist of the negative of the steering phases of an N point discrete Fourier transform (DFT) where the jth frequency coefficient is given by

$$F_j = \sum_{i=1}^{N} a_i \exp\left[-j\,(2\pi/N)\,(i-1)\,(j-1)\right] \tag{2}$$

where a_i is the ith complex input time sample. Thus each row of the Frank code phase matrix consists of a phase group which corresponds to the sampled phases of frequencies which are subharmonics of the sampling frequency. Also the Frank code phases correspond to the appropriately sampled phases of a step chirp waveform after converting to a baseband signal by synchronous demodulation using a coherent local oscillator. In this case, the local oscillator frequency is equal to that of the lowest frequency in the step chirp intermediate frequency (IF) waveform.

Upon reception the Frank code waveform is compressed as follows. The received signal is converted to baseband in-phase I and quadrature Q signals by synchronous detectors. These signals are then each sampled at the Nyquist rate to obtain one sample in I and Q every τ seconds. The I and Q analog samples are then converted to digital data and are processed by a digital matched filter. The output of this filter is a

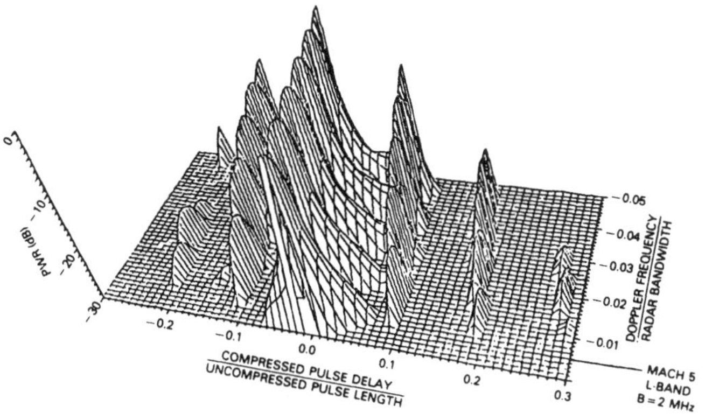

Fig. 1. Expanded Frank code ambiguity diagram.

waveform one sample period wide at the peak having an amplitude N^2 greater than that of any code element. Since the uncompressed waveform was N^2 samples in duration and compressed to one sample in duration, the pulse compression ratio is N^2.

The Doppler response or ambiguity diagram of a 100-element Frank coded waveform is shown in Fig. 1. This diagram shows the response of the pulse compressor when the return signal is Doppler shifted by any arbitrary amount. For reasonable velocity targets in a search radar application, the Doppler response is quite acceptable. For example, for a target whose velocity is less than Mach 5, the Doppler to bandwidth ratio for an *L*-band search radar having a 2 MHz bandwidth is less than 0.005. In this region the Frank code has good Doppler tolerance.

Excluding bandlimiting effects, which are discussed in the next section, the ease of implementation, relative Doppler tolerance, and low sidelobes of the Frank code make it a good candidate for digital processing in radar applications.

B. Bandlimiting Effects

The gross shape of the spectrum of an ideally generated Frank coded waveform is approximately given by $\sin(\pi f/B) \,/\, (\pi f/B)$, where B is approximately $1/\tau$, centered on the carrier frequency. In radar practice, transmitters are operated in saturation so that very abrupt phase transitions can be made from one code element to the next. The transmitted spectrum thus approximates the ideal. Radar receivers, however, generally have an approximately rectangular passband which is usually the same as the 3-dB bandwidth of the transmit spectrum. Thus the receiver response does not match the transmitted Frank code spectrum. This bandlimiting of the radar receiver is intentionally done to reject out-of-band interference and noise foldover caused by sampling in the analog to digital (A/D) conversion process. The limited receiver bandwidth results in an unfavorable mismatch to the Frank coded waveform which causes a degradation in the peak sidelobe ratio of the compressed pulse. A similar degradation also occurs for binary coded waveforms. By increasing the receiver bandwidth, sampling faster, and matching to this oversampled signal, the degradation can be reduced. However this requires faster circuitry and additional complexity. The reason for the degradation of the Frank coded waveform is that the smoothing or averaging effect caused by bandlimiting has the least effect on the ends of the Frank code, where the phase changes between adjacent code elements are the smallest, and the most effect in the middle of the code where the phase changes from code element to code element are the largest. That is, thinking of the phase groups as corresponding to frequencies, bandlimiting has the effect of attenuating the center frequencies of the Frank coded waveform the most, which amounts to an inverse of the normal amplitude weighting employed to reduce sidelobes. These considerations, in part, catalyzed the search for phase coded waveforms which have the largest phase increments at the ends of the waveform which represent the highest and lowest frequencies in the radar passband.

C. P1 and P2 Polyphase Codes

Recalling the analogy between the Frank code and the sampled phases of a baseband step chirp waveform, it is found that by conceptually changing the synchronous oscillator frequency, different phase codes may be generated which have the same ACF amplitude as the Frank code ACF, but different ACF phases. By placing the synchronous oscillators at the center frequency of the step chirp IF waveform and sampling the baseband waveform at the Nyquist rate, the polyphase code referred to as the P1 code may be derived which has the desired properties. Note that if an even number of frequencies compose the step chirp waveform, the synchronous oscillator frequency placed at the center of the band does not correspond to an actual frequency in the step chirp waveform. For an odd number of frequency steps in the chirp waveform, beating the synchronous oscillator with the center frequency produces the same sampled baseband phase groups or rows as the Frank code. However the rows of the resultant code matrix are rearranged such that the phases correspond to baseband frequencies that are symmetric about zero frequency or direct current. This allows the highest phase changes to occur at either end of the code.

The P1 code also consists of N^2 elements and the ith element of the jth group may be expressed as

$$\phi_{i,j} = -(\pi/N)\,[N-(2j-1)]\,[(j-1)\,N+(i-1)] \tag{3}$$

where i and j are integers ranging from 1 to N.

An $N = 3$ P1 code is given by the sequence

$\phi_{1,1}$	$\phi_{2,1}$	$\phi_{3,1}$	$\phi_{1,2}$	$\phi_{2,2}$	$\phi_{3,2}$	$\phi_{1,3}$	$\phi_{2,3}$	$\phi_{3,3}$
0	$-2\pi/3$	$-4\pi/3$	0	0	0	0	$2\pi/3$	$4\pi/3$

which can be seen to be a rearranged Frank code with the zero frequency group in the middle.

The P2 code, which also has the desired features, is similar to the Butler matrix steering phases used in antennas [4] to form orthogonal beams. The P2 code is valid for N even, and each group of the code is symmetric about 0 phase. The usual Butler matrix phase groups are not symmetric about 0 phase and result in higher sidelobes. For N even, the P1 code has the same phase increments within each phase group as the P2 code, except that the starting phases are different. The ith element of the jth group of the P2 code is given by

$$\phi_{i,j} = \{(\pi/2)\,[(N-1)/N] - (\pi/N)\,(i-j)\}\,[N+1-2j] \tag{4}$$

where i and j are integers ranging from 1 to N as before. The requirement for N to be even in this code stems from the desire for low autocorrelation sidelobes. An odd value for N results in high autocorrelation sidelobes.

An $N = 4$ P2 code example is given by the sequence

$\phi_{1,1}$	$\phi_{2,1}$	$\phi_{3,1}$	$\phi_{4,1}$	$\phi_{1,2}$	$\phi_{2,2}$	$\phi_{3,2}$	$\phi_{4,2}$
$+9\pi/8$	$+3\pi/8$	$-3\pi/8$	$-9\pi/8$	$+3\pi/8$	$+\pi/8$	$-\pi/8$	$-3\pi/8$
$\phi_{1,3}$	$\phi_{2,3}$	$\phi_{3,3}$	$\phi_{4,3}$	$\phi_{1,4}$	$\phi_{2,4}$	$\phi_{3,4}$	$\phi_{4,4}$
$-3\pi/8$	$-\pi/8$	$+\pi/8$	$+3\pi/8$	$-9\pi/8$	$-3\pi/8$	$+3\pi/8$	$+9\pi/8$

This code has the frequency symmetry of the P1 code and also has the property of being a palindromic code since the phases are symmetric about the center of the code.

The ambiguity diagram of the P1 code for N odd is identical to that of the Frank code. For N even, the ambiguity diagrams of the P1 and P2 codes are similar to each other and to that of the Frank code.

It should be noted that the P1 and P2 codes could have also been generated in reverse order by letting j index from N to 1. By proper matched filtering the results are the same.

III. New Code Utilization Techniques

The digital hardware similarities and simplifications permitted by the P1 and P2 codes can be demonstrated by comparing block diagrams with the simple one used in generating and compressing the Frank Code (Fig. 2). In Fig. 2 SR signifies clocked digital shift registers, and while clock pulses are assumed, they are not shown. The input shift registers

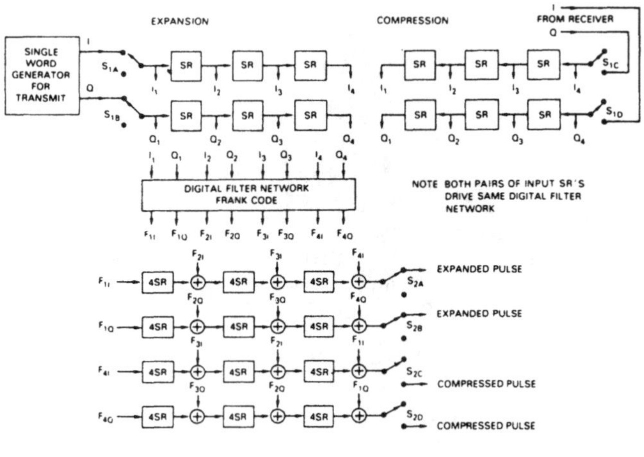

Fig. 2. DFT pulse expander-compressor with $N = 4$ (Frank code).

and digital filter network together compose a digital discrete Fourier transform circuit or a digital fast Fourier transform (FFT) circuit.

The signal code to be transmitted (pulse expansion) can be produced by clocking a single complex number with $Q = 0$ having sufficient I magnitude through switch S_1 into an N point DFT circuit or an FFT circuit. Although not required, the input Q channels of the expansion side are shown in Fig. 2 for generality. In response to this input number at each I time sample input to the filter network, in-phase (I) and quadrature (Q) numbers are simultaneously clocked out of each frequency port $F_1, F_2, \ldots, F_N$. Successive I and Q numbers from each filter, which occur as the input number is sequentially clocked to each I filter time sample input, represent complex numbers that define both the phase and magnitude of the steering weights used in the DFT to produce its various digital filter banks. The N successive samples of the output of the F_N port are sent to one terminal of a digital adder whose output passes through switch S_2 to a radar transmitter where each successive complex number defines the phase to be transmitted for a time interval equal to the clock period used in the DFT.

The output of the F_{N-1} port is sent to one input of a second digital adder whose output drives a chain of N shift registers operating at the clock rate used in the DFT. The output of this chain drives the second input of the adder driven by F_N. When the N complex numbers from F_N finally issue from S_2, the N complex numbers from F_{N-1} start to issue contiguously from the N shift register delay line. A similar process is used with each output from a frequency port as shown in Fig. 2 to produce a stream of N^2 digital complex numbers lasting for N^2 times the clocking period in time. Thus the duration of the total transmitted code is N^2 times as long as the duration of one element in the code. When this total code exits from S_2, S_1 and S_2 are thrown to their second position.

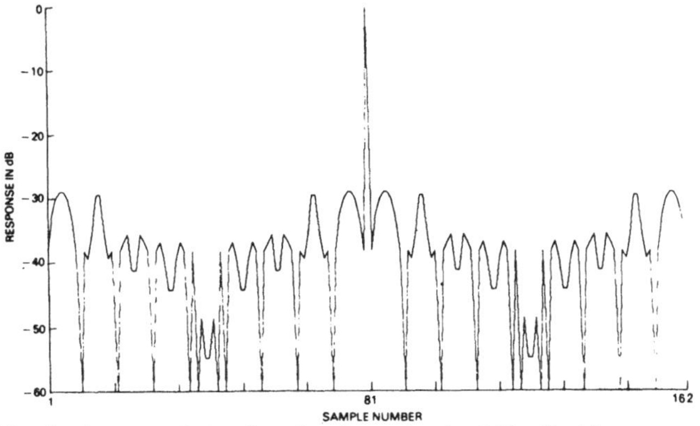

Fig. 3. Autocorrelation function from circuit of Fig. 2 with $N = 9$.

Fig. 4. Implementation of $N = 3$ P1 code expander-compressor.

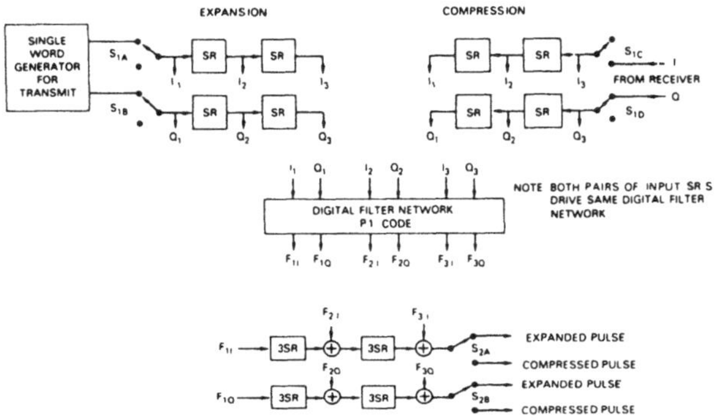

Pulse compression is achieved by coherently detecting echoes to baseband I and Q video, sampling this data at the DFT clock rate, converting these samples to digital numbers, changing the sign of the Q numbers (conjugating), and inserting these I and Q numbers into the compressor side of the DFT as illustrated in Fig. 2.

Since F_N was the first code group or frequency to be transmitted, it will be the first back in any echo. Thus, when the first N conjugated complex numbers of any echo index into the N points of the DFT, a complex number N times larger in magnitude than that of any single input number will issue from the F_N frequency port.

This F_N output will be connected to the input of an N shift register chain whose output is digitally added to the output from the F_{N-1} port. N sample periods after F_N indexed, F_{N-1} will index in the DFT points and produce an N times magnified output from the F_{N-1} port having the same phase as that from the F_N port. At this time the N times magnified number out of F_N will issue from its shift register delay to add coherently with the magnified signal out of F_{N-1}. This process continues until the differentially delayed magnified numbers out of each matched port sum to a number magnified by N^2 when F_1 indexes into the DFT points. This N^2 magnified number is the desired compressed pulse and is sent through switch S_2 to following processors.

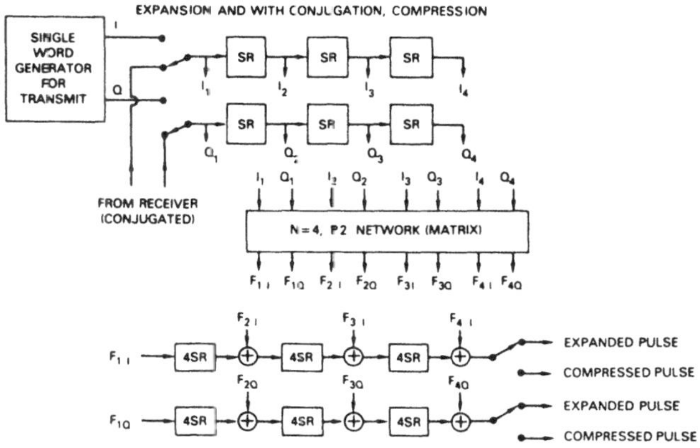

Fig. 5. P2 code implementation with $N = 4$.

Prior to and after the compressed pulse issues from S_2, mismatched responses (autocorrelation function sidelobes) will exit from S_2 starting with the first code element received and ending after the last code element received clears the DFT and the delay lines between the F_1 output and S_2.

Fig. 3 is a computer drawn plot of the ACF resulting from the circuit of Fig. 2 for $N = 9$ with no Doppler shift on the received signal. This ACF is that of an 81-element Frank code. Note that the highest sidelobe is down from the peak by $\pi^2 N^2$.

Fig. 4 illustrates the implementation of the P1 code for N equal to 3. Note (1) and (3) to identify the reordering of the frequencies transmitted. The same type of DFT is employed, but for odd N the code can be rearranged symmetrically about the radar's carrier frequency (the zero frequency code group). This reordering halves the phase increments that are employed in the high frequency groups and removes ambiguities in these frequencies. At no time is a phase increment between code elements allowed to equal or exceed π radians as can be seen from (3). This will be shown by computer simulation results to be important in precompression bandwidth limiting effects of IF amplifiers.

Note that in the implementation in Fig. 4 the echo signal does not need to be conjugated and two separate sets of differential shift register delay lines are not required for synthesis and compression as in Fig. 2.

If F_3 is transmitted first, it is received first and, without conjugating, it will match the F_1 filter since this filter's weights are the conjugates of those in F_3 due to the code symmetry. Thus its match will be delayed properly by the same differential shift register delay line that was used to form the code.

The autocorrelation function of a P1 code for $N = 9$ is identical in magnitude to that of the Frank code shown in Fig. 3. However the autocorrelation function phases differ from those corresponding to the Frank code.

Fig. 5 portrays the implementation of the P2 code whose element phases are given by (4). This code differs from the Frank or the P1 codes in that there is odd phase symmetry within each group corresponding to a frequency and there is even phase symmetry about the center of the complete code. These phase symmetries make it possible to input the conjugated received code for compression into the same input used to form the code. This eliminates the need for using two sets of *I, Q* shift registers as in Fig. 4 to obtain the *N* time samples that are used to drive the matrix filter networks. In other respects, this implementation works in a manner identical to that described for the P1 code. The autocorrelation function of a 100-element P2 code is shown in Fig. 6.

Frequency to frequency amplitude weighting can be used to control sidelobes, if desired, on all of the codes discussed. This weighting is applied at the output of the various frequency ports prior to summing in the differential delay lines and adders. Fig. 7 illustrates the results of amplitude weighting the frequency ports of a P2 code on receive with a cosine on a pedestal of 0.4.

IV. Simulation of Precompression Bandwidth Limitations

The effect of a restricted bandwidth in the IF amplifiers and the *I, Q* detectors preceding analog to digital conversion and compression of the phase codes was simulated on a digital computer. The various codes to be compressed were oversampled by 5 to 1 and sliding window averaged by 5, 7, and 10 to simulate the precompression bandwidth limitation. Only the results for the sliding window average of five samples are presented here. Similar results were obtained for the other cases. The compressor phases were matched to the input phases which existed prior to oversampling. The resultant oversampled and averaged waveform was then sampled every fifth sample beginning with the first sample and sent to the compressor. To take time of arrival variations into account, the sliding window average for the 5-sample case was taken starting four, three, two, one, and zero over-sample-periods ahead of the first received code element and sample correlation functions were developed for each case. Note that a match condition occurs for the latter case and otherwise a mismatch occurs.

The results of this study revealed that precompression bandwidth limitations were similar to amplitude weighting the frequency output ports of the digital filters in the compressor when the symmetrical P1 or P2 codes were employed and time of arrival variations were taken into consideration (Fig. 8). However this was not the case when the Frank code was processed (Fig. 9). For the Frank coded waveform, the bandwidth limitation did not affect the dc group and had little effect on the highest frequency code group since it is the conjugate of the frequency code group closest

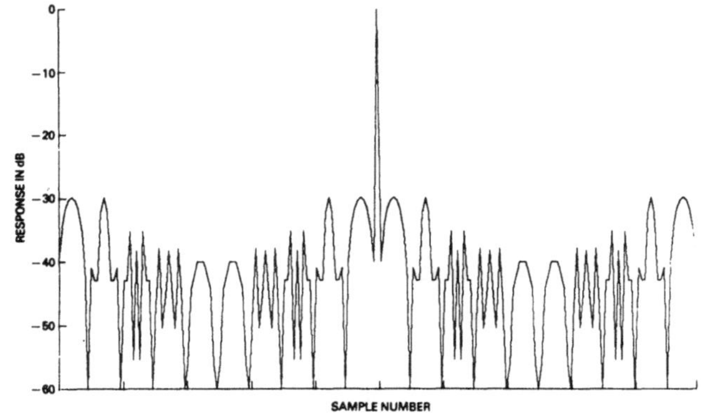

Fig. 6. Autocorrelation function of $N = 10$ P2 code.

Fig. 7. Effect of amplitude weighting, with cosine on pedestal of 0.4, frequency output ports of P2 code with $N = 10$.

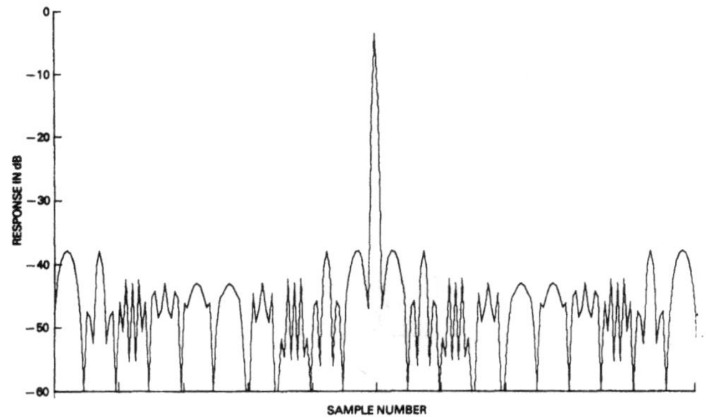

to the dc term. As a consequence, precompression bandwidth limitation did not drop the far out sidelobe caused by the dc group indexing into the highest frequency filter and vice versa.

Comparison of Figs. 8 and 9 show that for each corresponding time of arrival case, with the exception of the match condition shown in part (e) of each figure, the sidelobes of the P2 code are lower than the Frank code while the corresponding peak values are the same.

We mention at this point that although a sliding window average was used to simulate the band limitation effects in this paper, similar results would be expected for any other band-limiting filter since the sidelobe reduction is due to the smoothing effect of the filter.

V. Effect of Precompression Bandwidth Limitation on Signals and Noise

The phase coded waveforms discussed in this paper have $\sin(\pi f/B)\ /\ (\pi f/B)$ baseband spectrum envelopes as shown in Figs. 10 and 11 for the Frank and P2 codes. Note that the P2 (and P1) coded waveform has a more symmetrical spectrum than one coded with the Frank code due to its symmetry about the carrier.

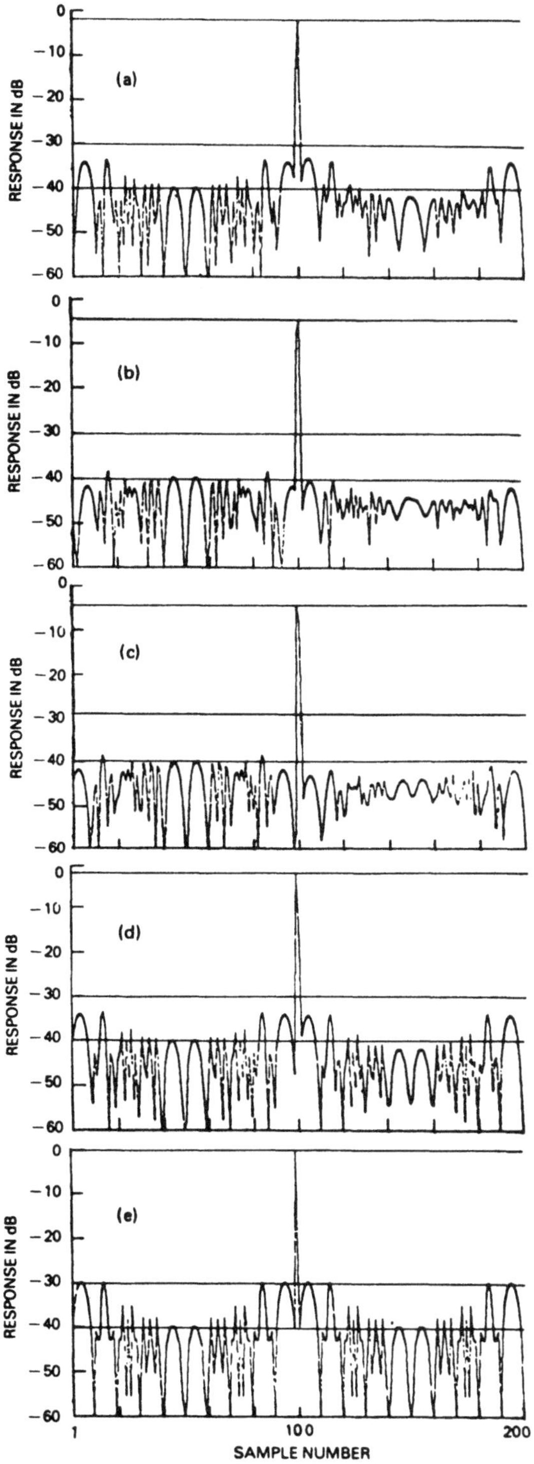

Fig. 8.

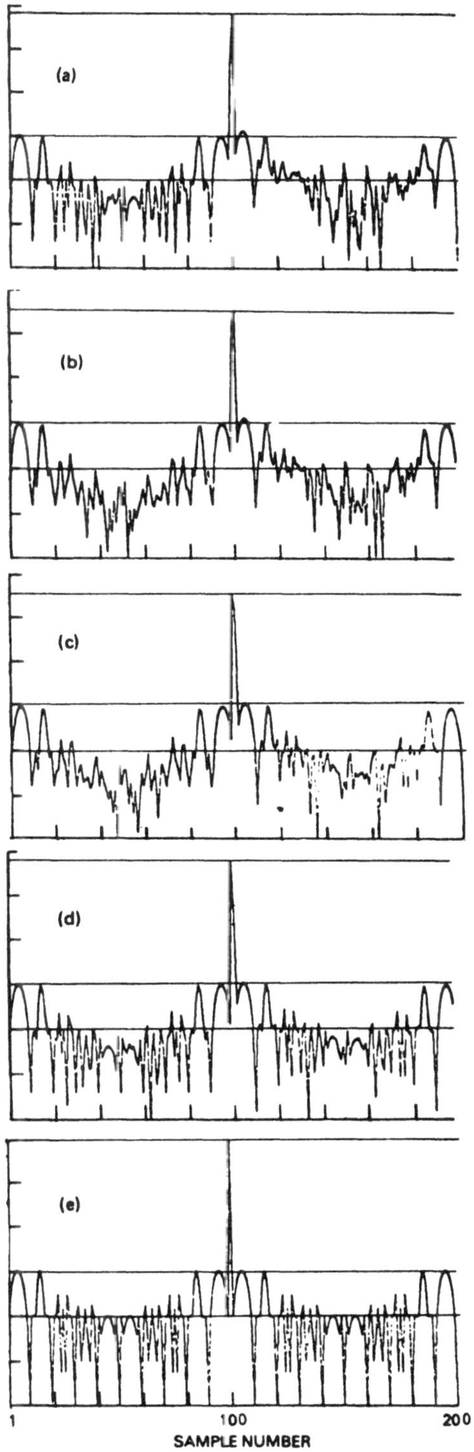

Fig. 9.

Figs. 8. and 9. Effect of precompression bandwidth limitations on N = 10, P2 code and Frank Code, respectively, with 5-sample average started (a) 4, (b) 3, (c) 2, (d) 1, and (e) 0 samples ahead of first point to be sent to compressor.

The average loss of the peak values shown in Figs. 8 and 9 was computed to be approximately 2.3 dB for both the Frank and P2 codes. Some of this loss can be attributed to the passband limitation while the remaining loss represents the loss due to time of arrival variation or range cusping. The passband limitation loss is due to the loss of the signal power contained in the sidelobes of the signal spectrum. The thermal noise contribution is also reduced by the bandlimiting and is the same for each code in Figs. 8 and 9 which account for signal only. It is important to note, however, that the symmetrical P1 or P2 code sidelobes drop more than the peak due to precompression bandwidth limitation while the sidelobes of the Frank code do not drop at all. This results in much lower sidelobes in the new codes for the same SNR loss due to the precompression bandwidth limitation.

VI. Summary

New polyphase codes have been described which share with the Frank code the advantages of compatibility with digital processing, low peak sidelobe ratios of approximately $\pi^2\varrho$, good Doppler tolerance in search radar applications, and ease of implementation. The new codes exhibit the additional advantage of being more tolerant than the Frank coded waveform to bandwidth limitations prior to pulse compression.

Acknowledgment

The helpful suggestions of J.P. Shelton of the Naval Research Laboratory are gratefully acknowledged.

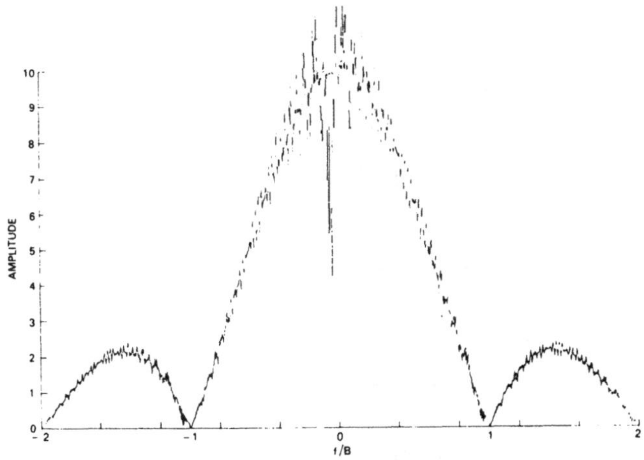

Fig. 10. Partial spectrum of N = 10 Frank-coded baseband waveform.

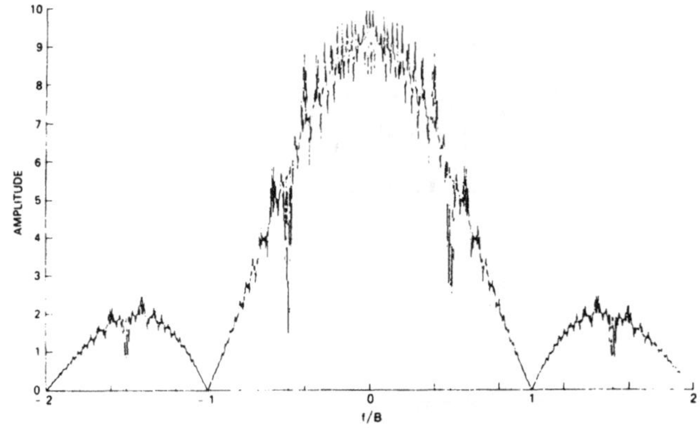

Fig. 11. Partial spectrum of N = 10 P2-coded baseband waveform.

References

[1] Frank, R.L. (1963)
Polyphase codes with good nonperiodic correlation properties.
IEEE Transactions on Information Theory, Jan. 1963, *IT-9,* 43-45.

[2] Nathanson, F.E. (1968)
Radar Design Principles.
New York: McGraw Hill, 1969.

[3] Cook, C. and Bernfeld, (1967)
Radar Signals, An Introduction to Theory and Application.
New York: Aeademic Press, 1967.

[4] Skolnik, M.I. (1970)
Radar Handbook.
New York: McGraw Hill, 1970.

[5] Lewis, B.L. and Kretschmer, F.F. (1980)
A new class of pulse compression codes and techniques.
Washington, DC, Naval Research Laboratory, NRL Report 8387, Mar. 26, 1980.

[6] Linder, J. (1975)
Binary sequences up to length 40 with best possible autocorrelation function.
Electronics Letters, 1975, *11,* 507

[7] Somaini, U. and Ackroyd, M.H. (1974)
Uniform complex codes with low autocorrelation sidelobes.
IEEE Transactions on Information Theory, Sept. 1974, *IT-20,* 689-691.

[8] Golay, M.J.E. (1961)
Complementary series.
IRE Transactions on Information Theory, Apr. 1961, *IT-7,* 82-87.

[9] Hollis, E.E. (1975)
Another type of complementary sequence.
IEEE Transactions on Aerospace and Electronic Systems, Sept. 1975, *AES-11,* 916-920.

NEW POLYPHASE PULSE COMPRESSION WAVEFORMS AND IMPLEMENTATION TECHNIQUES

Bernard L. Lewis and Frank F. Kretschmer, Jr.

Naval Research Laboratory, USA

INTRODUCTION

It is the purpose of this paper to introduce new polyphase pulse compression waveforms and efficient digital implementation techniques. The new waveforms provide previously unobtainable doppler and/or bandwidth limitation tolerance and the new implementation techniques permit systems to be implemented with cost effective available digital hardware.

Doppler tolerance in the paper will be taken to mean that the peak response and peak response to range-time-sidelobe ratio of compressed pulses do not degrade catastrophically with doppler.

Bandwidth limitation tolerance will be taken to mean that the peak response to range-time-sidelobe ratio of compressed pulses will not be degraded by receiver bandwidth limitations necessary to minimize the effect of receiver thermal noise.

POLYPHASE WAVEFORMS

The new waveforms to be discussed are carrier signals phase modulated with the P1, P2, P3, and P4 phase codes, Lewis and Kretschmer (1, 2). These codes are similar to the well known Frank polyphase code Frank (3), Cook and Bernfield (4) which will be labled PF in this paper. The PF code will be used as a standard for comparing the performances of the P codes.

The PF code can be derived as inphase I and quadrature Q samples taken at the Nyquist rate of a step approximation to a linear frequency modulation waveform coherently detected with a local oscillator of frequency equal to the lowest frequency of the waveform. This is illustrated by Fig. 1 in which a four frequency coherently detected step-chirp waveform that would yield a pulse compression ratio $\rho=4^2$ is sampled to obtain a 16 element PF code. A PF waveform is a carrier whose phase is changed every τ_c seconds by the phase difference between successive code elements of the PF code where τ_c is the uncompressed pulse length T divided by ρ. The ith code element of the jth frequency group of the PF code is defined by

$$\theta_{ij} = (2\pi/N)(i-1)(j-1) \qquad (1)$$

where i ranges from 1 to N for each j taken in ascending order from 1 to N.

It was noted Cantrell and Lewis (5) that (1) defines the complex conjugates of the steering phases used in an N point FFT to subdivide a bandwidth B into N subbands of width B/N where B is the I and Q sampling rate. This suggested that the PF code could be efficiently generated and compressed digitally as illustrated in Fig. 2. This implementation is the new form of implementation to be discussed in this paper.

Fig. 3 illustates the autocorrelation function of a 100 element PF code where sample number refers to resolvable time increments equal to the sampling period. Note the low range-time-sidelobes.

Experience with the PF waveform revealed that it was not doppler tolerant with high freqency carriers and that its peak to maximum range-time-sidelobe response ratio deteriorated with receiver bandwidth limitations. As a consequence, new codes and waveforms were sought. These statements will be justified in a following section of this paper.

NEW CODES AND IMPLEMENTATIONS

It was recognized that the intolerance of the PF waveform to bandwidth limitations was due to the fact that the code groups having the largest phase changes from code element to code element were in the center of the uncompressed pulse. This caused bandwidth limitations in a radar receiver to attenuate the center of the waveform more than the ends which adversely effected the peak to range-time-sidelobe ratio obtainable in the compressed pulse.

This suggested that the PF code groups representing the j frequencies be rearranged in time order of transmission to place the lowest phase increments from code element to code element in the center of the waveform. This triggered the invention of the P1 code whose code element phases are given by

$$\theta_{i,j} = -(\pi/N)[N-(2j-1)][(j-1)N+(i-1)] \qquad (2)$$

Note that when N is odd, the P1 code is the PF code rearranged to have conjugate symmetry about the D.C. term. The P1 code has an autocorrelation function nearly identical to that of the PF code (Fig. 3) with no bandwidth limitations.

Implementation of the P1 code for N odd simply requires changing the connections on the tapped delay line in Fig. 2 that delays the successive frequency groups. Implementation for N even is illustrated in Fig. 4.

In addition to the P1 code, the P2 code was developed whose successive code elements are defined by

$$\theta_{ij} = \{(\pi/2)[(N-1)/N]-(\pi/N)(i-1)\}\{N+1-2j\} \qquad (3)$$

This can be implemented as illustrated in Fig. 5 with N even. The P2 code is similar to the steering weights used on Butler matrices to steer antenna beams.

The P2 code has an autocorrelation function nearly identical in magnitude to that of the PF and P1 codes (Fig. 3) with no bandwidth limitations.

However, the P2 autocorrelation function is real rather than complex because the code is symmetrical.

The P1 and P2 waveforms were found to be much more bandwidth limitation tolerant than the PF waveform. However, they shared the PF doppler intolerance when used on high frequency carriers. This was recognized to be a characteristic of the analog step chirp waveform from which the phase codes were derived (4).

Recognizing that the linear chirp is much more doppler tolerant than the step chirp, a new phase code (the P3 code) was derived from the linear chirp waveform. This code consisted of Nyquist rate samples of a linear chirp waveform coherently detected using a local oscillator of frequency equal to the lowest frequency in the linear chirp waveform. The phases of successive code elements of the P3 code are given by

$$\theta_i = \pi(i-1)^2/\rho \qquad (4)$$

where i ranges from 1 to ρ and ρ is the pulse compression ratio. This code can also be implemented using a modified FFT as illustrated in Fig. 6.

The P3 code was found to be just as doppler tolerant as the analog linear chirp but it was found to be bandwidth limitation intolerant for the same reason as found for the PF code.

The same solution that worked for the PF code and transformed it to the P1 code was used on the P3 code to develop the P4 code which was implemented as illustrated in Fig. 7. The successive code elements of the P4 code have phases

$$\theta_i = [\pi(i-1)^2/\rho] - \pi(i-1) \qquad (5)$$

This code was found to be both doppler and bandwidth limitation tolerant.

BANDWIDTH LIMITATION EFFECTS

The effect of bandwidth limitations on the various polyphase coded compressors was investigated by bandlimiting the waveform to be compressed to the reciprocal of the length of a code element. The average response to waveforms with different times of arrival was then determined. Typical results are illustrated in Fig. 8 for the various polyphase codes. Note that the bandwidth limitation decreased the PF and P3 peak to range-time-sidelobe response ratios while it increased that of the P1, P2 and P4. Fig. 9 illustrates the effect of sliding window two sample adding the outputs of the various code compressors with ρ = 400.

DOPPLER EFFECTS

The effect of doppler on the various code peak to peak-sidelobe and image lobe responses is illustrated in Fig. 10. Note that grating lobes appear in the PF, P1 and P2 codes that are absent in the P3 and P4 codes. Note also the high peak to peak sidelobe ratio at zero doppler that these waveforms offer without amplitude weighting. This high ratio is characteristic of frequency derived polyphase coded compressors. It should be noted that bandlimiting suppresses the image lobe of the P1, P2 and P4 waveforms caused by doppler.

ADVANTAGES OF DIGITAL OVER ANALOG PROCESSING

Some advantages of digital over analog processing include the following:

1. Freedom from ringing produced by impedance mismatches or analog filters.
2. Low loss long delays.
3. Use of standard hardware.
4. Reproducibility of response.
5. High peak to peak-sidelobe ratios without weighting.
6. Ability to change bandwidth by changing clock frequency.
7. Ability to change pulse compression ratio by controlling number of time samples used and output tapped delay line delays.
8. Compatibility with other digital signal processors such as MTI.

CONCLUSIONS

At this point it can be concluded that doppler and bandwidth limitation tolerant digital pulse compressors are feasible for high carrier frequency radars. In addition, it can be concluded that such digital compressors have many advantages to offer over analog processors.

REFERENCES:

1. Lewis, B., and Kretschmer, F., May 1981, IEEE/AES, 17, 364-372.

2. Lewis, B., and Kretschmer, F., (scheduled for publication Sep, 1982), "Linear Frequency Modulation Derived Polyphase Pulse Compression Codes," IEEE/AES.

3. Frank, R., June 1963, IEEE/IT, 9, 43-45.

4. Cook, C., and Bernfield M., 1967, "Radar Signals, An Introduction to Theory and Applications," Academic Press, New York.

5. Cantrell, B., and Lewis, B., December 2, 1980, "High Speed Digital Pulse Compressor," U.S. Patent #4,237,461.

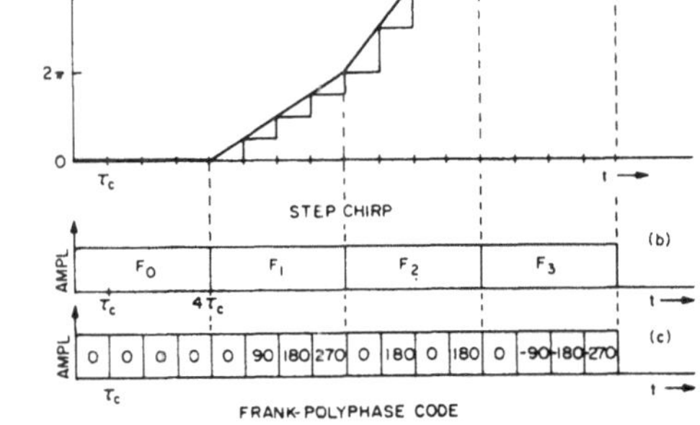

Figure 1 Step-chirp and Frank-polyphase-code relationships

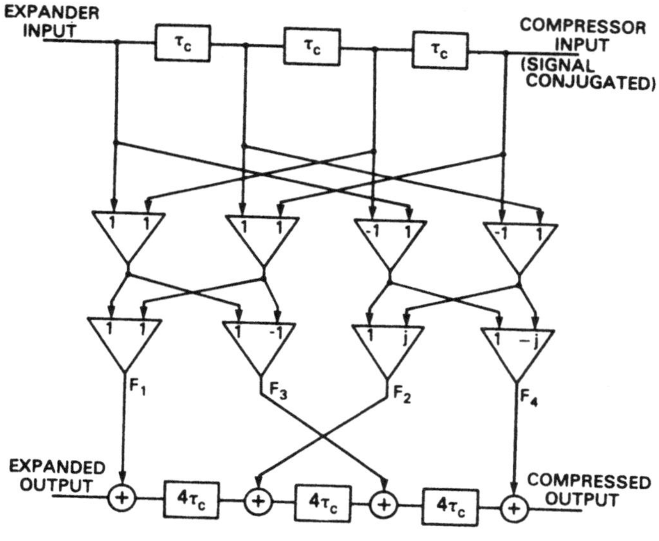

Figure 2 PF expander-compressor, $\rho=T/\tau_c=16$

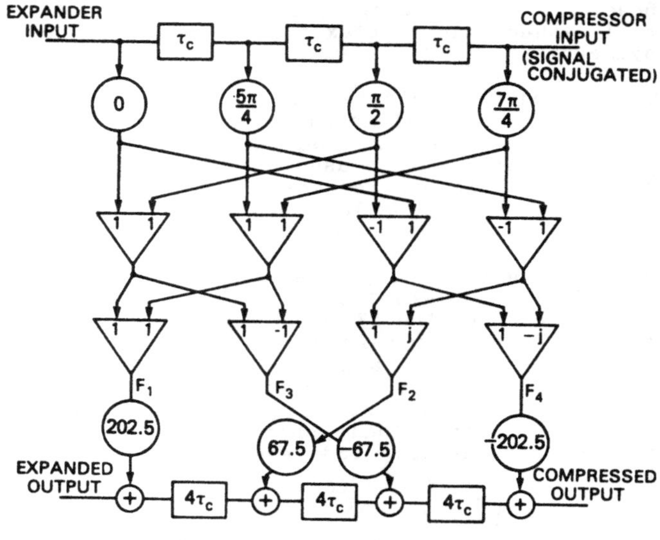

Figure 5 P2 expander-compressor, $\rho=T/\tau_c=16$

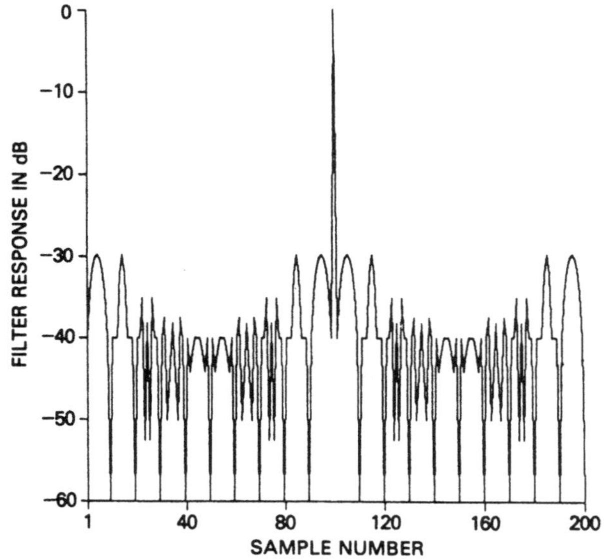

Figure 3 Compressed pulse of Frank code

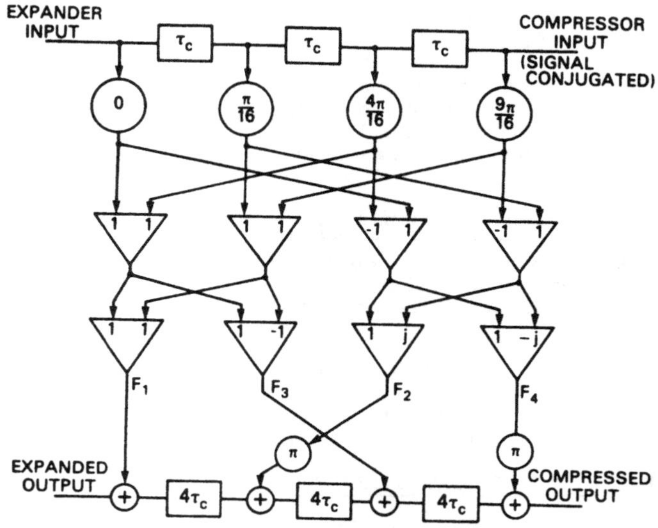

Figure 6 P3 expander-compressor, $\rho=T/\tau_c=16$

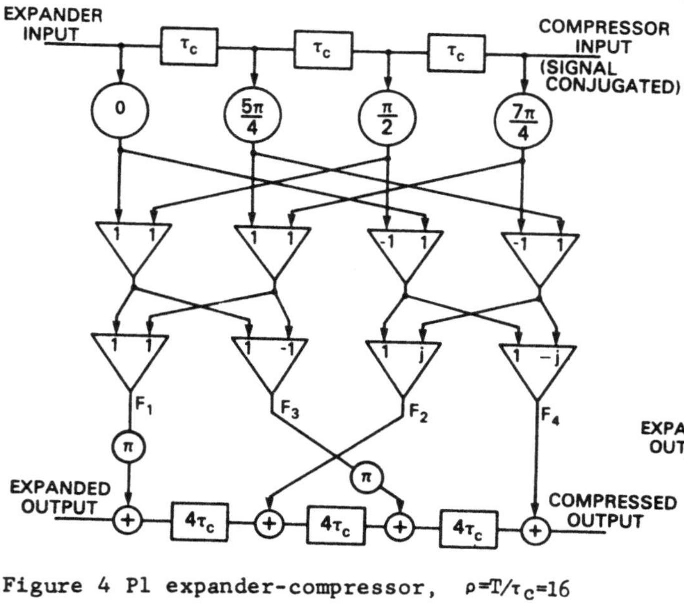

Figure 4 P1 expander-compressor, $\rho=T/\tau_c=16$

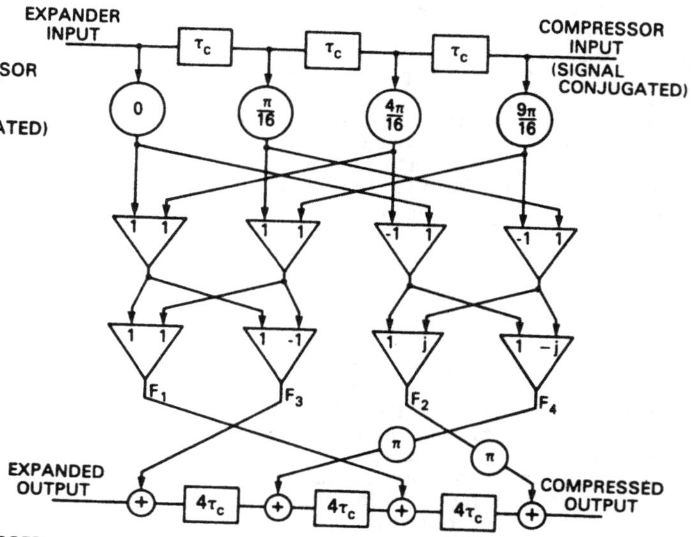

Figure 7 P4 expander-compressor, $\rho=T/\tau_c=16$

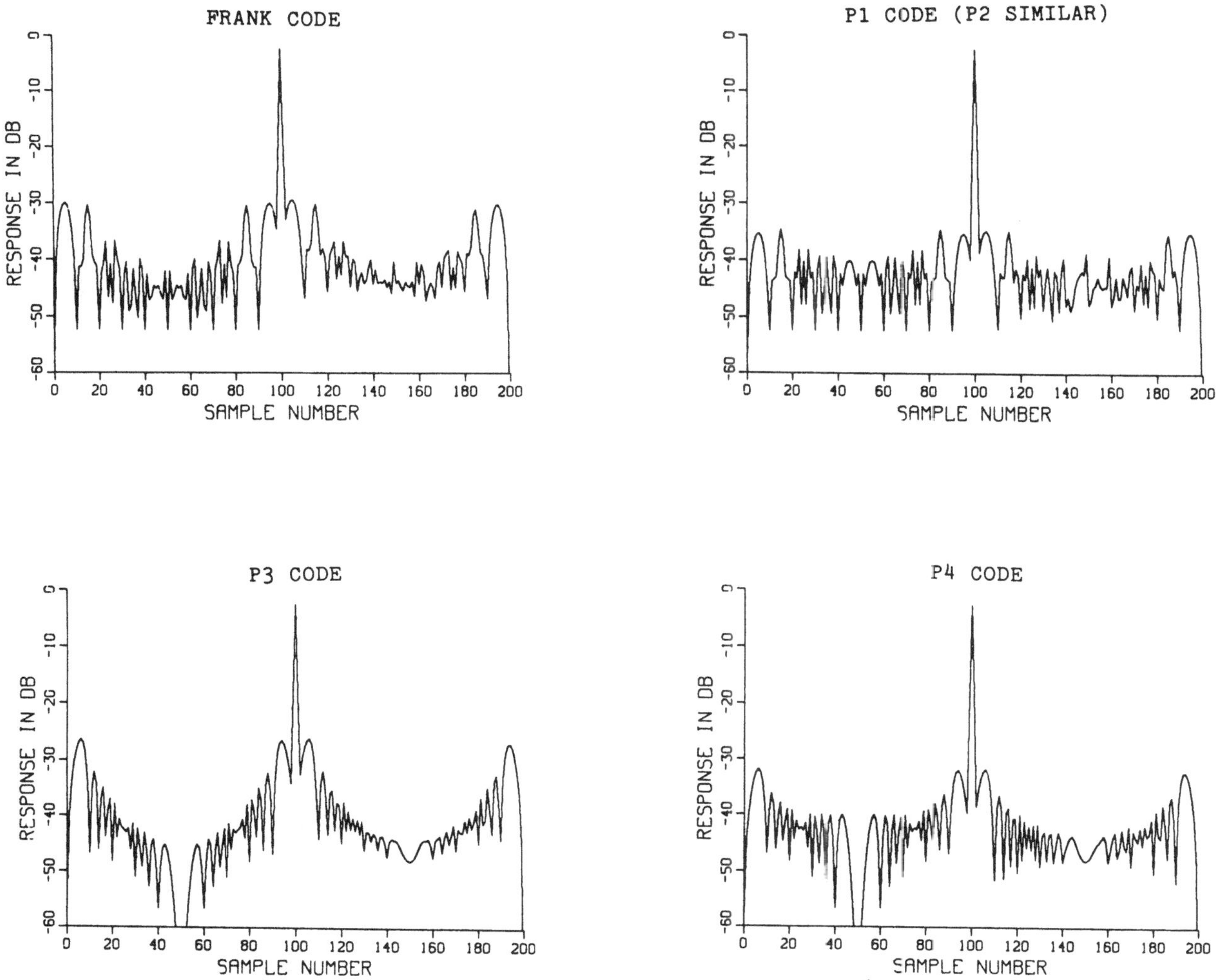

Figure 8 Effect of precompression bandlimiting on autocorrelation functions, ρ = 100. (Input oversampled by 5 and sliding window averaged by 5)

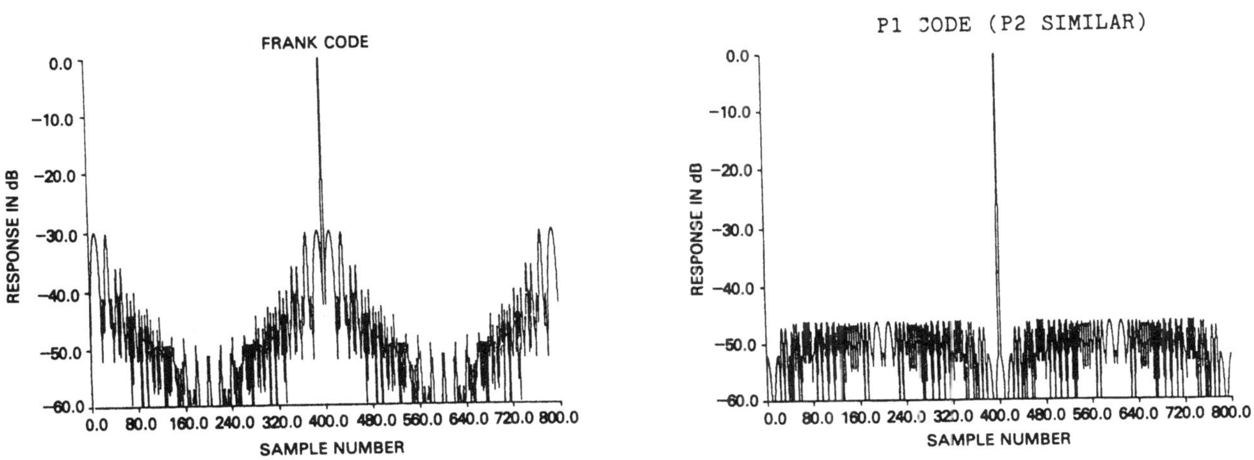

Figure 9(a) Effect of 2 sample adder on compressor output, ρ = 400

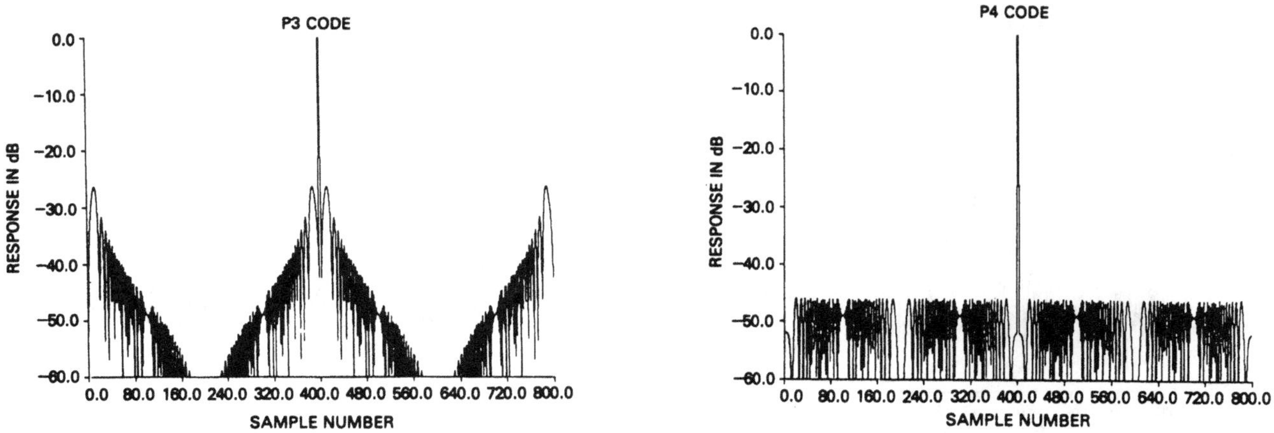

Figure 9(b) Effect of 2 sample adder on compressor output, ρ = 400

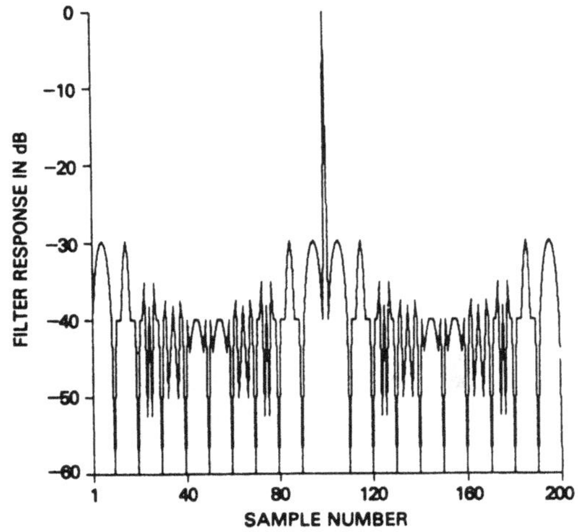

Figure 10(a) Frank code autocorrelation, ρ=100, zero doppler shift and no bandwidth limitation (P1 and P2 similar)

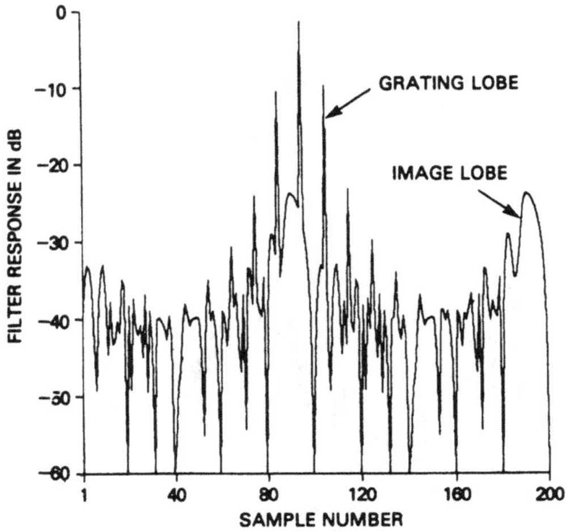

Figure 10(b) Frank code autocorrelation function, ρ=100, doppler =0.05B with no bandwidth limitation (P1 and P2 similar)

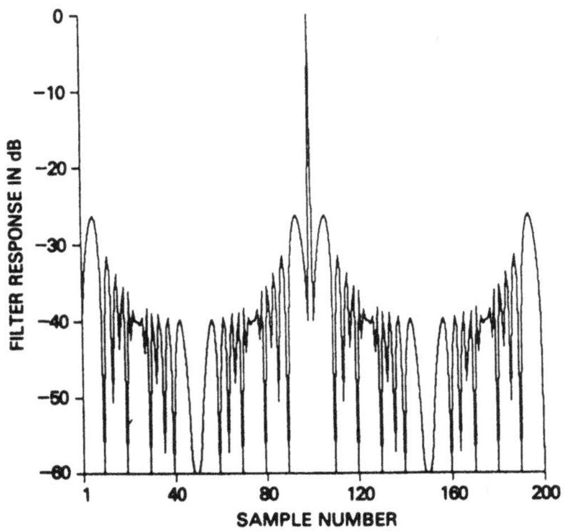

Figure 10(c) P3 or P4 autocorrelation function, ρ=100, zero doppler shift and no bandwidth limitation

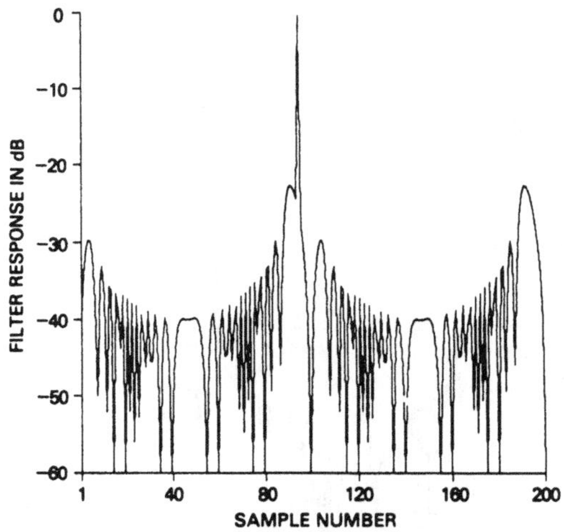

Figure 10(d) P3 and P4 autocorrelation function, ρ=100, doppler =0.05B with no bandwidth limitation

POLYPHASE PULSE COMPRESSION WAVEFORMS

INTRODUCTION

Pulse-compression techniques have been recognized for some time as a means of obtaining sufficient average power on targets for detection, while retaining a desired range resolution with peak-power limited radars. In radar practice, waveforms having a constant amplitude are usually generated to obtain maximum transmitted signal power. Under these conditions, a constant-amplitude pulse of length T can be compressed to a pulse of length τ by phase modulating the signal so that the spectral bandwidth is approximately equal to $1/\tau$. The resultant pulse compression ratio ρ is then equal to T/τ or TB, where B is the bandwidth which is equal to $1/\tau$. This phase modulation is commonly achieved by a linear *fm*, or chirp waveform where the phase varies quadratically with time so that the instantaneous frequency varies linearly with time. The frequency spectrum of the chirp signal is nearly rectangular with a width B, and the compressed pulse is approximately equal to the Fourier transform of the frequency spectrum.

The resultant sin t/t pulse has large time sidelobes which are capable of masking a nearby weak target and therefore a weighting, such as the Taylor weighting, is generally applied to reduce the sidelobe levels [1,2]. These weights symmetrically reduce the amplitude of the rectangular spectrum at the edges of the band and result in lower sidelobes. A weighting applied to the received waveform results in a mismatch which causes a loss typically on the order of 1 to 2 dB in the output peak-signal to noise ratio. Also the pulsewidth of the compressed pulse is widened.

Another common pulse-compression waveform is the binary-phase-coded waveform where the carrier is modulated by 0° and 180° phases. Pseudo-random binary sequences may be generated by using shift registers and, in general, the best binary pseudo-random sequences have a peak sidelobe level which is down from the main response by a factor of ρ. These codes are useful where a thumbtack ambiguity surface is desired. The doppler response of these codes is generally poor, and multiple doppler channels are required over the range of expected doppler returns.

Complementary codes generally consist of two binary sequences which are combined after pulse compression and result in low sidelobes. However, these codes likewise have a poor doppler response and are not generally useful in radar because of the need to separate the two codes in frequency, time, or polarization to permit them to be compressed separately. This separation can cause decorrelation by radar targets or distributed clutter and prevent cancellation of the sidelobes of the combined compressed pulses.

POLYPHASE CODES AND DIGITAL PULSE COMPRESSION

Advantages

The polyphase-coded waveforms discussed in this report offer many advantages over analog pulse-compression waveforms. These advantages include the ability to achieve low sidelobes without weighting, although weighting can be applied easily to achieve still lower sidelobes. Also, the polyphase coded waveforms are: (a) relatively doppler tolerant; (b) easily implemented; (c) have no

Manuscript submitted August 27, 1981.

reflections, as there may be in acoustic delay lines; (d) relatively insensitive to phase errors; and (e) enjoy the advantages of digital processing. These advantages include reliability, reproducibility and compatibility with other digital signal-processing functions, such as moving-target indicator (MTI) and pulse doppler. The use of digital pulse compression allows the digital MTI to precede the digital pulse compressor without requiring multiple A/D and D/A conversions. Also, placing the MTI before the pulse compressor reduces the dynamic range requirements of the MTI.

Equivalance to Analog Processing

In this section we denote by a_i the complex (I,Q) baseband samples of a received uncompressed pulse having an intermediate frequency (IF) bandwidth B. The complex video bandwidth is $B/2$ and it is assumed that samples are taken every $1/B$ s. Recalling that the optimization criterion leading to the matched filter maximizes the output peak signal to average noise power ratio S_p/N, which is given by $2E/N_0$, [2] it will be shown that digital processing achieves the same value for S_p/N.

For the digital case, the peak signal output power at the matchpoint is given by

$$S_p = k\left(\sum a_i a_i^*\right)^2 = k\left(\sum |a_i|^2\right)^2 \tag{1}$$

where k is a constant. In the above and following summations, the index ranges from 1 to ρ. The output noise voltage of the matched filter is

$$n = k \sum a_i^* n_i \tag{2}$$

where n_i is equal to the complex value of the ith noise sample. The average noise power in the signal envelope is then

$$\overline{|n|^2} = k\,\overline{\sum_i \sum_j a_i^* a_j n_i n_j^*}\ . \tag{3}$$

For complex, zero-mean, band-limited white noise, the coefficients are uncorrelated for a sampling interval T_s equal to $1/B$ and Eq. (3) becomes

$$\overline{|n|^2} = k\,\overline{|n_i|^2} \sum |a_i|^2\ . \tag{4}$$

The noise powers of interest may be computed by considering the narrowband representation for the IF noise waveform $x(t)$ given by

$$x(t) = n_I(t)\cos \omega_0 t + n_Q(t) \sin \omega_0 t$$

where n_I and n_Q are slowly varying independent Gaussian noise processes; n_I and n_Q have 0 means and equal variances σ_I^2 and σ_Q^2, which are also denoted by σ^2. The average noise power is

$$\sigma_x^2 = \left(\sigma_I^2 + \sigma_Q^2\right) / 2 = \sigma^2$$

$$= KT_0FB = N_0B \, , \tag{5}$$

where K is the Boltzman constant, T_0 is the standard noise temperature, and F is the noise figure. Also,

$$\overline{|n_i|^2} = \sigma_I^2 + \sigma_Q^2 = 2\sigma^2 = 2N_0B \, . \tag{6}$$

Substituting Eq. (6) in Eq. (4), the average envelope noise power is

$$\overline{|n|^2} = k2N_0B\sum|a_i|^2 \, , \tag{7}$$

and using the relation that N is equal to one half of the envelope noise power computed in Eq. (7), we have

$$\frac{S_p}{N} = \frac{\sum|a_i|^2}{N_0B} \, . \tag{8}$$

From Reference 3,

$$\sum|a_i|^2 = 2BE$$

so that

$$\frac{S_p}{N} = \frac{2E}{N_0} \tag{9}$$

is in agreement with the analog value.

Note that although the ratio S_p/N is the same for the analog and digital compressed pulses, the sidelobes are generally different.

THE FRANK-POLYPHASE-CODED WAVEFORM AND A COMPARISON WITH OTHER WAVEFORMS [4,5]

It will be shown that the phases of the Frank-coded waveforms are the same as the appropriately sampled phases of a step-chirp waveform. These phases are shown to be the same as the steering phases of a discrete Fourier transform (DFT), which means that this code can be generated efficiently or compressed by using a fast Fourier transform (FFT).

The doppler properties of the Frank code are similar to those of a step-chirp waveform and the Frank-coded waveform is more tolerant of doppler than the pseudo-random binary codes or the nonlinear chirp waveforms [1]. The doppler response of the compressed Frank-coded waveform is down approximately 4 dB, like the binary code, when the total accumulated phase shift due to doppler across the uncompressed pulse is π. The binary-code response continues to decrease with increasing doppler shift, while the Frank-code response increases to nearly full amplitude for a phase shift of 2π. The Frank-code response is cyclic with troughs occurring at odd multiples of π and with peaks occurring at multiples of 2π phase shift across the uncompressed pulse. This was not recognized in the publisher literature [1,6] since the doppler cuts were taken at much larger doppler intervals. The cyclic nature of the Frank code doppler response can be easily compensated to further improve the doppler response.

It is later shown that for a Frank code consisting of $N^2 = \rho$ phases, the peak sidelobe is down from the main response by a factor of $(\rho\pi^2)$. The best pseudo-random shift-register binary codes have peak sidelobes that are down by a factor which approaches ρ so that the Frank-code waveforms have lower peak and rms sidelobes than the binary codes. This means that, in a distributed clutter environment, the clutter received via the Frank-code-waveform sidelobes is less than that received via the binary-waveform sidelobes.

The sidelobe level of the Frank code decreases with increasing pulse-compression ratio and low sidelobes are achieved without weighting. However, a further reduction in the sidelobe level can be achieved easily by weighting. In contrast, the chirp signal is generally weighted and there is an attendant loss in S/N and widening of the pulsewidth. This section of this report concludes with a discussion of polyphase-code sidelobe reduction techniques and the sensitivity of polyphase codes to phase errors.

Frank-Polyphase-Coded Waveform

The Frank-polyphase-coded waveform may be described and generalized by considering a hypothetically sampled step-chirp waveform [4]. The Frank code was not originally described in this manner, but was given in terms of the elements of a matrix [7]. As an example, consider a four-frequency step-chirp waveform as shown in Fig. 1(b) where the F_i's denote frequency tones. In this waveform, the frequency steps are equal to the reciprocal of the tone duration $4\tau_c$, where τ_c denotes the compressed pulse width. Assuming this waveform has been beat to baseband I and Q using a synchronous oscillator having a frequency the same as the first tone frequency, the resultant phase-vs-time characteristic consists of four linear sections as shown on Fig. 1(a). The corresponding baseband frequencies are the subharmonics of the frequency $1/\tau_c$. If the baseband phases of the step-chirp waveforms are sampled every τ_c s and held for τ_c s, the phase sequence shown in Fig. 1(c) is obtained. This sequence of phases constitutes the phases of a Frank code for $N = 4$, corresponding to the four baseband frequencies of the hypothetical step-chirp waveform. The actual transmitted Frank-coded-waveform consists of a carrier whose phase is modulated according to the indicated baseband waveform sequence. For each frequency, or section, of the step-chirp phase characteristic, a phase group consisting of N phase samples is obtained and the total number of code phases is N^2 which is equal to the pulse-compression ratio. Note that the phase increments within the four phase groups are 0°, 90°, 180° and 270°. However, the phases of the last group are ambiguous (>180°) and appear as −90° phase steps or as the conjugate of the F_1 group of phases, which corresponds to the lower sideband of F_1. The last group of phases appears, because of the ambiguity, to complete one 360° counterclockwise rotation rather than the $(N - 1)$ rotations of the end frequency of the step-chirp waveform.

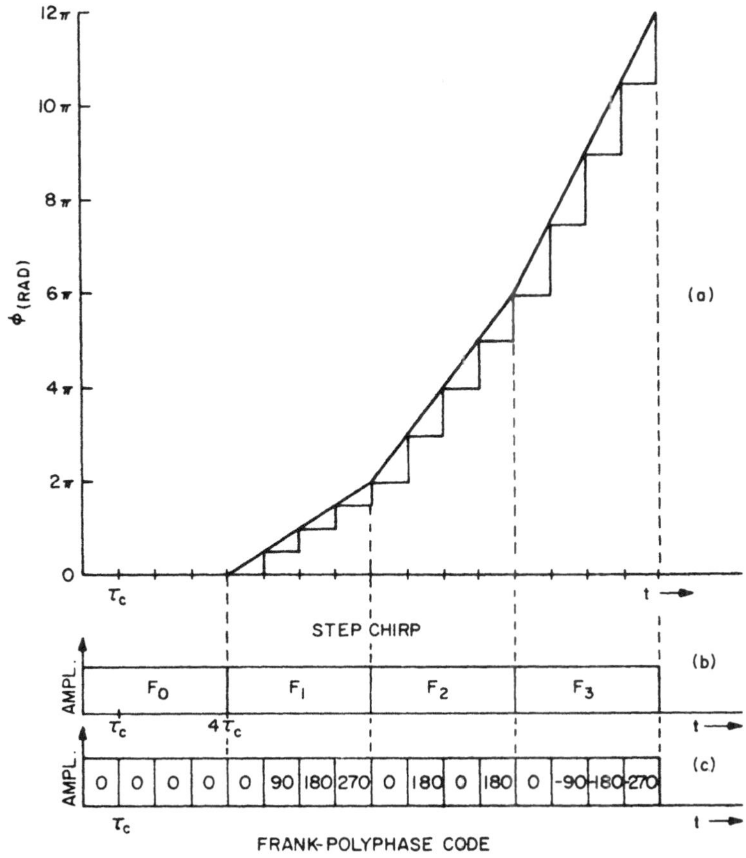

Fig. 1 — Step-chirp and Frank-polyphase-code relationships

The Frank-code phases may be stated mathematically as follows. The phase of the *i*th code element in the *j*th phase group, or baseband frequency, is

$$\phi_{i,j} = (2\pi/N)(i - 1)(j - 1) \tag{10}$$

where the index i ranges from 1 to N for each of the values of j ranging from 1 to N. An example of a Frank-code pulse generation for $N = 3$ is shown in Fig. 2. The Frank-code phases are the same as the negative of the steering phases of an N point DFT where the jth frequency coefficient is:

$$F_j = \sum_{i=1}^{N} a_i e^{-j\frac{2\pi}{N}(i-1)(j-1)}, \tag{11}$$

where a_i is the ith complex input time sample. This means that a considerable savings in hardware can be achieved by using the efficiency of an FFT.

The matched-filter output for an $N = 10$ or 100-element Frank code is shown in Fig. 3. This figure and the following figures showing the compressed pulse were obtained by sampling the input baseband waveform once per code element or per reciprocal bandwidth unless stated otherwise. Using a discrete-time matched filter the output signal is also a discrete-time sampled signal.

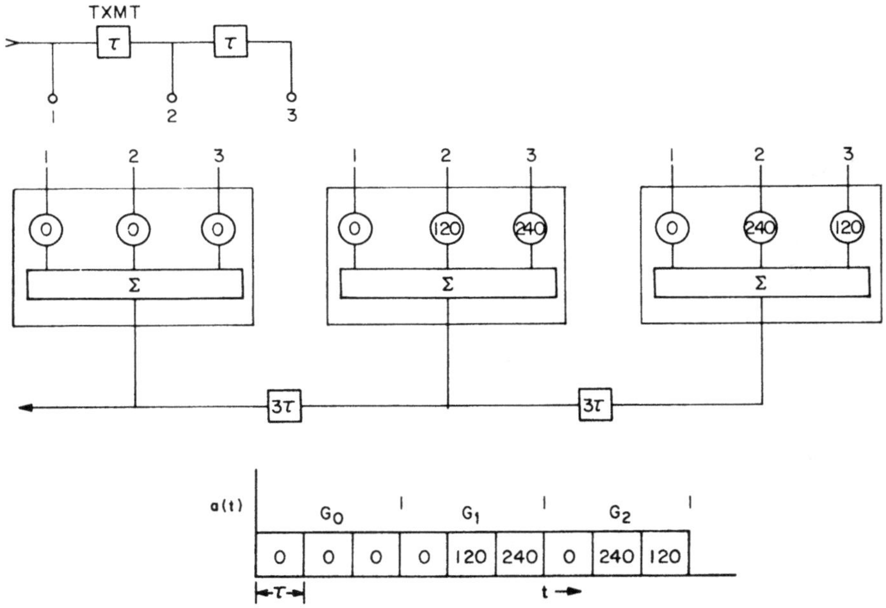

Fig. 2 — Simplified Frank-code generation

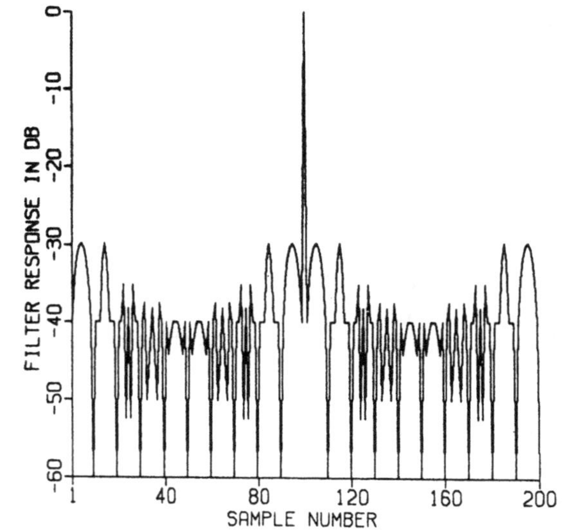

Fig. 3 — Compressed pulse of 100-element Frank-polyphase code

However, for ease of plotting and viewing, the points were connected by straight lines. The four sidelobe peaks on each half of the match point (peak response) are of equal magnitude. The first peak sidelobe at sample number 5 in Fig. 3 occurs as the last phase group having $-36°$ phase increments indexes halfway into the first phase group of zero phase vectors in the autocorrelation process. In general, at sample number $N/2$, there are $N/2$ vectors adding to complete a half circle. The end phase group indexing into the first phase group of $0°$ vectors makes an approximate circle since the phases of the last phase group make only one rotation as stated previously. The peak sidelobe amplitude may be approximated by the diameter D of the circle from the relation,

$$\text{Perimeter} = N = \pi D \tag{13}$$

$$\text{or} \qquad D = N/\pi \, . \tag{14}$$

At the match point the amplitude is N^2 so that the peak-sidelobe to peak-response power ratio R is

$$R = \frac{N^4}{(N/\pi)^2} = N^2\pi^2 = \rho\pi^2 \, . \tag{15}$$

For a 100-element Frank code, this ratio is approximately 30 dB as shown in Fig. 3.

Had the phases of the polyphase-coded waveform been generated by using the phases of step-chirp phase characteristics sampled at 1/5 of the interval used for the Frank code, the compressed code would appear as shown in Fig. 4. In this figure, five samples are equal in time to one sample in Fig. 3. Note in Fig. 4 that the near-in sidelobes are approximately 13 dB and that the envelope of the sidelobe peaks is approximately that of a sin x/x pulse. The 13-dB sidelobes also appear for an oversampling of 2:1. Also note that the compressed pulsewidth in Fig. 4 has not decreased since it is determined by the underlying bandwidth of the step-chirp waveform.

A comparison between the Frank code and a "good" binary code may be made by referring to Figs. 5 and 6 which have similar pulse-compression ratios. Fig. 7 shows a comparison of the

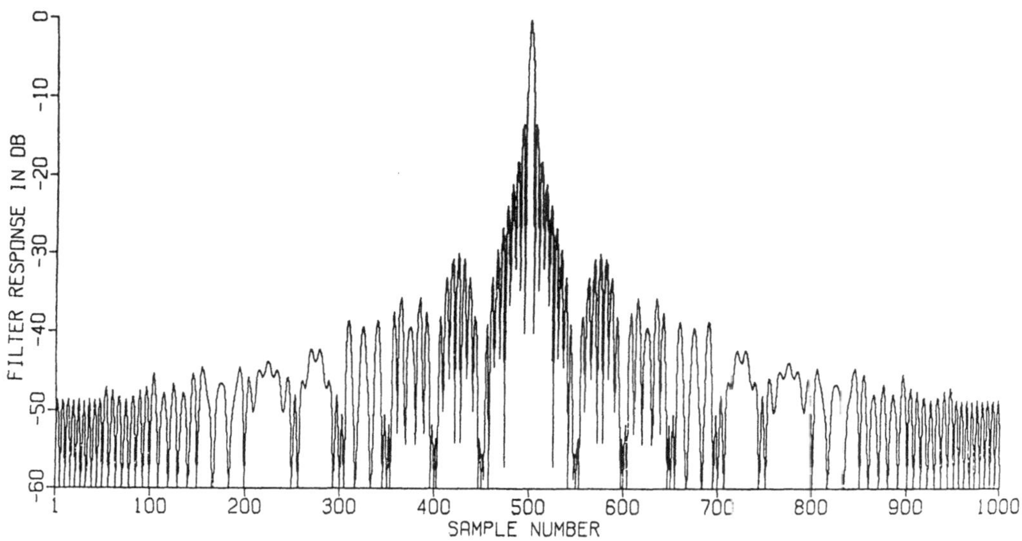

Fig. 4 — Compressed pulse of oversampled step-chirp (5:1)

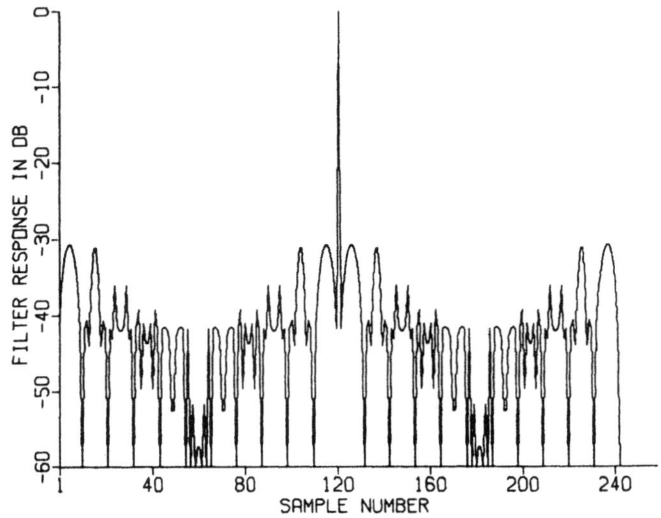

Fig. 5 — Compressed pulse of 121-element Frank code

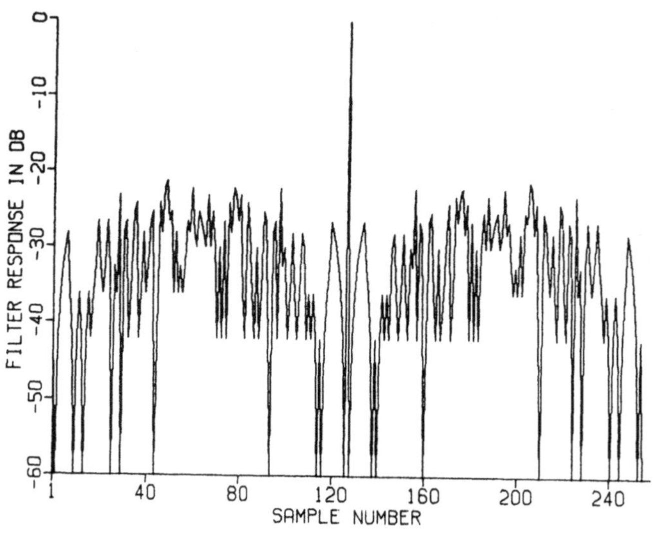

Fig. 6 — Compressed pulse of 127-element binary shift register code

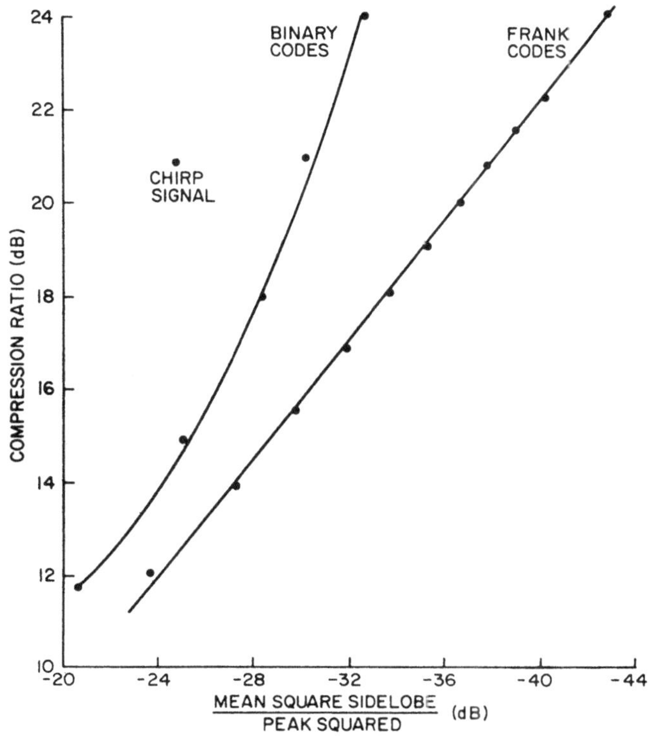

Fig. 7 - Comparison of mean-squared sidelobes

mean-square sidelobe power of the two codes. From this, one can see that better performance is achieved with the Frank code in a distributed clutter environment where clutter is introduced via the sidelobes of the compressed pulse. Also shown in Fig. 7 is a point, for the sake of comparison, for an unweighted chirp signal. The relatively high mean-square sidelobe level is due to the high near-in sidelobes.

Sidelobe Reduction Techniques

Various methods have been investigated to cause a further reduction of the Frank-code sidelobe levels. One method is based on a least-squares technique [8] whereby, for a given input waveform, the filter coefficients are found such that the output of the compression filter best approximates an idealized impulse function. This technique can also be applied to binary waveforms. It was found that this technique did not produce a symmetrical output waveform for a Frank-polyphase-coded input waveform. However, for the new P1 coded waveform, to be discussed in this report, a small sidelobe reduction was achieved.

Another method for reducing the sidelobes was investigated by Somaini and Ackroyd [9]. Their approach was to perturb the phases of a Frank code by search methods until an improved autocorrelation function was achieved. Using their resultant perturbed waveform, the peak sidelobes for a 100-element code were reduced from 30 to approximately 36 dB.

The shortcomings of the preceding techniques are that the doppler responses are not quite as good as the Frank code and that the filter cannot be implemented using FFT efficiency. The most effective method that has been found for reducing the sidelobes is achieved by simply weighting the output frequency ports of the FFT compression filter. Any of the recognized weightings can be used in this manner. Fig. 8 shows the results of using a cosine-on-a pedestal (of 0.4) weighting on a 100-element Frank-coded waveform. The peak signal is reduced as shown but the loss in signal-to-noise ratio (S/N) is small.

Doppler Response of Frank Code

A partial ambiguity function for a 100-element Frank code is shown in Fig. 9 which shows the amplitude in dB of a matched-filter output for given doppler shifts of the input. The doppler is normalized to the signal bandwidth and the delay axis is normalized to the uncompressed pulse length. The vertical scale ranges from 0 dB to −60 dB, and the −30-dB sidelobes for 0 doppler are evident. A front view is shown in Fig. 10 where the sidelobes are plotted down to the −30-dB level. The normalized doppler shift of −0.05 shown in this figure corresponds to a mach-50 target for an L-band radar having a signal bandwidth of 2 MHz. The first doppler cut shown in the literature [6] is taken at this normalized doppler and the resultant high-peak sidelobes have perhaps discouraged usage of the Frank code. The region shown between 0 and mach-5 doppler and a delay interval of ±0.3 is of interest, and it is shown on an expanded scale in Fig. 11. In this region the doppler response is good in terms of the sidelobe levels. At the doppler shift of −0.005, or more generally $1/(2\rho)$, the total phase shift across the uncompressed pulse is π and the peak response drops approximately 4 dB. At this doppler, there is a range-doppler coupling of 1/2 of a range cell with the result that the signal splits between two range cells. At a normalized doppler shift of −0.01, or in general $1/\rho$, there is a range-doppler coupling of one range cell resulting from a total phase shift of 2π radians across the uncompressed pulse, and the main peak response is nearly restored to full amplitude as shown in Fig. 9. This effect is cyclic and an approximate loss of nearly 4 dB is encountered when the total phase shift due to doppler is an odd multiple of 180°. This also occurs for the binary code except that the response is not cyclic and it monotonically decreases with frequency. Moreover, the troughs in the doppler response of the Frank code can be easily compensated by using an additional channel having a phase compensation of 180°. Also, it has been found that

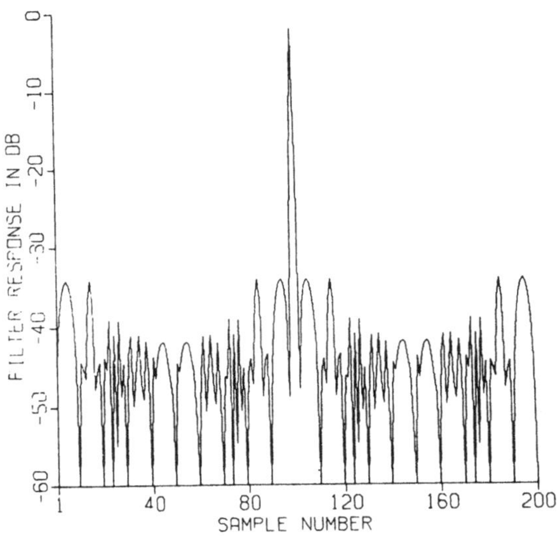

Fig. 8 — Compressed pulse of weighted Frank code

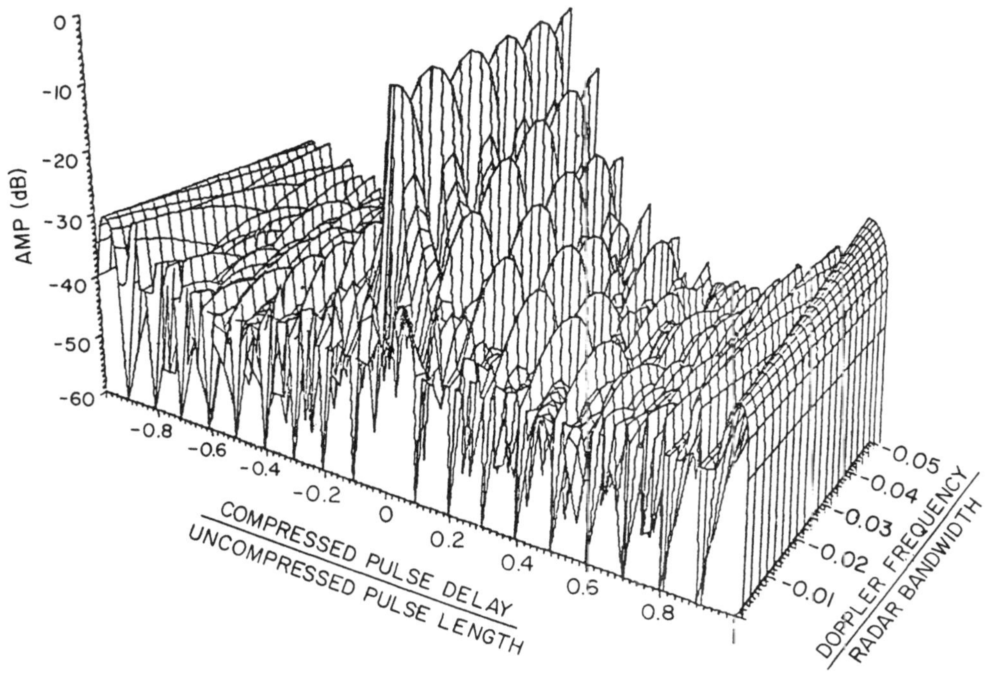

Fig. 9 — One hundred-element Frank code, partial ambiguity diagram

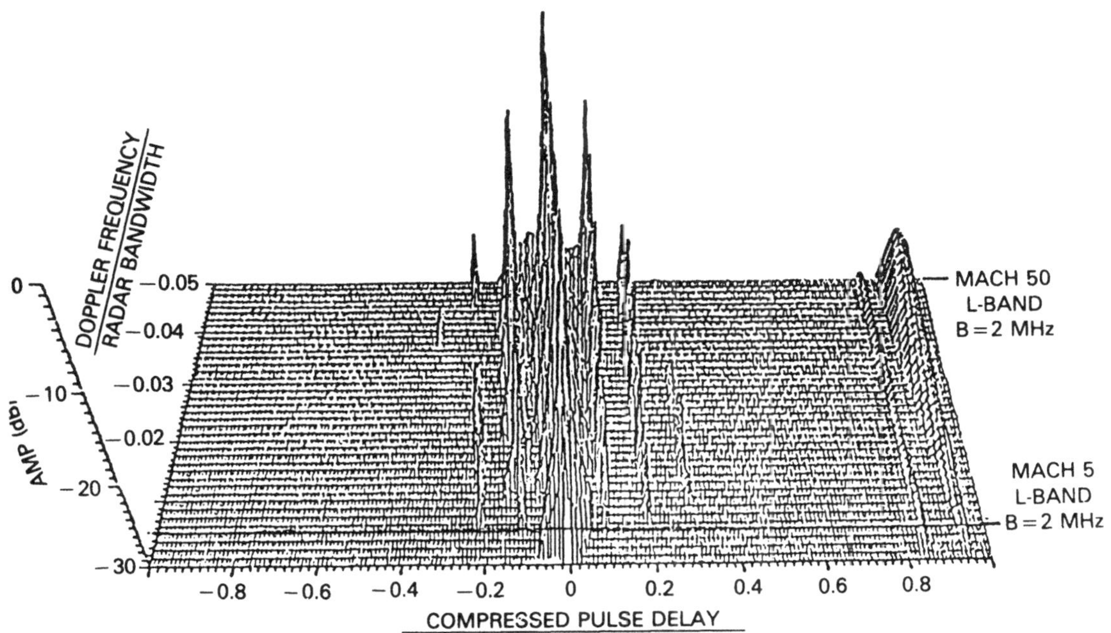

Fig. 10 — One hundred-element Frank code, partial ambiguity diagram

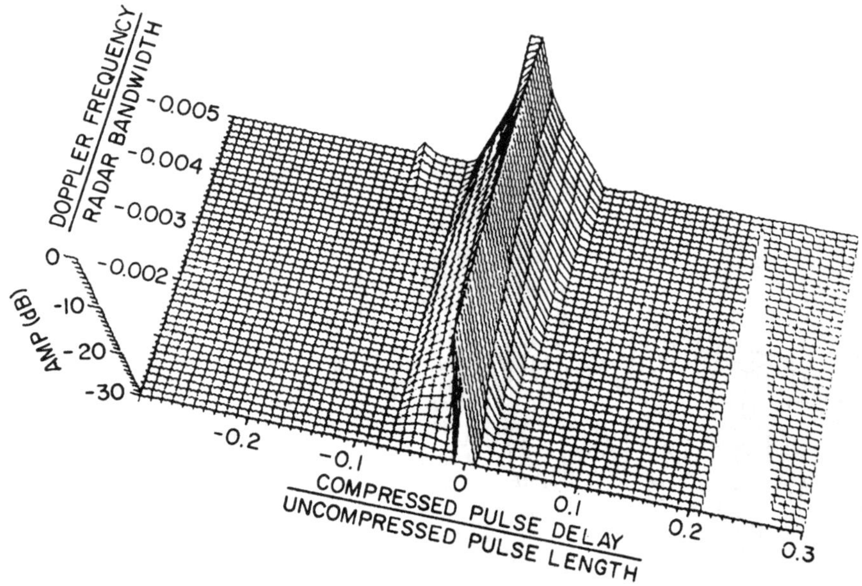

Fig. 11 — Magnified ambiguity diagram of 100-element Frank code

that the use of weighting improves the sidelobes and reduces the variation in the mainlobe peak in the presence of doppler.

Figure 12 shows the output pulse for a 100-element-Frank-coded waveform having a doppler shift of −0.005 or a total phase shift of π across the uncompressed pulse. Figure 13 shows the effect of weighting on receive. In addition to the reduction of the end sidelobes, the mainlobe width has been reduced. These aspects of doppler compensation techniques are discussed in more detail in Ref. 10.

Error Analysis

Computer simulations were performed to determine the sensitivity of the polyphase codes to phase and amplitude errors. The two types of errors considered were random errors in I and Q and quantization errors in I and Q which are encountered in A/D conversions.

Random Errors

Two types of random errors were considered as shown in Fig. 14. In each case independent, uniformly distributed errors in I and Q were generated over an interval $\pm x$. For the first type shown in Fig. 14(a), the error ϵ was determined by letting x be a given percentage of the nominal I or Q value for each code element phasor. The resultant vector is denoted as E_R. The other type of error ϵ', shown in Fig. 14(b) was generated as explained above, except that x was specified as a fixed error rather than a percentage of I or Q. In this case, the resultant vector is denoted by E_A. In determining E_R and E_A, the nominal signal amplitude is assumed to be unity.

Monte Carlo simulations were performed to determine the effect of the relative and absolute errors on the peak and average sidelobes of Frank codes with pulse compression ratios of 256 and 64. The results for $\rho = 256$ are shown in Fig. 15 with similar results obtaining for $\rho = 64$. Each point, other than for zero error, was obtained by taking an average of 100 compressed pulses to compute the indicated peak sidelobe and average sidelobe levels. The errors were assumed to occur on either transmission or reception but not both. The results of this simulation indicate that the

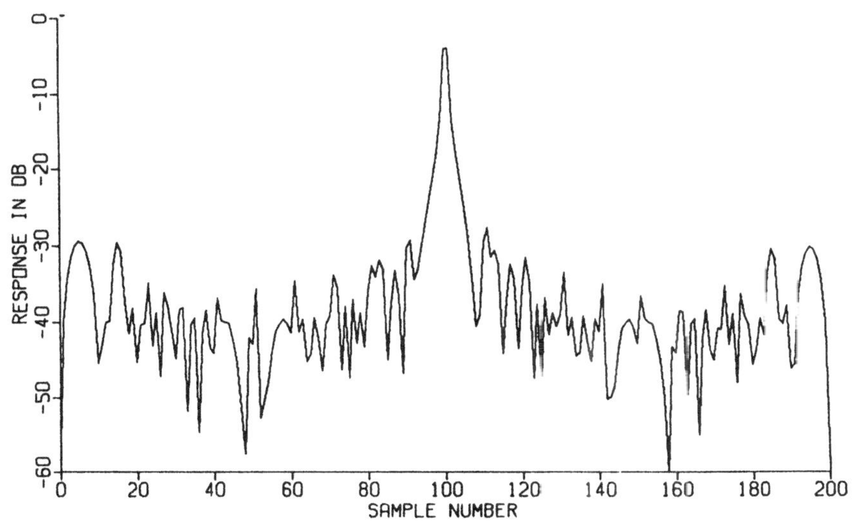

Fig. 12 — Compressed pulse of 100-element Frank code doppler shifted by 0.005

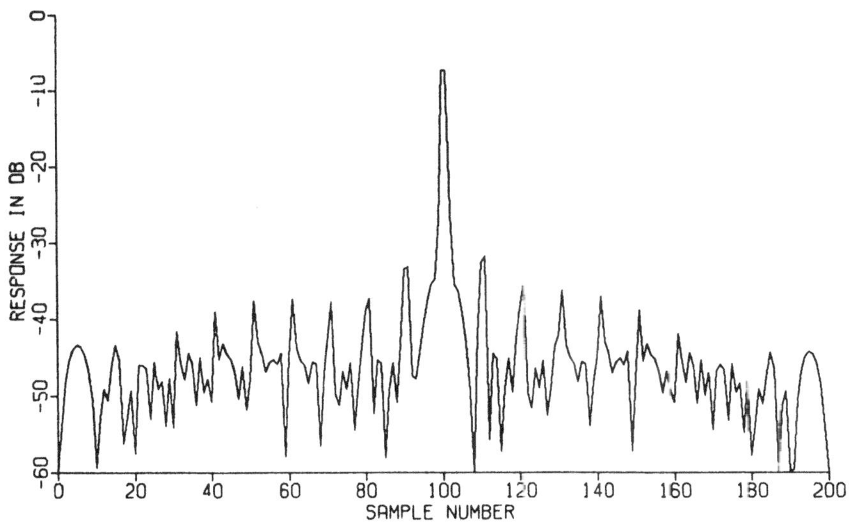

Fig. 13 — Effects of weighting on compressed pulse of 100-element Frank code doppler shifted by 0.005

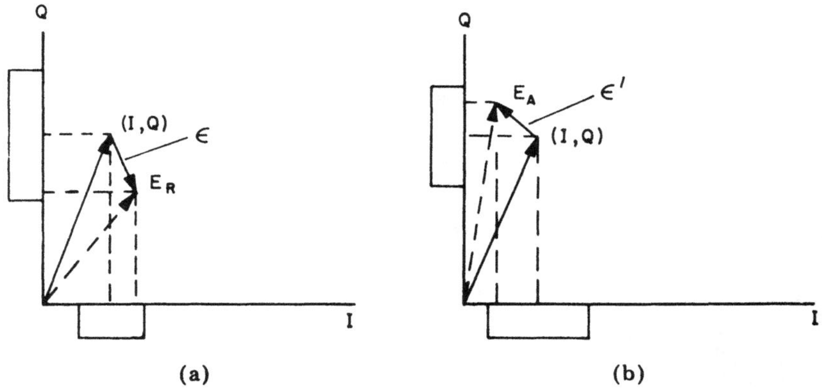

Fig. 14 — Random errors

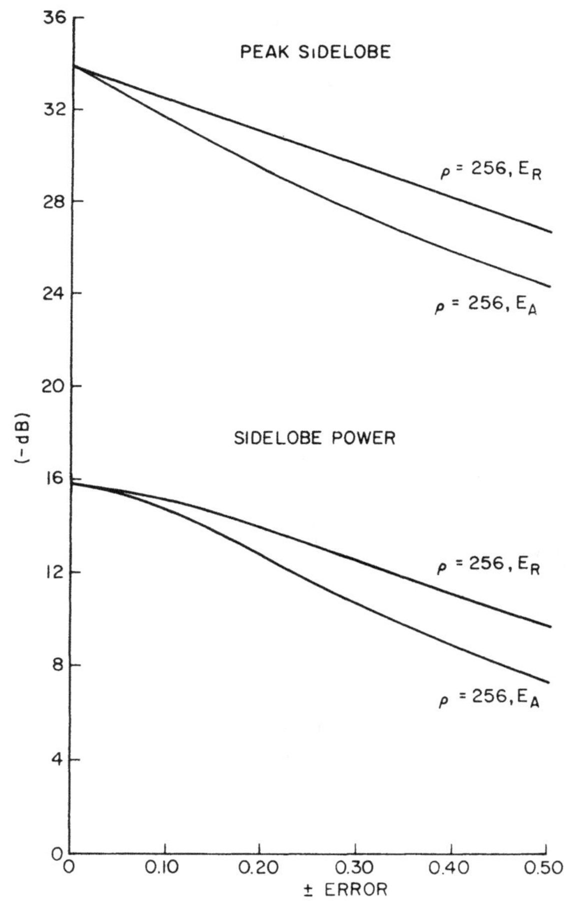

Fig. 15 — Frank-polyphase-code error analysis

sidelobes are not sensitive to the errors. For example, for the absolute error case, an error distribution of ±0.10 results in approximately a 2-dB average degradation in the peak sidelobe and a 1.2-dB degradation in the average sidelobe power.

Quantization Errors

The results of the Monte Carlo simulations previously described indicate the robustness of the polyphase codes to random errors. To quantify the effects of quantization errors, computer simulations were performed. The average and peak sidelobes were determined for a symmetric A/D characteristic having the phase and amplitude specified within the accuracy of the quantization levels determined by the number of bits (including sign). Compression ratios of 144 and 36 were considered in the simulations, which did not include noise. It was assumed that the errors were due only to the A/D converters and that the matched-filter phases and amplitude were perfect. The results are shown in Fig. 16 where each curve exhibits a knee. The knee location is seen to vary the most between the ρ = 36 and ρ = 144 peak sidelobes. The general conclusion reached from these results is that the polyphase code is relatively insensitive to the number of bits beyond a certain number. Other considerations, such as dynamic range, may dictate the use of more bits than indicated in Fig. 16.

NEW POLYPHASE CODES [4,5]

Effects of Bandlimiting Prior to Pulse Compression

A Frank-coded waveform is depicted in Fig. 17(a) where the G_K's denote the phase groups corresponding to the sampled phases of a step-chirp waveform as previously discussed. Each group

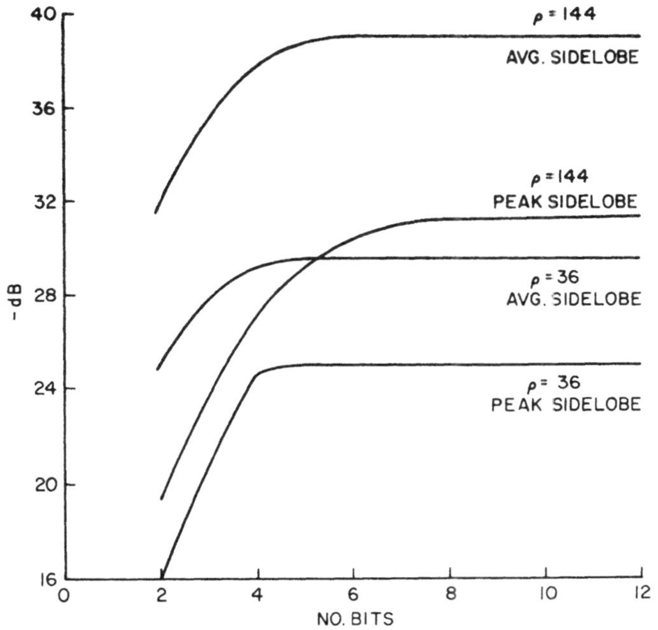

Fig. 16 — Frank-polyphase code, A/D quantization errors

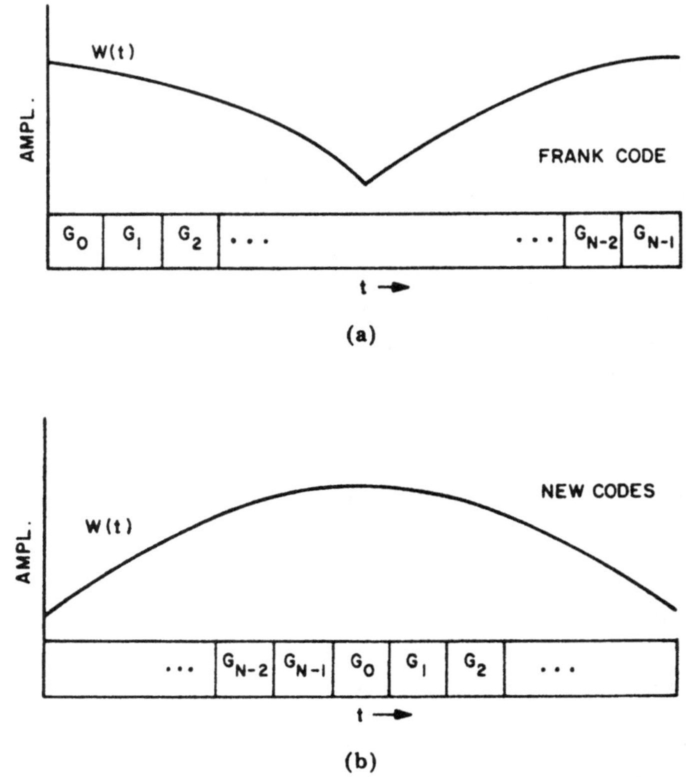

(a)

Fig. 17 — Effect of bandlimiting before pulse compression

consists of N vectors beginning with a vector at a phase angle of 0°. The phase increments within the Kth group are

$$\Delta\phi_k = K\frac{360°}{N}. \tag{16}$$

Thus G_0 consists of N vectors at 0°, G_1 has vectors separated by $360°/N$ until at the center of the coded waveform the phase increments approach or become 180° depending on whether N is odd or even. For phase increments greater than 180°, the phases are ambiguous with the result that the phasors of phase group G_{N-K} are the conjugates of the phasors of phase group G_K so that the vectors have the same increments but rotate in opposite directions. The result is that the phase increments are small at the ends of the code and become progressively larger toward the center of the code where the increments approach 180° from opposite directions.

If a receiver is designed so that it has an approximate rectangular bandwidth corresponding to the 3-dB bandwidth of the received waveform, the received waveform becomes bandlimited and a mismatch occurs with the compressor. This bandlimiting would normally occur prior to sampling in the A/D conversion process in order to prevent noise foldover and aliasing. The result of any bandlimiting is to average (or smooth) the vectors constituting the coded waveform, and for the Frank code, a weighting $W(t)$ such as illustrated in Fig. 17(a) takes place due to the larger phase increments toward the middle of the code. This weighting causes an unfavorable mismatch with the compressor which results in a degradation of the sidelobes relative to the peak response.

New symmetrical codes have been found which have the common property that the phase groups with the small phase increments are at the center of the code and the larger increment groups progress symmetrically toward the ends of the code. This is illustrated in Fig. 17(b) where a favorable amplitude weighting resulting from pre-pulse compression bandlimiting is shown.

P1 and P2 Polyphase Codes

The two new polyphase codes which tolerate bandlimiting are referred to as the P1 and P2 codes. The P1 code was derived from use of the previously described relationship between the Frank-code phases and those of a sampled step-chirp waveform. The desired symmetry, having the dc or small incremental phase group at the center of the code, can be achieved by determining the phases which result from placing the hypothetical synchronous oscillator at the center frequency of the step-chirp waveform. For an odd number of frequencies, the synchronous oscillator frequency corresponds to one of the waveform frequencies and the resultant phases are the same as the Frank code except the phase groups are rearranged as indicated in Fig. 17. If there is an even number of frequencies, the synchronous oscillator frequency placed at the center frequency does not correspond to one of the frequencies in the step-chirp signal. The phase of the ith element of the jth group is

$$\phi_{i,j} = -(\pi/N)[N - (2j - 1)]\,[(j - 1)N + (i - 1)] \,, \tag{17}$$

where i and j are integers ranging from 1 to N.

An $N = 3$, P1, code is given by the sequence

$\phi_{1,1}$	$\phi_{2,1}$	$\phi_{3,1}$	$\phi_{1,2}$	$\phi_{2,2}$	$\phi_{3,2}$	$\phi_{1,3}$	$\phi_{2,3}$	$\phi_{3,3}$
0	$-2\pi/3$	$-4\pi/3$	0	0	0	0	$2\pi/3$	$4\pi/3$

which can be seen to be a rearranged Frank code with the zero frequency group in the middle.

The P2 code, which also has the desired features, is similar to the Butler matrix steering phases used in antennas to form orthogonal beams. The P2 code is valid for N even, and each group of the code is symmetric about 0 phase. The usual Butler matrix phase groups are not symmetric about 0 phase and result in higher sidelobes. For N even, the P1 code has the same phase increments, within each phase group, as the P2 code except that the starting phases are different. The ith element of the jth group of the P2 code is

$$\phi_{i,j} = \left[(\pi/2)\,\frac{N-1}{N} - (\pi/N)(i-1)\right]\left[N + 1 - 2j\right] \,, \tag{18}$$

where i and j are integers ranging from 1 to N as before. The requirement for N to be even in this code stems from the desire for low autocorrelation sidelobes. An odd value for N results in high autocorrelation sidelobes.

An $N = 4$, P2, code example is given by the sequence

$\phi_{1,1}$	$\phi_{2,1}$	$\phi_{3,1}$	$\phi_{4,1}$	$\phi_{1,2}$	$\phi_{2,2}$	$\phi_{3,2}$	$\phi_{4.2}$	$\phi_{1,3}$	$\phi_{2,3}$
$+9\pi/8$	$+3\pi/8$	$-3\pi/8$	$-9\pi/8$	$+3\pi/8$	$+\pi/8$	$-\pi/8$	$-3\pi/8$	$-3\pi/8$	$-\pi/8$

$\phi_{3,3}$	$\phi_{4,3}$	$\phi_{1,4}$	$\phi_{2,4}$	$\phi_{3,4}$	$\phi_{4,4}$
$+\pi/8$	$+3\pi/8$	$-9\pi/8$	$-3\pi/8$	$+3\pi/8$	$+9\pi/8$

This code has the frequency symmetry of the P1 code and also has the property of being a palindromic code which is defined as a code having symmetry about the center.

The ambiguity diagram of the P1 code for N odd is identical to that of the Frank code. For N even, the ambiguity diagrams of the P1 and P2 codes are similar to each other and to that of the Frank code.

Simulation of Precompression Bandwidth Limitations

The effect of a restricted bandwidth in the IF amplifiers and the I,Q detectors preceding analog-to-digital conversion and compression of the phase codes was simulated on a digital computer. The various codes to be compressed were over-sampled by 5 to 1 and sliding-window-averaged by 5, 7, and 10 to simulate the precompression bandwidth limitation. Only the results for the sliding-window average of 5 samples are presented here. Similar results were obtained for the other cases. The compressor phases were matched to the input phases which existed prior to oversampling. The resultant oversampled and averaged waveform was then sampled every fifth sample beginning with the first sample and sent to the compressor. To account for take time-of-arrival variations, the sliding-window average for the five-sample case was taken starting 4, 3, 2, 1, and 0 over-sample-periods ahead of the first received code element and sample correlation functions were developed for each case. Note that a match condition occurs for the latter case and otherwise a mismatch occurs.

The results of this study revealed that precompression bandwidth limitations were similar to amplitude weighting the frequency output ports of the digital filters in the compressor when the symmetrical P1 and P2 codes were employed and time-of-arrival variations were taken into consideration (Fig. 18). However, this was not the case when the Frank code was processed (Fig. 19). For the Frank coded waveform, the bandwidth limitation did not affect the dc group and had little effect on the highest frequency code group since it is the conjugate of the frequency code group closest to the dc term. As a consequence, precompression bandwidth limitation did not drop the far-out sidelobe caused by the dc group indexing into the highest frequency filter and vice versa.

Comparison of the sidelobes between the −30 and −40-dB lines of Figs. 18 and 19 shows that, for each corresponding time-of-arrival case, with the exception of the match condition shown in Part (e) of each figure, the sidelobes of the P2 code are lower than the Frank code, while the corresponding peak values are the same.

We mention at this point that although a sliding-window average was used to simulate the band limitation effects in this report, similar results would be expected for any other bandlimiting filter since the sidelobe reduction is due to the smoothing effect of the filter.

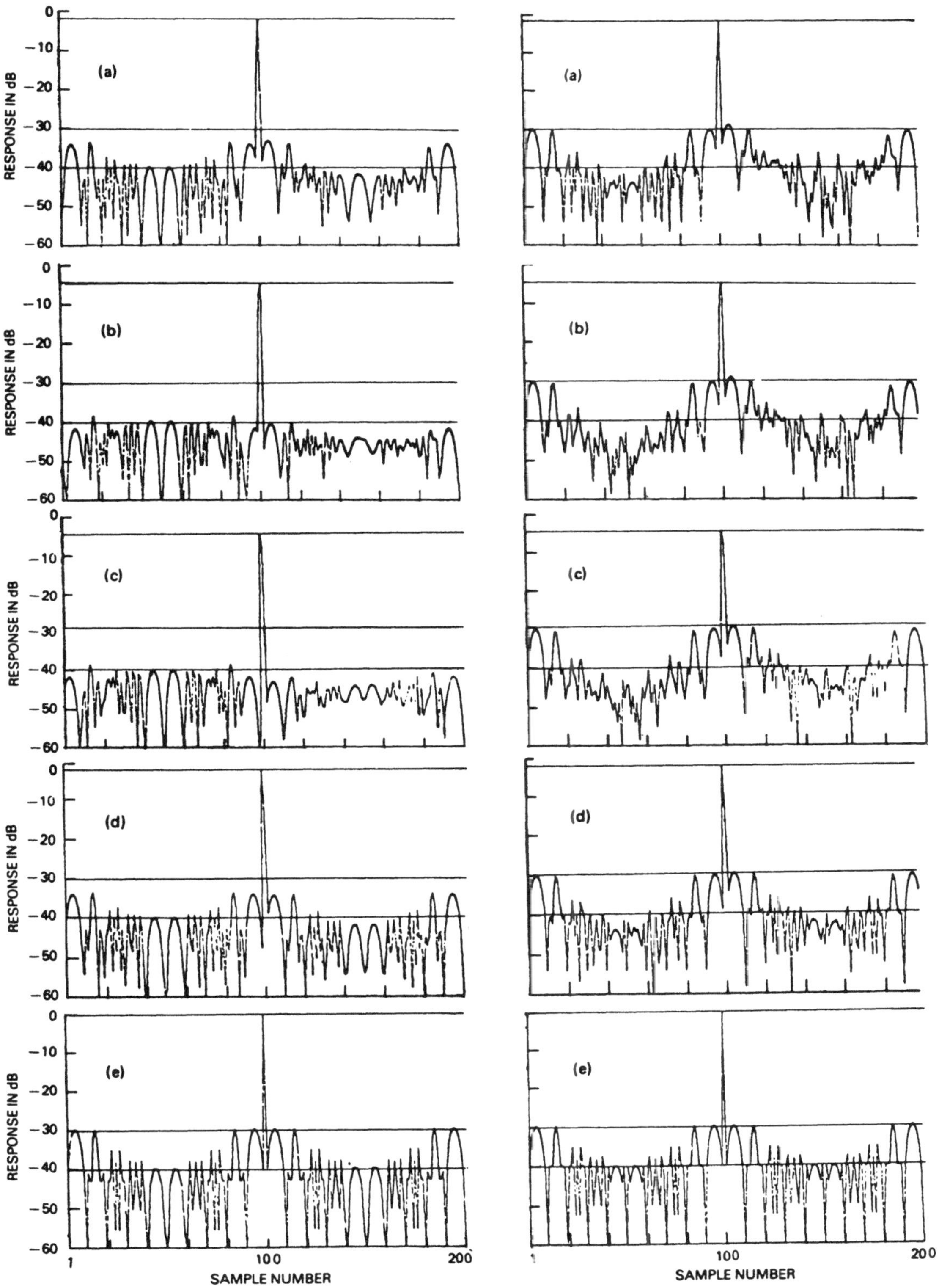

Fig. 18 and 19 — Effect of precompression band limitations on N = 10, P2 code and Frank code respectively, with 5-sample average started (a) 4, (b) 3, (c) 2, (d) 1, and (e) 0 samples ahead of first point to be sent to compressor

The average loss of the peak signal values shown in Figs. 18 and 19 is the same for both the Frank and P2 codes. Some of this loss can be attributed to the passband limitation while the remaining loss represents the loss due to time-of-arrival variation or range cusping. The passband limitation loss is due to the loss of the signal power contained in the sidelobes of the signal spectrum. The thermal-noise contribution is also reduced by the bandlimiting and is the same for each code in Figs. 18 and 19 which account for signal only. It is important to note, however, that the symmetrical P1 and P2 code sidelobes drop more than the peak due to precompression bandwidth limitation the sidelobes of the Frank code do not drop at all. This results in lower sidelobes in the new codes for the same signal-to-noise ratio loss due to the precompression bandwidth limitation.

APPLICATION OF POLYPHASE CODES

The polyphase codes discussed in this report may be used wherever pulse compression is needed and where the anticipated doppler-to-bandwidth ratio is less than approximately $1/(2\rho)$ corresponding to a range-doppler coupling of 1/2 of a range cell. This doppler extent would apply to many search-radar and radar-mapping applications. These polyphase codes have much better doppler tolerance than the binary codes and have lower sidelobe levels.

The polyphase codes may be efficiently implemented to provide large pulse compression ratios, with normalized peak sidelobes given by $1/(\rho\pi^2)$. The achievable compression ratio is primarily limited by the signal bandwidth, which impacts on the A/D sampling rates and the digital circuit speeds. The polyphase pulse compressor does not become less efficient for long-duration waveforms as the analog acoustic delay-line compressors do.

For odd N, the P1 code, which is tolerant of precompression bandlimiting, can be implemented using FFT technology. This results in a considerable hardware savings for large ρ and allows the compression of different pulsewidths using the same processor.

The use of digital processing to compress the polyphase codes is compatible with digital MTI and pulse-doppler processing. As mentioned previously, the digital MTI can precede the digital pulse compressor to reduce the dynamic range requirements of the MTI without the need for multiple A/D and D/A conversions.

SUMMARY

The properties of Frank polyphase codes have been investigated in detail and extended. It was shown how the Frank code can be conceptually derived by appropriately sampling a step-chirp waveform and how the Frank and new polyphase codes are useful for doppler-to-bandwidth ratios less than approximately $1/(2\rho)$. Doppler compensation techniques were presented to improve the performance of the polyphase codes. Also it was found that the polyphase codes are not very sensitive to amplitude and phase errors.

New polyphase codes were described which have more tolerance to precompression bandlimiting than the Frank codes. The precompression bandlimiting acts as a weighting on the Frank codes, which increases the sidelobe levels relative to the peak. The normalized sidelobe levels of the new codes are reduced by the effective weighting caused by precompression bandlimiting.

REFERENCES

1. F. E. Nathanson, "Radar Design Principles," New York: McGraw-Hill, 1969.
2. M. I. Skolnik, "Introduction to Radar Systems," New York: McGraw-Hill, 1980.
3. R. S. Berkowitz, "Modern Radar Analysis, Evaluation, and System Design," New York: Wiley and Sons, Inc., 1965 p. 98.
4. B. L. Lewis and F. F. Kretschmer, Jr., "A New Class of Pulse Compression Codes and Techniques," NRL Report #8387, March 26, 1980.
5. B. L. Lewis and F. F. Kretschmer, Jr., "A New Class of Polyphase Pulse Compression Codes and Techniques," IEEE Transactions on Aerospace and Electronic System," May 1981, AES-17, pp. 364-372.
6. C. Cook and M. Bernfield, "Radar Signals, An Introduction to Theory and Applications," New York: Academic Press, 1967.
7. R. L. Frank, "Polyphase Codes With Good Nonperiodic Correlation Properties," IEEE Transactions on Information Theory, June 1963, IT-9, pp. 43-45.
8. M. H. Ackroyd and F. Ghani, "Optimum Mismatched Filters for Sidelobe Suppression," IEEE Transactions on Aerospace and Electronic Systems, March 1973, AES-9, No. 2, pp. 214-218.
9. U. Somaini and M. H. Ackroyd, "Uniform Complex Codes With Low Autocorrelation Sidelobes," IEEE Transactions on Information Theory, Sept. 1974, IT-20, pp. 689-691.
10. F. F. Kretschmer, Jr. and B. L. Lewis, "Doppler Induced Losses in a Frank Code Pulse Compression System and Correction Techniques," Submitted to IEEE Transactions on Aerospace and Electronic Systems.

EFFECTS OF BANDWIDTH LIMITATION ON POLYPHASE CODED PULSE COMPRESSION SYSTEMS

INTRODUCTION

Bandwidth limitations are found in all well-designed radar systems. Such bandwidth limitations are the results of attempts to maximize the signal to thermal noise ratio in the receiver.

Radar receiver bandwidth limitations are very detrimental to some phase coded pulse compression systems but actually improve the performance of compressors that employ recently developed polyphase codes [1]. The purpose of this report is to document the effects of both pre- and postcompression bandwidth limitations on the performance of digitally implemented polyphase coded pulse compressors.

It is assumed that the radar transmits the phase codes as a train of contiguous constant-amplitude pulses with discrete phase changes from pulse to pulse in the train.

POLYPHASE PULSE COMPRESSION CODES TO BE CONSIDERED

The codes to be considered in detail are limited to what have been called frequency derived polyphase codes [1-6], i.e., the Frank, P1, P2, P3, and P4 codes. These codes are the phase weights or their conjugates taken in succession that would be used in a digital filter to separate the resolvable frequency components of analog frequency modulation waveforms sampled at a rate equal to the bandwidth over which the frequency is varied (herein called the Nyquist rate). In this case, resolvable frequencies are separated by integer multiples of the reciprocal of the duration of the signal being processed in the digital filter.

For a pulse compression ratio $\rho = N^2$, where N is the number of resolvable frequencies, the Frank code is defined by

$$\phi_{i,j} = (2\pi/N)(i-1)(j-1) \tag{1}$$

where $i = 1,2,3, \ldots N$ and $j = 1,2,3, \ldots N$. The index i designates the ith steering weight of the jth frequency filter. In forming the code, i ranges from 1 to N for each value of j. For example, with $\rho = 16$ and $N = 4$, the Frank code would consist of 16 code elements $\phi_{1,1}$, $\phi_{2,1}$, $\phi_{3,1}$, $\phi_{4,1}$, $\phi_{1,2}$, $\phi_{2,2}$, $\ldots$, $\phi_{4,4}$ where $\phi_{2,2} = (2\pi/N)(2-1)(2-1) = \pi/2$. Note that the Frank code would be obtained by inphase "I" and quadrature "Q" detecting a step-approximation-to-a-linear-modulation-waveform (SALFMW) with a coherent local oscillator of frequency equal to the first frequency step of the SALFMW and sampling at the Nyquist rate starting at the leading edge of the waveform.

The P1 code is similar to the Frank code in being derived from a SALFMW. However, the local oscillator used in the I, Q detectors in deriving the P1 code would have a frequency equal to the average frequency of the SALFMW. Because of this difference, the Frank code can be thought of as the result of a single sideband detection while the P1 code would be the result of a double sideband detection. The phase of the ith element of the jth frequency of the P1 code is defined by

$$\phi_{i,j} = -(\pi/N)[N-(2j-1)][(j-1)N+(i-1)]. \tag{2}$$

Manuscript submitted May 25, 1982.

It should be noted that, for N odd, the DC term is in the middle of the P1 code instead of at the beginning as in the Frank code. The P1 code has frequency symmetry about its center while the Frank code is unsymmetrical.

The P2 code differs from both the Frank and P1 codes by being derived from a Butler matrix such as used in phased array antennas. It is palindromic in that it has conjugate symmetry across each frequency or beam and even symmetry about the center of the code. The ith code element of the jth beam or frequency of the P2 code is defined by

$$\phi_{i,j} = \{(\pi/2)[(N-1)/N] - (\pi/N)(i-1)\}\{N+1-2j\}. \tag{3}$$

The P3 code is derived from an I, Q detected and sampled linear-frequency-modulation-waveform (LFMW) where the local oscillator frequency is the lowest frequency of the LFMW. The ith code element is defined by

$$\phi_i = \pi(i-1)^2/N^2 = \pi(i-1)^2/\rho. \tag{4}$$

The P4 code is obtained by moving the local oscillator to the center frequency of the LFMW and sampling at the Nyquist rate starting at the leading edge of the LFMW. The P4 code is defined by code element phases of

$$\phi_i = [\pi(i-1)^2/\rho] - \pi(i-1). \tag{5}$$

The P4 code differs from the P3 code by having the largest code element to code element phase changes on the ends of the code instead of the middle as in the P3 code. In this way, the P4 code differs from the P3 code in the same way as the P1 code differs from the Frank code.

Note that the P1 code can be made completely symmetrical (palindromic) by subtracting $\phi_{N,j}$ from each code element in the lower sideband frequencies when N is odd. This makes the autocorrelation function real rather than complex. Similarly, the P3 and P4 codes can be made palindromic by taking the first sample of the LFMW 1/2 period of a code element after the leading edge of the waveform while still sampling at the Nyquist rate, i.e., I and Q sampling rates equal to the waveform bandwidth.

PHASE CODED WAVEFORM SPECTRA

The spectrum of a Frank coded waveform with 100 code elements is illustrated in Fig. 1 out to the second nulls in the spectrum. The envelope is $(\sin X)/X$, but there is unsymmetrical fine structure. The abscissa is normalized to a frequency equal to the reciprocal of the duration of a code element.

Figure 2 represents the spectrum of a P1 coded waveform. It also has a $(\sin X)/X$ envelope, but its fine structure is symmetrical, unlike that of the Frank code. This difference is attributed to the differences of the time order of the frequencies represented by the code groups in the two waveforms.

The spectra of the P2 and P4 codes are very similar to that of the P1 code, and that of the P3 code resembles that of the Frank code.

Note that if the I, Q detected LFMW that was used to derive the P3 and P4 codes is sampled faster than at the Nyquist rate, the spectrum of the resultant phase code on a waveform changes dramatically from that of the P3 and P4 codes. Figures 3 and 4 show the spectra obtained by sampling at 2 and 5 times faster than the Nyquist rate. These spectra are nearly rectangular instead of $(\sin X)/X$. This explains the difference in the peak range-time-sidelobes of the Frank and P codes compared to an analog LFMW compressor since a $(\sin X)/X$ spectrum has a time function with zero sidelobes while a rectangular spectrum produces a $(\sin X)/X$ time function. The phase codes derived by sampling at the Nyquist rate have peak sidelobe levels more than the pulse compression ratio below the match point

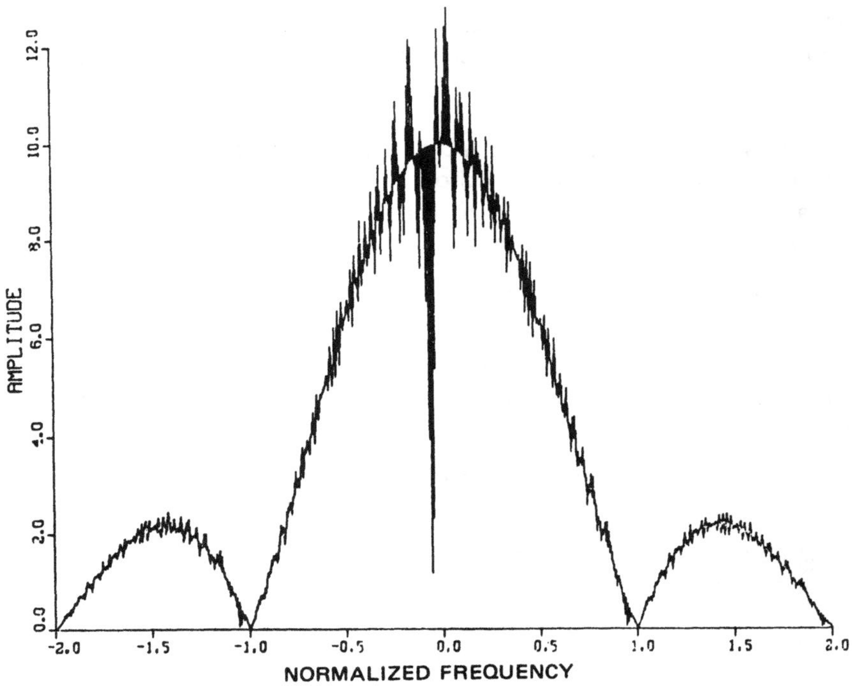

Fig. 1 — Spectrum of a Frank coded waveform

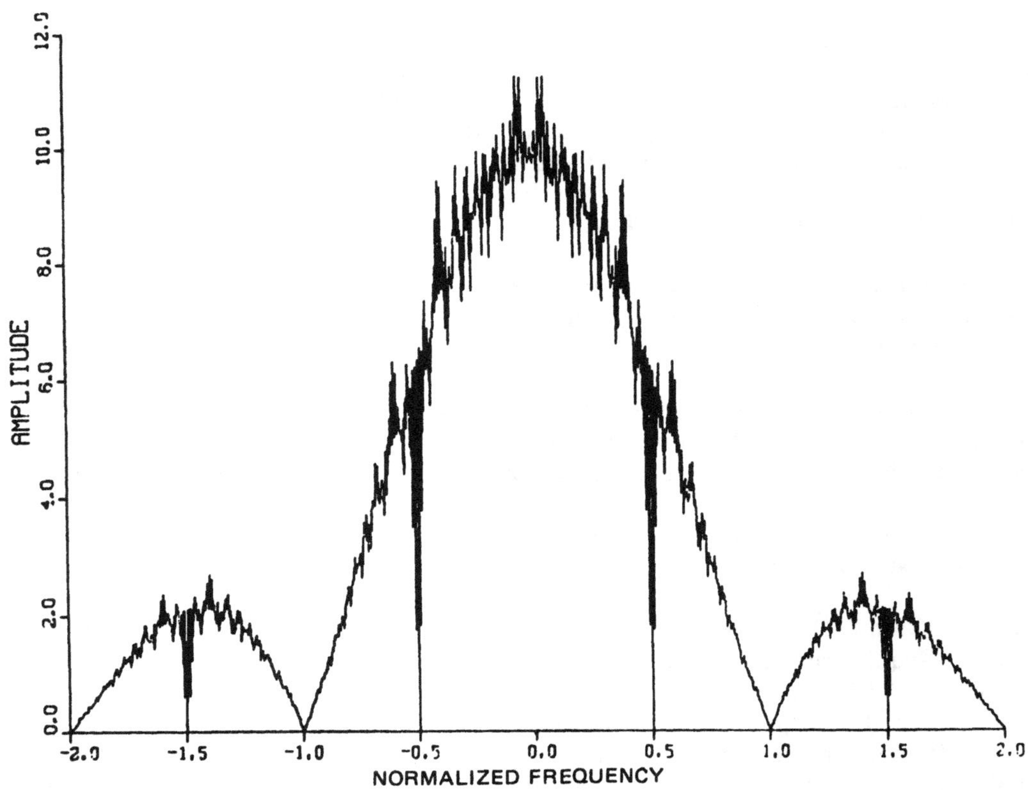

Fig. 2 — Spectrum of a P1 coded waveform

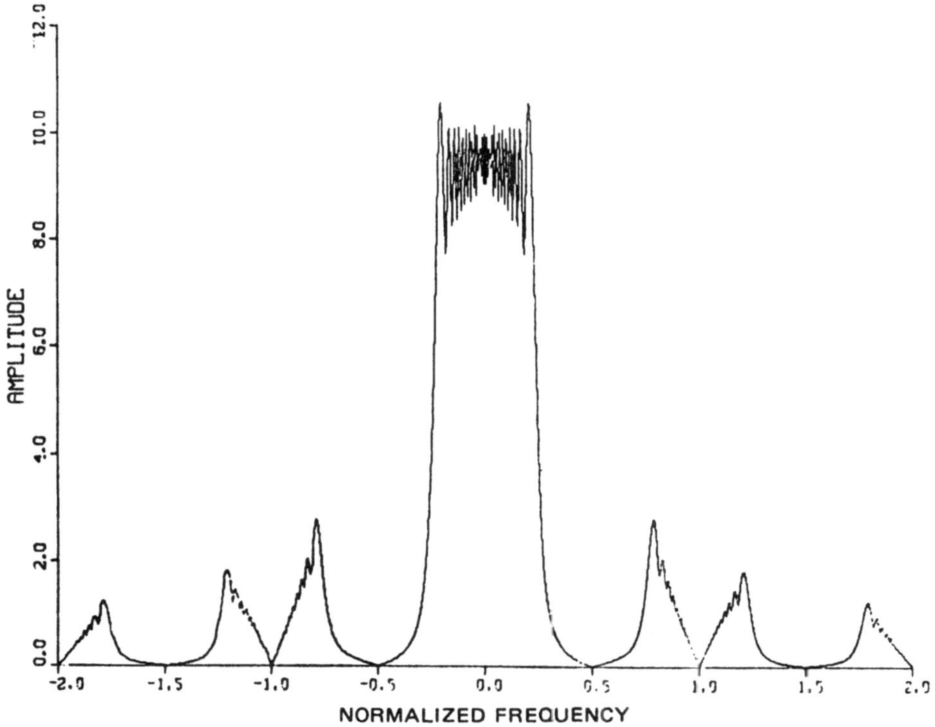

Fig. 3 — Spectrum of a linear frequency modulated waveform baseband sampled at twice the Nyquist rate

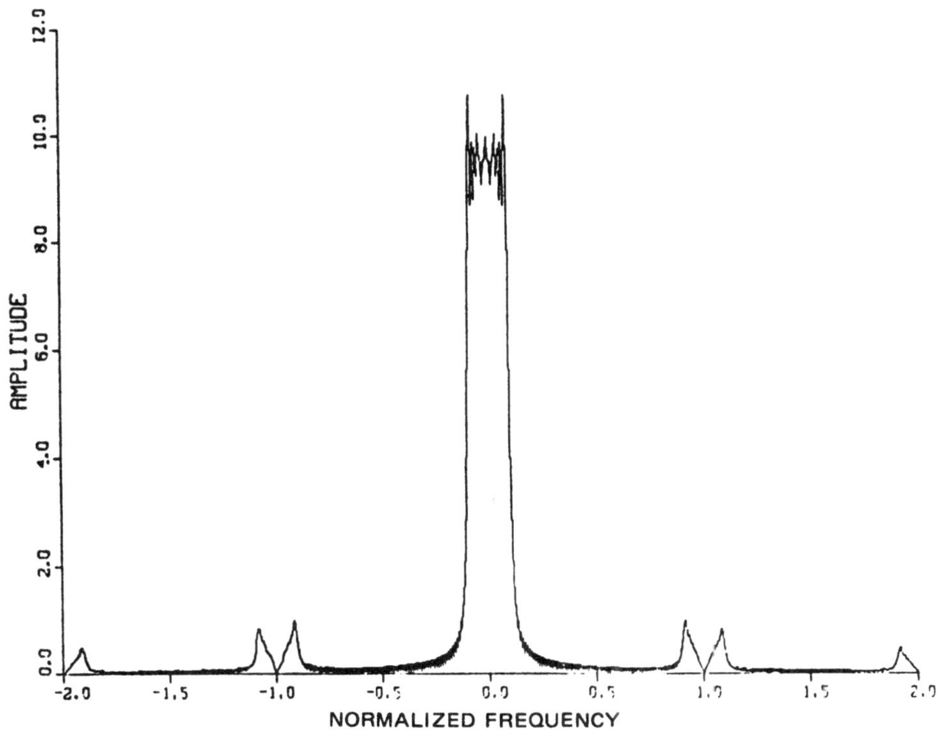

Fig. 4 — Spectrum of a linear frequency modulated waveform baseband sampled at 5 times the Nyquist rate

while the analog waveform has a $(\sin X)/X$ autocorrelation function independent of the pulse compression ratio. The finite sidelobes in the time functions of the Nyquist rate phase coded compressor outputs are due to the fine structure in the phase coded waveform spectra.

PRECOMPRESSION BANDLIMITING EFFECTS

The effects of precompression band-limitations on the response of polyphase coded pulse-compressors were evaluated by using two different techniques. In one technique, a fourth order Butterworth filter was placed ahead of the compressor and autocorrelation functions were determined for several different code leading edge arrival times with respect to a sampling pulse. In a second technique, the received code was sampled at 5 times the Nyquist rate and sliding window 5 sample averages (Fig. 5) were taken digitally to reduce the signal bandwidth. The resultant signal was then sampled at the Nyquist rate and compressed in a compressor matched to the unbandlimited code.

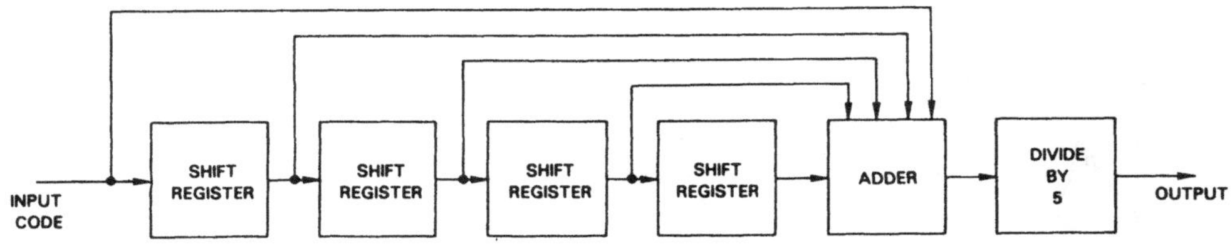

Fig. 5 — Sliding window 5 sample averager

Figures 6(a) and 7(a) show the real part or I values of a 144-element complex number Frank and modified P1 code [1] respectively. The modified P1 code was a rearranged Frank code representing 12 frequencies. The code group using the π phase increments from code element to code element was placed on the left-hand side of the uncompressed waveform and was followed by the frequency groups having progressively smaller negative phase increments from code element to code element. This modification was employed to permit a 12-point fast Fourier transform (FFT) circuit to be used to generate and compress both codes. In this case, each code element of the waveform generated for the test was 0.5 μs in duration.

Figures 6(b) and 7(b) show the effect of a 2 MHz bandwidth fourth order Butterworth filter on the I parts of the codes. Figures 6(c), 6(d), 7(c), and 7(d) show the corresponding imaginary parts or Q values of the complex numbers specifying the code elements of the two codes without and with filtering, respectively,

Figures 6(e) and 7(e) show the autocorrelation functions (compressor outputs) of the two codes sampled 0.1 sampling period prior to the leading edge of the waveform. Note that the peak range-time-sidelobes of both compressor outputs are $\pi^2\rho = 144\ \pi^2$ below the peak response.

Figures 6(f), 6(g), 7(f), and 7(g) show the effect of the 2 MHz filter on the compressor outputs using the two different codes sampled 0.3 and 0.5 sampling periods ahead of the leading edge of the waveforms. In both cases, the average peak gain loss was 1.5 dB. However, the filter reduced the P1 levels. This occurred because the filter effectively amplitude reduced the edges of the P1 code waveform more than the center while it reduced the amplitude of the center of the Frank code waveform more than its edges. In this way, the filter decreased the average Frank code peak response to peak sidelobe ratio by 1.5 dB but increased the P1 average ratio by 1.5 dB.

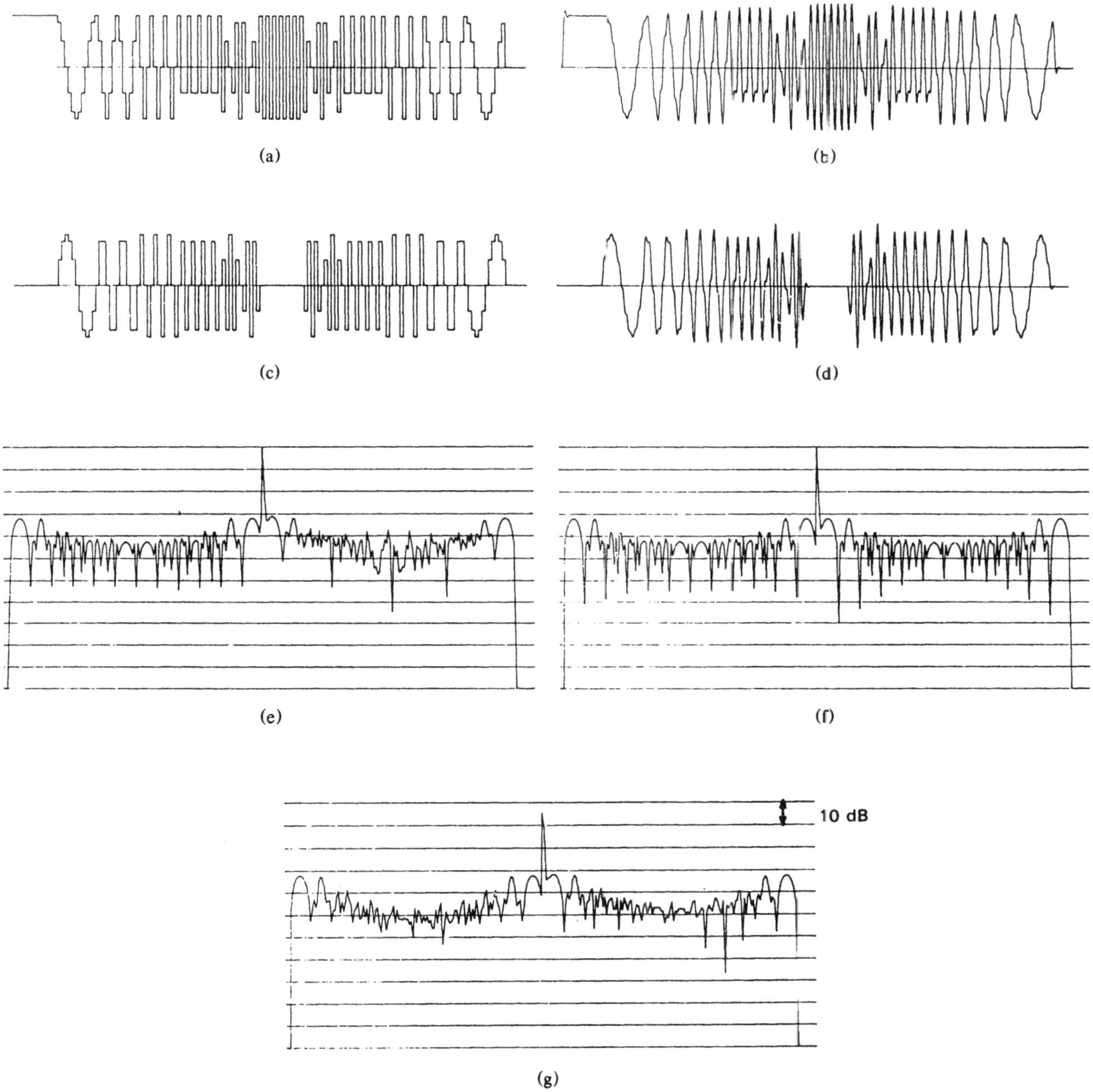

Fig. 6 — Frank code, $\rho = 144$. (a) Unfiltered I; (b) filtered I, 2 MHz; (c) unfiltered Q; (d) filtered Q, 2 MHz; (e) autocorrelation function sampled 0.1 sampling period early; (f) sampled 0.3 sampling period early; (g) sampled 0.5 sampling period early.

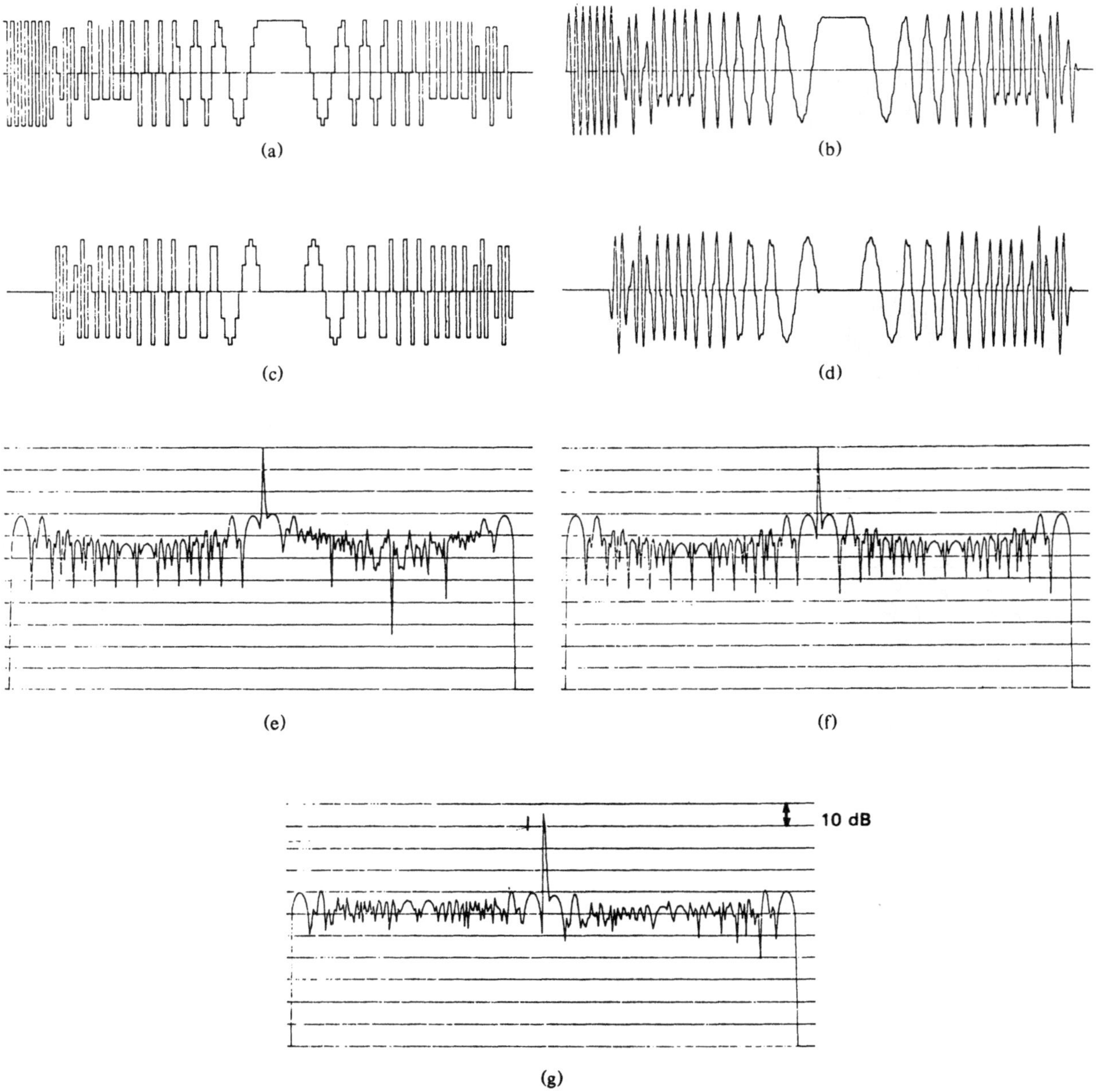

Fig. 7 — P1 code, $\rho = 144$. (a) Unfiltered I; (b) filtered I, 2 MHz; (c) unfiltered Q; (d) filtered Q, 2 MHz; (e) autocorrelation function sampled 0.1 sampling period early; (f) sampled 0.3 sampling period early; (g) sampled 0.5 sampling period early.

Figures 8 and 9 show the effect of reducing the filter bandwidth to 1 MHz without changing the input signal bandwidth of 2 MHz. Note that the filter drops both peaks by the same amounts without significantly altering the Frank code peak sidelobes. However, the P1 code peak sidelobes are reduced.

The P3 code yielded results similar to those shown in Figs. 6 and 8 for the Frank code, and the P2 and P4 codes yielded results similar to those shown in Figs. 7 and 9 for the P1 codes.

Figure 10 illustrates the effect (averaged over the possible arrival times) of oversampling the polyphase codes by 5 to 1, averaging with a 5 sample sliding window and pulse compressing with a compressor matched to the unfiltered phase code. Again, each code had 100 code elements. Note that the peak sidelobes of the Frank and P3 codes were unaffected while the peak sidelobes of the P1, P2, and P4 codes were reduced.

POSTCOMPRESSION BANDLIMITING EFFECTS

Postcompression bandlimiting effects were investigated by using a two-sample sliding window adder on the output of a digital pulse compressor with pulse compression ratics of 100 and 400.

Figure 11 illustrates the effect of a two-sample sliding window adder after pulse compression on the various polyphase codes with a pulse compression ratio ρ of 100. Figure 12 is similar data for a pulse compression ratio ρ of 400. Note that the peak response is not changed because the output of the adder was not divided by 2 to normalize the data. Note also that the Frank and P3 code peak sidelobes increase by 6 dB due to the adder so the adder would not have changed them if the data had been normalized by dividing by 2 after addition. The most important thing to note, however, is that the peak sidelobes of the P1, P2, and P4 codes are actually reduced by a significant amount, about 5 dB for $\rho = 100$ and 11 dB for $\rho = 400$. Thus, if the data had been normalized, the peak response would have dropped by 6 dB and the peak sidelobes by 11 dB and 17 dB for $\rho = 100$ and $\rho = 400$ respectively with the P1, P2, and P4 codes.

It is interesting to note that with $\rho = 100$ or $\rho = 400$ and the sliding window two sample adder, the peak sidelobes of the P1, P2, and P4 codes are only about 5 dB above those that would have been obtained with a so-called perfect code like the Barker code [5] if one could be found with these high pulse compression ratios. The Barker codes have peak sidelobes down by the square of the pulse compression ratio.

CONCLUSIONS

The symmetrical P1, P2, and P4 code peak to range-time-sidelobe ratios are increased by either pre- or postcompression bandlimiting. Such bandlimiting, however, reduces this ratio using the Frank or P3 code.

Bandlimiting the P1, P2, or P4 code results in peak to maximum range time sidelobes within 6 dB of those of a so-called perfect code like the Barker codes. However, the P1, P2, and P4 codes can provide any desired pulse compression ratio in contrast to the Barker codes that are limited to pulse compression ratios of 13.

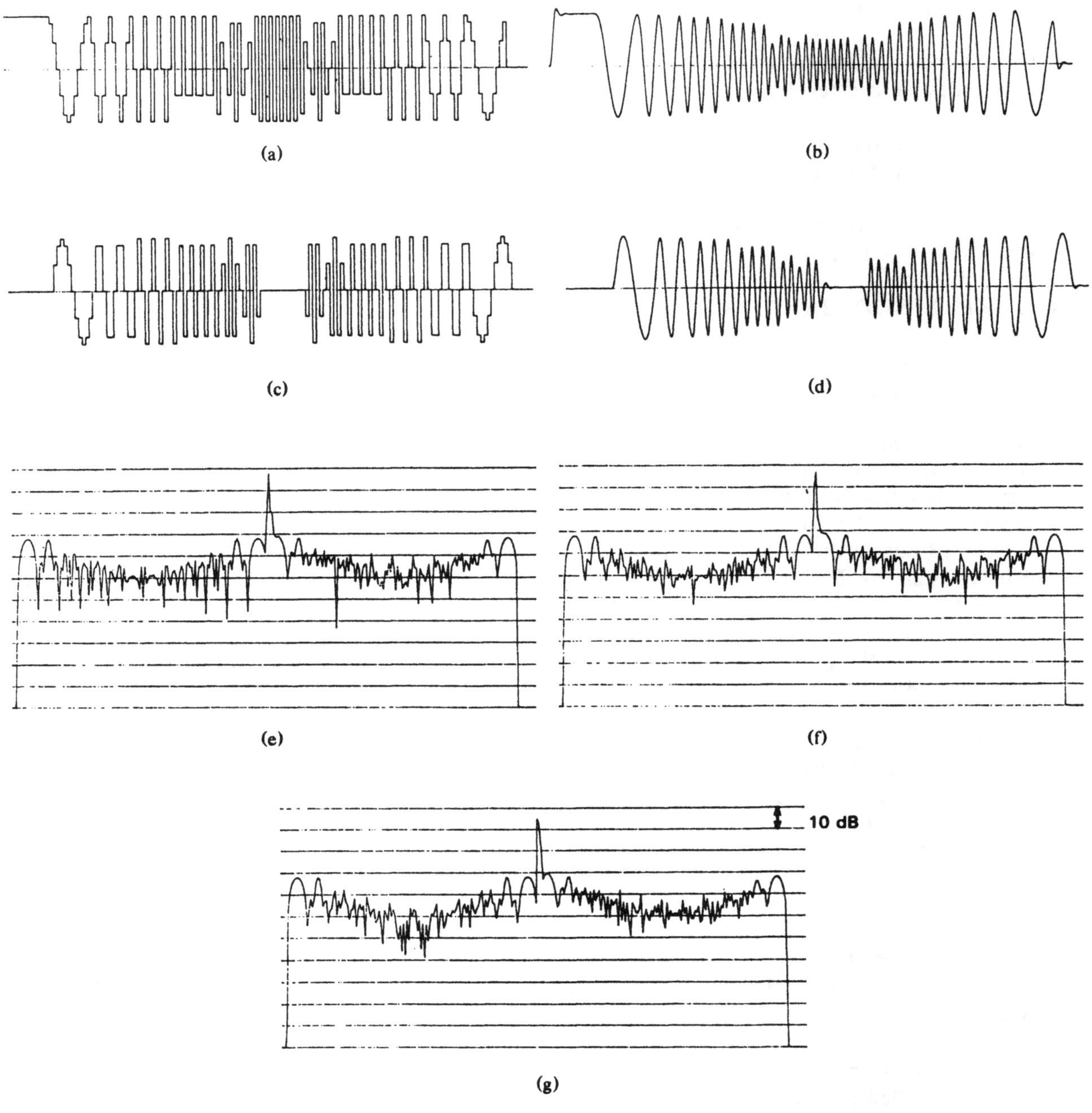

Fig. 8 — Frank code, $\rho = 144$. (a) Unfiltered I; (b) filtered I, 1 MHz; (c) unfiltered Q; (d) filtered Q, 1 MHz; (e) autocorrelation function sampled 0.1 sampling period early; (f) sampled 0.3 sampling period early; (g) sampled 0.5 sampling period early.

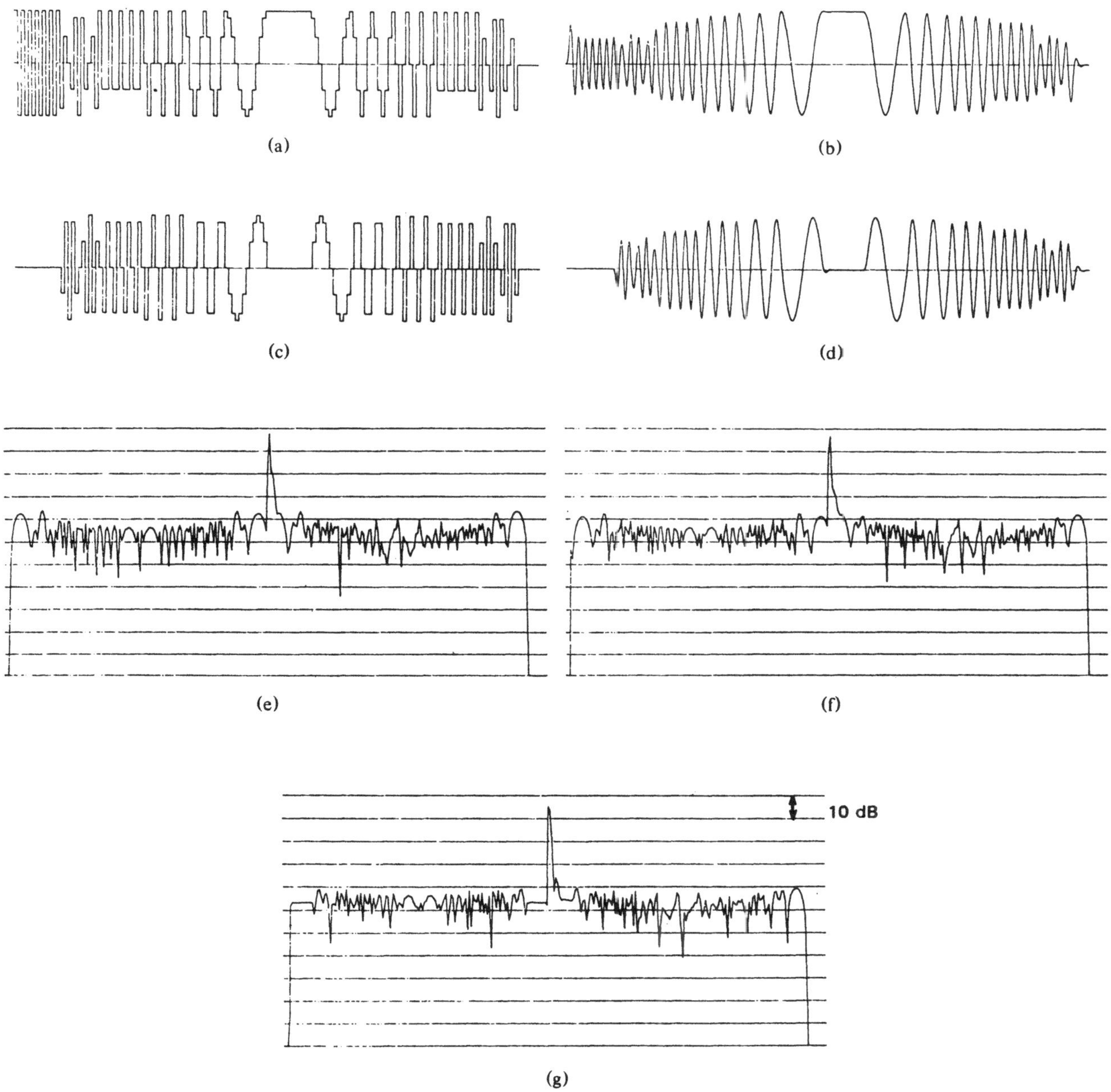

Fig. 9 — P1 code, $\rho = 144$. (a) Unfiltered I; (b) filtered I, 1 MHz; (c) unfiltered Q; (d) filtered Q, 1 MHz; (e) autocorrelation function sampled 0.1 sampling period early; (f) sampled 0.3 sampling period early; (g) sampled 0.5 sampling period early.

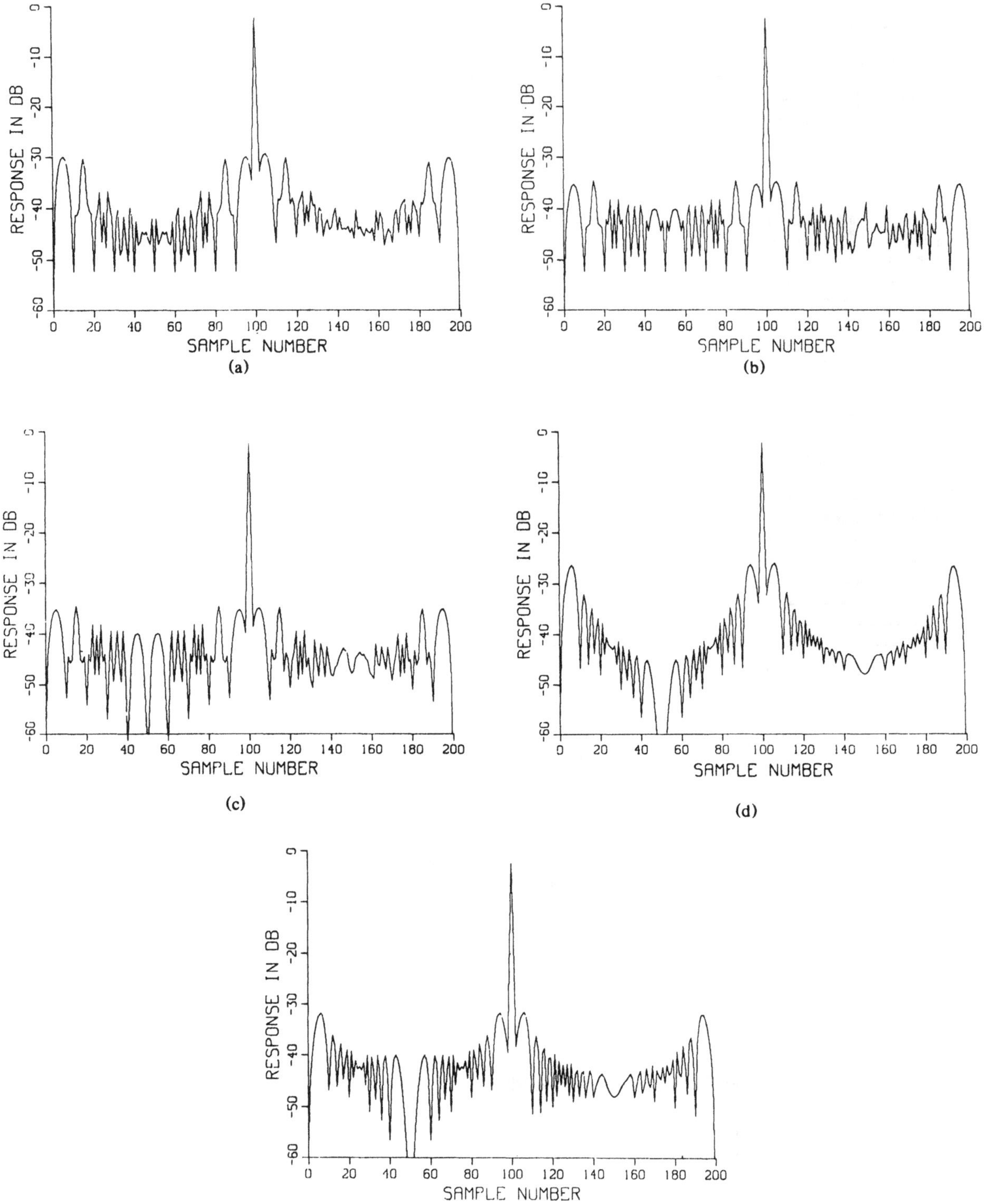

Fig. 10 — Average response over arrival times of various code compressors to waveforms over sampled by 5, averaged by 5 and resampled at the Nyquist rate for the unbandlimited waveform with $\rho = 100$. (a) Frank code; (b) P1 code; (c) P2 code; (d) P3 code; (e) P4 code.

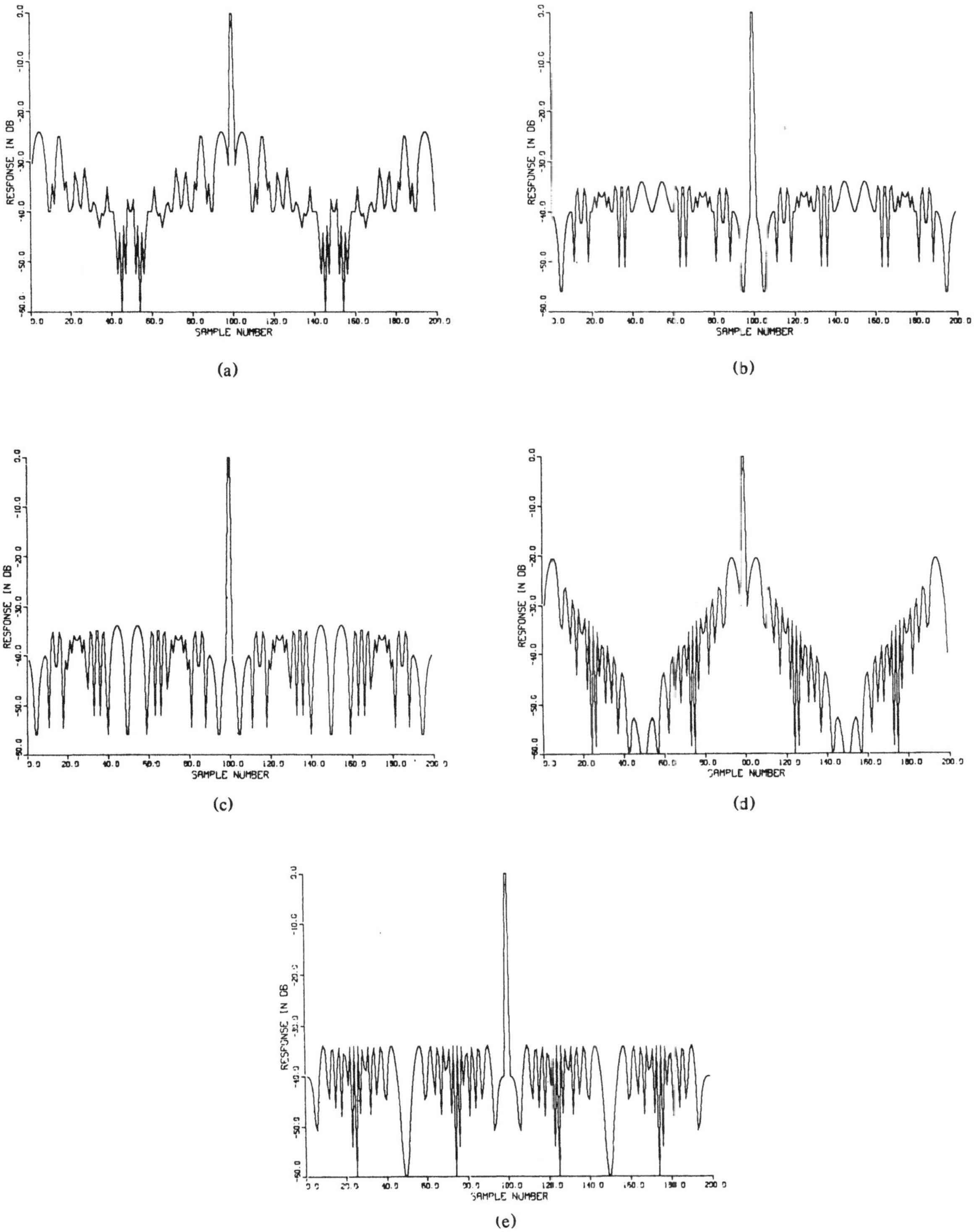

Fig. 11 — Compressed pulse waveforms of polyphase codes sampled at the Nyquist rate with $\rho = 100$ and with sliding window 2 sample adder at output. (a) Frank code; (b) P1 code; (c) P2 code; (d) P3 code; (e) P4 code.

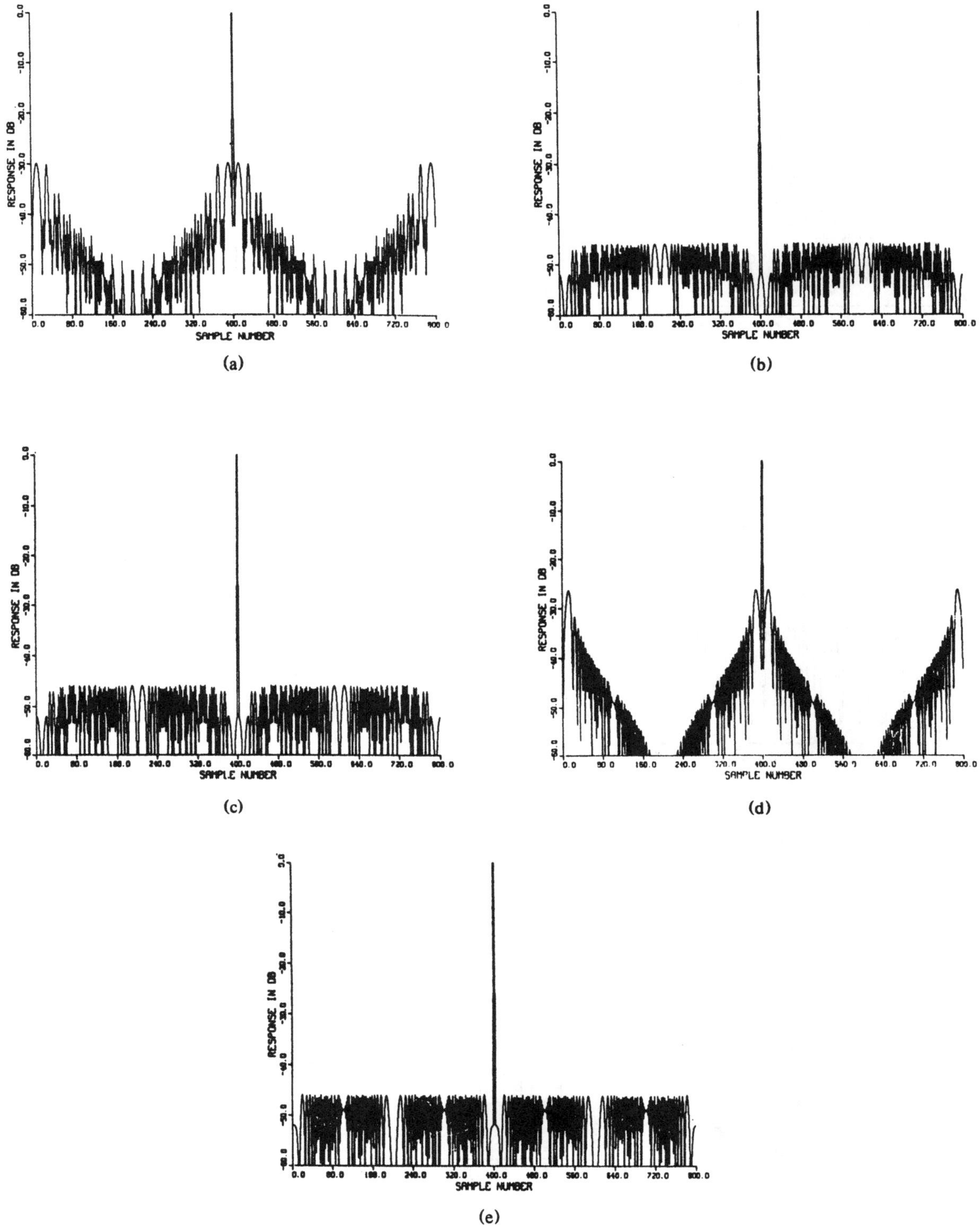

Fig. 12 — Compressed pulse waveforms of polyphase codes sampled at the Nyquist rate with $\rho = 400$ and with sliding window 2 sample adder at output. (a) Frank code; (b) P1 code; (c) P2 code; (d) P3 code; (e) P4 code.

REFERENCES

1. B.L. Lewis and F.F. Kretschmer, Jr., "A New Class of Polyphase Pulse Compression Codes and Techniques," IEEE Trans. Aerospace and Electronic Systems **AES-17**(3), 364-372, May 1981.

2. B.L. Lewis and F.F. Kretschmer, Jr., "Linear Frequency Modulation Derived Polyphase Pulse-Compression Codes and an Efficient Digital Implementation," NRL Report 8541, November 1981.

3. F.F. Kretschmer, Jr., and B.L. Lewis, "Polyphase Pulse Compression Waveforms," NRL Report 8540, January 1982.

4. R.L. Frank, "Polyphase Codes With Good Nonperiodic Correlation Properties," IEEE Trans. Information Theory **IT-9**, 43-45, January 1963.

5. F.E. Nathanson, *Radar Design Principles: Signal Processing and the Environment,* McGraw-Hill, 1969.

6. C.E. Cook and M. Bernfeld, *Radar Signals: An Introduction to Theory and Application*, Academic Press, 1967.

Doppler Properties of Polyphase Coded Pulse Compression Waveforms

F.F. KRETSCHMER, JR.

B.L. LEWIS

Naval Research Laboratory

Doppler properties of the Frank polyphase code and the recently derived *P*1, *P*2, *P*3, and *P*4 polyphase codes are investigated and compared. An approximate 4 dB cyclic variation of the peak compressed signal is shown to occur as the Doppler frequency increases. The troughs in the peak-signal response occur whenever the total phase shift across the uncompressed pulse, due to Doppler, is an odd multiple of π radians.

It is shown that while the *P*3 and *P*4 codes have larger zero-Doppler peak sidelobes than the other codes, the *P*3 and *P*4 codes degrade less as the Doppler frequency increases. Also, the effects of amplitude weighting and receiver bandlimiting for both zero and nonzero Doppler are investigated.

Manuscript received May 12, 1982; revised August 12, 1982.

Authors' address: Target Characteristics Branch, Radar Division, Naval Research Laboratory, Code 5340.1, Washington, D.C. 20375.

0018-9251/83/0700-0521 $00.75

I. INTRODUCTION

The Frank polyphase coded waveform, and the *P*1, *P*2, *P*3, and *P*4 polyphase pulse-compression waveforms previously described by the authors [1–3], provide a class of frequency-derived phase coded waveforms that can be sampled upon reception and processed digitally.

These waveforms are derivable from the chirp and step-chirp analog waveforms and are therefore similar in certain respects. There are some important differences, however, which include differences in sidelobe levels, implementation techniques, and Doppler characteristics.

The compressed pulse of the polyphase coded waveforms has sidelobes which decrease as the pulse compression ratio ρ is increased. For ρ equal to 100, the peak sidelobes range from 26 to 30 dB below the peak signal response depending on the particular code. In contrast, the compressed chirp or step-chirp pulse has approximately 13 dB sidelobes, independent of the pulse compression ratio, which can mask a relatively weak nearby target. An amplitude weighting is generally applied to reduce the sidelobes and the resulting mismatch reduces the output *S/N* by 1 to 2 dB.

The polyphase coded waveforms are capable of large pulse compression ratios which may be efficiently implemented using the phase shifts provided by a fast Fourier transform (FFT). Thus the FFT can be used directly as the pulse compressor. These waveforms can also be efficiently compressed with another pulse compression technique where the FFT is used to convert to the frequency domain where the matched filtering and weighting are performed. This processing is followed by an inverse FFT to restore the signal to the time domain.

This paper first reviews the properties of the polyphase coded waveforms, then focuses on the Doppler characteristics of these waveforms. A cyclic loss of approximately 4 dB is discussed which is characteristic of frequency derived polyphase coded waveforms having low sidelobes. This cyclic variation was not recognized in the prior literature dealing with Frank codes [4, 5]. A method of compensating for this loss is described.

The Doppler characteristics of the various polyphase codes are investigated in detail. Also, the effects of weighting on the Doppler performance of the codes is presented. This weighting may be due to an applied amplitude weighting and/or it may be caused by bandlimiting in the receiver.

II. PROPERTIES OF POLYPHASE CODED WAVEFORMS

A. Frank-Polyphase-Coded Waveforms

The Frank-polyphase-coded waveform may be described and generalized by considering a hypothetically sampled step-chirp waveform [1, 2]. The Frank code was not originally described in this manner, but was given in terms of the elements of a matrix [6]. As an example, consider a four-frequency step-chirp waveform as shown in Fig. 1(b) where the F_i denote frequency tones. In this waveform, the frequency steps are equal to the reciprocal

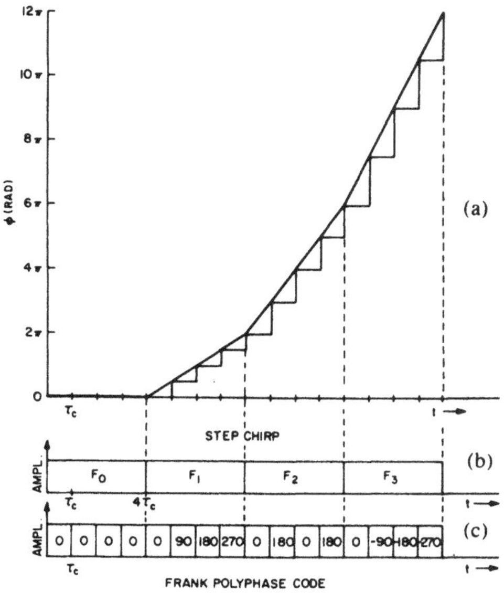

Fig. 1. Step-chirp and Frank polyphase code relationships.

of the tone duration $4\tau_c$, where τ_c denotes the compressed pulsewidth. Assuming this waveform has been beat to baseband I and Q using a synchronous oscillator having a frequency the same as the first tone frequency, the resultant phase-versus-time characteristic consists of four linear sections, as shown in Fig. 1(a). The corresponding baseband frequencies are the subharmonics of the frequency $1/\tau_c$. If the baseband phases of the step-chirp waveforms are sampled every τ_c seconds and held for τ_c seconds, the phase sequence shown in Fig. 1(c) is obtained. This sequence of phases constitutes the phases of a Frank code for $N = 4$, corresponding to the four baseband frequencies of the hypothetical step-chirp waveform. The actual transmitted Frank-coded waveform consists of a carrier whose phase is modulated according to the indicated baseband waveform sequence. For each frequency, or section, of the step-chirp phase characteristic, a phase group consisting of N phase samples is obtained and the total number of code phases is N^2 which is equal to the pulse-compression ratio. Note that the phase increments within the four phase groups are 0°, 90°, 180°, and 270°. However, the phases of the last group are ambiguous (>180°) and appear as −90° phase steps or as the conjugate of the F_1 group of phases, which corresponds to the lower sideband of F_1. The last group of phases appears, because of the ambiguity, to complete one 360° counter-clockwise rotation rather than the $(N-1)$ counter-clockwise rotations of the end frequency of the step-chirp waveform.

The Frank-code phases may be stated mathematically as follows. The phase of the ith code element in the jth phase group, or baseband frequency, is

$$\phi_{i,j} = (2\pi/N)(i-1)(j-1) \tag{1}$$

where the index i ranges from 1 to N for each of the values of j ranging from 1 to N. An example of a brute force Frank-code pulse generation for $N = 3$ is shown in Fig. 2. The Frank-code phases are the same as the nega-

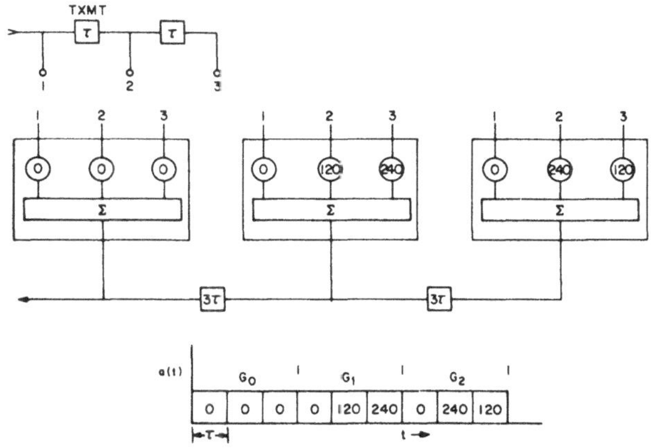

Fig. 2. Simplified Frank code generation.

tive of the steering phases of an N point discrete Fourier transform where the jth frequency coefficient is

$$F_j = \sum_{i=1}^{n} a_i \exp\left[(-j2\pi/N)(i-1)(j-1)\right] \tag{2}$$

where a_i is the ith complex input time sample. This means that a considerable savings in hardware can be achieved by using the efficiency of an FFT.

The matched-filter output for an $N = 10$ (100-element) Frank code is shown in Fig. 3. This figure and the

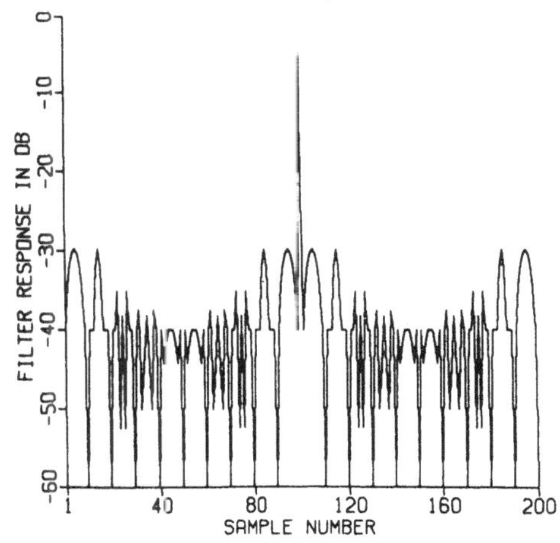

Fig. 3. Compressed pulse of 100-element Frank code.

following figures showing the compressed pulse were obtained by sampling the input baseband waveform once per code element or per reciprocal bandwidth, unless stated otherwise. Using a discrete-time matched filter, the output signal is also a discrete-time sampled signal. However, for ease of plotting and viewing, the points were connected by straight lines. The four sidelobe peaks on each half of the match point (peak response) are of equal magnitude. The first peak sidelobe at sample number 5 in Fig. 3 occurs as the last phase group having −36° phase increments indexes halfway into the first phase group of zero phase vectors in the autocorrelation process. In general, at sample number $N/2$, there are $N/2$ vectors adding to complete a half circle. The end phase group indexing into the first phase group of 0° vectors makes an approximate circle since the phases of the last phase group make only one rotation as stated previously. The peak sidelobe amplitude may be approximated by the diameter D of the circle from the relation,

$$\text{perimeter} = N = \pi D \tag{3}$$

or

$$D = N/\pi. \tag{4}$$

At the match point the amplitude is N^2 so that the peak response to peak-sidelobe power ratio R is

$$R = N^4/(N/\pi)^2 = N^2\pi^2 = \rho\pi^2. \tag{5}$$

For a 100-element Frank code, this ratio is approximately 30 dB, as shown in Fig. 3.

Had the phases of the polyphase coded waveform been generated by using the phases of the step-chirp phase characteristic sampled at 1/5 of the interval used for the Frank code, the compressed code would appear as shown in Fig. 4. In this figure, five samples are equal in

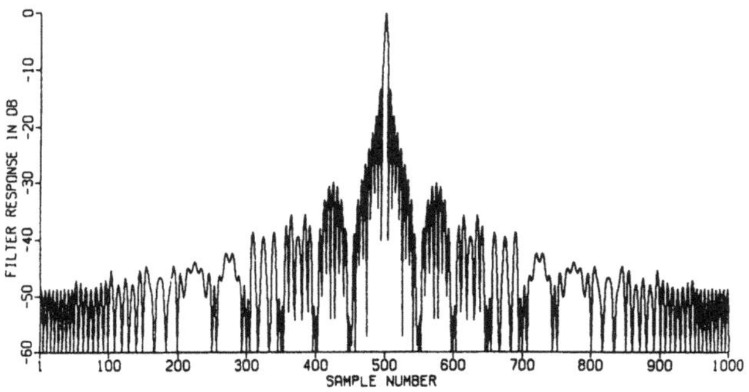

Fig. 4. Compressed pulse of oversampled (5:1) step chirp.

time to one sample in Fig. 3. Note in Fig. 4 that the near-in sidelobes are approximately 13 dB and that the envelope of the sidelobe peaks is approximately that of a sin x/x pulse. The 13 dB sidelobes also appear for an oversampling of 2:1. These comments also apply to the other polyphase codes described in this paper. Also note that the compressed pulsewidth in Fig. 4 has not decreased since it is determined by the underlying bandwidth of the step-chirp waveform.

(1) Effects of Bandlimiting Prior to Pulse Compression: A Frank-coded waveform is depicted in Fig. 5(a) where the G_K denote the phase groups corresponding to the sampled phases of a step-chirp waveform as previously discussed. Each group consists of N vectors beginning with a vector at a phase angle of 0°. The phase increments within the Kth group are

$$\Delta\phi_K = K\,\frac{360°}{N}. \tag{6}$$

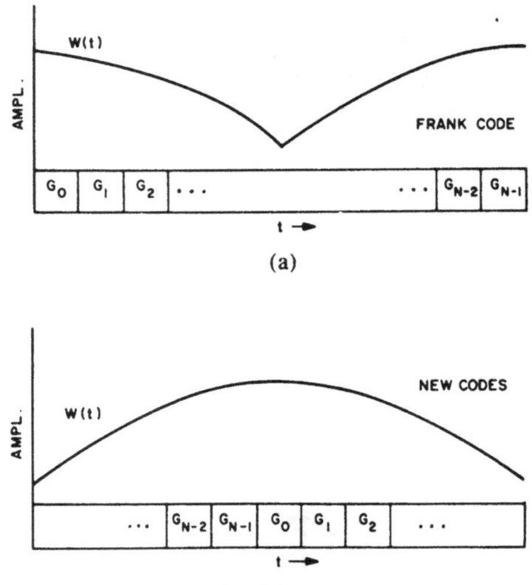

Fig. 5. Effect of bandlimiting before pulse compression.

Thus G_0 consists of N vectors at 0°, G_1 has vectors separated by 360°/N until at the center of the coded waveform the phase increments approach or become 180° depending on whether N is odd or even. For phase increments greater than 180°, the phases are ambiguous with the result that the phasors of phase group G_{N-K} are the conjugates of the phasors of phase group G_K so that the vectors have the same increments but rotate in opposite directions. The result is that the phase increments are small at the ends of the code and become progressively larger toward the center of the code where the increments approach 180° from opposite directions.

If a receiver is designed so that it has an approximately rectangular bandwidth corresponding to the 3 dB bandwidth of the received waveform, the received waveform becomes bandlimited and a mismatch occurs with the compressor. This bandlimiting would normally occur prior to sampling in the A/D conversion process in order to prevent noise foldover and aliasing. The result of any bandlimiting is to average (or smooth) the vectors constituting the coded waveform, and for the Frank code, a weighting $W(t)$ such as illustrated in Fig. 5(a) takes place due to the larger phase increments toward the middle of the code. This weighting causes an unfavorable mismatch with the compressor which results in a degradation of the sidelobes relative to the peak response.

The new symmetrical codes found by the authors have the common property that the phase groups with the small phase increments are at the center of the code and the larger increment groups progress symmetrically toward the ends of the code. This is illustrated in Fig. 5(b) where a favorable amplitude weighting resulting from bandlimiting prior to pulse compression is shown.

B. *P*1, *P*2, *P*3, and *P*4 Polyphase Codes

The new polyphase codes which tolerate bandlimiting are referred to as the $P1$, $P2$, and $P4$ codes. The $P1$ code was derived from use of the previously described relationship between the Frank-code phases and those of a sampled step-chirp waveform. The desired symmetry, having the dc or small incremental phase group at the center of the code, can be achieved by determining the phases which result from placing the hypothetical synchronous oscillator at the band center of the step-chirp waveform. For an odd number of frequencies, the synchronous oscillator frequency corresponds to one of the waveform frequencies and the resulting phases are the same as the Frank code except the phase groups are rearranged as indicated in Fig. 5(b). If there is an even number of frequencies, the synchronous oscillator frequency placed at the band center does not correspond to one of the frequencies in the step-chirp signal. The phase of the ith element of the jth group is given in degrees by

$$\phi_{i,j} = -(180/N)\,[N-(2j-1)]\,[(j-1)N + (i-1)] \quad (7)$$

where i and j are integers ranging from 1 to N.

The $P2$ code, which also has the desired features, is similar to the Butler matrix steering phases used in antennas to form orthogonal beams. The $P2$ code is valid for N even, and each group of the code is symmetric about zero phase. The usual Butler matrix phase groups are not symmetric about zero phase and result in higher sidelobes. For N even, the $P1$ code has the same phase increments, within each phase group, as the $P2$ code except that the starting phases are different. The ith element of the jth group of the $P2$ code is given in degrees by

$$\phi_{i,j} = (90/N)\,[N + 1 - 2i]\,[N + 2 - 2j] \quad (8)$$

where i and j are integers ranging from 1 to N as before. The requirement for N to be even in this code stems from the desire for low autocorrelation sidelobes. An odd value for N results in high autocorrelation sidelobes. This code has the frequency symmetry of the $P1$ code and also has the property of being a palindromic code which is defined as a code having symmetry about the center.

The $P4$ code is similar to the $P1$ code except that the phase samples are those of a sampled chirp waveform rather than step-chirp waveform. In each case, the synchronous oscillator is placed at the band center with the result that the codes are symmetrical. The $P3$ code is also derived from a chirp waveform and is the counterpart of the Frank code where the synchronous oscillator is put at the lowest frequency to determine the phases. The $P3$ code is therefore intolerant of bandlimiting like the Frank code.

The phases of a modified $P4$ code[1] are given by

$$\phi_i = (45/\rho)\,(2i - 1)^2 - 90(2i - 1), \qquad 1 \le i \le \rho \quad (9)$$

and the phases of the $P3$ code are

$$\phi = (180/\rho)\,(i - 1)^2, \qquad 1 \le i \le \rho. \quad (10)$$

[1]This code varies slightly from the one given in [3]. In effect, the first sample of the Nyquist sampling of a chirp signal has been shifted by 1/2 of a sample period to produce a palindromic code.

The compressed pulse for a zero-Doppler, 100-element $P3$ or $P4$ code is shown in Fig. 6. It is similar to the Frank, $P1$, and $P2$ 100-element codes except that the peak sidelobes are approximately 4 dB higher.

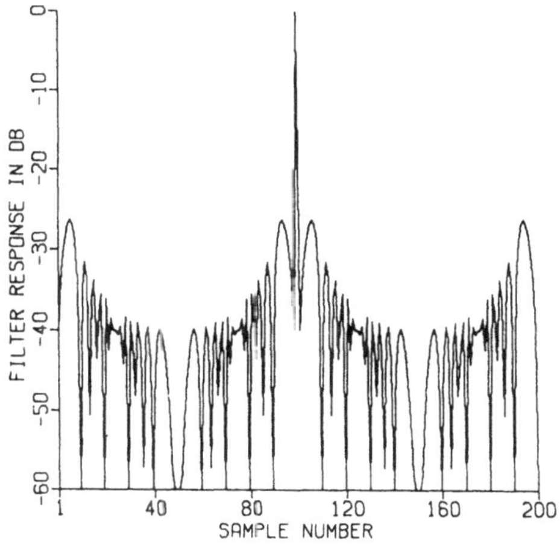

Fig. 6. Compressed pulse of 100-element $P3$ or $P4$ code.

III. DOPPLER PROPERTIES OF POLYPHASE CODES

A. Ambiguity Functions and Cyclic Losses

A partial ambiguity function for a 100-element Frank code is shown in Fig. 7 which shows the amplitude in dB

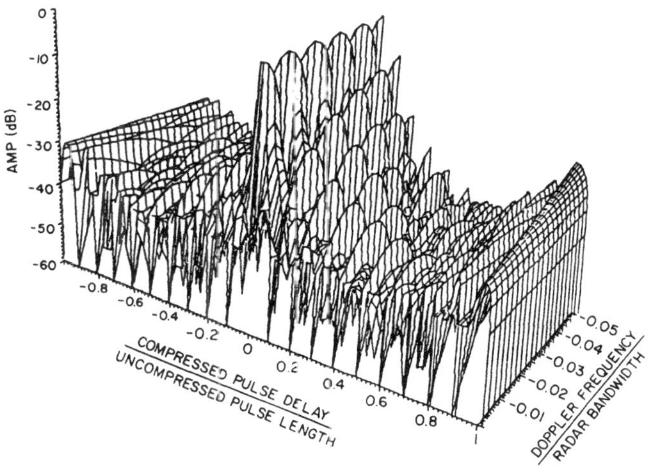

Fig. 7. Partial ambiguity function for 100-element Frank code.

of a matched-filter output for given Doppler shifts of the input. The Doppler is normalized to the signal bandwidth and the delay axis is normalized to the uncompressed pulse length. The cut through this diagram at zero Doppler shows the output of a perfectly matched receiver. In this case, the output pulse is the same as the autocorrelation function of the input waveform. A cut along any other Doppler axis shows the output of the receiver for an input waveform having a Doppler which is mismatched to the receiver by the stated amount. The vertical scale ranges from 0 dB to −60 dB, and the −30 dB sidelobes for zero Doppler are evident. The normalized Doppler shift of -0.05 shown in this figure corresponds to a mach-50 target for an L-band radar having a signal bandwidth of 2 MHz. The first Doppler cut shown in the literature [4] is taken at this normalized Doppler and the resultant high-peak sidelobes shown in Fig. 8 have perhaps discouraged usage of the Frank code. The region shown between zero and mach-5 Doppler, and a delay interval of $\pm$ 0.3 is of interest, and it is shown on an expanded scale in Fig. 9. In this region, the Doppler response is good in terms of the sidelobe levels. (The blank spot on the plot was caused by a computer plotting glitch.) The corresponding ambiguity function for the 100-element $P4$ code is shown in Fig. 10 where the peak response is seen to have the same cyclic variation as the Frank code. The differences between these ambiguity functions are discussed in later sections and it will be shown that amplitude weighting or bandlimiting will reduce the image lobes at the ends of the compressed pulse. At the Doppler shift of -0.005, or more generally $\pm$ 1/(2ρ), the total phase shift across the uncompressed pulse is π and the peak response drops approximately 4 dB. At this Doppler, there is a range-Doppler coupling of 1/2 of a range cell with the result that the signal splits between two range cells. This is illustrated in Fig. 11 for the Frank code. At a normalized Doppler shift of -0.01, or in general $\pm$ 1/ρ, there is a range-Doppler coupling of one range cell resulting from a total phase shift of 2π radians across the uncompressed pulse, and the main peak response is restored to nearly full amplitude as shown in Figs. 7 and 10. This effect is cyclic and a loss of approximately 4 dB is encountered when the total phase shift due to Doppler is an odd multiple of 180°.

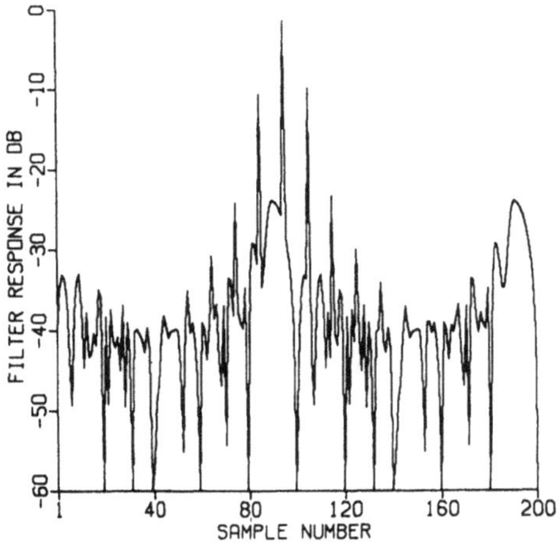

Fig. 8. Compressed pulse of 100-element Frank code, Doppler = -0.05.

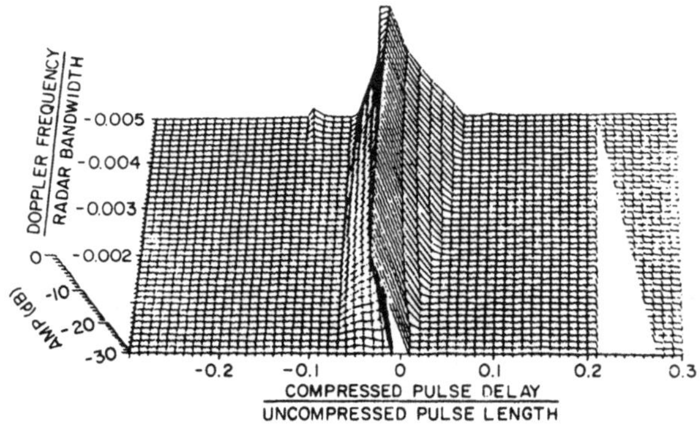

Fig. 9. Magnified ambiguity function of 100-element Frank code.

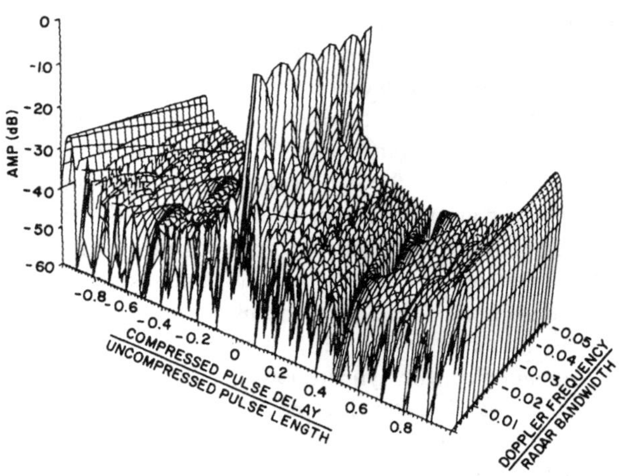

Fig. 10. Partial ambiguity function for 100-element $P4$ code.

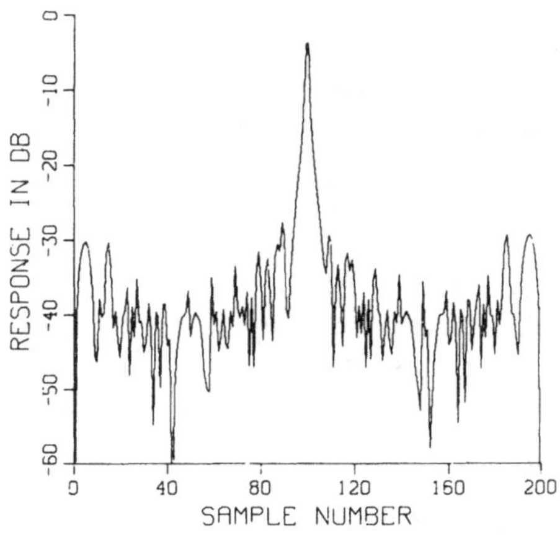

Fig. 11. Compressed pulse of 100-element Frank code, Doppler = -0.005.

The loss may also be shown by considering the misalignment of the vectors at the pulse compressor output due to a Doppler shift which results in phase shifts across the uncompressed pulse duration T. The loss occurs because the matched filter does not desteer the phases due to Doppler. For a Doppler frequency f_d, the phase increments from subpulse to subpulse are

$$\Delta\theta = 2\pi f_d T/\rho \tag{11}$$

and the resultant unit-normalized signal is

$$S = (1/\rho) \sum_{n=0}^{\rho-1} \exp(-jn\Delta\theta)$$

$$= (1/\rho) \frac{\sin \rho(\Delta\theta/2)}{\sin(\Delta\theta/2)}. \tag{12}$$

For $\Delta\theta = 0$, the maximum normalized output of one is obtained. When the total phase shift across the uncompressed pulse is π, one finds that

$$S = 2/\pi. \tag{13}$$

This is equivalent to an approximate 4 dB loss and corresponds to a range-Doppler coupling of 1/2 of a range cell. As mentioned previously, for a range Doppler coupling of 1 range cell, the total phase shift across the uncompressed pulse is 2π and the peak amplitude is nearly restored. The trough following each peak is down approximately 4 dB from the peak.

It should be noted that this 4 dB loss also occurs in a pseudorandom binary shift register code when the Doppler phase shift across the uncompressed pulse is π. However, the response is not cyclic and monotonically decreases as the Doppler frequency is increased. For this reason, a Doppler filter bank is usually instrumented to cover the Doppler band of interest.

The cyclic loss indicated for the polyphase code is a consequence of deriving the phases of the polyphase code from those of a step-chirp or chirp waveform which is sampled at the Nyquist rate. Had the phases been sampled faster, the cyclic loss would decrease and the peak sidelobes of the compressed pulse would increase to approximately 13 dB as previously described. This is a general property of the polyphase codes described in this paper.

B. FFT Implementations and Doppler Compensation

Another property of the polyphase codes described in this paper is that they can be implemented using a modified FFT phase structure. An example is shown in Fig. 12 for a $P1$ code. Each code can be generated or compressed using the same standard FFT phase filter shown in Fig. 13. The phase shifts used before and after the FFT phase filter depend on the particular code.

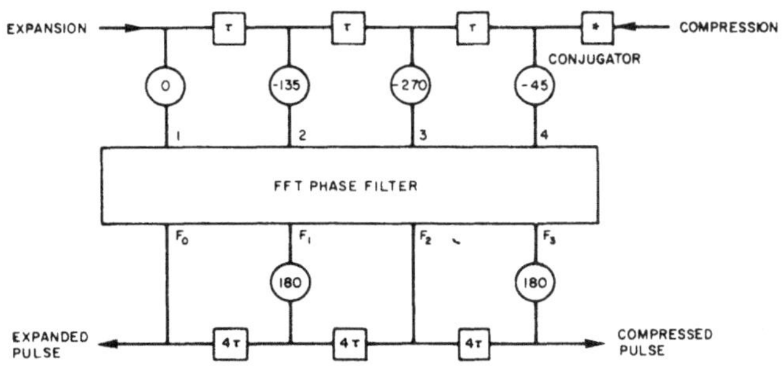

Fig. 12. $P1$ code generation and expansion using FFT.

One way to reduce the 4 dB cyclic variation of the polyphase codes is to provide an additional output port for the compressed pulse which provides an approximate phase compensation of π. This could be achieved by using additional phase shifters and delay lines in the F_i output ports of the FFT phase filter shown in Fig. 12.

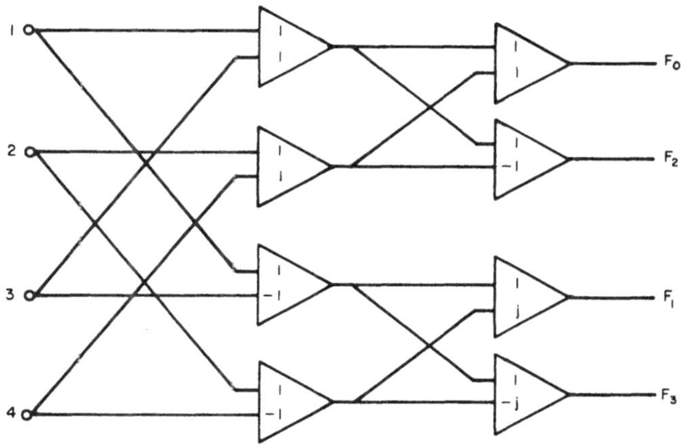

Fig. 13. FFT phase filter.

C. Comparison of Doppler Responses and Effects of Weighting

In this section we compare the Doppler responses of the various 100-element codes by examining the Doppler variation of the peak compressed signal, the peak sidelobes or secondary maxima, and the image signal. The image signal is due to the polyphase codes being derived from a Nyquist rate sampling of the step-chirp or linear-chirp phase characteristics. These are illustrated in Fig. 14 for a Frank-coded receive waveform having a normal ized Doppler frequency of -0.05.

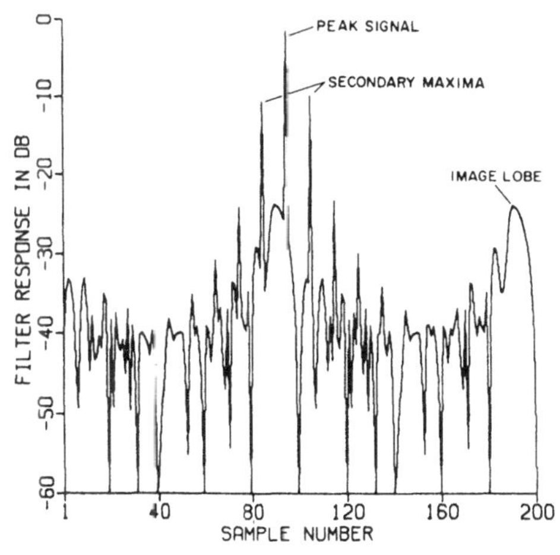

Fig. 14. Compressed pulse of 100-element Frank code, Doppler $= -0.05$.

The Doppler behavior of the Frank, $P1$, and $P2$ codes is the same and is shown in Fig. 15 where the cyclic variation of the peak amplitude is evident. There is also a cyclic behavior of the secondary maxima and the envelope of the peak signal response every 0.1 in normalized Doppler. This Doppler corresponds to a range-Doppler coupling of 10 range cells for the 100-element code, which is the equivalent duration of one phase or frequency group. These secondary maxima are nearly the same as those which occur for an analog step-chirp compressed pulse having the same Doppler shifted input waveform [4]. Fig. 15, except for the rapid cyclic behav-

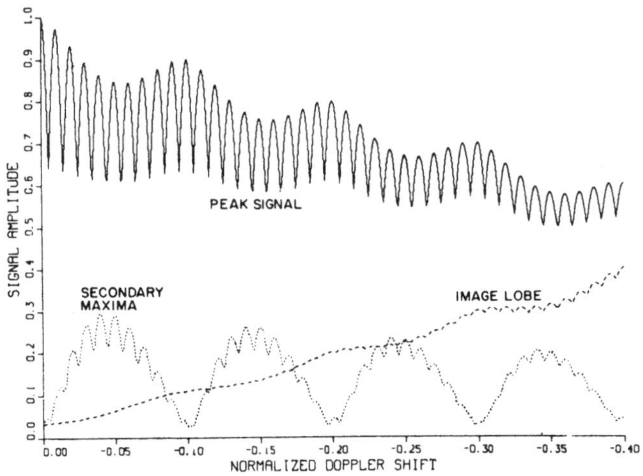

Fig. 15. Doppler properties of 100-element Frank, $P1$, and $P2$ codes.

ior, is similar to [4, fig. 8.28]. However, the peak response in Fig. 15 does not fall off as fast with Doppler. This is attributed to our sampling the received Frank code once per code element. Also, our secondary maxima and image signal include the zero Doppler values.

The $P3$ and $P4$ codes also have the same Doppler behavior which is shown in Fig. 16, where it is evident that

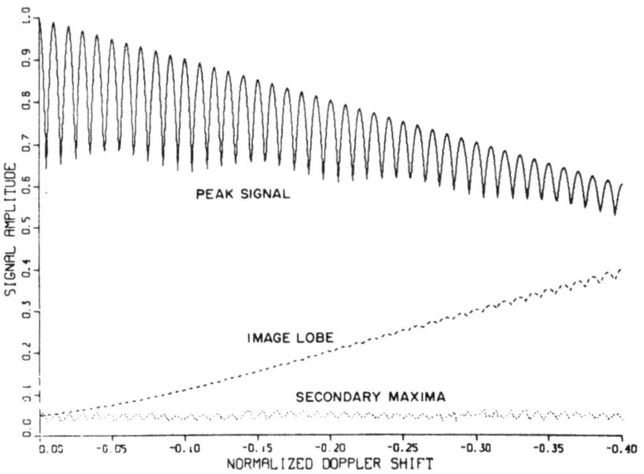

Fig. 16. Doppler properties of 100-element $P3$ and $P4$ codes.

the secondary maxima sidelobes are generally much less than for the Frank, $P1$, and $P2$ codes shown in Fig. 15. At zero Doppler, the $P3$ and $P4$ codes have sidelobes that are approximately 4 dB higher than the Frank, $P1$, and $P2$ codes. The compressed pulse for the $P3$ or $P4$ code having a Doppler shift of -0.05 is shown in Fig. 17, which should be compared with the Frank, $P1$, and $P2$ compressed pulse in Fig. 14.

The effects of weighting, using a cosine on a pedestal of 0.2, are depicted in Fig. 18 for the Frank, $P1$, and $P2$ codes (Case 1) and in Fig. 19 for the $P3$, $P4$ codes (Case 2). Figs. 20 and 21 show the compressed pulses for the weighted Case 1 and Case 2 codes. Figs. 20(a) and 21(a) and 20(b) and 21(b) show normalized Doppler frequencies of 0.0 and -0.05, respectively. A comparison of these figures with the corresponding unweighted responses

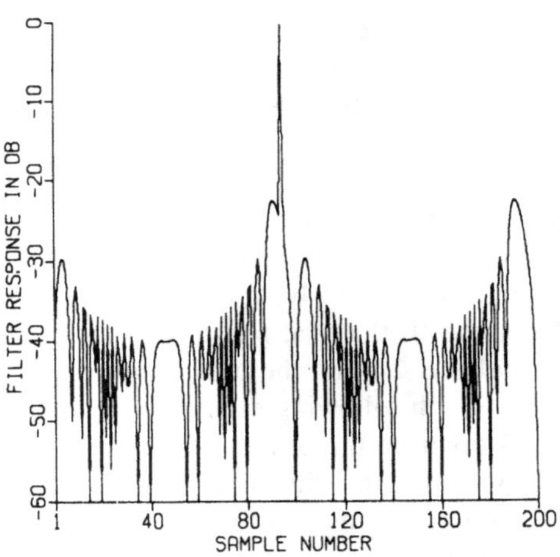

Fig. 17. Compressed pulse of 100-element $P4$ code, Doppler $= -0.05$.

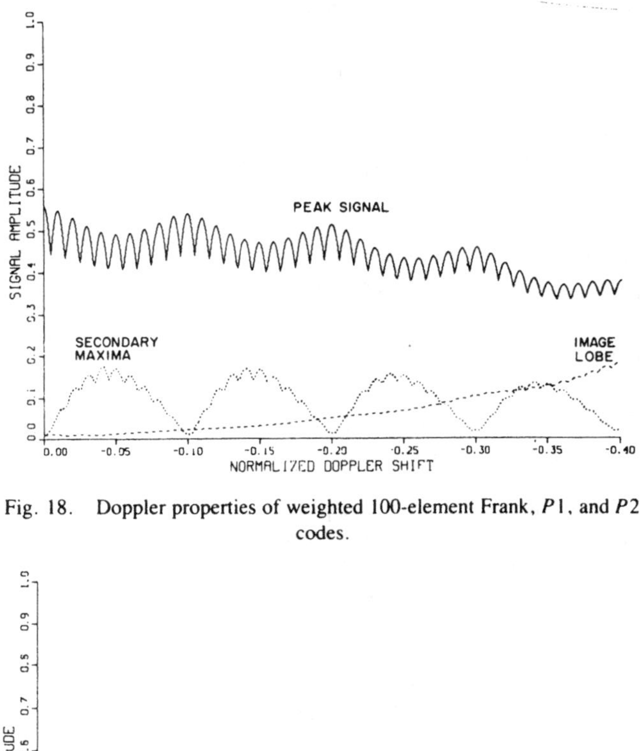

Fig. 18. Doppler properties of weighted 100-element Frank, $P1$, and $P2$ codes.

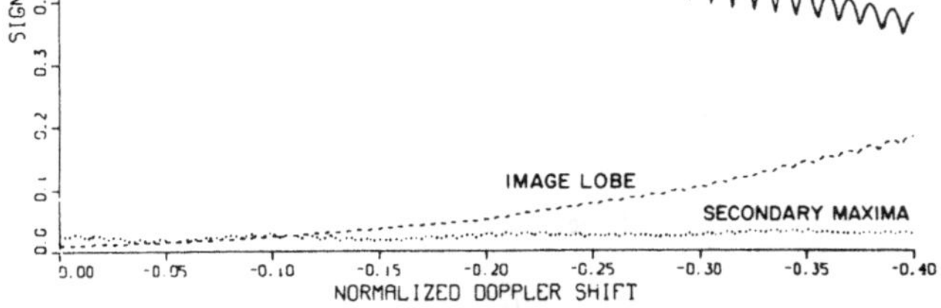

Fig. 19. Doppler properties of weighted 100-element $P3$ and $P4$ codes.

shown in Figs. 3, 14, 6, and 17, respectively, allows several observations. First, weighting reduces the percentage cyclic variation of the compressed pulse peak with Doppler. Second, weighting increases the ratios of the peak signal to the image lobe, the mean-squared sidelobes, and to a lesser extent, the secondary maxima. An-

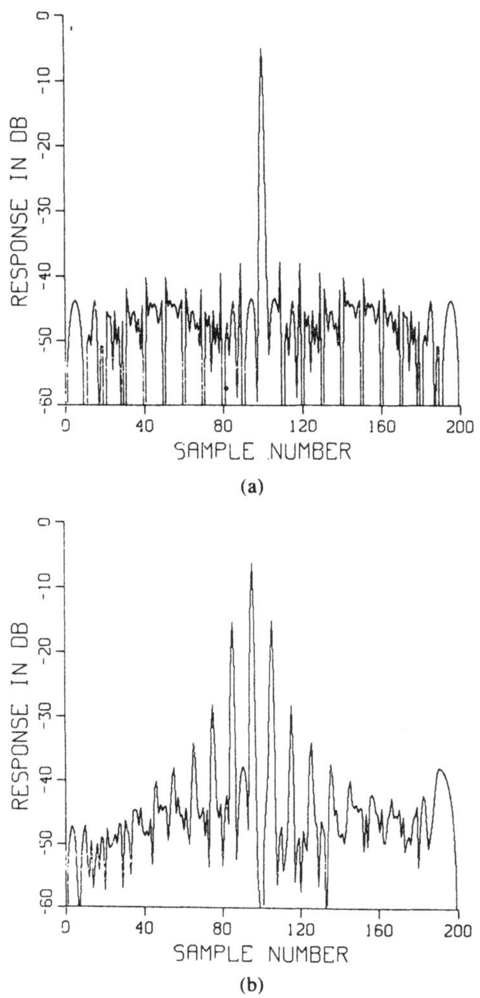

Fig. 20. Compressed pulse of weighted 100-element Frank, $P1$, and $P2$ codes. (a) Doppler = 0.0. (b) Doppler = − 0.05.

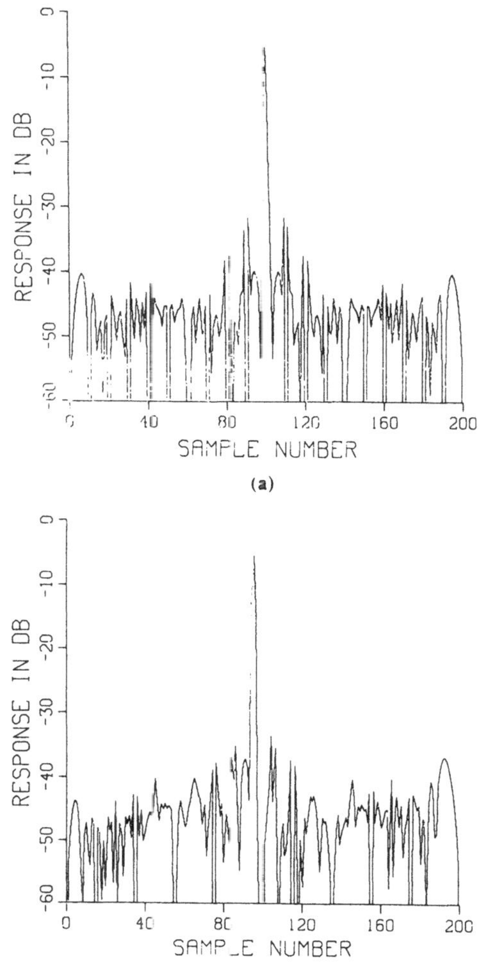

Fig. 21. Compressed pulse of weighted 100-element $P3$ and $P4$ codes. (a) Doppler = 0.0. (b) Doppler = −0.05.

other aspect of weighting is that it can be shown to reduce the fluctuation of the compressed pulsewidth as the Doppler varies.

D. Bandlimitation Effects on Doppler Responses

In this section we describe the effects of bandlimitation prior to pulse compression. It was previously described how bandlimitation acts as an adverse amplitude weighting on the Frank and $P3$ codes but improves the $P1$, $P2$, and $P4$ symmetrical codes.

The behavior of the polyphase codes was compared in the presence of bandlimiting. The bandlimiting effect was simulated by oversampling the received coded waveform by a factor of 5 to approximate the analog received waveform and then filtering this waveform with a 5-sample sliding window average. The inputs to the pulse compressor were then taken using every fifth sample. The particular set of inputs is therefore dependent on the starting time. The outputs of the compression filter were computed for the five sets of input data, corresponding to different sampling times within the subpulsewidth of the coded element. This data was averaged and is shown in Figs. 22, 23, and 24 for the Frank, $P2$, and $P4$ codes, respectively, for normalized Doppler frequencies of 0 and −0.05. The results for the $P3$ and $P1$ codes were similar to the Frank and $P2$ codes, respectively, and are not shown. In Figs. 22(a), 23(a), and 24(a) showing the zero Doppler cases, the peak responses are each reduced approximately 2.4 dB. However, for the Frank code, the secondary maxima and the image sidelobes are not reduced. Thus, it is seen that in the presence of bandlimiting, the Frank code compressed pulse degrades. Referring to Figs. 23(a) and 24(a), it is seen that the secondary maxima and image sidelobes are approximately 5 dB lower with the result that the peak signal to sidelobe ratio is improved. In Figs. 22(b), 23(b), and 24(b), we show the same codes having a normalized Doppler shift of −0.05. Comparing 22(b) with Fig. 14 for the Frank code, it may be seen that the sidelobe levels are approximately the same although the peak signal is reduced by nearly 4 dB. This again shows that the Frank code is degraded in the presence of bandlimiting. For the $P2$ code in Fig. 23(b), the ratio of the peak signal to secondary maxima is the same as for no bandlimiting (this ratio is the same as for the Frank code shown in Fig. 14); how-

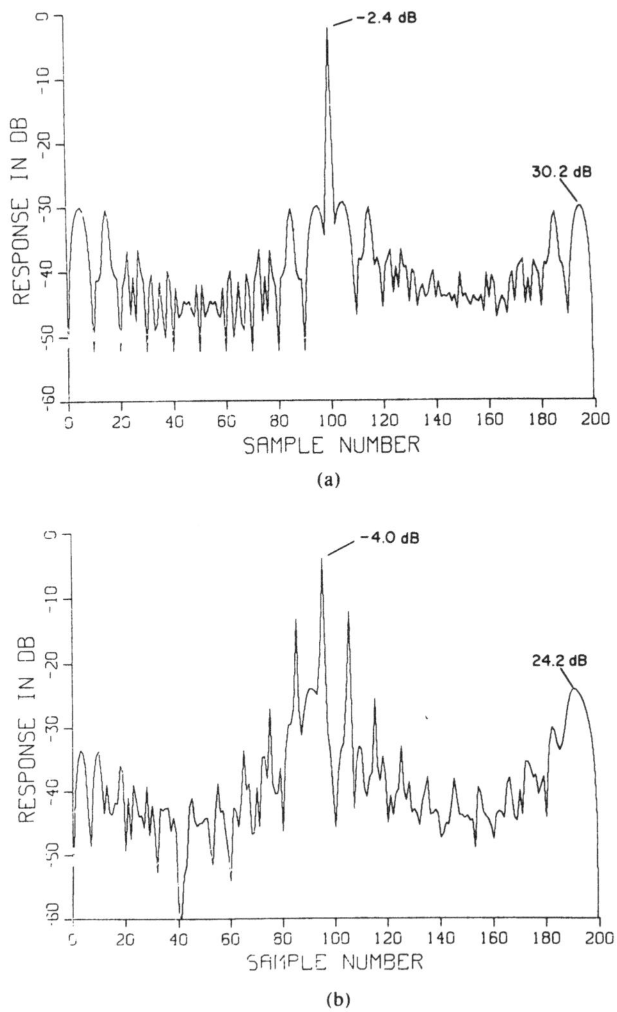

Fig. 22. Average compressed pulse of 100-element Frank code after bandlimiting. (a) Doppler = 0.0. (b) Doppler = −0.05.

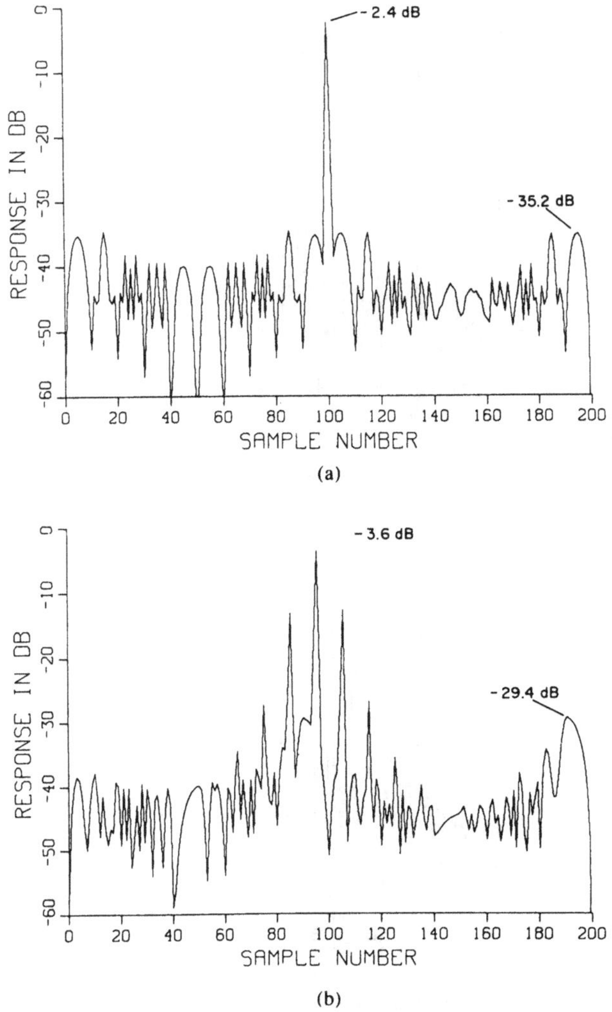

Fig. 23. Average compressed pulse of 100-element $P2$ code after bandlimiting. (a) Doppler = 0.0. (b) Doppler = −0.05.

ever, the ratio of the peak signal to the image lobe has improved by approximately 3 dB. In Fig. 24(b), showing the bandlimited $P4$ coded waveform having a Doppler shift of −0.05, the ratio of the peak signal to image lobe is also improved by approximately 3 dB compared with the case shown in Fig. 17 for no bandlimiting.

From these results and prior comments, it is seen that for zero and nonzero Doppler shifts, the Frank and $P3$ codes degrade in the presence of bandlimiting. On the other hand, the symmetrical $P1$, $P2$, and $P4$ codes improve mainly in terms of the ratios of the peak signal to the image lobe and the mean-squared sidelobes. The peak signal to secondary maxima ratio is improved by several dB for zero Doppler, and for higher Dopplers the ratio is approximately the same. The large secondary maxima of the Frank, $P1$, and $P2$ codes which occur at the higher Doppler frequencies are not present with the $P3$ or $P4$ codes.

IV. SUMMARY

The basic properties of the Frank and the $P1$, $P2$, $P3$, and $P4$ polyphase codes were presented. It was shown how these codes may be obtained from considering the sampled phases of the step-chirp and chirp baseband waveforms. These codes can be digitally compressed by using FFTs directly or by a fast convolution technique.

The Doppler properties of these waveforms were investigated in detail. It was shown that these waveforms have a cyclic loss of approximately 4 dB which is attributed to the ρ discrete phases that are used. As the number of phase samples increases, this loss diminishes and the sidelobe levels approach the 13 dB level of the chirp waveform. It was shown that this loss occurs when the total phase shift across the uncompressed pulse is an odd multiple of π radians. This loss can therefore be reduced by providing a phase compensated channel which has approximately a π phase shift across the uncompressed pulse. Then, for example, the channel having the largest signal is selected. Also, it was shown or stated that weighting reduced the cyclic variation of the peak response and the variation of the pulsewidth with Doppler.

The Doppler responses of the $P3$ and $P4$ codes were shown to have much lower secondary maxima than the Frank, $P1$, and $P2$ codes and to have comparable image lobes. However, the effects of amplitude weighting were

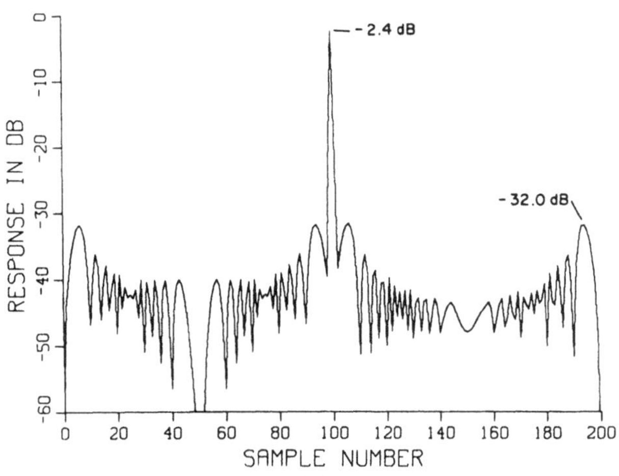

(a)

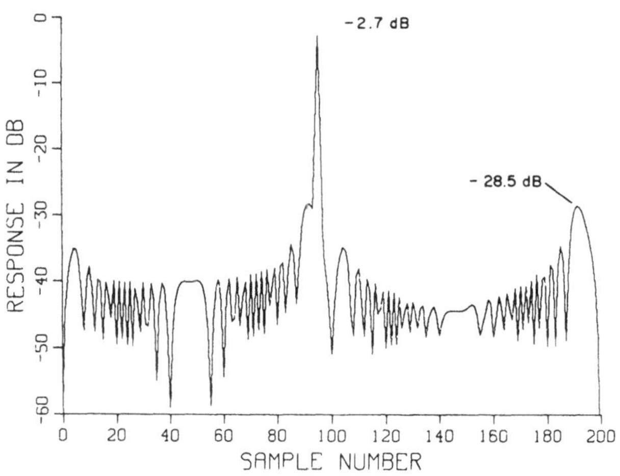

(b)

Fig. 24. Average compressed pulse of 100-element $P4$ code after bandlimiting. (a) Doppler = 0.0. (b) Doppler = −0.05.

shown to primarily increase the ratios of the peak signal to the image lobe, the mean-squared sidelobes, and, to a smaller extent, the ratio of the peak signal to the secondary maxima.

The effects of bandlimiting were investigated for zero and nonzero Doppler-shifted waveforms; for the symmetrical $P1$, $P2$, and $P4$ codes, the results were found to be similar to amplitude weighting. For these codes, it was found that the ratios of the peak signal to the image lobe, the mean-squared sidelobes and, to a lesser extent, the secondary maxima are improved. However, for the unsymmetrical Frank and $P3$ codes, these ratios are degraded.

The $P4$ code, in addition to being tolerant of bandlimiting, was shown to provide better Doppler tolerance than the other codes in the presence of relatively large Doppler shifts. However, for small normalized Doppler shifts of less than approximately $1/(2\rho)$, the $P1$ or $P2$ codes have lower peak sidelobes. The preferred code therefore depends on the expected Doppler shifts.

ACKNOWLEDGMENTS

The assistance of Mrs. F.C. Lin in obtaining the computer plots is gratefully acknowledged.

REFERENCES

[1] Lewis, B.L., and Kretschmer, F.F., Jr. (1981)
A new class of polyphase pulse compression codes and techniques.
IEEE Transactions on Aerospace and Electronic Systems, AES-17 (May 1981), 364–372.

[2] Kretschmer, F.F., Jr., and Lewis, B.L. (1982)
Polyphase pulse compression waveforms.
NRL Report 8540, Naval Research Laboratory, Washington, D.C., Jan. 5, 1982.

[3] Lewis, B.L., and Kretschmer, F.F., Jr. (1981)
Linear frequency modulation derived polyphase pulse-compression codes and an efficient digital implementation.
NRL Report 8541, Naval Research Laboratory, Washington, D.C., Nov. 2, 1981.

[4] Cook, C., and Bernfield, M. (1967)
Radar Signals, An Introduction to Theory and Applications.
New York: Academic Press, 1967.

[5] Nathanson, F.E. (1969)
Radar Design Principles.
New York: McGraw-Hill, 1969.

[6] Frank, R.L. (1963)
Polyphase codes with good nonperiodic correlation properties.
IEEE Transactions on Information Theory, IT-9 (June 1963), 43–45.

Linear Frequency Modulation Derived Polyphase Pulse Compression Codes

BERNARD L. LEWIS, Senior Member, IEEE

FRANK F. KRETSCHMER, JR., Senior Member, IEEE
Naval Research Laboratory

Two new polyphase pulse compression codes and efficient digital implementation techniques are presented that are very Doppler tolerant and that can provide large pulse compression ratios. One of these codes is tolerant of precompression bandwidth limitations.

Manuscript received September 1, 1981

Authors' address: Naval Research Laboratory, Code 5340.1, Washington, DC 20375.

INTRODUCTION

In previous publications [1, 2], the authors introduced a new class of polyphase pulse compression codes and techniques for use in digitally coded radars. Such codes and compressors can be employed to obtain much larger time-bandwidth products (pulse compression ratios) than are feasible with analog dispersive delay lines.

It is the purpose of this paper to extend this class to include two new codes, one of which is tolerant of precompression bandwidth limitation that appears in radar receivers [2]. The availability of several different codes provides a radar designer with more flexibility.

These new phase codes are conceptually derived from a linear frequency modulation waveform (LFMW) and are more Doppler tolerant than other phase codes derived from a step approximation to a LFMW. By Doppler tolerant, we mean that the compressed pulse does not degrade significantly with relatively large Doppler shifts on echoes. These new phase codes also have low range-time sidelobes without amplitude weighting. Also described is an efficient technique for implementing these new phase codes in a digital pulse-expander-compressor, and performance data is presented.

It should be noted that these phase codes are designed to be used both on transmission and reception to insure that the receive filter matches the transmitted waveform independent of time differences between the leading edge of the echo and a sampling pulse, i.e., independent of target range.

NEW PHASE CODES

The two new phase codes will be called the P3 and P4 codes to distinguish them from the P1 and P2 codes discussed in [2]. The P3 code is not precompression bandwidth limitation tolerant but is much more Doppler tolerant than the Frank [3] or P1 and P2 codes. The P4 code is a rearranged P3 code with the same Doppler tolerance and with better precompression bandwidth limitation tolerance.

P3 CODE

The P3 code is conceptually derived by converting a linear frequency modulation waveform to baseband using a local oscillator on one end of the frequency sweep and sampling the inphase I and quadrature Q video at the Nyquist rate. Letting the waveform to be coherently detected have a pulse length T and frequency

$$f = f_0 + kt \tag{1}$$

where k is a constant, the bandwidth of the signal will be approximately

$$B = kT. \tag{2}$$

This bandwidth will support a compressed pulse length of approximately

$$t_c = 1/B \tag{3}$$

and the waveform will provide a pulse compression ratio of

$$\varrho = T/t_c = BT. \tag{4}$$

Assuming that the first sample of I and Q is taken at the leading edge of waveform, the phases of successive samples taken t_c apart is

$$\phi_i^{(P3)} = 2\pi \int_0^{(i-1)t_c} [(f_0 + kt) - f_0]dt$$

$$= \pi k(i - 1)^2 t_c^2 \tag{5}$$

where $i = 1, 2, \ldots, \partial$. From (2), $k = B/T$ and from (3), $t_c = 1/B$; therefore (5) can be written as

$$\phi_i^{(P3)} = \pi(i - 1)^2/BT = \pi(i - 1)^2/\varrho. \tag{6}$$

With $\varrho = 16$, the P3 code modulo 2π is as shown in Table I.

TABLE I
P3 Code Modulo 2π, $\varrho = 16$

$i =$	1	2	3	4	5	6	7	8
$\phi_i^{(P3)} =$	0	$\pi/16$	$4\pi/16$	$9\pi/16$	π	$25\pi/16$	$4\pi/16$	$17\pi/16$
$i =$	9	10	11	12	13	14	15	16
$\phi_i^{(P3)} =$	0	$17\pi/16$	$4\pi/16$	$25\pi/16$	π	$9\pi/16$	$4\pi/16$	$\pi/16$

P4 CODE

The P4 code is conceptually derived from the same waveform as the P3 code. However, in this case, the local oscillator frequency is set equal to $f_0 + kT/2$ in the I, Q detectors. With this frequency, the phases of successive samples taken t_c apart are

$$\phi_i^{(P4)} = 2\pi \int_0^{(i-1)t_c} [(f_0 + kt) - (f_0 + kT/2)]dt$$

$$= 2\pi \int_0^{(i-1)t_c} k(t - T/2)dt \tag{7}$$

or

$$\phi_i^{(P4)} = \pi k(i - 1)^2 t_c^2 - \pi kT(i - 1)t_c$$

$$= [\pi(i - 1)^2/\varrho] - \pi(i - 1). \tag{8}$$

With $\partial = 16$, the P4 code modulo 2π is as shown in Table II.

TABLE II
P4 Code Modulo 2π, $\varrho = 16$

$i =$	1	2	3	4	5	6	7	8
$\phi_i^{(P4)} =$	0	$17\pi/16$	$4\pi/16$	$25\pi/16$	π	$9\pi/16$	$4\pi/16$	$\pi/16$
$i =$	9	10	11	12	13	14	15	16
$\phi_i^{(P4)} =$	0	$\pi/16$	$4\pi/16$	$9\pi/16$	π	$25\pi/16$	$4\pi/16$	$17\pi/16$

It should be noted that the largest phase increments from code element to code element are on the two ends of the P4 code but are in the middle of the P3 code. Thus the P4 code is more precompression bandwidth limitation tolerant than the P3 code. This follows since precompression bandwidth limitations average the code phase increments and would attenuate the P4 code on the ends and the P3 code in the middle as discussed in [2] and in later discussions in this paper. The former increases the peak-to-sidelobe ratio of the compressed pulse while the latter decreases it. Precompression bandlimiting will be found in any well designed system to prevent out of band noise foldover and signal aliasing due to the time sampling required in the conversion from analog-to-digital data format.

EFFICIENT DIGITAL PULSE EXPANDER-COMPRESSOR IMPLEMENTATION

The P3 and P4 codes can be implemented in a pulse expander-compressor employing digital fast Fourier transform circuits (FFT [4]) similar to those discussed for the Frank and P1 codes in [2]. These expander-compressors take advantage of the fact that the P3 code only differs from the Frank code by π phase shifts every $\varrho^{1/2}$ code elements and by added phase increments from code element to code element that repeat every $\varrho^{1/2}$ samples. These added phase shifts are caused by the linear frequency shift during the time the equivalent Frank code frequency is constant. This difference between the P3 and Frank codes is illustrated in Tables III-VI for $\varrho = 16$. The extra phase increments can be added to the $\varrho^{1/2}$ individual time samples in the FFT and their accumulated value π in the time that Frank code frequency exists can be added to the frequency port outputs of FFT as illustrated in Figs. 1 and 2. (Note that even number multiples of π need not be added since they are equivalent to zero phase shift.)

TABLE III
P3 Code Matrix, $\pi = 16$

0	$\pi/16$	$4\pi/16$	$9\pi/16$
π	$25\pi/16$	$4\pi/16$	$17\pi/16$
0	$17\pi/16$	$4\pi/16$	$25\pi/16$
π	$9\pi/16$	$4\pi/16$	$\pi/16$

TABLE IV
Adding π to Every Other Row of Table III and Subtracting Even Multiples of π

0	$\pi/16$	$4\pi/16$	$9\pi/16$
0	$9\pi/16$	$20\pi/16$	$\pi/16$
0	$17\pi/16$	$4\pi/16$	$25\pi/16$
0	$25\pi/16$	$20\pi/16$	$17\pi/16$

TABLE V
Frank Code Matrix, $\varrho = 16$

0	0	0	0
0	$8\pi/16$	$16\pi/16$	$24\pi/16$
0	$16\pi/16$	0	$16\pi/16$
0	$24\pi/16$	$16\pi/16$	$8\pi/16$

TABLE VI
Subtracting Frank Code Phases (Table V) from Phases in Table IV

0	$\pi/16$	$4\pi/16$	$9\pi/16$
0	$\pi/16$	$4\pi/16$	$9\pi/16$
0	$\pi/16$	$4\pi/16$	$9\pi/16$
0	$\pi/16$	$4\pi/16$	$9\pi/16$

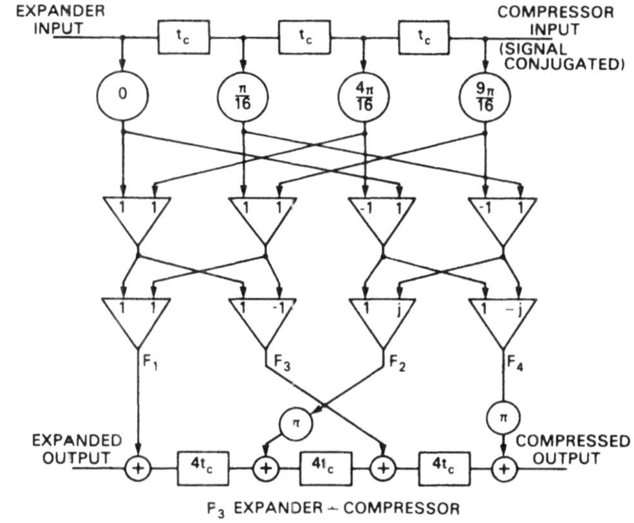

Fig. 1. P3 expander-compressor, $\varrho = T/t_c = 16$.

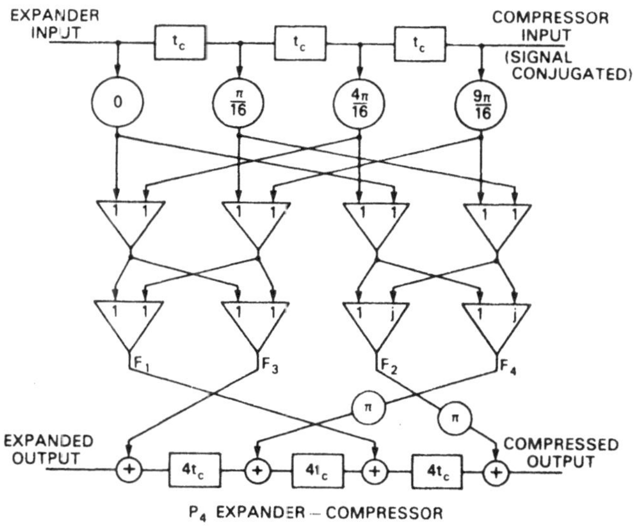

Fig. 2. P4 expander-compressor, $\varrho = T/t_c = 16$.

PERFORMANCE DATA

Fig. 3 illustrates the autocorrelation function or compressed pulse waveform that is obtained with the digital pulse-expander-compressors illustrated in Figs. 1 and 2 with no Doppler and no bandwidth limitation. The sample number corresponds to a range cell in time in a radar. In this case the pulse compression ratio is $\varrho = 100$ and the highest range time sidelobe is 4ϱ below the peak response. For comparison purposes, Fig. 4 illustrates the autocorrelation function of the Frank or P1 codes with zero Doppler and no bandwidth limitation. With zero Doppler and no bandwidth limitations the Frank and P1 codes have 4 dB lower peak range-time sidelobes than the P3 or P4 codes.

Fig. 5 illustrates the effect of a Doppler shift equal to 5 percent of the bandwidth on the P3 or P4 code compressors. Fig. 6 (for comparison purposes) shows the compressed pulse that would result from using the Frank or P1 codes with this Doppler shift. Note that the large grating lobes that appear with Doppler in the Frank or P1 code autocorrelation functions are absent in the P3 and P4 results.

The effect of precompression bandlimiting on the P3 and P4 codes was found to be the same as that illustrated in [2] for the Frank code and the P2 code. The precompression bandlimiting weights the P2 code to reduce sidelobes but does not change the Frank code sidelobes. The same phenomenon reduces the P4 sidelobes but not the P3 sidelobes.

The P3 and P4 code peak gain cycles with increasing Doppler like the Frank (and P1 and P2) code as illustrated in [2, Fig. 1]. This cycling repeats at Doppler shifts equivalent to an odd multiple of a half range cell range Doppler coupling, i.e., when the Doppler causes π phase shift across the uncompressed pulse. These gain changes and the accompanying peak broadening can be controlled by amplitude weighting the FFT frequency ports in the compressor on receive as discussed in [1].

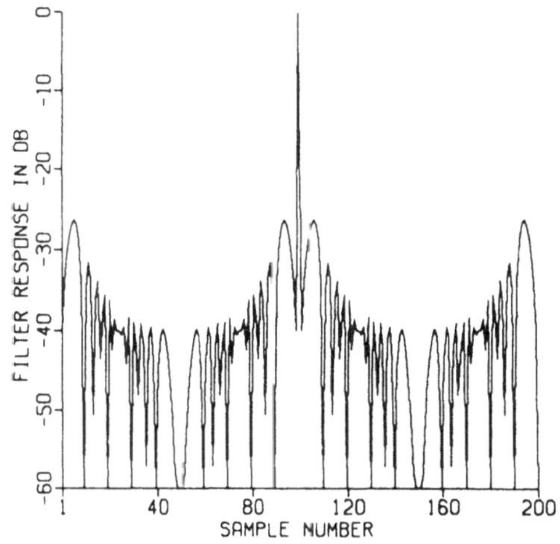

Fig. 3. P3 or P4 autocorrelation function, $\varrho = 100$, zero Doppler shift, and no bandwidth limitation.

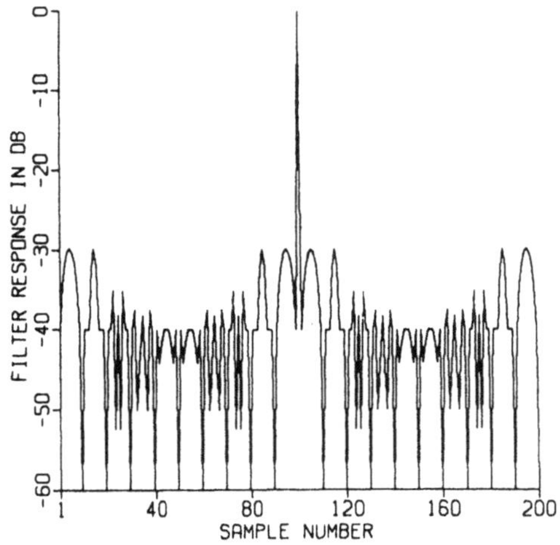

Fig. 4. Frank code autocorrelation function, $\varrho = 100$, zero Doppler shift, and no bandwidth limitation.

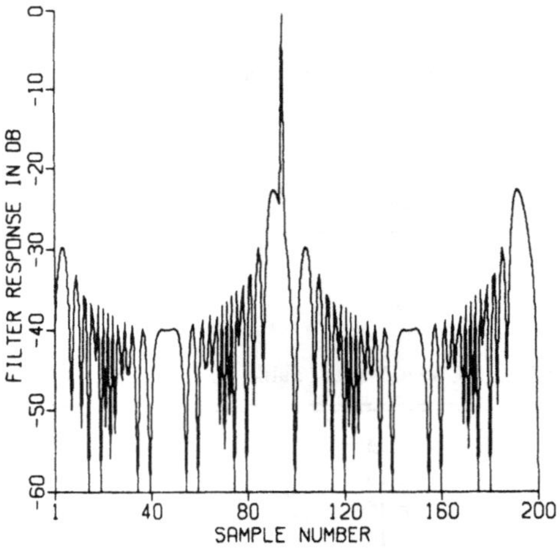

Fig. 5. P3 or P4 autocorrelation function, $\varrho = 100$, Doppler = 0.05B with no bandwidth limitation.

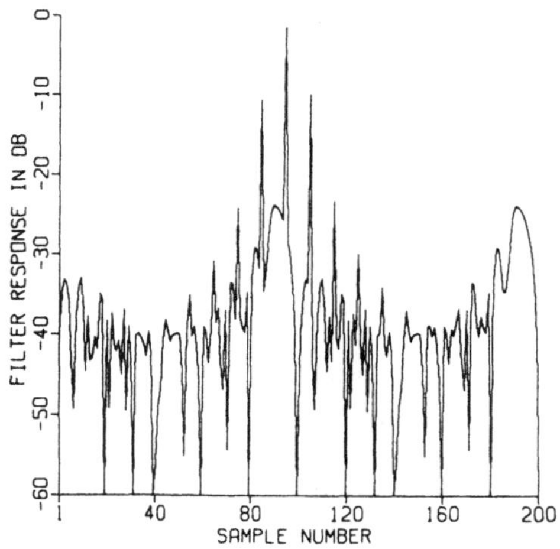

Fig. 6. Frank code autocorrelation function, $\varrho = 100$, Doppler = 0.05B with no bandwidth limitation.

With large Doppler shifts, as illustrated in Fig. 5, all frequency derived polyphase pulse compression code autocorrelation functions are folded in frequency due to the time sampling. This folding produces the asymmetry in the far out sidelobes. An up Doppler makes the highest frequency part of the code match the lowest frequency part of the matched filter while down Doppler produces a reverse effect. However, precompression bandlimiting will prevent this foldover and control the sidelobes furthest away.

CONCLUSIONS

The P3 and P4 polyphase pulse compression codes do not produce the large range-time grating lobes with large Doppler shifts that are characteristic of the Frank, P1, and P2 codes.

Both the P3 and P4 codes can be compressed with digital FFT circuits with additive phase shifts in the FFT time samples and π phase shifts in every other FFT output (frequency) port.

The P4 code is much more tolerant of precompression bandwidth limiting than the P3 code.

The digitally implemented phase coding and compressing makes larger pulse compression ratios possible than would be feasible with dispersive delay line analog systems. In addition, the better Doppler tolerance of the P3 and P4 codes permit these large time-bandwidths to be effective in the presence of large Doppler shifts on echo pulses.

REFERENCES

[1] Lewis, B.L., and Kretschmer, F.F., Jr. (1980)
A new class of pulse compression codes and techniques. NRL Report 8387, Naval Research Laboratory, Washington, D.C., Mar. 26, 1980.

[2] Lewis, B.L., and Kretschmer, F.F., Jr. (1981)
A new class of polyphase pulse compression codes and techniques.
IEEE Transactions on Aerospace and Electronic Systems, May 1981, *AES-17*, 364-371.

[3] Frank, R.L. (1963)
Polyphase codes with good nonperiodic correlation properties.
IEEE Transactions on Information Theory, Jan. 1963, *IT*, 43-45.

[4] Skolnik, M.I. (1970)
Radar Handbook.
New York: McGraw-Hill, 1970, 35-15–35-16.

HUFFMAN-CODED PULSE COMPRESSION WAVEFORMS

INTRODUCTION

Modern radars generally incorporate pulse compression waveforms to avoid transmitting a pulse having a large peak power which can result in waveguide arcing. Pulse compression waveforms enable one to transmit a long pulse to obtain sufficient energy on a target for detection and to simultaneously obtain the desired range resolution. This is achieved by modifying the time-bandwidth product (TB) of the transmitted waveform. A larger transmit time duration T allows sufficient energy on the target for detection, while $1/B$ determines the resolution of the compressed pulse if no mismatch occurs. The desired signal bandwidth is generally obtained by modulating the signal's phase or frequency while maintaining a constant maximum pulse amplitude. This is illustrated by the linear chirp signal, pseudorandom phase codes, and polyphase pulse compression waveforms.

A desirable property of the compressed pulse is that it have low sidelobes to prevent a weak target from being masked in the time sidelobes of a nearby stronger target. It is generally also desired that the compressed pulse does not significantly degrade when the return signal has been doppler shifted due to target motion. For the commonly used chirp signal, the sidelobes are reduced by weighting the received signal which results in a mismatch loss of approximately 1 dB, and a broadening of the pulsewidth due to the band-limiting associated with the weighting. The polyphase codes and the binary codes may also be weighted, if desired, to increase peak-to-sidelobe levels.

The Huffman-coded waveform [1 to 5] results in an ideal compressed pulse having no sidelobes except for the unavoidable sidelobe at either end of the compressed pulse. However, the Huffman waveform consists of code elements which vary in amplitude as well as in phase. Because of the amplitude fluctuations, the waveforms were not previously very practical. However, with the increased utilization of solid-state transmitters and the ability to switch the transmitters or the power transistors on and off of a transmitter bus, there has been a recent interest in investigating the properties and performance capabilities of the Huffman codes.

In this report we describe a general waveform synthesis procedure, and the synthesis of Huffman codes for a desired compressed pulse. For an N-element code, there are 2^{N-1} different uncompressed pulses having the same compressed pulse. Procedures are described for selecting an efficient Huffman code in terms of transmitted power, and the Huffman codes are compared with a polyphase code and a binary shift-register code. The effects of errors on the Huffman code sidelobes are also investigated.

WAVEFORM SYNTHESIS AND HUFFMAN CODES

A. Waveform Synthesis from a Known Compressed Pulse

We consider in the following discussion a coded transmit waveform consisting of N-subpulses which can vary in both phase and amplitude from subpulse to subpulse. This is exemplified by Barker-coded waveforms, pseudorandom binary shift-register codes, and polyphase codes. It is assumed that on reception the waveform is digitized so that an inphase I and a quadrature Q sample, or equivalently an amplitude and a phase sample, are obtained for each subpulse. This simplifies the description of the waveforms and allows us to conveniently describe the time sequence by using z-transforms. Accordingly, we denote the return sequence by $E(z)$ which is given by

Manuscript approved February 7, 1985.

$$E(z) = a_0 + a_1 z^{-1} + a_2 z^{-2} + \ldots + a_{N-1} z^{-(N-1)}, \tag{1}$$

where a_i is the complex value which specifies the amplitude and phase of the ith subpulse.

A filter matched to this sequence, $M(z)$, is then given by

$$M(z) = a_{N-1}^* + a_{N-2}^* z^{-1} + a_{N-3}^* z^{-2} + \ldots + a_0^* z^{-(N-1)}. \tag{2}$$

The output of the matched filter is given by

$$G(z) = E(z) M(z) \tag{3}$$

which can be represented by

$$G(z) = g_0 + g_1 z^{-1} + g_2 z^{-2} + \ldots + g_{N-1} z^{-(N-1)} + \ldots + g_{2(N-1)} z^{-2(N-1)}, \tag{4}$$

where g_i denotes the samples of the compressed waveform.

Substituting Eqs. (1) and (2) in Eq. (3), we obtain

$$G(z) = (a_0 + a_1 z^{-1} + \ldots + a_{N-1} z^{-(N-1)})(a_{N-1}^* + a_{N-2}^* z^{-1} + \ldots + a_0^* z^{-(N-1)}). \tag{5}$$

Equation (5) can be written as

$$\begin{aligned} G(z) &= z^{-(N-1)} (a_0 + a_1 z^{-1} + \ldots + a_{N-1} z^{-(N-1)}) (a_0^* + a_1^* z + \ldots + a_{N-1}^* z^{(N-1)}) \\ &= z^{-(N-1)} E(z) E^*(1/z^*). \end{aligned} \tag{6}$$

Next, we express the z-plane zeros of $E(z)$ in factored form as

$$\prod_{i=1}^{N-1} (z - z_i), \tag{7}$$

and note from Eqs. (3) and (6) that the factored zeros of $M(z)$ may be expressed as

$$\prod_{i=1}^{N-1} (z - 1/z_i^*). \tag{8}$$

The z-plane is related to the complex S-plane by the relation

$$z = e^{ST} = e^{(\alpha + j\omega)T} = e^{\alpha T} e^{j\omega T}, \tag{9}$$

where T denotes the sampling interval or the time duration of each subpulse. From Eq. (9) it is observed that the roots of $M(z)$ given by $1/z_i^*$ are equal to the reciprocal amplitude of the roots z_i of $E(z)$, and the phase angles of the roots z_i and $1/z_i^*$ are the same. In summary, the zeros of the output function $G(z)$ occur in related pairs which are in the z-plane at the same angle but are reciprocal in amplitude. For an input N-pulse sequence there are $N-1$ such pairs of zeros.

The upshot of these relations is that the generation of the specified compressed pulse is not uniquely defined but can be accomplished by selection of either zero of each zero pair. There are 2^{N-1} possible choices of z-plane zeros that can be selected to produce the same compressed pulse. From each given choice of $(N - 1)$ zeros, a different input sequence $E(z)$ can be determined from

$$E(z) = kz^{-(N-1)} \prod_{i=1}^{N-1} (z - z_i), \tag{10}$$

where k is an energy normalization constant. The resulting coefficients of the polynomial in z^{-1} specify the amplitude and the phase of the uncompressed input waveform subpulses.

These procedures are next illustrated for the well-known Barker codes having a peak-to-sidelobe power ratio that is equal to or greater than N^2. The compressed pulse for the 13-element Barker code is shown in Fig. 1(a) on a voltage scale. From the z-transform of this response, the zero locations were determined and are shown in Fig. 1(b). It is seen that there are 12 pairs of zeros and thus 2^{12} different input waveforms which, when match-filtered, produce the same compressed pulse shown in Fig. 1(a). Each of the zeros comprising a pair are related to each other as being conjugate reciprocal zeros. For a given selection of zeros the input waveform is found from the coefficients of the polynomial resulting from performing the multiplication in Eq. (10).

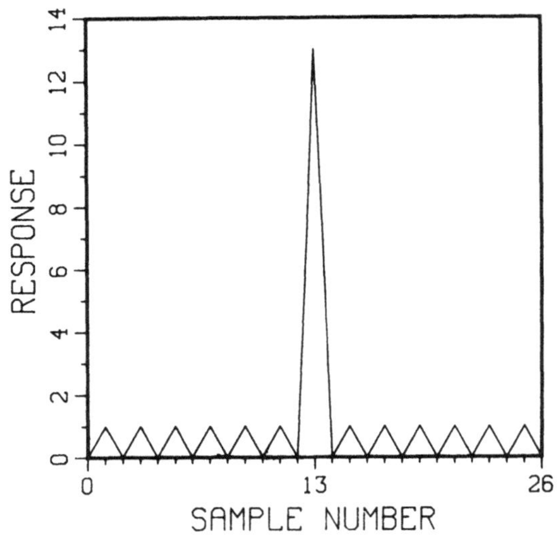

Fig. 1(a) — Compressed pulse for a 13-element Barker code

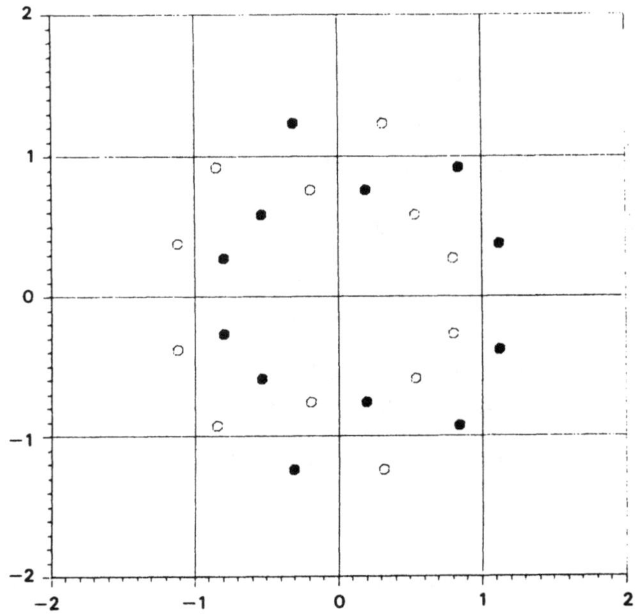

Fig. 1(b) — z-plane zero locations for a compressed 13-element Barker code

This can be efficiently and more accurately achieved using fast Fourier transform (FFT) algorithms when N becomes very large [6]. The particular set of zeros indicated by the filled-in circles corresponds to a constant amplitude binary waveform which is the well-known 13-element Barker code. In general, any other selection of zeros would result in an input waveform which varies in both amplitude and phase from subpulse to subpulse.

In Fig. 2 we show, as another example, the zeros of a compressed pulse for a four-element generalized Barker code making use of a sextic alphabet [7] whose elements consist of powers of $\exp(j\pi/3)$.

B. Huffman Codes

Huffman considered the idealized compressed pulse which contains no sidelobes except the unavoidable sidelobes at either end of the compressed pulse. The end sidelobe-level is a design parameter, and a compressed 64-element, zero-doppler Huffman code is illustrated in Fig. 3 for a relative sidelobe amplitude level of 0.1/64 V in voltage or −56 dB.

From Eq. (4) the z-transform of the compressed Huffman-coded waveform, which has been normalized to unity at the peak, takes the simple quadratic form

$$G(z) = s + z^{-(N-1)} + sz^{-2(N-1)} = s(z^{-2(N-1)} + z^{-(N-1)}/s + 1), \tag{11}$$

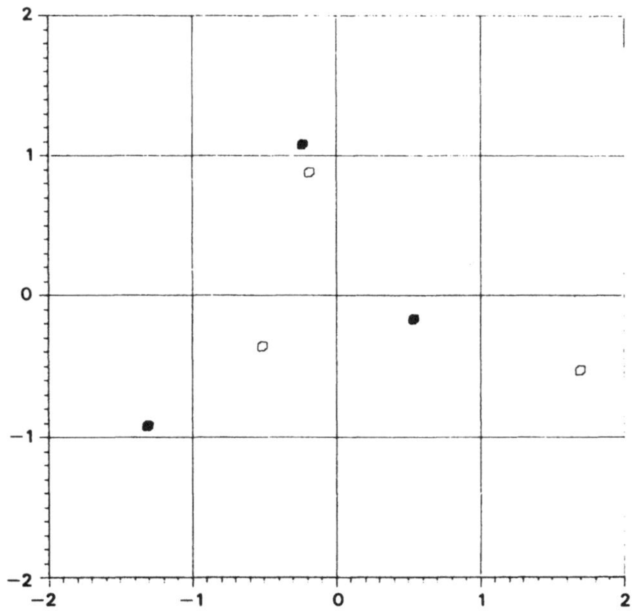

Fig. 2 — z-plane zero locations for a compressed four-element sextic code

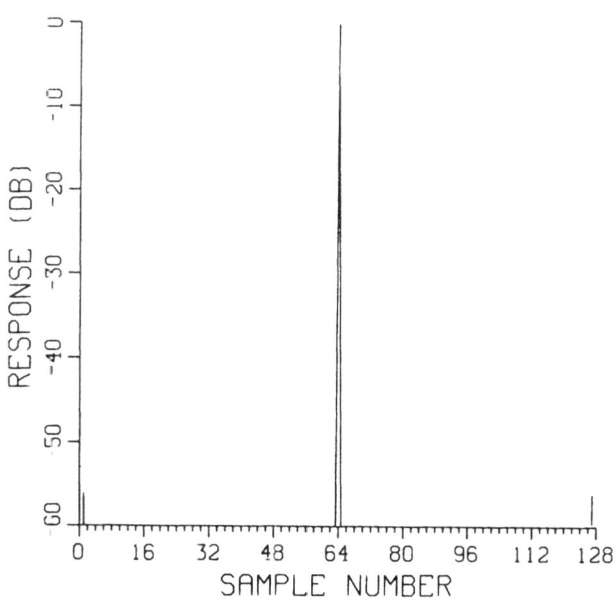

Fig. 3 — Compressed 64-element Huffman-coded waveform

where s is the normalized sidelobe voltage (that can be positive or negative). The roots of this equation lie in the z-plane at intervals of $2\pi/(N-1)$ on two circles whose radii R and R^{-1} are given by

$$\left[\left|\frac{1}{2s}\right| \pm \left(\frac{1}{4s^2} - 1\right)^{1/2}\right]^{\frac{1}{(N-1)}} \tag{12}$$

Thus, the distinguishing feature of the zeros of the compressed Huffman code, in contrast to the other codes, is that they are on two circles whose radii are reciprocal of each other at regularly spaced intervals of $2\pi/(N-1)$ radians.

The design of a Huffman code consists of specifying the number of code elements N, the sidelobe level s, and the particular choice of z-plane zeros which results from choosing one zero from each of the 2^{N-1} pairs. In general, for given N and s, the amplitude as well as the phase of each resulting uncompressed code subpulse varies with the different choice of z-plane zeros.

DESIGN OF EFFICIENT HUFFMAN CODES

Selecting the zero pattern of $G(z)$ in a random manner to determine the input-coded waveform generally results in codes which vary considerably in amplitude from subpulse to subpulse. This represents a loss in terms of the power that could be transmitted at the maximum level. We define an efficiency factor E as the ratio of the power represented by a given Huffman-coded waveform to the power that results from transmitting a constant envelope waveform having a value equal to the largest subpulse value. This can be written as

$$E = \frac{\sum_{i=0}^{N-1} |a_i|^2}{N|a_i|^2_{\max}} . \tag{13}$$

Although stated differently, this definition is equivalent to the definition of Huffman and Ackroyd except that the maximum value of E is normalized to unity.

Ackroyd [5] proposed a method for improving the efficiency of a Huffman code based on a paper by Schroeder [8] whereby the efficiency of a waveform is improved by modifying the phase spectrum of the waveform. In particular it was found that waveforms having a high FM content tended to be efficient. Accordingly, Ackroyd cleverly determined which zero to use in each pair by noting that the pulse spectrum due to the zeros changes by plus or minus π, depending on whether the zero is inside or outside the unit circle, as one traverses the unit circle in the vicinity of the zero-pair. Ackroyd suggested that the desired zero selection could be determined by using the desired phase-spectral function as an input to a delta-modulator which provides a staircase approximation to the phase spectrum in step sizes of plus or minus π. By noting the polarity of the staircase function steps, one could identify the appropriate zero of each zero pair.

The desired phase spectrum is given by Ackroyd [5], in our notation, as

$$\arg C_n = \arg C_0 - \pi n^2/N + \pi n \qquad n = 0, 1, \ldots N-1, \tag{14}$$

where the C_n represents the Fourier coefficients of the waveform. Ackroyd [5] states that C_0 is arbitrary. However, it will be shown later that C_0 is not arbitrary in terms of the achievable efficiency. We mention at this juncture that the desirable phase characteristic given by Schroeder is very similar to the phase characteristic used in the P3 and P4 polyphase-coded waveforms devised by Lewis and Kretschmer [9,10].

To illustrate these concepts we randomly selected the zeros from the zero pairs of a 64-element Huffman code having a sidelobe voltage of −0.1 V. This was done with a computer random-number generator having equally likely plus and minus ones which were associated with a zero location inside or outside the unit circle. From these zero locations, the input waveform was determined as previously described. The amplitude distribution of this waveform is shown in Figs. 4(a) and 4(b) for different trials. The efficiencies of these waveforms are 11% and 16%. Using the method of imparting a large FM content to the waveform resulted in the waveform shown in Fig. 4(c) whose efficiency is 39%. The waveforms shown in Fig. 4 were normalized to have a compressed pulse peak equal to 64.

Ackroyd noted that the efficiency could be improved by varying the design sidelobe level. We confirmed this and in addition found that the value of C_0 in the desired characteristic given by Eq. (14) could be altered to result in a different zero selection and hence a different efficiency. The variation of the efficiency with the initial phase and the sidelobe level is shown in Fig. 5 for the 64-element Huffman code.

COMPARISON OF HUFFMAN CODES WITH OTHER CODES

We first describe and compare the 64-element Huffman code with a 64-element polyphase P4 code and a 63-element pseudorandom shift-register binary code in terms of doppler sensitivity. Next, the sensitivity of the Huffman-code sidelobes due to tolerance errors and due to finite quantization levels is presented.

The ambiguity function of an efficient 64-element Huffman code (as shown in Fig. 4(c)) having an initial phase angle of $-\pi$ and a sidelobe level of −56 dB is shown in Fig. 6. The delay is normalized to the uncompressed pulse length T, and the y-axis is the product of the doppler frequency and T. This product can also be interpreted as the number of 2π phase shifts across the uncompressed pulse due to doppler. A blowup of this figure is shown in Fig. 7 where the product of the doppler frequency and T ranges from 0 to 1. It is seen in Figs. 6 and 7 that the sidelobes of the Huffman code grow rapidly with an increase in doppler. For comparison we show the ambiguity function of a P4 code in

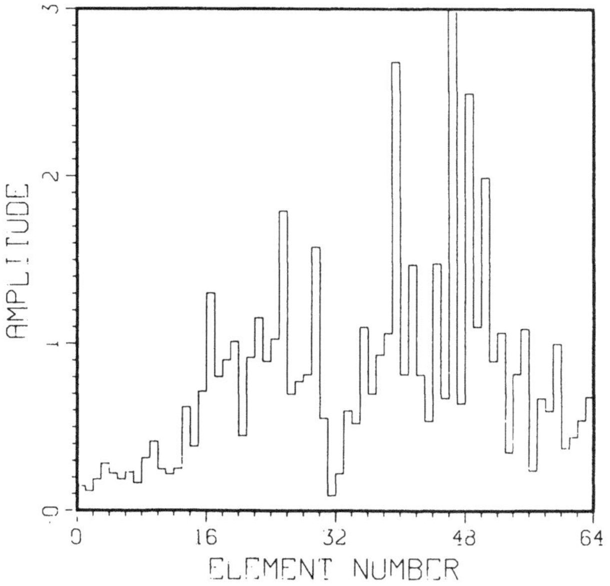

(a) efficiency is 11%

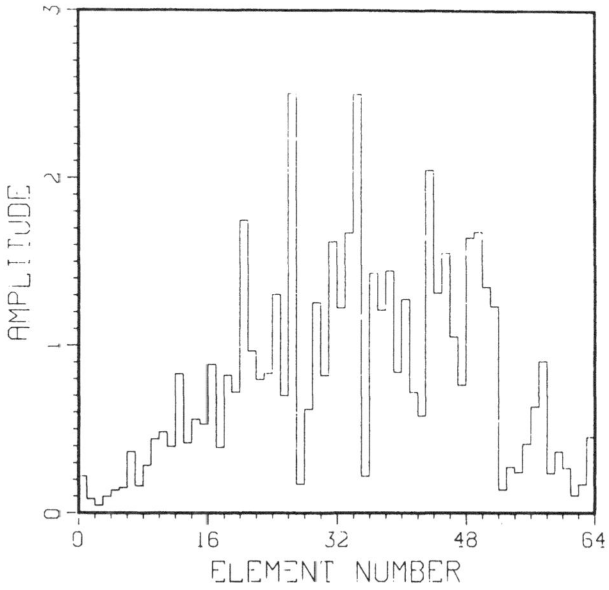

(b) efficiency is 16%

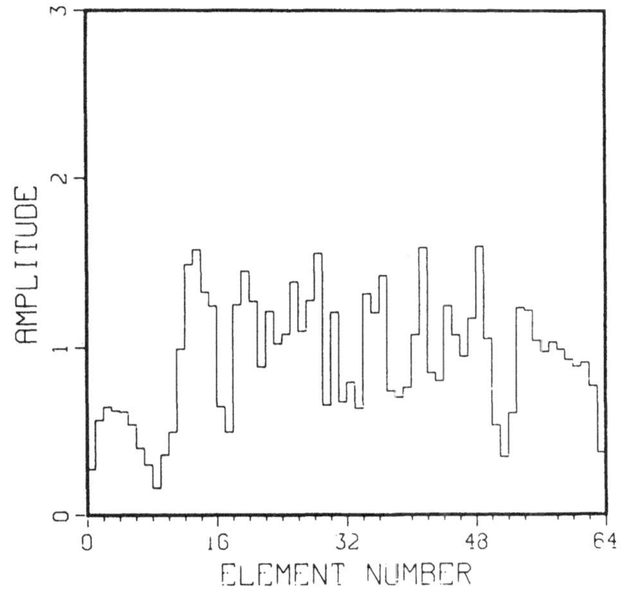

(c) efficiency is 39% due to imparting an FM phase characteristic on waveform

Fig. 4 — Amplitudes of different 64-element Huffman-coded waveforms having the same compressed pulse as shown in Fig. 3

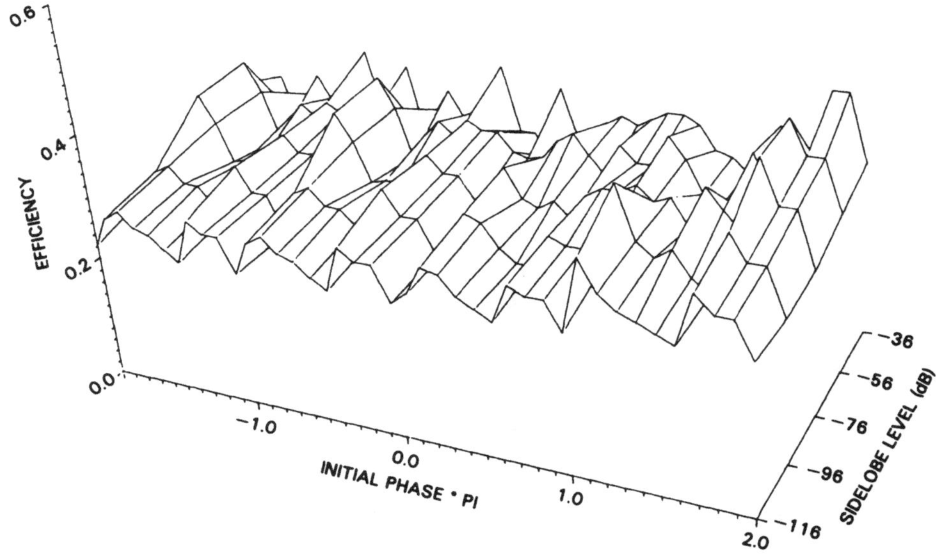

Fig. 5 — Variation of 64-element Huffman-coded waveform efficiency with the design sidelobe level and the initial phase angle of the FM pulse characteristic

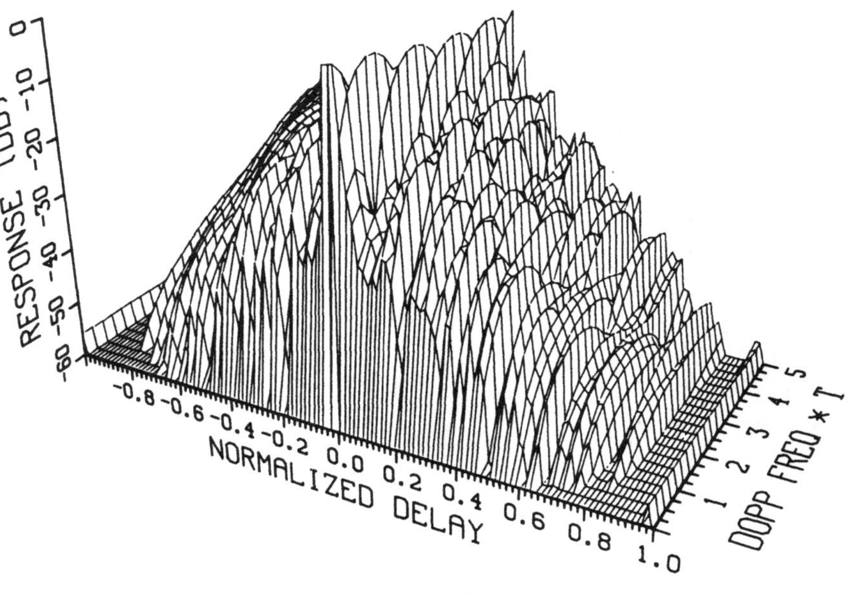

Fig. 6 — Partial ambiguity surface for an efficient 64-element Huffman-coded waveform

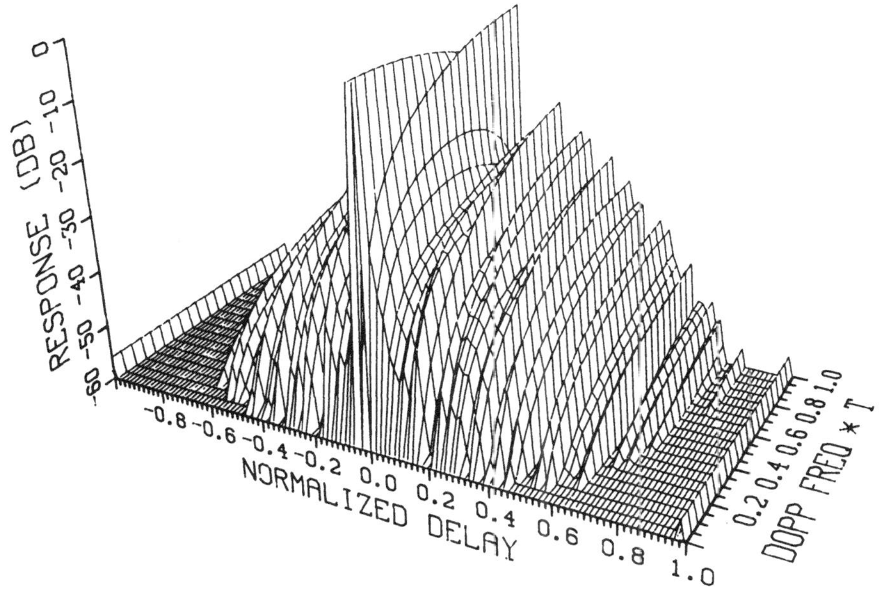

Fig. 7 — Expanded ambiguity surface for an efficient 64-element Huffman-coded waveform

Fig 8. This code, which is derived from a linear chirp signal, is seen to be relatively insensitive to doppler. Figures 9 and 10 show the ambiguity function of a 63-element pseudorandom binary shift-register code which has a thumbtack ambiguity function. Although this is an excellent waveform for simultaneous determination of range and doppler, it has a relatively poor doppler response. The response in doppler for 0-delay is approximately a $(\sin x/x)^2$ response with the first zero occurring at a normalized doppler shift of 1.0.

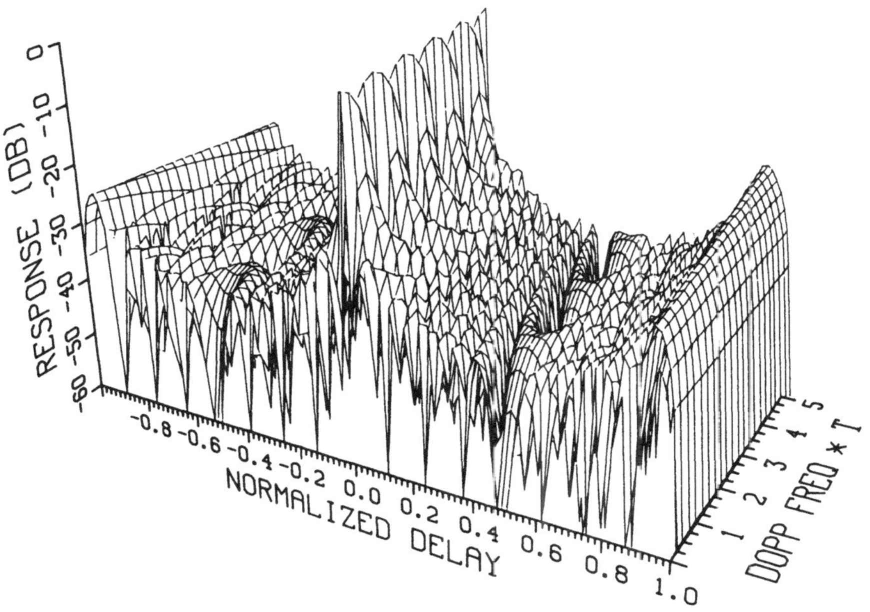

Fig. 8 — Partial ambiguity surface for a 64-element P4 polyphase-coded waveform

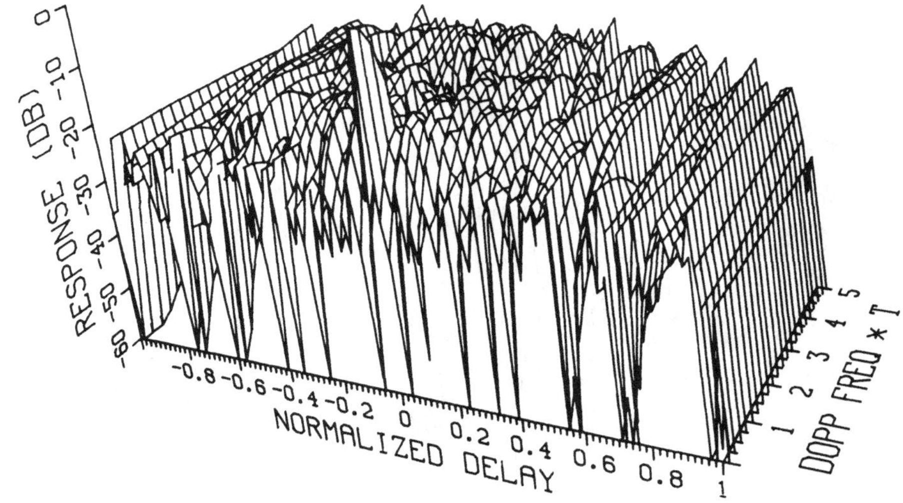

Fig. 9 — Partial ambiguity surface for a 63-element binary shift-register code

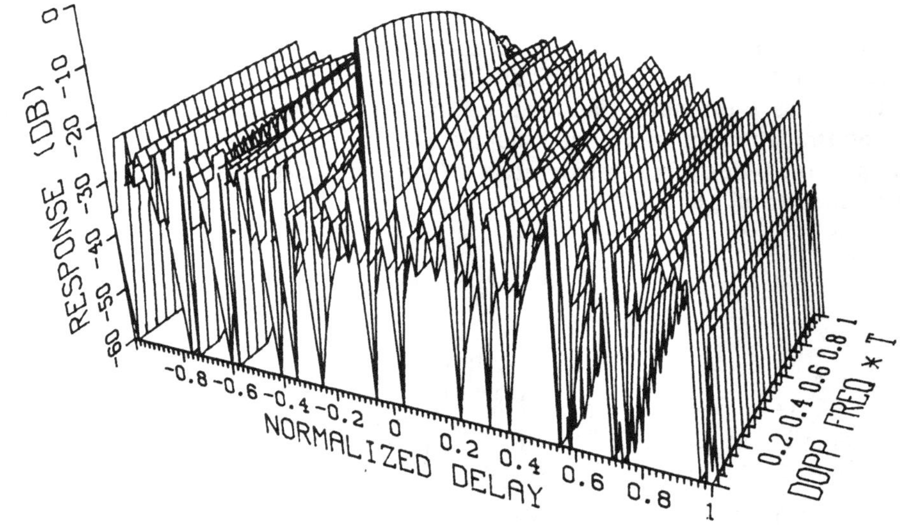

Fig. 10 — Expanded ambiguity surface for a 63-element binary shift-register code

In Fig. 11 we show a cut through the ambiguity surfaces of the three codes at a normalized doppler shift of 0.25, corresponding to a total phase shift across the uncompressed waveform of $\pi/2$ rad. The sample number in the abscissa corresponds to the number of range cells or time intervals whose duration is equal to the subpulse width. By comparing Figs. 11(a) and 11(b) it is seen that for this doppler, the peak sidelobe of the Huffman code is approximately equal to that of the P4 code at approximately −25 dB. For higher doppler, the Huffman-peak sidelobe is larger than the P4 code. The peak sidelobes of the binary code shown in Fig. 11(c) are approximately −19 dB and are not very sensitive to doppler. As described above however, the peak signal in the compressed binary pulse degrades rapidly with increasing doppler as a $(\sin x/x)^2$ function as shown in Figs. 9 and 10.

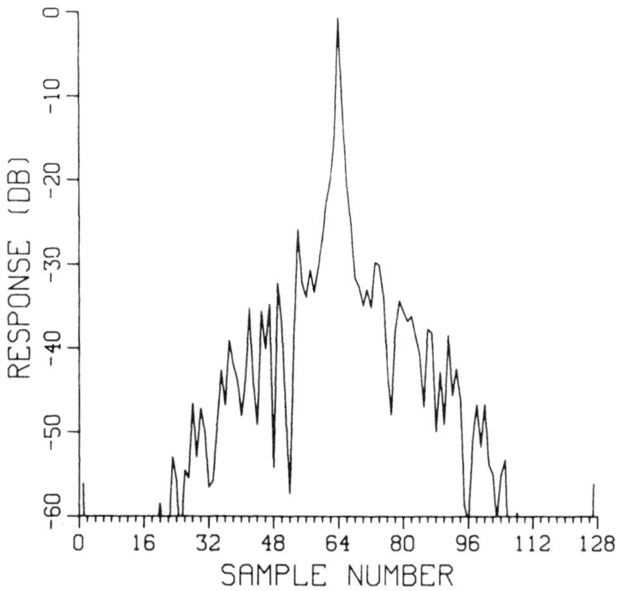

(a) efficient 64-element Huffman code

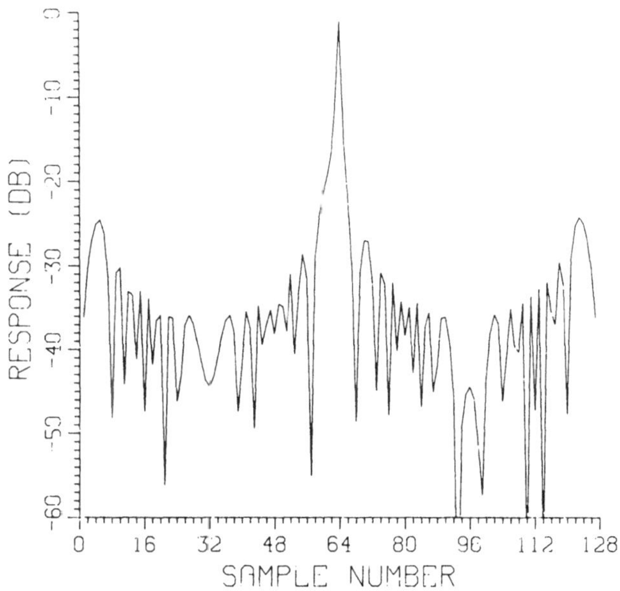

(b) 64-element P4 polyphase code

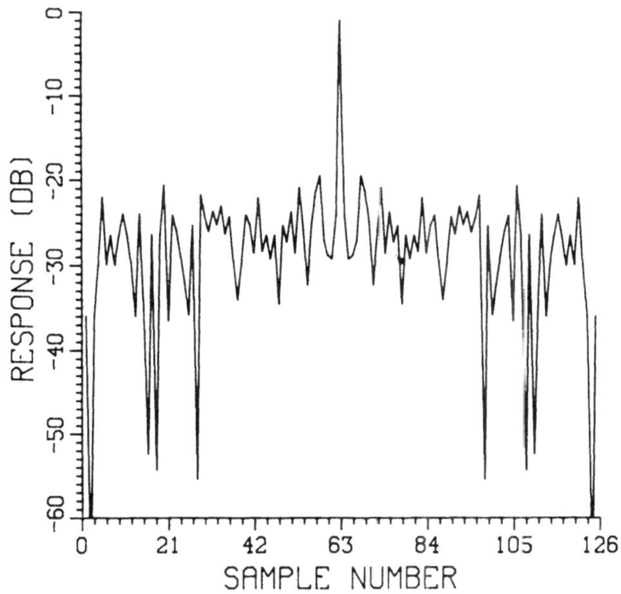

(c) 63-element binary shift-register code

Fig. 11 — Compressed pulses for a normalized doppler shift of 0.25

Another cause of degradation of the Huffman code is errors in generating and compressing the code. The zero sidelobe level, except for the end sidelobes, is an idealization which is not achievable in practice. To assess the sensitivity of the Huffman code to errors, two types of errors were investigated. The first is a random error due to tolerances and the second is due to A/D quantization.

In the first case, an independent random error is added to the inphase I and quadrature Q nominal values for each subpulse. The error is determined from a uniformly distributed sample having a maximum error of ±5% of the nominal value of I or Q. The new I and Q values for each subpulse, denoted by primes, are given by

$$I' = I(1 + e_i)$$

$$Q' = Q(1 + e_q),$$

where e_i and e_q are independent samples from a uniform probability distribution over the range of ±0.05. An error-free compressed pulse for the efficient 64-element Huffman code is shown in Fig. 12(a), and in Fig. 12(b) a realization is shown for a maximum 5% error imposed on the transmit waveform. It is assumed here that the matched-filter is matched to the transmitted imperfect Huffman waveform. It is seen that the peak sidelobe levels increase to −40 dB over the region where all the sidelobes are zero in the error-free case. An error-free compressed P4 code is shown in Fig. 13(a), and the compressed pulse containing a maximum random error of ±5% is shown in Fig. 13(b). This compressed pulse is seen to be only slightly different from the error-free compressed pulse. Similar results were obtained when the errors were only one way, that is, on either transmit or in the receiver filter.

The effect of quantization errors incurred with the utilization of A/D converters was also simulated for the −36 and −56 dB sidelobe level cases. The results are shown in Fig. 14 for the two-way average and peak sidelobe level vs the number of bits, including the sign bit. From this figure one sees that, as expected, it requires more bits to achieve lower sidelobe levels. Again, the results were nearly the same for the case of a one-way error.

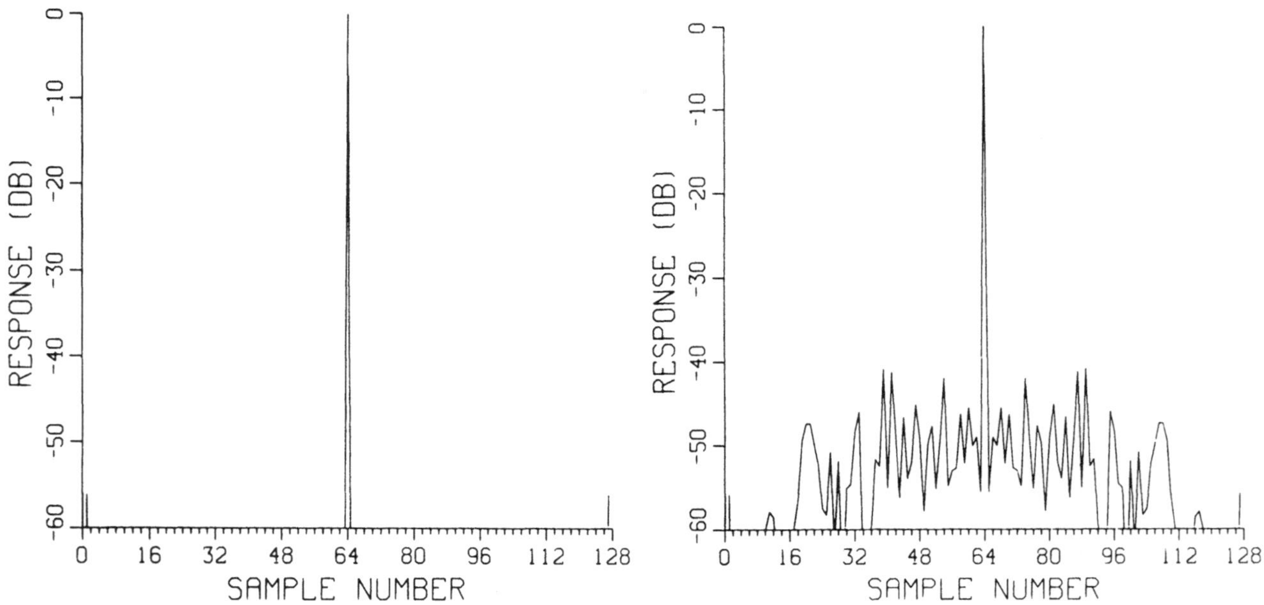

(a) no errors

(b) uniformly distributed random error on each subpulse I and Q values

Fig. 12 — Zero-doppler compressed pulse for an efficient 64-element Huffman code

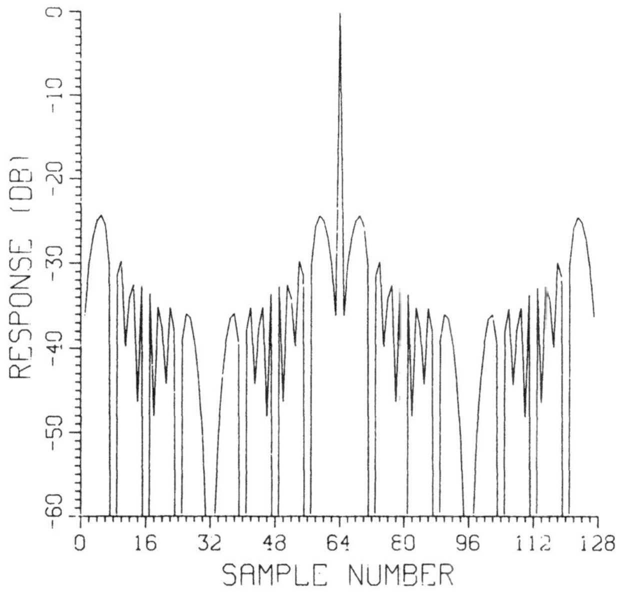

(a) no errors

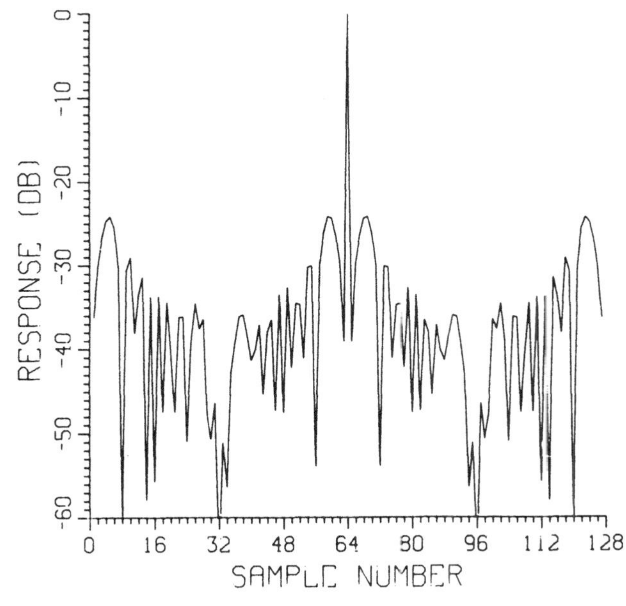

(b) uniformly distributed random error on each subpulse I and Q values

Fig. 13 — Zero-doppler compressed pulse for 64-element P4 polyphase code

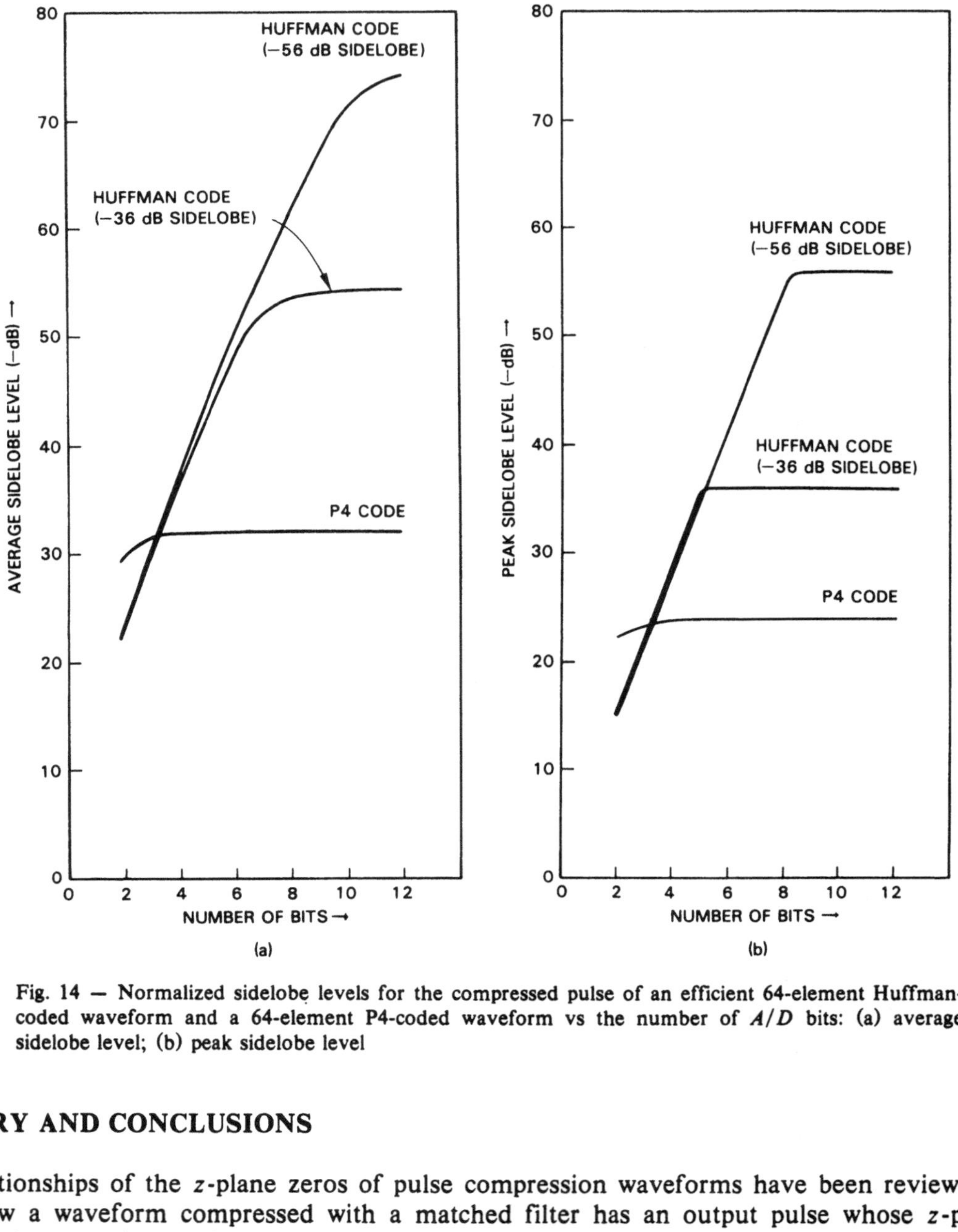

Fig. 14 — Normalized sidelobe levels for the compressed pulse of an efficient 64-element Huffman-coded waveform and a 64-element P4-coded waveform vs the number of A/D bits: (a) average sidelobe level; (b) peak sidelobe level

SUMMARY AND CONCLUSIONS

Relationships of the z-plane zeros of pulse compression waveforms have been reviewed. It was shown how a waveform compressed with a matched filter has an output pulse whose z-plane zeros occur in pairs occurring at the same angle and which are reciprocally related in amplitude. For an N-element input waveform, there are $N-1$ such zero pairs. A synthesis procedure then consists of selecting one zero from each pair. The remaining zeros then correspond to the zeros of the matched filter.

In general, determination of the zeros of the compressed pulse requires factorization of a polynomial of degree $2(N-1)$ in z. However, for the idealized Huffman-coded waveform, the compressed pulse consists of only the central peak and zero sidelobes except for those at either end of the compressed pulse. The consequence is that the z-plane polynomial is quadratic in the variable $z^{-(N-1)}$ so that it can be easily factored. The result is that the zeros lie on two circles, which are reciprocal in amplitude and whose radii depend on the specified sidelobe level, and these zeros also lie on radial lines at regularly spaced intervals of $2\pi/(N-1)$ rad. Random selection of the zeros of the compressed pulse to determine the input waveform usually results in an inefficient waveform in terms of the of the waveform energy compared to the maximum that could be transmitted. Techniques for improving the efficiency by imparting a large FM content of the waveform were discussed. These methods, based on the work of Schroeder and Ackroyd, were shown to result in a much more efficient waveform. Ackroyd suggested an additional improvement in the waveform efficiency by varying the specified sidelobe design level for a given fixed initial phase angle. We have shown that the efficiency could be further improved by also varying the initial phase angle of the FM-phase characteristic used in determining the zero locations.

The effects of doppler shifts were investigated and it was shown that the zero sidelobes of the Huffman-compressed-pulse increase rapidly with doppler. Ambiguity diagrams were presented for the exemplar 64-element Huffman code, a P4 polyphase code, and a binary shift-register code. Above a normalized doppler shift of approximately 0.25, corresponding to a total phase shift across the uncompressed pulse of $\pi/2$ rad, the peak sidelobes of the 64-element Huffman code exceed those of the 64-element polyphase code. A doppler shift of 0.25 corresponds, for example, to a Mach 1.7 target for a 1 GHz radar having an uncompressed pulse width equal to 64 μs. The sidelobes of the binary code are higher than those of the P4 code and they, as well as those of the P4 code, remain relatively constant with increasing doppler shift. However, the central peak of the binary-coded waveform falls off as a $(\sin \pi fT/\pi fT)^2$ function with the normalized doppler shift fT. For higher pulse compression ratios, it is expected that the sidelobes of the Huffman code would exceed those of the polyphase code for smaller fT because of the lower sidelobes of the P4 code.

The sensitivity of the 64-element Huffman code to random errors and to A/D quantization errors was investigated. It was found that the idealized zero-level sidelobes increased to approximately the -40 dB level in the presence of independent random errors which were a maximum of $\pm 5\%$ of the nominal I and Q channel values for each subpulse. An investigation of the effects of the number of bits used in an A/D converter showed that, to maintain the low sidelobe levels, it is necessary to utilize additional bits. The sidelobe level was found to diminish approximately 7 dB per additional bit.

In conclusion, it is found that the Huffman codes, which are theoretically idealized waveforms in terms of zero sidelobes except for the end sidelobes, degrade rapidly in the presence of doppler shifts. The sidelobes also degrade in the presence of tolerance errors and an insufficient number of A/D bits. The efficiency factor of the Huffman codes may also be an important consideration. For example, for a Huffman code having an efficiency of 50%, there is an approximately 15% reduction in the detection range of the radar.

In view of the foregoing comments, the Huffman waveforms appear to have the most appeal in sufficiently low, or compensated doppler applications where low sidelobes are achievable. The tradeoffs are generally a reduction in range performance, depending on the code efficiency, increased transmitter complexity, and providing the necessary number of A/D bits to support the sidelobe levels. Depending on the code and radar, this number of bits may not need to be greater than what is normally required for proper radar operation.

REFERENCES

1. D.A. Huffman, "The Generation of Impulse-Equivalent Pulse Trains," *IRE Trans. on Information Theory* **IT-8**, S10-S16 (Sept. 1962).

2. C.E. Cook and M. Bernfeld, *Radar Signals* (Academic Press, New York, 1967).

3. M.H. Ackroyd, "The Design of Huffman Sequences," *IEEE Trans. Aerospace and Electronic Systems* **AES-6** (6), 790-796 (Nov. 1970).

4. M.H. Ackroyd, "Amplitude and Phase Modulated Pulse Trains for Radar," *The Radio and Electronic Engineer* **41** (12), 541-552 (1971).

5. M.H. Ackroyd, "Synthesis of Efficient Huffman Sequences," *IEEE Trans. Aerospace and Electronic Systems* **AES-8** (1), 1-8 (Jan. 1972).

6. M.H. Ackroyd, "Computing the Coefficients of High-Order Polynomials," *Electronics Letters* **6**, 715-717 (Oct. 1970).

7. S.W. Golomb and R.R. Scholtz, "Generalized Barker Sequences," *IEEE Trans. Information Theory* **IT-11** (4), 533-537 (Oct. 1965).

8. M.R. Schroeder, "Synthesis of Low-Peak-Factor Signals and Binary Sequences with Low Autocorrelation," *IEEE Trans. Information Theory* **IT-16**, 85-89 (Jan. 1970).

9. B.L. Lewis and F.F. Kretschmer, Jr., "Linear Frequency Modulation Derived Polyphase Pulse Compression Codes," *IEEE Trans. Aerospace and Electronic Systems* **AES-18** (5), 637-641 (Sept. 1982).

10. F.F. Kretschmer, Jr. and B.L. Lewis, "Doppler Properties of Polyphase Coded Pulse Compression Waveforms," *IEEE Trans. Aerospace and Electronic Systems* **AES-19** (4), 521-531 (July 1983).

PARAMETER ESTIMATION AND TARGET DETECTION IN A DISTRIBUTED-CLUTTER ENVIRONMENT

INTRODUCTION

The three basic purposes of a radar are target detection, target resolution, and parameter estimation. For a single target, the tasks of target detection and parameter estimation are in principle quite simple if sufficient signal energy guarantees precise measurements of target range, velocity, and other parameters. The real test of a radar is its ability to resolve the desired targets from clutter and to detect targets and estimate parameters simultaneously in the presence of clutter. The mutual interference caused by other scatterers will in general make target resolution difficult.

Any response of a practical receiver always has a main peak surrounded by several time sidelobes or slowly decreasing tails. In an environment with distributed clutter or multiple targets, the sum of all these low-level responses may build up to a level sufficiently high so as to mask even relatively strong targets. It has been stated [1] that parameter estimation is one of the fundamental applications of large time-bandwidth signals. It is generally believed that the optimum implementation for parameter estimation is a bank of matched filters obtained by doppler-correcting a nonrange-doppler-coupled pulse compressor [1,2]. Although this is true in the absence of clutter, this report demonstrates that the doppler filter bank described above cannot provide subclutter visibility in a distributed-clutter environment. In such an environment the clutter will mask the targets, preventing both target detection and parameter estimation.

The implementation of a moving-target indicator (MTI) at each output port of a doppler-steered pulse compressor can eliminate the excess clutter (to be defined later), but the radar will require at least two transmitted pulses and will operate with blind-speed limitations. As will be described, for two or more pulses when the proper waveforms are employed, radar performance can be improved significantly in distributed nonmoving clutter. Both subclutter visibility and accurate parameter estimation can be achieved at the same time.

CODES WITH A THUMBTACK-TYPE AMBIGUITY FUNCTION

It has been stated [1] that the simultaneous measurement of the range and velocity of a radar target (parameter estimation) is accomplished with minimum error when the waveform employed does not range-doppler couple. The thumbtack-type ambiguity function, with a narrow spike surrounded by a uniformly low pedestal, is usually considered for parameter-estimation applications.

The aperiodic maximum-length binary shift-register codes have an ambiguity function which approximates a thumbtack type of characteristic. These codes are derived from recurrence formulas which are suitable for shift-register implementation [3]. The coefficients of the primitive polynomial of degree n specify the stages used in the feedback path. Codes of length 2^n-1 are generated by sensing predetermined stages of the n-bit shift register and summing modulo 2, with the result applied to the input of the shift register. When code generation starts, different initial conditions with binary elements in the shift register will yield cyclic permutation of the code. Among these permutations, there are codes with either the lowest peak sidelobes or the lowest RMS sidelobes. In addition to the maximum-length binary codes, Barker codes also have thumbtack-type ambiguity functions. However, no known Barker code exists with length greater than 13.

Manuscript approved December 2, 1982.

In the matched-filter-bank implementation for parameter estimation, each filter in the bank is tuned to a different center frequency (doppler-steered pulse compressor). For a maximum-length-sequence and Barker-coded waveforms, a doppler shift of m/τ (m = integer, τ = code duration) reduces all the matched-filter peak responses to the sidelobe level, except the F_m filter, which is the filter matched to the doppler shift of the received signal. Under the latter condition a peak response is obtained; its location indicates the true range, and the filter number identifies the target velocity. The responses of the F_0,F_1,F_2, and F_3 doppler filters with a zero-doppler target are shown in Fig. 1 for one of the 127-element-maximum-length pseudorandom binary waveforms and in Fig. 2 for the 13-element Barker code. The responses of the F_0,F_1,F_2, and F_3 filters with an F_2-doppler target (with a doppler shift of $2/\tau$) are shown in Fig. 3 for the same maximum-length binary code as in Fig. 1.

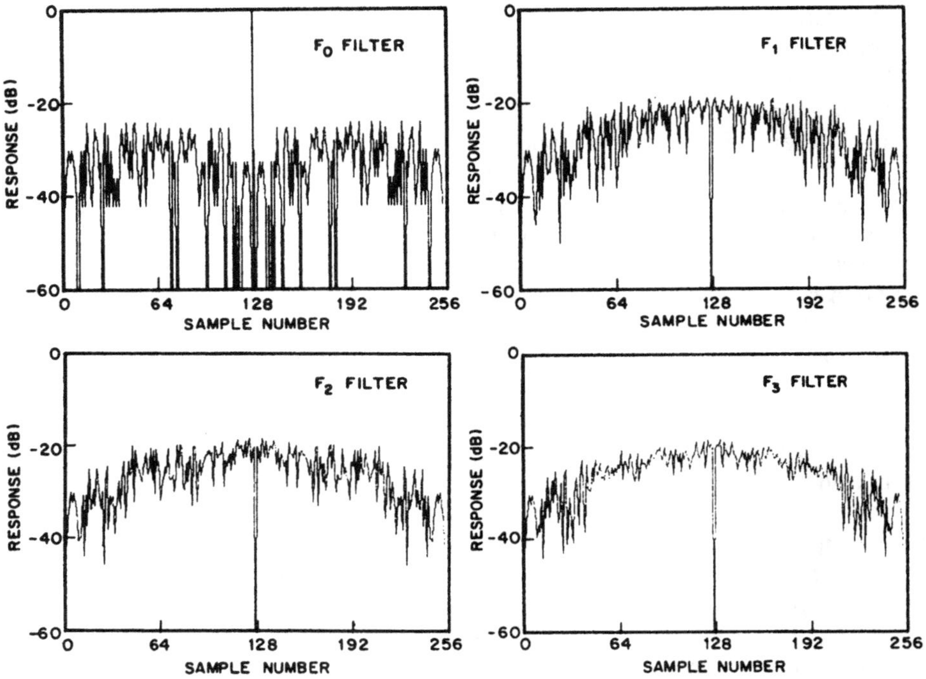

Fig. 1 — Responses of the doppler filters with a zero-doppler target for a 127-element-maximum-length binary code: (217_8, 127)

DISTRIBUTED-NONMOVING-CLUTTER ENVIRONMENT

In a distributed-clutter environment, the range extent of the clutter is assumed to be much greater than the transmitted pulse duration. For the distributed clutter, it is assumed that the I and Q components of clutter cells are independently Gaussian distributed with zero mean and equal variance. Therefore, the amplitudes of clutter cells are Rayleigh distributed, and the phases are uniformly distributed. The responses for one realization of the nonmoving-distributed-clutter model were obtained at outputs of doppler-steered pulse compressors when several maximum-length binary-sequence and Barker codes were adopted for input waveforms. The mean-squared output clutter powers from the doppler-steered pulse compressors employing these waveforms are listed in Table 1 for the F_0,F_1,F_2, and F_3 filters. The results consistently show that the clutter power effectively comes through all the doppler-steered passbands on the range-time sidelobes with nearly as much power as that at the matchpoint in the zero-frequency passband. This integrated time-sidelobe power has been called excess clutter and gives rise to what has been called processing loss or degradation [4]. Work to date has revealed that no codes with a thumbtack-type ambiguity function in the implementation of doppler-corrected pulse compressors will provide subclutter visibility in the presence of nonmoving distributed clutter.

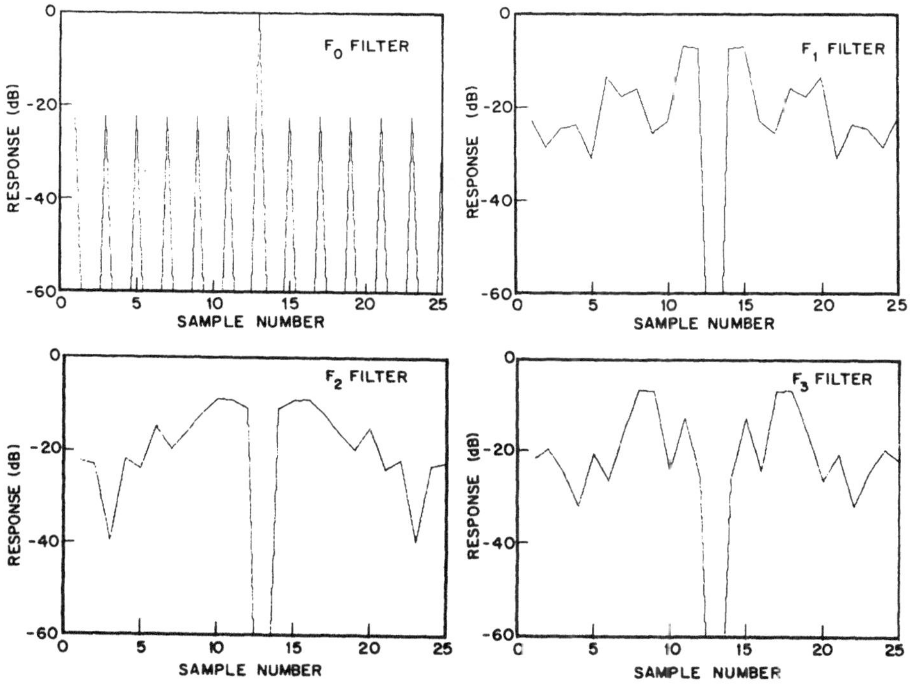

Fig. 2 — Responses of the doppler filters with a zero-doppler target for the 13-element Barker code

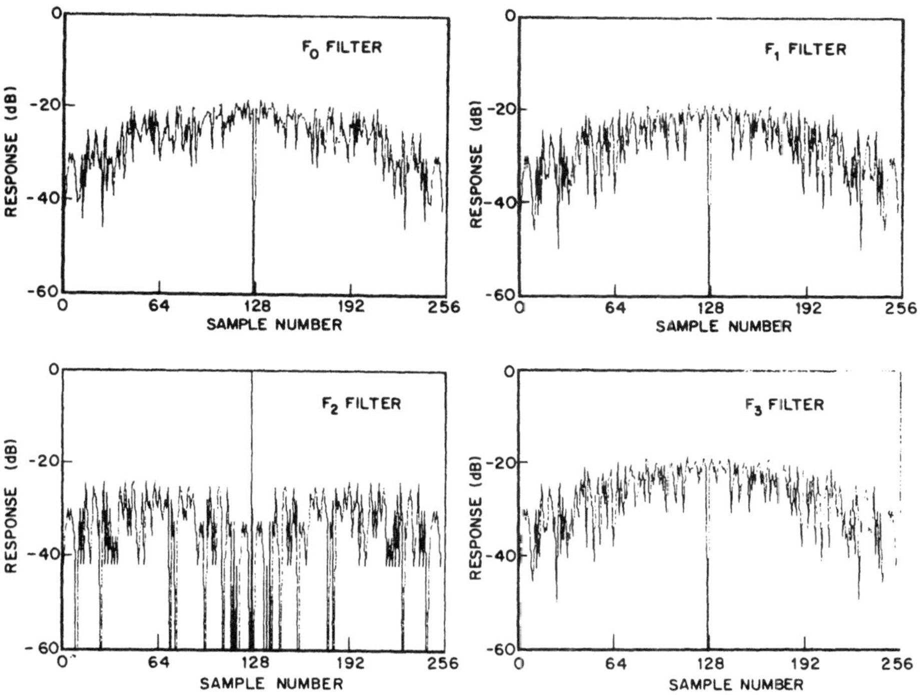

Fig. 3 — Responses of the doppler filters with an F_2-doppler target for the same 127-element-maximum-length binary code as in Fig. 1

Table 1 — Mean-Squared Clutter Level Output from Doppler-Steered Pulse Compressors

Waveform	Mean-Squared Clutter Level (dB)			
	F_0	F_1	F_2	F_3
13-Element Barker code	0.471	0.164	0.087	0.260
15-Element-maximum-length binary code*				
$(23_8, 1)^{\dagger\ddagger}$	1.501	0.482	-0.119	0.217
31-Element-maximum-length binary code				
$(45_8, 20)^{\ddagger}$	1.103	0.135	-0.058	-0.029
$(45_8, 4)^{\dagger}$	1.379	0.143	-0.059	0.099
63-Element-maximum-length binary code				
$(103_8, 31)^{\dagger\ddagger}$	1.186	0.103	0.168	0.008
127-Element-maximum-length binary code				
$(211_8, 39)^{\ddagger}$	0.875	-0.138	-0.099	0.131
$(217_8, 127)^{\dagger}$	1.211	0.088	0.172	0.225
$(203_8, 64)$	1.385	-0.516	0.143	0.122
$(203_8, 1)$	1.494	-0.510	0.093	0.207

*The maximum length binary codes are represented by (α_8, β), where α is the coefficient of primitive polynomial in octal notation and β is the initial condition in binary notation.
†Code with the lowest peak sidelobe.
‡Code with the lowest RMS sidelobe.

Note: The initial conditions derived for those codes wth either the lowest peak sidelobes or the lowest RMS sidelobes are different from those obtained by Taylor and MacArthur [3].

MTI DELAY-LINE CANCELERS

To eliminate the excess clutter, two or more successive echoes from each doppler passband must be subtracted in an MTI (Fig. 4). This prevents the system from parameter estimation with a single pulse. The delay-line canceler acts as a filter which not only rejects the *DC* component but also eliminates any moving target whose doppler frequency is the same as the pulse-repetition frequency (PRF) or an integer multiple therof. This gives rise to a blind-speed problem as in a conventional MTI radar.

RANGE-DOPPLER-COUPLED WAVEFORM SYSTEM

When at least two pulses are dedicated to the MTI, it is no longer necessary to employ waveforms with thumbtack ambiguity diagrams for accurate parameter estimation. For example, a variation of an MTI technique that takes advantage of the range-doppler-coupling effect could be used. This effect causes the range of the echo to vary with doppler frequency in a direction depending on the target velocity and the direction of the radar frequency sweep. By alternating the radar-frequency-sweep direction on successive transmissions, a moving target will appear at different ranges on successive pulses, while a stationary target will appear at same range. The echo from one pulse subtracted coherently or noncoherently from the echo of the next pulse (Fig. 5) will effectively cancel the nonmoving-target response but not the moving-target response. In this type of MTI, true range and velocity of the moving target can be estimated from the output. In addition, blind-speed problems are also eliminated except at zero doppler, where the blind speed is desired. It is emphasized that the implementation of the scheme shown in Fig. 5 does not require any doppler-steered pulse compressor and that only one delay-line canceler is needed.

The relatively doppler-tolerant low-sidelobe polyphase-coded waveforms [5-7] are particularly suitable for range-doppler-coupled-MTI applications. Some of these codes could become palindromic with

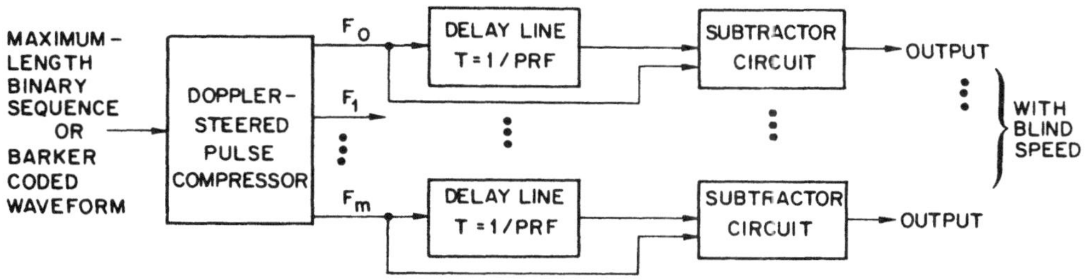

Fig. 4 — Regular delay-line canceler, with no range-doppler coupling

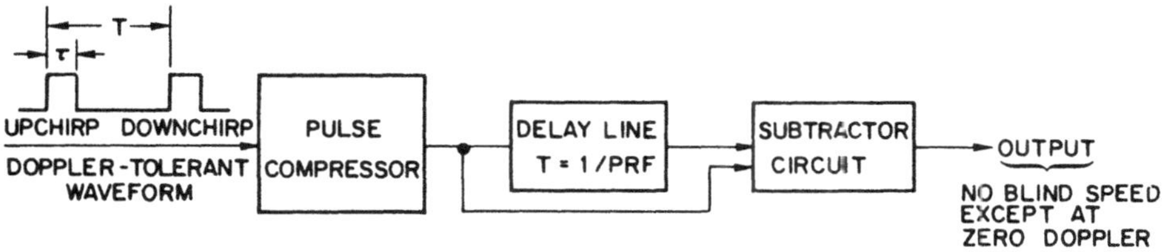

Fig. 5 — Range-doppler-coupled MTI

real autocorrelation functions by slightly modifying the existing waveforms. For example, the PP4 code was obtained from the P4 code by taking the first sample of the waveform half a code-element duration after the leading edge while still sampling at the Nyquist rate. The frequency response of the single-delay-line canceler in a range-doppler-coupled MTI employing a PP4-coded waveform is shown in Fig. 6. It is evident that blind-speed problems do not occur except at zero doppler.

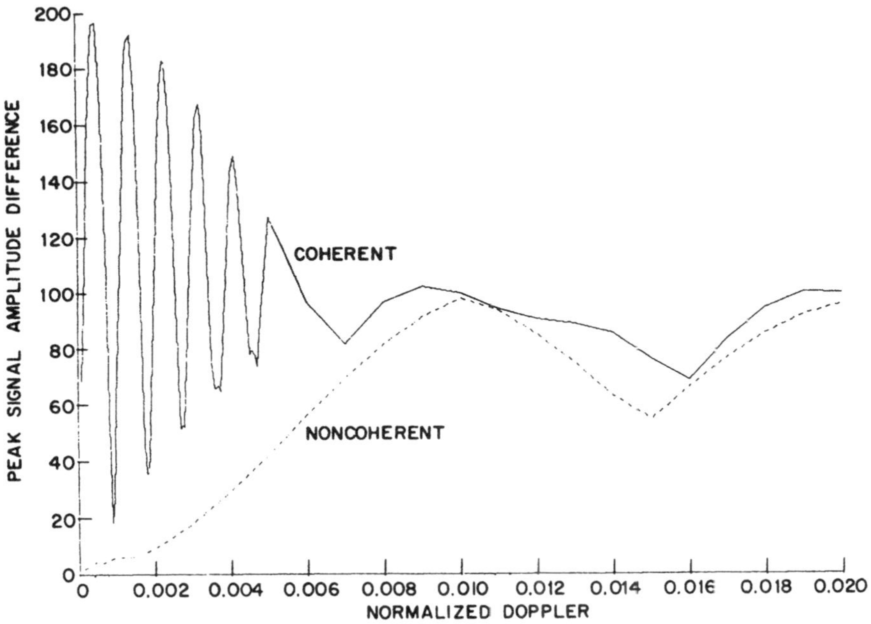

Fig. 6 — Frequency response of a two-pulse range-doppler-coupled MTI using a PP4 code

In the event of nonmoving distributed clutter, simulation of a range-doppler-coupled MTI output for those palindromic codes showed that the returns canceled perfectly and that no clutter residue was obtained. It was demonstrated in the simulation that if there were moving targets in addition to the nonmoving distributed clutter, only the target responses were presented at the MTI output. As an example, Figs. 7 and 8 show the responses of a target with doppler frequency 0.01B, where B is the radar bandwidth. Figure 7 shows that the target is embedded in the clutter before cancellation, and Fig. 8 shows that the target is visible after cancellation using the PP4-coded waveform. The pulse-compression ratio was 100 in this case. At the MTI input, the clutter-to-target power ratio was 20 dB. At the MTI output, the oppositely range-doppler-coupled target responses were separated by two range cells and were resolved in range. The separation of these two target responses indicates that the target velocity and the true target range lies halfway between the two responses. Consequently, parameter estimation and subclutter visibility can both be accomplished by implementing such a range-doppler-coupled MTI with a palindromic polyphase-coded waveform.

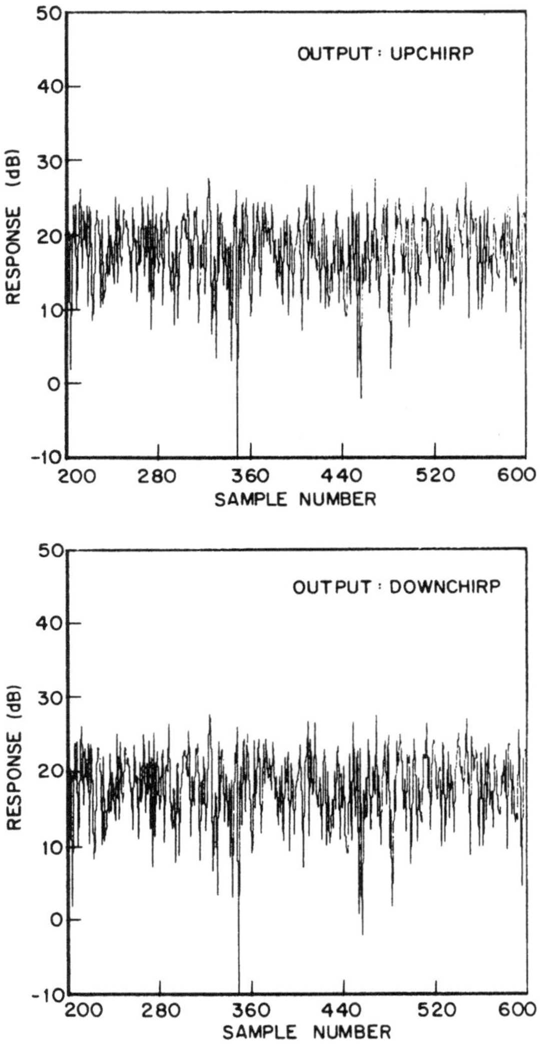

Fig. 7 — Responses of a pulse compressor before cancellation by a single-delay-line canceler

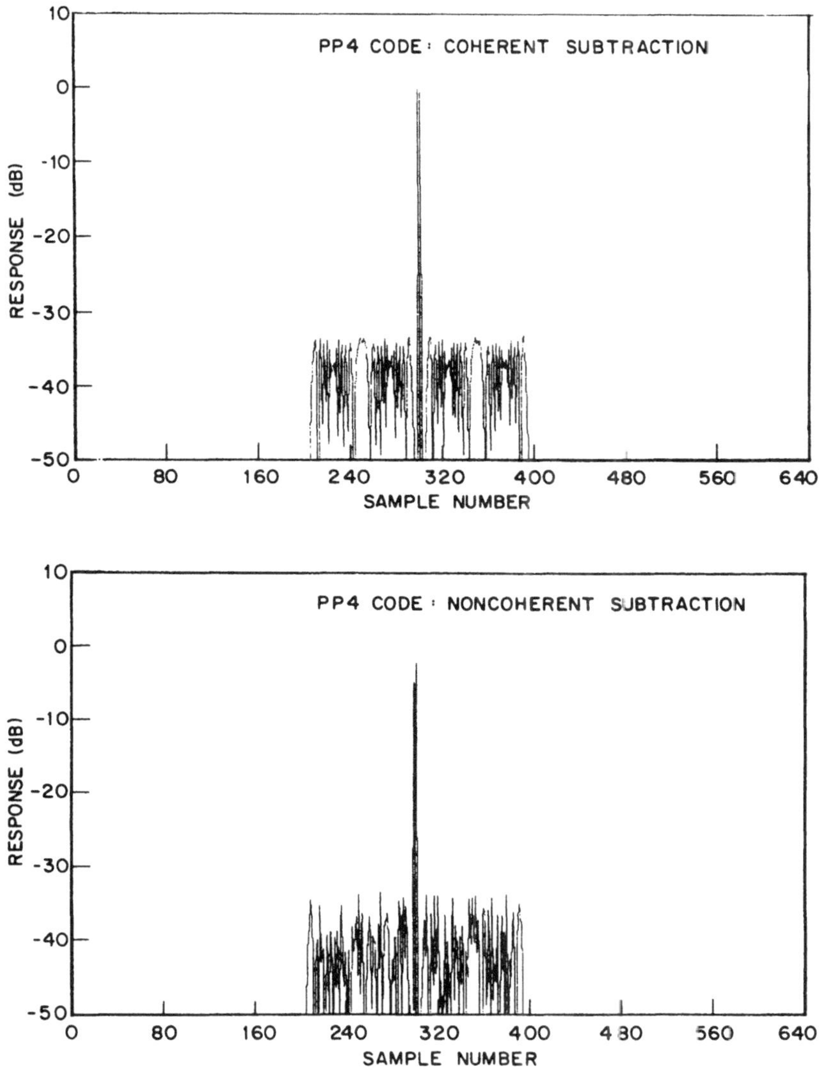

Fig. 8 — MTI output after cancellation by a single-delay-line canceler

CONCLUSIONS

It is demonstrated in the simulation that doppler resolution and clutter attenuation are not simultaneously available from doppler-steered pulse compressors using waveforms with thumbtack ambiguity diagrams in a nonmoving-distributed-clutter environment. As a consequence the implementation of a matched filter bank does not provide the capability of parameter estimation in a clutter environment. Subsequent conventional MTI processing at each filter output port can eliminate excess clutter with more than one transmitted pulse but has blind-speed problems.

In an MTI system, if the waveform with a thumbtack ambiguity diagram is replaced with one that range-doppler couples and is palindromic, both parameter estimation and subclutter visibility can be obtained without implementation of any matched filter bank.

REFERENCES

1. C.E. Cook and M. Bernfeld, *Radar Signals: An Introduction to Theory and Applications,* Academic Press, 1967.

2. E. Brookner, *Radar Technology,* Chap. 8, Artech House, 1978.

3. S.A. Taylor and J.L. MacArthur, "Digital Pulse Compression Radar Receiver," APL Tech. Digest **6** (No. 4), 2-10 (1967).

4. F.E. Nathanson, *Radar Design Principles,* McGraw-Hill, 1969.

5. B.L. Lewis and F.F. Kretschmer, Jr., "A New Class of Polyphase Pulse Compression Codes and Techniques," IEEE Trans. Aerospace and Electronic Systems **AES-17**, 364 (May 1981).

6. B.L. Lewis and F.F. Kretschmer, Jr., "Linear Frequency Modulation Derived Polyphase Pulse Compression Codes," IEEE Trans. Aerospace and Electronic Systems **AES-18**, 637 (Sept. 1982).

7. F.F. Kretschmer, Jr., and B.L. Lewis, "Polyphase Pulse-Compression Waveforms," NRL Report 8540, Jan. 1982.

Chapter 3
Interference Suppression

3.1 TYPES OF INTERFERENCE CONSIDERED

As noted in Chapter 1, the interference considered in this chapter is defined to be externally generated. This interference can have either a *high duty cycle* (continuous or nearly continuous in time) or a *low duty cycle* (short pulses of interference, such as another radar's transmissions, false target repeaters, *et cetera*). The interference may be either *synchronized* or *unsynchronized* with the radar's pulse repetition rate. It may enter the radar antenna's *main lobe* or its *sidelobes* and be *matched* or *mismatched* to the radar's receiver. We also assume that the interference originates from point sources at different azimuth or elevation angles from the radar.

Differences between desired signals and externally generated interference can be used to suppress the interference and to augment the desired signals.

Typical interference suppressors are known as defruiters, sidelobe blankers, coherent sidelobe cancelers, mainlobe notchers and adaptive arrays.

3.2 LOW-DUTY-CYCLE INTERFERENCE BLANKERS

Low-duty-cycle *interference blankers* are known as *defruiters* and *sidelobe blankers*. Defruiters blank the impulsive type of interference, which is not synchronized with the radar's pulse repetition frequency. Sidelobe blankers, by contrast, blank interference entering the radar receiver with less amplitude than it has when entering an auxiliary receiver employing an auxiliary antenna.

Defruiters are common on civilian FAA radars that detect and track aircraft near airports, where there are usually several radars in operation simultaneously. Defruiters operate on interference entering either the radar main lobe or any of its sidelobes. They employ interpulse period delay lines and look for coincidence of echoes from successive radar transmissions. Defruiters pass coincident echoes while blocking noncoincident signals such as impulse interference unsynchronized with the radar's PRF.

Sidelobe blankers employ omnidirectional auxiliary antennas having gains that are greater than the highest sidelobe gain of the radar antenna, but less than that of the radar antenna's main lobe. The auxiliary antenna feeds a receiver identical to that used by the radar, and the detected output of this receiver is compared to that of the radar receiver. If the detected output of the auxiliary receiver is larger than that of the radar receiver, then the latter is blanked in all range cells where this occurs. Signals entering the radar main lobe are not blanked because they produce larger radar outputs than the auxiliary system.

3.3 COHERENT SIDELOBE CANCELERS

Coherent sidelobe cancelers were invented by Paul Howells [1] to cancel high-duty-cycle interference entering the sidelobes of a radar antenna. Sidney Applebaum was the first to analyze such systems mathematically and both names have been associated with coherent sidelobe cancelers. The first cancelers were analog circuits that were subject to drift with time, temperature, *et cetera*. As a consequence, they were implemented as closed loops in order to allow the loop gain to oppose any drift. Recently, digital systems [2, 3] have been invented that perform much better than the analog closed loops. Many future systems will be implemented digitally to take advantage of the improved performance available from digital techniques.

3.3.1 Analog Coherent Sidelobe Cancelers

The principle of operation of a Howells coherent sidelobe canceler implemented in analog form is as outlined below.

An auxiliary antenna with a gain less than the main lobe gain of a radar antenna but greater than the highest sidelobe gain of the radar antenna is used to obtain samples of any high-duty-cycle interference that may be present. The auxiliary antenna is placed sufficiently close to the phase center of the radar antenna to ensure that the samples of the interference which it obtains will be correlated with the interference received in the radar antenna sidelobes. This requires that the separation X between the radar antenna's phase center and the auxiliary antenna's phase center divided by the velocity of light c be much less than the reciprocal of the smaller of the radar bandwidth B_r or the interference bandwidth.

The auxiliary antenna feeds a receiver which is identical to that of the radar and the two receivers share the same local oscillator as illustrated in Fig. 3–1.

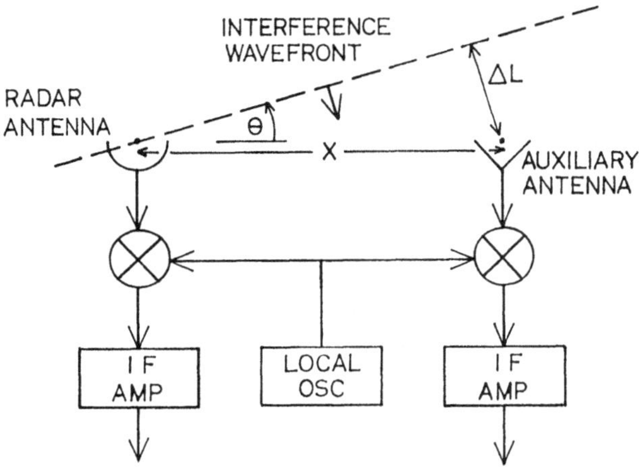

Fig. 3–1 Radar and Auxiliary Receiving Systems.

It is obvious from Fig. 3–1 that the maximum amplitude of difference between the time of arrival of the interference at the radar and at an auxiliary antenna, separated by a distance X, is X/c, and it is encountered at angles of arrival of $\theta = \pm 90°$. The time-of-arrival difference for any value of θ is given by

$$T_a = \Delta L/c = (x \sin \theta)/c \quad (3\text{–}1)$$

For proper sidelobe canceler operation, T_a must be limited so that its amplitude is much less than $1/B_r$ in order to maintain correlation between the radar and auxiliary interference signals.

The outputs of the radar and auxiliary receivers at intermediate frequency (IF) feed an analog loop as illustrated in Fig. 3–2.

The interference voltages V_m and V_a in Fig. 3–2 are assumed to be functions of time $A(t)$ modulating an IF carrier that is much higher in frequency ($\omega_1 = 2\pi f_1$rad/s) than the bandwidth B of the modulating function $A(t)$. In addition, V_a is assumed to be correlated with V_m, (i.e., V_a only differs from V_m by a constant amplitude and by a constant phase shift of the carrier frequency). This difference is assumed to result from the difference in gain and location of the phase centers of the radar and auxiliary antennas as illustrated in Fig. 3–1.

The Howells adaptive loop employs a local oscillator with an angular frequency ω_2 to translate the radar receiver's output IF to $\omega_1 - \omega_2$. The resulting signal

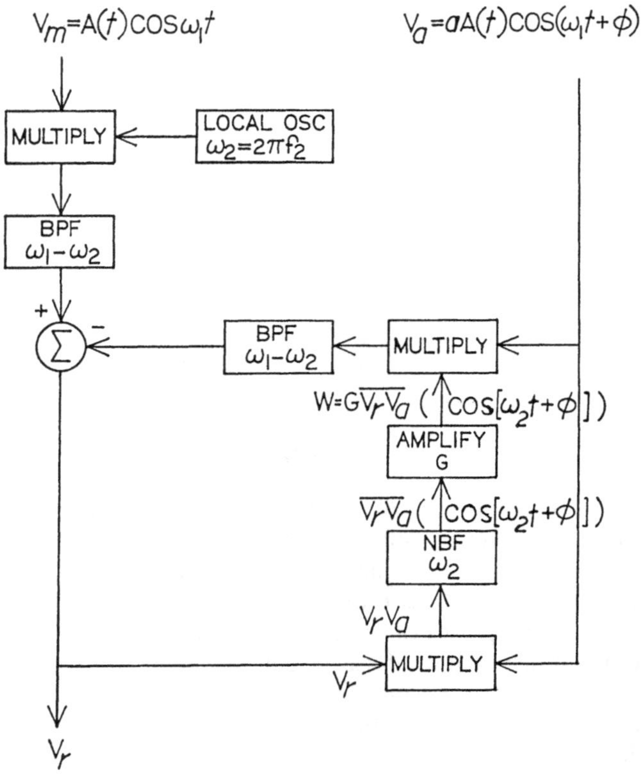

Fig. 3–2 Howells Analog Adaptive Loop.

V_m enters the positive input of a subtractor, the output of which is both the adaptive-loop residue V_r and the input to a multiplier, where it translates the auxiliary signal down to a center frequency ω_2 with phase ϕ. The resulting product passes through a narrowband filter centered on frequency ω_2, is amplified by a gain G, and translates the auxiliary signal to $\omega_1 - \omega_2$ and zero phase in a multiplier and bandpass filter centered on $\omega_1 - \omega_2$. The resulting signal of zero phase is then subtracted from the frequency-translated radar signal of phase zero to form the loop residue V_r.

The output of the amplifier (gain G) in the loop is a weighting signal that controls the amplitude and phase of V_a so that when the weighted V_a is subtracted from V_m, the residue V_r approaches zero as the loop gain approaches infinity, if V_a is perfectly correlated with V_m.

To facilitate analysis of the Howells loop, the loop of Fig. 3–2 can be redrawn in equivalent form for baseband signals using complex notation, where V_m, V_a, and V_r in Fig. 3–3 now denote complex signals consisting of I and Q components.

We can see from Fig. 3–3 that the residue signal may be expressed as

$$V_r = V_m - WV_a \quad (3\text{–}2)$$

where W is the weight used by the loop, identified in Fig. 3–3 as

$$W = G\overline{V_r V_a^*} \quad (3\text{–}3)$$

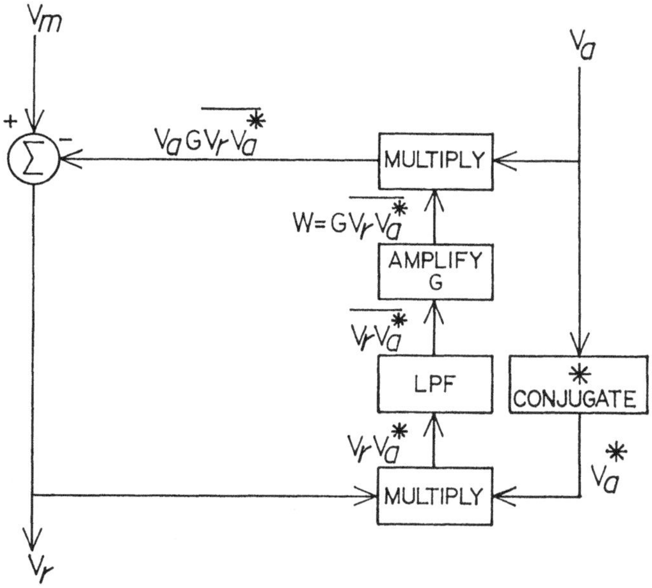

Fig. 3–3 Equivalent Circuit of Howells Adaptive Loop.

and where the bar indicates averaging (low-pass filtering). Substitution of (3–2) into (3–3) yields

$$W = G\overline{V_m V_a^*} - G\overline{W V_a V_a^*} \tag{3–4}$$

Solution of (3–4) for W with $\overline{V_a V_a^*} = \overline{|V_a|^2}$ reveals

$$W = \left[\frac{\overline{V_m V_a^*}}{\overline{|V_a|^2}}\right]\left[\frac{G\overline{|V_a|^2}}{1 + G\overline{|V_a|^2}}\right] \tag{3–5}$$

Now, W in (3–5) may be expressed as

$$W = W_{\text{opt}}\left[\frac{G\overline{|V_a|^2}}{1 + G\overline{|V_a|^2}}\right] \tag{3–6}$$

where

$$W_{\text{opt}} = \frac{\overline{V_m V_a^*}}{\overline{|V_a|^2}} \tag{3–7}$$

We can readily show that the minimum residue power is achieved for weighting W_{opt} as follows. From (3–2), we obtain

$$\overline{|V_r|^2} = \overline{|V_m|^2} - W\overline{V_m^* V_a} - W^*\overline{V_m V_a^*} + |W|^2\,\overline{|V_a|^2} \tag{3–8}$$

Letting $W = W_{\text{opt}} + \Delta$, where Δ is an arbitrary complex value, and substituting (3–6) in (3–8) results in

$$\overline{|V_r|^2} = \overline{|V_m|^2} - W_{\text{opt}}\overline{V_m^* V_a} - W_{\text{opt}}^*\overline{V_m V_a^*} + |W_{\text{opt}}|^2\overline{|V_a|^2} + |\Delta|^2\overline{|V_a|^2} \tag{3–9}$$

From (3–9), we can see that the average residue power is a minimum for $\Delta = 0$ so that $W = W_{\text{opt}}$, which is the Wiener weight.

The *cancellation ratio* of an adaptive loop, defined as

$$CR = \overline{|V_m|^2}/\overline{|V_r|^2} \tag{3–10}$$

can be derived in terms of the *normalized correlation coefficient* ρ, given by

$$\rho = \overline{V_m V_a^*}/[(\overline{|V_m|^2})^{1/2}(\overline{|V_a|^2})^{1/2}] \tag{3–11}$$

If we let

$$k = (G\overline{|V_a|^2})/(1 + G\overline{|V_a|^2}) \tag{3–12}$$

then the weight of the adaptive loop given by (3–5) can be expressed in terms of ρ in (3–11) as

$$W = \rho[(\overline{|V_m|^2})^{1/2}/(\overline{|V_a|^2})^{1/2}]k \tag{3–13}$$

From (3–8), (3–11), and (3–13), we can show that

$$\overline{|V_r|^2} = \overline{|V_m|^2}(1 - 2|\rho|^2 k + |\rho|^2 k^2) \tag{3–14}$$

From (3–10) and (3–14), the cancellation ratio is given by

$$CR = (1 - 2|\rho|^2 k + |\rho|^2 k^2)^{-1} \tag{3–15}$$

For the case where $|\rho|^2 = 1$, we obtain from (3–12) and (3–13):

$$CR = (1 - G\overline{|V_a|^2})^2 \tag{3–16}$$

It should be noted that the effective loop gain in a Howells analog adaptive loop is

$$G_e = G\overline{|V_a|^2} \tag{3–17}$$

which is proportional to the auxiliary signal power. This proportionality to the auxiliary signal is one of the disadvantages of the closed-loop circuit. If the auxiliary signal level is too low, the cancellation ratio is poor. If the auxiliary signal level is too high, the loop may oscillate. These considerations limit the dynamic range over which a closed loop can operate.

The above-mentioned auxiliary signal level problem has been solved to some extent by incorporating what has been called an "inner limiter" in the loop. This limiter is placed between the auxiliary signal input and the multiplier that forms the product of V_a and V_r in Fig. 3–2. The limiter makes the loop gain proportional to V_a rather than to $\overline{|V_a|^2}$ and increases the loop's dynamic range. Unfortunately, however, the limiter exacts a price for the increased dynamic range, which is a reduced cancellation of weak interference in the presence of strong interference from separate sources at different locations in space. This problem becomes important in systems that have multiple degrees of freedom, which employ several auxiliary antennas, receivers, and adaptive loops.

The closed-loop bandwidth of a Howells adaptive loop is proportional to the product of the loop gain and the bandwidth of the narrowband filter, which follows the V_r, V_a multiplier in Fig. 3–2. This bandwidth cannot be allowed to become so large that the loop begins to cancel desired signals. This is another reason why a loop gain proportional to $\overline{|V_a|^2}$ is undesirable.

Cancellation of the desired signal in an adaptive loop is caused by the cancellation not being limited to correlated components between V_a and V_r. When the loop bandwidth becomes too large, the loop may force the weighted V_a to be correlated with V_r by stripping the modulation from V_r to be used in the modulation of V_a.

Multiple Degrees of Freedom

At this point, we should note that if there are two independent and uncorrelated sources of interference exciting the radar and auxiliary antenna simultaneously from different directions, a single system of auxiliary antenna, receiver, and loop cannot cancel both interfering signals. This is due to the fact that a loop weight only represents one amplitude and phase change on V_a, and two interference sources in different directions require that different weights be used in each source's signal.

Howells solved the problem of cancelling interference from multiple, independent, and uncorrelated sources at different locations about a radar by inventing the parallel-loop configuration illustrated in Fig. 3–4. To simplify the figure, the radar and auxiliary receivers are not shown, but are understood to be located between the antennas and the loop inputs.

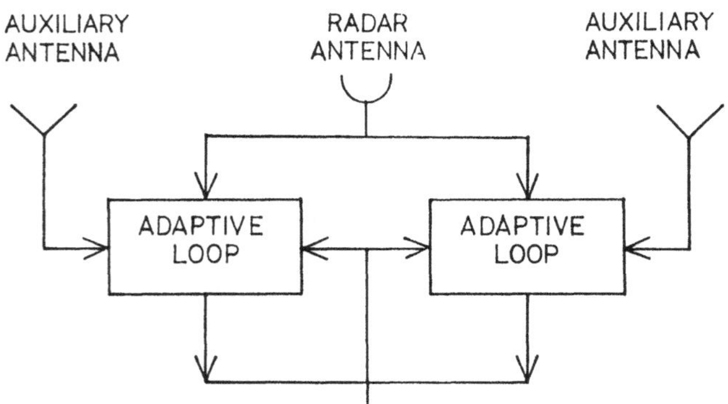

Fig. 3–4 Parallel-Loop Configuration for Canceling Two Interference Signals Arriving from Different Directions.

Figure 3–5 shows a more detailed block diagram of the system with two-degrees of freedom shown in Fig. 3–4.

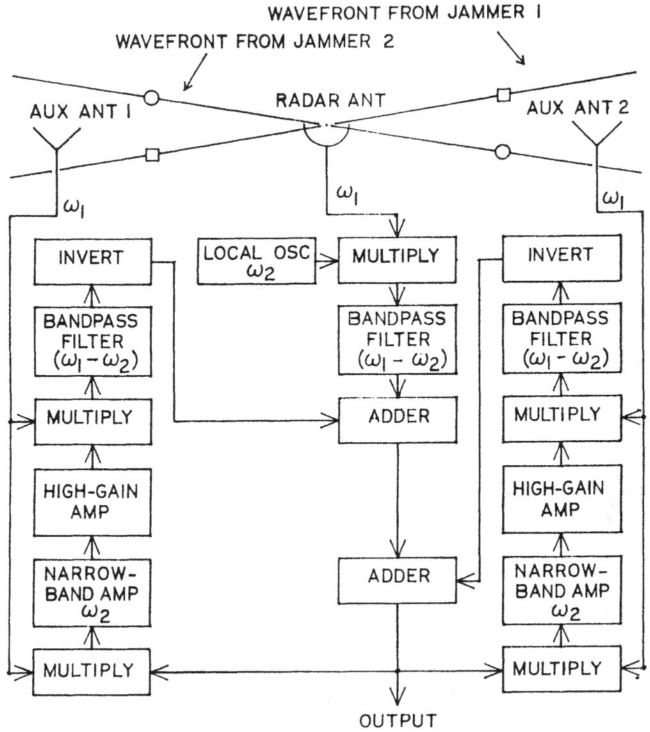

Fig. 3–5 Functional Block Diagram of a Two-Degree-of-Freedom Coherent Sidelobe Canceler (After Howells).

In this figure, wavefronts from interference sources, called *jammers,* are shown arriving at the radar and two auxiliary antennas. One wavefront is marked by a circle, the other by a square. Both wavefronts pass through the phase center of the radar antenna. In this way, each interference signal at the radar antenna's phase center can be considered to be a phase reference (zero phase) for describing the phase of counterpart signals at the various auxiliary antennas. Any differences in amplitude or phase among the various signals entering each antenna can thus be indicated as illustrated in Fig. 3–6 by using vectors marked with a circle or a square next to the antenna in question.

This *vector technique* for defining the interference components at the various antennas can be used to explain parallel-loop operation [4]. In the case of Fig. 3–6, the two vectors shown in auxiliary antenna 1 would produce equal and opposite phase signals of frequency ω_2 at the output of the narrowband amplifier following multiplier 2. In this process, cross products between circle and square vectors can be ignored because the two signals have been assumed to be uncorrelated (i.e., the average value of their product has been assumed to be zero). The two out-of-phase signals from the narrowband amplifier will cancel and the loop using aux-

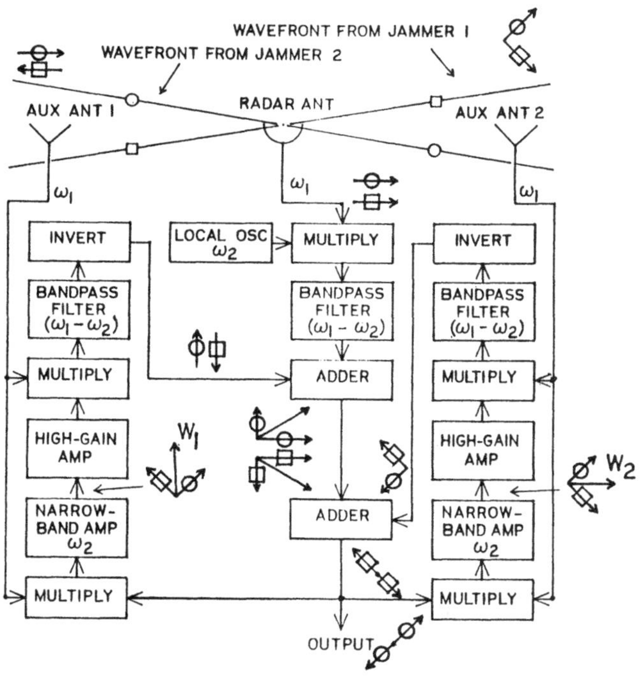

Fig. 3–6 Definition of Phase and Amplitude of the Vectors at the Antennas of a Radar and Sidelobe Canceler System.

iliary antenna 1 will not be able to form a weight. Were this the only loop employed, the output of adder 2 would be the radar input to adder 1.

The presence of loop 2 with auxiliary antenna 2, however, permits the system to operate. The product of the radar and auxiliary 2 circle signals will produce a circle signal out of the narrowband amplifier that follows multiplier 4 having a frequency ω_2 and a 45° phase, as shown in Fig. 3–6. The products of the square signals will produce an equivalent signal with the same frequency, but a −45° phase angle. These two signals will add vectorially to produce the weight W_1 as illustrated in Fig. 3–6. This weight will be amplified and multiply the auxiliary 2 signals in multiplier 5. This will translate them down to a center frequency $\omega_1 - \omega_2$ without changing the respective phases. The resulting signals after bandpass filtering will be inverted and added to the output of adder 1 in adder 2. In addition, the horizontal components of the two weighted auxiliary signals will cancel some of the equivalent signals out of adder 1, and the weighted auxiliary signals will add some vertical components to the resulting signals in opposite directions. These oppositely directed circle and square components out of adder 2 will correlate with their counterparts on auxiliary antenna 1 to produce a weight W_2 at the output of the narrowband amplifier following multiplier 2. W_2 will translate the frequency and shift the phase of the signals from auxiliary antenna 1 in multiplier 5, the output of which will be inverted and appear at the weighted auxiliary input to adder 1 as illustrated in Fig. 3–6. The vertical components added to the radar input to adder 1 will then cancel the vertical components added to the input of adder 2 by loop 2.

From this example, we can see that two uncorrelated interference signals can be cancelled by two adaptive loops if the two auxiliary signals employed by the loops correlate differently with the interference in the radar. In this case, loop 2 removed the horizontal components of the two interference signals in the radar and loop 1 removed the two vertical components.

It is interesting to note at this point that if the angles defining the phases of the interference signals at the auxiliary 2 antenna of Fig. 3–6 were larger, the weights used by the two loops would become larger and approach infinity as the angles approach 90°. This results from the fact that the projections of the auxiliary 2 signals on the signals in the radar receiver will decrease with the described angular increase until the projections reduce to zero when the angles reach 90°.

This increase in weighting magnitude as the auxiliary signals on the auxiliary antennas become correlated requires the loops to have wide dynamic range. Also, large magnitude weights translate great amounts of thermal noise from the auxiliary channel into the radar channel to reduce the signal to thermal noise ratio in the radar.

Quantitative Analysis of Multiple Degree-of-Freedom Systems

Quantitative analysis of Howells' multiple degree-of-freedom systems becomes more difficult when the number of degrees of freedom exceeds two. A well known algorithm applicable to the analysis of multiple degree-of-freedom systems is the *Widrow-Hoff least mean square (LMS) algorithm* [5–7]. Although it is widely used in simulations, this algorithm is *conditionally stable* [8] and should be used with caution to simulate fast convergence systems. The Widrow-Hoff algorithm for simulating a single Howells loop can be explained as follows:

Discrete data, normally sampled at the Nyquist rate, is used with Howells' loop logic illustrated in Fig. 3–3. In addition, a one-sample delay is placed in the weight prior to its application in the loops used by the system. The $(j + 1)$th weight in this algorithm is given by

$$W(j+1) = W(j) + 2uV_r(j)V_a^*(j) \qquad (3\text{–}18)$$

where

$$V_r(j) = V_m(j) - W(j)V_a(j) \qquad (3\text{–}19)$$

with $W(j)$, $V_r(j)$, $V_m(j)$, and $V_a(j)$ denoting the jth sample of the complex modulation functions respectively corresponding to the weighting signal, the error or output signal, the primary input or desired signal, and the reference input or auxiliary signal. The parameter u is a gain factor that controls stability and convergence rate. The primary input signal V_m consists of the desired signal plus interference, while V_a is assumed to consist only of interference that is correlated with the interference contained in V_m. In general, said interference may consist of correlated components plus uncorrelated components, such as uncorrelated thermal noise in each channel. The LMS criterion determines the weight that minimizes the mean squared value of the error or output signal given in (3–19), which, in turn, corresponds to minimizing the average output noise power. Substitution of (3–19) in (3–18) results in

$$W(j+1) = W(j)(1 - 2u|V_a(j)|^2) + 2uV_m(j)V_a^*(j) \qquad (3\text{–}20)$$

Substitution of (3–20) into (3–19) yields

$$V_r(j+1) = V_m(j) - [W(j)(1 - 2u|V_a(j)|^2) + 2uV_m(j)V_a^*(j)]V_a(j) \qquad (3\text{–}21)$$

Equation (3–21) can be used with a single loop or multiple loops in parallel. In the multiloop case, V_r is the summed output of the parallel loops, V_m is the main input to all loops, and (3–21) is written as

$$V_r(j+1) = V_m(j) - \sum_{i=1}^{n} [W_i(j)(1 - 2u|V_{ai}(j)|^2) + 2uV_m(j)V_{ai}^*(j)]V_{ai}(j) \qquad (3\text{–}22)$$

where $V_{ai}(j)$ is the jth sample of the ith auxiliary signal.

The attractive feature of the Widrow-Hoff algorithm is that it is well adapted to computer analysis, but, as previously mentioned, it does not work well with loop bandwidths greater than one-tenth of the input interference bandwidth, unless it is modified. This limitation is due to the fact that a one-sample delay equal to the reciprocal of the Nyquist sampling rate is normally used in the derivation and application of the weight in the Widrow-Hoff algorithm. With data sampled at the Nyquist rate, the algorithm becomes completely unstable with a loop bandwidth equal to the input signal bandwidth (i.e., when the loop gain is unity at the frequency where a 180° phase shift exists around the closed loop). One remedy for this problem is to sample the data at a rate higher than the Nyquist rate. Another solution was investigated by Kretschmer and Lewis [8] whereby the *sample-delay offset* is removed from between the weight and the data it multiplies. Their investigation indicated that the modified algorithm was *unconditionally stable* ([8], *see also* Readings, Chapter 3). However, the equations became more complex than the LMS solution for a multiple degree-of-freedom system. A recursive algorithm to determine the derived weights [8] in a multiple degree-of-freedom system was published by Johnson [9]. In a later publication, Nitzberg [10] investigated an algorithm which effectively normalizes the loop gain and showed that his solution was similar to that of Kretschmer and Lewis.

The potentially slow convergence speeds of Howells' parallel loops have encouraged researchers to seek better solutions to multiple degree-of-freedom systems. One solution that has been investigated extensively by Lewis and Kretschmer is referred to as the *Gram-Schmidt approach* because the auxiliary signals are preprocessed so as to be mutually orthogonal to each other in a manner that is similar in principle to the mathematical method of Gram-Schmidt for orthogonalization of dependent vectors.

3.3.2 Gram-Schmidt Approach

The Gram-Schmidt approach uses series-connected cancelers (adaptive loops) employing decorrelated auxiliary signals as illustrated in Fig. 3–7 [3, 11].

In Fig. 3–7, an adaptive loop decorrelates the radar signal from interference picked up on auxiliary antenna 1. Another loop decorrelates the signal picked up on auxiliary antenna 2 from that used in the first loop and sends the results to be used by a second series loop in the radar channel. Similarly, the signal of auxiliary antenna 3 is decorrelated from both of the previously used auxiliary signals in two adaptive loops, and the result is used in a third series loop in the radar channel. In this process, none of the series loops in the radar channel can reinsert interference that has been removed by a previous loop because all auxiliary signals employed in the radar channel have been decorrelated prior to use.

The Gram-Schmidt configuration is substantially better than the parallel-loop configuration in terms of stability, speed of convergence, and ease of analysis. Because no loop is in another's feedback path, the loops do not interact to complicate either the performance or analysis.

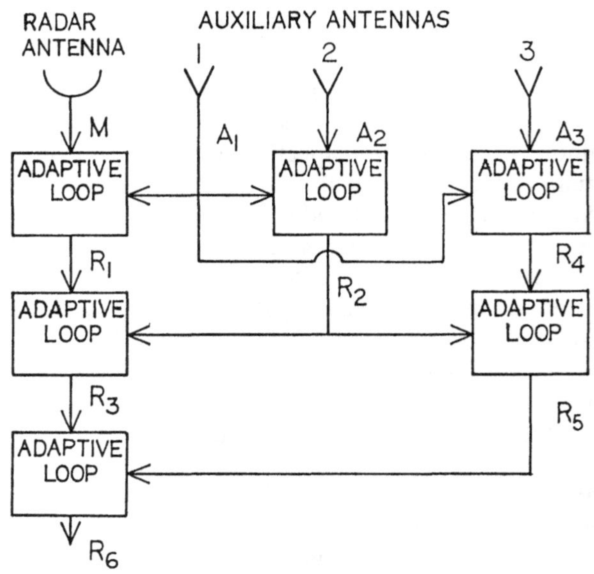

Fig. 3–7 Gram-Schmidt Configuration.

Let us represent the complex modulation functions on the radar and auxiliary antennas by M and A, respectively. Thus, the residues R of the loops can be calculated by using (3–2) and (3–5). For example,

$$R_1 = M - \left[\frac{\overline{MA_1^*}}{\overline{A_1A_1^*}}\right]\left[\frac{G\overline{|A_1|^2}}{(1 + G\overline{|A_1|^2})}\right]A_1 \quad (3\text{–}23)$$

and

$$R_2 = A_2 - \left[\frac{\overline{A_2A_1^*}}{\overline{A_1A_1^*}}\right]\left[\frac{G\overline{|A_1|^2}}{(1 + G\overline{|A_1|^2})}\right]A_1 \quad (3\text{–}24)$$

et cetera up through R_6. The convergence time of the system illustrated in Fig. 3–7 has been shown to be the settling time of each loop as determined by its local auxiliary inputs, which, in turn, determines the loop bandwidth. The settling time is not effected by the number of degrees of freedom, as is the case in the Howells parallel-loop configuration.

The availability of the Gram-Schmidt configuration for implementing multiple degree-of-freedom coherent sidelobe-canceler systems without requiring feedback thus opened the door to digital systems [2, 12]. A digital open-loop adaptive processor was developed, which used the optimum weight (3–7) to obtain optimum cancellation and a cancellation ratio independent of auxiliary signal power level.

3.3.3 Digital Open-Loop Adaptive Processor [2, 12]

We note from (3–5) that the Howells adaptive loop is limited by the loop gain in achieving the minimum residue provided by W_{opt}. The feedback is important in analog circuitry to compensate for component drifts. In digital implementation, however, this problem is not present and the optimum weight can be determined directly in an open-loop configuration from the input signals. For optimum weighting, k in (3–15) is effectively unity and the cancellation ratio is a maximum.

In digital application, the ensemble average indicated in (3–5) is approximated by a finite time sample average. The simulation results, which are subsequently discussed in this book, were derived by computing W_{opt} from sliding-window averages. Thus, the jth sampled residue was computed from

$$V_r(j) = V_m(j) - W_{\text{opt}}(j)V_a(j) \quad (3\text{–}25)$$

where, for an n-sample window size, we have

$$W_{\text{opt}} = \sum_{k=j-n+1}^{j} V_m(k)V_a^*(k) \Big/ \sum_{k=j-n+1}^{j} |V_a(k)|^2 \quad (3\text{–}26)$$

We can see from (3–26) that in the case of a single sample, W_{opt} becomes V_m/V_a. Substitution of this into (3–25) results in perfect cancellation, irrespective of V_m or V_a. This suggests that more samples be averaged to prevent cancellation of a valid target signal by an uncorrelated auxiliary signal.

An illustrative implementation is shown in Fig. 3–8 for a two-sample average. The delay time shown in the figure corresponds to one sampling interval, where it is assumed that successive samples are uncorrelated and each sample consists of both the in-phase (I) and quadrature (Q) components.

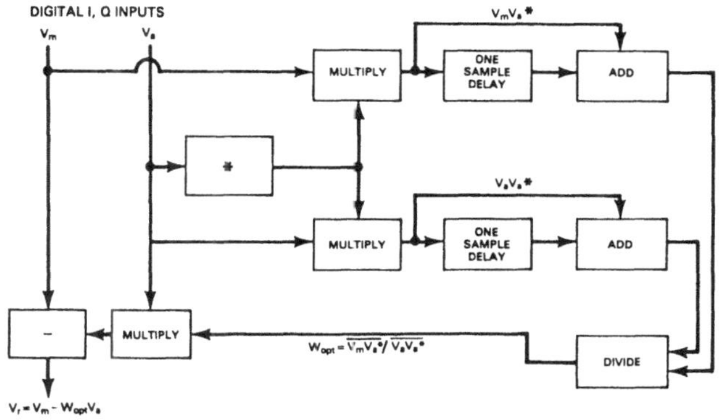

Fig. 3–8 Digital Open-Loop Adaptive Processor.

Performance Simulation

Computer simulations were run to evaluate the logic of Fig. 3–8. The input jamming signal was simulated by a step input of Gaussian samples, corresponding to random noise modulation. Each sample generated in the simulated radar channel was phase shifted by an arbitrary 90° for the corresponding auxiliary channel sample. The amplitude of the auxiliary channel sample was selected to be the same as, or 20 dB higher than, the radar signal amplitude.

In practice, the gain of the auxiliary channel antenna is equal to or greater than the highest sidelobe level of the radar antenna to prevent excessive translation of thermal noise from the auxiliary channel into the radar channel as well as to reduce the dynamic range requirements. In the simulations, successive samples were statistically independent so that each sample corresponds to a sample from a different radar range-resolution cell. Also, Gaussian samples, independent of the jamming samples, were generated to represent thermal noise in the radar and auxiliary channels, which were independent of one another. Successive noise samples in each channel were also independent. Target signals were simulated by injecting a constant signal value into the radar or into both channels at selected sample numbers.

In the simulations, various numbers of pulses were averaged (NAVG) to form the weight, and average residue powers were computed. The simulation results were expressed in terms of the ratio (in decibels) of the average input jamming power to the sampled residue power. Thus, the 0 dB level in the simulation results corresponds to the average input jamming power level, and the residue sample power shown by the ensuing plots is given in terms of the number of decibels below the input jamming power level. The first 200 input samples in the simulations contain jamming and thermal noise, while targets were injected at preselected points throughout the next 300 points. The normalized residue power was averaged over the first region, excluding transients, and over the second region at the points where targets were injected. These averages, expressed in decibels, are denoted by *CJ* and *CS*, respectively. The quantity *CJ* also corresponds to the cancellation ratio as defined in (3–10).

Also calculated was a figure of merit, similar to the improvement factor used in moving target indication (MTI). The figure of merit is defined here as the ratio of the average signal-to-interference ratio at the canceler output to that at the input. This improvement factor, I_j may be expressed in dB as

$$I_j = \frac{(S/I)_{\text{out}}}{(S/I)_{\text{in}}} = CJ - CS - [(S/I)_{\text{in}}]_{\text{dB}} \qquad (3\text{–}27)$$

The interference consists of jamming and thermal noise. The thermal noise is insignificant compared to the input jamming level in the simulations, so $(S/I)_{\text{in}}$ is approximated by $(S/J)_{\text{in}}$.

Results are presented in Figs. 3–9, 3–10, and 3–11 for an input thermal noise level, N, 50 dB below the jamming level. In these figures, a target signal S_m was injected in the radar channel at a level 20 dB below the average input jamming level, beginning at sample number 200 and recurring every 10 samples thereafter. Prior to sample number 200, jamming and thermal noise alone are present. The jamming levels in the radar channel J_m and the auxiliary channel J_a were the same, and the target signal in the auxiliary channel S_a was zero.

Several general observations can be made from these three figures. First, we observe that the adaptive digital canceler reduces the jamming down to the thermal noise level except in the regions where targets were injected. This is because of the effective infinite gain associated with the canceler, which enables it to completely cancel correlated signals between the radar and auxiliary channels. Because the thermal noise is uncorrelated between these channels, the thermal noise level represents the floor of the residue.

Next, we observe that, in the first part of the plot on the first sample, the cancellation ratio increases to 120 dB, limited only by the single precision of the computer used. At this point, the weight is V_m/V_a, which multiplied by V_a equals V_m and subtracted from V_m leaves a residue of zero. On the second sample, some averaging has taken place on the weight and the residue becomes thermal-noise limited.

Another general observation that can be made is that as NAVG increases, the target becomes less likely to be reduced by the canceler. If no cancellation of the target signals takes place, the target points are at the 20 dB level in the plots. We can see from Figs. 3–9, 3–10, and 3–11 that the amount of the target cancellation, and the variability thereof, reduces as NAVG increases. More averaging of the weight causes it to approach the correlation coefficient between the radar and auxiliary signals, which is defined as a long time average.

In these simulations, computations of the average residues *CJ* and *CS* were made in the first region, where

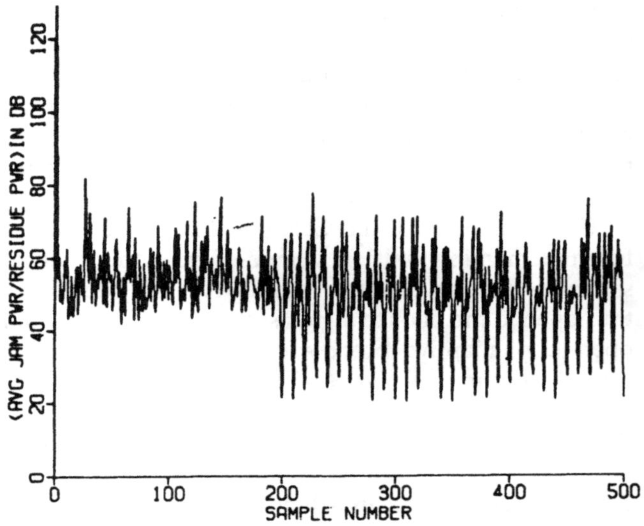

Fig. 3–9 Digital Canceler Performance [$W = W_{\text{opt}}$, NAVG = 2, (J_m/N) = 50 dB, (J_m/S_m) = 20 dB, $J_a = J_m$, $S_a = 0$].

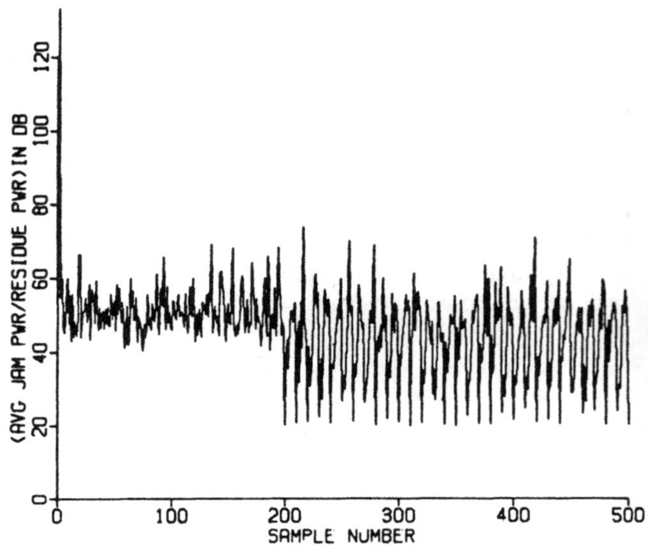

Fig. 3–10 Digital Canceler Performance [$W = W_{\text{opt}}$, NAVG = 4, (J_m/N) = 50 dB, (J_m/S_m) = 20 dB, $J_a = J_m$, $S_a = 0$].

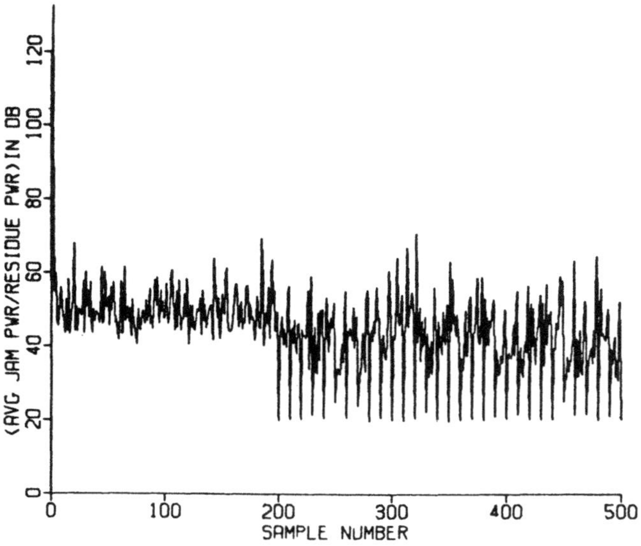

Fig. 3–11 Digital Canceler Performance [$W = W_{opt}$, NAVG = 8, (J_m/N) = 50 dB, (J_m/S_m) = 20 dB, $J_a = J_m$, $S_a = 0$].

no targets were present, and in the second region on the points at which the targets were injected. The first average was obtained from 175 points, which excluded the transients of the response. The target average was determined from 30 samples. The results are tabulated in decibels in Table 3–1. The cancellation improvement factor I_j is also shown in the table. Therein, we can see that the jamming residue power increases as more pulses are averaged, but the target cancellation decreases so that the improvement factor remains fairly constant.

Parameter	NAVG = 2	NAVG = 4	NAVG = 8
CJ	50.5	48.2	47.5
CS	23.7	21.8	20.9
I_J	46.8	46.4	46.6

Table 3–1 Performance Results Using W_{opt}.

The transient response arising after application of a target signal in the radar channel is addressed next. Again, referring to Figs. 3–9, 3–10, and 3–11, we see that following the application of each target signal there is transient recovery time in jamming cancellation. The transient may be observed in greater detail in Figs. 3–12, 3–13, 3–14, and 3–15, wherein it is plotted for NAVG equal to 2, 4, 8, and 32, respectively. In these simulations the thermal noise was at the 100 dB level and the target signals were injected at intervals of 50 samples, beginning with sample number 200, for better illustration of the transient effects. Also, these figures again demonstrate cancellation down to the thermal noise level. The canceler residue is essentially equal to that in Figs. 3–9, 3–10, and 3–11, except for a translation of the residue level.

From Figs. 3–12 through 3–15, we observe that the total recovery span following a target input point is the same as the averaging window or NAVG samples. Although the target is present for only one sample at a given point of injection, the influence of the target is retained in the sliding window for the window duration. The transient, or *back porch* response is shown to diminish in level as NAVG increases. The weight perturbation of W_{opt}, caused by the presence of the target signal in the window, is of the form:

$$\delta W = \frac{SV_{ai}}{\sum_{k=1}^{\mathrm{NAVG}} |V_{ak}|^2}, \quad 1 \leq i \leq \mathrm{NAVG} \tag{3–28}$$

where

V_{ak} = kth auxiliary channel sample in the averaging window;
S = target signal sample corresponding to the ith auxiliary channel sample V_{ai}.

The value of $|\delta W|^2$ reduces as the square of NAVG, which is shown in the simulations to be a reduction in the porch level by the same amount as the increase in NAVG. The back porch not only diminishes target resolution, but is also capable of masking a weaker target located in the window. This phenomenon is also present in conventional analog loops.

Results of a simulation are shown in Fig. 3–16 for the case in which the auxiliary channel jamming level

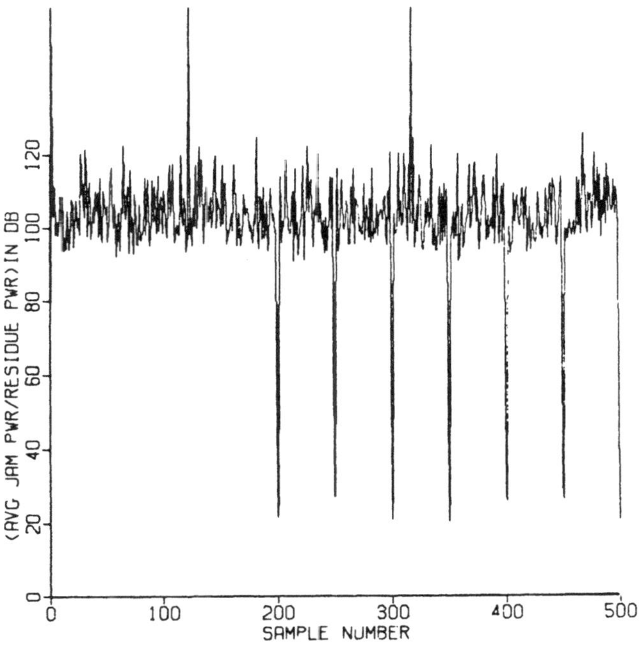

Fig. 3–12 Digital Canceler Performance [$W = W_{opt}$, NAVG = 2, (J_m/N) = 100 dB, (J_m/S_m) = 20 dB, $J_a = J_m$, $S_a = 0$].

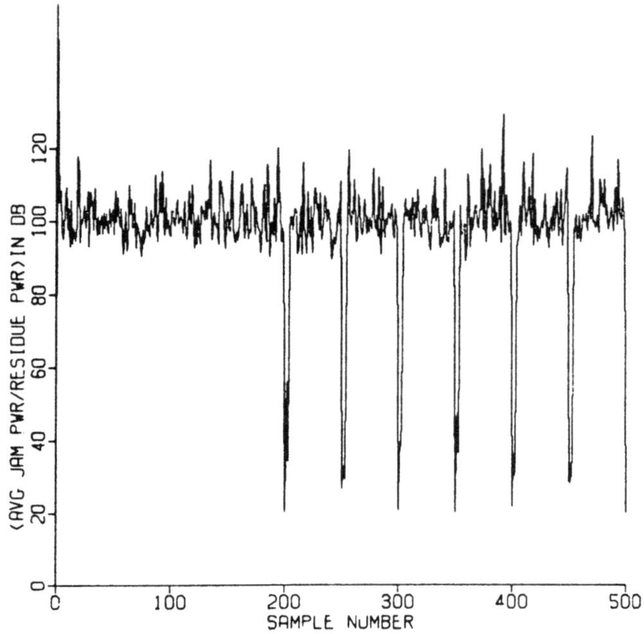

Fig. 3–13 Digital Canceler Performance [$W = W_{opt}$, NAVG = 4, (J_m/N) = 100 dB, (J_m/S_m) = 20 dB, $J_a = J_m$, $S_a = 0$].

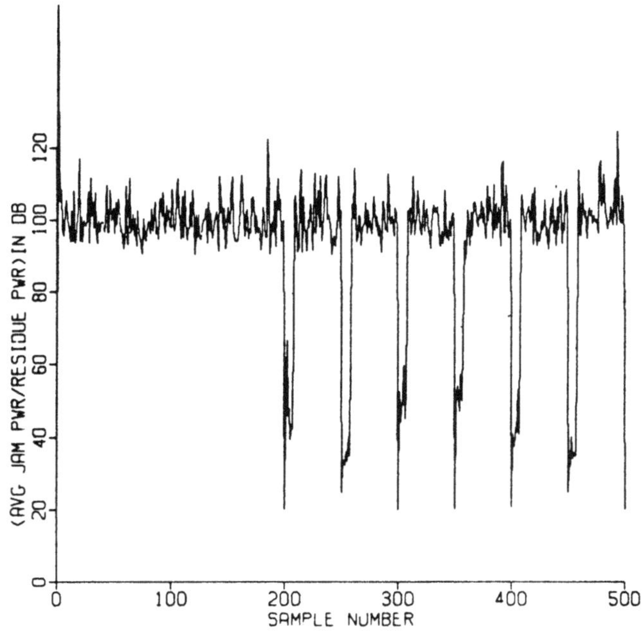

Fig. 3–14 Digital Canceler Performance [$W = W_{opt}$, NAVG = 8, (J_m/N) = 100 dB, (J_m/S_m) = 20 dB, $J_a = J_m$, $S_a = 0$].

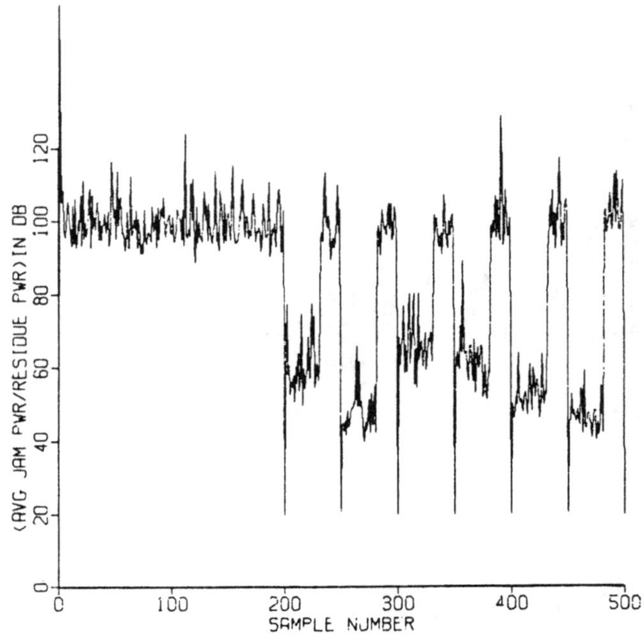

Fig. 3–15 Digital Canceler Performance [$W = W_{opt}$, NAVG = 32, (J_m/N) = 100 dB, (J_m/S_m) = 20 dB, $J_a = J_m$, $S_a = 0$].

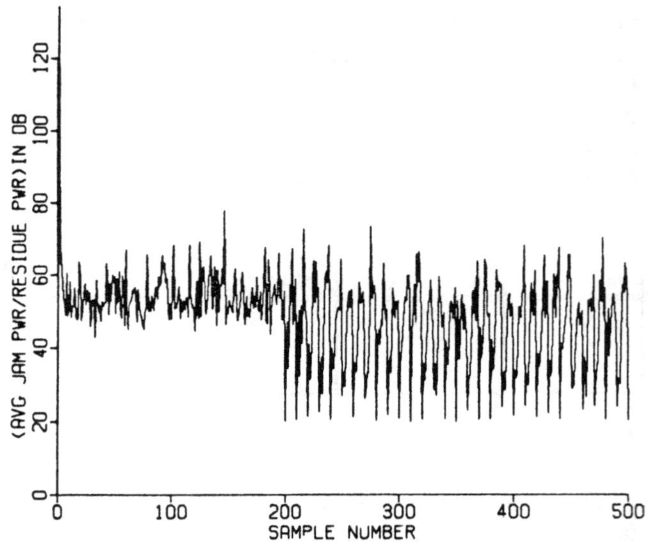

Fig. 3–16 Digital Canceler Performance [$W = W_{opt}$, NAVG = 4, (J_m/N) = 50 dB, (J_m/S_m) = 20 dB, J_a/J_m = 20 dB, $S_a = 0$].

was 20 dB higher than that in the radar channel. The results are basically the same as those shown in Fig. 3–10, which demonstrates that cancellation of correlated signals is independent of the power level in the auxiliary channel, in contrast to the situation of the closed-loop canceler. Calculation of *CJ* indicated a 3 dB decrease of the residue as compared to the case in which the jamming signal is the same in the radar and auxiliary channels. This was due to the required weighting being 20 dB lower, which precluded adding thermal noise from the auxiliary channel to the radar channel.

Figure 3–17 shows the results for the case in which the signal was also present in the auxiliary channel, 10 dB above the noise level, or 40 dB below the jamming level. For this simulation, NAVG = 4 and the thermal

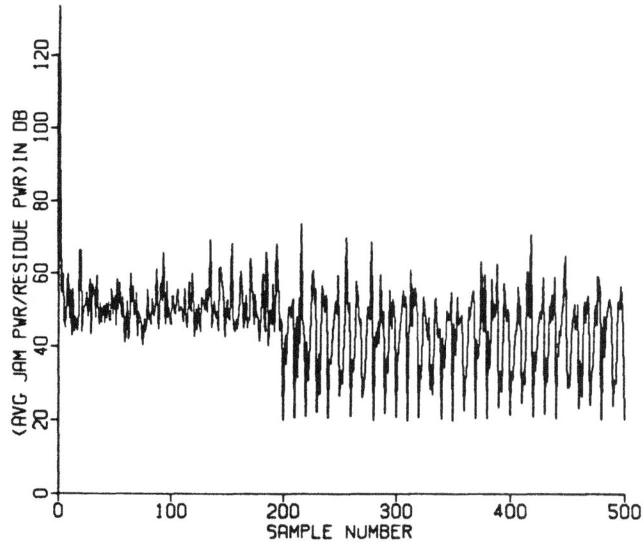

Fig. 3–17 Digital Canceler Performance [$W = W_{opt}$, NAVG = 4, (J_m/N) = 50 dB, (J_m/S_a) = 20 dB, $J_a = J_m$, (J_m/S_a) = 40 dB].

noise level was 50 dB below the input jamming level, which was the same in both channels. Because the jamming power level was much greater than the signal level, the canceler weight was not altered by the signal in the auxiliary channel. The results for different values of NAVG gave results identical to those in Table 3–1.

From the simulation results presented here, we see that the new digital cancellation technique using the optimum weight derived from the input signals in an open-loop configuration provides improved cancellation performance as compared to conventional analog sidelobe-canceler loops that rely on closed-loop adaptation.

3.3.4 System Requirements for Good Cancellation

In order to obtain good cancellation from coherent sidelobe cancelers, correlation must be maintained between the radar and auxiliary channels. All receivers must be matched in terms of phase and amplitude response across the radar passband, and no internal or external multipath should be allowed to introduce differential time delays between the radar and auxiliary signals. This is required because the adaptive cancelers can only use one weight on all frequency components in the signals to be cancelled.

The effect of differential time delays between the radar and auxiliary signals can be appreciated from Fig. 3–18.

The differential time delay causes a differential phase shift $\Delta\phi = 2\pi B\tau$, across the band B of the input signal, which cannot be removed by a canceler that is only able to use one phase shift on all frequencies. If the delayed and undelayed signals are rectangular

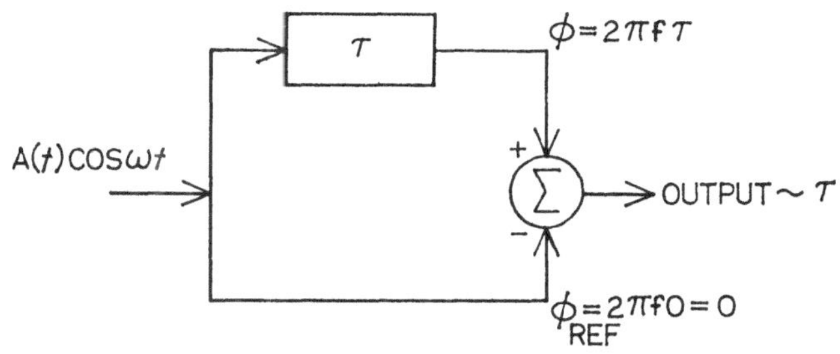

Fig. 3–18 Effect of Differential Time Delay τ on Finite Bandwidth Interference.

band-limited white noise, they become uncorrelated when τ is a multiple of $1/B$. This means that the average value of the product of the two signals is zero when τ becomes a multiple of $1/B$.

3.4 MAIN-LOBE NOTCHERS [13, 14]

Main-lobe notchers are used to cancel high-duty-cycle interference entering the main lobe of a radar from a single direction in space. Main-lobe notchers are implemented in the same way as coherent sidelobe cancelers, except that the auxiliary antenna for the main-lobe notcher must have a much higher gain than that of the auxiliary antenna for a sidelobe canceler. This requirement is due to the fact that the radar antenna's main lobe has a much higher gain than its sidelobes. Thus, to keep the notcher weight near unity, the auxiliary antenna must have a gain close to that of the radar antenna's main lobe. If the weight is kept near unity, the canceler is prevented from amplifying the thermal noise in the auxiliary receiver and adding it to the radar signals. Were the weight allowed to become large, the canceler would effectively jam the radar with thermal noise by reducing the effective radar antenna gain to that of the auxiliary antenna.

3.5 SIMULATED DOPPLER

A new signal processing technique for reducing the response of a radar to signals entering the radar antenna's sidelobes [15] is described next. This technique simulates rapid antenna motion normal to the radar boresight axis while the radar is in transmission. This motion causes Doppler shifts to any signals radiated into the sidelobes so that they are out of the radar receiver's passband. Motion is simulated by switching the excited area of a phased array antenna through several adjacent locations during the radar's transmission time. This effectively moves the radar antenna's phase center during transmission, as if the actual antenna had moved. The simulated Doppler technique increases the angular resolving power of the radar by allowing it to ignore large reflectors in the radar antenna sidelobes.

In operation, only a fraction of the available elements of an antenna array are excited at any one time upon transmission and the excited group of elements is moved across the array in steps of short duration as compared to the reciprocal of the radar signal bandwidth. For example, if a 1 μs pulse were transmitted from 10 of 20 available elements, then the 10 excited elements would be allowed to radiate for 0.1 μs and would be switched one element at a time every 0.1 μs.

3.6 ADAPTIVE ANTENNA ARRAYS AND SIGNAL PROCESSING SYSTEMS

Many papers, reports, and books (*see* [3, 7, 11] for background material and references) have been written in recent years dealing with adaptive array systems, which adapt the antenna pattern to nullify (called *nulling*) interference sources in the sidelobes, and other *adaptive processing systems* that use similar techniques. In many practical applications, the attainable interference cancellation is limited by multipath reflections and mismatches between channels. In this situation, improved cancellation is obtained by using a *tapped delay line* [3, 16]. Also, improved performance is obtained by carrying out cancellation after FFT processing [7]. *Spectral analysis techniques* have been extensively investigated as a means of nulling, locating, and estimating the strength of interference sources (for example, see [17]).

PROBLEMS

Problem 3–1 Given $V_m = \exp(j\omega t)$, $V_a = 2\exp[j(\omega t + \pi/2)]$, and $G = 10$ in a Howells loop, answer the following questions:
(a) What weight would the loop develop?
(b) What would be the loop residue V_r?
(c) What would be the effective gain G_e of the Howells loop?
(d) What complex number would represent the relative amplitude and phase of V_a and V_m?

Problem 3–2 What is the maximum number of independent interference signals that the system illustrated in Fig. 3–7 would cancel?

Problem 3–3 Draw a Gram-Schmidt configuration that would cancel a maximum of four independent interference signals.

Problem 3–4 Draw a Gram-Schmidt configuration that would cancel a maximum number of two independent interference signals.

Problem 3–5 By using the configuration drawn in problem 3–4 and the Howells adaptive-loop canceler, let $M = \exp(j\omega_1 t) + \exp(j\omega_2 t)$, $A_1 = \exp(j\omega_1 t) + \exp[2(j\omega_2 t + \pi)]$, $A_2 = \exp(j\omega_1 t) + \exp[j(\omega_2 t + \pi/2)]$, and $G = 10$, and answer the following questions:
(a) What are the weight W_1 and the residue R_1 of the canceler using M and A_1 as its inputs?
(b) What are the weight W_2 and the residue R_2 of the canceler using A_2 and A_1 as its main and auxiliary inputs, respectively?
(c) What are the weight W_3 and the residue R_3 of the canceler using R_1 and R_2 as its main and auxiliary inputs, respectively?

Problem 3–6 Repeat problem 3–5 with $G = 100$.

Problem 3–7 Repeat problem 3–5 with digital open-loop adaptive processors.

Problem 3–8 Given $V_m = \exp(j\omega_1 t) + 0.3\exp(j\omega_2 t)$ and $V_a = \exp(j\omega_1 t) + 0.3\exp(j[\omega_2 t + \pi])$ in Howells loop with $G = 10$, calculate:
(a) the weight W;
(b) the loop residue V_r;
(c) the cancellation ratio CR.

Problem 3–9 Repeat problem 3–8 with $G = 100$.

Problem 3–10 With $V_m = 0.01\exp(j\omega_1 t) + 10\exp(j\omega_2 t)$, $V_a = 0.01\exp(j\omega_3 t) + 10\exp(j\omega_2 t)$, and $G = 1000$ in a Howells loop, find:
(a) the weight W;
(b) the loop residue V_r;
(c) the cancellation ratio CR.

Problem 3–11 Find the effective loop gain G_e in problem 3–9.

Problem 3–12 With $V_a = 2\exp(j[\omega_1 t + \pi/4]) + 2\exp(j[\omega_2 t - \pi/4])$, $V_m = \exp(j\omega_1 t) + \exp(j\omega_2 t)$, and $G = 100$ in a Howells loop, find:
(a) the weight W;
(b) the loop residue V_r;
(c) the cancellation ratio CR.

Problem 3–13 Assume that the loop using auxiliary antenna 2 (loop 2) in Fig. 3–6 is open. Let $V_m = \exp(j\omega_1 t) + \exp(j\omega_2 t)$ and the signal of auxiliary antenna 1 be $V_{a1} = \exp(j\omega_1 t) - \exp(j\omega_2 t)$. With a loop gain $G = 100$ in loop 1, calculate the weight W, the loop gain V_r, and the cancellation ratio CR due to loop 1.

Problem 3–14 Repeat problem 3–13 with loop 2 using an auxiliary 2 signal of $V_{a2} = \exp(j[\omega_1 t + \pi/4]) + \exp(j[\omega_2 t - \pi/4])$, assuming loop 1 to be open and $G = 100$ in loop 2.

Problem 3–15 With the weight determined in 3–14 fixed, repeat 3–13.

Problem 3–16 With the weight determined in 3–15 fixed, repeat 3–14.

Problem 3–17 Describe what happens to the cancellation ratio *CR* as the steps involved in 3–13 through 3–16 are repeated.

REFERENCES

1. P. W. Howells, Intermediate Frequency Sidelobe Canceler, U.S. Patent No. 3,202,990, Aug. 1965.
2.* F. F. Kretschmer, Jr., and B. L. Lewis, "A Digital Open-Loop Adaptive Processor," *IEEE Trans. on Aerospace and Electronic Systems,* Vol. AES-14, No. 1, Jan. 1978, pp. 165–171.
3. R. A. Monzingo and T. W. Miller, *Introduction to Adaptive Arrays,* Wiley-Interscience Publication, New York, 1980.
4.* B. L. Lewis and J. P. Hansen, "Understanding and Optimizing Multiple Sidelobe Canceler Operation," NRL Report 7610, Oct. 1973.
5. B. Widrow, R. Mantey, L. Griffiths, and B. Good, "Adaptive Antenna Systems," *Proc. IEEE,* Vol. 55, Dec. 1967, pp. 2143–2158.
6. B. Widrow, *et al.*, "Adaptive Noise Canceling: Principles and Applications," *Proc. IEEE,* Vol. 69, Dec. 1975, pp. 1692–1719.
7. B. Widrow and S. D. Stearns, "Adaptive Signal Processing," Prentice-Hall, Englewood Cliffs, NJ, 1985.
8.* F. F. Kretschmer, Jr., and B. L. Lewis, "An Improved Algorithm For Adaptive Processing," *IEEE Trans. on Aerospace and Electronic Systems,* Vol. AES-14, No. 1, Jan. 1978, pp. 172–177.
9. C. R. Johnson, Jr., "An Implementation of the Multidimensional Modified LMS Algorithm," *IEEE Trans. on Aerospace and Electronic Systems,* May 1980, pp. 398–399.
10. R. Nitzberg, "Application of the Normalized LMS Algorithm to MSLC," *IEEE Trans. on Aerospace and Electronic Systems,* Vol. AES-21, No. 1, Jan. 1985, pp. 79–91.
11. J. E. Hudson, *Adaptive Array Principles,* Peter Peregrinus, NY, 1981.
12. F. F. Kretschmer, Jr., and B. L. Lewis, "Digital Open-Loop Sidelobe Canceler Techniques," NRL Report 8100, April 1977.
13. W. F. Gabriel, "Using Spectral Estimation Techniques in Adaptive Array Systems," *Phased Array Symposium Proceedings,* RADC TR 85171, 1985.
14. E. Brookner and J. M. Howell, "Adaptive Array Processing," *Phased Array Symposium Proceedings,* RADC TR 85177, 1985.
15.* B. L. Lewis and J. B. Evans, "A New Technique for Reducing Radar Response to Signals Entering Antenna Sidelobes," *IEEE Trans. on Antennas and Propagation,* Vol. AP-31, No. 6, Nov. 1983, pp. 993–996.
16.* K. Gerlach, "Adaptive Canceller Limitations due to Frequency Mismatch Errors," NRL Report 8949, Jan. 2, 1986.
17.* W. F. Gabriel, "Using Spectral Estimation Techniques in Adaptive Processing Antenna Systems," NRL Report 8920, Oct. 9, 1985.

*Reference reprinted herein. *See* Readings, Chapter 3.

A Digital Open-Loop Adaptive Processor

FRANK F. KRETSCHMER, JR., Senior Member, IEEE
BERNARD L. LEWIS, Senior Member, IEEE
Naval Research Laboratory

Abstract

A new technique for adaptive processing applications, which is superior to the conventional Applebaum-Howells adaptive loop, is presented. The new technique is based on open-loop digital processing and does not have the limitations of the conventional closed-loop analog processor. In contrast with the conventional adaptive loop, the open-loop processor has effectively infinite gain, is unconditionally stable, and does not depend on the power level of the auxiliary signal.

Manuscript received January 17, 1977; revised September 13, 1977.

Authors' address: Radar Division, Code 5345, Naval Research Laboratory, Washington, DC 20375.

0018-9251/78/0100-0165 $00.75 © 1978 IEEE

I. Introduction

Removal of correlated interference signals by adaptive processing has applications in many areas such as adaptive antennas, medical electronics, pattern recognition, and parameter estimation [1], [2]. The conventional Applebaum-Howells analog adaptive loop, which is often used to remove correlated interference signals from two input channels, requires the weight to be derived in a closed-loop form to accommodate component drifts in the loop. This closed-loop approach does not provide the optimum weight and makes the loop performance and stability dependent upon internal gain and the power level of the auxiliary channel signal. It is the purpose of this paper to describe a new approach based on driftless digital techniques that do not depend on a closed-loop weight derivation. This new approach takes advantage of the fact that the optimum weight is derivable in terms of the input signals.

Conventional analog closed-loop theory is summarized here and the theory of the digital open-loop approach is presented and verified via computer simulations. The results indicate that the open-loop processor is superior to the conventional closed-loop form.

II. Conventional Adaptive Loops

The conventional analog adaptive processor loop is shown in Fig. 1. In this figure G denotes the internal gain, V_m denotes the input signal from the main or primary channel, V_a is the input signal from the auxiliary or reference channel, W is the weighting signal, and V_r denotes the residue signal. These signals represent the complex modulation functions of the actual narrow-band signals. The basic principle of operation of the loop is that a weighting signal W is developed, in a closed-loop fashion, which causes the residue of correlated signal components to be a minimum depending on the value of the loop gain given by $G\overline{|V_a|^2}$, where the overbar is used to denote an ensemble average. From Fig. 1 it is seen that the residue signal may be expressed as

$$V_r = V_m - WV_a. \tag{1}$$

Making the usual assumptions of a slow loop and ergodicity, the steady state weight approaches

$$W = G\overline{V_r V_a^*}. \tag{2}$$

Substituting (1) into (2) yields

$$W = G\overline{V_m V_a^*} - GW\overline{|V_a|^2}$$

which results in

$$W = (\overline{V_m V_a^*}\, G)/(1 + G\,\overline{|V_a|^2}). \tag{3}$$

Now, W in (3) may be expressed as

$$W = W_{\text{opt}}[(G\overline{|V_a|^2})/(1 + G\overline{|V_a|^2})] \tag{4a}$$

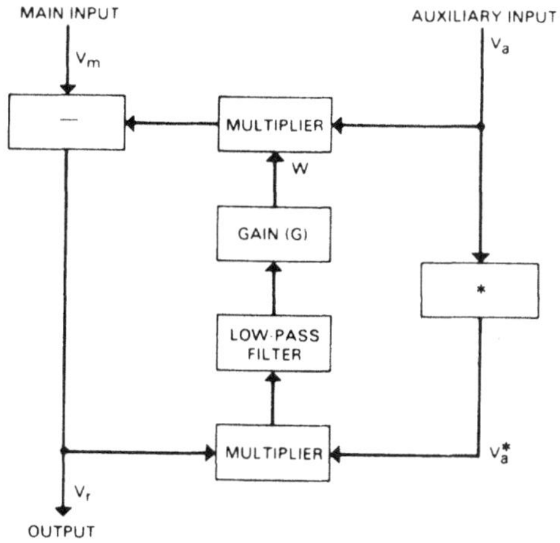

Fig. 1. Basic analog processor.

where

$$W_{opt} = \overline{V_m V_a^*} / \overline{|V_a|^2}. \quad (4b)$$

It is readily shown that the minimum residue power is achieved for the weighting W_{opt} as follows. From (1) we obtain

$$\overline{|V_r|^2} = \overline{(V_m - WV_a)(V_m^* - W^* V_a^*)}$$
$$= \overline{|V_m|^2} - W\overline{V_m^* V_a} - W^* \overline{V_m V_a^*} + |W|^2 \overline{|V_a|^2}. \quad (5)$$

Letting $W = W_{opt} + \Delta$, where Δ is an arbitrary complex value, and substituting (4b) in (5) results in

$$\overline{|V_r|^2} = \overline{|V_m|^2} - W_{opt} \overline{V_m^* V_a} - W_{opt}^* \overline{V_m V_a^*}$$
$$+ |W_{opt}|^2 \overline{|V_a|^2} + |\Delta|^2 \overline{|V_a|^2}. \quad (6)$$

From (6) it is seen that the average residue power is a minimum for Δ equal to zero so that W equals W_{opt}, which is the Wiener weight.

The cancellation ratio of an adaptive loop, defined as

$$\text{CR} = \overline{|V_m|^2} / \overline{|V_r|^2} \quad (7)$$

is next derived in terms of the normalized correlation coefficient ρ given by

$$\rho = \overline{V_m V_a^*} / [(\overline{|V_m|^2})^{1/2} (\overline{|V_a|^2})^{1/2}]. \quad (8)$$

Letting

$$k = (G\overline{|V_a|^2})/(1 + G\overline{|V_a|^2}) \quad (9)$$

the weight of an adaptive loop given by (3) can be expressed in terms of ρ in (8) as

$$W = \rho[(\overline{|V_m|^2})^{1/2}/(\overline{|V_a|^2})^{1/2}]\ k. \quad (10)$$

From (5), (8), and (10) it can be shown that

$$\overline{|V_r|^2} = \overline{|V_m|^2}\ (1 - 2|\rho|^2 k + |\rho|^2 k^2). \quad (11)$$

From (7) and (11) the cancellation ratio is given by

$$\text{CR} = (1 - 2|\rho|^2\ k + |\rho|^2\ k^2)^{-1}. \quad (12)$$

For the case where $|\rho|^2 = 1$, corresponding to perfectly correlated signals, one obtains from (12) and (9)

$$\text{CR} = (1 + G\overline{|V_a|^2})^2. \quad (13)$$

III. Digital Open-Loop Adaptive Processor

From (4a) it is noted that the conventional adaptive loop is limited by the loop gain in achieving the minimum residue provided by W_{opt}. The feedback is important in analog circuitry to compensate for component drifts. In digital implementations, however, this problem is not present and the optimum weight can be determine directly in an open-loop manner from the input signals. For optimum weighting, k in (12) is effectively unity and the cancellation ratio is a maximum.

In the digital application, the ensemble average indicated in (4b) and (5) is taken to mean a finite time sample average. It is then seen from (5) that W_{opt} minimizes the finite sample average residue for any V_m and V_a. The assumption of a fixed W_{opt} over the averaging window in (5) corresponds to batch processing. The simulation results which are subsequently discussed in this paper were derived by computing W_{opt} from sliding window averages. Thus the jth sampled residue was computed from

$$V_r(j) = V_m(j) - W_{opt}(j)\ V_a(j) \quad (14)$$

where, for an n-sample window size,

$$W_{opt}(j) = \sum_{k=j-n+1}^{j} V_m(k)\ V_a^*(k) / \sum_{k=j-n+1}^{j} |V_a(k)|^2. \quad (15)$$

Several simulations were run to compare the batch and sliding window methods and little difference was observed in the residue power averaged over a sample size much larger than the window or batch size.

From (15) it is noted that in the case of a single sample, W_{opt} becomes V_m/V_a. Substituting this in (14) results in perfect cancellation irrespective of V_m and V_a. This suggests that more samples be averaged to prevent cancellation of a valid target signal. This aspect is pursued in Section IV.

An illustrative implementation is shown in Fig. 2 for a two pulse average. The delay time shown in the figure corresponds to one sampling interval where it is assumed that successive samples are uncorrelated and each sample consists of both the inphase (I) and quadrature (Q) components.

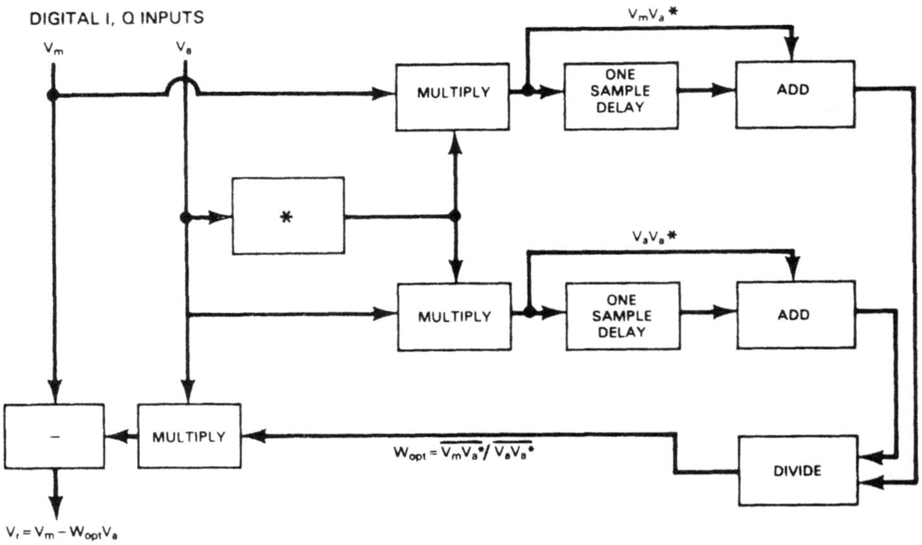

Fig. 2. Basic digital open-loop processor.

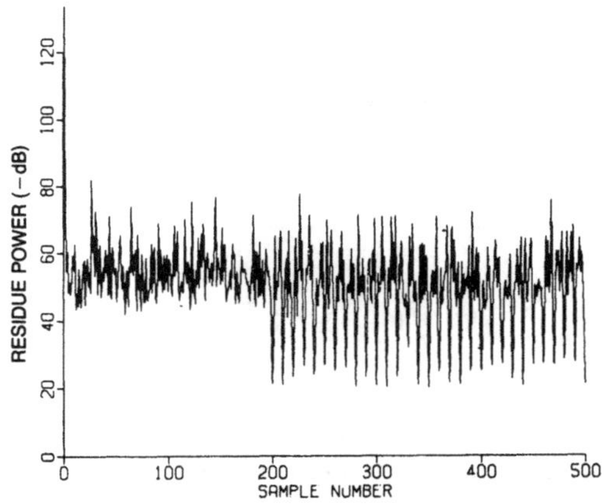

Fig. 3. Digital processor performance (NAVG = 2, N = -50 dB, S_m = -20 dB, $N_a = N_m$, S_a = 0).

IV. Performance Simulation

Computer simulations were run for the optimum weight previously described using different averaging window sizes. The input correlated interference signal was simulated by a step input of Gaussian samples corresponding to random noise modulation. Each sample generated in the main channel was phase-shifted by an arbitrary 90° for the corresponding auxiliary channel sample. The amplitude of the auxiliary channel sample was taken the same as, or 20 dB higher than, the main channel counterpart. In the simulations, successive samples were statistically independent in each channel but correlated from channel to channel as described above. Also, Gaussian samples were generated which were independent of the interference samples to represent uncorrelated thermal noise in the main and auxiliary channels. Successive noise samples in each channel were also independent. Desired signals were simulated by injecting a constant signal value into the main or both channels at selected sample numbers.

In the simulations, various numbers of pulses were averaged (NAVG) for W_{opt} and average residue powers were computed. The simulation results are expressed in terms of the sampled residue power normalized to the average input correlated noise power level N_m in the main channel. Thus the 0-dB level in the simulation results corresponds to N_m and the residue sample power shown in the ensuing plots is in terms of the number of decibels below N_m. The first 200 input samples in the simulations contain only correlated interference and thermal noise, while desired signals were injected at preselected points throughout the next 300 points. The normalized residue power was averaged over the first region excluding the transient part and over the second region at the points where desired signals were injected. These averages, expressed in decibels, are denoted by C and CS, respectively. The negative of C also corresponds to the cancellation ratio as defined in (7). A figure of merit, similar to the improvement factor used in moving target indication (MTI), was also calculated and is defined herein as the ratio of the average signal-to-interference ratio at the adaptive processor output and input. This improvement factor Q may be expressed as

$$Q = (S/I)_{out}/(S/I)_{in} = -C + CS - [(S/I)_{in}] \text{ dB}. \quad (16)$$

The interference consists of correlated noise and uncorrelated thermal noise. The thermal noise is insignificant compared with the input correlated noise level in the simulations so that $(S/I)_{in}$ is approximated by $(S/N_m)_{in}$.

Results are presented in Figs. 3, 4, and 5 for an input thermal noise level N, 50 dB below N_m and for NAVG equal to 2, 4, and 8, respectively In these figures, a desired signal S_m

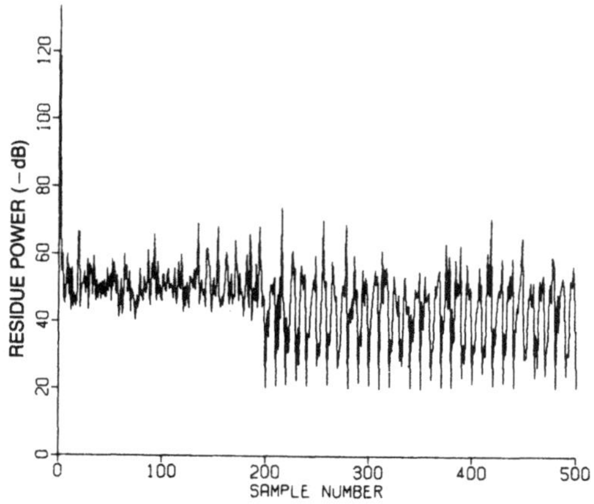

Fig. 4. Digital processor performance (NAVG = 4, N = -50 dB, S_m = -20 dB, $N_a = N_m$, S_a = 0).

Fig. 5. Digital processor performance (NAVG = 8, N = -50 dB, S_m = -20 dB, $N_a = N_m$, S_a = 0).

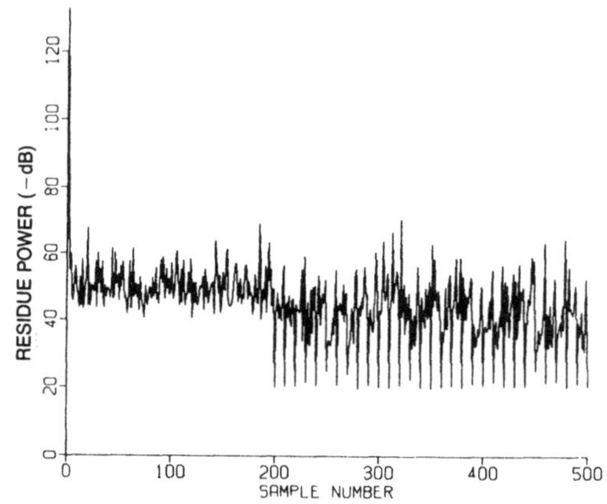

was injected in the main channel at a level 20 dB below N_m beginning at sample number 200 and recurring every 10 samples thereafter. Prior to sample number 200, correlated and thermal noise alone are present. The correlated interference levels N_m and N_a in the main channel and the auxiliary channel were the same and the desired signal S_a in the auxiliary channel was zero.

Several general observations can be made from these figures. First, it is observed that the adaptive digital processor cancels the correlated interference down to the thermal noise level except in the regions where desired signals were injected. This is because of the effective infinite gain associated with the processor which enables it to completely cancel correlated signals between the main and auxiliary channels. Since the thermal noise is uncorrelated between these channels, the thermal noise level represents the floor of the residue except where the desired signal (also an uncorrelated signal) appears. Another general observation is that as NAVG increases, the desired signal is less likely to be reduced by the processor. If no cancellation of the desired signals took place, the signal points would be at the -20-dB level in the plots. It is seen in Figs.

TABLE I

Performance Results Using W_{opt}

	NAVG		
	2	4	8
C	-50.5	-48.2	-47.5
CS	-23.7	-21.8	-20.9
Q	46.8	46.4	46.6

3, 4, and 5 that the amount of a signal cancellation reduces as NAVG increases. In these simulations, computations of the average residues C and CS were made in the first region where no desired signals were present and in the second region at the points where the desired signals were injected. The first average was obtained from 175 points which excluded the transient part of the response. The desired signal average was determined from 30 samples. The results are tabulated in decibels in Table I along with the processing improvement factor Q.

From Table I it is observed that the interference residue power increases as more pulses are averaged but that the target cancellation decreases so that the improvement factor remains fairly constant.

Referring to Figs. 3, 4, and 5, it is observed that very high cancellation occurs for the first sample of the input step of interference occurring at sample number 1 on the abscissa. This aspect is discussed next. The averaging for W_{opt} consists of a sliding window average as may be seen from the implementation shown in Fig. 2. In general, the average is taken over NAVG input samples including the present value except during the transient condition. Referring to Fig. 2, it is noted that the first value of W_{opt} for a step input is V_m/V_a, which is the exact weight for perfect cancellation. The next weight consists of a sum of two successive values of $V_m V_a^*$ divided by the sum of two successive values of $|V_a|^2$. This process continues until NAVG samples are contained within the window. Thus the optimum weight processor is able to immediately cancel a correlated interference source turning on without experiencing a transient which results in a fringing effect in the conventional closed-loop processor.

The transient response arising after the application of a desired signal in the main channel is next addressed. Referring again to Figs. 3, 4, and 5 it may be seen that following the application of each desired signal there is a transient recovery time in cancellation of the correlated interference. The transient may be observed in greater detail in Figs. 6, 7, and 8 corresponding to NAVG equal to 2, 8, and 32, respectively. In these simulations the thermal noise was at the -100-dB level and the desired signals were injected at intervals of 50 samples beginning with sample number 200 to better illustrate the transient effects. These figures also demonstrate again the cancellation down to the thermal noise level. From Figs. 6, 7, and 8 it is observed that the total recovery span following a desired signal input point is the same as the averaging window or NAVG samples. Although the desired signal is only present for one sample at

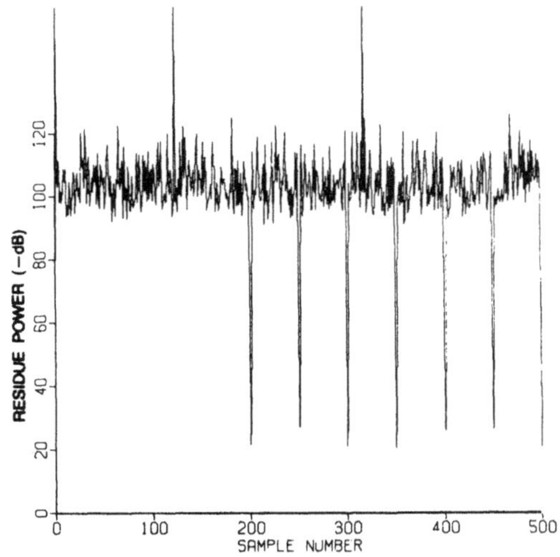

Fig. 6. Digital processor performance (NAVG = 2, N = -100 dB, S_m = -20 dB, $N_a = N_m$, S_a = 0).

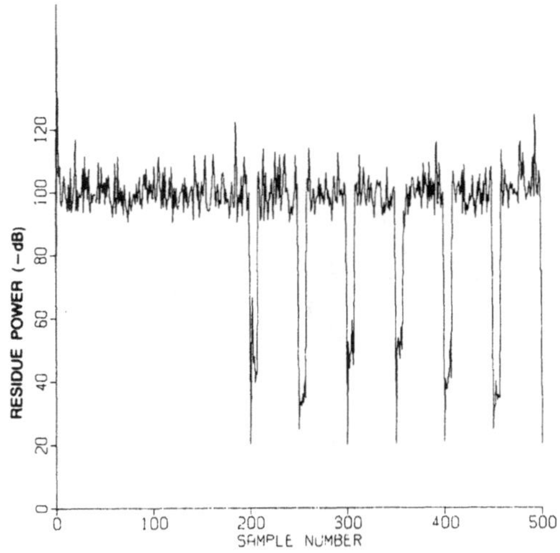

Fig. 7. Digital processor performance (NAVG = 8, N = -100 dB, S_m = -20 dB, $N_a = N_m$, S_a = 0).

Fig. 8. Digital processor performance (NAVG = 32, N = -100 dB, S_m = -20 dB, $N_a = N_m$, S_a = 0).

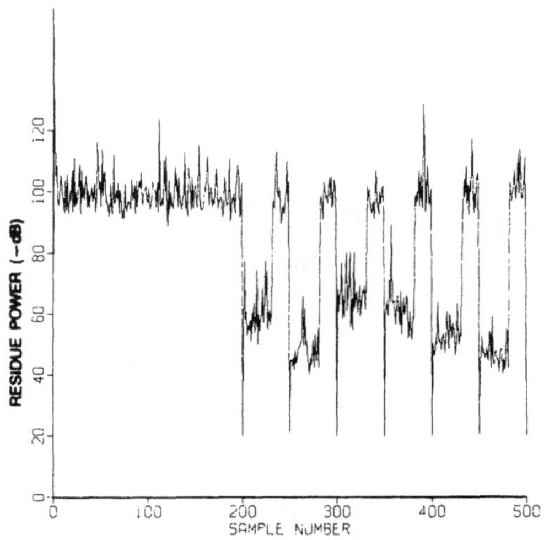

a given point of injection, the influence of the desired signal is retained in the sliding window for the window duration. The back porch or transient response is seen to diminish in level as NAVG increases. The weight perturbation of W_{opt}, caused by the presence of the desired signal in the window, is of the form

$$\delta W = S_m V_{ai} / \sum_{k=1}^{NAVG} |V_{ak}|^2, \qquad 1 \leqslant i < NAVG$$

where V_{ak} is the kth auxiliary channel sample in the averaging window and S_m is the main channel signal sample corresponding to the ith sample V_{ai} in the auxiliary channel. The value of $|\delta W|^2$ reduces as the square of NAVG which is shown in the simulations as a reduction in the porch level by the same amount as NAVG increases. The back porch is seen to be capable of masking a weaker signal located within the window. For this reason, and because of improvement factor and implementation considerations, a digital processor having a shorter averaging time of 2 or 4 pulses is preferable. It is mentioned here that the porch can be removed if desired by additional digital processing and that the porch effect is also present in conventional analog loops.

In Fig. 9, results of a simulation are shown for the case where the auxiliary channel correlated noise level was 20 dB higher than in the main channel and NAVG equaled 4. The results are basically the same as those shown in Fig. 4, which demonstrates that the cancellation of correlated signals is independent of the power level in the auxiliary channel. It should be noted that this differs from the conventional closed-loop processor where the cancellation ratio given by (13) is dependent on the auxiliary signal power level. Calculation of C indicated a 3-dB decrease in the residue as compared to the case where the correlated interference is the same in the main and auxiliary channels. This was due to the required weighting being 20 dB lower, which precluded adding thermal noise from the auxiliary to the main channel.

Simulations were run for the case where the signal S_a was also present in the auxiliary channel 10 dB above the noise level or 40 dB below N_a. Since N_a was much greater than the signal level, the processor weight was not altered by the signal in the auxiliary channel. The results were nearly identical to those where S_a equaled zero.

It has been found for the case of M auxiliary inputs that good results are obtained by cascading M open-loop processors. Each processor uses one of the auxiliary signal inputs and uses the residue of the previous processor as its main input. This processing is repeated by cascading another M processors and reusing the original M auxiliary signals. Several iterations are generally adequate for good cancellation. In low bandwidth applications, the number of processors can be substantially reduced by commutation and storage of intermediate results.

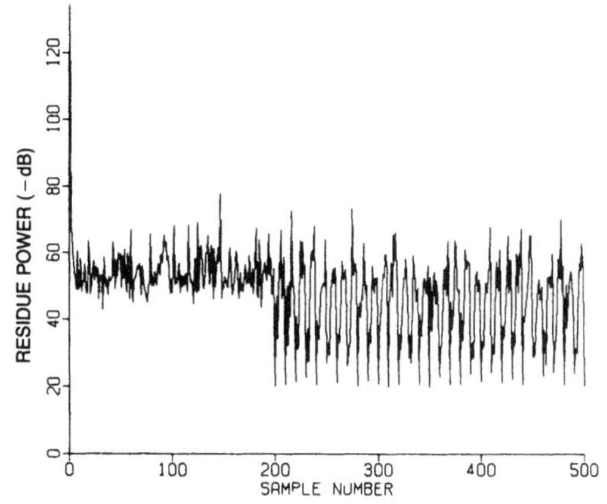

Fig. 9. Digital processor performance (NAVG = 4, N = -50 dB, S_m = -20 dB, N_a/N_m = 20 dB, S_a = 0).

V. Summary

A new technique of open-loop digital processing has been described. It was indicated that the closed-loop operation found in conventional analog adaptive processors, to compensate for component drifts, is unnecessary in the digital implementation. It was also shown that the conventional analog processor is restricted by the loop gain and instability in developing the optimum weight. In contrast, the digital processor technique determines the optimum weight from the input signals in an open-loop manner. The effective loop gain of the digital processor is infinite, and stability considerations are completely eliminated. It was shown that the optimum weighting used by the digital processor allows cancellation of correlated interference to the thermal noise level. In practice this is achievable with a sufficient number of bits in the A/D converters.

The performance for the optimum weighting as a function of NAVG was compared by a figure of merit referred to as the processing improvement factor Q, which was defined as the ratio of the average signal-to-interference ratios at the processor output and input. The improvement factor calculations indicated the excellent performance obtainable with averaging only a few pulses in the optimum weight processor. This in turn impacts on the ease of implementation and on the desired signal resolution. A transient recovery time lasting over the averaging window was shown to exist following a desired signal which permits masking of small signals colocated within the window. This also occurs in conventional analog adaptive processors, but the ability to achieve good cancellation with a small window in the digital canceller mitigates this effect. Moreover, it was shown that the digital processor using the optimum weight is capable of immediately cancelling a step input of correlated interference. On the other hand, a transient build-up time is required for the conventional analog processor.

It is concluded that the new digital adaptive technique using the optimum weight derived in an open-loop manner from the input signals provides a means of obtaining improved performance in reducing correlated interference compared to conventional analog processors which rely on closed-loop adaptation.

References

[1] B. Widrow et al., "Adaptive noise cancelling: Principles and applications," *Proc. IEEE,* vol. 63, pp. 1692-1716, Dec. 1975.

[2] (Special Issue on Adaptive Antennas), *IEEE Trans. Antennas Propagat.* vol. AP-24. Sept. 1976.

UNDERSTANDING AND OPTIMIZING MULTIPLE SIDE-LOBE CANCELER OPERATION

INTRODUCTION

The conventional mathematical technique of analyzing the operation of multiple side-lobe cancelers on multiple jammers does not provide a clear insight into the actual processes involved, nor does it automatically suggest methods for optimizing the canceler performance in practical situations. As a consequence, a simple vector technique for analyzing such loops has been developed that provides a lucid picture of the processes involved and offers help in optimization.

This technique allows an analyst to determine easily what the residue of multiple loops will be under any given geometry of jammers and auxiliary antennas through a complete 360° scan of the radar antenna if desired. It then allows him to determine the effect of repositioning the auxiliary antennas or tells him that he must add more antennas and loops to obtain the desired performance.

The technique is based on the fact that side-lobe canceler loops are linear, and when operating in groups the operation of any loop is independent of the operation of all other loops to a first approximation. This allows an analyst to pick any loop to start his analysis and to ignore the operation of all others while he completes his analysis of the first loop. In this way he can analyze one loop after another, with each successive loop operating on the residue of those already analyzed, and thus determine an initial approximation of the final residue produced by the group operating together. The approximation can be made much more accurate by repeating the process, using this initial result as the radar input to the cancelers.

With practice, the analyst can obtain a feel for the problem that will allow him to design an optimum arrangement of auxiliary antennas to protect adequately a radar from any given number of jammers.

This technique has been tested experimentally and found to be valid and valuable. It reveals that the common belief that *n* loops can adequately cancel the signals from *n* jammers is false when the jammer geometry is variable and the auxiliary antennas are omnidirectional and fixed in space. It explains why poor cancellation can be obtained from a given system with certain jammer geometries and good cancellation can be obtained with other jammer geometries without invoking multipath effects as an excuse for the poor performance.

It is hoped that, because the technique is so easy to use, it will prove valuable both to engineers designing systems and to customers buying them.

BACKGROUND

Side-lobe canceler loops are usually composed of an auxiliary omnidirectional antenna and a feedback loop that uses a sample of the jamming environment to subtract jamming signals from a radar receiver. In this process, it is assumed that the spacing between the omnidirectional antenna and the radar antenna is small compared to the autocorrelation distance of any jamming signal to be considered. In this way, any jamming signal picked up by the auxiliary antenna can be considered to be correlated with its counterpart picked up by the radar antenna. This permits the relative radio-frequency phase and amplitude of the auxiliary antennas' signal to be specified with respect to its counterpart picked up by the radar antenna, independently of the spectral characteristics of the jamming signal.

DEMONSTRATION OF VECTOR TECHNIQUE

The vector method of analyzing the operation of multiple canceler loops on the signals from multiple jammers can be demonstrated as follows.

Figure 1 shows a scanning serach radar antenna with two omnidirectional auxiliary antennas placed symmetrically about the radar antenna. The two auxiliary antennas are shown driving two coherent side-lobe canceler loops. The phase center of the radar antenna is the chosen reference point for defining the phase angle of all signals incident at the three antennas. The wavefronts from two jammers are shown incident upon the antennas from two different directions. One signal is represented by a circle on the line and the other is represented by a square.

The average rms magnitudes and the phases of the two jamming signals received at the auxiliary antennas with respect to their counterparts received by the radar antenna can be graphically represented by vectors as illustrated in Fig. 2. It is obvious that "circle" signals will lag and "square" signals will lead their radar counterparts in phase at auxiliary antenna 1, since the wavefront of the circle has not arrived at antenna 1, whereas that of square has already passed antenna 1 when the wavefronts arrive at the radar antenna phase center. Similarly, circle will lead and square will lag their counterparts in the radar antenna in auxiliary 2 as illustrated.

Radar antenna scan can be accounted for by controlling the magnitudes of the vectors in the radar antenna and inverting one or the other as they move from side lobe to side lobe. In Fig. 2, the two jamming signals were arbitrarily set equal in magnitude at the auxiliary antennas. They could just as easily have been made unequal. However, their resultants in the radar antenna were made unequal to demonstrate the directive nature of the radar antenna.

The signals picked up by the radar antenna are frequency translated by ω_1 from a local oscillator and pass through adder 1 and adder 2. The output of adder 2 enters multiplier 2 where it is multiplied by the jamming signals picked up by auxiliary antenna 1. The difference frequencies formed in this process are passed through a narrowband amplifier centered on ω_1. The multiplier and narrowband amplifier act like a correlator and produce outputs only when the inputs to the multiplier are correlated.

Mathematically, this operation can be demonstrated by letting circle be $A \cos(\omega t + \theta)$ in auxiliary 1 and $B \cos \omega t$ in the radar antenna. The output of adder 2 will then be

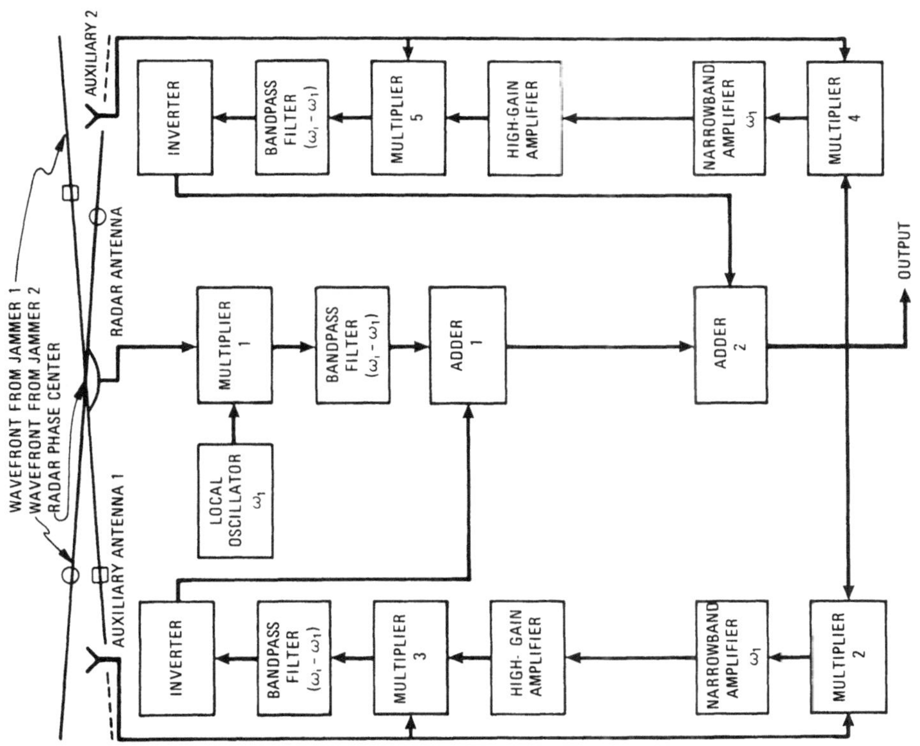

Fig. 1—Starting point in applying the vector technique

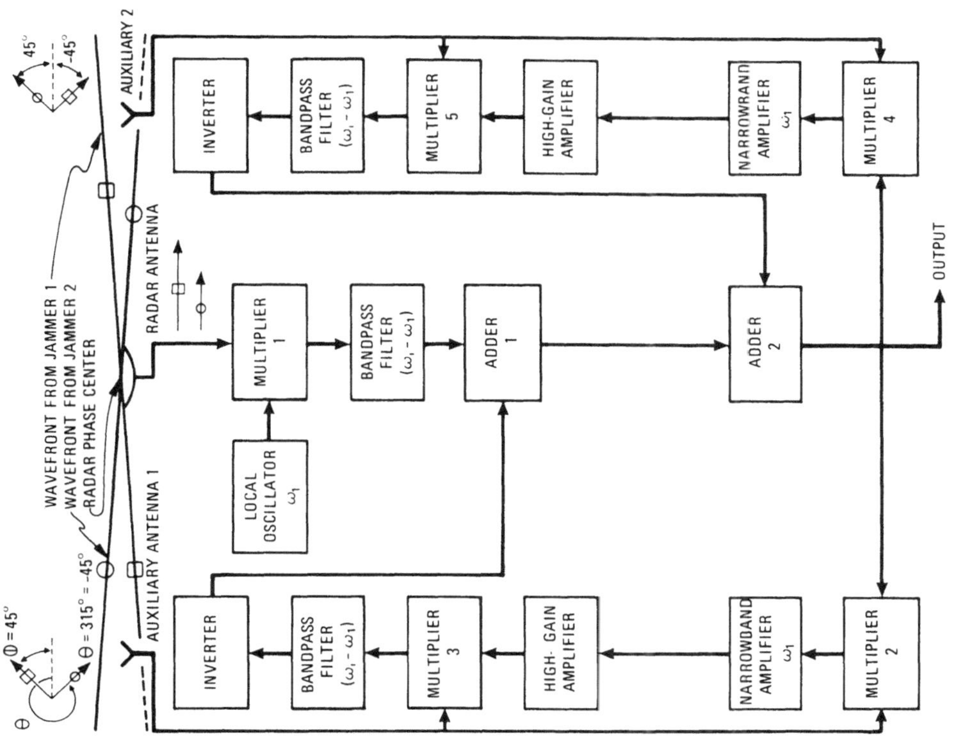

Fig. 2—Definition of phase and amplitude of the vectors at the antennas

$KB \cos(\omega - \omega_1)t$, and the output of the narrowband amplifier following multiplier 2 will be $GKAB \cos(\omega_1 t + \theta)$, where G is a gain function greater than unity.

The output of the narrowband amplifier for each correlated signal can be represented vectorially as illustrated by the dotted vectors in Fig. 3. The magnitude k and phase α of the resultant can also be portrayed graphically since all vectors are correlated at this point. This initial resultant will be labeled F_1^1 and has been called a weighting function. F_1^1 enters multiplier 3 where it multiplies the jamming signals picked up on auxiliary antenna 1. The resultant lower sideband is passed by a bandpass filter closely matched to that following multiplier 1.

The result of this process will be the auxiliary jamming signals rotated in phase by $-\alpha$ and having magnitudes proportional to k (Fig. 4). These vectors are inverted and added to the radar input to adder 1 to form the output of canceler 1 at the output of adder 1 as illustrated in Fig. 5. These vectors now become the input to multiplier 2 by passing adder 2, and the new output of the narrowband amplifier is portrayed by solid vectors. Note that the resulting phase of F_1, the new resultant, is rotated toward zero and its magnitude is greatly reduced through cancellation of the two correlated components. This cancellation permits high loop gain to be employed without saturating the narrowband amplifier when the loop is closed.

It should be noted that, in this case, the amplitudes and phase relationships of the residues with respect to their counterparts in auxiliary antenna 1 are specified by the necessity for reducing F_1 by the amount required by the loop gain. If this gain is high, the phase of square signals out of the narrowband amplifier $(\phi - \gamma)$ must approach 180° with respect to circle whose phase will be $\theta - \delta$. Thus, the phase angle between the residue vectors must be such that $\phi - \gamma + 180° = \theta - \delta$ or $\gamma - \delta = \phi - \theta + 180°$. Putting in the assumed values for ϕ and θ yields $\gamma - \delta = 45° - 315° + 180° = -90° = 270°$. Since the phase angle between the two residue vectors is $360° - (\gamma - \delta)$, this phase angle can be seen to be $360° - 270° = 90°$.

The necessity for reducing F_1 to a small value also implies that the magnitudes of circle and square out of the narrowband amplifier must be nearly equal as well as nearly 180° out of phase. Since the signals in auxiliary antenna 1 were chosen equal, this means that circle and square residues out of adder 1 must be nearly equal.

These residues can now be considered the inputs to the second loop at multiplier 4 (Fig. 6). They will produce an initial weighting function F_2^1 as shown by the dotted vectors. This weighting function multiplied by the jamming signals in auxiliary antenna 2 will produce the vectors shown at the output of the bandpass filter following multiplier 5. These vectors are inverted and added to the residue of adder 1 in adder 2 to cancel the residue in the assumed case, since the angle between the residue vectors from loop 1 equals that between the vectors on antenna 2. Note that the residue decrease in the second loop reduces the output of the narrowband amplifier to produce the final weighting function F_2 shown by the solid vector. It should be noted at this point that cancelers can only rotate all vectors from their auxiliary antenna by a given angle α, control all magnitudes by the same gain function k, and subtract the resultant vectors from the radar antenna's signals.

The utility and power of this technique can now be illustrated by showing that two loops cannot always cancel two jamming signals. This is done in Fig. 7 where the phase angle of one jamming signal picked up on one auxiliary leads in phase by 90° its

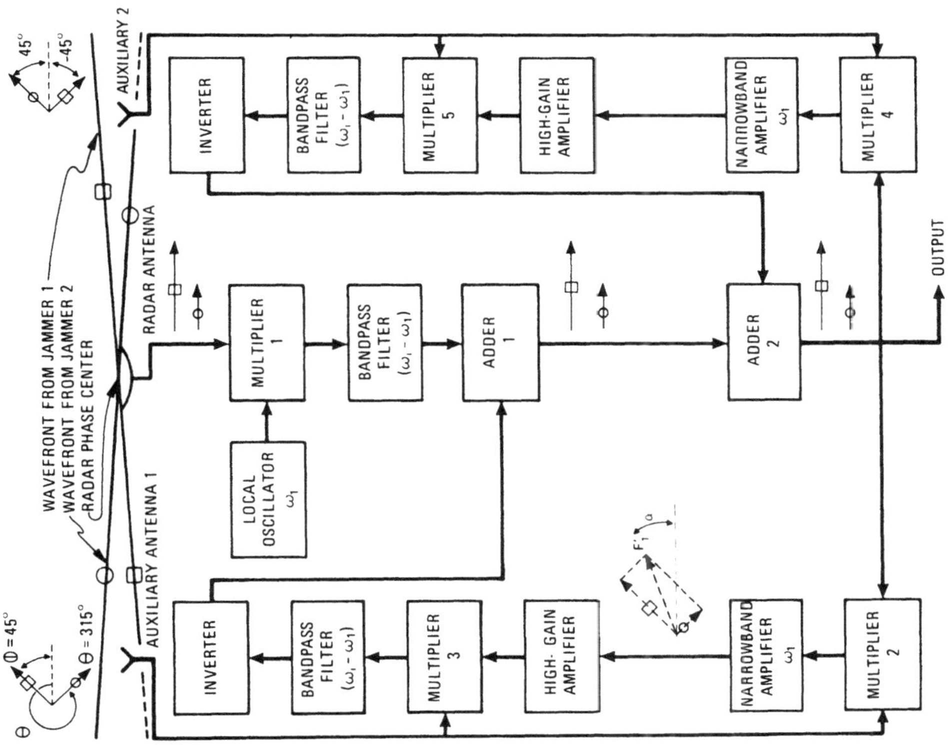

Fig. 3—Derivation of initial weighting function in loop 1

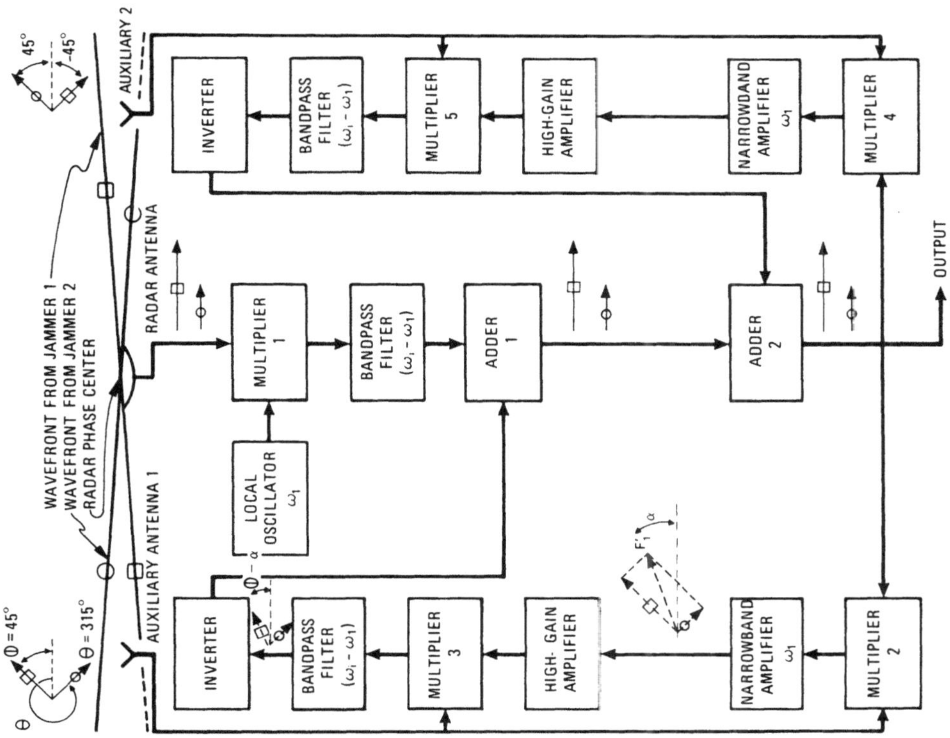

Fig. 4—Translation of auxiliary signals

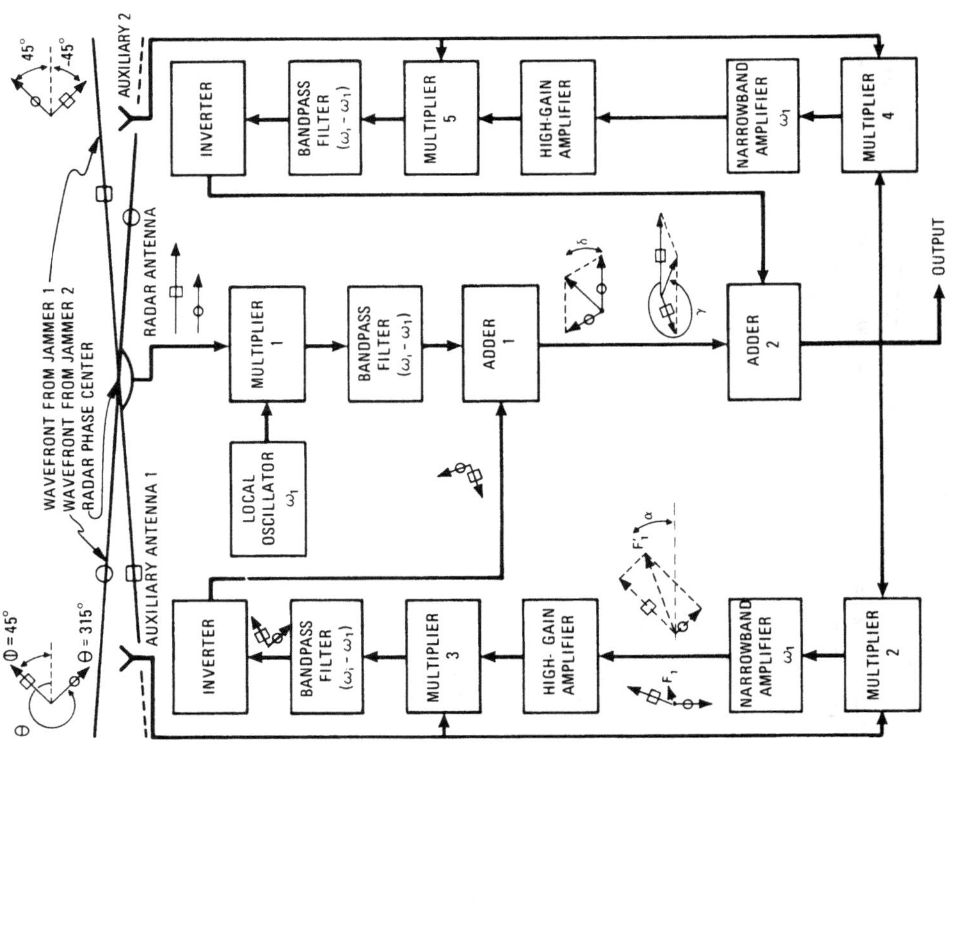

Fig. 5—Derivation of final weighting function in loop 1

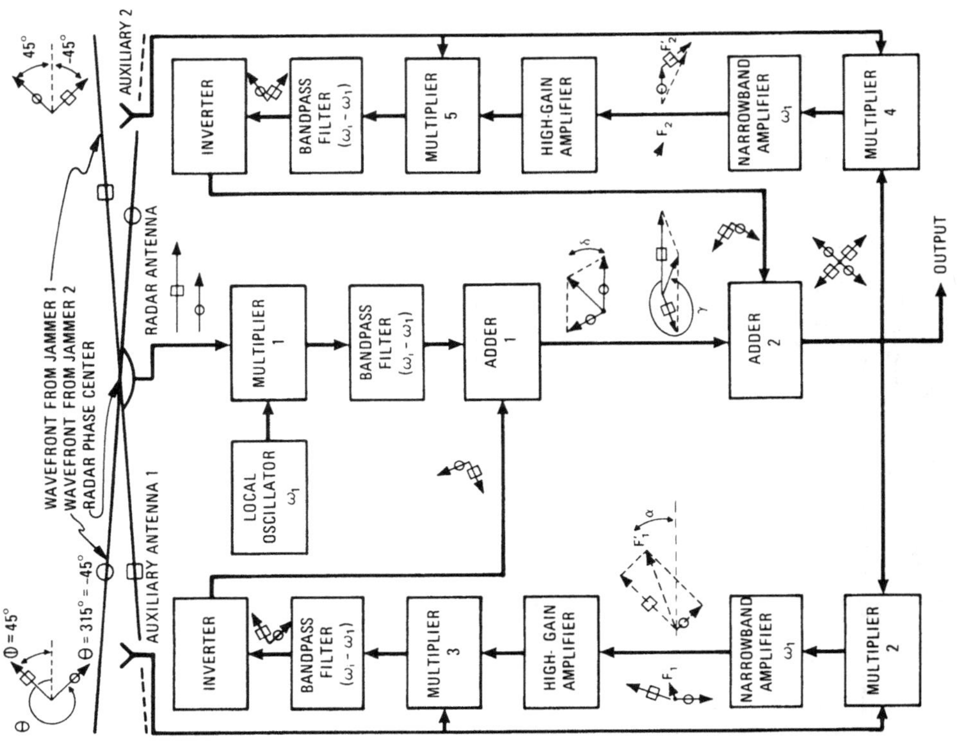

Fig. 6—Demonstration of vector technique for analyzing multiple canceler operation on the signals from multiple jammers

counterpart in the radar and the other lags its counterpart by 90°. With symmetrical auxiliary spacing, the inverse will hold true on the other auxiliary. In this case, the initial weighting function F_1^1 has a phase angle of 90°. This subtracts from the phase angles of the signals in auxiliary 1 to produce a square with 0° phase and a circle with 180° phase with respect to their counterparts in the radar antenna. Inverting these signals and adding them to the radar signals in adder 1 results in square canceling and circle adding. This produces a final weighting function F_1 such that with high loop gain, the residue consists of nearly equal parts of square and circle with magnitudes intermediate between square and circle in the radar antenna (the loop adds circle to circle and subtracts square from square). The second loop will form a weighting function of the same magnitude but opposite polarity as the first, but it cannot be shown since it would be so small. (Note that function F_1 is shown larger than actual size for demonstration purposes only.) This will translate the signals from auxiliary 2 so that they aid the first loop in canceling square and adding to circle in adder 2. In trying to help loop 1, loop 2 effectively increases its gain but cannot significantly change the residue, since infinite gain would only make the residue of square and circle equal and opposite in phase at the outputs of the narrowband amplifiers.

Since circle is not correlated with square at the output of adder 2, however, the jammer power output of adder 2 will be the sum of the powers of the two residues. In this way, the vector technique reveals that two loops cannot always handle two jammers and it provides information about the magnitude of the residue.

It should be noted that radar antenna scan might invert the sense of one of the vectors in the radar antenna in the situation illustrated in Fig. 7. If this happens, better cancellation will be obtained as illustrated in Fig. 8. Again, in this case, the second loop cannot contribute significantly.

It should also be noted that, if the jamming signals in the radar antenna in Fig. 7 had been equal, neither loop would have been able to form a weighting function, and no cancellation would have been obtained. This condition is illustrated in Fig. 9.

Experimental Confirmation

An analog simulator of the block diagram illustrated in Fig. 1 was constructed. The antennas were adders, and two uncorrelated thermal noise sources (balanced, modulating, variable-phase, uncorrelated carriers) were used as the jamming signal sources. The carrier phase and the total signal average rms amplitude of each signal input to each adder simulating the antennas were variable to permit any desired condition to be established. The operation of this simulator confirmed the validity of the vector technique for estimating multiple canceler operation on multiple jammers.

Uses of Vector Technique

The simulator verification of the vector technique encourages its use in obtaining an understanding of multiple-loop operation in expected conditions. For example, the effect of the jammer location, the effect of radar antenna scan, the effect of varying the spacing of the auxiliary antennas, and the effect of rotation of the radar antenna about a point not located at its phase center can be determined. This knowledge can then be used to specify how many and what locations should be filled with auxiliary antennas in order to optimize the canceler

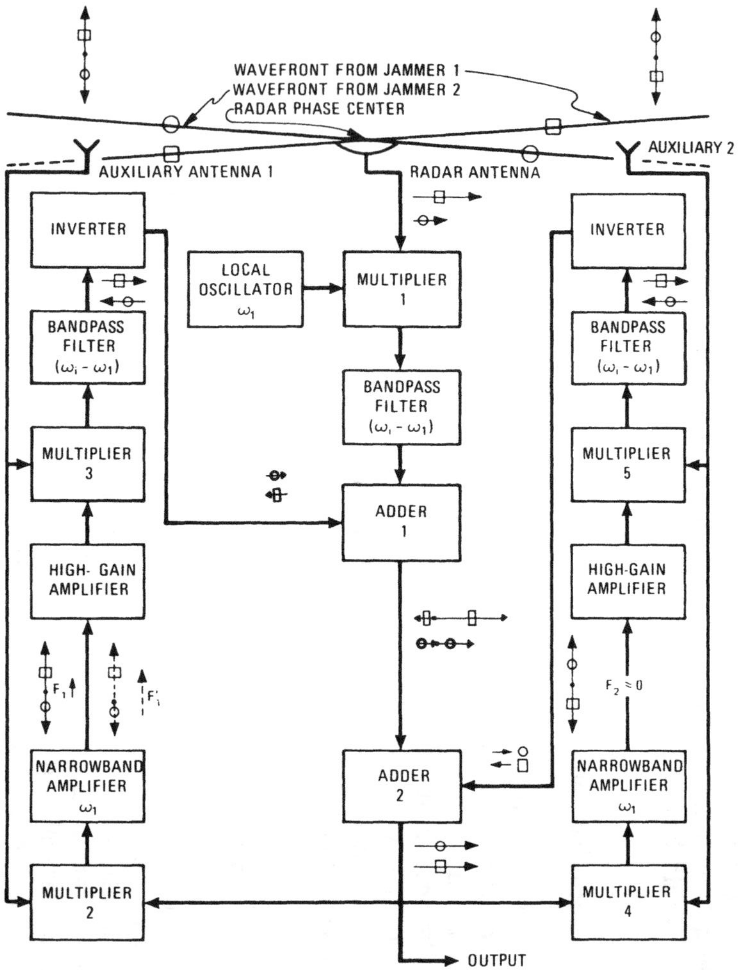

Fig. 7—Vector technique used to show a condition in which two cancelers offer poor cancellation of the signals from two jammers caused by antenna scan

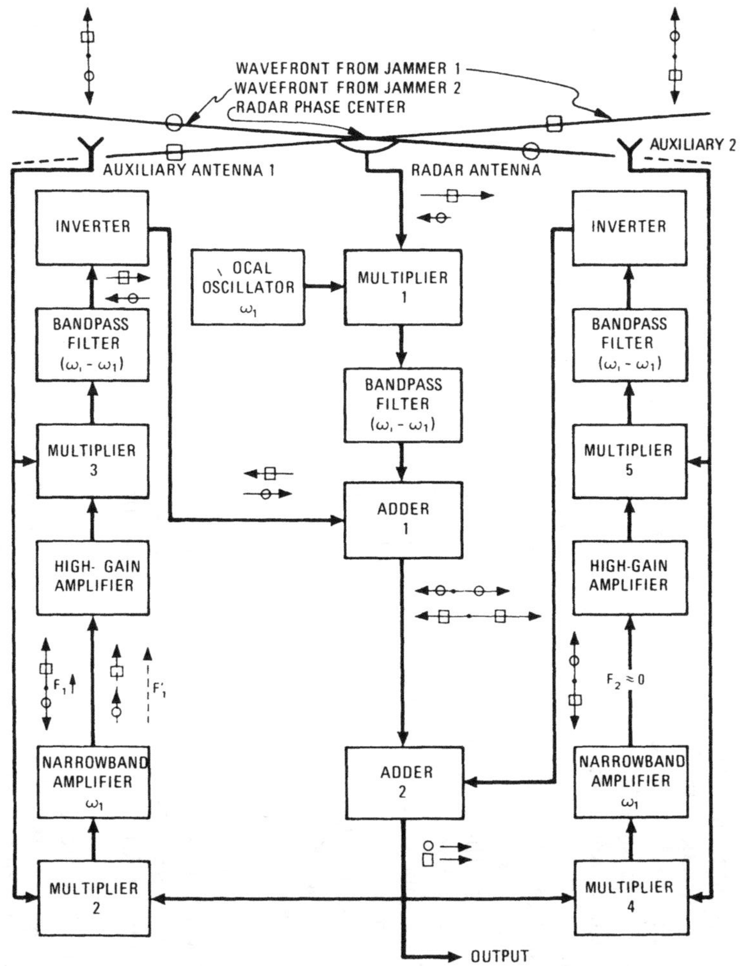

Fig. 8—Effect of phase inversion of one of the radar components in Fig. 7

operation. The effect of radar antenna scan about its phase center with symmetrical auxiliary antenna locations is analyzed in the next section as an example.

EFFECT OF RADAR ANTENNA SCAN

Radar antenna scan about its phase center will cause the relative magnitudes of the signals from two jammers at different locations to vary independently and will invert the sense of each jamming signal when the jammer moves from one side lobe to the next. An idea of the effect of this variation on multiple cancelers can be obtained as follows.

FIgure 10 illustrates one case in which one jammer (circle) is near a null of the radar antenna pattern and the signals on the auxiliary antennas are in phase. In this case, F_1^1 is produced by square alone. However, F_1^1 multiplies both square and circle from auxiliary antenna 1. The inverter thus provides two signals to adder 1 while the radar only provides one. The output of adder 1 becoming the new input to multiplier 2 produces an F_1 that is the difference of the two correlating components as illustrated. With high loop gain, the output of adder 1 will approach equal values of circle and square with a total power equal to half the power of the single component in the radar antenna. The half-power condition is evident by virtue of the residue being composed of two uncorrelated vectors, each having one-half the magnitude of the vector in the radar antenna. Each component thus has one-fourth of the power, which results in their sum having one-half the power. The output of adder 1 is now the input to multiplier 4 in the second loop. However, since these vectors are nearly equal and opposite, the second loop cannot form more than a token weighting function. As a consequence, it cannot contribute significantly to adder 2 and the output of adder 2 will remain that out of adder 1 with loop 1 gain increased by 2. In this case, the two loops will provide only a 3-dB cancellation ratio.

Figure 11 illustrates the action of the loop when the signals on the auxiliary antennas are in phase and equal in magnitude, while the signals in the radar antenna are 180° out of phase due to the jammers' being on side lobes oppositely sensed but equal in magnitude. Note that this condition prevents both loops from developing a weighting function, and no cancellation is obtained.

Figure 12 demonstrates the equal-amplitude, equal-phase case in which good cancellation is obtained by one loop, and the other does nothing but double the gain of the first.

The trend that falls between equality of signals and one signal missing in the radar antenna is illustrated for the in-phase condition in Fig. 13 and the out-of-phase condition in Fig. 14. In the in-phase case, Fig. 13, the residue of each signal has a magnitude equal to one-half the difference in magnitude of the signals in the radar antenna, and the second loop does little. In the out-of-phase case, Fig. 14, the residue of each signal has a magnitude equal to the magnitude of the smaller signal plus one-half the difference between the two signal amplitudes in the radar channel. In this case, also, the second loop does little.

It should be noted that jammer geometries that allow one jammer to enter a null on the radar antenna while the other jammer is near a side-lobe peak are bad cases if the signals from each jammer on the two auxiliaries are in phase. It should also be noted that jammer geometries in which the jamming signals arrive at both auxiliary antennas with 90° lags and leads are also bad. Good jammer geometries are those in which the jamming signals in the auxiliary antennas lead and lag by 45° in one auxiliary and lag and lead by 45° in the other. The latter case is illustrated in Fig. 6 and Figs. 15 and 16 which illustrate

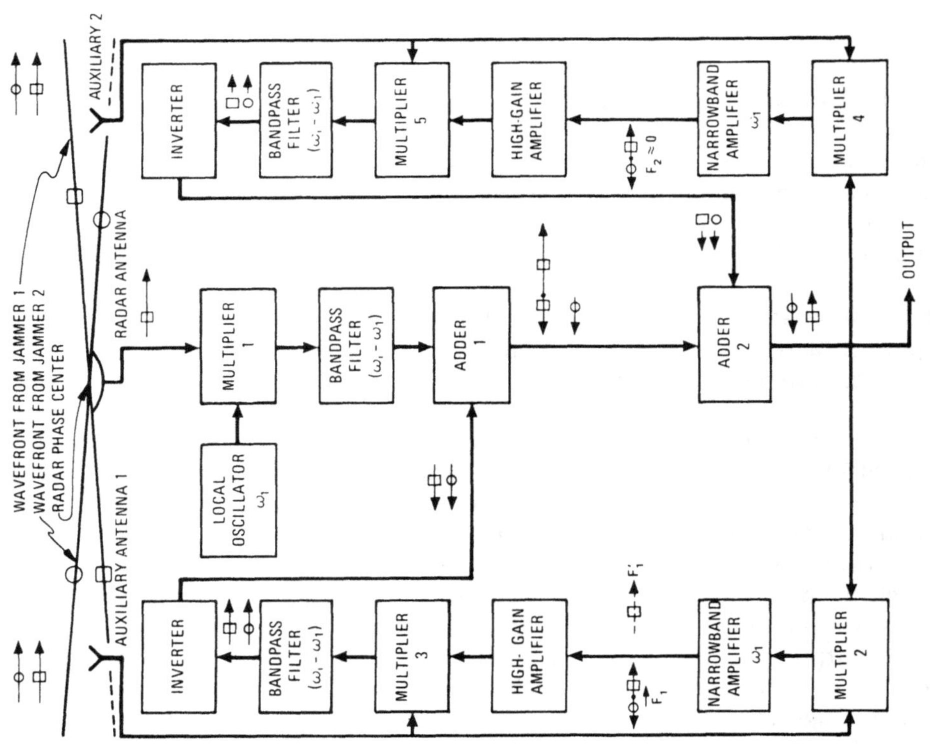

Fig. 10—One jammer in a radar antenna pattern null

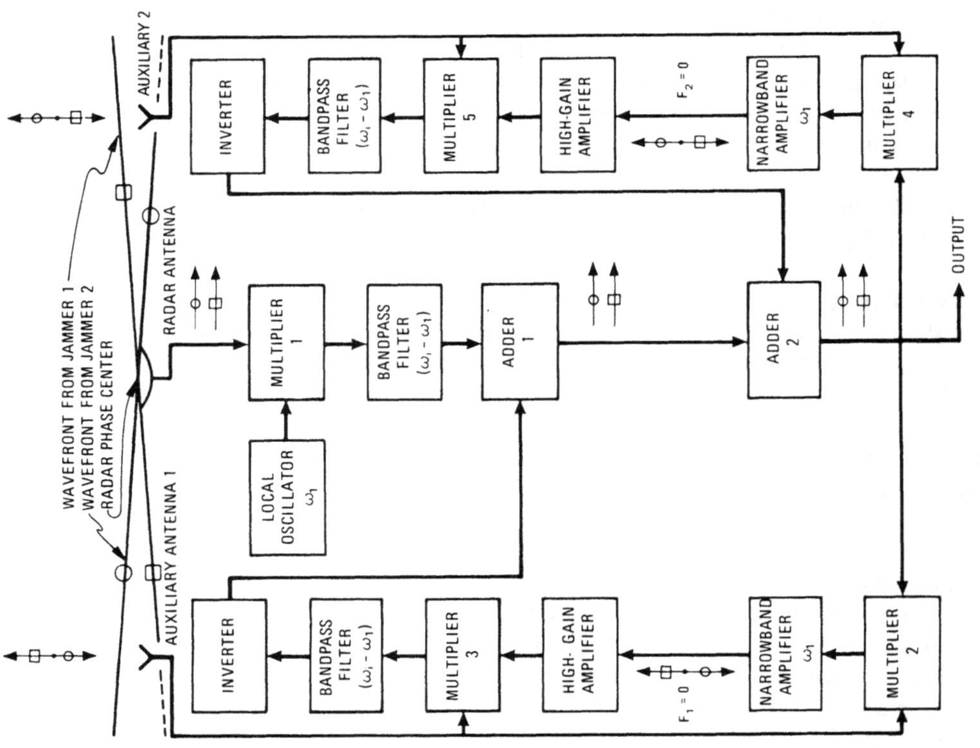

Fig. 9—The case of Fig. 7 with equal radar components

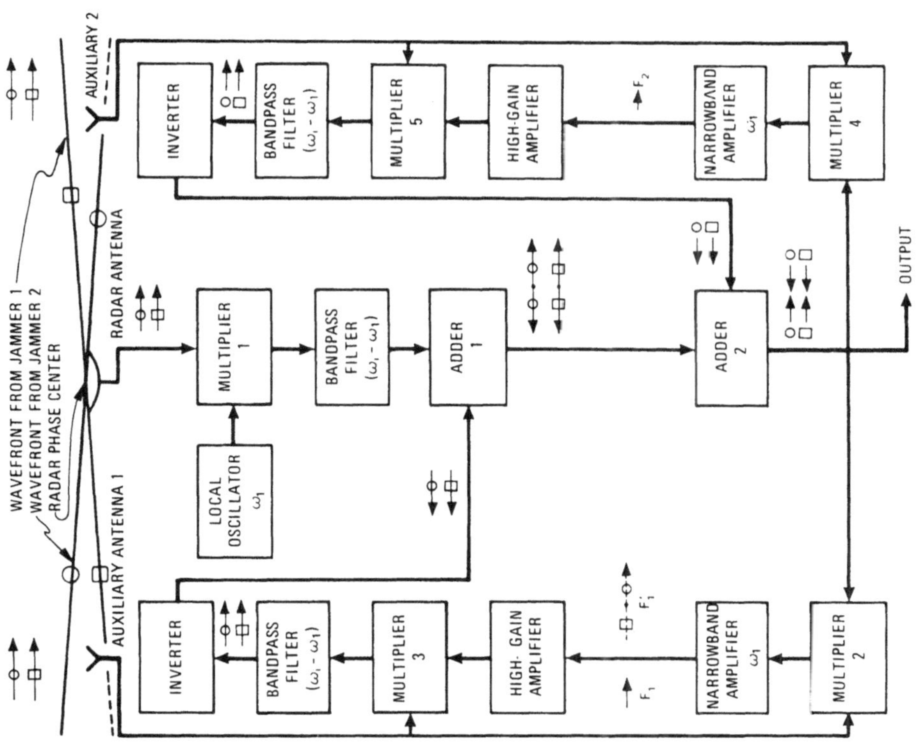

Fig. 12—Equal-amplitude components all in phase

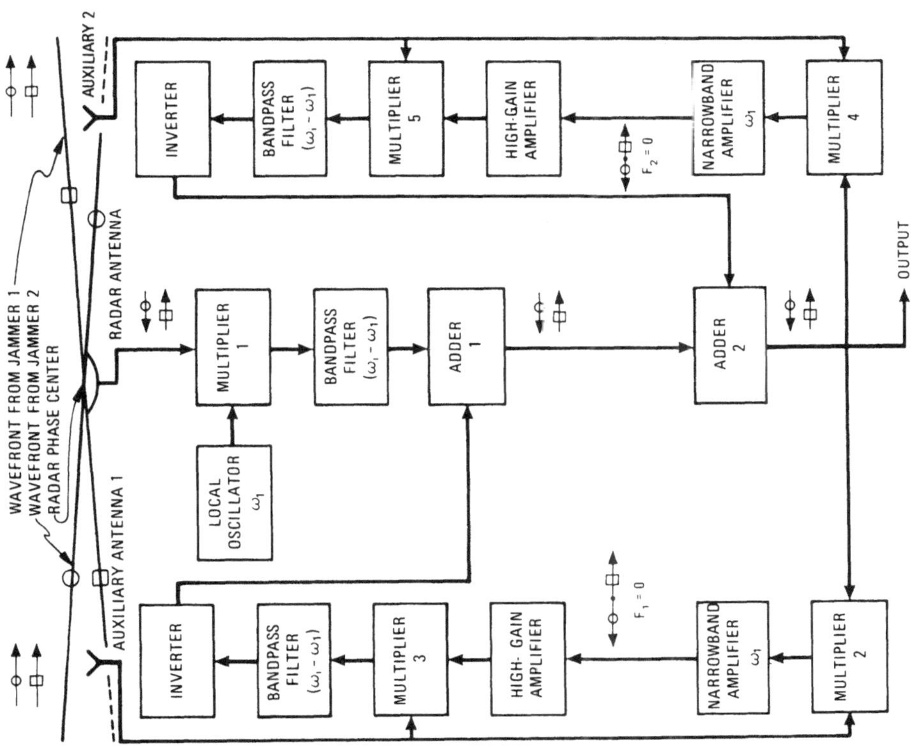

Fig. 11—Equal vectors with radar components 180° out of phase

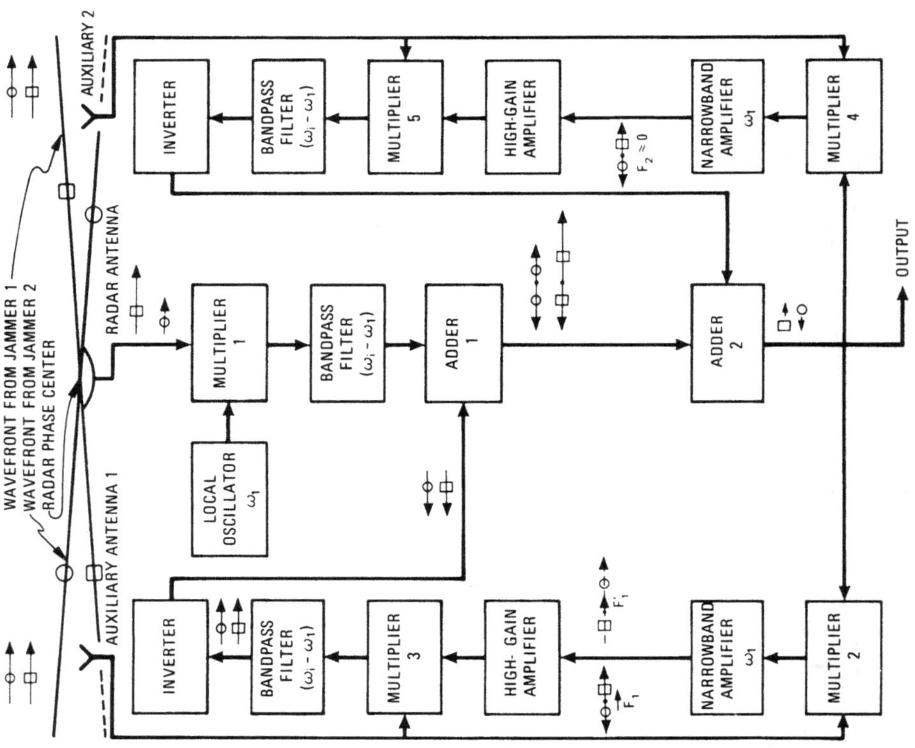

Fig. 13—Unequal radar components, in phase

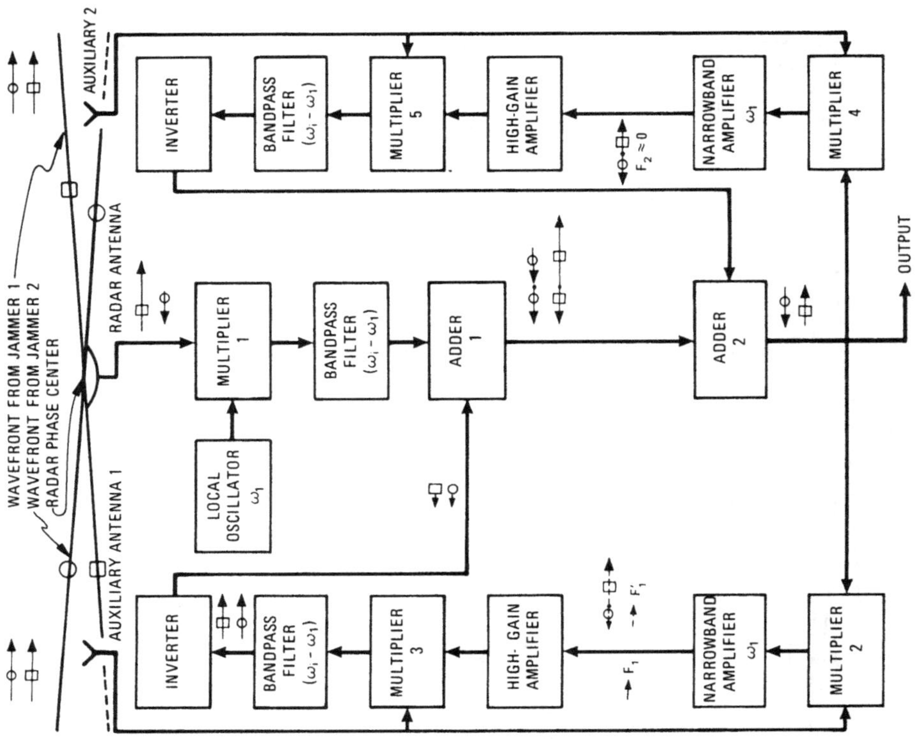

Fig. 14—Unequal amplitudes, 180° out of phase

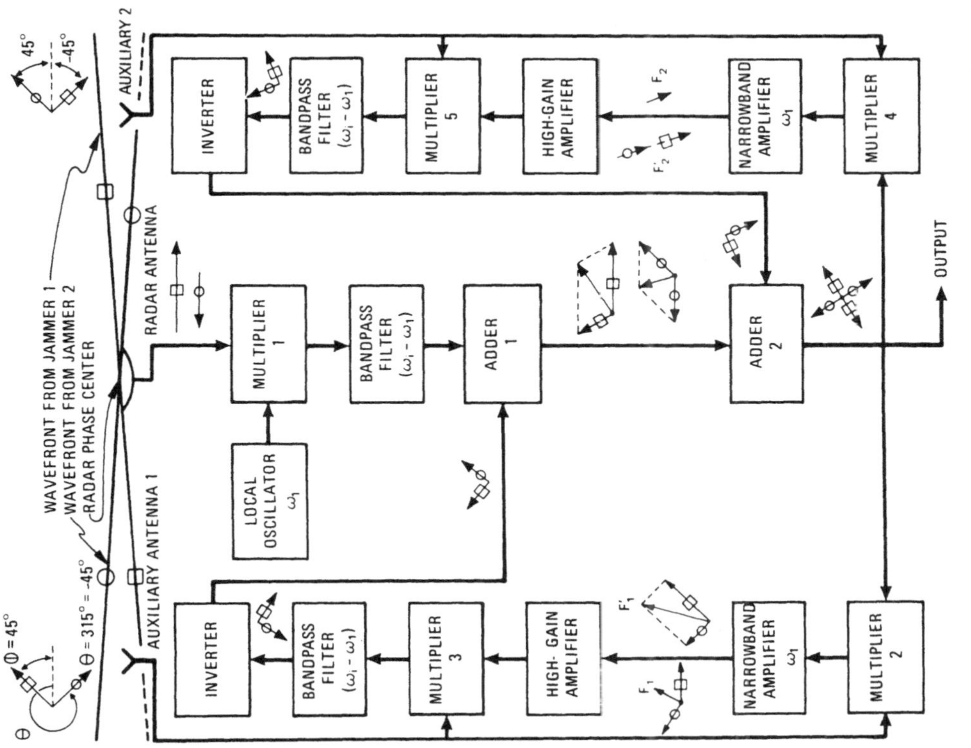

Fig. 15—Best-case auxiliary phasing, one vector flipped by radar antenna

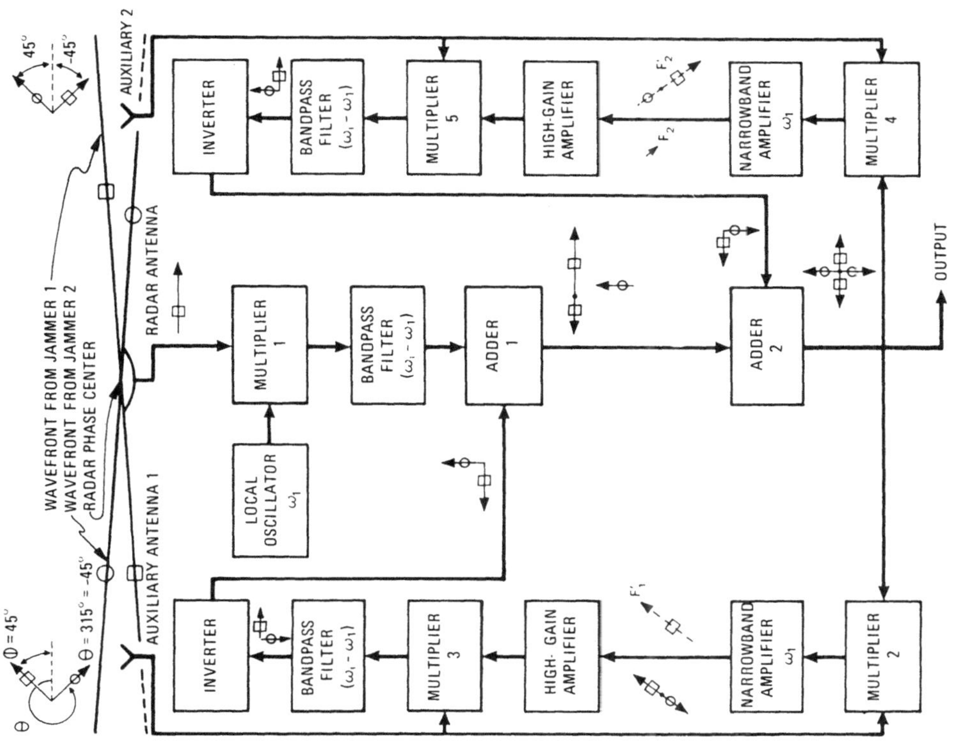

Fig. 16—Best-case auxiliary phasing, one jammer in radar antenna null

typical variations that could occur due to radar scan and show good cancellation in all cases.

Loop Interaction Effects

A case in which the operation of one loop allows another loop to function is illustrated in Fig. 17. In this case, the order of loop analysis would be important or two successive analyses would be required.

Figure 17 portrays two equal and zero-phase signals in the radar with 180° and 90° relationships in the auxiliaries. If the analysis starts with auxiliary antenna 1, it is obvious that no weighting function can be developed by loop 1 and an analyst would proceed to loop 2. Loop 2 would be found to produce a residue out of adder 2 composed of equal parts of circle and square with phase angles of +45° and -45° with respect to their counterparts in the radar antenna. The residue magnitudes would be smaller than their radar counterparts by $(1/2)^{1/2}$. Loop 2 would then be said to produce a 3-dB cancellation ratio (ratio of power out of adder 2 with canceler loops open and closed).

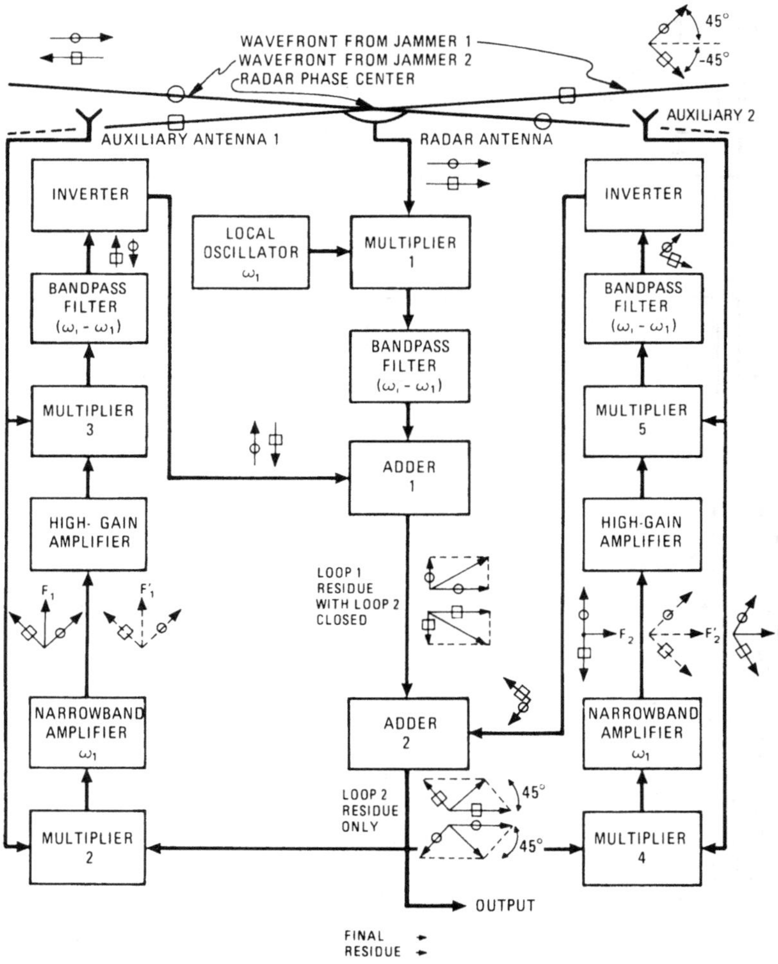

Fig. 17—Demonstration of loop interaction

At this point, the analyst should note that loop 1 develops a weighting function F_1^1. Loop 1 attempts to cancel the vertical components of the residue of loop 2. In so doing, it reduces the magnitudes and phase angles of the two components of the residue. The reduction in the phase angles is now opposed by loop 2 developing a new weighting function F_2^{11} that attempts to hold the phase angles at +45° and −45°. Stability is reached with loop 1 canceling the vertical components of the residue of loop 2 as loop 2 cancels the horizontal radar components. The cancellation ratio of both loops operating in this case would only be limited by the gain of the loops.

Loop interaction, when it exists, is evident as soon as the first approximation to the multiple loop residue is obtained. The angles and relative magnitudes of the residue components can be compared to the angles and relative magnitudes of the signals on the auxiliary antennas to determine if any loop or loops can further reduce the residue.

Analysis of Three-Loop Systems

A three-loop system can be obtained by adding a third adder following adder 2 and a third auxiliary antenna driving a third canceler that transmits inputs to adder 3. The addition of a third auxiliary antenna permits an extra sample of the jamming environment to be obtained. However, its effectiveness is still dependent upon the assumed jammer geometry and the assumed locations of the auxiliary antennas. For example, in the case illustrated in Fig. 6, the addition of a third loop would not significantly improve jamming since two loops are adequate. However, in Fig. 7, the third loop would be extremely useful if it were positioned to receive both jamming signals in phase with their radar counterparts, and nearly any phases other than those shown in Fig. 7 would yield some cancellation.

It should be noted that the best location for auxiliary antenna 3 is that in which the phases of the jamming signals are such that the loop can rotate them to be out of phase with the residues of the first two loops. In Fig. 7, this would require both the jamming signals on auxiliary 3 to be in or 180° out of phase with their radar counterparts. In Fig. 10, it would require one of the jamming signals on auxiliary 3 to be in phase with its radar counterpart and the other to be 180° out of phase.

It should be noted that the good cancellation obtained in Fig. 1, 15, and 16 would not have been obtained if auxiliary 2 had received a circle with lag and a square with lead, since loop 2 could not have rotated them to have each component out of phase with its counterpart in the residue of loop 1.

More than three loops can be obtained by adding more loops as done to obtain three and the analysis can be continued loop by loop. In all cases, the optimum position for the next auxiliary antenna can be determined by determining the phases of the residues it has to work on.

Triple Jammer Analysis

Three jammers and multiple loops can be easily accommodated by drawing a third phase front through the radar phase center and denoting it by a different symbol. The analysis with three jammers then proceeds as it did with two. However, three vectors contribute to the weighting functions and one, two, or three vectors may appear in the residue.

It is important to remember that, regardless of the number of loops and jammers, each loop can only rotate all of the vectors on its auxiliary antenna by the same angle, control all of their magnitudes by the same gain constant, and subtract the resultant vectors from the radar signals.

CONCLUSION

The vector technique of analyzing multiple side-lobe canceler loop operation on multiple jammers is a valid check of performance and permits optimum auxiliary antenna locations to be determined for any given jammer geometry. With practice, many loops and many jammers can be analyzed in a short time. In addition, the use of this technique and the simulator results reveal that the popular belief that n loops can effectively cancel n jammers in all condition is completely false when the auxiliary antennas are omnidirectional so that they cannot resolve the various jammers in angle.

☆ U. S. GOVERNMENT PRINTING OFFICE : 1973—542-166/Z-13

An Improved Algorithm for Adaptive Processing

FRANK F. KRETSCHMER, JR., Senior Member, IEEE
BERNARD L. LEWIS, Senior Member, IEEE
Naval Research Laboratory

Abstract

The Widrow-Hoff least mean square (LMS) algorithm based on the method of steepest descent is conditionally stable. A modified algorithm is given which is unconditionally stable, capable of better performance when used in adaptive filter processing, and provides a more realistic means for simulating the Applebaum-Howells adaptive loop.

Manuscript received July 15, 1976; revised August 29, 1977.

Authors' address: Radar Division, Code 5345, Naval Research Laboratory, Washington, DC 20375.

0018-9251/78/0100-0172 $00.75

I. Introduction

A discrete form of the Widrow-Hoff least mean square (LMS) error algorithm based on the method of steepest descent is commonly used in adaptive array and adaptive filter work [1]-[3]. However, this algorithm is conditionally stable with the result that the loop gain and bandwidth are constrained. This in turn impacts on the convergence rate and performance of the filter. By changing the indexing of the LMS algorithm, a modified LMS algorithm (MLMS) is obtained which is unconditionally stable and achieves the same mean steady state weight as the LMS algorithm. In this paper we compare the performance of the LMS and MLMS algorithms, which for illustrative purposes are applied to noise cancellation filters, for the two cases of ideal and leaky integration. The later case is illustrated by the Applebaum-Howells adaptive loop for which simulation results are presented for the two algorithms.

II. Ideal Integration

A. LMS Algorithm

We first consider, without loss in generality, a one-dimensional form of the Widrow-Hoff LMS algorithm for the noise cancellation application [2] given by

$$W(j+1) = W(j) + 2\mu\, E(j)\, V_a^*(j) \tag{1a}$$

where

$$E(j) = V_m(j) - W(j)\, V_a(j) \tag{1b}$$

and where $W(j)$, $E(j)$, $V_m(j)$, and $V_a(j)$ denote the jth sample of the complex modulation functions corresponding, respectively, to the weighting signal, the error or output signal, the primary input or desired signal, and the reference input signal. The parameter μ is a gain factor that controls stability and convergence rate. The desired signal V_m consists of the actual desired signal plus interference while V_a is assumed to consist only of interference which is correlated with the interference comprising V_m. In general, the interference may consist of correlated components plus uncorrelated components such as uncorrelated thermal noise in each channel. The LMS criterion determines the weight which minimizes the mean squared value of the error or output signal given in (lb), which in turn corresponds to minimizing the average output noise power. Substituting (lb) in (1a) results in

$$W(j+1) = W(j)\,(1 - 2\mu|V_a(j)|^2) + 2\mu\, V_m(j)\, V_a^*(j). \tag{2}$$

The solution of this equation for a step input of V_m and V_a is readily determined by iterating (2), using the initial condition $W(1)$ equals zero, and by determining the sum of j terms of the resulting geometric series. The results are

$$W(j) = B[1 - A^{j-1}]/(1 - A), \qquad j \geqslant 2 \tag{3}$$

where $W(1) = 0$, $A = 1 - 2\mu|V_a|^2$, and $B = 2\mu V_m V_a^*$. For stability it is required that

$$|A| < 1 \tag{4}$$

which reduces to the requirement

$$0 < \mu < 1/|V_a|^2. \tag{5}$$

For the general N-dimensional or N-channel case, conventional analysis reduces the problem to an orthogonal coordinate system where the covariance matrix of the input signal is diagonalized. The result is that each weight can be expressed independent of the other weights by equations having the same form as (1). The stability criterion is then determined as in (5) except the maximum eigenvalue is used in place of $|V_a|^2$. In our one-dimensional analysis, $|V_a|^2$ corresponds to the unique eigenvalue so that a direct comparison can be made between the LMS and MLMS algorithms without incurring a loss in generality.

The mean steady state weight, for any input, is obtained from averaging (2) and making the usual assumption that

$$\overline{W(j)\, V_a(j)} = \overline{W(j)}\; \overline{V_a(j)}. \tag{6}$$

The result is

$$W_\infty = \overline{V_m V_a^*}/\overline{|V_a|^2} \tag{7}$$

which may be shown to be the Wiener weight, which minimizes the average interference residue power.

Next we investigate the time constant of the LMS algorithm for a step input of V_m and V_a. The time constant is defined as the time taken for the weight to get to $1 - \exp(-1)$ of the steady state value. From (3) we obtain

$$1 - A^{j-1} = 1 - \exp(-1) \tag{8}$$

which results in the time constant, normalized by the sampling interval,

$$T = -1/\ln|A| + 1 \tag{9}$$

$$T = -1/\ln|1 - 2\mu|V_a|^2| + 1. \tag{10}$$

The magnitude signs are necessary since it is possible for A to become negative, which corresponds to a damped oscillation of $W(j)$ provided $|A| < 1$. From (10) the minimum time constant occurs for

$$1 - 2\mu|V_a|^2 = 0 \tag{11a}$$

in which case

$$T_{\min} = 1. \tag{11b}$$

B. MLMS Algorithm

Next we investigate the MLMS algorithm which is obtained by modifying the indexing of the LMS algorithm. The MLMS algorithm is determined from the feedback control-loop equation

$$W(j) = W(j-1) + 2\mu\, E(j)\, V_a^*(j) \tag{12a}$$

where

$$E(j) = V_m(j) - W(j)\, V_a(j). \tag{12b}$$

Here, the present weight $W(j)$ is obtained from the present values of V_a and E in contrast to the LMS algorithm of (1) which generates the next weight from the present V_a and E. Substituting (12b) in (12a) and solving for $W(j)$ results in the MLMS algorithm

$$W(j) = [W(j-1) + 2\mu\, V_m(j)\, V_a^*(j)]/[1 + 2\mu|V_a(j)|^2]. \tag{13}$$

This equation is of the form $W(j) = CW(j-1) + D$, which, for an output step of V_m and V_a, has the solution

$$W(j) = [D(1 - C^j)]/(1 - C), \quad j \geqslant 1 \tag{14}$$

where $C = 1/(1 + 2\mu|V_a|^2$ and $D = 2\mu V_m V_a^*/[1 + 2\mu|V_a|^2]$. For stability it is required that $|C| < 1$ or

$$1/(1 + 2\mu|V_a|^2) \leqslant 1 \tag{15}$$

which is always satisfied when μ and $|V_a|^2$ are nonzero. The mean steady state weight is obtained from (13) as

$$W_\infty = W_{\mathrm{opt}} = \overline{V_m V_a^*}/\overline{|V_a|^2}$$

so that the weight of the MLMS algorithm is also unbiased. The time constant for the MLMS algorithm is

$$T = -1/\ln[1/(1 + 2\mu|V_a|^2)]$$

$$= 1/\ln[1 + 2\mu|V_a|^2]. \tag{16}$$

The minimum time constant case for the LMS criterion requires from (11a) that

$$2\mu|V_a|^2 = 1. \tag{17}$$

Substituting this value in (16) results in the MLMS algorithm time constant, for the same parameters, being

$$T = 1/\ln 2 = 1.44. \tag{18}$$

Hence, for the same parameters the MLMS algorithm is slightly slower; however, this is somewhat deceiving since the time constant of the MLMS algorithm can be

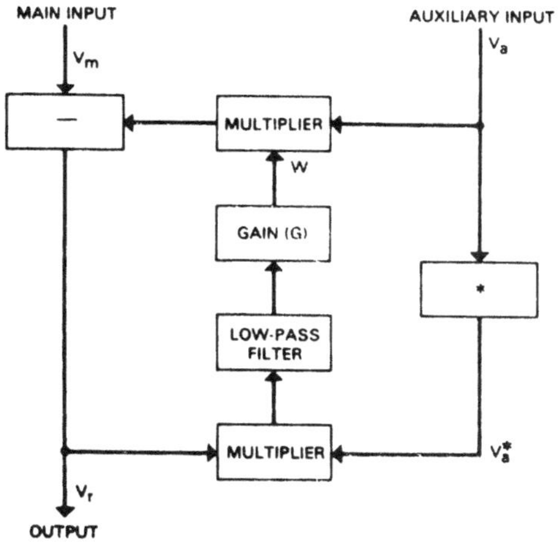

Fig. 1. Basic canceller loop.

reduced by simply increasing the gain term without becoming unstable.

III. Leaky Integration

We next consider the case of nonideal or leaky integration such as found in a low-pass filter. This situation is typified by the Applebaum-Howells adaptive loop [4] which is shown in Fig. 1 in its basic form. The residue signal V_r in Fig. 1 corresponds to the error signal in (1b) which the adaptive loop attempts to minimize in the least squares sense by subtracting the weighted auxiliary or reference signal from the main or primary input. The analysis parallels that for the ideal integration and only the results are given.

A. LMS Algorithm

The LMS algorithm applied to this loop is

$$W(j+1) = KW(j) + G(1-K)\, V_r(j)\, V_a(j^*) \tag{19a}$$

where

$$V_r(j) = V_m(j) - W(j)\, V_a(j)$$

$$K = 1 - 1/\tau = 1 - 2\pi f_{3\,dB} \tag{19b}$$

with τ the normalized time constant of the low-pass filter and $f_{3\,dB}$ the normalized bandwidth of the low-pass filter. τ and $f_{3\,dB}$ are normalized to the sampling interval and sampling frequency, respectively. Substituting (19b) in (19a) and solving for $W(j)$, using the initial condition $W(1)$ equals zero, results in the solution, for step inputs given by

$$W(j) = b(1 - a^{j-1})/(1-a), \qquad j \geq 2 \tag{20}$$

where $W(1) = 0$, $a = K - G(1-K)|V_a|^2$, and $b = G(1-K)|V_a|^2$.

For stability it is required that

$$|K - G(1-K)|V_a|^2| < 1. \tag{21}$$

Substituting the value $K = 1 - 1/\tau = 1 - 2\pi f_{3\,dB}$ in (21) leads to

$$(1+G)|V_a|^2 \pi f_{3\,dB} < 1. \tag{22}$$

Also, the weight will not ring in amplitude for a less than unity, which implies positive b. The mean steady state weight is given by

$$W_\infty = [G\overline{|V_a|^2}/(1 + G\overline{|V_a|^2})]\, W_{opt} \tag{23}$$

where $W_{opt} = \overline{V_m V_a^*}/\overline{|V_a|^2}$. Thus, the weight is biased from the Wiener weight and is the weight obtained by a gain-limited Applebaum-Howells adaptive loop [4]. The time constant is given by

$$T = -1/\ln|K - G(1-K)|V_a|^2| + 1. \tag{24}$$

T is a minimum when $K - G(1-K)|V_a|^2 = 0$. This corresponds to $f_{3\,dB} = 1/2\pi(1 + G|V_a|^2)$ which is approximately equal to one-half of the critical value of $f_{3\,dB}$ given by (22). It should be noted that T decreases with increasing $f_{3\,dB}$ until the minimum occurs and that increasing $f_{3\,dB}$ beyond this point results in larger values of T.

B. MLMS Algorithm

The MLMS algorithm applied to Fig. 1 is determined from (also see [4])

$$W(j) = KW(j-1) + G(1-K)\, V_r(j)\, V_a^*(j) \tag{25a}$$

where

$$V_r(j) = V_m(j) - W(j)\, V_a(j). \tag{25b}$$

Note that these equations more realistically represent the behavior of the actual adaptive loop since the present weight is determined from the present data. The LMS algorithm corresponds to using a delayed value of the weight which introduces a phase shift that is not actually present. Substituting (25b) in (25a), the solution for a step input of V_a and V_m is given by

$$W(j) = d(1 - c^j)/(1-c), \qquad j \geq 1 \tag{26}$$

where

$$c = K/[1 + G(1-K)|V_a|^2]$$

$$d = G(1-K)|V_a|^2/[1 + G(1-K)|V_a|^2].$$

Since $|c| < 1$, the algorithm is always stable. The mean steady state weight may again be shown to be the same as

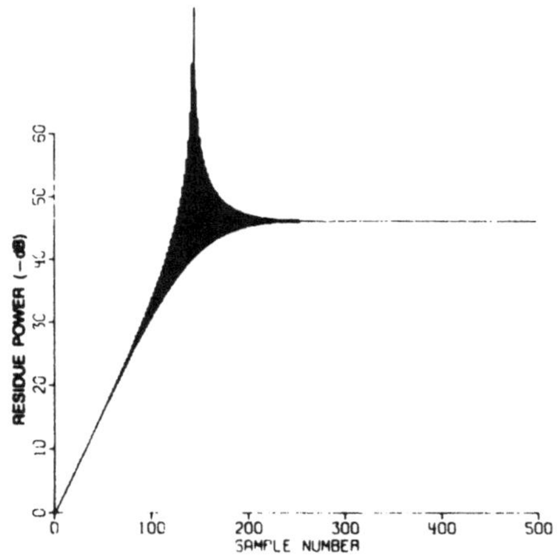

Fig. 2. Response of LMS algorithm (f_{3dB} = 0.00155).

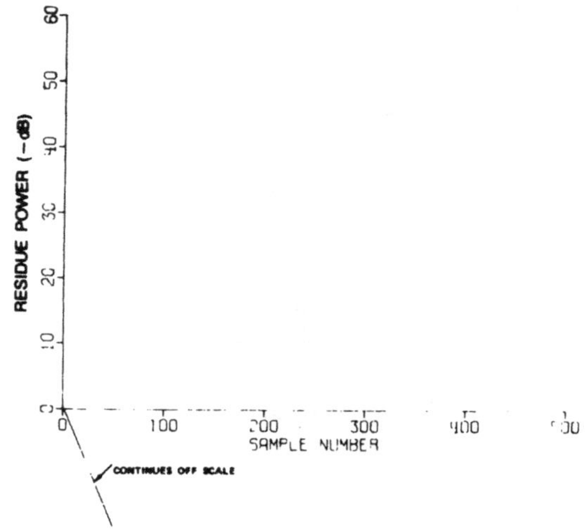

Fig. 3. Response of LMS algorithm (f_{3dB} = 0.00162).

Fig. 4. Response of MLMS algorithm (f_{3dB} = 0.00162).

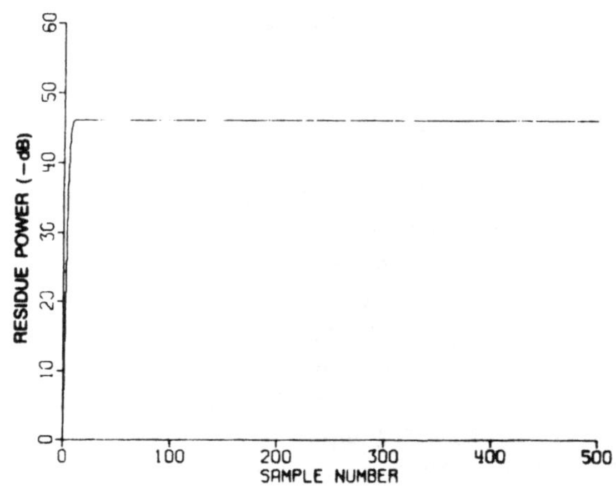

that for the LMS algorithm given by (23). The time constant for the MLMS algorithm for a step input of V_a and V_m is given by

$$T = -\ \{\ln(K/[1 + G\,(1 - K)|\,V_a|^2\,])\}^{-1}. \tag{27}$$

IV. Computer Simulations of Adaptive Loops

A. Nonrandom Interference: Step Input

Computer simulations were run to demonstrate the instability associated with use of the LMS algorithm given by (19) and are shown in Figs. 2 and 3. Residue power plotted in the figures is the residue sample power normalized by the average input interference power. In Figs. 2 and 3, G is equal to 100, $|V_a|^2$ and $|V_m|^2$ are equal to 2, and several values of f_{3dB} are plotted.

From (22) the stability condition becomes

$$f_{3dB} < 0.00158 \tag{28}$$

for the specified values of G and $|V_a|^2$. Fig. 2 shows a damped oscillation occurring for f_{3dB} equal to 0.00155. For this value of f_{3dB}, (24) yields a time constant of 20 samples which is the time required for the weight to reach $1 - \exp(-1)$ of the steady state value and corresponds to a normalized residue of −8.7 dB. In Fig. 3, f_{3dB} is equal to 0.00162 and instability occurs with the weight phase alternating between 0 and 180 degrees and the weight magnitude growing unbounded.

In Fig. 4 the response to a step input is plotted using the MLMS algorithm determined from (25) for the same value of f_{3dB} (0.00162) which caused unstable operation of the steepest descent algorithm shown in Fig. 3. It should be noted that there is no overshoot or ringing in Fig. 4 since the response is unconditionally stable.

B. Random Interference: Step Input

Computer simulations were run using independent samples of a Gaussian random process having a mean of zero and a variance equal to 2. Successive samples were correlated by taking a sliding window average of two samples and renormalizing so that the resultant power remained equal to 2. The same samples were then applied to the main and auxiliary channels of the sidelobe canceller. A comparison between the LMS and the MLMS algorithms is shown in Figs. 5 to 9 for an input step of random values which are the same for each simulation. In these simulations a constant target signal was introduced in the main channel at sample number 250 at an interference-to-signal level of 20 dB. However, in this paper we wish to focus our attention in the region where the signal is absent.

The LMS and MLMS algorithm results are shown in Figs. 5 and 6 for f_{3dB} equal to 0.00025 which corresponds to an effective loop bandwidth B_E to interference bandwidth B_J ratio of 0.1. B_E is shown in [4] to be given by

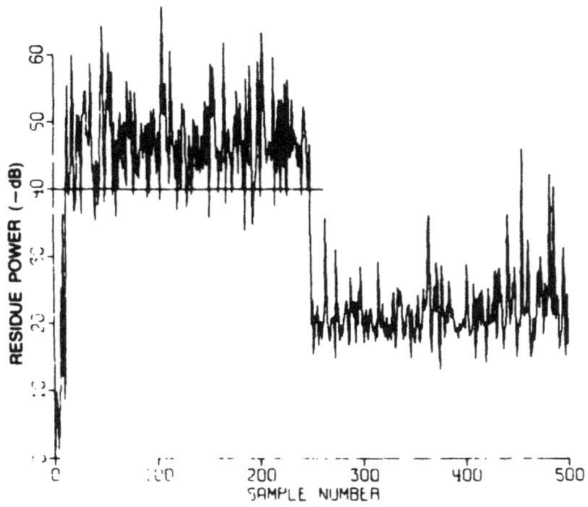

Fig. 5. Response of LMS algorithm (f_{3dB} = 0.00025).

Fig. 6. Response of MLMS algorithm (f_{3dB} = 0.00025).

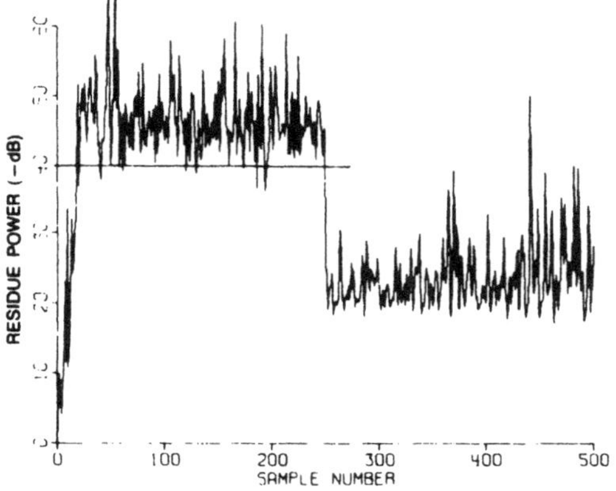

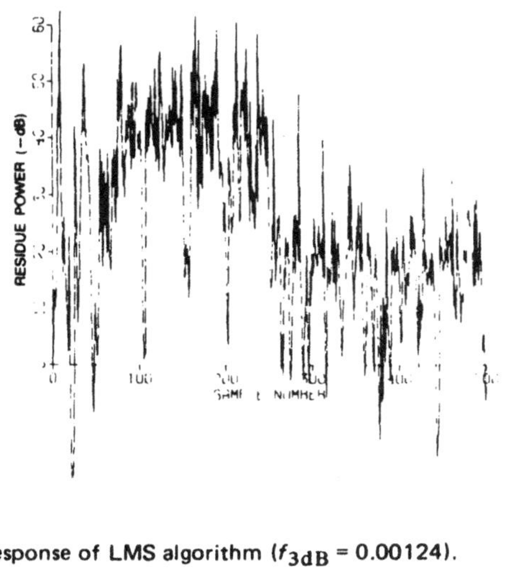

Fig. 7. Response of LMS algorithm (f_{3dB} = 0.00124).

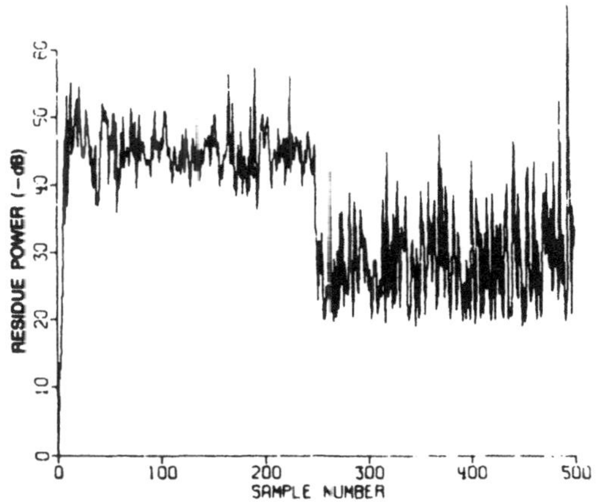

Fig. 8. Response of MLMS algorithm (f_{3dB} = 0.00124).

Fig. 9. Response of MLMS algorithm (f_{3dB} = 0.0025).

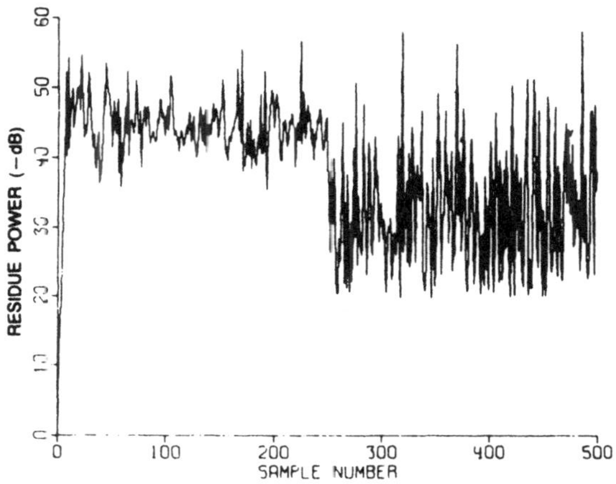

$$B_E = (1 + G\overline{|V_a|^2}) f_{3dB} \tag{29}$$

which, for sampled data, is normalized by the sampling frequency. Comparing Figs. 5 and 6 it is seen that the LMS algorithm gives more points of lower cancellation (e.g., under the 40-dB line) than the MLMS algorithm. This is attributed to the ringing effect in the steepest descent algorithm which is present even for the slower loop adaptation.

In Figs. 7 and 8, the LMS and MLMS algorithm results are shown for f_{3dB} = 0.00124, or correspondingly B_E/B_J = 0.5, and poor performance is seen to result from the LMS method while good performance is obtained with the MLMS algorithm. For the case of f_{3dB} = 0.0025 or B_E/B_J = 1, the LMS algorithm gives unstable loop performance while the MLMS algorithm gives stable performance as shown in Fig. 9.

The degradation of the interference cancellation for faster adaptation using the LMS algorithm is primarily attributed to the ringing effect caused by the excitation of the random inputs. The results for the MLMS algorithm shown in Figs. 6, 8, and 9 do not reveal any appreciable degradation of the interference cancellation (shown in the first 250 points of each figure) with faster adaptation.

It should be noted from (22) and the dependence of f_{3dB} on the sampling rate that any tendency toward ringing or instability in the LMS algorithm can be reduced or eliminated by sampling at a higher rate. This, in effect, reduces the delay or the corresponding phase shift which is introduced by the LMS algorithm.

V. Summary

It has been shown that the LMS algorithm is conditionally stable and that by modifying the indexing of the LMS algorithm a modified MLMS algorithm is obtained which is unconditionally stable. Both algorithms were shown to converge in the mean to the same weight which is the Wiener weight for the ideal integration case and a biased Wiener weight for the leaky integration case. The biased weight corresponds to the weight actually obtained in a gain-limited Applebaum-Howells adaptive loop. For the leaky integration case it was shown that the allowable loop gain and effective bandwidth of the LMS algorithm are limited by stability conditions. It was shown that even for slower loop adaptation the MLMS algorithm gives better cancellation of interference. The superior cancellation of the MLMS algorithm became more evident as B_E/B_J was increased. As B_E/B_J was increased, the simulations indicated degraded cancellation for the LMS algorithm. This is attributed to the ringing effect of the weight in the LMS algorithm when driven by random noise. The faster the adaptation time of the LMS algorithm, the poorer the cancellation became. The simulations for the MLMS algorithm did not reveal any significant degradation as the adaptation time was decreased. Moreover, it was pointed out that the MLMS algorithm provides a more realistic algorithm for simulating actual loop behavior.

It was indicated for the leaky integrator case that the tendency toward instability of the LMS algorithm could be reduced by increasing the sampling rate. Thus, to prevent performance degradation of the LMS algorithm for faster adaptation, a sampling rate higher than the Nyquist rate is necessitated. For the ideal integrator LMS case, increasing the sampling rate reduces the settling time for the same gain but does not affect stability or cancellation performance.

A comparison of algorithms was shown for the one-dimensional noise canceller application for illustrative purposes. However, the concept generalizes to any adaptive processing which minimizes the mean square error. The generalization of (25) for the M-dimensional MLMS case is given for the ith weight by

$$W_i(j) = KW_i(j-1) + G(1-K)\, V_r(j)\, V_i^*(j) \tag{30a}$$

where

$$V_r(j) = V_m(j) - \sum_{n=1}^{M} W_n(j)\, V_n(j) \tag{30b}$$

which requires the solution of simultaneous equations. The generalization of the LMS algorithm of (19) is given by

$$W_i(j+1) = KW_i(j) + G(1-K)\, V_r(j)\, V_i^*(j) \tag{31a}$$

where

$$V_r(j) = V_m(j) - \sum_{i=1}^{M} W_i(j)\, V_i(j). \tag{31b}$$

The multidimensional LMS algorithm is simpler to use than the multidimensional MLMS algorithm since each new weight can be directly computed from all present data. Simplification of the MLMS equations (30) is the subject of a future investigation. However, a possible simplification, which retains its basic features, is given by

$$W_i(j) = KW_i(j-1) + G(1-K)\, V_r(j)\, V_i^*(j) \tag{32a}$$

$$V_r(j) = V_m(j) - W_i(j)\, V_i(j) - \sum_{n=1, n\neq i}^{M} W_n(j-1)\, V_n(j). \tag{32b}$$

In this simplified MLMS case each weight is found in a closed loop fashion, as in the single loop case, while the other weights are frozen. The actual residue signal resulting from this procedure is then taken to be

$$V_r(j) = V_m(j) - \sum_{n=1}^{M} W_n(j)\, V_n(j) \tag{33}$$

where the weights are derived as explained above.

References

[1] B. Widrow, R. Mantey, L. Griffiths, and B. Goode, "Adaptive antenna systems," *Proc. IEEE*, vol. 55, pp. 2143-2158, Dec. 1967.

[2] B. Widrow et al., "Adaptive noise cancelling: Principles and applications," *Proc. IEEE*, vol. 69, pp. 1692-1719, Dec. 1975.

[3] I.S. Reed, J.D. Mallett, and L.W. Brennan, "Rapid convergence rate in adaptive arrays," *IEEE Trans. Aerosp. Electron. Syst.*, vol. AES-10, pp. 853-863, Nov. 1974.

[4] F.F. Kretschmer, "Effects of cascading sidelobe canceller loops," *Rec. IEEE 1975 Internat. Radar Conf.*, pp. 181-185.

Biographies and photographs of **F.F. Kretschmer** and **B.L. Lewis** may be found on page 171 of this issue.

A New Technique for Reducing Radar Response to Signals Entering Antenna Sidelobes

BERNARD L. LEWIS, SENIOR MEMBER, IEEE, AND
JAMES B. EVINS

***Abstract*—A new technique for reducing a radar's response to undesired signals entering a radar's sidelobes is described and analyzed theoretically. This technique involves moving a phased array antenna's phase center to Doppler shift sidelobe signals out of the radar receiver's passband.**

INTRODUCTION

The purpose of this communication is to describe and evaluate a new technique for suppressing undesired echoes that enter the sidelobes of a radar antenna on receive. The technique also changes the frequency of signals radiated by the radar through the sidelobes of the radar antenna on transmit.

The concepts employed in this technique include but are not limited to those discussed by Bickmore [1] and Crotty *et al.* [2]. This communication notes the value of limiting the spectral sidelobes of the transmitted signal which was not considered by [1] and [2].

CONCEPT

The concept involves moving the phase center of a phased array antenna in the plane of the aperture to Doppler shift signals radiated and received on the antenna sidelobes out of the passband

Manuscript received December 8, 1982; revised May 23, 1983.

The authors are with the Naval Research Laboratory, 4555 Overlook Avenue, Washington, DC 20375.

of the radar receiver. The phase center motion is achieved by illuminating only part of an available phased array aperture and moving the illuminated part across the aperture while the antenna is transmitting.

The technique can be implemented by switches in the feed lines of the antenna elements. If N elements are available in the desired dimension, the left hand M elements ($M < N$) can be turned on for a time τ equal to the reciprocal of the radar bandwidth B divided by $(N - M + 1)$. The first or left hand element is then turned off, elements from 2 through M are left on and the $M + 1$ element is turned on. Elements 2 through $M + 1$ then radiate for another time τ before changing to radiation from elements 3 through $M + 2$, etc.

If the antenna elements are separated by 1/2 wavelength λ of the radar and, if the M excited elements are stepped through $N - M$ positions in the time the radar transmits its pulse of length $T = 1/B$, the apparent phase center velocity will be

$$V = (N - M)\lambda/2T. \tag{1}$$

This velocity will produce a Doppler shift on any signal radiated along any line in the plane defined by the normal to the aperture and the direction of motion of the phase center except a line normal to the aperture. With λ = 30 cm, T = 1 μs and $N - M$ = 20, $V = 3 \times 10^6$ m/s which is much faster than any target can move. A mach 3 target moves approximately 10^3 m/s.

The magnitude and sign of the Doppler shift in a direction making an angle θ with the normal to the aperture in the plane defined by the normal to the aperture and the direction of motion of the phase center will be

$$F_d = V \sin\theta/\lambda = [(N - M)\lambda \sin\theta]/2\lambda T. \tag{2}$$

Since $1/T = B$ the radar bandwidth, (2) can be written as

$$F_d = [(N - M)B \sin\theta]/2 \tag{3}$$

and, by the proper choice of $N - M$, F_d can be made to exceed B at any angle θ_d or greater.

THEORETICAL STUDY

A computer simulation of an M out of N element line array was developed to calculate the far field response of the antenna in phase and amplitude as the M excited elements were stepped across the array. The response of the array as a function of angle θ was then processed with a fast Fourier transform (FFT) package and three-dimensional plots of antenna response and signal spectrum were made as a function of θ.

Fig. 1 illustrates an example plot of the one-way response of a system where N was 40, M was 20, and no weighting on the aperture or the transmit pulse was used. In this case, the antenna and spectral responses were sin x/x. The peak to first null in the antenna response was arc sin 0.1 and the peak to first null in the spectral response was the radar bandwidth B or $1/T$. Note that the peak spectral response moves off zero Doppler as θ moves away from zero and approaches ±10 MHz at θ = ±85° as given by (3). Note also the grating lobes in the signal spectrum caused by the time sampling of the moving illuminated portion of the antenna which occurred every 1/21st of the transmitted pulse length T. Fig. 2 is an expanded view of one quadrant of Fig. 1 to permit detail to be observed. An optimum radar receiver passband is indicated about zero on the frequency scale in Fig. 2.

Fig. 3 illustrates the effect of weighting the 20 excited elements with a Hamming weighting without weighting the transmit pulse.

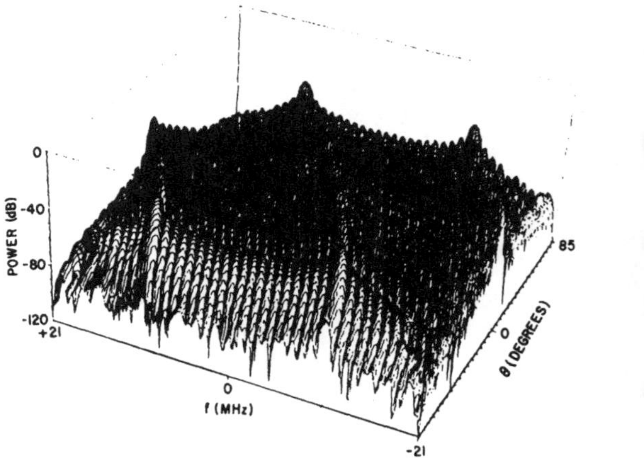

Fig. 1. Space (θ) and frequency (F) response of an unweighted 20-element array swept through 20 elements spaced $\lambda/2$ apart while radiating a 1 μs rectangular pulse.

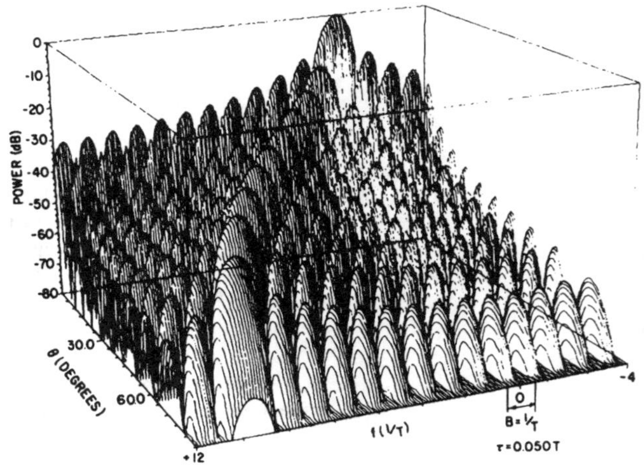

Fig. 2. Expanded space and frequency response plot of 20 element swept through 20 elements with unweighted antenna and 1 μs rectangular pulse.

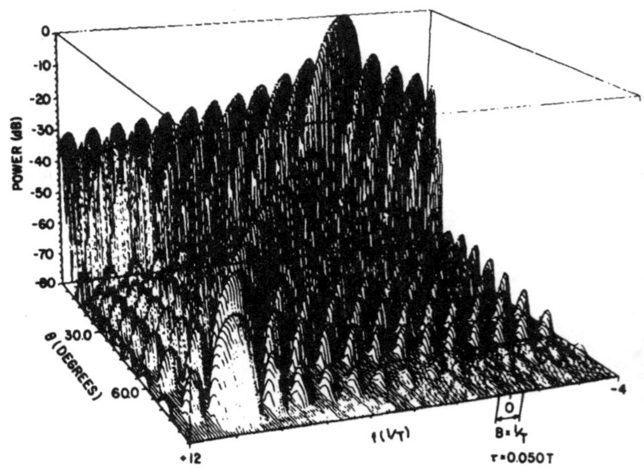

Fig. 3. Expanded space and frequency response plot of 20 Hamming weighted elements swept through 20 elements while radiating a μs rectangular pulse.

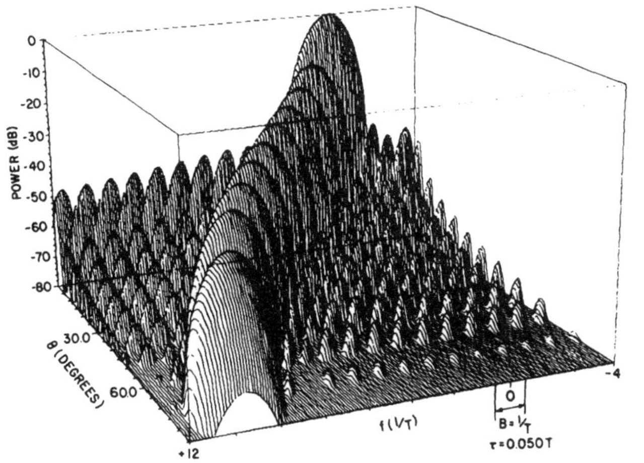

Fig. 4. Expanded space and frequency response plot of 20 unweighted elements swept through 20 elements while radiating a Hamming weighted 1 μs pulse.

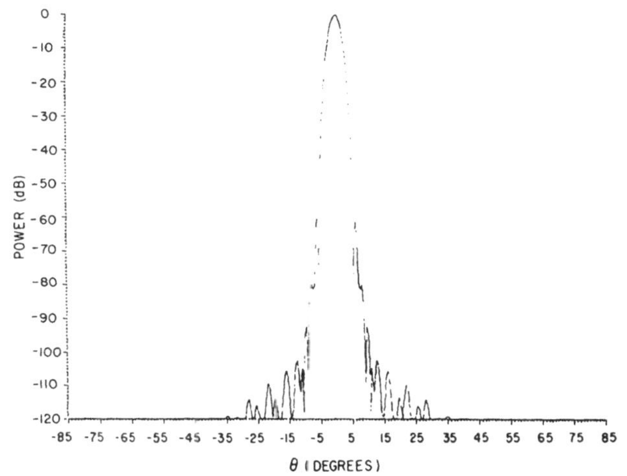

Fig. 6. Two-way antenna and receive filter response of 20 Hamming weighted elements swept through 20 elements on transmit during a 1 μs Hamming weighted pulse with 40 Hamming weighted unswept elements used on receive.

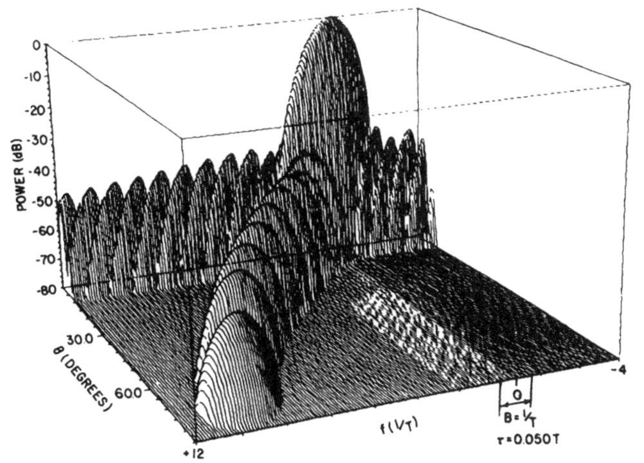

Fig. 5. Expanded space and frequency response of 20 Hamming weighted elements swept through 20 elements while radiating a Hamming weighted 1 μs pulse.

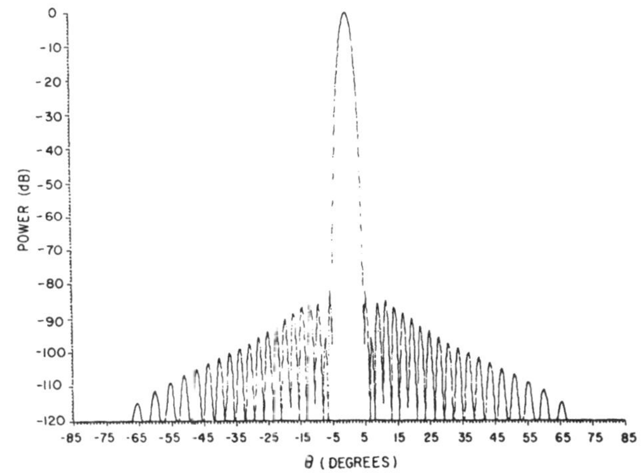

Fig. 7. Two-way antenna pattern of 40 unswept Hamming weighted elements.

In this case, the highest sidelobe in the antenna response is down 42 dB one way.

Fig. 4 is a one-way plot illustrating the effect of weighting the transmit pulse with a Hamming weighting without weighting the 20 excited elements of the array.

Fig. 5 illustrates the effect of a Hamming weighting of both the transmitted pulse and the 20 excited elements on the one-way pattern of the aperture. Note that nearly all sidelobes in the radar passband are below 80 dB down one way.

ANTENNA PATTERN SIMULATIONS

The antenna patterns that would result from a combination of a stepped aperture and a band-limiting filter on receive were obtained by squaring and adding the FFT coefficients in the assumed radar passband and normalizing to the peak response.

Fig. 6 is an example of the use of 20 Hamming weighted excited array elements stepped through 20 elements separated by $\lambda/2$ during the transmit pulse with all 40 available elements with Hamming weighting used on receive. Fig. 7 is a two-way pattern of an array with 40 stationary Hamming weighted elements used both on transmit and on receive for comparison with Fig. 6. Note the sidelobe reduction possible with the stepped aperture.

ANTENNA PHASE CENTER MOTION ON RECEIVE

Theoretically, the antenna phase center can be swept both on transmit and on receive which would double the Doppler shifts of signals received from directions not normal to the line of motion. This could be used to reduce the distance traveled by the phase center in the transmit time if desired to improve the antenna efficiency by making M/N closer to unity.

Unfortunately, however, the arrival time of echoes to be detected in radar will not be known. As a consequence, the antenna phase center sweep cannot be synchronized with all echo arrival times and any timing errors would widen the received signal spectra. This spectral width increase would increase the radars response to signals entering its antenna's sidelobes and its receiving bandpass filter.

In addition to the synchronization problem discussed above, the grating lobes introduced in the transmit pulse spectrum (Fig. 1) by the switches that move the phase center would be a problem. Phase center motion both on transmit and receive would shift the spectral grating lobes closer to the center frequency of the radar receiver than would motion only on transmit. As a result, the radar's response to signals entering the sidelobes of its antenna would increase. This problem, however, could be solved by an implementation of the phase center motion that would produce continuous rather than discrete step motion of the phase center.

CONCLUSION

On the basis of the preceding analysis, it appears that moving phase centers on phased arrays provide significant benefits in the form of unwanted sidelobe suppression. The available Doppler shift provides an added dimension to antenna designers that may have many uses.

REFERENCES

[1] R. C. Hansen, "Microwave scanning antennas," in *Array Theory and Practice*, vol. 2. New York: Academic, 1966.

[2] R. E. Crotty and C. G. Goss, "Side lobe response reducing system," U.S. Patent 3 412 405, Nov. 19, 1968.

ADAPTIVE CANCELLER LIMITATIONS DUE TO FREQUENCY MISMATCH ERRORS

I. INTRODUCTION

An adaptive canceller combines auxiliary channels of data with a main channel of data in such a way so as to minimize the main channel output noise power residue. Hence, this is an effective way of eliminating unwanted data (or noise) from a main channel (the information channel) by inputting correlated data from auxiliary channels. Mismatch errors of any kind between channels of an adaptive canceller can cause a reduction in the achievable cancellation ratio. These mismatch errors can include small time delay differences, in-phase (I) and quadature-phase (Q) imbalances, strobing errors, and frequency mismatch errors among the various channels. For a radar or communications digital canceller, many of these errors occur due to the radio frequency (RF)-to-intermediate frequency (IF)-to-baseband-to-sample and hold (S+H)-to-analog-to-digital (A/D) chain which is present in each channel. If any link of this chain is not identical among the channels, there are mismatch errors which cause the canceller performance to degrade.

In this report, we concern ourselves with just frequency mismatch errors. Other research in this area can be found in Ref. 1. To compensate for frequency mismatch errors, often adaptive digital transversal filters are inserted into the auxiliary channels. Figure 1.1 illustrates a two channel compensated adaptive canceller. Here, we have two signals $y_M(t)$ and $y_A(t)$ inputted into the main and auxiliary channels, respectively. These signals will normally pass through bandpass filters in each channel. The bandwidth of these filters is set equal to the bandwidth of the desired signal. Let the frequency transfer functions (FTF) of the main and auxiliary channels be $H_M(j\omega)$ and $H_A(j\omega)$, respectively. Normally, both of these FTFs are designed to be equal to some desired FTF: $H(j\omega)$. However, because of inaccuracies in the filter synthesis process, $H_M(j\omega)$ and $H_A(j\omega)$ may not be equal.

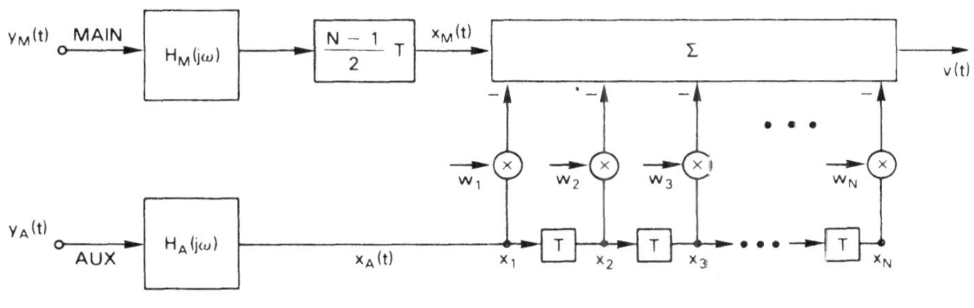

Fig. 1.1 — Two channel model of a compensated adaptive processor

To compensate for this mismatch, a transversal filter (or a tapped delay line) is inserted into the auxiliary channel, and weights $w_n, n = 1,2,\ldots,N$ on these taps are adjusted so that the output noise power residue of $v(t)$ (see Fig. 1.1) is minimized. Note that the time delay, T, normally approximates the Nyquist sampling interval: $1/B$ where B is the input signal's bandwidth. In addition, the main channel is delayed such that the auxiliary samples are time-centered

Manuscript approved August 26, 1985.

If we define $\mathbf{w} = (w_1, w_2, \ldots, w_N)^T$ to be the optimal complex valued weighting vector where T denotes the transpose operating, then it can be shown [1] that $\mathbf{w}$ is the solution of the following vector equation:

$$\mathbf{R}\,\mathbf{w} = \mathbf{r} \tag{1.1}$$

where $\mathbf{R}$ is the covariance matrix of the time delayed taps in the auxiliary channel and $\mathbf{r}$ is the cross covariance vector between the auxiliary taps and the time-centered main channel. More formally

$$\mathbf{R} = E\{\mathbf{x}^*\,\mathbf{x}^T\} \tag{1.2}$$

and

$$\mathbf{r} = E\{\mathbf{x}^*\,x_M\} \tag{1.3}$$

where $E\{\cdot\}$ denotes the expected value, * denotes the complex conjugate, and $\mathbf{x} = (x_1, x_2, \ldots, x_N)^T$ is the vector of tapped time delayed signals in the auxiliary channels.

To completely understand the effects of the frequency mismatch errors, the statistical characteristics of the input signals must be known. However, in many instances this may not be possible. We have chosen to characterize and investigate the effects of the frequency mismatch errors on cancellation when the adaptive canceller is in the self-cancellation mode as illustrated in Fig 1.2. Here, we have tied the main and auxiliary inputs together and have inputted a wideband signal, $y(t)$. In this mode the optimal weights, $\mathbf{w}$, are chosen such that the product of $H_A(j\omega)$ and the transversal FTF match as closely as possible, $H_M(j\omega)$. We then calculate the output cancellation power residue. In a sense, the self-canceller mode yields best case (or an upper bound on) cancellation performance.

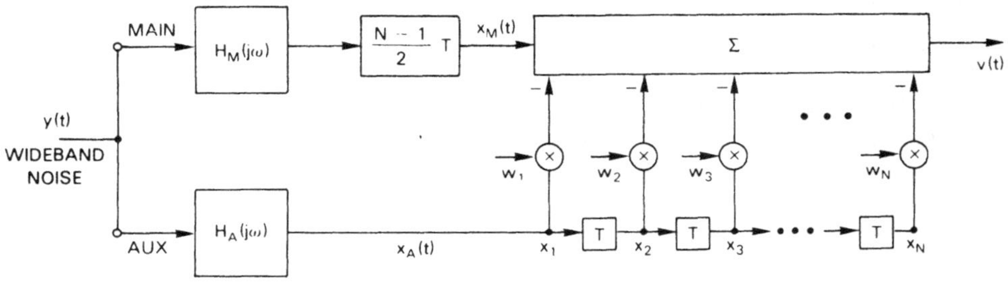

Fig. 1.2 — Self cancellation mode of a two channel compensated adaptive processor

This report is laid out as follows: Section II describes the pole/zero error model which is used to characterize the errors in filter fabrication. In Section III, a formula is derived for the output noise power residue of the self-canceller. In Section IV, a formulation is given for the special case of when the desired FTF is a Butterworth filter. Results generated from this analysis are presented in Section V.

II. POLE/ZERO ERROR MODEL

In this section, we develop a first order pole/zero error model for the frequency transfer functions (FTF) of the main and auxiliary channels: $H_M(j\omega)$ and $H_A(j\omega)$, respectively. This error model will allow us to derive a closed form solution for the cancellation residue as a function of the adaptive canceller system parameters. We assume that both these FTFs are designed to be some desired FTF: $H(j\omega)$. However, because of errors in the synthesis process, the poles and zeroes of $H(j\omega)$ will not be as designed and will have small perturbations around the desired poles and zeroes. This is illustrated in Fig. 2.1. There perturbations are assumed small enough so that we can use first order approximations for the filter responses in the main and auxiliary channels.

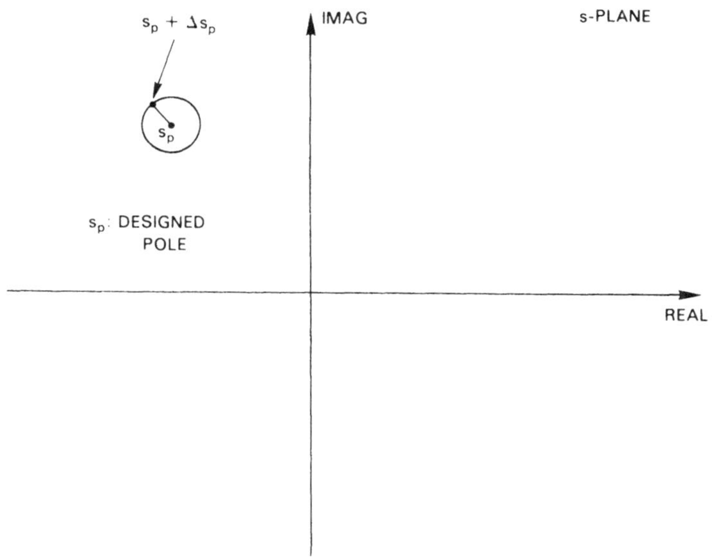

Fig. 2.1 — Pole with perturbation

We assume that real and imaginary parts of each perturbation are statistically independent and identically distributed zero mean random real variables. Also, for convenience we assume that the perturbations are statisically independent and identically distributed random complex variables. The variance of the magnitude of each perturbation is denoted by σ_F^2. Later, we show that the identically distributed limitation can be deleted. We note at this time that this variance may be a function of the order of filter and moreover each perturbation of a pole or zero may have a different variance.

We assume that the desired FTF is a ratio of polynomials such that

$$H(j\omega) = \frac{P(j\omega)}{Q(j\omega)} \tag{2.1}$$

where $P(\cdot)$ and $Q(\cdot)$ are polynominals of order m and n, respectively. Consider the Laplace transform representations of $P(j\omega)$ and $Q(j\omega)$: $P(s)$ and $Q(s)$. Let $s_1^{(p)}, s_2^{(p)}, \ldots, s_m^{(p)}$ be the roots of $P(s)$ and $s_1^{(q)}, s_2^{(q)}, \ldots, s_n^{(q)}$ be the roots of $Q(s)$. Therefore $P(j\omega)$ and $Q(j\omega)$ can be expressed as

$$P(j\omega) = (j\omega - s_1^{(p)}) \ldots (j\omega - s_m^{(p)}), \tag{2.2}$$

$$Q(j\omega) = (j\omega - s_1^{(q)}) \ldots (j\omega - s_n^{(q)}). \tag{2.3}$$

Consider just $P(j\omega)$. Let each root, $s_k^{(p)}, k = 1,2,\ldots,m$ be perturbed by a small amount, $\Delta s_k^{(p)}$. Then the numerator polynomial is actually $\tilde{P}(j\omega)$, where

$$\tilde{P}(j\omega) = (j\omega - s_1^{(p)} - \Delta s_1^{(p)}) \ldots (j\omega - s_m^{(p)} - \Delta s_m^{(p)}) . \tag{2.4}$$

We assume that no roots of $P(s)$ and $Q(s)$ lie on or are arbitrarily close to the $j\omega$ axis. This assumption allows us to write an expansion of $\tilde{P}(j\omega)$ and $\tilde{Q}(j\omega)$ which does not have any singular points. If we expand Eq. (2.4) and retain only the lower order terms, then

$$\tilde{P}(j\omega) = (j\omega - s_1^{(p)})\ldots(j\omega - s_m^{(p)}) - \sum_{k=1}^{m} (j\omega - s_1^{(p)})\ldots(j\omega - s_{k-1}^{(p)})\,(j\omega - s_{k+1}^{(p)})\ldots(j\omega - s_m^{(p)})\,\Delta s_k^{(p)} \tag{2.5}$$

$$= (j\omega - s_1^{(p)})\ldots(j\omega - s_m^{(p)})\left[1 - \sum_{k=1}^{m} \frac{\Delta s_k^{(p)}}{j\omega - s_k^{(p)}}\right]$$

$$= P(j\omega)\left[1 - \sum_{k=1}^{m} \frac{\Delta s_k^{(p)}}{j\omega - s_k^{(p)}}\right].$$

Similarly, we can show that the denominator polynomial when perturbed has the form

$$\tilde{Q}\,(j\omega) = Q(j\omega)\left[1 - \sum_{k=1}^{n} \frac{\Delta s_k^{(q)}}{j\omega - s_k^{(q)}}\right\}. \tag{2.6}$$

Therefore, the perturbed FTF has the form

$$\tilde{H}(j\omega) = \frac{P(j\omega)}{Q(j\omega)} \cdot \frac{1 - \sum_{k=1}^{m} \frac{\Delta s_k^{(p)}}{j\omega - s_k^{(p)}}}{1 - \sum_{k=1}^{n} \frac{\Delta s_k^{(q)}}{j\omega - s_k^{(q)}}} \tag{2.7}$$

or

$$\tilde{H}(j\omega) = H(j\omega)\left[1 + \sum_{k=1}^{n} \frac{\Delta s_k^{(q)}}{j\omega - s_k^{(q)}} - \sum_{k=1}^{m} \frac{\Delta s_k^{(p)}}{j\omega - s_k^{(p)}}\right] \tag{2.8}$$

where we have retained only the lower order terms.

We rewrite Eq. (28) as

$$\tilde{H}(j\omega) = H(j\omega)\left[1 + \sum_{k=1}^{n+m} \frac{\Delta s_k}{j\omega - s_k}\right] \tag{2.9}$$

where we have set

$$\left.\begin{aligned} \Delta s_k &= \Delta s_k^{(p)} \\ s_k &= s_k^{(p)} \end{aligned}\right\} \quad k = 1,2,\ldots,m \tag{2.10}$$

and

$$\left.\begin{aligned} \Delta s_{m+k} &= \Delta s_k^{(q)} \\ s_{m+k} &= \Delta s_k^{(q)} \end{aligned}\right\} \quad k = 1,2,\ldots,n\,. \tag{2.11}$$

As it was previously mentioned, we assume that H_M and H_A are designed to be matched to $H(j\omega)$, but because of inaccuracies are not equal to $H(j\omega)$. We use the pole/zero error model to express

$$H_M(j\omega) = H(j\omega)\left[1 + \sum_{m=1}^{M} \frac{\Delta s_m^{(M)}}{j\omega - s_m}\right] \tag{2.12}$$

and

$$H_A(j\omega) = H(j\omega)\left[1 + \sum_{m=1}^{M} \frac{\Delta s_m^{(A)}}{j\omega - s_m}\right] \tag{2.13}$$

where M is the number of poles and zeros of $H(j\omega)$ and m is now an index. The parameters $s_m, m = 1,2,\ldots M$ are some ordering of the poles and zeros of $H(j\omega)$, and $\Delta\ s_m^{(A)}, \Delta s_m^{(M)}$ are the perturbations of the poles and zeros of $H_A(j\omega)$ and $H_M(j\omega)$, respectively. These perturbations are assumed to be independent and identically distributed for both $H_A(j\omega)$ and $H_M(j\omega)$.

We set

$$\Delta H_A\ (j\omega) = \sum_{m=1}^{M} \frac{\Delta s_m^{(A)}}{j\omega - s_m} \tag{2.14}$$

and

$$\Delta H_M\ (j\omega) = \sum_{m=1}^{M} \frac{\Delta s_m^{(M)}}{j\omega - s_m}\ . \tag{2.15}$$

Therefore, the first order approximations of the perturbed main and auxiliary channel's FTF are

$$H_M(j\omega) = H(j\omega)\ (1 + \Delta H_M(j\omega)) \tag{2.16}$$

$$H_A(j\omega) = H(j\omega)\ (1 + \Delta H_A(j\omega))\ . \tag{2.17}$$

III. RESIDUE DERIVATION

In this section, we derive an expression for the output residue of the compensated canceller seen in Fig. 1.2. From this figure, we see that the output voltage, $v(t)$, can be expressed as

$$v(t) = x_M(t) - \mathbf{w}^T\mathbf{x}(t)\ . \tag{3.1}$$

If we set

$$P_{\text{out}} = E\{|v(t)|^2\} \tag{3.2}$$

and

$$P_{\text{in}} = E\{|x_M(t)|^2\} \tag{3.3}$$

where P_{out} and P_{in} are the output and input noise powers, respectively, then we can show [1] that

$$P_{\text{out}} = P_{\text{in}} - \mathbf{w}^t\mathbf{R}\mathbf{w} \tag{3.4}$$

where $\mathbf{R}$ is defined by Eq. (1.2) and $\mathbf{w}$ is the vector solution of Eq. (1.1). In fact, by using Eq. (1.1), we can show that

$$P_{\text{out}} = P_{\text{in}} - \mathbf{r}^t\mathbf{R}^{-1}\mathbf{r} \tag{3.5}$$

where t denotes the complex conjugate transpose operation. The output cancellation (or noise attenuation factor) $P_{\text{out}}/P_{\text{in}}$ can then be expressed by

$$\frac{P_{\text{out}}}{P_{\text{in}}} = \frac{P_{\text{in}} - \mathbf{r}^t\mathbf{R}^{-1}\mathbf{r}}{P_{\text{in}}}\ . \tag{3.6}$$

Note for the self-canceller that if $H_A(jw) = H_M(jw)$, that $P_{out}/P_{in} = 0$. We can show this as follows. Under the previous assumption of this analysis that the main and auxiliary inputs are identical, the optimal weighting, $\mathbf{w}_0$, for the self canceller is

$$\mathbf{w}_0 = (0\ 0 \ldots \underset{\substack{\uparrow \\ \frac{N+1}{2}\text{ position}}}{1}\ 0\ 0 \ldots 0)^T. \tag{3.7}$$

This is due to the fact that

$$x_{N_2}(t) = x_M(t) \tag{3.8}$$

where we have set

$$N_2 = \frac{N+1}{2}. \tag{3.9}$$

Hence, we simply subtract the N_2th output of the transversal filter seen in Fig. 1.2 from the output of the time delay element in the main channel to yield zero output noise power residue. As a result, if $\mathbf{r}_0$ and $\mathbf{R}_0$ are the cross covariance vector and covariance matrix under these ideal conditions (perfect matched filters), then

$$\mathbf{R}_0^{-1}\mathbf{r}_0 = \mathbf{w}_0 = (0\ 0 \ldots \underset{\substack{\uparrow \\ \frac{N+1}{2}\text{ position}}}{1}\ 0 \ldots 0)^T \tag{3.10}$$

We use the result of Eq. (3.10) quite often in the upcoming derivations to simplify many of our expressions.

The noise power spectrum $S_{yy}(\omega)$, of $y(t)$ is assumed to be white so that $S_{yy}(\omega) = 1$ for all ω. Expressions for the elements of $\mathbf{R}_0$ and $\mathbf{r}_0$ are easily derivable. It can be shown that if $R_{0,nm}$ is the nmth element of the matrix $\mathbf{R}_0$, then

$$R_{0,nm} = a\int_{-\infty}^{\infty} |H(j\omega)|^2 e^{j\omega\pi BT(n-m)}\,d\omega \quad n, m = 1, 2, \ldots, N \tag{3.11}$$

where a is some nonzero proportionality constant. In fact, in the following discussions we arbitrarily set $a = 1$ because we will be dealing with ratios of powers which implies that none of the outputs calculated will be a function of a. Note that we have normalized the angular frequency to the desired angular bandwidth πB where B is the frequency bandwidth of the desired FTF, $H(j\omega)$. Similarly, if $r_{0,n}$ is the nth element of $\mathbf{r}_0$, then

$$\mathbf{r}_{0,n} = \int_{-\infty}^{\infty} |H(j\omega)|^2 e^{j\omega\pi BT(n-N_2)}\,d\omega \quad n = 1, 2, \ldots, N. \tag{3.12}$$

We define the elements of the inverse of $\mathbf{R}_0$ as

$$\mathbf{R}_0^{-1} = (R_0^{(nm)}) \quad n, m = 1, 2, \ldots, N. \tag{3.13}$$

Expressions for the elements of $\mathbf{R}$ and $\mathbf{r}$ are given by

$$R_{nm} = \int_{-\infty}^{\infty} |H_A(j\omega)|^2\, e^{j\omega\pi BT(n-m)}\,d\omega \quad n, m = 1, 2, \ldots, N \tag{3.14}$$

and

$$r_n = \int_{-\infty}^{\infty} H_A^*(j\omega) H_M(j\omega) e^{j\omega\pi BT(n-N_2)}\,d\omega \quad n = 1, 2, \ldots, N. \tag{3.15}$$

If we use the first order approximations of $H_M(j\omega)$ and $H_A(j\omega)$ given by Eqs. (2.16) and (2.17) respectively, we can show by using Eqs. (3.14) and (3.15) that

$$R_{nm} = R_{0,nm} + \Delta R_{nm} \qquad n,m = 1,2,\ldots,N \tag{3.16}$$

$$r_n = r_{0,n} + \Delta r_n \qquad n = 1,2,\ldots,N \tag{3.17}$$

where

$$\Delta R_{nm} = \int_{-\infty}^{\infty} |H|^2(\Delta H_A + \Delta H_A^*) e^{j\omega\pi BT(n-m)}\, d\omega + \int_{-\infty}^{\infty} |H|^2\, |\Delta H_A|^2\, e^{j\omega\pi BT(n-m)}\, d\omega$$
$$n,\ m = 1,\ 2,\ \ldots,\ N \tag{3.18}$$

and

$$\Delta r_n = \int_{-\infty}^{\infty} |H|^2\, (\Delta H_A^* + \Delta H_M^*) e^{j\omega\pi BT(n-N_2)}\, d\omega + O(\Delta H_A^* \Delta H_M) \quad n = 1,\ 2,\ \ldots,\ N. \tag{3.19}$$

Furthermore, if we define

$$P_{\text{in}} = P_{\text{in}}^{(0)} + \Delta P_{\text{in}} \tag{3.20}$$

where $P_{\text{in}}^{(0)}$ is the input power when there are no perturbations, then we can show that

$$\Delta P_{\text{in}} = \int_{-\infty}^{\infty} |H|^2\, (\Delta H_M + \Delta H_M^*) d\omega + \int_{-\infty}^{\infty} |H|^2\, |\Delta H_M|^2 d\omega. \tag{3.21}$$

We rewrite the output power residue given by Eq. (3.5) in terms of the perturbations given by Eqs. (3.16), (3.17), and (3.20):

$$P_{\text{out}} = P_{\text{in}}^{(0)} + \Delta P_{\text{in}} - (\mathbf{r}_0 + \Delta\mathbf{r})'\, (\mathbf{R}_0 + \Delta\mathbf{R})^{-1}\, (\mathbf{r}_0 + \Delta\mathbf{r}) \tag{3.22}$$

where

$$\Delta\mathbf{r} = (\Delta r_1, \Delta r_2, \ldots, \Delta r_N)^T \tag{3.23}$$

and

$$\Delta\mathbf{R} = (\Delta R_{nm}) \quad n,\ m = 1,\ 2,\ \ldots,\ N. \tag{3.24}$$

Note that the $\Delta\mathbf{R}$ matrix is hermitian Toeplitz. We use a second order approximation of $(\mathbf{R}_0 + \Delta\mathbf{R})^{-1}$ This can be shown to be

$$(\mathbf{R}_0 + \Delta\mathbf{R})^{-1} = \mathbf{R}_0^{-1} - \mathbf{R}_0^{-1}\Delta\mathbf{R}\mathbf{R}_0^{-1} + \mathbf{R}_0^{-1}\Delta\mathbf{R}\mathbf{R}_0^{-1}\Delta\mathbf{R}\mathbf{R}_0^{-1}. \tag{3.25}$$

If Eq. (3.22) is expanded and only the second order and below perturbation terms are retained, then

$$\begin{aligned} P_{\text{out}} = P_{\text{in}}^{(0)} &+ \Delta P_{\text{in}} - \mathbf{r}_0'\mathbf{R}_0^{-1}\mathbf{r}_0 - \Delta\mathbf{r}'\mathbf{R}_0^{-1}\mathbf{r}_0 - \mathbf{r}_0'\mathbf{R}_0^{-1}\Delta\mathbf{r} \\ &- \Delta\mathbf{r}'\mathbf{R}_0^{-1}\Delta\mathbf{r} + \mathbf{r}_0'\mathbf{R}_0^{-1}\Delta\mathbf{R}\mathbf{R}_0^{-1}\mathbf{r}_0 + \Delta\mathbf{r}'\mathbf{R}_0^{-1}\Delta\mathbf{R}\mathbf{R}_0^{-1}\mathbf{r}_0 \\ &+ \mathbf{r}_0'\mathbf{R}_0^{-1}\Delta\mathbf{R}\mathbf{R}_0^{-1}\Delta\mathbf{r} - \mathbf{r}_0'\mathbf{R}_0^{-1}\Delta\mathbf{R}\mathbf{R}_0^{-1}\Delta\mathbf{R}\mathbf{R}_0^{-1}\mathbf{r}_0. \end{aligned} \tag{3.26}$$

Note that an immediate simplification of Eq. (3.26) results because

$$P_{\text{in}}^{(0)} - \mathbf{r}_0'\mathbf{R}_0^{-1}\mathbf{r}_0 = 0. \tag{3.27}$$

We average P_{out} over the identical zero mean probability density functions (p.d.f.'s) of the pole and zero perturbations in order to obtain an average cancellation residue. Since the p.d.f.'s are zero mean, it follows immediately that

$$E\{\Delta\mathbf{r}'\mathbf{R}_0^{-1}\mathbf{r}_0\} = 0 \tag{3.28}$$

and

$$E\{\mathbf{r}_0'\mathbf{R}_0^{-1}\Delta\mathbf{r}\} = 0. \tag{3.29}$$

Further simplifications are possible due to Eq. (3.10). Derivations of the rest of the terms in Eq. (3.26) averaged over the pole/zero perturbations are given in Appendix A. We merely state the results. If σ_F^2 is the variance of a pole/zero perturbation, then

$$E\{\Delta P_{\text{in}}\} = \sigma_F^2\Gamma_1 \tag{3.30}$$

$$E\{\Delta\mathbf{r}'\mathbf{R}_0^{-1}\Delta\mathbf{r}\} = 2\sigma_F^2\Gamma_2 \tag{3.31}$$

$$E\{\mathbf{r}_0'\mathbf{R}_0^{-1}\Delta\mathbf{R}\mathbf{R}_0^{-1}\Delta\mathbf{R}\} = E\{\Delta\mathbf{r}'\mathbf{R}_0^{-1}\Delta\mathbf{R}\mathbf{R}_0^{-1}\mathbf{r}_0\} = \sigma_F^2\Gamma_2 \tag{3.32}$$

$$E\{\mathbf{r}_0'\mathbf{R}_0^{-1}\Delta\mathbf{R}\mathbf{R}_0^{-1}\mathbf{r}_0\} = \sigma_F^2\Gamma_1 \tag{3.33}$$

$$E\{\mathbf{r}_0'\mathbf{R}_0^{-1}\Delta\mathbf{R}\mathbf{R}_0^{-1}\Delta\mathbf{R}\mathbf{R}_0^{-1}\mathbf{r}_0\} = 2\sigma_F^2\Gamma_2 \tag{3.34}$$

where

$$\Gamma_1 = \sum_{i=1}^{M}\int_{-\infty}^{\infty}\frac{|H|^2 d\omega}{|j\omega - s_i|^2} \tag{3.35}$$

and

$$\Gamma_2 = \sum_{i=1}^{M}\sum_{k=1}^{N}\sum_{m=1}^{N}\left[R_0^{(km)}\cdot\int_{-\infty}^{\infty}\frac{|H|^2 e^{j\omega\pi BT(N_2-k)}d\omega}{j\omega - s_i}\cdot\int_{-\infty}^{\infty}\frac{|H|^2 e^{j\omega\pi BT(m-N_2)}d\omega}{-j\omega - s_i^*}\right]. \tag{3.36}$$

We define $P_{\text{out}}^{(\text{ave})}$ to be equal to P_{out} average over all the perturbations. Examining Eq. (3.6) and Eqs. (3.30) to (3.34), we see by use of the above results that $P_{\text{out}}^{(\text{ave})}/P_{\text{in}}$ is proportional to σ_F^2 and that the constant of proportionality in the first order approximation does not change if we set $P_{\text{in}} = P_{\text{in}}^{(0)}$. By adjusting the gain of $|H(j\omega)|^2$ we can arbitrarily set $P_{\text{in}}^{(0)} = 1$; i.e.,

$$\int_{-\infty}^{\infty}|H(j\omega)|^2 d\omega = 1. \tag{3.37}$$

Thus by substituting Eqs. (3.30) to (3.36) into Eq. (3.26), and by simplifying and normalizing we can show that

$$\frac{P_{\text{out}}^{(\text{ave})}/P_{\text{in}}}{\sigma_F^2} = 2\sum_{i=1}^{M}\left[\int_{-\infty}^{\infty}\frac{|H|^2 d\omega}{|j\omega - s_i|^2} - \sum_{k=1}^{N}\sum_{m=1}^{N}R_0^{(km)}\int_{-\infty}^{\infty}\frac{|H|^2 e^{j\omega\pi BT(N_2-k)}d\omega}{j\omega - s_i}\cdot\int_{-\infty}^{\infty}\frac{|H|^2 e^{j\omega\pi BT(m-N_2)}d\omega}{-j\omega - s_i^*}\right]. \tag{3.38}$$

We call $[P_{\text{out}}^{(\text{ave})}/P_{\text{in}}]/\sigma_F^2$ the cancellation-filter mismatch ratio, CFMR.

We note that the restriction that all the poles and zeroes have the same variance, σ_F^2, can be easily removed. If $\sigma_{F_m}^2$ is the variance of the perturbation on the mth pole or zero of $H(s)$, then we could have simply expanded $P_{\text{out}}^{(\text{ave})}$ in terms of $\sigma_{F_m}^2$, $m = 1,2, \ldots, M$. The resultant formulation would have exactly the same form as Eq. (3.38) except that σ_F^2 is replaced by $\sigma_{F_{\text{ave}}}^2$ where

$$\sigma_{F_{\text{ave}}}^2 = \frac{1}{M} \sum_{m=1}^{M} \sigma_{F_m}^2 . \tag{3.39}$$

IV. A SPECIAL CASE: THE BUTTERWORTH FILTER

In this section we evaluate the cancellation-filter mismatch ratio, CFMR, for the case when the desired transfer function is a Butterworth filter. This filter is of much interest because it is easily synthesized and is a bandpass filter with the attenuation of the skirts controlled by the order of the filter.

This filter has the following magnitude squared angular frequency response:

$$|H(j\omega)|^2 = \frac{c_0}{1 + \omega^{2M}} \tag{4.1}$$

where M defines the order of the filter, the angular frequency has been normalized to the desired angular bandwidth, πB, and

$$c_0 = \frac{M}{\pi} \sin \frac{\pi}{2M} . \tag{4.2}$$

The constant c_0 has been chosen so that Eq. (3.37) is satisfied. Curves of the Butterworth filter response are shown in Fig. 4.1 for various values of M. Note, that by increasing the order of the Butterworth filter, M, that the skirts of the bandpass filter become more attenuated.

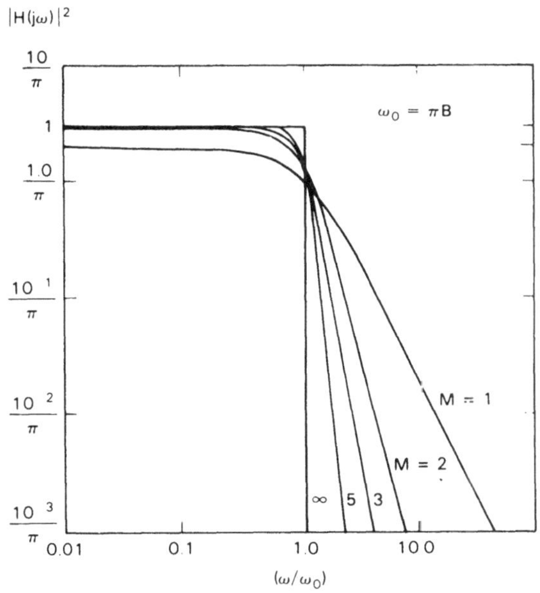

Fig. 4.1 — Butterworth filter response

The filter is synthesized by finding an $H(s)$ function whose poles are in the left-hand side of the s-plane such that

$$H(s)H(-s)|_{s=j\omega} = |H(j\omega)|^2 . \tag{4.3}$$

Now the poles of $|H(j\omega)|^2$ can be shown to lie on the unit circle and are spaced equally in angle as illustrated in Fig. 4.2 for $M = 3$. Hence in order to find $H(s)$, the M left-hand plane poles of $|H(j\omega)|^2$ are identified and used to form the polynomial, $H(s)$; i.e., if s_i, $i = 1, 2, \ldots, M$ are the left-handed poles, then

$$H(s) = \sqrt{c_0}[(s - s_1)(s - s_2) \ldots (s - s_M)]^{-1}. \tag{4.4}$$

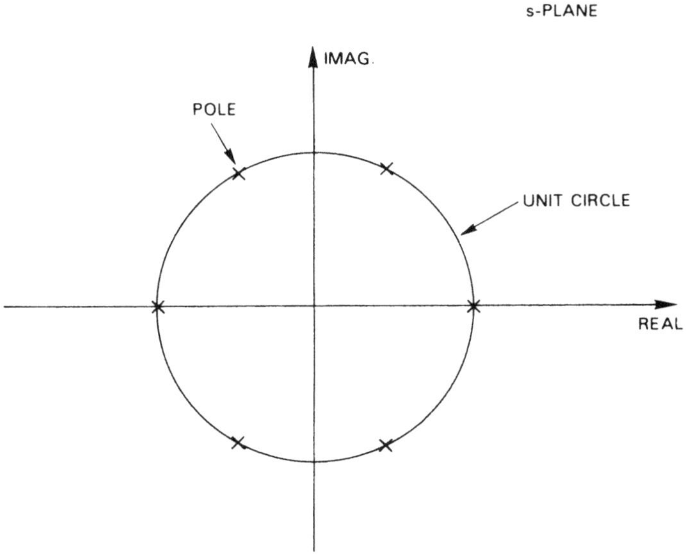

Fig. 4.2 — Pole plot of Butterworth filter of order, 3

To evaluate the CFMR, the integrals of Eqs. (3.35) and (3.36) must be evaluated. This can be done by using the Theory of Residues [2] and is outlined in Appendix B where expressions for these integrals are obtained. From Eq. (3.38), we also need an exact expression for the elements of $\mathbf{R}_0$ from which the elements of $\mathbf{R}_0^{-1}$ can be obtained through matrix inversion. Summarizing the results of Appendix B:

$$R_{nm} = \frac{\pi c_0}{M} \sum_{l=1}^{M} e^{\pi BT(n-m)\sin\frac{\pi}{2M}(2l-1))} e^{j\pi(\frac{1}{2}-\frac{1}{2M}(2l-1)+BT(n-m)\cos\frac{\pi}{2M}(2l-1))} \quad \text{for } m > n \tag{4.5a}$$

$$R_{nm} = \frac{\pi c_0}{M} \sum_{l=1}^{M} e^{-\pi BT(n-m)\sin\frac{\pi}{2M}(2l-1)} e^{j\pi(-\frac{1}{2}+\frac{1}{2M}(2l+1)+BT(n-m)\cos\frac{\pi}{2M}(2l-1))} \quad \text{for } m \leqslant n \tag{4.5b}$$

$$\int_{-\infty}^{\infty} \frac{|H|^2 d\omega}{|j\omega - s_i|^2} = \frac{\pi c_0}{2} \frac{1}{\sin \frac{\pi}{2M}(2i - 1)} \tag{4.6}$$

$$\int_{-\infty}^{\infty} \frac{|H|^2 e^{j\omega\pi BT(N_2-n)} d\omega}{j\omega - s_i}$$

$$= \frac{\pi c_0}{M} \sum_{m=1}^{M} \frac{e^{(N_2-n)\pi BT\sin\frac{\pi}{2m}(2m-1)} e^{j(N_2-n)\pi BT\cos\frac{\pi}{2M}(2m-1)}}{1 - e^{j\frac{\pi}{M}(i + m - 1)}} \quad \text{for } N_2 \leqslant n \tag{4.7a}$$

$$\int_{-\infty}^{\infty} \frac{|H|^2 e^{j\omega \pi BT (N_2 - n)} d\omega}{j\omega - s_i}$$

$$= \frac{\pi c_0}{M} \sum_{m=1}^{M} \left[(M - .5 - (N_2 - n)\, \pi BTje^{j\frac{\pi}{2M}(2i-1)})\delta_{im} - \frac{1 - \delta_{im}}{1 - e^{j\frac{\pi}{M}(i-m)}} \right]$$

$$\cdot\, e^{-(N_2-n)\pi BT \sin\frac{\pi}{2M}(2m-1)}\, e^{j(N_2 - n)\pi BT \cos\frac{\pi}{2M}(2m-1)} \quad \text{for } N_2 > n \tag{4.7b}$$

and

$$\int_{-\infty}^{\infty} \frac{|H|^2 e^{j\omega\pi BT(n-N_2)} d\omega}{-j\omega - s_i^*} = \left[\int_{-\infty}^{\infty} \frac{|H|^2 e^{j\omega\pi BT(N_2-n)} d\omega}{j\omega - s_i} \right]^* \tag{4.8}$$

where

$$\delta_{im} = \begin{cases} 1 & \text{if } i = m \\ 0 & \text{otherwise} \end{cases} . \tag{4.9}$$

Note that Eq. (4.8) follows by the definition of the complex conjugate operation.

V. RESULTS

A. Introduction

In this section, we present some results of the effects of frequency mismatch errors on self-cancellation. Specifically, curves of the Cancellation-Frequency Mismatch Ratio, CFMR, versus the various input parameters,

- M, the order of the Butterworth filter
- N, the number of time delay taps, and
- BT, the filter bandwidth-tapped time delay product

are plotted and discussed. Also, a procedure for choosing the "optimum" M, N, and BT is outlined in the Discussion subsection.

Note that in some cases the numerical calculations involved in obtaining CFMR were very sensitive to the eigenvalues of $\mathbf{R}_0$, the ideal covariance matrix (even using double precision complex arithmetic on a VAX-750). Because of this sensitivity it was necessary to add statistically independent internal noise to each of the tapped delay outputs. In regards to $\mathbf{R}_0$, this added an internal noise power term, σ_n^2, to each of the diagonal elements of $\mathbf{R}_0$. It was found that an external jamming (the self-cancellation input signal) to internal noise ratio, J/N, of 50 dB resulted in excellent numerical stability in calculating CFMR. Furthermore, if J/N were increased to 70 dB, the results changed minutely. Hence for the curves to be presented, we list the J/N under which the CFMR was calculated.

B. The Butterworth Filter Order

Figures 5.1 through 5.10 present plots of CFMR versus the order of the Butterworth filter, M. For each figure, BT is held constant and contours of CFMR with N, the number of time delay taps are plotted. We note that cancellation improves as CFMR decreases. From these figures, we observe that

- cancellation (or CFMR) obviously improves by increasing the number of time delay taps;
- the CFMR in many cases has a minimum as a function of the order of the Butterworth filter, M; and
- the minimum also depends on the bandwidth-tapped time delay product, BT.

We discuss the first and third observations in more detail in subsections D and C, respectively.

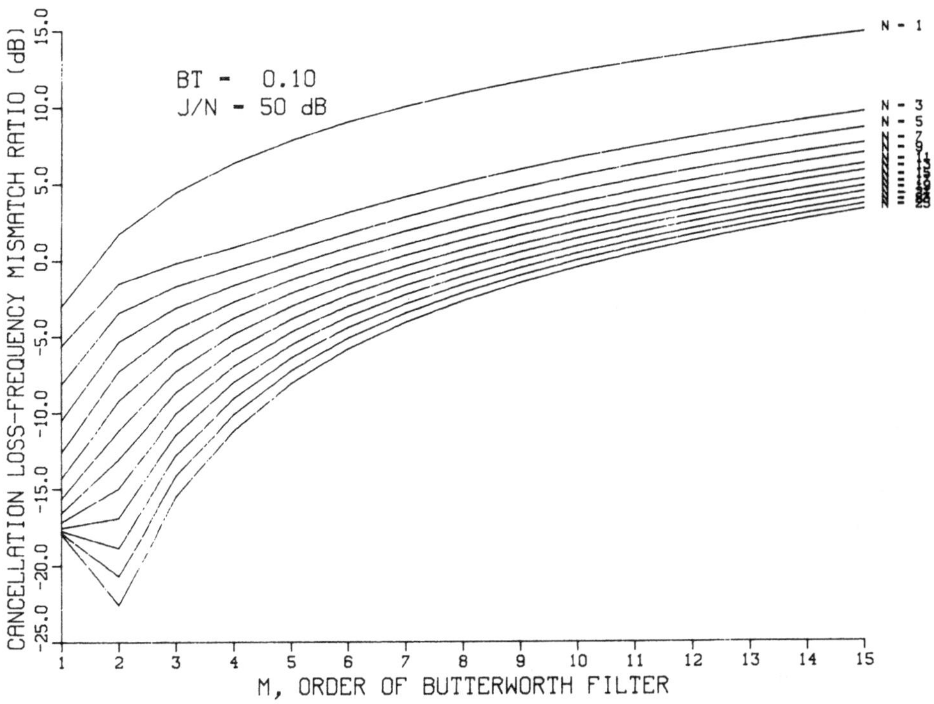

Fig. 5.1 — CFMR vs M, BT = .1

The occasional minimum in the curves as function of M can be explained as follows. There is a certain amount of aliasing which occurs due to the transversal filter in the auxiliary channel; i.e., the transversal filter has a periodic frequency response. Hence the tails of the main channel's perturbed Butterworth filter frequency response near the sampling rate frequency (which is assumed constant for each figure and related to BT) are not accurately matched. As the order of the Butterworth filter increases, these tails decrease in magnitude about the sampling rate frequency and hence auxiliary channel aliasing effects decrease. As a result there is better matching up to a point. The fact that we are degrees of freedom (DOF) limited (recall that N is assumed a constant), results in the CFMR increasing after reaching some minimum; i.e., the more poles in the Butterworth filter, the more irregularities in the frequency response which must be matched.

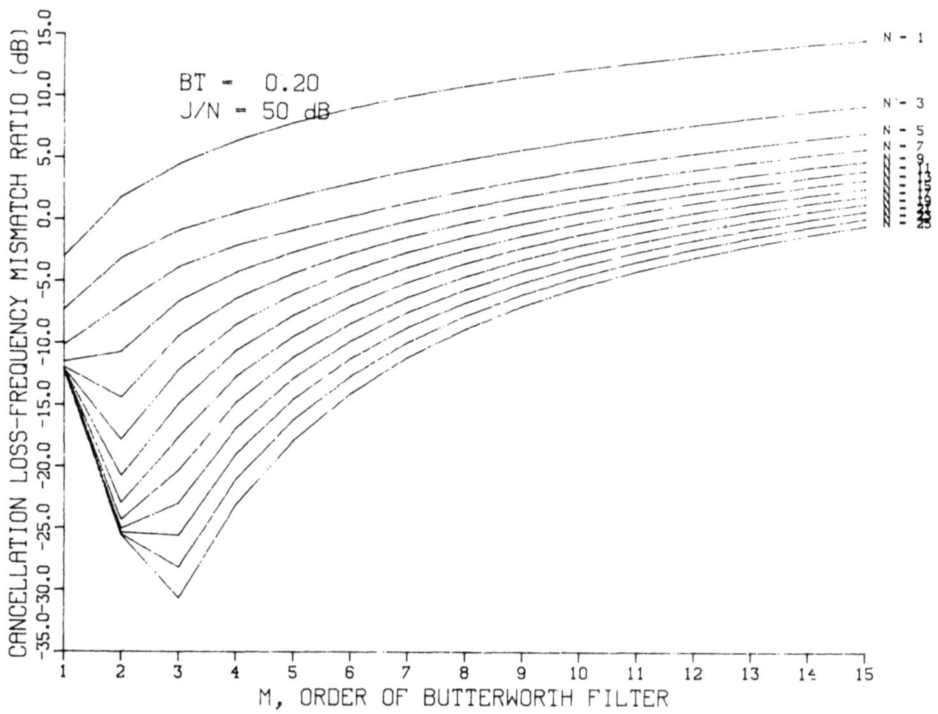

Fig. 5.2 — CFMR vs M, BT = .2

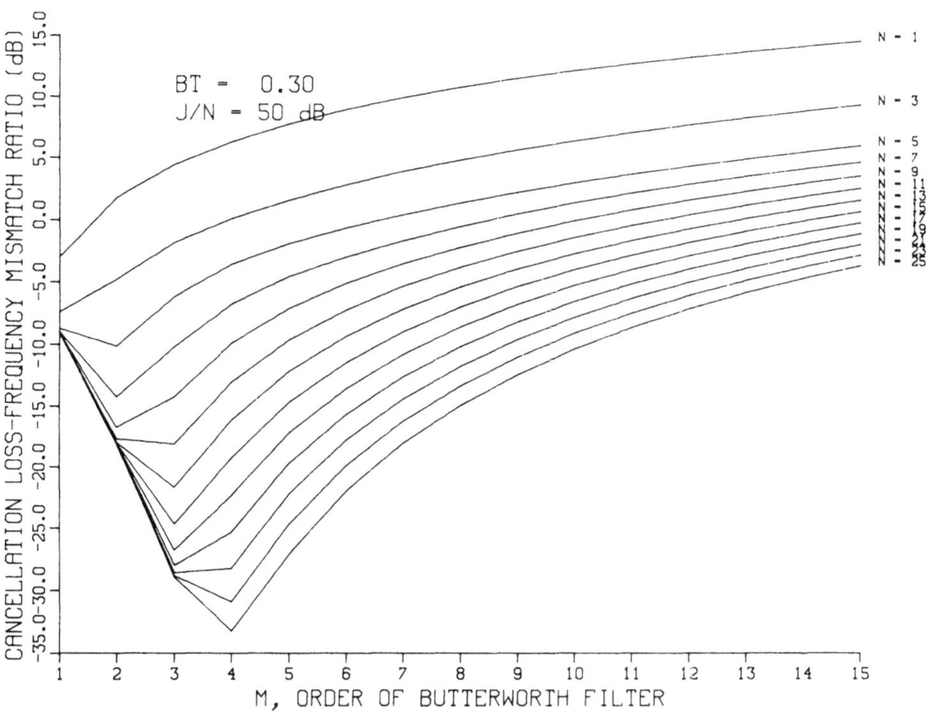

Fig. 5.3 — CFMR vs M, BT = .3

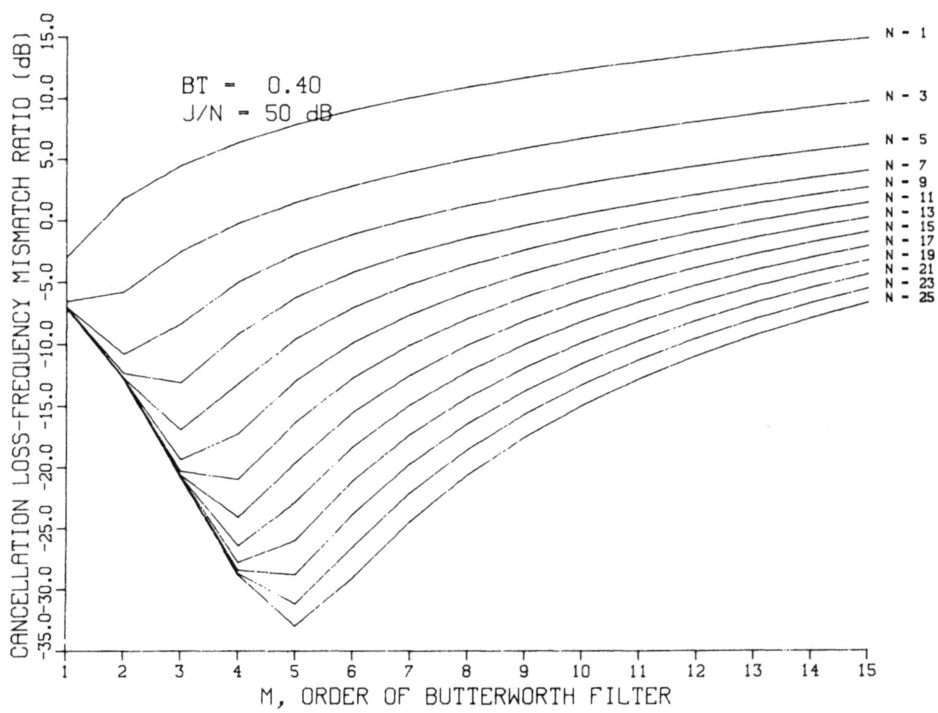

Fig. 5.4 — CFMR vs M, BT = .4

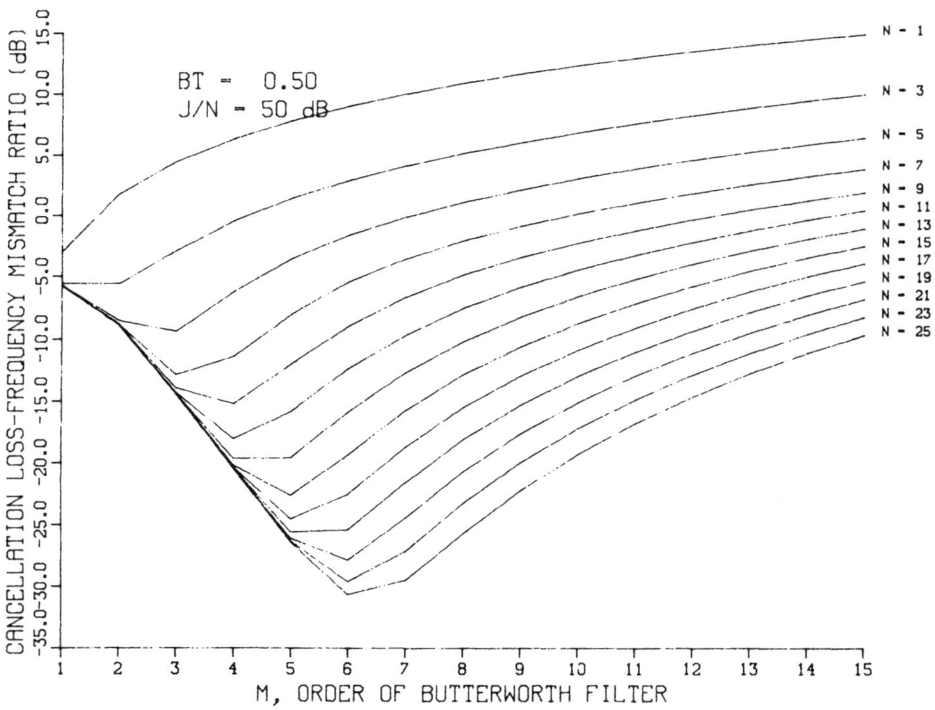

Fig. 5.5 — CFMR vs M, BT = .5

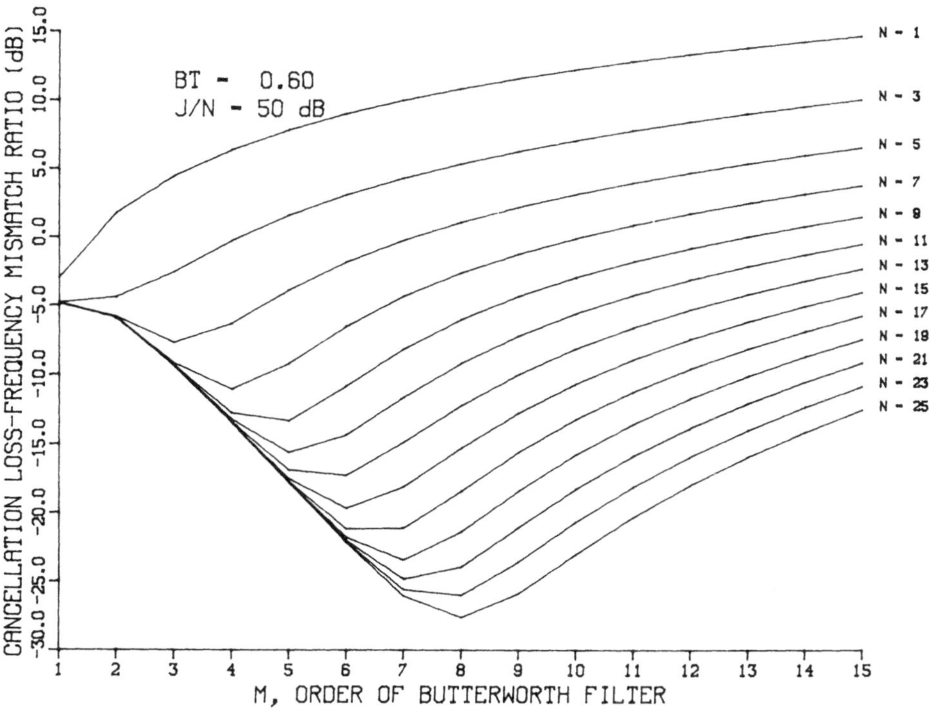

Fig. 5.6 — CFMR vs M, BT = .6

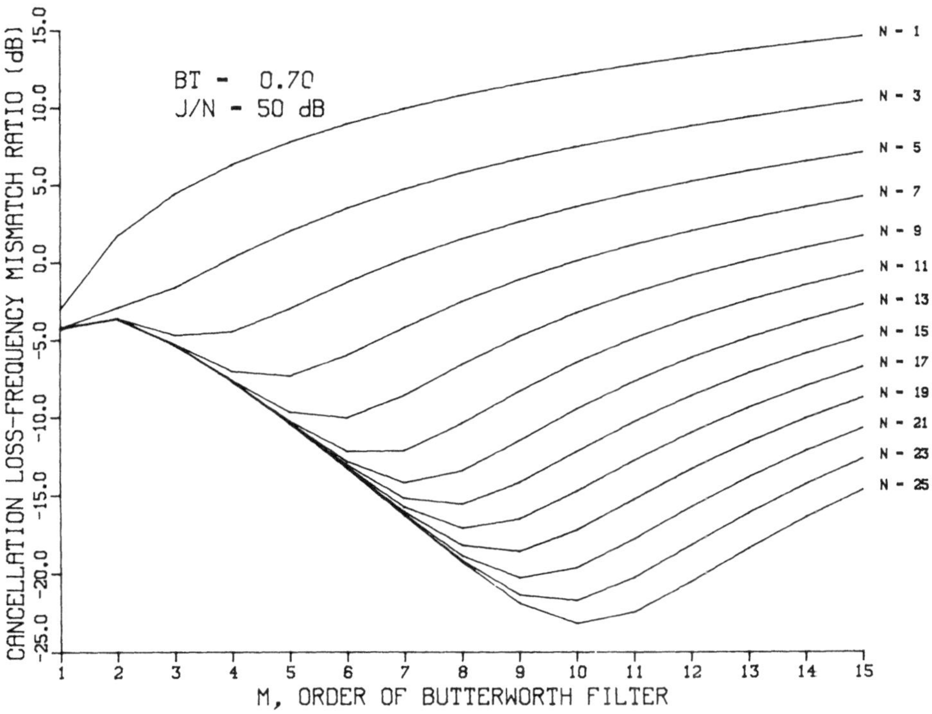

Fig. 5.7 — CFMR vs M, BT = .7

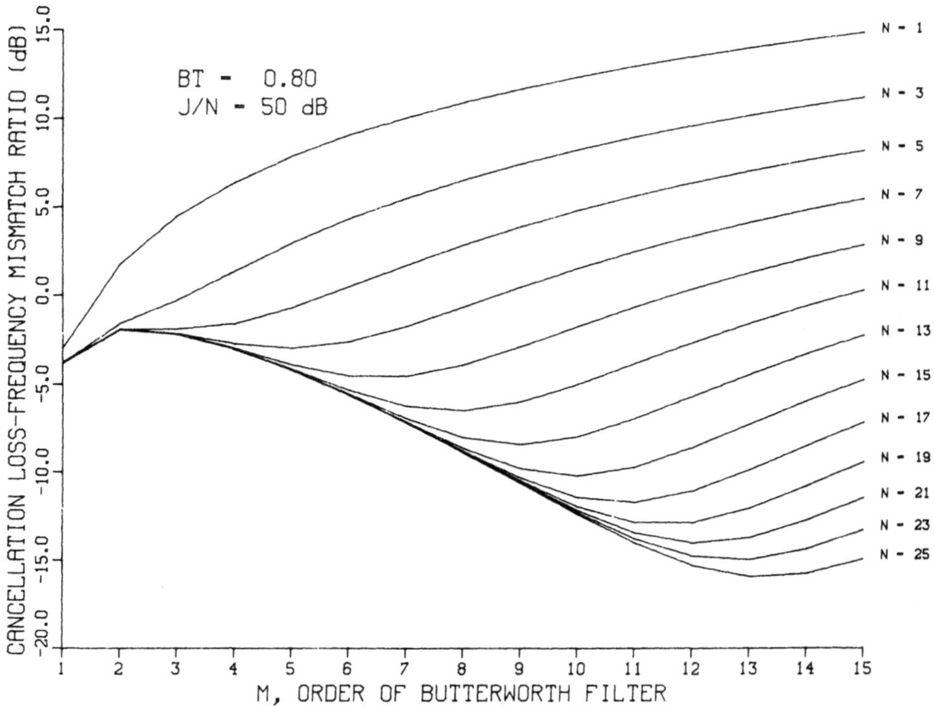

Fig. 5.8 — CFMR vs M, $BT = .8$

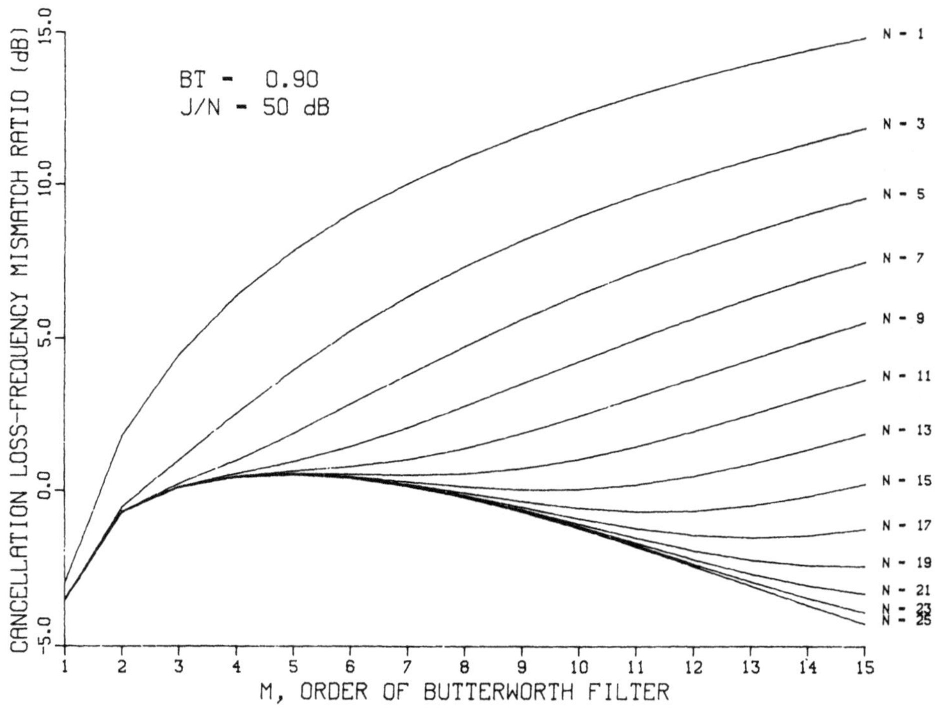

Fig. 5.9 — CFMR vs M, $BT = .9$

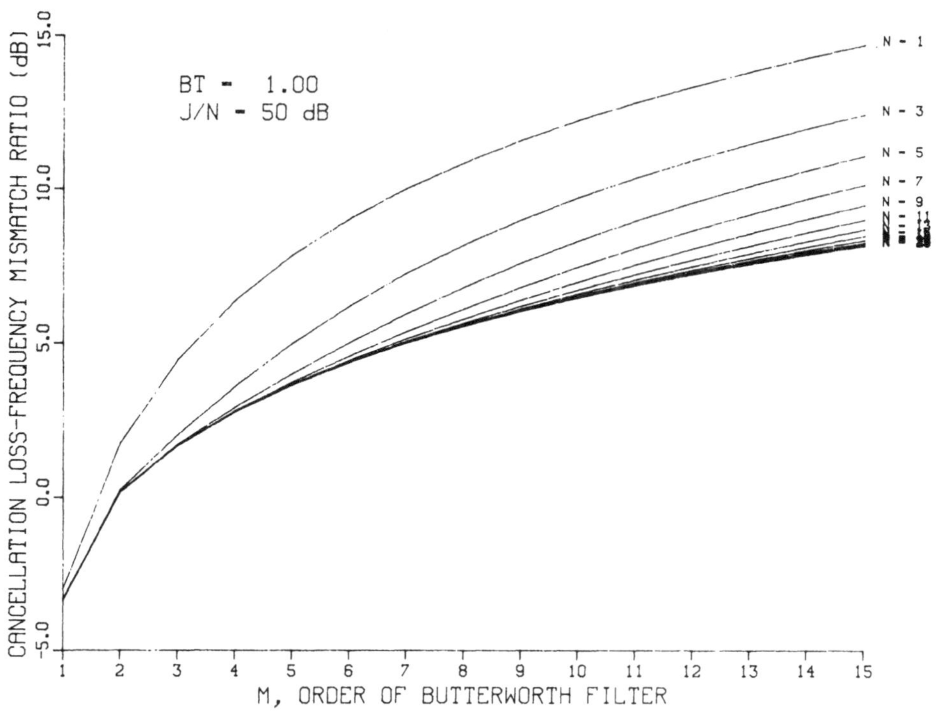

Fig. 5.10 — CFMR vs M, BT = 1

C. The Bandwidth-Tapped Time Delay Product

The minimum values of CFMR as seen in Figs. 5.1 to 5.9 are also functions of the bandwidth-tapped time delay product, BT. From these figures, we see that for a constant N and decreasing BT, that the minimal CFMR occurs at decreasing values of M. This can again be explained by considering aliasing effects: a higher sampling rate (or smaller BT) can tolerate a lower order of Butterworth filter while maintaining a constant aliasing degradation.

Note that by comparing Fig. 5.10 where $BT = 1$ with Figs. 5.1 to 5.9 where $BT < 1$, that sampling at the filter bandwidth (or the information bandwidth) results in very poor cancellation performance. This is again due to aliasing, which is caused by the periodic frequency response of the compensating transversal filter. Hence, for cancellation systems using a bandpass filter, one should never sample at the Nyquist rate if possible.

In Fig. 5.11, we have plotted CFMR versus BT for various values of N and $M = 6$. Here, we observe that the CFMR has a minimum with respect to BT (for this example, the minimums over all curves occurs when $BT \approx 0.6$). The value of CFMR decreases as BT decreases from one because the negative effects of aliasing are being reduced. However, as BT becomes smaller, the tapped time delay decreases and because there is a fixed number of taps, the transversal filter cannot accurately match the main channel's irregularities. In essence, the tails of the time correlation function associated with the power spectrum of the transversal filter degrades in matching the tails of the main channel's time correlation function (associated with its spectrum) because NT decreases as BT decreases. In the limit of course as BT or $T \rightarrow 0$, the transversal filter with N taps is equivalent to a transversal filter with just one tap (and no time delays). Hence, CFMR must increase as BT becomes very small.

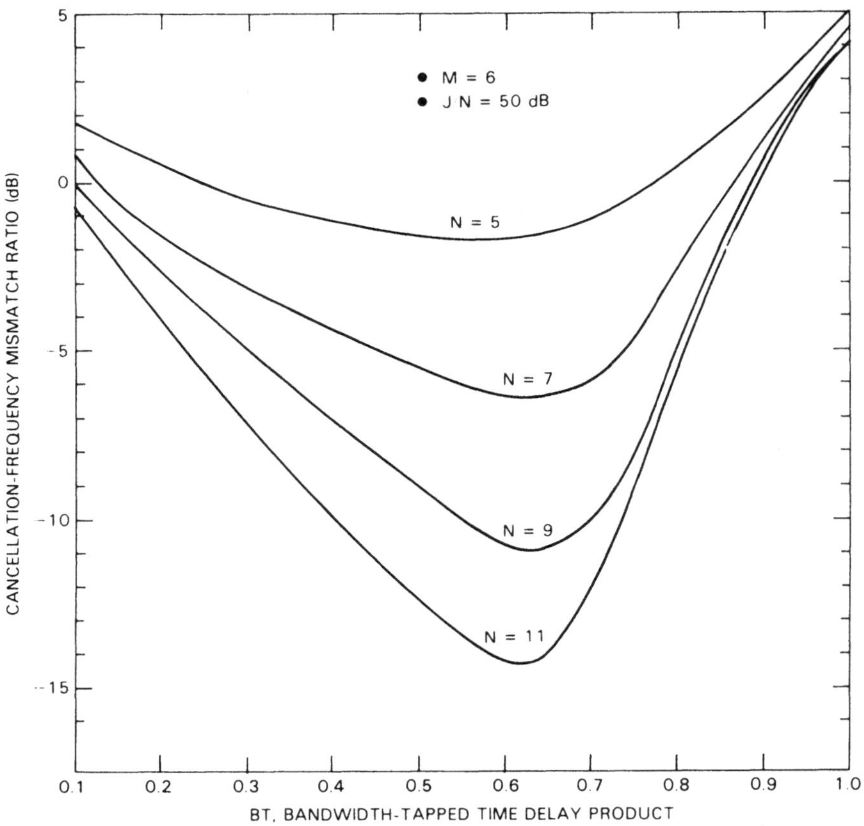

Fig. 5.11 — CFMR vs BT, $M = 6$

D. The Number of Time Delay Taps

In Fig. 5.12, curves of CFMR are plotted versus N, the number of time delay taps for various values of M with $BT = 1$. It was previously observed that CFMR decreases as N increases for a constant M and BT. However, note that the cancellation goes to some lower bound as $N \rightarrow \infty$. This results because of the adaptive transversal filter, regardless of its order, is not a good match to the main channel beyond the sampling rate frequency. The main channel's frequency spectrum has tails which extend beyond the sampling rate frequency. These tails are poorly matched by the transversal filter, and a finite noise power residue results which is independent of N but dependent on BT and M. Hence, we see that merely increasing the number of time taps does *not* result in the cancelled noise residue going to zero or some arbitrarily small number. We also note that the CFMR is not necessarily monotonic with the order of the Butterworth filter, M, as depicted in Fig. 5.12 for when $BT = 1$. In fact for all cases, $BT < 1$, it is not.

E. Discussion

In this subsection, we briefly outline a design procedure for choosing BT, M, and N such that the cancellation is "optimized" in some sense. In our previous analysis, we have only considered one auxiliary channel. In many applications there are multiple auxiliary channels which are used to cancel the noise in the main channel. Each of these channels has a bandpass filter and an adaptive compensating transversal filter, as was shown for the case of the single auxiliary as seen in Fig. 1.1. For the self-canceller, because the adaptive transversal filter in each auxiliary channel attempts to match each of the

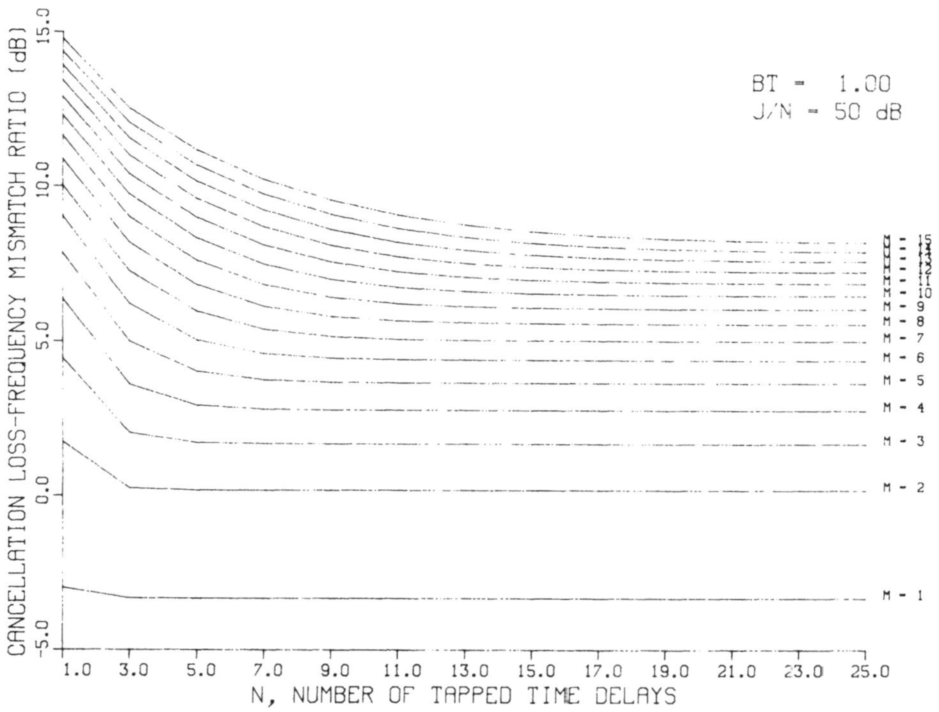

Fig. 5.12 — CFMR vs N, BT = 1

auxiliary channels to the main channel, we see that choosing the best BT, M, and N for one auxiliary channel is essentially equivalent to choosing the best BT, M, and N for any number of auxiliary channels. Hence, the design procedure, which will be outlined, is applicable to a canceller with any number of auxiliary channels.

Let us set

$$\frac{C}{\sigma_F^2} = \text{CFMR} \tag{5.1}$$

where

$$C = \frac{P_{\text{out}}^{(\text{ave})}}{P_{\text{in}}}. \tag{5.2}$$

Here we have defined C to be the average cancellation. We can readily discern two types of optimization problems:

1. maximize (in a negative sense) the cancellation, C, under a given set of constraints or,

2. given a required C, maximize σ_F^2 under a given set of constraints.

For many kinds of antijamming design problems, the second type of problem is more realistic. In this case, we calculate how much cancellation is needed in a given jamming scenario, and try to relax to system specification, σ_F^2, as much as possible.

Both types of design problems are subject to a number of constraints on M, N, and BT. We briefly list the constraints and some of the causes of these constraints.

- $N_{min} \leqslant N$: set by the matching requirements among channels for the external signals, multipath, antennas, RF front ends, etc.
- $N \leqslant N_{max}$: set by computational load limits, finite sampling window
- $BT_{min} \leqslant BT$: information bandwidth requirement
- $BT \leqslant BT_{max}$: system limitations
- $M \leqslant M_{max}$: synthesis limitations, cost, response time
- $\sigma_F^2 = f(M)$: error is a function of Butterworth filter order.

Note in the optimizations procedure for most cases the optimal N will equal N_{max}. The last item mentioned above would have to be determined by a statistical analysis of the rms error of the poles as a function of the Butterworth filter order and the errors in the synthesis process.

Hence we see that given that the constraints on BT, M, and N are defined, and given the curves of CFMR (or Eq. 3.38), a computer search program could be developed which finds the BT, M, and N which either maximizes the cancellation (problem 1) or maximizes σ_F^2 (problem 2, given C required).

We should point out, however, that one rarely goes to a filter designer with a specification such as the variance of the pole perturbation. Normally one specifies the ripple across the passband and the rms difference between the synthesized filter and the ideal filter across the passband. Hence the canceller designer must make a conversion from σ_F^2 to these filter design parameters if some meaningful specification is to be made. After the filter is fabricated, the canceller designer can have the filter designer test the filter and see that the poles fall within the variance, σ_F^2. (Note that if one specifics $\sigma_{F_{ave}}^2$ over all poles to less than σ_F^2, the filter specifications are satisifed.)

For example, let us assume that σ_F^2 is not a function of M and that BT, M, and N have the following constraints: $1 \leqslant BT \leqslant 1$, $5 \leqslant N \leqslant 11$, $1 \leqslant M \leqslant 15$. We desire -30 dB of cancellation. Under these conditions, we would find that the optimal parameters are $BT = 0.4$, $N = 11$, and $M = 4$ with $\text{CFMR}_{opt} = -19$ dB. Hence $\sigma_F^2 = -11$dB, or the relative rms perturbation error on the poles is approximately 30%. Note that if we desired 50 dB of cancellation, this relative rms error would be approximately 3%.

VI. SUMMARY AND CONCLUSIONS

The effects of frequency mismatch errors on adaptive cancellers have been investigated. The frequency mismatch errors occur because of errors in the synthesis process of supposedly identical bandpass filters which are in each of the input channels. These frequency mismatches among the channels result in cancellation degradation. Tapped delay line transversal filters can be used to compensate for these frequency mismatches and thus improve cancellation performance.

A pole/zero error model of the filters has been developed whereby closed form solutions of the maximum achievable cancellation are obtained. This cancellation is a function of the order of the ideally matched frequency filters, the number of time delay taps in the compensating transversal filter, the bandwidth-tapped time delay product, and the constraints on these parameters. A design procedure was outlined for "optimizing" the canceller with respect to these parameters and their constraints. Specifically, results were presented for when the input filters are the Butterworth type. It was shown that an arbitrarily low output noise residue *cannot* be achieved by arbitrarily increasing the number of time delay taps.

VII. REFERENCES

1. R.A. Monzingo and T.W. Miller, *Introduction to Adaptive Arrays* (John Wiley and Sons, 1980).

2. R.V. Churchill, *Complex Variables and Applications* (McGraw-Hill, 1960).

Appendix A
DERIVATION OF EQUATIONS (3.30) TO (3.34)

1. The $E\{\Delta P_{in}\}$ Term:

We showed in the text (see Eq. 3.21) that

$$\Delta P_{\text{in}} = \int_{-\infty}^{\infty} |H|^2 \, (\Delta H_M + \Delta H_M^*) \, d\omega + \int_{-\infty}^{\infty} |H|^2 \, |\Delta H_M|^2 \, d\omega. \tag{A1}$$

Now

$$\Delta H_M(j\omega) = \sum_{i=1}^{M} \frac{\Delta s_i^{(M)}}{j\omega - s_i}. \tag{A2}$$

Let $\Delta s_i^{(M)}, i = 1,2, \ldots, M$, be identically distributed independent zero mean random variables with covariance: σ_F^2. Also let the real and imaginary parts of $\Delta s_i^{(M)}$ be identically distributed and independent. If we substitute Eq. (A2) into Eq. (A1) and take the expected value, the first term of Eq. (A1) is zero because the Δs_i is zero mean and the following expression results:

$$E\{\Delta P_{\text{in}}\} = \int_{-\infty}^{\infty} |H|^2 \, E\left\{\left| \sum_{i=1}^{M} \frac{\Delta s_i^{(M)}}{j\omega - s_i} \right|^2\right\} d\omega. \tag{A3}$$

If we rewrite the summation by expanding the magnitude in Eq. (A3) and evaluate the expected values of the cross terms (many are equal to zero), then Eq. (A3) can be simplified to

$$E\{\Delta P_{\text{in}}\} = \sigma_F^2 \sum_{i=1}^{M} \int_{-\infty}^{\infty} \frac{|H|^2 d\omega}{|j\omega - s_i|^2}. \tag{A4}$$

2. The $E\{\mathbf{r}_0^t \, \mathbf{R}_0^{-1} \, \Delta\mathbf{R} \, \mathbf{R}_0^{-1} \, \mathbf{r}_0\}$ Term:

Because $\mathbf{R}_0^{-1} \, \mathbf{r}_0 = (0,0,\ldots,1,0,\ldots 0)^T$ where the 1 is in the N_2 position (the middle), the evaluation of this term reduces to just finding the expected value of the center element of $\Delta\mathbf{R}$. From Eq. (3.18), it follows that

$$\Delta R_{N_2 \, N_2} = \int_{-\infty}^{\infty} |H|^2 \, (\Delta H_A + \Delta H_A{}^*) \, d\omega + \int_{-\infty}^{\infty} |H|^2 |\Delta H_A|^2 \, d\omega. \tag{A5}$$

Now

$$\Delta H_A \, (j\omega) = \sum_{i=1}^{M} \frac{\Delta s_i^{(A)}}{j\omega - s_i}. \tag{A6}$$

The $\Delta s_i^{(A)}$ have the same kind of statistics as $\Delta s_i^{(M)}$. The forms of Eqs. (A5) and (A6) are identical to Eqs. (A1) and (A2), so that it follows from our previous derivation that

$$E\{\mathbf{r}_0^t \, \mathbf{R}_0^{-1} \, \Delta\mathbf{R}\mathbf{R}_0^{-1} \, \mathbf{r}_0\} = E\{\Delta P_{\text{in}}\}. \tag{A7}$$

The expression for $E\{\Delta P_{\text{in}}\}$ is given by Eq. (A4).

3. The $E\{\mathbf{r}_0^t \mathbf{R}_0^{-1} \Delta\mathbf{R}\mathbf{R}_0^{-1} \Delta\mathbf{R}\, \mathbf{R}_0^{-1}\mathbf{r}_0\}$ *Term:*

We can write

$$\mathbf{r}_0^t\mathbf{R}_0^{-1} \Delta\mathbf{R}\mathbf{R}_0^{-1} \Delta\mathbf{R}\mathbf{R}_0^{-1} \mathbf{r}_0 = (\mathbf{R}_0^{-1}\mathbf{r}_0)^t \Delta\mathbf{R}\mathbf{R}_0^{-1} \Delta\mathbf{R}\, (\mathbf{R}_0^{-1} \mathbf{r}_0) . \tag{A8}$$

Since $\mathbf{R}_0^{-1}\mathbf{r}_0 = (0,0,\ldots,1,0,\ldots0)^T$, the expected value of the above is equal to the expected value of the center element of the matrix, $\Delta\mathbf{R}\mathbf{R}_0^{-1}\Delta\mathbf{R}$. It is straightforward to show that

$$E\{(\mathbf{R}_0^{-1}\mathbf{r}_0)^t \Delta\mathbf{R}\mathbf{R}_0^{-1} \Delta\mathbf{R}(\mathbf{R}_0^{-1} \mathbf{r}_0)\} = \sum_{k=1}^{N} \sum_{m=1}^{N} R^{(km)}\, E\{\Delta R_{N_2k}\, \Delta R_{mN_2}\} . \tag{A9}$$

If we use expressions for ΔR_{N_2k} and ΔR_{mN_2} by using Eq. (3.18), we find that

$$E\{\Delta R_{N_2k}\, \Delta R_{mN_2}\} = E\left\{\int_{-\infty}^{\infty} |H|^2\, (\Delta H_A + \Delta H_A{}^*) e^{j\omega\pi BT(N_2 - k)} d\omega \right.$$
$$\left. \cdot \int_{-\infty}^{\infty} |H|^2\, (\Delta H_A + \Delta H_A{}^*) e^{j\omega\pi BT(m-N_2)} d\omega \right\} + E\{O(\Delta H_A^3)\}. \tag{A10}$$

If we evaluate only the first term of Eq. (A10) by substituting Eq. (A6) into Eq. (A10), and use the fact that

$$E\{\Delta s_i^2\} = 0 , \tag{A11}$$

then it can be shown that

$$E\{\Delta R_{N_2k}\, \Delta R_{mN_2}\} = 2\sigma_F^2 \sum_{i=1}^{M}\left[\int_{-\infty}^{\infty} \frac{|H|^2\, e^{j\omega\pi BT(N_2-k)}}{j\omega - s_i} d\omega \cdot \int_{-\infty}^{\infty} \frac{|H|^2\, e^{j\omega\pi BT(m-N_2)}}{-j\omega - s_i{}^*} dw\right] . \tag{A12}$$

Hence by substituting Eq. (A12) into Eq. (A9), Eq. (3.39) results.

4. The $E\{\mathbf{r}_o^t\, \mathbf{R}_0^{-1} \Delta\mathbf{R}\mathbf{R}_0^{-1} \Delta\mathbf{r}\}$ *Term:*

We write

$$\mathbf{r}_0^t\mathbf{R}_0^{-1} \Delta\mathbf{R}\mathbf{R}_0^{-1} \Delta\mathbf{r} = (\mathbf{R}_0^{-1} \mathbf{r}_0)^t \Delta\mathbf{R}\mathbf{R}_0^{-1} \Delta\mathbf{r} . \tag{A13}$$

The term $\Delta\mathbf{R}\mathbf{R}_0^{-1}\Delta\mathbf{r}$ is a vector and because of Eq. (3.10), the above expected value is simply the expected value of the center element of this vector. Thus, we can show that

$$E\{\mathbf{r}_0^t\mathbf{R}_0^{-1} \Delta\mathbf{R}\mathbf{R}_0^{-1} \Delta\mathbf{r}\} = \sum_{m=1}^{N} \sum_{k=1}^{N} \mathbf{R}_0^{(km)} E\{\Delta R_{N_2k}\, \Delta r_m\} . \tag{A14}$$

If we substitute the forms of ΔR_{N_2k} and Δr_m given by Eqs. (3.18) and (3.19) respectively, we find that

$$E\{\Delta R_{N_2k}\, \Delta r_m\} = E\left\{\int_{-\infty}^{\infty} |H|^2(\Delta H_A + \Delta H_A{}^*)\; e^{j\omega\pi BT\,(N_2-k)} d\omega \right.$$
$$\left. \cdot \int |H|^2\, \Delta H_A{}^*\; e^{j\omega\pi BT(m-N_2)}\, d\omega\right\} + E\{O(\Delta H_A\, \Delta H_M)\} . \tag{A15}$$

The last term in Eq. (A15) is equal to zero, and the first term can be evaluated to be

$$E\{\Delta R_{N_2k}\, \Delta r_m\} = \sigma_F^2 \sum_{i=1}^{M}\left[\int_{-\infty}^{\infty} \frac{|H|^2 e^{j\omega\pi BT(N_2-k)}}{j\omega - s_i} d\omega \cdot \int_{-\infty}^{\infty} \frac{|H|^2 e^{j\omega\pi BT(m-N_2)}}{-j\omega - s_i{}^*} d\omega\right] . \tag{A16}$$

Thus Eq. (A16) is substituted into Eq. (A14) and Eq. (3.32) follows.

5. The $E\{\Delta \mathbf{r}' \mathbf{R}_0^{-1} \, \Delta \mathbf{R} \, \mathbf{R}_0^{-1} \, \mathbf{r}_0\}$ Term:

By examination of Eqs. (A14) and (A16), and the fact that $R_0^{(mk)} = R_0^{(km)*}$, we can show that

$$R_0^{(mk)} \; E\{\Delta R_{N_2 m} \, \Delta_{r_k}\} = \left[R_0^{(km)} \; E\{\Delta R_{N_2 k} \, \Delta r_m\}\right]^* . \tag{A17}$$

Hence, it follows that the expression given in Eq. (3.32) is real because every term of Eq. (A14) is either real or has an associated complex conjugate term. Hence, $E\{\Delta \mathbf{r}' \, \mathbf{R}_0^{-1} \, \Delta \mathbf{R} \, \mathbf{R}_0^{-1} \mathbf{r}_0\}$ can be found by using the expression for $E\{\mathbf{r}_0' \, \mathbf{R}_0^{-1} \, \Delta \mathbf{R} \mathbf{R}_0^{-1} \, \Delta \mathbf{r}\}$ previously given.

6. The $E\{\Delta \mathbf{r}' \, \mathbf{R}_0^{-1} \, \Delta \mathbf{r}\}$ Term:

We can show that

$$E\{\Delta \mathbf{r}' \, \mathbf{R}^{-1} \, \Delta \mathbf{r}\} = \sum_{k=1}^{N} \sum_{m=1}^{N} \mathbf{R}_0^{(km)} \; E\{\Delta r_k{}^* \, \Delta r_m\} . \tag{A18}$$

Using the definitions of the $\Delta \mathbf{r}$ elements given by Eq. (3.19) and multiplying out these expressions we obtain

$$\begin{aligned} E\{\Delta r_k{}^* \, \Delta r_m\} = {} & E\left\{\int_{-\infty}^{\infty} |H|^2 \, \Delta H_A \, e^{j\omega\pi BT\,(N_2-k)} \, d\omega \right. \\ & \left. \cdot \int_{-\infty}^{\infty} |H|^2 \, \Delta H_A{}^* e^{j\omega\pi BT(m-N_2)} \, d\omega\right\} \\ & + E\left\{\int_{-\infty}^{\infty} |H|^2 \, \Delta H_M{}^* e^{\,j\omega\pi BT(N_2-k)} \, d\omega \right. \\ & \left. \cdot \int_{-\infty}^{\infty} |H|^2 \, \Delta H_M \, e^{\,j\omega\pi BT(m-N_2)} d\omega\right\} + E\{\, O(\Delta H_A \, \Delta H_M)\}. \end{aligned} \tag{A19}$$

The last term in Eq. (A19) equals zero. Because of the identical statistics of ΔH_A and ΔH_M, the first two terms are equal. Thus, we can evaluate the above as

$$E\{\Delta r_k{}^* \, \Delta r_m\} = 2\sigma_F^2 \sum_{i=1}^{M} \left[\int_{-\infty}^{\infty} \frac{|H|^2 e^{j\omega\pi BT(N_2-k)} d\omega}{j\omega - s_i} \int_{-\infty}^{\infty} \frac{|H|^2 e^{j\omega\pi BT\,(m-N_2)} d\omega}{-j\omega - s_i^*} \right] . \tag{A20}$$

Appendix B
EVALUATION OF INTEGRALS

In this appendix, we evaluate the integrals seen in Eqs. (3.35) and (3.36) for when the FTF is a Butterworth filter. Expressions for these integrals were listed in Eqs. (4.5) to (4.7). These integrals can be derived by using the Theory of Residues [2].

First, we find an expression for $H(s)H(-s)$ which is consistent with Eqs. (4.1) and (4.3). It is easily shown that

$$H(s)H(-s) = \frac{1}{1 + (-1)^M \, s^{2M}} . \tag{B1}$$

Next, the poles of $H(s)H(-s)$ are identified. These poles lie on the unit circle and are equally spaced in angle (see Fig. 4.2). We can show that the right-hand plane poles are given by the expression

$$p_i^{(R)} = j \, e^{-j\frac{\pi}{2M}(2i-1)}, \qquad i = 1,2,\ldots, M \tag{B2}$$

and the left-hand plane poles are given by the expression

$$p_i^{(L)} = j \, e^{j\frac{\pi}{2M}(2i-1)}, \qquad i = 1,2,\ldots, M. \tag{B3}$$

Note that the expressions do not depend on whether M is even or odd as does $H(s)H(-s)$ and that there are no poles on the imaginary axis (the $j\omega$ axis). We set the poles of $H(s)$ equal to the left-hand plane poles:

$$s_i = p_i^{(L)}, \qquad i = 1,2,\ldots, M. \tag{B4}$$

In Fig. B1, we designate the path of integration of the integrals seen in Eqs. (3.35) and (3.36) as contour C_1; i.e., $-j\infty < j\omega < j\infty$. In addition, we have shown two other contours: C_2 and C_3. Let $K(s)$ be the kernal of any of the integrals to be evaluated. We can then write

$$\int_{C_1} K(s)\, ds = \int_{C_1 + C_2} K(s)\, ds - \int_{C_2} K(s)\, ds \tag{B5}$$

or

$$\int_{C_1} K(s)\, ds = \int_{C_1 + C_3} K(s)\, ds - \int_{C_3} K(s)\, ds . \tag{B6}$$

Depending on the kernal, we can show that as $R_c \rightarrow \infty$ then either both the C_2 and C_3 contour integrals go to zero, or one goes to zero and the other goes to infinity. In our evaluation, we always choose the one that goes to zero so that we can form a closed contour about either the right-side plane poles (contour $C_1 + C_2$) or left-side plane poles (contour $C_1 + C_3$) and thus use the Theory of Residues.

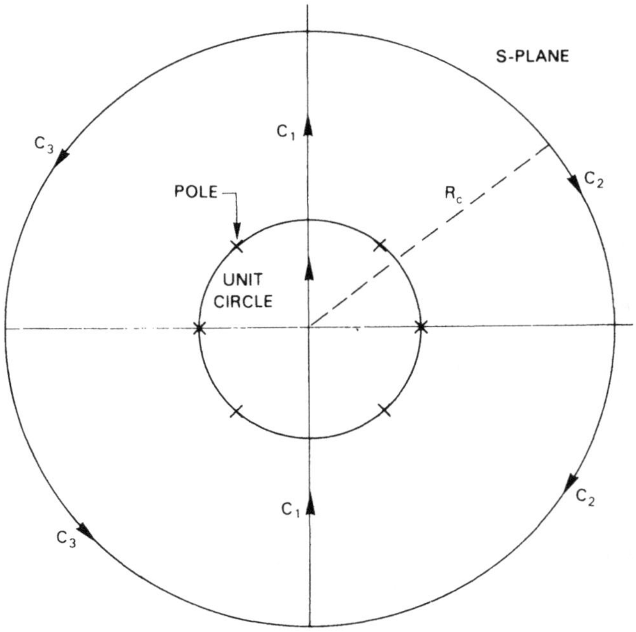

Fig. B1 — Contours of integration

1. Derivation of Eq. (4.5)

We can show for $m \leqslant n$

$$\int_{-\infty}^{\infty} |H(j\omega)|^2 \, e^{j\omega\pi BT(n-m)} \, d\omega = \frac{1}{j} \oint_{C_1 + C_3} \frac{c_0 \, e^{s\pi BT(n-m)} \, ds}{1 + (-1)^M \, s^{2M}}$$

$$= 2\pi \sum_{i=1}^{M} \mathrm{Res}_i \tag{B7}$$

where Res_i are the M residues of the above kernal about the poles in the left-hand plane. Note we have used the fact that the integral along C_3 as seen in Eq. (B6) goes to zero as $R_c \rightarrow \infty$. All the poles are single poles. It can be shown by use of the Theory of Residues that

$$\mathrm{Res}_i = \lim_{s \rightarrow p_i^{(L)}} \left\{ \frac{c_0 \, e^{s\pi BT(n-m)} \, (s - p_i^{(L)})}{1 + (-1)^M \, s^{2M}} \right\} \tag{B8}$$

or evaluating the above limit

$$\mathrm{Res}_i = -\frac{c_0}{2M} \, p_i^{(L)} \, e^{p_i^{(L)} \pi BT(n-m)} . \tag{B9}$$

Substituting Eq. (B3) into Eq. (B9) and then substituting Eq. (B9) into Eq. (B7) results in Eq. (4.5b). Note we have substituted l for i in Eq. (4.5b).

We can also show for $m > n$ that

$$\int_{-\infty}^{\infty} |H(j\omega)|^2 \, e^{j\omega\pi BT(n-m)} \, d\omega = \frac{1}{j} \oint_{C_1 + C_2} \frac{c_0 \, e^{s\pi BT(n-m)} \, ds}{1 + (-1)^M \, s^{2M}} = -2\pi \sum_{i=1}^{M} \mathrm{Res}_i \tag{B10}$$

where Res_i are the M residues of the above kernal. We used the fact that the integral along C_2 seen in Eq. (B5) goes to zero as $R_c \to \infty$.

It can be shown that

$$\text{Res}_i = \lim_{s \to p_i^{(R)}} \left\{ \frac{c_0\, e^{s\pi BT(n-m)}\,(s - p_i^{(R)})}{1 + (-1)^M s^{2M}} \right\} \tag{B11}$$

or evaluating the above limit

$$\text{Res}_i = -\frac{c_0}{2M}\, p_i^{(R)}\, e^{p_i^{(R)}\pi BT(n-m)}\,. \tag{B12}$$

Substituting Eq. (B2) into Eq. (B12) and then substituting Eq. (B12) into Eq. (B10) results in Eq. (4.5a). Again we have set $l = i$ in Eq. (4.5a).

2. Derivation of Eq. (4.6)

We can show that

$$\int_{-\infty}^{\infty} \frac{|H(j\omega)|^2\, d\omega}{|j\omega - s_i|^2} = \frac{1}{j} \oint_{C_1 + C_2} \frac{c_0\, ds}{(1 + (-1)^M s^{2M})(s - s_i)(-s - s_i^*)}$$

$$= -2\pi \sum_{m=1}^{M} \text{Res}_m \tag{B13}$$

where Res_m $m = 1,2,\ldots, M$ are the residues of the above kernal. Again we have used the fact that the integral along C_2 seen in Eq. (B5) goes to zero as $R_c \to \infty$. All of the poles (in the right-hand plane) of this kernal are single poles except when $i = m$ or

$$p_i^{(R)} = -s_i \tag{B14}$$

which results in a double pole.

For the single poles, $m \neq i$, we can show

$$\text{Res}_m = \lim_{s \to p_m^{(R)}} \left\{ c_0 \frac{s - p_m^{(R)}}{(1 + (-1)^M s^{2M})(s - s_i)(-s - s_i^*)} \right\}. \tag{B15}$$

We can show using l'Hopital rule that

$$\text{Res}_m = \frac{c_0}{2M} \frac{1}{p_m^{(R)} - p_m^{(R)*} + s_i^* - s_i}$$

$$= -\frac{j\, c_0}{4M} \frac{1}{\cos \dfrac{\pi}{2M}(2m - 1) - \cos \dfrac{\pi}{2M}(2i - 1)}\,. \tag{B16}$$

Since all of the Res_m, for $m \neq i$, are purely imaginary and we know that the integral to be evaluated is real, these terms will cancel out with the imaginary part of Res_i. We can show that

$$\text{Res}_i = \lim_{s \to -s_i^*} \left\{ \frac{d}{ds} \left[-c_0 \frac{s + s_i^*}{(1 + (-1)^M s^{2M})(s - s_i)} \right] \right\}. \tag{B17}$$

After taking the derivative and applying l'Hopital rule twice, we find that

$$\mathrm{Res}_i = \frac{c_0}{2}\,\frac{c_0}{s_i + s_i^*} + \frac{1}{4M}\,\frac{s_i^* - s_i}{(s_i + s_i^*)^2} \tag{B18}$$

$$= -\frac{c_0}{4}\,\frac{1}{\sin\frac{\pi}{2M}(2i-1)} - j\,\frac{c_0}{8M}\,\frac{\cos\frac{\pi}{2M}(2i-1)\cdot}{\sin^2\frac{\pi}{2M}(2i-1)}.$$

Adding all the residues and evaluating Eq. (B13) results in Eq. (4.6).

3. Derivation of Eq. (4.7)

It can be shown that for $k \leqslant n$

$$\int_{-\infty}^{\infty} \frac{|H(j\omega)|^2 e^{j\omega\pi BT(k-n)} d\omega}{j\omega - s_i} = \frac{1}{j}\oint_{C_1+C_2} \frac{c_0\, e^{s\pi BT(k-n)}\, ds}{(1 + (-1)^M s^{2M})(s - s_i)}$$

$$= -2\pi \sum_{m=1}^{M} \mathrm{Res}_m \tag{B19}$$

where $\mathrm{Res}_m, m = 1,2,\ldots, M$ denotes the residues of the above kernal. All of the poles are single poles. We can show that

$$\mathrm{Res}_m = \lim_{s \to p_m^{(R)}} \left\{ c_0 \frac{(s - p_m^{(R)})\, e^{s\pi BT(k-n)}}{(1 + (-1)^M s^{2M})(s - s_i)} \right\} \tag{B20}$$

$$= -\frac{c_0}{2M}\,\frac{p_m^{(R)}}{p_m^{(R)} - s_i}\, e^{p_m^{(R)}(k-n)\pi BT} \tag{B21}$$

If Eq. (B2) is substituted into Eq. (B21) and then Eq. (B21) is substituted into Eq. (B19), then Eq. (4.7a) results with $k = N_2$.

For $k > n$, we can show that

$$\int_{-\infty}^{\infty} \frac{|H(j\omega)|^2 e^{j\omega\pi BT(k-n)} d\omega}{j\omega - s_i} = \frac{1}{j}\oint_{C_1+C_3} \frac{c_0\, e^{s\pi BT(k-n)}\, ds}{(1 + (-1)^M s^{2M})(s - s_i)}$$

$$= 2\pi \sum_{m=1}^{M} \mathrm{Res}_m \tag{B22}$$

where Res_m, $m = 1,2,\ldots, M$ are the residues of the above kernal. This kernal has $M - 1$ single poles and one double pole for when $m = i$. For $m \neq i$, we can show that

$$\mathrm{Res}_m = \lim_{s \to p_m^{(L)}} \left\{ c_0 \frac{(s - p_m^{(L)})\, e^{s\pi BT(k-n)}}{(1 + (-1)^M s^{2M})(s - s_i)} \right\}$$

$$= -\frac{c_0}{2M}\,\frac{e^{\pi BT(k-n)\sin\frac{\pi}{2M}(2m-1)}\; e^{j\pi BT(k-n)\cos\frac{\pi}{2M}(2m-1)}}{1 - e^{j\frac{\pi}{M}(i-m)}}. \tag{B23}$$

For $i = m$, we can show that

$$\text{Res}_i = \lim_{s \to s_i} \frac{d}{ds} \left\{ \frac{(s - s_i)\, e^{s\pi BT(k-n)}}{1 + (-1)^N\, s^{2M}} \right\}. \tag{B24}$$

After taking the derivative and applying l'Hopital rule twice, the following expression results:

$$\text{Res}_i = \frac{1}{2M} \left\{ M - .5 - BT(k-n)\, je^{j\frac{\pi}{2M}(2i-1)} \right\} e^{-\pi BT(k-n) \sin \frac{\pi}{2M}(2i-1)} \cdot e^{j\pi BT(k-n) \cos \frac{\pi}{2M}(2i-1)}. \tag{B25}$$

Substituting Eqs. (B23) and (B25) in Eq. (B22) results in Eq. (4.7b) with $k = N_2$.

USING SPECTRAL ESTIMATION TECHNIQUES IN ADAPTIVE PROCESSING ANTENNA SYSTEMS

INTRODUCTION

Improved spectral estimation techniques are an emerging technology which derives largely from modern spectral estimation theory of the past decade and adaptive array processing techniques [1,2,3]. Coupled with the phenomenal advances in digital processing, these techniques are becoming a valuable asset for adaptive array antenna systems. Their value lies in the considerable amount of additional useful information which they can provide about the environment while using only a relatively small number of degrees-of-freedom (DOF). For example, current spectral estimation algorithms can provide asymptotically unbiased estimates of the number of interference sources, source directions, source strengths, and any cross-correlations (coherence) between sources [4,5]. Such information can then be used to track and "catalogue" interference sources, hence assign adaptive DOF.

These newer techniques are not viewed as a "superresolution" replacement for more conventional estimation methods such as mainbeam search, analogue beamformers, or spatial discreet Fourier transforms (DFT); but rather, the new technology is considered complementary to the other methods and best used in tandem. For example, "superresolution" techniques cannot compete with the speed of a DFT. Some comparisons of the various methods may be found described in the literature [3,5,6].

The purpose of this report is to present two conceptual application areas that use spectral estimation techniques, partially-adaptive low-sidelobe antennas, and fully-adaptive tracking arrays. In a partially adaptive array only a part of the DOF, array elements or beams, is individually controlled adaptively [7,8,9]. Obviously, the fully adaptive configuration is preferred since it offers the most control over the response of the antenna system. But, when the number of elements or beams becomes moderately large (hundreds), the fully adaptive processor implementation can become prohibitive in cost, size, and weight.

This report is divided into three principal parts. The second section of this report discusses partially-adaptive, low-sidelobe antennas with the focus upon a constrained beamspace system. In the third section of the report, the source estimation and beam assignment from *superresolution* techniques are considered; and in the fourth section, an all-digital, fully-adaptive tracking array concept is discussed. Several Appendices are also used for referral of the more tedious details.

PARTIALLY-ADAPTIVE LOW-SIDELOBE ANTENNAS

The antenna system, addressed in this section, is assumed to be a moderately large aperture array of low-sidelobe design wherein the investment is already considerable and one simply could not afford to make it fully adaptive. The assumption of low-sidelobes (30 dB or better) is intended to give us good initial protection against modest interference sources and to reduce the problems from strong sources, i.e., in regard to the number of adaptive DOF required and the adaptive dynamic range of the processor. Thus, retention of the low sidelobes is considered a major goal in our adaptive system. In the discussion that follows it is shown that using improved spectral estimation techniques in such a system can result in the following benefits over a fully adaptive array system:

Manuscript approved May 17, 1985.

a. Reduction in overall cost because relatively few adaptive DOF are implemented.

b. Simple adaptive weight constraints permit minimal degradation of the mainbeam and sidelobe levels.

c. Reduction in computation burden.

d. Considerably faster adaptive response.

e. Compatibility with a larger number of adaptive algorithms, including analogue versions.

f. Greater flexibility in achieving a "tailored" response due to greater information available.

On the negative side, a partially-adaptive system can never be guaranteed a cancellation performance equal to that of a fully adaptive array, and will deteriorate abruptly in performance when the interference situation exceeds its adaptive DOF. These risks are an inherent part of the package and must be carefully weighed for any specific system application.

Low-Sidelobe Eigenvector Constraint

In this section we review how unconstrained adaptive arrays can experience very "noisy" sidelobe fluctuations and mainbeam perturbations when the data observation/integration time is not long enough, even though the quiescent mainbeam weights are chosen for low sidelobes. Consider the simple schematic shown in Fig. 1, and let us compute the complex adaptive element weights W_k from the well-known Sample Matrix Inverse (SMI) algorithm [9,10] as expressed in the following matrix notation,

$$\mathbf{W} = \mu \hat{\mathbf{R}}^{-1} \mathbf{S}^* \tag{1}$$

where:

$\mathbf{W}$ is the adaptive weights vector,
$\hat{\mathbf{R}}$ is the sample covariance matrix,
$\mathbf{S}^*$ is the quiescent mainbeam weights vector, and μ is a constant.

* denotes the conjugate of a complex vector or matrix.

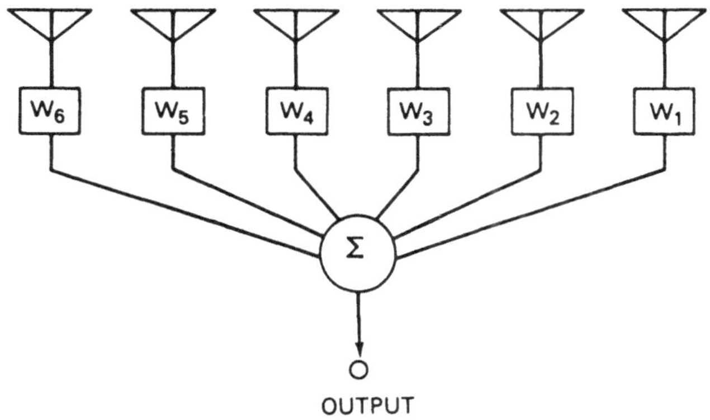

Fig. 1 — Schematic of adaptive array

Furthermore, the sample covariance matrix is computed via the simple "block" average taken over N snapshots,

$$\hat{\mathbf{R}} = \frac{1}{N} \sum_{n=1}^{N} \left[\mathbf{E}(n)\mathbf{E}(n)^{*t} \right], \tag{2}$$

where $\mathbf{E}(n)$ is the element signal data vector received at the nth time sampling. (See Appendix A for the description of snapshot signal model.) The data observation/integration time in (2) is the parameter **N**. If $\hat{\mathbf{R}}$ is estimated over a lengthy observation time, like thousands of snapshots, then the sidelobe fluctuations from **W** updates will be relatively small. However, practical system usage often demands short observation times on the order of hundreds of snapshots or even less.

Figure 2 illustrates a typical adapted pattern behavior for independent estimates of $\hat{\mathbf{R}}$ using $N = 256$ snapshots per update for the case of three 30 dB noncoherent sources located at 14, 18 and 22°. The antenna aperture chosen for this example is a 16-element linear array with half-wavelength element spacing and a 30 dB Taylor illumination incorporated in $\mathbf{S}^*$. The adaptive algorithm maintains the mainbeam region and successfully nulls out the interference sources, but it also raises the sidelobe levels elsewhere. The adaptive patterns are in continual fluctuation in the sidelobe regions and may exceed the quiescent sidelobe level by a considerable margin. Also, the mainbeam suffers a significant modulation which would degrade tracking performance. These effects worsen as the value of N decreases.

To understand the reason for this undulating pattern behavior, it is helpful to analyze the optimum weights in terms of eigenvalue/eigenvector decomposition. Appendix B contains a derivation of such a decomposition for Eq. (1), and we reproduce Eq. (B-18) below:

$$\mathbf{W} = \mu' \left[\mathbf{S}^* - \sum_{i=1}^{K} \left(\frac{\beta_i^2 - \beta_0^2}{\beta_i^2} \right) \alpha_i \mathbf{e}_i \right] \tag{3}$$

where:

$$\alpha_i = \mathbf{e}_i^{*t} S^*$$

$$\mu' = \mu / \beta_0^2$$

t denotes the transpose of a vector or matrix. The β_i^2 and $\mathbf{e}_i$ are the eigenvalues and eigenvectors, respectively, of the sample covariance matrix, and β_0^2 is equal to receiver channel noise power level. Equation (3) shows that **W** consists of two parts: the first part is the quiescent mainbeam weight $\mathbf{S}^*$; the second part, which is subtracted from $\mathbf{S}^*$, is the summation of weighted orthogonal eigenvectors. This is a clear expression of the fundamental principle of pattern subtraction which applies in adaptive array analysis [9,11].

We introduce the term *principal eigenvectors* (PE) to mean those eigenvectors which correspond to unique eigenvalues generated by the spatial source distribution; and the term *noise eigenvectors* to mean those eigenvectors which correspond to the small noise eigenvalues generated by the receiver channel noise contained in the finite $\hat{\mathbf{R}}$ estimates. The PE are generally rather robust and tend to remain relatively stable from one data trial to the next, whereas the noise eigenvectors tend to fluctuate considerably because of the inherent random behavior of noise. This difference in behavior is illustrated in Fig. 3 for the three source case described above; wherein there are three PE and thirteen noise eigenvectors associated with each $\hat{\mathbf{R}}$ estimate. Figure 3(a) shows the stability of the three PE for nine trials, and Fig. 3(b) shows the random behavior of typical noise eigenvectors for the exact same trials. Thus, we would expect that the sidelobe undulations in Fig. 2(b) are associated primarily with the noise eigenvectors. This thesis is verified in Fig. 4, which illustrates the adapted patterns resulting from Eq. (3) when only the PE are subtracted.

The above adaptive array pattern behavior leads to the following observations for source distributions which do not encroach upon the mainbeam and involve a small number of the available degrees-of-freedom (DOF):

a. It is possible to retain low sidelobes in the adapted patterns, even with short observation times, by constraining our algorithm, Eq. (3), to utilize only the PE. The weight solution is unique and therefore stable.

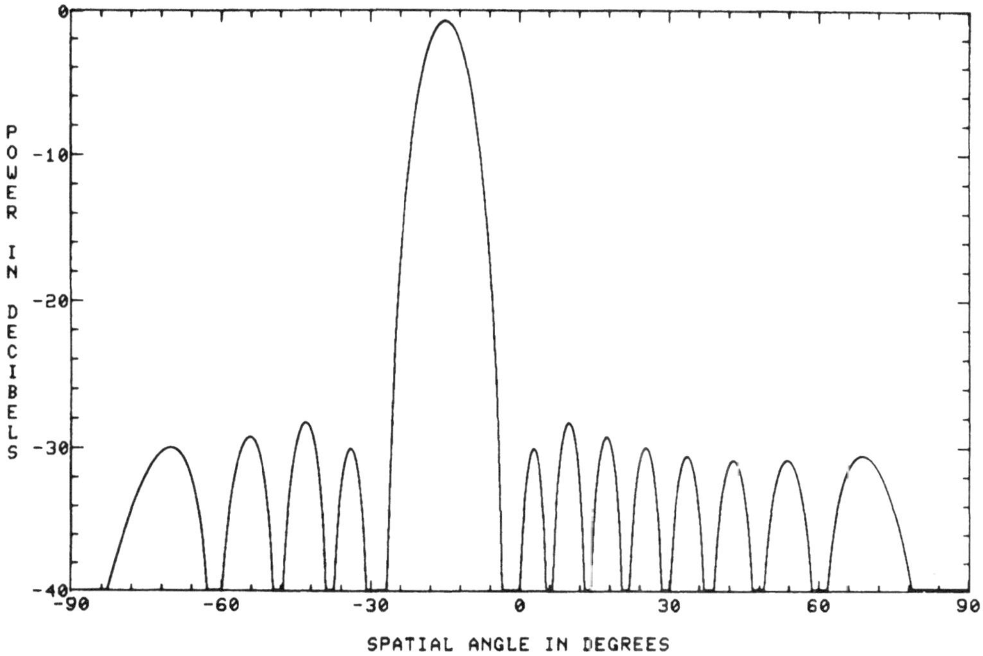

(a) Quiescent mainbeam pattern, 30 dB Taylor weighting

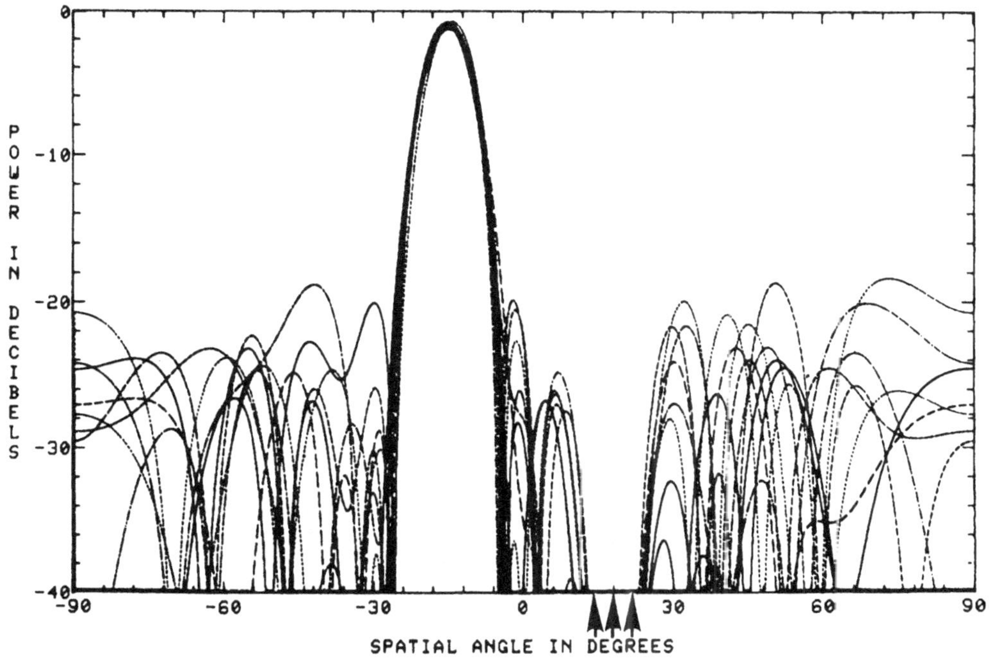

(b) Typical adapted patterns, nine update trials plotted

Fig. 2 — Fully adaptive 16-element linear array, SMI algorithm with $\hat{\mathbf{R}}$ estimated from 256 snapshots per update, three 30 dB non-coherent sources located at 14, 18, and 22°

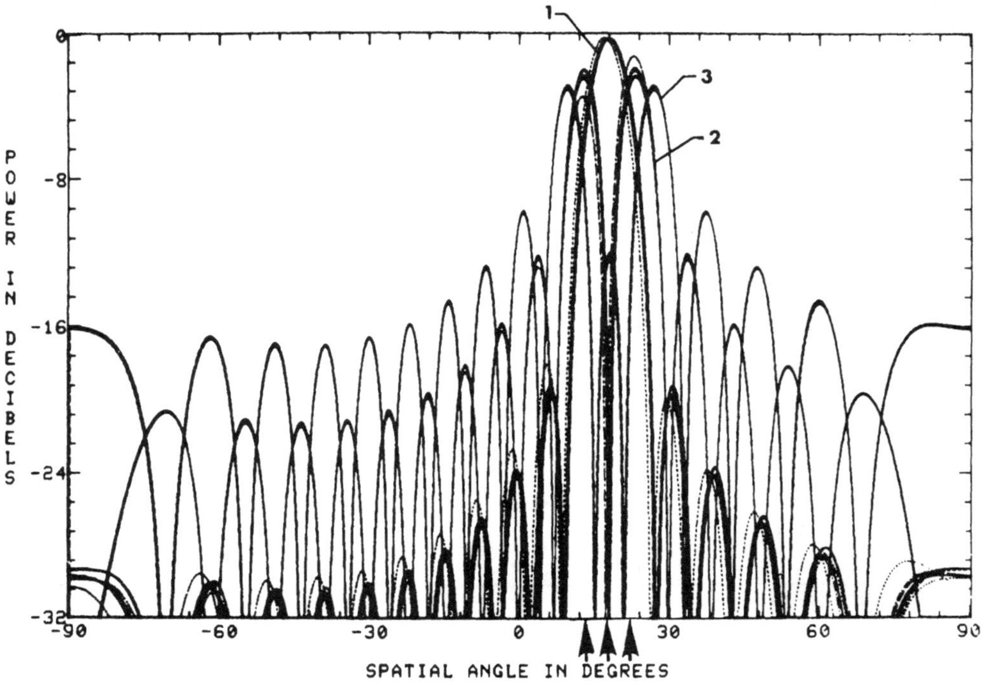

(a) Principal eigenvectors (PE) Nos. 1, 2, and 3

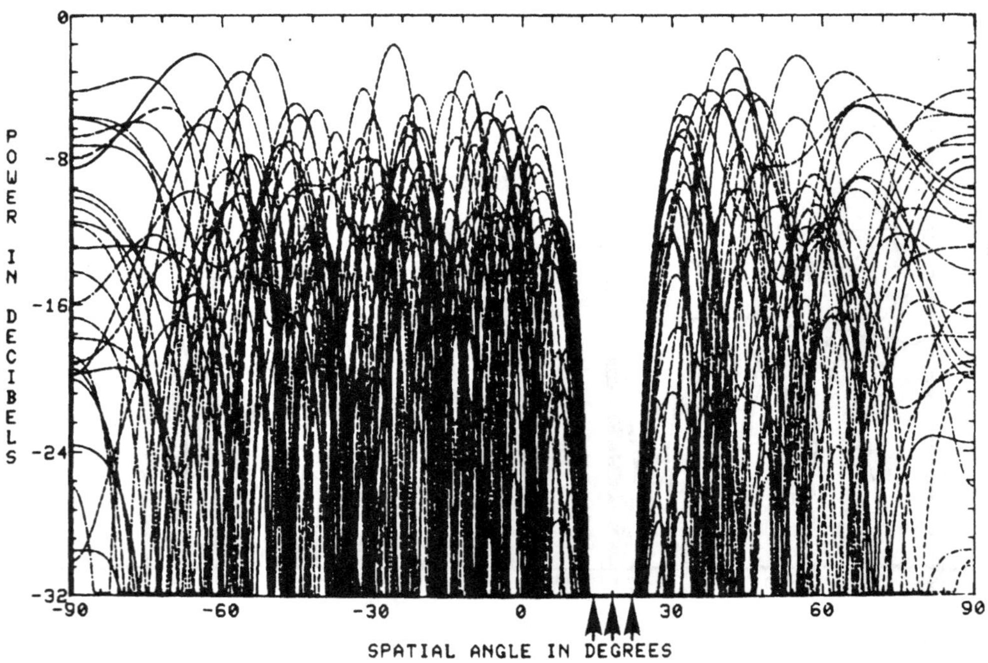

(b) Typical noise eigenvectors Nos. 4, 10, and 16

Fig. 3 — Plots of principal eigenvectors (PE) and noise eigenvectors computed from the $\hat{\mathbf{R}}$ estimates associated with the three-source case of Fig. 2, nine update trials

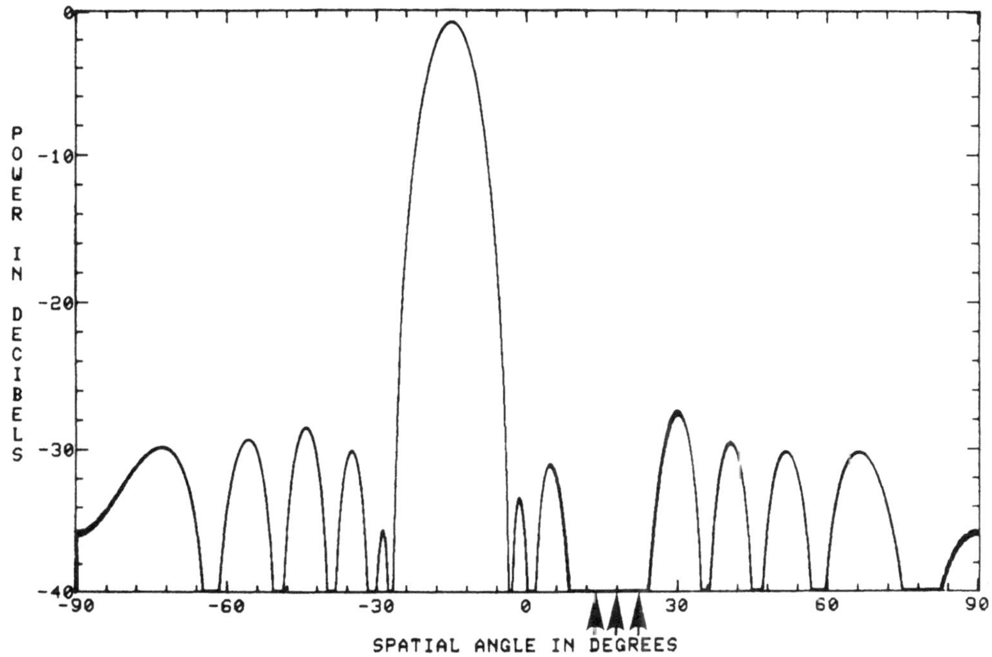

Fig. 4 — Typical adapted patterns resulting from the constraint of utilizing only the PE, three-source case of Fig. 2

b. Utilizing only the PE is tantamount to operating our adaptive system in beamspace (as opposed to element space) with a set of weighted orthogonal canceller beams.

c. The fully adaptive array automatically forms and "assigns" its PE canceller beams to cover the interference source distribution, with one beam per each DOF needed.

Therefore, we have set forth a low-sidelobe eigenvector constraint algorithm for this type of restricted interference situation.

Low-Sidelobe Constraints for a General Beamformer

Consider next a more interesting configuration which is shown by the schematic diagram in Fig. 5, where we represent an adaptive array system operating in beamspace so as to have available some pre-adaptation spatial filtering. Applebaum and Chapman [8,9,12] were the first to describe beamspace systems of this type, using a Butler matrix [13] beamformer, wherein the vector of the beamformer outputs, $\hat{\mathbf{E}}$, may be expressed as follows:

$$\hat{\mathbf{E}} = \mathbf{B}'\mathbf{E} \tag{4}$$

where **B** is a KxK matrix containing the beamformer element weights (see Appendix C). Other descriptions of beamspace systems are also available in the literature [9,14,15,16], of which Adams et al. [15], is particularly germane to our discussion. Chapman [8] pointed out that when used in a partially adaptive configuration, such beamspace systems are susceptible to aperture element errors and cannot arbitrarily compensate the random error component of their sidelobe structure. This makes it necessary to control element errors in accordance with the quiescent mainbeam sidelobe level desired, and fits into our initial assumption of low-sidelobe design as mentioned earlier. A separate weighted

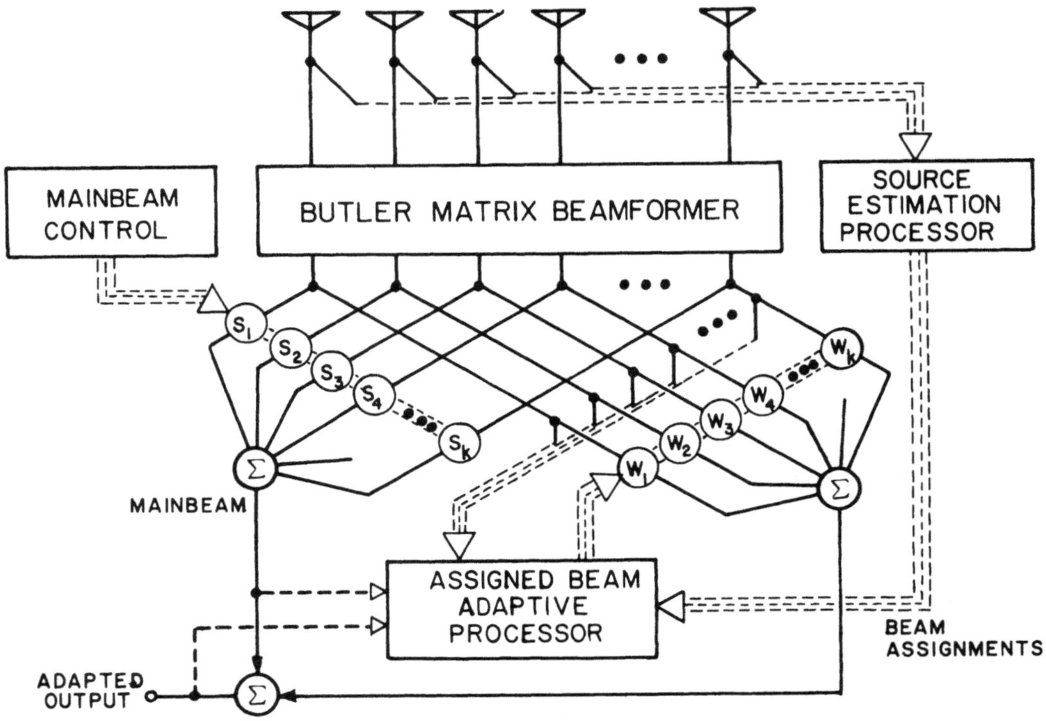

Fig. 5 — Beamspace adaptive array with a separately weighted mainbeam

mainbeam summing is indicated which may be obtained either by coupling into the beamformer outputs as shown, or by coupling off from the elements and providing suitable phase shifters for steering plus a corporate feed network. Our purpose here is to examine the sidelobe performance of such a partially-adaptive beamspace system in which element errors are kept low and beamformer beams are subjected to simple constraints.

Spatial estimation data on the interference source distribution shall determine which beamformer beams are to be adaptively controlled. Such beams are defined herein as *assigned* beams, and the idea is to assign only enough beams to accommodate the DOF required by the source distribution. Whenever the two are equal, the adaptive weight solution is unique and we avoid adding any extra "noisy" weight perturbations. The reader will recognize that we are attempting to replace the PE beams of the previous section, Low-Sidelobe Eigenvector Constraint, with assigned beams from our general beamformer. Thus, we are defining a partially-adaptive array which will utilize only a relatively small number of its available DOF. In addition to this assigned beam constraint, we seek to limit the adaptive weights of assigned beams to a maximum level, γ, which was chosen to exceed the mainbeam sidelobe level by only a few dB. This prevents an excessive rise in adaptive sidelobe level, including the condition where the number of assigned beams exceeds the DOF required.

γ represents the product of assigned beam gain and adaptive weight magnitude, so that we have the option of working with beamformer beams which are considerably decoupled/attenuated.

An equation formulation may be expressed in terms of the same pattern subtraction principle as used in Eq. (3) for K beams,

$$\mathbf{W}_0 = \mathbf{S}^* - \sum_{k=1}^{K} W_k \mathbf{b}_k \tag{5}$$

where:

$|W_k| \leqslant \gamma$ for J assigned beams

$W_k = 0$ for all other beams

and $\mathbf{b}_k$ is the kth Butler matrix beam element-weight vector. When $W_k = 0$, that beam port is essentially disconnected from the output summation and it is much to our advantage to reduce the DOF of the adaptive weight processor accordingly, i.e., this processor reduction relates directly to the computational burden, response time, sidelobe degradation, and overall cost mentioned earlier. For example, utilizing the SMI technique described in Eqs. (1) and (2), we would now have the advantage that our sample covariance matrix of signal inputs, $\hat{\mathbf{R}}$, involves only the J assigned beams and its dimensions reduced from KxK to JxJ, thereby greatly easing the computation burden involved in obtaining its inverse [9]. The equivalent "steering vector", Λ, per Applebaum [7] is also reduced to dimension J and consists of the cross-correlation between the mainbeam signal V and the J assigned beam outputs, $\mathbf{Y}$,

$$\Lambda = \frac{1}{N} \sum_{n=1}^{N} V(n)\mathbf{Y}^*(n) . \tag{6}$$

The jth assigned beam output for the nth snapshot signal sample is simply

$$Y_j(n) = \mathbf{E}^t(n)\mathbf{b}_k , \quad k \text{ set by } j \tag{7}$$

where the particular value of k must be selected for the jth assigned beam. Our J dimension adaptive weight solution thus becomes,

$$\mathbf{W} = \hat{\mathbf{R}}^{-1}\Lambda . \tag{8}$$

Equation (8) gives us the J assigned beam weights required in Eq. (5). The proposed constraint $|W_k| \leqslant \gamma$ can be applied directly to the solution from (8), but be aware that this is a "hard" constraint and the results will not be optimal when the limit is exceeded.

A softer, more flexible constraint for our purposes is one suggested by Brennan* based upon Owsley [17], where weights are selected which simultaneously minimize both the output and the sum of the weight amplitudes squared, i.e.,

$$\text{minimize} \left\{ \overline{|V - \mathbf{W}^t\mathbf{Y}|^2} + \alpha \mathbf{W}^t\mathbf{W}^* \right\}$$

where the overbar denotes averaging over N snaps. The solution is a simple modification to Eq. (8) wherein

$$\mathbf{W} = \left[\hat{\mathbf{R}} + \alpha \mathbf{I} \right]^{-1} \Lambda \tag{9}$$

where:

$$\alpha = \left(\frac{\gamma^2}{J} \right) \text{Trace}[\hat{\mathbf{R}}].$$

We note here that Eq. (9) adds a small percentage of the average assigned beam power to the diagonal terms of $\hat{\mathbf{R}}$. We also recall that γ was selected to be close to the mainbeam sidelobe level. Although α is a small percentage of the Trace $[\hat{\mathbf{R}}]$, it is generally much larger than the receiver noise level, β_0^2; this domination over receiver noise by a constant will tend to severely dampen weight fluctuations due to noise. Equation (9) deviates from the optimum Weiner weights and will result in a slightly larger output residue, however, the cost is negligible compared to the remarkably stable results achieved from

*Private communication, L.E. Brennan, Adaptive Sensors, Inc.

this rather simple constraint. It essentially permits the number of assigned beams to exceed the DOF required, and yet retain low sidelobe levels.

Equations (5) through (9) were used in computing the adaptive pattern examples which follow. The reader should recognize that the J dimension adaptive weight solution may be arrived at via any of the current adaptive processing algorithms such as Howells-Applebaum [7], Gram-Schmidt [9], Sample Matrix Inverse Update [18], etc.

When applying these constraints to our three-source case of Fig. 2, we would assign beamformer beams Nos. 10, 11, and 12 to cover the sources as illustrated in Fig. 6(a). These assigned beams are then given a maximum gain level of about 5 dB above the -30 dB mainbeam sidelobes; thus, the assigned beam weights are constrained to $|W_k| \leqslant 0.055$. All other W_k are set to zero. Typical resultant adapted patterns are shown in Fig. 6(b), where nine trials of 160 snapshots each are plotted. The pattern stability is near-perfect for a unique solution like this, and note that the three sources have been nulled with very little perturbation of the mainbeam sidelobes except in the immediate vicinity of the sources. Since we are inverting a matrix of only 3x3 dimension in Eq. (8), for this case, it follows that the number of snapshots processed per trial could be reduced by an order of magnitude [10] and still obtain excellent results.

Figure 7 demonstrates how the adaptive weights will become *noisy* if we include even one extra DOF beyond the unique solution. Beamformer beam No. 16 was deliberately added for the same case as in Fig. 6, and we may note the consequent sidelobe fluctuations. However, if we use the "soft" constraint of Eq. (9) in solving for the weights, stable performance is again restored despite the extra DOF. It may be of interest to the reader that for this particular example:

$$J = 4$$
$$\gamma = 0.045 \text{ (-27 dB)}$$
$$\text{Trace } [\hat{\mathbf{R}}]/J \approx 10{,}350\, \beta_0^2$$
$$\alpha = 21\beta_0^2 .$$

Although not shown here, another example of interest is the case of using a two-beam cluster (Nos. 11 and 12) to cancel a single 40 dB broadband source located at 22°. It was found that the source could be adequately cancelled at bandwidths up to 15%.

Many other combinations of source distributions and assigned beams were tested to further verify the technique, and the partially-adaptive performance was satisfactory *provided that the assigned beams were sufficient to cover the DOF demanded by the source distribution.*

Interference Sources in the Mainbeam Region

Extension of the foregoing partially-adaptive array technique for mainbeam interference is straightforward, provided we relax the constraint upon the value of γ in Eq. (5). Obviously, the low-sidelobe stratagem becomes secondary to the greater menace of an interference source coming in through our high-gain mainbeam. Low sidelobes could still be retained, if necessary, by implementing a beamformer which is capable of producing a family of low-sidelobe assigned beams [15].

SOURCE ESTIMATION AND BEAM ASSIGNMENT

Modern spectral estimation techniques are a welcome addition to the conventional methods for tracking and cataloging interference sources. They do not interfere with any functions of the mainbeam, and are capable of providing superior source resolution from fewer elements. The latter advantage is obtained in part because we assumed low sidelobes for the mainbeam, i.e., the only sources that require estimation are those few which are of sufficiently high SNR to get through the mainbeam sidelobes. Resolution performance is always directly related to SNR, [2,5,6].

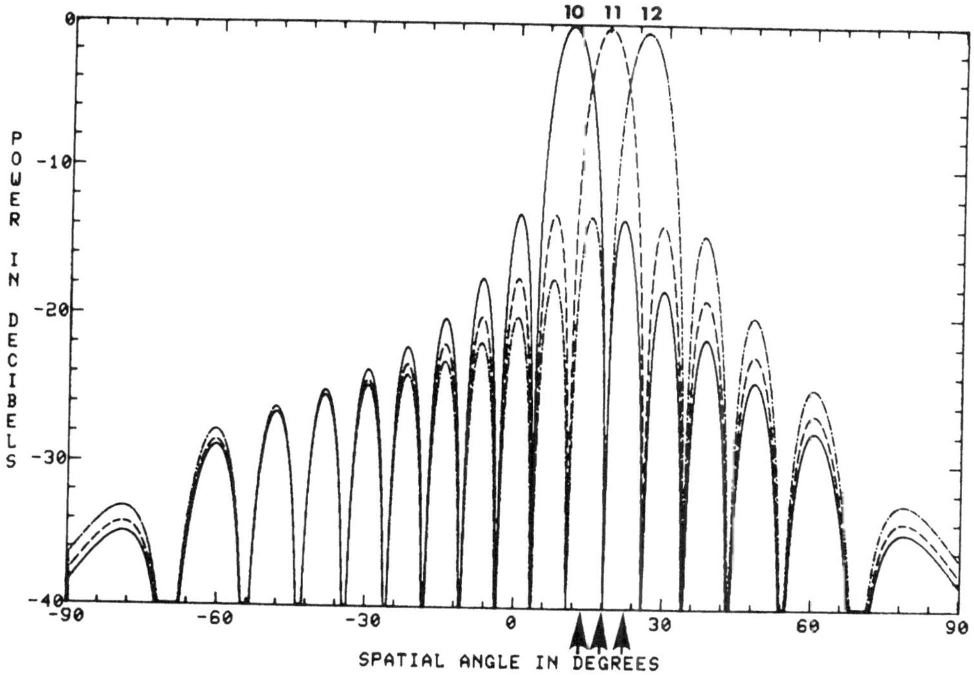

(a) Beamformer beams Nos. 10, 11, and 12

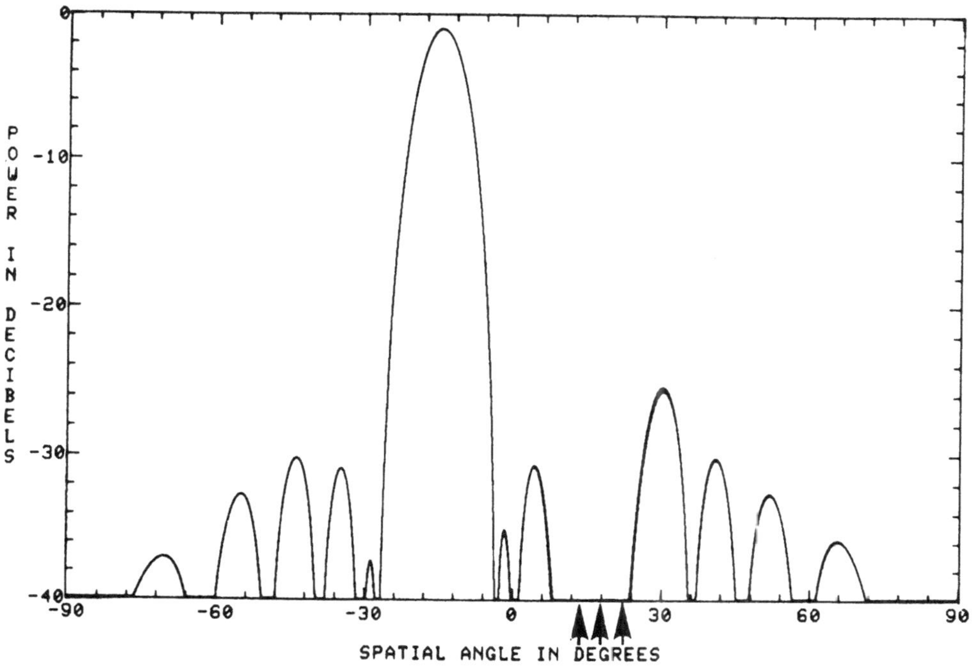

(b) Typical adapted patterns, nine trials of 160 snapshots

Fig. 6 — Partially adaptive linear array of 16 elements, using three assigned beams for the three source case of Fig. 2

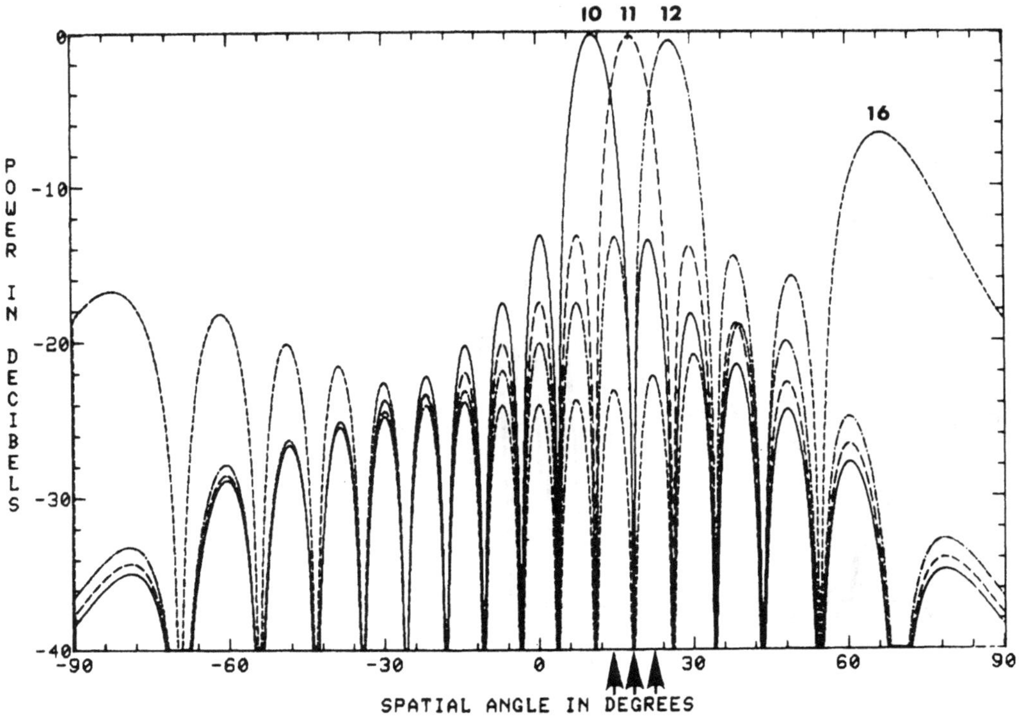

(a) Beamformer beams Nos. 10, 11, 12, and 16

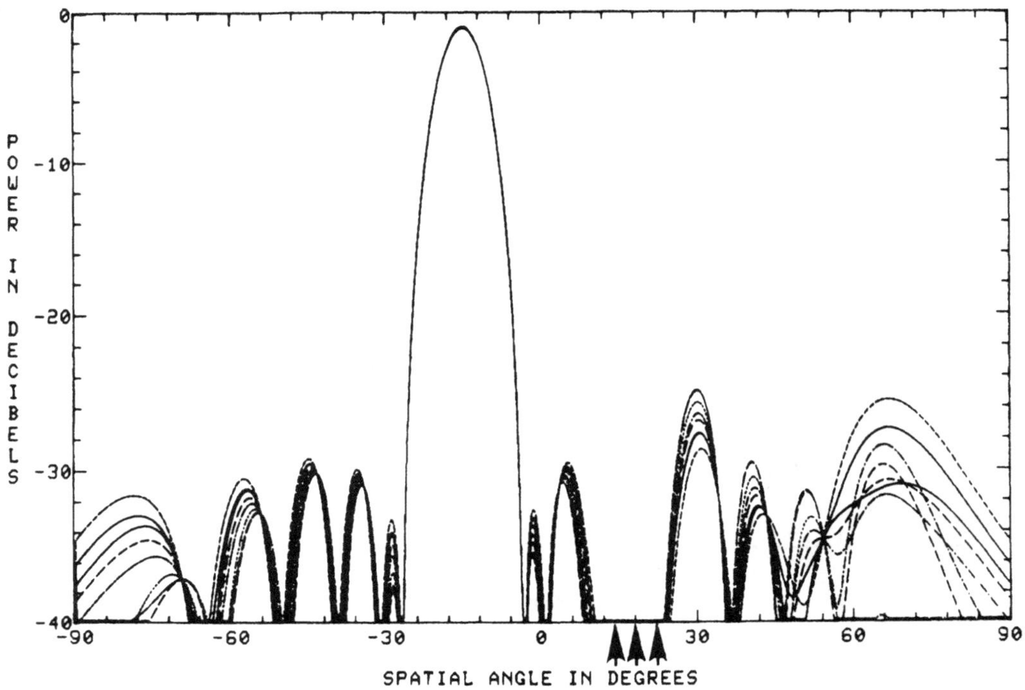

(b) Typical adapted patterns, nine trials of 160 snapshots

Fig. 7 — Partially adaptive linear array of 16 elements, using four assigned beams for the three-source case of Fig. 2

The principle for achieving source estimation from a small fraction of the aperture DOF has been demonstrated via many techniques, both conventional and optimal [1,3,19]. It is not within the scope of this report to attempt a comprehensive comparison of such techniques, but the point is important to our concept so that an example of a half-aperture linear array estimator is given in this section. The type of application envisioned is illustrated in Fig. 8, where we represent a KxK element aperture system in which the adaptive beam DOF are to be assigned on the basis of estimates derived from two orthogonal linear arrays of $K/2$ elements each. An extension of the 2D (two-dimension) beamspace adaptive array system of Fig. 5 to the 3D system suggested by Fig. 8 permits several beamformer options, including:

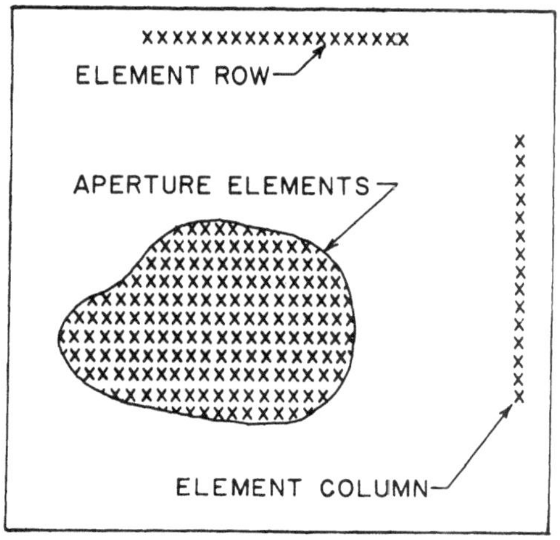

Fig. 8 — (K x K) element aperture within which row/column linear arrays couple into source estimation processors

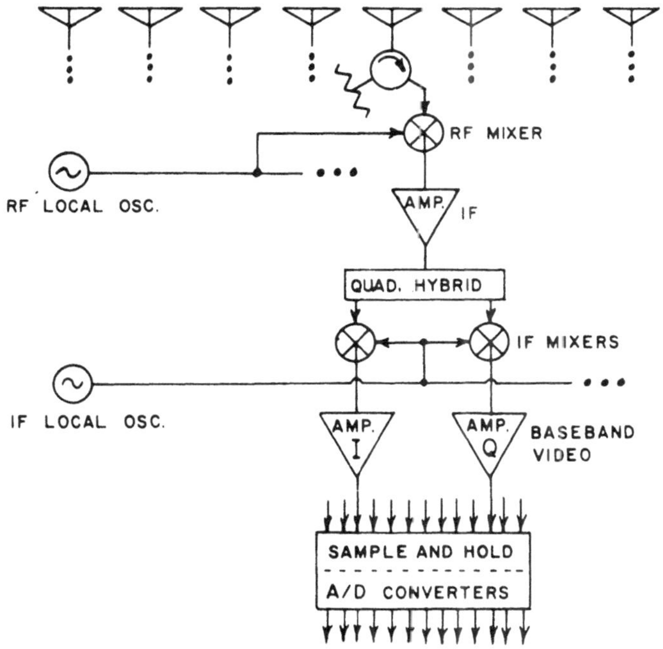

Fig. 9 — Typical RF receiver techniques associated with A/D complex data sampling

a. Two orthogonal 2D beamformers of which one is coupled into a row and the other coupled into a column of elements.

b. A complete 3D beamformer [20] coupled into the aperture elements, perhaps on a thinned basis.

The separate mainbeam must be summed from all K^2 elements to attain the desired low sidelobes.

Although they involve relatively few elements from the aperture, the linear array estimators represent a significant increase in system expense, because they are all-digital processing subsystems. Typical RF receiver components required prior to the signal analogue-to-digital (A/D) converters are shown in Fig. 9. The processing of the digital signals to estimate the sources may be carried out in accordance with a number of spectral estimation algorithms available in the literature [1-6]. Appendix B discusses several algorithms that were used in the simulations conducted for this report. For example, Fig. 10 illustrates a comparison plot of our mainbeam search scan vs half-aperture eigenanalysis processing results for the 16-element linear array case of Fig. 2.

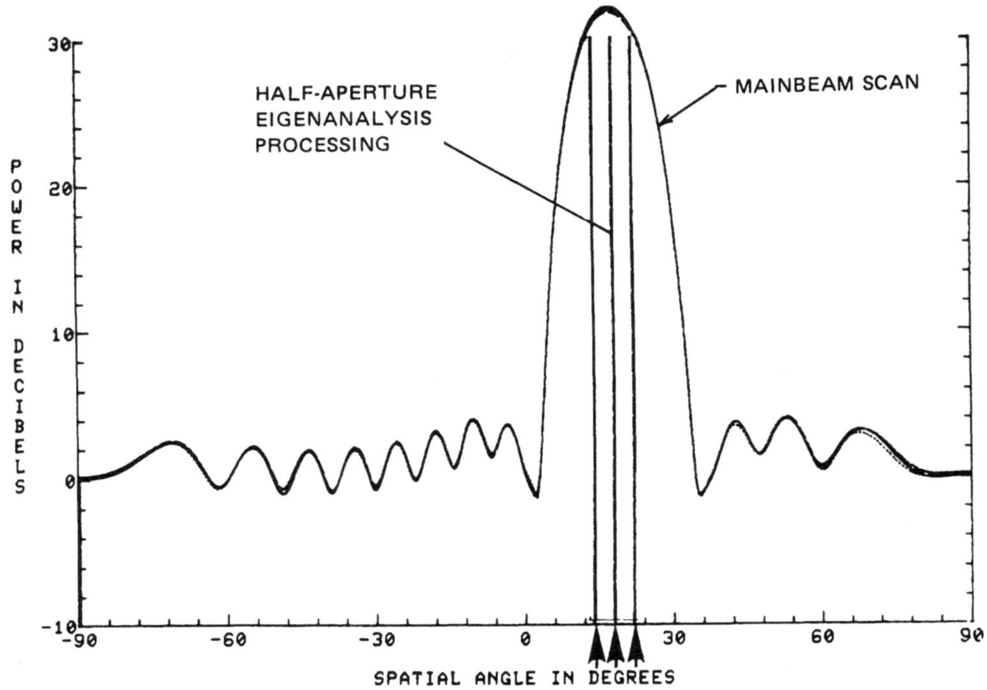

Fig. 10 — Comparison of mainbeam scan vs half-aperture eigenanalysis source estimation for three source case of Fig. 2

Once the source estimation information is available, we can proceed by assigning beamformer beams via a computer logic program. For the simulations reported in this report, a Fortran IV computer code named "BEAMASSIGN" was developed which accepts source information updates, compares the new data against a source directory kept in memory, computes track updates for sources already in memory, determines priority ranking, and assigns beams to cover the sources of highest priority. An important point to note is that beam assignment does not require great accuracy, i.e., a half-beamwidth is usually close enough. Also, clusters of two or three adjacent beams may be assigned for doubtful cases.

A demonstration of beam assignment was conducted with a moving source simulation involving the 16-element linear array of Fig. 2. Four sources of unequal strength were set up in the farfield, traveling in criss-crossing patterns. Two of the sources are of 30 dB strength with start angles of 3.0 and 39.0°, and two are of 43 dB strength with start-angles of 5.0 and 70.0°. The estimation of the scanned mainbeam for this example is shown in Fig. 11(a). Each time-unit plot cut is computed from $\hat{\mathbf{R}}$ averaged over 160 snapshots,

$$P_0 = \mathbf{S}^t \hat{\mathbf{R}} \mathbf{S}^* \tag{10}$$

where $\mathbf{S}^*$ is the mainbeam steering vector used to generate the display plot. As expected, this simple Fourier output is dominated by the two stronger sources. In contrast, Fig. 11(b) shows the source estimation derived from eigenanalysis processing using only half of the aperture (8 elements). Note that the "superresolution" characteristics of this type of optimal estimation produces excellent source tracking, even in the vicinity of cross-over of three of the sources.

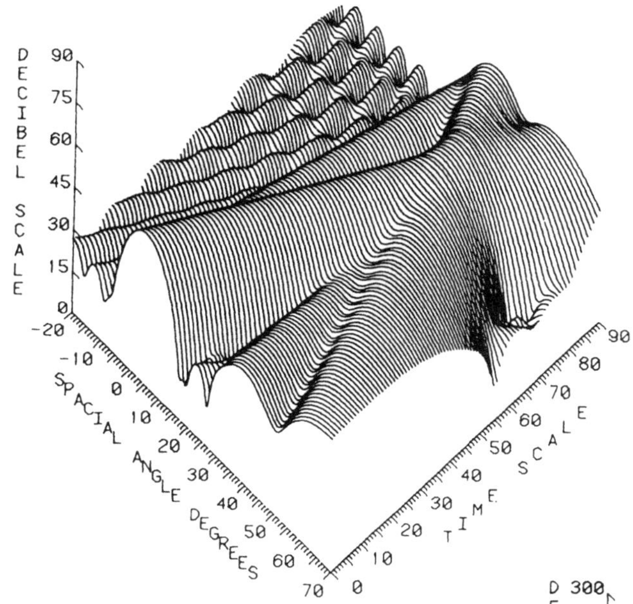

(a) Conventional mainbeam scanning, 16 elements, 30 dB Taylor illumination

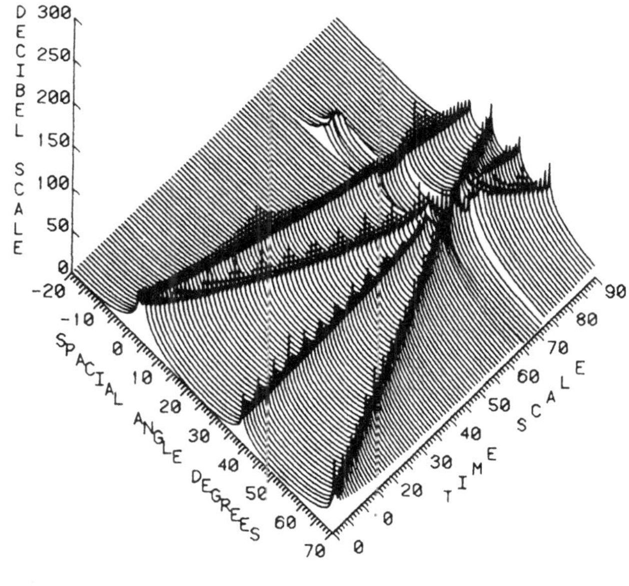

(b) Half aperture eigenanalysis source estimation

Fig. 11 — Estimation of four moving sources via mainbeam scan and half aperture eigenanalysis algorithm

The results from the use of the source information data contained in Fig. 11(b) to continuously update beam assignments are illustrated in the adapted pattern cuts shown in Fig. 12(a). Note that the mainbeam remains steady, and the sidelobes seldom exceed their quiescent 30 dB peak level, despite the drastic shifting of the nulls as the moving sources criss-cross in the sidelobe region. In contrast, Fig. 12(b) illustrates the adapted pattern cuts obtained when we utilize the SMI algorithm weights with the array fully adaptive. Although the source cancellation is excellent, the mainbeam suffers significant modulation and the peak sidelobe levels rise considerably.

(a) Partially adaptive, constrained, assigned beams, 32 snapshots processed per plot cut

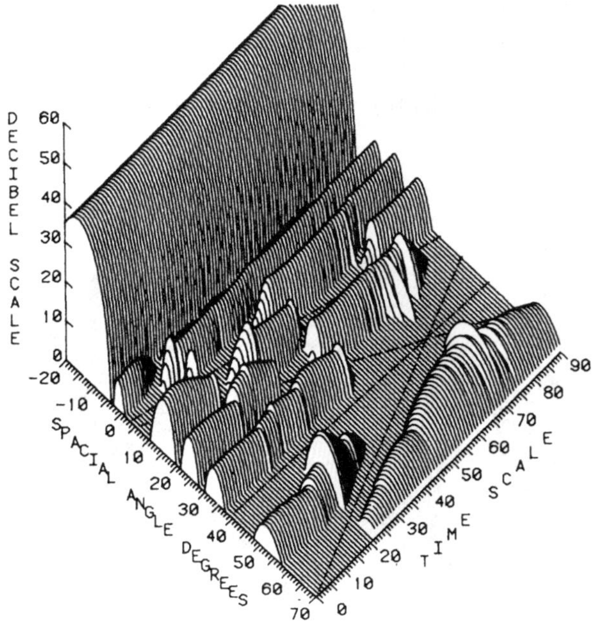

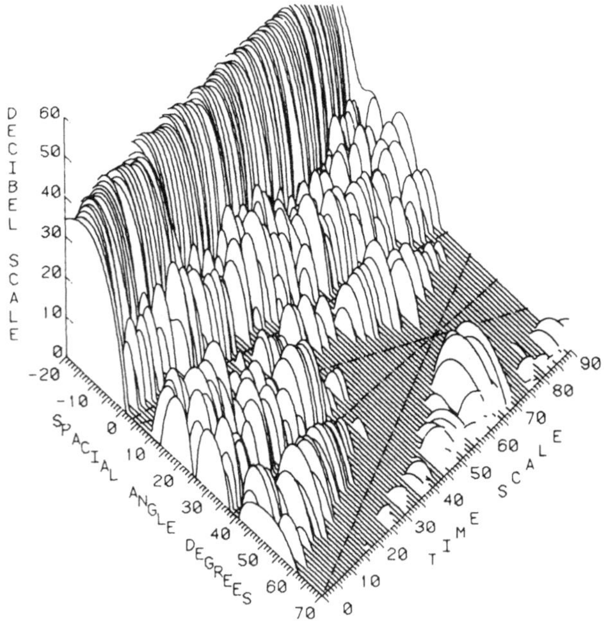

(b) Fully adaptive array, no constraints, 300 snapshots processed per plot cut

Fig. 12 — Adaptive patterns for 16 element linear array, SMI algorithm, four moving sources case of Fig. 11

AN ADAPTIVE ARRAY TRACKING APPLICATION

A second area where spectral estimation techniques can provide valuable assistance is that of adaptive array tracking systems. Here we are dealing with the problem of attempting to track targets under the condition of having interference sources present in the mainbeam region. Some early proposed solutions in this area evolved from the growing adaptive array technology of the 1970's. For example, a paper by White [21] discusses the radar problem of tracking targets in the low-angle regime, where conventional tracking radars encounter much difficulty because of the presence of a strong surface-reflected ray.

The first extension of fully adaptive arrays to angle estimation in external noise fields is the contribution of Davis et al. [22], who developed an algorithm based on the outputs of adaptively distorted sum and difference beams. The adaptive beams filter (null) the external noise sources, and distortion correction is then applied in the resultant monopulse output angle estimate. Their work is particularly appropriate as a starting point for this section, where we discuss the advantages of using spectral estimation techniques in an all-digital, fully adaptive, array tracking system. Reference [15] is also pertinent.

Coherent Spatial Interference Sources

The existence of significant coherence between spatial sources as, for example, in multipath situations involving a specular reflection, continues to represent a serious problem area even for a fully adaptive tracking array.

Reasons include:

a. Coherent signals in space are not stationary [2,5,23].

b. Adaptive systems may perform cancellation via weight phasing rather than null steering [5,23,24,25,26].

c. Adaptive tracking beam distortion is highly sensitive to coherent signal phasing.

d. Signal fading under anti-phase conditions.

To demonstrate these reasons, adaptive characteristics were computed for a 16-element linear array for an interference case in which there are two 13 dB coherent sources in the mainbeam region. There is also a third source, non-coherent, in the nearby sidelobe region that acts as a stable null comparison point.

In Fig. 13(a), we illustrate the severe changes in our mainbeam caused by variation of the phase shift between the two coherent sources. The quiescent mainbeam has the same Taylor weighting as that in Fig. 2(a). Figure 13(b) illustrates the spatial insertion loss associated with the three adaptive weightings involved. Note that for source phasing of 0° and 180°, the adaptive weights do not achieve cancellation by steering nulls onto the coherent sources, but rather by the weight phasing itself. The array output was driven down to receiver noise level in all three cases. The plots for 90° phase are very similar to what one would obtain if all three sources were non-coherent, i.e., cancellation is achieved by adaptive null steering, in this instance.

Such severe sensitivity to coherent source phasing in the mainbeam region produces different distortions in tracking estimates from adaptive Σ (sum) and Δ (difference) patterns, as shown in Fig. 14. Appendix C contains the equation development for this type of plot; however, the main point here is to show the considerable changes in track angle estimates due to phase variation. Once again, if all three sources were non-coherent, the distortion plot would be stable and very similar to the one shown for the 90° phase.

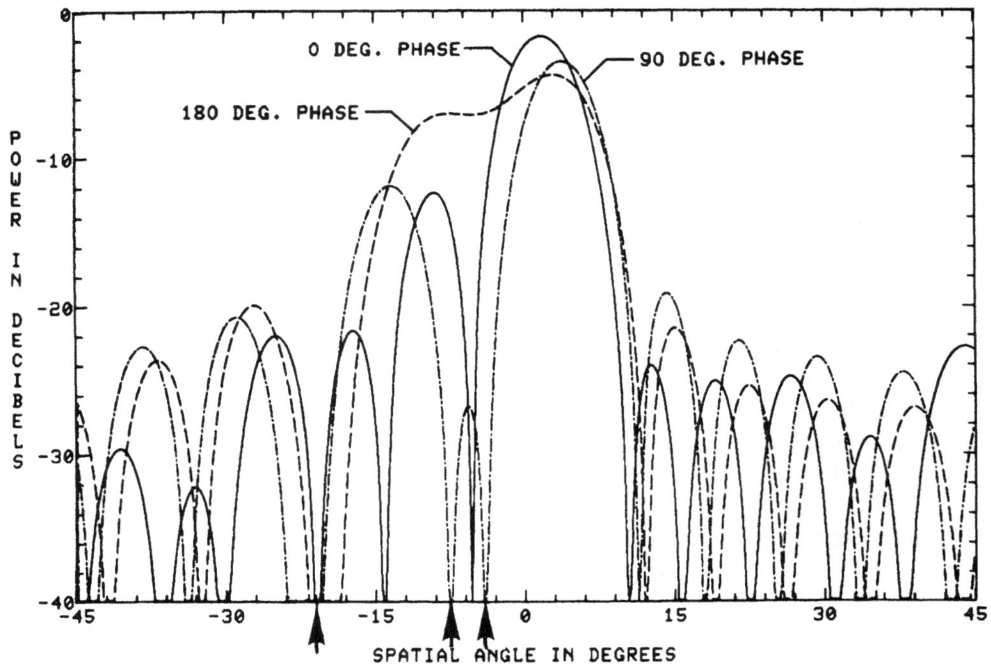

(a) Typical adaptive patterns for three coherent phases

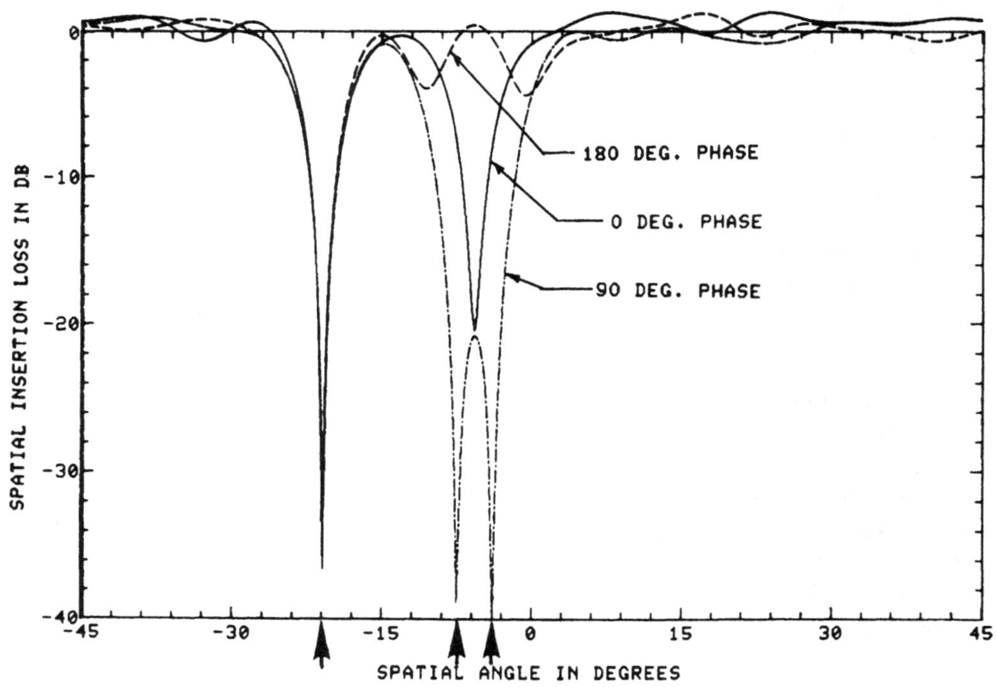

(b) Insertion loss associated with adaptive weighting

Fig. 13 — Mainbeam interference adaptive characteristics for 16 element linear array, SMI algorithm, 256 snapshots, three interference sources: one 10 dB non-coherent at −21°, and two 13 dB coherent at −7.6 and −4.0°

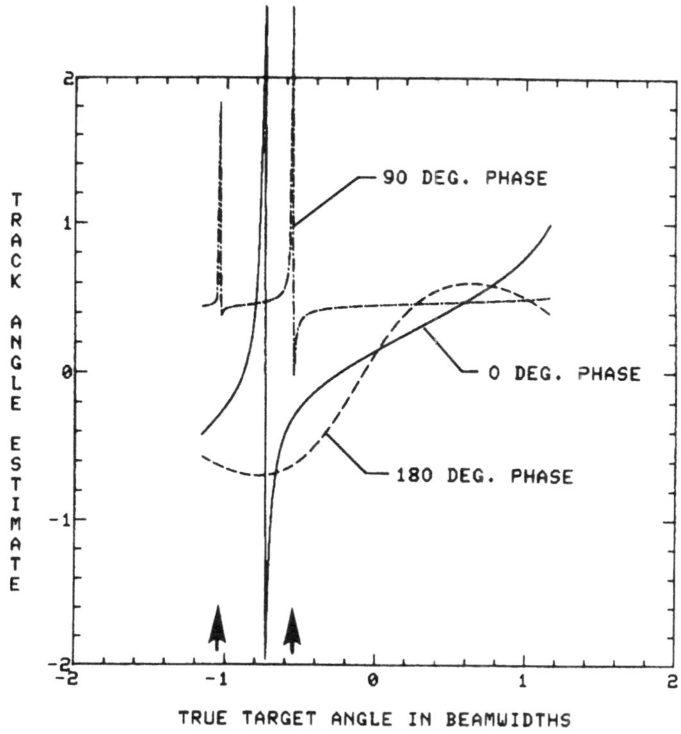

Fig. 14 — Track estimate distortion resulting from adaptive Σ and Δ tracking beams for coherent interference in the mainbeam region, same case as Fig. 13

All-Digital Tracking System Concept

The separate estimation of interference source data (total number, power levels, location angles, coherence) and its utilization for improving the output SNR of desired signal detections is a mode of system operation that has been addressed in the literature a number of times for various applications [5,6,15,16]. In this section, we briefly review such a system wherein the estimated data is used to drive a fully adaptive tracking processor [27]. The concept is illustrated in Fig. 15. Starting on the left-hand side, the system continuously computes/updates a sample covariance matrix $\hat{\mathbf{R}}$. Of particular significance is that $\hat{\mathbf{R}}$ may be dimensioned either equal to or less than the total number of array elements, i.e., the model order of the estimate is selectable per subaperture averaging option choice. Off-line processing on $\hat{\mathbf{R}}$ is then conducted at periodic intervals to estimate the locations and relative power levels of interference sources via the most appropriate spectral estimation algorithms. The central processor unit (CPU) then applies these data to the computation of optimized adaptive spatial filter weights for the right-hand side of Fig. 15. Separation of source estimation from adaptive filter weight computation can be done accurately only in an all-digital processing system, but it permits the following benefits:

a. Estimation of coherent interference source locations for deliberate adaptive null filter placement.

b. Remembering slow-changing or time-gated sources, and *colored-noise* distributions.

c. Anticipating sources from apriori data inputs.

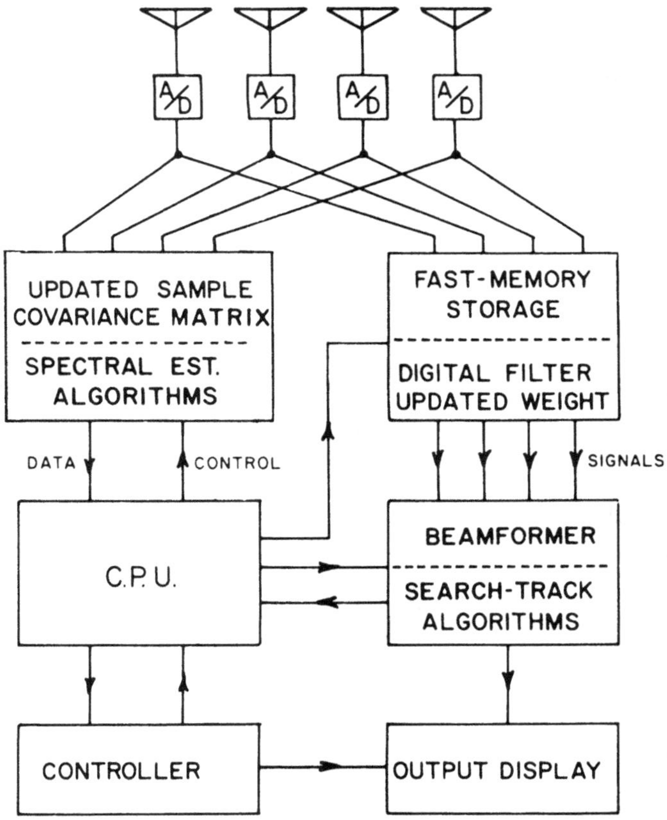

Fig. 15 — All-digital adaptive array tracking system concept

d. Flexibility in time-domain control of the filtering to counter interference time strategies.

e. Tracking/cataloging/ranking sources.

f. Efficient assignment of available DOF.

g. Compatible with fast-response adaptive algorithms, i.e., parallel algorithm processing.

The right-hand side of Fig. 15 indicates a fast-memory storage capability that is intended to permit selected time delays of the snapshots for feeding into the filter weights. The idea is to synchronize selected snapshots with their filter weight updates, if possible.

Finally, the filtered signal output residue is fed into a beamformer which is weighted to produce the desired search and monopulse track beams for target detection and tracking. The algorithms of Davis et al., [22], may be applied for estimating the target signal angle of arrival, based upon the outputs of adaptively distorted sum and difference beams. Appendix C discusses the equivalence of such beams to the concept shown in Fig. 15.

As an example, let us apply this concept to the coherent source case used in Figs. 13 and 14 wherein we would use a 16-element linear array feeding into our all-digital processor. An appropriate estimation algorithm is that of forward-backward subaperture spatial smoothing [5,28,29] combined with eigenanalysis. Appendix D describes the rudiments of this algorithm, and the results are plotted in Fig. 16 in comparison with a scanned mainbeam output. From this source estimation data, we can

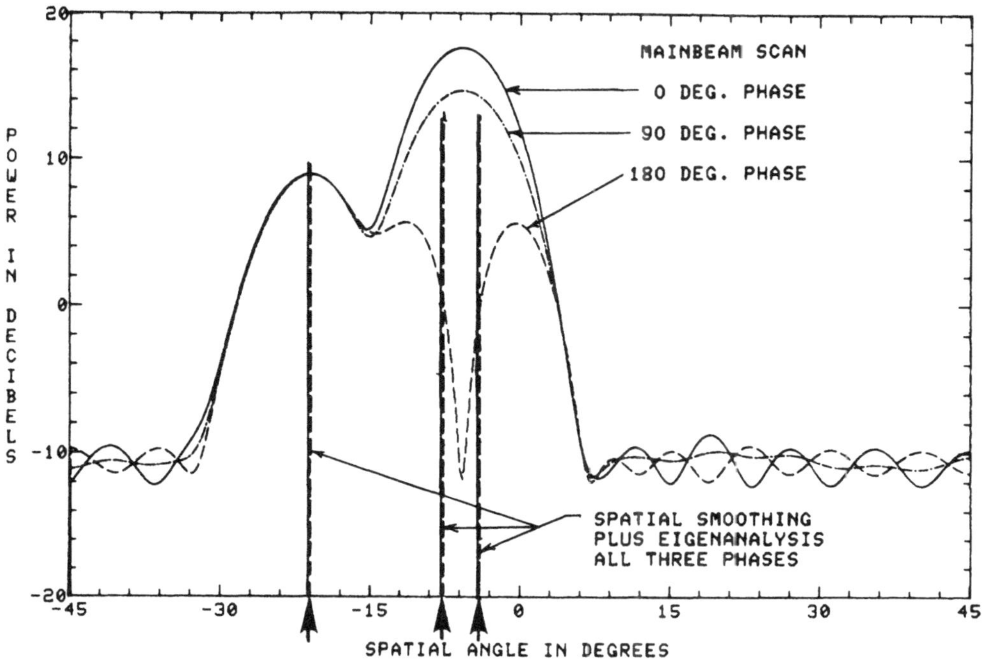

Fig. 16 — Comparison of mainbeam scan vs spatial smoothing processing for coherent source case of Fig. 13, PEGS eigenanalysis, 256 snapshots per trial

construct an equivalent covariance matrix dimensioned for the full aperture per the procedure given in Appendix A, and we can compute its inverse for obtaining the adaptive filtering. If we define the constructed covariance matrix as **M**, then its inverse may be viewed as a matrix set of adaptive *beamformer* filter weights to give us the filtered output nth snapshot vector $\mathbf{E}_f(n)$,

$$\mathbf{E}_f^t(n) = \mathbf{E}^t(n)\ \mathbf{M}^{-1} \tag{11}$$

Conventional beam weighting $\mathbf{S}^*$ can then be applied to the filtered output residue to obtain the final output for the nth snapshot,

$$Y_0(n) = \mathbf{E}_f^t(n)\mathbf{S}^* = \mathbf{E}^t(n)\mathbf{M}^{-1}\mathbf{S}^* \tag{12}$$

or

$$Y_0(n) = \mathbf{E}^t(n)\mathbf{W}_0$$

where $\mathbf{W}_0$ is the familiar optimum Wiener filter weight.

Note that the constructed covariance matrix, **M**, permits options such as adding synthetic sources or changing power levels. Furthermore, since it is always Toeplitz, solutions may be simplified somewhat.

For the current example, the computed adaptive characteristics would be very similar to those plotted in Figs. 13 and 14 for the 90° phase angle. Other examples along with a more detailed discussion of the processing may be found in [27].

CONCLUSIONS

Two conceptual application areas have been presented for using spectral estimation techniques; partially-adaptive low-sidelobe arrays, and fully-adaptive tracking arrays. In both cases, improved spectral estimation techniques are used separately to acquire information about the interference environment which is beyond that ordinarily available in a conventional adaptive array. Examples discussed included *superresolution* effects, relative power level determination, estimation of coherent sources, and the tracking/cataloging/ranking of sources. For the partially-adaptive area, the information was used for efficient assignment of a limited number of DOF in a beamspace constrained adaptive system to obtain the following benefits (as compared to a fully adaptive array): retention of low sidelobes plus a stable mainbeam; considerably faster adaptive response; reduction in overall cost; and greater flexibility. On the negative side, we incur the risk of possible inferior cancellation performance if the interference source situation is not adequately covered by the assigned DOF.

For the fully adaptive tracking array area, the information is used in an all-digital processing system to obtain the benefits of stable nulling of coherent interference sources in the mainbeam region, efficient assignment of the available DOF, and a far greater flexibility in the time-domain control of adaptive filtering strategy.

REFERENCES

1. D.G. Childers, ed., *Modern Spectrum Analysis*, IEEE Press, New York, N.Y. 1978.

2. W.F. Gabriel, "Spectral Analysis and Adaptive Array Superresolution Techniques," Proc. IEEE **68**, Jun 1980, pp. 654-666.

3. Special Issue on Spectral Estimation, Proc. IEEE **70**, Sep 1982.

4. R. Schmidt, "Multiple Emitter Location and Signal Parameter Estimation," Proc. of the RADC Spectrum Estimation Workshop, RADC-TR-79-63, Rome Air Development Center, Rome, NY, Oct 1979, p. 243.

5. J.E. Evans, J.R. Johnson, and D.F. Sun, "Application of Advanced Signal Processing Techniques to Angle of Arrival Estimation in ATC Navigation and Surveillance Systems," MIT Lincoln Laboratory Tech. Report 582, (FAA-RD-82-42), Jun 1982.

6. A.J. Barabell et al., "Performance Comparison of Superresolution Array Processing Algorithms," MIT Lincoln Laboratory Report TST-72, May 1984.

7. S.P. Applebaum, "Adaptive Arrays," IEEE Trans. Antennas Propagat., AP-**24**, Sep 1976, pp. 585-598.

8. D.J. Chapman, "Partial Adaptivity for the Large Array," IEEE Trans. Antenna Propag., AP-**24**, Sep 1976, pp. 685-696.

9. R.A. Mozingo and T.W. Miller, *Introduction to Adaptive Arrays*, (John Wiley and Sons, New York, 1980).

10. I.S. Reed, J.D. Mallett, and L.E. Brennan, "Rapid Covergence Rate in Adaptive Arrays," IEEE Trans. Aerosp. Electron, Syst., AES-**10**, Nov 1974, pp. 853-863.

11. W.F. Gabriel, "Adaptive Arrays — An Introduction," Proc. IEEE, **64**, Feb 1976, pp. 239-272.

12. S.P. Applebaum and D.J. Chapman, "Adaptive Arrays with Main Beam Constraints," IEEE Trans. Antennas Propagat. AP-**24**, Sep 1976, pp. 650-662.

13. J. Butler, "Multiple Beam Antennas," Sanders Assoc. Internal Memo RF 3849, Jan 1960.

14. J.T. Mayhan, "Adaptive Nulling with Multiple-Beam Antennas," IEEE Trans. Antennas Propag., AP-**26**, Mar 1978, pp. 267-273.

15. R.N. Adams, L.L. Horowitz, and K.D. Senne, "Adaptive Main-Beam Nulling for Narrow-Beam Antenna Arrays," IEEE Trans. Aerosp. Electron. Syst., AES-**16**, Jul 1980, pp. 509-516.

16. E.C. DuFort, "An Adaptive Low-Angle Tracking System," IEEE Trans. Antennas Propag., AP-**29**, Sep 1981, pp. 766-772.

17. N.L. Owsley, "Constrained Adaption," Array Processing Applications to Radar, Academic Press, 1980.

18. E. Brennan, J.D. Mallett, and I.S. Reed, "Adaptive Arrays in Airborne MTI Radar," IEEE Trans. Antennas and Propagation, AP-**24**, Sep 1976, pp. 607-615.

19. B.M. Leiner, "An Analysis and Comparison of Energy Direction Finding Systems," IEEE Trans. AES, AES-**15**, Nov 1979, pp. 861-873.

20. J.P. Shelton, "Focusing Characteristics of Symmetrically Configured Bootlace Lenses," IEEE Trans. Antennas and Propagation, AP-**26**, Jul 1978, pp. 513-518.

21. W.D. White, "Low-Angle Radar Tracking in the Presence of Multipath," IEEE Trans. AES, AES-**10**, Nov 1974, pp. 835-853.

22. R.C. Davis, L.E. Brennan and L.S. Reed, "Angle Estimation with Adaptive Arrays in External Noise Fields," IEEE Trans, AES, AES-**12**, Mar 1976, pp. 179-186.

23. W.D. White, "Angular Spectra in Radar Applications," IEEE Trans. Aerosp. Electron. Syst., AES-**15**, Nov 1979, pp. 895-899.

24. A. Cantoni and L. C. Godara, "Resolving the Directions of Sources in a Correlated Field Incident on an Array," Dept. of Electrical Engng., Univ. of Newcastle, Shortland, New South Wales, Australia, J. Acoustic. Soc. Am. (USA), **67**, No. 4, Apr 1980, pp. 1247-1255.

25. B. Widrow, et al., "Signal Cancellation Phenomena in Adaptive Antennas: Causes and Cures," IEEE Trans. Antennas and Prop., AP-**30**, May 1982, pp. 469-478.

26. T.J. Shan and T. Kailath, "Adaptive Beamforming for Coherent Signals and Interference," IEEE Trans. Acoust., Speech, Signal Processing, ASSP-**33**, pp. 527-536 (1985).

27. W.F. Gabriel, "A High-Resolution Target-Tracking Concept Using Spectral Estimation Techniques," NRL Report 8797, May 1984.

28. A.H. Nuttall, "Spectral Analysis of a Univariate Process with Bad Data Points, via Maximum Entropy and Linear Predictive Techniques," NUSC-TR-5303, Naval Underwater Systems Center, New London, CT, Mar 1976.

29. L. Marple, "A New Autoregressive Spectrum Analysis Algorithm," IEEE Trans. Acoust., Speech, Signal Process. ASSP-**28**, Aug 1980, pp. 441-454.

Appendix A
SNAPSHOT SIGNAL MODEL

Consider a simple linear array of K elements as shown in Fig. A1. The received signal samples are correlated in both space and time, giving rise to a two-dimensional data problem, but we convert this to spatial domain only by assuming that narrow-band filtering precedes our spatial domain processing. Bandwidth can be handled when necessary via a spectral line approach [A1] or tapped delay lines at each element [A2], but we did not consider such extra complication essential to the basic purposes of this analysis. Thus, the postulated signal environment on any given observation consists of I narrow-band plane waves arriving from distinct directions θ_i. The RF phase at the kth antenna element as a result of the ith source would be the product $\omega_i X_k$, where X_k is the location of the element phase center with respect to the midpoint of the array in wavelengths, and ω_i is defined as

$$\omega_i = 2\pi \sin \theta_i \,. \tag{A1}$$

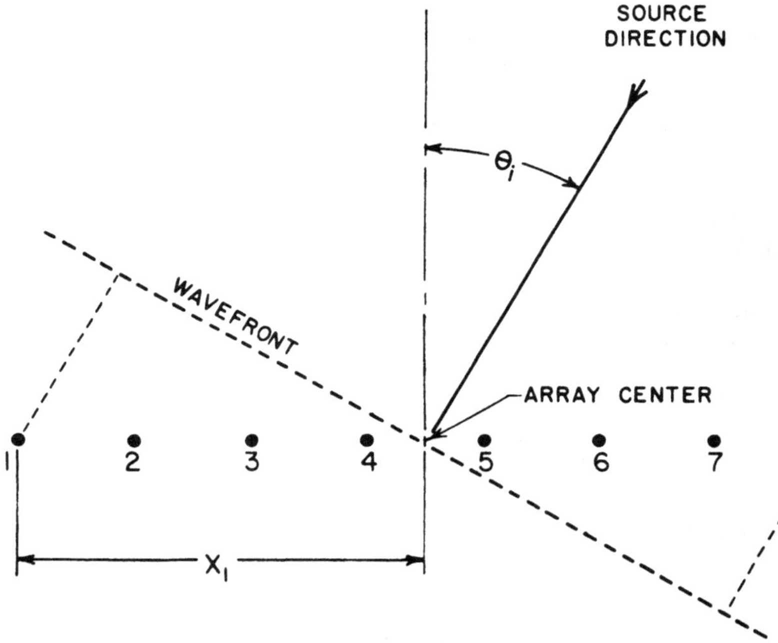

Fig. A1 — Geometry of linear array and signal wavefront

This notation is deliberately chosen to have the spatial domain dual of sampling in the time domain, so that the reader may readily relate to the more familiar spectral analysis variables. Sin θ_i is the dual of a sinusoid frequency f_i, and the X_k locations are the dual of time sampling instants t_k. Note that if our elements are equally spaced by a distance d, then X_k may be written,

$$X_k = \left(\frac{d}{\lambda}\right)\left(k - \left(\frac{K+1}{2}\right)\right) \tag{A2}$$

where λ is the common RF wavelength. The ratio d/λ becomes the dual of the sampling time T with the cutoff frequency equal to half of the reciprocal.

The complex amplitude of the ith source *at the array midpoint phase center* is p_i, such that we can now express the *n th* time-sampled signal at the kth element as,

$$E_k(n) = \eta_k(n) + \sum_{i=1}^{I} p_i(n) g_k(\theta_i) \exp(j\omega_i X_k) \tag{A3}$$

where $g_k(\theta_i)$ is the element pattern response in the direction θ_i, and $\eta_k(n)$ is the nth sample from the kth element independent Gaussian receiver noise. (The receiver noise component is assumed to be a random process with respect to both the time index n and the element index k.) Equation (A3) permits us to construct a convenient column vector of observed data in the form,

$$\mathbf{E}(n) = \mathbf{V}\mathbf{p}(n) + \boldsymbol{\eta}(n) \tag{A4}$$

$$\begin{bmatrix} E_1(n) \\ E_2(n) \\ E_3(n) \\ \cdot \\ \cdot \\ \cdot \\ \\ E_K(n) \end{bmatrix} = \begin{bmatrix} v_{11} & v_{21} & & v_{I1} \\ v_{12} & v_{22} & & v_{I2} \\ v_{13} & v_{23} & & v_{I3} \\ \cdot & \cdot & \cdots & \cdot \\ \cdot & \cdot & & \cdot \\ \cdot & \cdot & & \cdot \\ \\ v_{1K} & v_{2K} & & v_{IK} \end{bmatrix} \begin{bmatrix} p_1(n) \\ p_2(n) \\ \cdot \\ \cdot \\ \cdot \\ p_I(n) \end{bmatrix} + \begin{bmatrix} \eta_1(n) \\ \eta_2(n) \\ \eta_3(n) \\ \cdot \\ \cdot \\ \cdot \\ \\ \eta_K(n) \end{bmatrix}$$

where $\mathbf{V}$ is a $K \times I$ matrix containing a column vector $\mathbf{v}_i$ for each of the I source directions; i.e.,

$$v_{ik} = g_k(\theta_i) \exp(j\omega_i X_k)\,. \tag{A5}$$

Note that Eq. (A4) separates out the basic variables of source direction in the direction matrix $\mathbf{V}$, source baseband signal in the column vector $\mathbf{p}(n)$, and element receiver channel noise in the column vector $\boldsymbol{\eta}(n)$. The vector $\mathbf{E}(n)$ is defined as the nth snapshot, i.e., a simultaneous signal sampling across all K-array elements at the nth time instant. These snapshots would nominally occur at the Nyquist sampling rate corresponding to our receiver bandwidth [A3], so that a radar-oriented person may view them as range bin time samplings. However, for source estimation purposes, they need not necessarily be chosen from contiguous range bins, and for most applications it would be highly desirable to selectively time gate the snapshots used for source estimation. For this simple analysis, let us postulate that the snapshots are gated at more or less arbitrary instants of time.

Over typical processing intervals, the directions of arrival will not change significantly, so that $\mathbf{V}$ is a slowly changing matrix. In contrast, the signals $p_i(n)$ will generally vary rapidly with time, often unpredictably, such that we must work with their statistical descriptions. It is assumed that the signals are uncorrelated with receiver noise. Proceeding then from Eq. (A4), we can obtain the covariance matrix $\mathbf{R}$ via application of the expected value operator, $\mathscr{E}$, or ensemble average,

$$\mathbf{R} = \mathscr{E}[\mathbf{E}(n)\mathbf{E}^{*t}(n)] \tag{A6}$$

$$\mathbf{R} = \mathbf{V}\mathbf{P}\mathbf{V}^{*t} + \mathbf{N} \tag{A7}$$

where $\mathbf{N} = \mathscr{E}[\boldsymbol{\eta}(n)\boldsymbol{\eta}(n)^{*t}]$, $\mathbf{P} = \mathscr{E}[\mathbf{p}(n)\mathbf{p}(n)^{*t}]$, $*$ is the complex conjugate, and t is the transpose. $\mathbf{N}$ is a simple diagonal matrix consisting of the receiver channel noise power levels. The diagonal elements of $\mathbf{P}$ represent the ensemble average power levels of the various signal sources, and off-diagonal elements can be nonzero if any correlation exists between the sources. Note that correlated far-field signals can easily arise if significant specular reflection or diffraction multipath is present.

REFERENCES

A1. W.F. Gabriel, "Adaptive Arrays — An Introduction," Proc. IEEE, **64**, Feb 1976, pp. 239-272.

A2. E. Brennan, J.D. Mallett, and I.S. Reed, "Adaptive Arrays in Airborne MTI Radar," IEEE Trans. Antennas and Propagation, AP-**24**, Sep 1976, pp. 607-615.

A3. O.V. Oppenheim and R.W. Schafer, *Digital Signal Processing* (Prentice Hall, Englewood Cliffs, NJ, 1975).

Appendix B
EIGENVALUE/EIGENVECTOR DECOMPOSITION

When a signal is known to consist of pure sinusoids in white noise, an appropriate procedure to find the unknown frequencies and powers is the Pisarenko spectral-decomposition procedure [B1]. Although Pisarenko's method per se has not found widespread use, it has provided a fundamental eigenanalysis basis for several closely related techniques which have demonstrated excellent performance. Among these are the algorithms described by Reddi [B2], the MUSIC algorithm of Schmidt [B3], the work of Bienvenu and Kopp [B4], the singular value decomposition (or principal eigenvector) methods of Kumaresan and Tufts [B5,B6], the eigenassisted method of Evans et al. [B7], and the algebraic approach of Bronez and Cadzow [B8].

A key principle in all of these techniques is the geometric vector space relationships between the spatial source vectors and the eigenvectors of the sample covariance matrix; so we begin our discussion on this point. From the theory of matrices, we know that a positive definite Hermitian matrix such as **R** of Eq. (A7), can be diagonalized by a nonsingular orthonormal modal matrix transformation which shall be defined as the matrix **Q**. Furthermore, we know that the resulting diagonal components are the eigenvalues of matrix **R**. In accordance with the usual eigenvalue problem statements,

$$|\mathbf{R} - \beta_i^2 \mathbf{I}| = 0 \text{ and } \mathbf{R}\mathbf{e}_i = \beta_i^2 \mathbf{e}_i\,, \tag{B1}$$

the β_i^2 are the eigenvalues (real positive numbers) of **R**, **I** is the identity matrix, and $\mathbf{e}_i$ are the associated eigenvectors. These eigenvectors, which are normalized to unit Hermitian length and are orthogonal to one another, make up the columns of the **Q** matrix,

$$\mathbf{Q} = \begin{bmatrix} | & | & | & & | \\ \mathbf{e}_1 & \mathbf{e}_2 & \mathbf{e}_3 & \cdots & \mathbf{e}_K \\ | & | & | & & | \end{bmatrix} \text{ where } \mathbf{e}_i = \begin{bmatrix} e_{i1} \\ e_{i2} \\ e_{i3} \\ \cdot \\ \cdot \\ \cdot \\ e_{iK} \end{bmatrix} \tag{B2}$$

Diagonalization of **R** by the **Q** matrix transformation per Eq. (B1) may be written,

$$[\mathbf{Q}^{*t}\mathbf{R}\mathbf{Q}] = [\beta_i^2\delta_{ij}] = \begin{bmatrix} \beta_1^2 & 0 & 0 & . & . & . \\ 0 & \beta_2^2 & 0 & . & . & . \\ 0 & 0 & \beta_3^2 & . & . & . \\ . & . & . & . & & \\ & & & & . & \\ . & . & . & & & . \\ . & . & . & & & \beta_K^2 \end{bmatrix} \tag{B3}$$

where δ_{ij} is the Kronecker delta symbol. One can readily show a construction of **R** from its orthonormal components,

$$\mathbf{R} = \mathbf{Q}[\beta_i^2\delta_{ij}]\mathbf{Q}^{*t} = \sum_{k=1}^{K} \beta_k^2 \mathbf{e}_k \mathbf{e}_k^{*t}. \tag{B4}$$

Next, we introduce the term "principal eigenvector" (PE) to mean those eigenvectors which correspond to the unique eigenvalues generated by the spatial source distribution; and the term "noise eigenvector" to mean those eigenvectors which correspond to the small noise eigenvalues generated by the receiver channel noise in Eq. (A7). Under ideal conditions, the noise eigenvalues are all identical and equal to receiver channel noise power level β_0^2, such that we can factor Eq. (B4) to emphasize the PE,

$$\mathbf{R} = \sum_{i=1}^{q} (\beta_i^2 - \beta_0^2)\mathbf{e}_i\mathbf{e}_i^{*t} + \beta_0^2\mathbf{I} \tag{B5}$$

where q is the number of PE. Comparing Eq. (B5) with Eq. (A7) we note that the noise diagonal matrices are equal; i.e.,

$$\beta_0^2\mathbf{I} = \mathbf{N} \tag{B6}$$

so that one may equate the source direction vectors with the PE,

$$\mathbf{VPV}^{*t} = \sum_{i=1}^{I} |\overline{p}_i|^2\mathbf{v}_i\mathbf{v}_i^{*t} = \sum_{i=1}^{q} (\beta_i^2 - \beta_0^2)\mathbf{e}_i\mathbf{e}_i^{*t} \tag{B7}$$

where the $|\overline{p}_i|^2$ represents the expected power levels of uncorrelated sources. Equation (B7) embodies the key principle that the PE are linear combinations of the source direction vectors and vice-versa. In geometrical language, the $\mathbf{v}_i$ define an I dimensional source vector space, and the principal $\mathbf{e}_i$ span that same vector space. Furthermore, since the noise eigenvectors are always orthogonal to the PE, then it follows that the noise eigenvectors must occupy a subspace which is orthogonal to the source vector space. To put it another way, if the noise eigenvectors are viewed as antenna array element weights, then they should have pattern nulls at source direction angles because of their orthogonality. (Note the vivid demonstration of this point in Fig. 3(b).) Despite the fact that Eq. (B7) is based on ideal assumptions, it turns out to be a valuable concept for formulating algorithms, perhaps because it is inherently a noise-subtracted relationship, and the estimates of the PE are rather robust.

When working with finite sets of data snapshots which are not ideal, a nontrivial problem area arises in determining which eigenvectors to designate as principal, and which ones result from noise. This important problem will be addressed after discussing the associated algorithms.

Eigenanalysis of Three Algorithms

Eigenvalue/eigenvector decomposition is now applied to three different spatial spectrum estimation algorithms:

- the MLM (Maximum Likelihood Method),
- the MUSIC (Multiple Signal Classification), and
- the PEGS (Principal Eigenvector Gram-Schmidt).

Since these algorithms are dealt with in a very abbreviated manner in the report, the reader is encouraged to consult the references given for a better description and understanding of the techniques involved.

The MLM (Maximum Likelihood Method)

The maximum likelihood spectral estimate is defined as a filter designed to pass the power in a narrow-band about the signal frequency of interest and to minimize or reject all other frequency components in an optimal manner [B9,B10]. This is identical to the use of a zero-order mainbeam directional gain constraint in adaptive arrays [B11], where the spatial spectrum would be estimated by the output residual power P_0 from the optimized adapted array weights,

$$P_0(\theta) = \mathbf{W}_0^{*t}\mathbf{R}\mathbf{W}_0 . \tag{B8}$$

Where $\mathbf{W}_0$ is the optimum adaptive Wiener filter weight, and

$$\mathbf{W}_0 = \mu\mathbf{R}^{-1}\mathbf{S}^* \tag{B9}$$

where $\mathbf{S}^*$ is the usual mainbeam weight vector for steering angle θ, and μ is a complex number. Under the zero-order gain constraint, we require $\mathbf{S}^t\mathbf{W}_0 = 1$, whereupon μ becomes

$$\mu = [\mathbf{S}^t\mathbf{R}^{-1}\mathbf{S}^*]^{-1} . \tag{B10}$$

Substituting μ and $\mathbf{W}_0$ into Eq. (B8) results in

$$P_0(\theta) = \left(\frac{1}{\mathbf{S}^t\mathbf{R}^{-1}\mathbf{S}^*}\right) . \tag{B11}$$

Upon sweeping the steering vector $\mathbf{S}^*$ for a given covariance matrix inverse, $P_0(\theta)$ estimates the spatial spectrum.

In terms of eigenvalue/eigenvector decomposition, we can take the inverse of Eq. (B3) and express $\mathbf{R}^{-1}$ in the form,

$$\mathbf{R}^{-1} = \mathbf{Q}\left[\frac{\delta_{ij}}{\beta_i^2}\right]\mathbf{Q}^{*t} = \sum_{k=1}^{K}\left(\frac{1}{\beta_k^2}\right)\mathbf{e}_k\mathbf{e}_k^{*t} . \tag{B12}$$

Here, we see that this older algorithm simply uses all of the eigenvalues/eigenvectors. One advantage in this decomposition is that Eq. (B12) can be substituted into Eq. (B11) to form a simple summation of eigenvector beams referenced to the receiver noise power level,

$$\left(\frac{P_0(\theta)}{\beta_0^2}\right) = \frac{1}{\sum_{k=1}^{K}\left(\frac{\beta_0^2}{\beta_k^2}\right) g_k^2(\theta)} \tag{B13}$$

where $g_k(\theta) = \mathbf{S}^t\mathbf{e}^k$. This permits an insight into the peak values which occur at the nulls of the noise eigenvector beams; i.e., we get an evaluation of relative source power level if the source location is resolved.

The MUSIC (Multiple Signal Classification)

This algorithm was suggested by Schmidt [B3] to provide asymptotically unbiased estimates of the number of signal sources, directions of arrival, strengths and cross correlations among the directional waveforms, polarizations, and strength of noise/interference. His geometrical vector space description and interpretation is clearly presented and was used as the basis for the one above. Essentially, this MUSIC algorithm selects and uses only the noise eigenvectors to solve for the directions of arrival. This is tantamount to approximating $\mathbf{R}^{-1}$ in Eq. (B11) by the noise eigenvectors only; i.e.,

$$\text{let } \mathbf{R}^{-1} \approx \sum_{i=q+1}^{K} \left(\frac{1}{\beta_i^2} \right) \mathbf{e}_i \mathbf{e}_i^{*t} \tag{B14}$$

where q is the number of principal eigenvectors. The same indexing would apply as noise eigenvector beams in Eq. (B13), where we note that the ratio of eigenvalues would now become unity (or close to it).

This algorithm does indeed produce very large peaks in $P_0(\theta)$ for good covariance matrix estimates, because of the aforementioned orthogonality of the noise eigenvectors to the source vector space. Its performance is usually far superior to the older MLM algorithm in resolving closely spaced source directions. In addition, Schmidt points out that once the directions of arrival have been found, the direction matrix $\mathbf{V}$, in Eq. (A4) and (A7) becomes available and may be used to compute the source power matrix $\mathbf{P}$. We form the special matrix $\mathbf{U}$,

$$\mathbf{U} = [\mathbf{V}^{*t}\mathbf{V}]^{-1}\mathbf{V}^{*t} \tag{B15}$$

and

$$\mathbf{U}\mathbf{V}\mathbf{P}\mathbf{V}^{*t}\mathbf{U}^{*t} = \mathbf{P}.$$

If the matrix $\mathbf{U}$ exists or can be closely approximated, then we can apply it to the noise-subtracted covariance matrix of Eq. (A7) to solve for $\mathbf{P}$; i.e.,

$$\mathbf{P} = \mathbf{U}(\mathbf{R} - \mathbf{N})\mathbf{U}^{*t} \tag{B16}$$

$$\mathbf{P} = \begin{bmatrix} \overline{p_1{}^*p_1} & \overline{p_1{}^*p_2} & \cdots & \overline{p_1{}^*p_l} \\ \overline{p_2{}^*p_1} & \overline{p_2{}^*p_2} & \cdots & \overline{p_2{}^*p_l} \\ \cdot & \cdot & \cdot & \cdot \\ \cdot & \cdot & \cdot & \cdot \\ \cdot & \cdot & \cdot & \cdot \\ \overline{p_l{}^*p_1} & \overline{p_l{}^*p_2} & \cdots & \overline{p_l{}^*p_l} \end{bmatrix}$$

Note that the diagonal elements of $\mathbf{P}$ represent power estimates of the sources, and that the nonzero off-diagonal elements represent estimates of the correlations existing between partially coherent sources.

The ability to solve for the power estimates is of great importance in distinguishing "false alarms" and in selecting the sources of interest. We recommend References B7, B6, and B8 which are either related to or give a comparative analysis of the MUSIC algorithm.

The PEGS (Principal Eigenvector Gram-Schmidt)

Several eigenvalue/eigenvector decomposition techniques described in References B7, B2, B6, and B8 are based on the principal eigenvectors (PE) with some type of constraint imposed on the optimum weight vector.

This subgroup of PE methods is of interest in the current work because of their generally superior performance characteristics. An intuitive reasoning behind their use is that the estimates of the PE are robust; i.e., they tend to remain relatively stable from one data record to the next, whereas the noise eigenvectors tend to fluctuate because of noise perturbations. In addition, the PE methods are inherently a noise subtraction technique similar to noise power cancellation algorithms which attempt to remove the noise bias term that appears along the main diagonal of the covariance matrix.

Let us begin development of our PEGS algorithm by decomposing the inverse of the covariance matrix as given in Eq. (B12), normalized by receiver noise power,

$$\beta_0^2 \mathbf{R}^{-1} = \sum_{k=1}^{K} \left\{ 1 - \left(\frac{\beta_k^2 - \beta_0^2}{\beta_k^2} \right) \right\} \mathbf{e}_k \mathbf{e}_k^{*t}$$

$$= \mathbf{I} - \sum_{i=1}^{q} \left(\frac{\beta_i^2 - \beta_0^2}{\beta_i^2} \right) \mathbf{e}_i \mathbf{e}_i^{*t}. \qquad \text{(B17)}$$

Substituting Eq. (B17) into Eq. (B9) results in the optimum Wiener weight,

$$\mathbf{W}_0 = \frac{\mu}{\beta_0^2} \left[\mathbf{I} - \sum_{i=1}^{q} \left(\frac{\beta_i^2 - \beta_0^2}{\beta_i^2} \right) \mathbf{e}_i \mathbf{e}_i^{*t} \right] \mathbf{S}^*$$

$$= \mu' \left[\mathbf{S}^* - \sum_{i=1}^{q} \left(\frac{\beta_i^2 - \beta_0^2}{\beta_i^2} \right) \alpha_i \mathbf{e}_i \right] \qquad \text{(B18)}$$

where:

$\alpha_i = \mathbf{e}_i^{*t} \mathbf{S}^*$,

$\mu' = \mu / \beta_0^2$, and

$\mathbf{S}^*$ = a quiescent array weight vector.

In the limit of noise-free data, Eq. (B18) is suggestive of a simple Gram-Schmidt vector subtraction from $\mathbf{S}^*$ in which we would form an optimum weight that would be orthogonal to the PE, and therefore, orthogonal to the source direction vectors per Eq. (B7). Thus, let us formulate a PEGS algorithm by defining the optimum weight $\mathbf{W}_e$ from Eq. (B18) as

$$\mathbf{W}_e = \mathbf{S}^* - \sum_{i=1}^{q} \alpha_i e_i \qquad \text{(B19)}$$

where:

$\alpha_i = \mathbf{e}_i^{*t} \mathbf{S}^*$.

$\mathbf{W}_e$ possesses the necessary orthogonality to noise-free source direction vectors,

$$\langle \mathbf{W}_e, \mathbf{v}_i \rangle = 0 \qquad \text{(B20)}$$

and may readily incorporate the option of unit Hermitian length if desired,

$$|\mathbf{W}_e|^2 = 1 . \tag{B21}$$

The PEGS algorithm as applied in this report used an end-element weighting for $\mathbf{S}^*$; i.e.,

$$\mathbf{S}^* = [0\ 0,\ ...,\ 0,\ 1]' . \tag{B22}$$

Culling Principal Eigenvalues

The number of principal eigenvalues is usually directly related to the number of sources which, in practice, are not known and must be estimated. One of the early estimation techniques which has often been used is the AIC (Akaike Information Criterion) [B12,B13]. This criterion has been successfully applied to many model identification problems in engineering and statistics, including the well known problem of determining the order of an autoregressive (AR) process [B13]. Recent work reported by Wax and Kailath [B14] presents a new approach based on the AIC, which eliminates the need for any subjective judgment in the decision process; i.e., the procedure does not require any subjectively chosen threshold. This new approach was implemented during the current investigation and was found to be very effective for most of the examples tested. In addition to the Wax-Kailath AIC approach, we also used a second effective technique which is based on the following three processing operations:

- An initial decreasing-magnitude sort,
- Culling per coarse magnitude threshold, and
- Culling per sensitive threshold based on a quadratic curvefit predictor.

The entire procedure is listed in Ref. [B15] as a Fortran IV computer code.

REFERENCES

B1. V.F. Pisarenko, "The Retrieval of Harmonics from a Covariance Function," Geophys. J. (Royal Astron. Soc.) **33**, 1973, pp. 347-366.

B2. S.S. Reddi, "Multiple Source Location — A Digital Approach," IEEE Trans. Aerospace & Elect. Sys., AES-**15**, Jan 1979, pp. 95-105.

B3. R. Schmidt, "Multiple Emitter Location and Signal Parameter Estimation," Proc. of the RADC Spectrum Estimation Workshop, RADC-TR-79-63, Rome Air Development Center, Rome, NY, Oct 1979, p. 243.

B4. E. Bienvenu and L. Kopp, *Adaptive High Resolution Spatial Discrimination of Passive Sources,* Underwater Acoustics and Signal Processing (D. Reidel Publishing Co., Boston, 1981), pp. 509-515.

B5. R. Kumaresan and D. Tufts, "Singular Value Decomposition and Spectral Analysis," Proc. of the First ASSP Workshop on Spectral Estimation, McMaster University, Hamilton, Ontario, Canada, **2**, Aug 1981, pp. 6.4.1-6.4.12.

B6. R. Kumaresan and D.W. Tufts, "Estimating the Angles of Arrival of Multiple Plane Waves," IEEE Trans. Aerospace & Elect. Sys., AES-**19**, Jan 1983, pp. 134-139.

B7. J.E. Evans, J.R. Johnson, and D.F. Sun, "Application of Advanced Signal Processing Techniques to Angle of Arrival Estimation in ATC Navigation and Surveillance Systems," MIT Lincoln Laboratory Tech. Report 582, (FAA-RD-82-42), Jun 1982.

B8. T.P. Bronez and J.A. Cadzow, "An Algebraic Approach to Superresolution Array Processing," IEEE Trans. Aerospace & Elect. Sys., AES-**19**, Jan 1983, pp. 123-133.

B9. D.G. Childers, ed., Modern Spectrum Analysis, IEEE Press, New York, N.Y. 1978.

B10. W.F. Gabriel, "Spectral Analysis and Adaptive Array Superresolution Techniques," Proc. IEEE **68**, Jun 1980, pp. 654-666.

B11. S.P. Applebaum and D.J. Chapman, "Adaptive Array with Main Beam Constraints," IEEE Trans. Antenna Propagat. AP-**24**, Sep 1976, pp. 650-662.

B12. H. Akaike, *Statistical Predictor Identification*, (Ann. Inst. Statis. Math. **22**, p. 205, 1970).

B13. H. Akaike, "A New Look at the Statistical Model Identification," IEEE Trans. Autom. Contr. AC-**19**, 1974, pp. 716-723.

B14. M. Wax and T. Kaliath, "Determining the Number of Signals by Information Theoretic Criteria," Proc. of the IEEE/ASSP Spectrum Estimation Workshop II, Tampa, FL, Nov 1983 pp. 192-193.

B15. W.F. Gabriel, "A High-Resolution Target-Tracking Concept Using Spectral Estimation Techniques," NRL Report 8797, May 1984.

Appendix C
TRACKING BEAMS AND ADAPTIVE DISTORTION

The tracking beams used in Section, An Adaptive Array Tracking Application, are based on the selection of an adjacent pair of orthogonal uniform illumination beams generated by a Butler matrix beamformer [C1] transformation. The transformation matrix **B** for a linear array with half-wavelength element spacing will have individual matrix components of the form,

$$b_{km} = \frac{1}{\sqrt{K}} \exp\left\{\frac{2\pi}{K}\left(k - \frac{K+1}{2}\right)\left(m - \frac{K+1}{2}\right)\right\} \tag{C1}$$

where:

m = beam index

k = element index

K = total number of elements.

$$\mathbf{B} = \begin{bmatrix} b_{11} & b_{12} & & b_{1K} \\ b_{21} & b_{22} & & b_{2K} \\ b_{31} & b_{32} & & b_{3K} \\ \cdot & \cdot & \cdots & \cdot \\ \cdot & \cdot & & \cdot \\ \cdot & \cdot & & \cdot \\ b_{K1} & b_{K2} & & b_{KK} \end{bmatrix} \tag{C2}$$

The beamformer output vector $\hat{\mathbf{E}}$ is expressed,

$$\hat{\mathbf{E}} = \mathbf{B}'\mathbf{E} \tag{C3}$$

Figure C1(a) illustrates typical sinx/x patterns of an adjacent pair of beams for our 16-element array example. The familiar sum (Σ) and difference (Δ) tracking beam outputs are then obtained from this adjacent pair via the 3 dB hybrid junction as shown in Fig. C2, where we note that the Δ beam is in quadrature phase relationship to the Σ beam. Expressed in terms of element weights, the Σ beam weight vector **S** and the Δ beam weight vector **D** may be written via Eq. (C2),

$$\mathbf{S} = \frac{1}{2}\left[\mathbf{b}_{m+1} + \mathbf{b}_m\right]$$

$$\mathbf{D} = \frac{1}{2}\left[\mathbf{b}_{m+1} - \mathbf{b}_m\right]. \tag{C4}$$

The uniform illumination vectors $\mathbf{b}_m$ and $\mathbf{b}_{m+1}$ result in cosine illuminations for **S** and **D**, which are shown plotted in Fig. C1(b) for our 16-element linear array example.

Monopulse tracking [C2] involves an angle estimate for each pulse (snapshot) containing the target, and it is computed from the ratio of Δ/Σ. From the cosine illumination beams of Fig. C1, note that we can form the approximation,

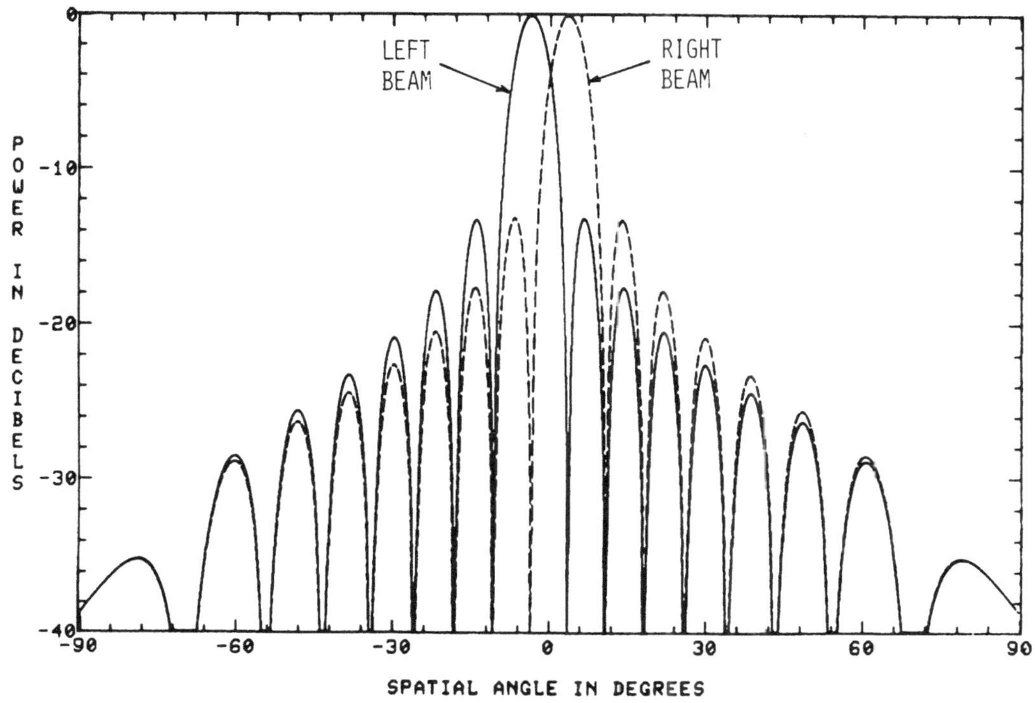

(a) Adjacent pair of Butler matrix beams

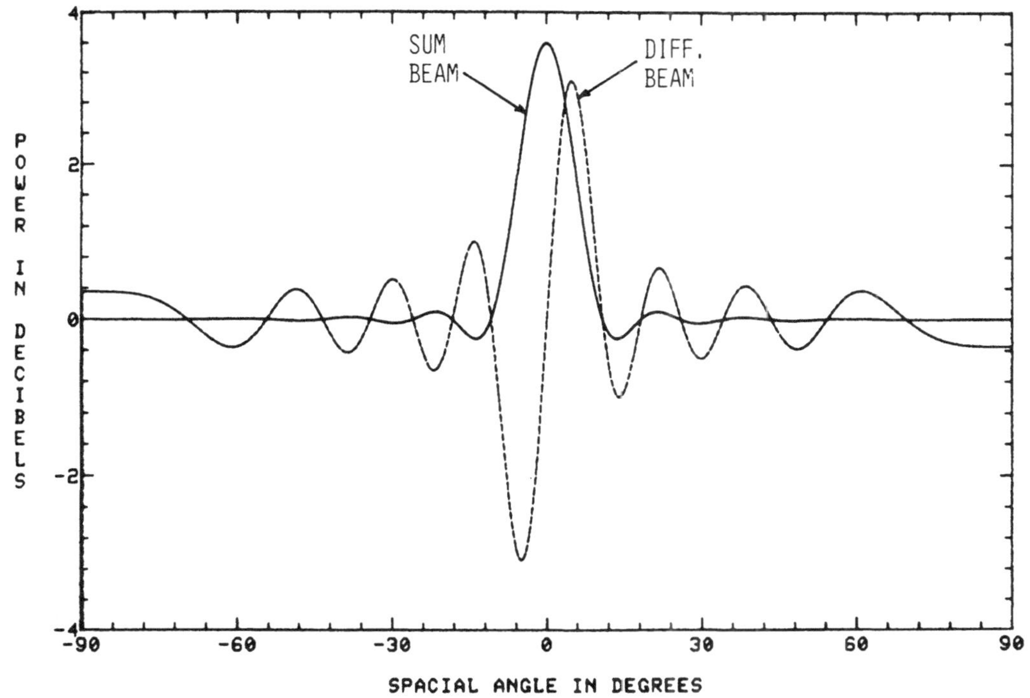

(b) Resultant sum and difference beams

Fig. C1 — Tracking beams formed for 16 element linear array

$$\left(\frac{\Delta}{\Sigma}\right) \approx C \left(\frac{\sin\psi}{1 + \cos\psi}\right) = C \tan\left(\frac{\psi}{2}\right) \tag{C5}$$

where

$$\psi = \left(\frac{\pi}{2}\right)\left(\frac{\delta}{B}\right),$$

δ is the angle of the target from boresight,

B is the angle to the first null of the sum pattern, and

C is a constant which depends upon the particular illuminations.

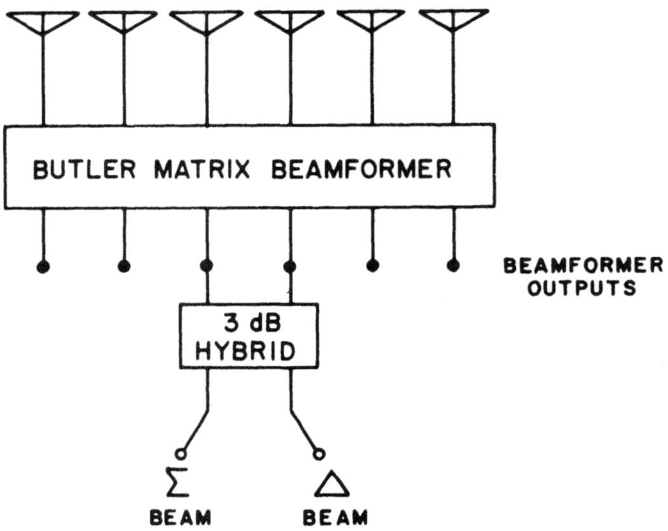

Fig. C2 — Block diagram of tracking beam formation via Butler matrix beamformer

For our example, $C = 3.8°$ and $B = 10.8°$. Thus, given values of Δ/Σ, we can solve for track angle estimates, δ/B, from Eq. (C5).

Next, let us address track estimate distortion. Whenever one performs spatial filtering as described in Section, All-Digital Tracking System Concept, a distortion of received plane wavefronts occurs because the spatial insertion loss generally varies as a function of sin θ.

This problem was first addressed in the literature by Davis, Brennan, and Reed [C3] who proposed an algorithm for estimating the angle of arrival, based on the outputs of adaptively distorted sum and difference beams. They used approximations to the optimum angle estimator which permitted correction of distortion at the tracking beam boresight position, and they demonstrated good performance via simulation for sidelobe and/or mainbeam interference.

In an all-digital system, we know our adaptive filter weights. Therefore, we can compute the resultant distortion error throughout the tracking beam region. In Section, All-Digital Tracking System Concept, above, we showed that it makes no difference whether we apply our quiescent beam weights to the spatially filtered signals, or the equivalent adaptive weights to the unfiltered signals. Thus, we may apply the monopulse sum and difference weights of Eq. (C4) to the spatially filtered output residue signal vector of Eqs. (11, 12) in Section, All-Digital Tracking System Concept, and obtain the equivalent forms,

$$\Sigma = \mathbf{S}'\mathbf{E}_f = \mathbf{E}'\mathbf{M}^{-1}\mathbf{S} = \mathbf{E}'\mathbf{W}_s \tag{C6}$$

and

$$\Delta = \mathbf{D}'\mathbf{E}_f = \mathbf{E}'\mathbf{M}^{-1}\mathbf{D} = \mathbf{E}'\mathbf{W}_d\,, \tag{C7}$$

where $\mathbf{W}_s$ and $\mathbf{W}_d$ are now the equivalent adapted (and distorted) sum and difference beam weights. The distorted ratio Δ/Σ can be computed for any direction vector $\mathbf{E}$, thus giving us the distortion curve across the entire tracking angle region.

REFERENCES

C1. J. Butler, "Multiple Beam Antennas," Sanders Assoc. Internal Memo RF 3849, Jan 1960.

C2. Merrill Skolnik, *Radar Handbook* (McGraw-Hill Book Co., NY, 1970).

C3. R.C. Davis, L.E. Brennan and L.S. Reed, "Angle Estimation with Adaptive Arrays in External Noise Fields," IEEE Trans, AES, AES-**12**, Mar 1976, pp. 179-186.

Appendix D
FORWARD-BACKWARD SUBAPERTURE AVERAGING

This is an excellent technique for increasing the effective averaging of our sample covariance matrix SCM when needed, and it may readily be implemented if the antenna array elements are identical and equally spaced. Figure D1 illustrates this technique. We form a reduced dimension subaperture of L elements, where L must be less than the total number of array elements K. Starting from the left-hand side, the subaperture samples its first snapshot as elements 1 through L, then bumps to the right by 1 and samples its second snapshot as elements 2 through $(L + 1)$, then bumps to the right by 1 and samples its third snapshot as elements 3 through $(L + 2)$, etc. After bumping across to the Kth element, we will have accumulated $(K - L + 1)$ subaperture snapshots from one overall array data snapshot, such that we can increase our SCM averaging by that same factor. This subaperture motion from left to right produces what is generally termed "forward averaging." The technique may be applied to any SCM method. For example, the simple block averaging of Eq. (2) in section, Low-Sidelobe Eigenvector Constraint becomes

$$\hat{\mathbf{R}}_f = \frac{1}{N(K-L+1)} \sum_{n=1}^{N} \sum_{i=1}^{(K-L+1)} [\mathbf{E}(n,i)\mathbf{E}(n,i)^{*t}] \qquad \text{(D1)}$$

where $\mathbf{E}(n,i)^t = [E_i(n), E_{i+1}(n), \ldots, E_{i+L-1}(n)]$ and $\hat{\mathbf{R}}_f$ is the new reduced L x L dimension SCM. Note that $\mathbf{E}(n,i)$ may be expressed as the matrix product,

$$\mathbf{E}(n,i) = \mathbf{I}_i \mathbf{E}(n) \qquad \text{(D2)}$$

where $\mathbf{I}_i$ is a special L x K rectangular sampling matrix in which the i index denotes the first column where the identity matrix $\mathbf{I}$ begins. For example, the $\mathbf{I}_i$ matrix for $L = 3$, $K = 6$, and $i = 2$ would be

$$\mathbf{I}_i = [\mathbf{O} \,|\, \mathbf{I} \,|\, \mathbf{O}] \qquad \text{(D3)}$$

$$\mathbf{I}_2 = \begin{bmatrix} 0 & 1 & 0 & 0 & 0 & 0 \\ 0 & 0 & 1 & 0 & 0 & 0 \\ 0 & 0 & 0 & 1 & 0 & 0 \end{bmatrix}$$

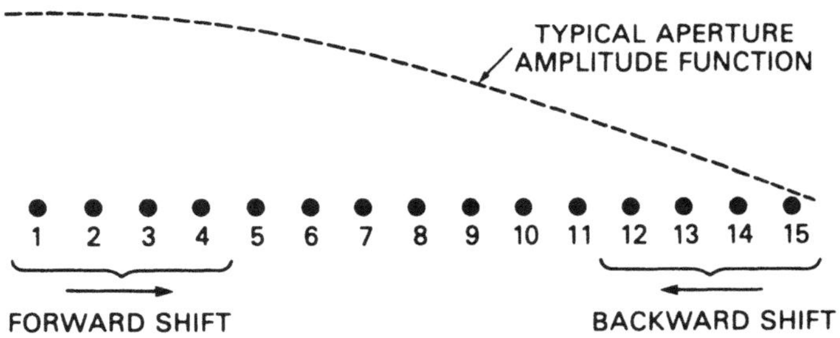

Fig. D1 — Forward-backward shift movement for a reduced dimension sampling subaperture where $L = 4$, along a linear antenna array of $K = 15$ elements

An $\mathbf{I}_i$ matrix may be used to reduce the number of computations by multiplying the SCM to obtain

$$\hat{\mathbf{R}}_f = \frac{1}{(K-L+1)} \sum_{i=1}^{(K-L+1)} [\mathbf{I}_i \hat{\mathbf{R}} \mathbf{I}_i^t] . \qquad \text{(D4)}$$

Equations (D4) and (D1) give similar results, and both are mathematical expressions of the additional spatial averaging or "smoothing" that is obtained via the moving subaperture technique.

Furthermore, the averaging can be doubled again by reversing our subaperture at the right-hand side and bumping across to the left-hand side in similar fashion; however, it requires conjugating the subaperture snapshots. This subaperture motion from right to left produces what is generally termed "backward-averaging."

If we define the reversed and conjugated vector, $\mathbf{E}^{\ddagger}$, in terms of our array element data samples of equation (A3), then the kth element signal sample becomes

$$E_k^{\ddagger} = E^*_{K-k+1} \; ; \qquad k = 1, 2, 3, \ldots, K \tag{D5}$$

The reader can verify that the resulting SCM will be an index-exchanged $\hat{\mathbf{R}}_f^{\ddagger}$, and that we can combine the two into a final forward-backward average SCM which is denoted as the reduced L x L matrix $\hat{\mathbf{R}}_{fb}$:

$$\hat{\mathbf{R}}_{fb} = 1/2\ [\hat{\mathbf{R}}_f + \hat{\mathbf{R}}_f^{\ddagger}]\,. \tag{D6}$$

Note that $\hat{\mathbf{R}}_{fb}$ is a symmetric matrix, but is generally not Toeplitz. References D1 and D2 are recommended for further detailed discussion of the technique.

Although forward-backward subaperture averaging is a very simple concept, it usually produces remarkable improvements in output estimates, and it becomes crucial to processing in the following situations:

- When only a few data snapshots are available per SCM computation. Note that the method can be used even under the extreme condition of only a single snapshot.
- When significant coherence exists between spatial sources as for example in multipath situations involving a specular reflection. For this particular condition, the fields arriving at the aperture are nonstationary in space and the SCM is not Toeplitz [D1,D3,D4,D5].

A caveat concerning this averaging technique is that, as the dimension L of the subaperture becomes smaller, the subaperture antenna gain, resolution, and degrees of freedom decrease. Thus, the advantage of increased averaging must always be balanced against these factors, and it is usually prudent to process with as large a subaperture dimension as possible.

REFERENCES

D1. J.E. Evans, J.R. Johnson, and D.F. Sun, "Application of Advanced Signal Processing Techniques to Angle of Arrival Estimation in ATC Navigation and Surveillance Systems," MIT Lincoln Laboratory Tech. Report 582, (FAA-RD-82-42), Jun 1982.

D2. A.H. Nuttall, "Spectral Analysis of a Univariate Process with Bad Data Points, via Maximum Entropy and Linear Predictive Techniques," NUSC-TR-5303, Naval Underwater Systems Center, New London, CT, Mar 1976.

D3. T.J. Shan and T. Kailath, "Adaptive Beamforming for Coherent Signals and Interference," IEEE Trans. Acoust., Speech, Signal Processing, Vol. ASSP-**33**, pp. 527-536, (1985).

D4. W.D. White, "Angular Spectra in Radar Applications," IEEE Trans. Aerosp. and Electron. Syst., AES-**15**, Nov 1979, pp. 895-899.

D5. W.F. Gabriel, "Spectral Analysis and Adaptive Array Superresolution Techniques," Proc. IEEE **68**, Jun 1980, pp. 654-666.

Chapter 4
Clutter Suppression

Radar *clutter echoes* consist of radar returns from reflectors that are not of interest and often obscure the signals from targets that are of interest. Radar *clutter signals* are typically caused by things such as rain, chaff, sea, woods, and mountains. Examples of radar targets are ships, aircraft, and satellites. What is clutter in some applications might be reflectors or targets of interest in other applications. This is illustrated by a ground-mapping radar where reflections from the ground are of interest.

Often, clutter returns are much stronger than target returns and radar processing is required to improve the signal-to-clutter ratio, *S/C*. Several techniques are available to achieve this goal. One method is to reduce the resolution cell size in angle and range in order to decrease the amount of clutter return that competes with the target. This technique is effective because the clutter is usually distributed over a larger area of volume than the target. Another method is to take advantage of differences between polarization characteristics of the target and the clutter. A third method is to filter the signals to improve *S/C* based on the differences in Doppler frequency because echoes from desired targets are generally moving at a higher radial velocity than clutter signals.

Radial velocity of a target with respect to the radar causes the target's range to change with time. This range rate produces a phase rate in the echo that appears as a Doppler frequency shift. The lack of phase shift from pulse to pulse in the echo from clutter reflectors is used in the basic form of *moving target indicators* (MTI) to cancel echoes from nonmoving reflectors by subtracting one pulse from the next pulse occuring at the radar repetition interval τ_r. In this process, the phase shift from pulse to pulse on moving reflector echoes causes successive pulses to add vectorially, instead of subtracting. This addition enhances or accentuates echoes from rapidly moving targets as desired, while canceling clutter. Digital filters called fast Fourier transform (FFT) circuits, such as those discussed in Chapter 2, can also be used to form a contiguous Doppler filter bank in order to separate clutter from Doppler signals of the desired target.

Under certain conditions, clutter can vary more rapidly in amplitude and phase than echoes from desired targets. This occurs when high range resolution radars attempt to detect floating debris on a disturbed sea surface. In this case, nonmoving target detectors are employed to suppress clutter. Because the characteristics of clutter and desired target echoes are so important in the design of clutter suppression and target-enhancement systems, these characteristics will be discussed prior to going into detail on clutter suppression concepts.

4.1 CLUTTER CHARACTERISTICS

The principal clutter characteristics of interest to radar designers are the following:

σ_0 is the *equivalent scattering cross section* in square meters per square meter of illuminated land or sea. In the case of volumetric clutter, like rain or chaff, η_0 is the equivalent scattering cross section in square meters per cubic meter.

σ_v is the *standard deviation of the clutter radial velocity* in meters per second. This quantity defines the width of the spectrum (generally assumed to be Gaussian) of the clutter component that is directed toward or away from the radar.

$\overline{V}_c$ is the *mean radial velocity* of the clutter with respect to the radar.

In addition to the above values, the *spectral shape* of the clutter is of interest when it is not adequately described by a Gaussian function. Also, the *polarization* properties of the target and clutter are of interest and have been discussed in most standard radar texts. The effects of polarization on high resolution sea clutter are presented later in this chapter.

4.1.1 Published Clutter Data

Often quoted sources of clutter data in order of their original publication date are Skolnik, 1962 [1]; Kerr, 1964 [2]; Barton, 1964 [3]; Nathanson, 1969 [4]; and Shrader, 1970 [5]. Shrader summarizes some of the available clutter information in terms of reflectivity per unit area (σ_0) or per unit volume (η_0) and the standard deviation of the clutter velocity (σ_v). However, the various sources are in disagreement about the expected range of σ_v for rain and chaff. Barton states ([3], p. 110):

> The spectral spread of chaff will be approximately the same as that for precipitation in the same atmosphere. However, generally smaller values for chaff are shown since chaff is often used in a clear atmosphere without turbulent winds. Rain, on the other hand, is usually accompanied by gusty winds. The ratio of wind speed to rms spectral spread will be approximately five to one for both chaff and rain.

Also, a table of clutter spectral spreads appearing in Barton, and reproduced by Shrader, is based on some early clutter measurements by Barlow [6] and Goldstein [7]. The data indicate that the spectral spread of chaff is not only less than rain, but also comparable to that of sea clutter. It should be noted that some of the data were taken under unknown conditions.

Nathanson ([4], pp. 206, 227), on the other hand, states that chaff or rain clutter spectral spreads are dominated by wind shear rather than turbulence in long-range, ground-based radars. The former can be on the order of 6 m/s and may occur on clear days, while the turbulence is limited to about 1 m/s, even on rainy days. More recently, a detailed analysis of wind shear was performed by Trizna and Pilon [8], making use of radiosonde data. Based on certain assumptions, Trizna and Pilon generally concluded [8] that a median wind-shear index of 2 m/s/km is a more appropriate value than 4 m/s/km as stated by Nathanson [4]. However, they pointed out [8] that this wind shear index cannot be applied over the entire elevation extent of rain or chaff, and wind shear is limited to values that are roughly in agreement with Nathanson (approximately 6 m/s). Nathanson's results are also supported by data (Fig. 4–1) taken from the *Handbook of Geophysics and Space Environment* [9].

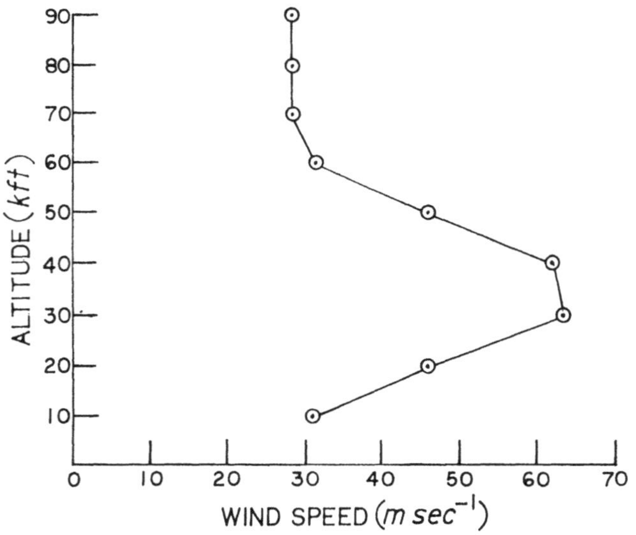

Fig. 4–1 Wind Shear Data.

In addition to disagreements on σ_v for chaff, there are issues on the Doppler spectral shape for both chaff and rain. A Gaussian spectrum is often used for ease of analysis and also because in some situations this may be appropriate. The clutter may appear Gaussian due to long-term averaging, or can be approximated by a Gaussian distribution because of the shape of the radar beam in the vicinity of the peak. However, for wider spectral widths, the tails of the Gaussian distribution become significant and thus may not accurately represent the actual conditions. Rain, for example, is approximately limited to an upper elevation between 15 or 20 thousand feet and a lower elevation determined by the radar horizon. Also, bright spectral bands are associated with rain, thus dominating the spectrum over a relatively narrow band [4]. Moreover, the returns from rain analyzed on a short-term basis can be quite different than would be expected from a Gaussian spectrum [10, 11].

With regard to sea clutter, all sources are in relatively good agreement when the radar range and azimuth resolution cells are large (i.e., the radar pulses are long). However, they disagree with recently acquired data when the range-azimuth resolution cells are small. Nathanson ([4], pp. 249) reports that the spectral spread of sea clutter diminishes with decreasing radar pulse lengths, for pulse lengths less than 0.1 μs. However, in recently acquired sea clutter data, described in the next section, this was not the case. None of the references provide sea clutter information in a form that would permit a radar designer to differentiate between sea clutter and echoes from slowly moving rigid targets. With regard to land clutter, all sources are in relatively good agreement. At this point, we can state that problems arise when heavy data smoothing is employed to provide single numbers or ranges of numbers to radar designers.

The next section gives a specific example of the utility of nonaveraged clutter data.

4.1.2 Recently Acquired Sea Clutter Data

Recently acquired high resolution radar data on sea clutter is now available and promises to be very valuable to both radar designers and sea scatter theorists. The data were taken on single carrier frequencies and simultaneously on two widely different carrier frequencies (8.6 and 9.2 GHz, and 9.2 and 3 GHz) with synchronized radar and optical views of the scattering surface. Data were acquired in the Chesapeake Bay, offshore at Boca Raton, Florida, and from a tower 12 miles offshore near Panama City, Florida. Measurements were made for sea states from 0 to 5 with the radar looking at different angles to the wind using both vertical and horizontal polarization [4, 12–16].

These data reveal that sea scatter originates in widely separated scattering centers which have dimensions that are small compared to the range resolution offered by a 40 ns transmitted pulse and a 30-foot cross-range antenna resolution. Figure 4–2 illustrates the resolvability of the scattering centers in range and the radar's polarization dependence. Note that relatively long times are available with no sea return in any given range cell using horizontal polarization.

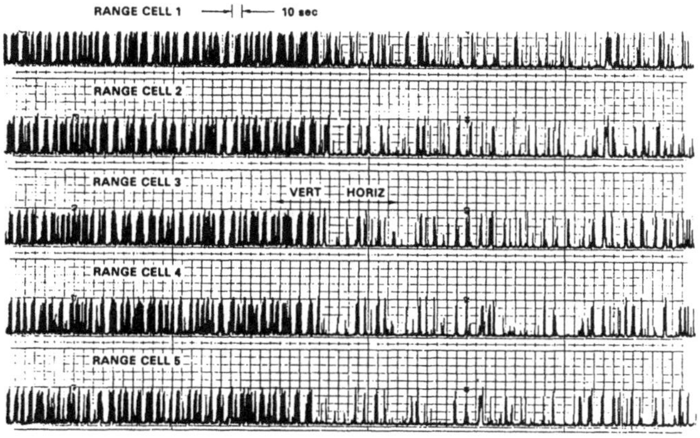

Fig. 4–2 Time Functions of Sea Return in Gates with 10-Foot Waves 12 Miles Offshore at 100-Foot Water Depth.

Figure 4–3 is an example of wave staff measurements taken with the sea scatter data in Fig. 4–2, showing wave heights greater than 10 feet. Note that the wave period corresponds to the period between low returns in Fig. 4–2 with vertical polarization. This indicates that these low return regions are due to shadowing by wave peaks. Note also that vertically polarized returns are obtained on nearly every wave peak, while horizontally polarized returns are not obtained from every wave peak. Thus, targets would be detected with horizontal polarization if they could be resolved from clutter.

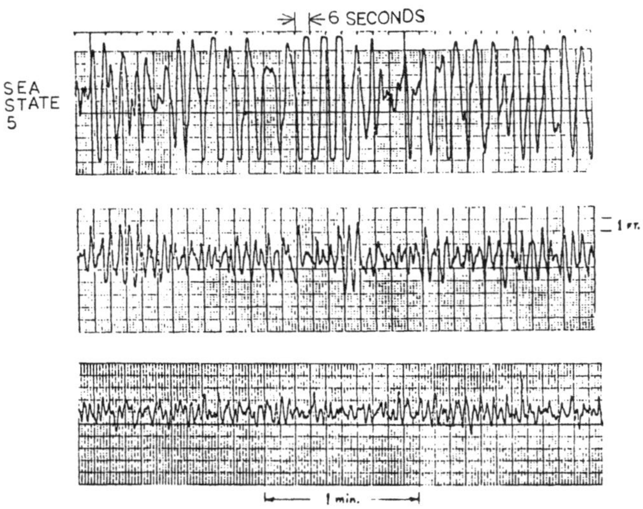

Fig. 4–3 Wave Staff Recordings of Different Sea States.

Figure 4–4 illustrates the spectral energy distribution of the 9.2 GHz echo envelope from a single range gate as a function of time. To obtain this data, successive 0.1 s samples of the envelope data were used as inputs to a spectrum analyzer. The absence of spectra for finite periods of time was due to the absence of the an echo from this range cell during that time. Note that spectra are nearly flat to about 200 Hz, which would imply a σ_v on the order of 3 m/s using Barton's expression $\sigma_v = 0.85 f_{3dB}\lambda/2$.

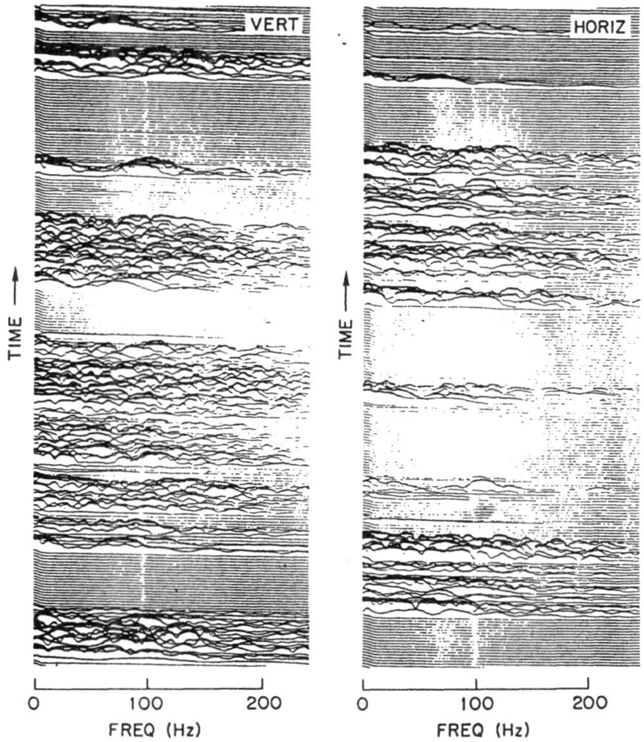

Fig. 4–4 Sea Clutter Spectra *versus* Time.

Data taken in crosswind show even higher frequency components due to the high growth rate in range (up to infinity) of white caps breaking across the line of sight. This phenomena can present higher frequency sea clutter fluctuations than those exhibited by rain or wind-driven chaff.

Whenever sea echoes appeared, they were always found to be heavily amplitude modulated by a relatively wideband modulating waveform. The bandwidth of this waveform was found to be much wider than that modulating echoes from rigid floating objects (Fig. 4–5), and the persistence of a return in any range cell was found to be much less than the echo from rigid floating objects.

This difference between echoes from the disturbed sea surface and floating rigid debris suggested the clutter suppressor illustrated in Fig. 31 of Lewis *et al* ([12]; *see* Readings, Chapter 4). We might call this device a nondecorrelating target detector or a nonmoving target detector.

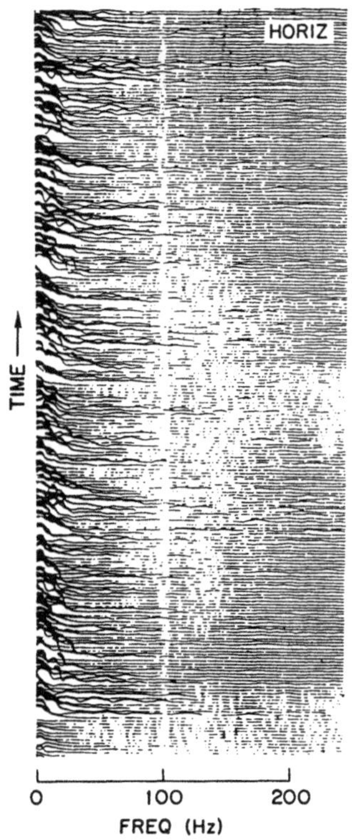

Fig. 4–5 Spectra of Floating Target *versus* Time.

4.2 MOVING TARGET INDICATORS

In general, radar targets of interest produce echoes that have different characteristics than clutter. As previously mentioned, an important difference is the radial velocity with respect to the radar. The radial velocity is usually higher for targets of interest than for clutter, causing the desired echo phase to change more rapidly with time than the phases of clutter echoes. The *phase rate of change with time* gives rise to a Doppler frequency, which can be filtered in a moving target indicator (MTI) to cancel clutter and enhance moving target echoes [1–5, 17–20].

The basic concept of an MTI is illustrated in Fig. 4–6.

This figure shows a two-pulse MTI that subtracts present echoes from previous echoes of the radar's transmissions. This circuit cancels echoes from reflectors that change neither phase nor amplitude from pulse to pulse, but does not cancel those which do change from pulse to pulse.

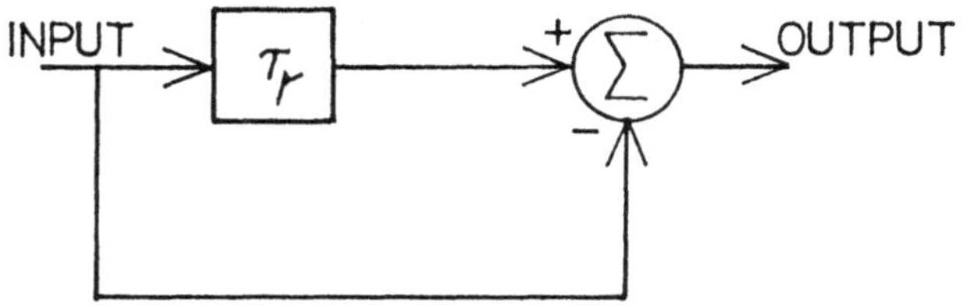

Fig. 4–6 Two-Pulse MTI.

Let $a(t)e^{j\omega_d t}$ be the input to the MTI, where $\omega_d = 2\pi f_d$ and f_d is the Doppler frequency of the signal of amplitude $a(t)$. Therefore, the output will be

$$S_o = a(t - \tau_r)\, e^{j\omega_d(t-\tau_r)} - a(t)\, e^{j\omega_d t} \tag{4–1}$$

With $a(t - \tau_r) = a(t)$, the amplitude of S_o in (4–1) can be written as

$$|S_o| = |a(t)|2(1 - \cos \omega_d\tau_r) \tag{4–2}$$

Note that $|S_o| = 0$ when $\omega_d = 0$, and that $|S_o|$ has a peak of $2|a(t)|$ when ω_d is an odd integer multiple of π/τ_r. Also note that $|S_o|$ returns to zero whenever $\omega_d\tau_r$ is an integer multiple of 2π. This cycling between a zero, a peak, and a zero is caused by the radar's sampling rate (pulse repetition rate) being a subharmonic of the Doppler frequency so that phase shifts by multiples of π occur across the delay line in the MTI.

If $a(t - \tau_r)$ is not equal to $a(t)$ in (4–1), then $|S_o|$ does not equal zero for any ω_d. This phenomenon is encountered with search radars where the antennas scan in angle with time so that the antenna gain $G(\theta) = G(t)$ varies from pulse to pulse due to scan and a finite pulse repetition frequency equal to $1/\tau_r$ as illustrated in Fig. 4–7.

The effect of the antenna scanning-induced variation of $G(t)$ can be reduced in the MTI by using cascaded cancelers as shown in Fig. 4–8.

We observe that, with nonmoving target echoes entering the MTI in Fig. 4–8, the output of the first subtractor will be proportional to the gain change in one

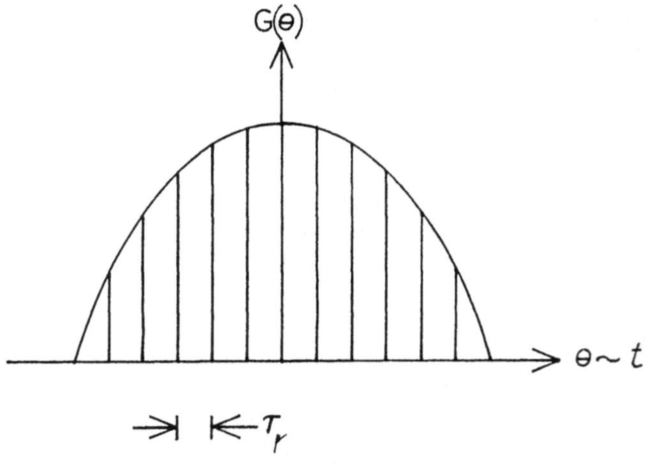

Fig. 4–7 $G(t)$ Variation with Antenna Scan in Angle θ as a Function of Time.

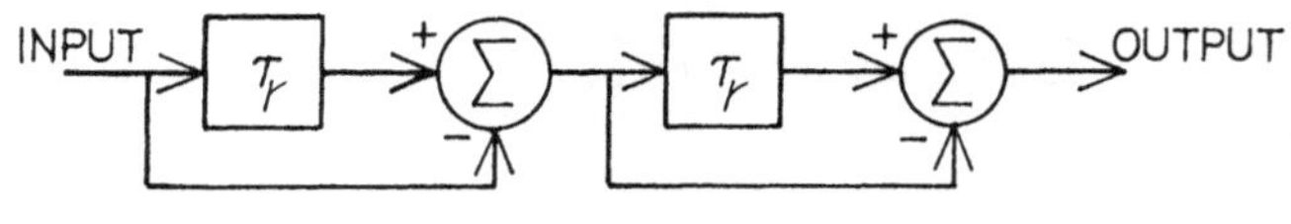

Fig. 4–8 Three-Pulse MTI.

interpulse period due to the antenna scanning with time. The output of the second subtractor, therefore, will be the difference between two successive residues caused by antenna scanning. This second difference will reduce to zero if the change in gain with time is a linear function of time. In this way, the use of a three-pulse canceler can compensate for a first derivative of $G(t)$ with respect to time t. Similarly, a four-pulse canceler can compensate for finite first and second derivatives with respect to time, *et cetera*. We should point out that the system illustrated in Fig. 4–8 is equivalent to the one illustrated in Fig. 4–9.

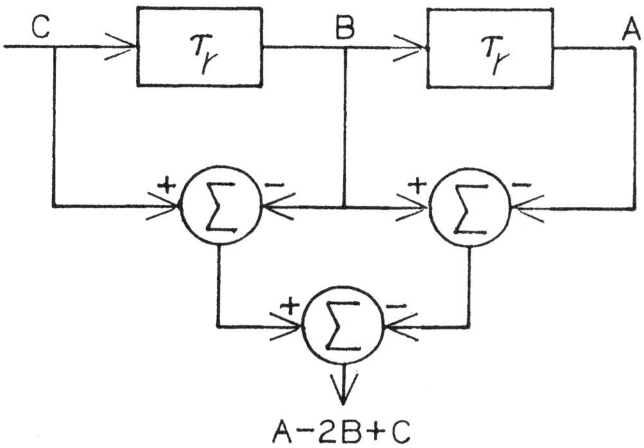

Fig. 4–9 Three-Pulse MTI with A = the Echo from the First Pulse Transmitted, B = the Echo from the Second Pulse Transmitted, and C = the Echo from the Third Pulse Transmitted.

The system of Fig. 4–9 can be implemented as shown in Fig. 4–10 to save a subtractor if the implementation is digital. Note that a multiplication by 2 can be done as a hard-wire shift so that the only hardware in Fig. 4–10 are two interpulse period delays, a subtractor, and an adder.

We can see that if the subtractor of Fig. 4–10 were changed to an adder and if B were multiplied by -2 instead of $+2$, the output would have remained the same as in Fig. 4–10. Effectively, A would have been multiplied by $+1$, B by -2, and C by $+1$ prior to adding in order to form $(1A - 2B + 1C)$. These multipliers are the coefficients of a polynominal obtained from $(x - y)^2$. These multipliers are called *weights* and higher-order MTIs employ higher-order polynomial coefficients. For example, a four-pulse MTI employs weights of $1, -3, +3, -1$, which are the coefficients of $(x - y)^3$. This is illustrated in Fig. 4–11.

As previously mentioned, an N-delay, binomially weighted MTI is able to cancel amplitude samples of an $(N-1)$th-order polynomial. It is also interesting to note that the transfer function and its $(N-1)$ derivatives are equal to zero at zero frequency.

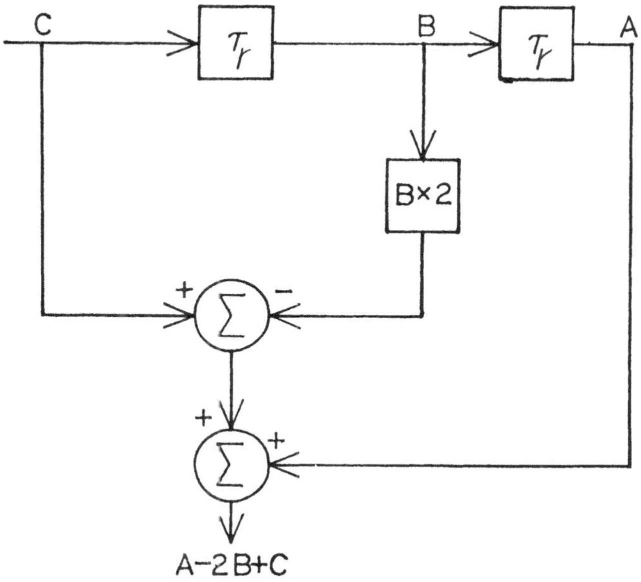

Fig. 4–10 Binomial-Weighted Three-Pulse MTI.

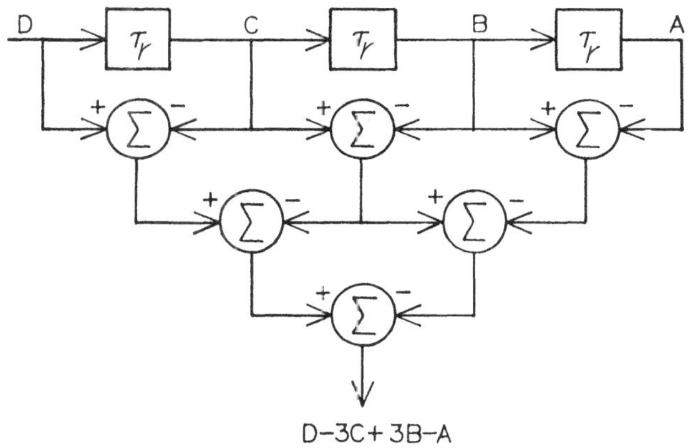

Fig. 4–11 Four-Pulse MTI Showing Binomial Weights.

We can see that the addition of another sample or pulse to the system in Fig. 4–11 through the addition of another interpulse delay adds another layer to the tree of subtractors. Observe that the top layer must contain as many subtractors as delay lines and that there are as many layers as delays. This large number of subtractors can be reduced using serial processing as shown in Fig. 4–12, where there are only as many subtractors as interpulse delays.

We should note that the output of an MTI does not provide proper cancellation until enough echo pulses have been received to occupy all the taps on the delay line. At this time, and for successive intervals, the

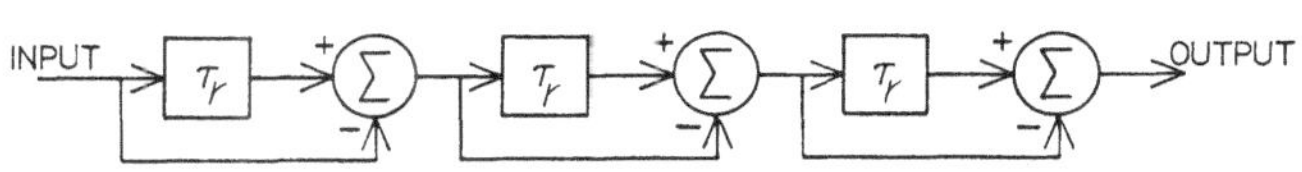

Fig. 4–12 Serial MTI Configuration.

weighting is correct so as to provide the desired cancellation. However, for an N-delay MTI, successive output samples are correlated and independent output pulses can only be obtained by sampling the output after every $(N + 1)$ or more transmitted pulses. Thus, an MTI reduces the effective number of independent pulses available for detection (to be discussed later) and for azimuth location within the beam of a scanning radar using beam-splitting methods. The latter affects the scan rate in that a sufficient number of pulses are required for the antenna's change of gain with time due to antenna scanning to be recognizable so that an estimate can be made of the time at which the target is in the center of the beam. This estimate and a knowledge of the pointing direction of the beam with time establish the azimuth position for a target of interest. In addition, MTI cancellation for an MTI of a given order depends on there being enough pulses per beam position to restrict the pulse-to-pulse amplitude change, which is due to antenna scanning modulation, in the return echoes from a point target.

It is apparent that there are several conflicting requirements placed on a search radar in order to suppress clutter and to enhance echoes from moving targets when using an MTI. As noted in Chapter 1, the radar pulse repetition frequency (PRF) must be low enough to provide the required unambiguous range. Nevertheless, the PRF must be high enough to provide a sufficient number of echoes per scan for clutter to be suppressed and targets located accurately in azimuth angle. Because the number of echoes per scan from each reflector illuminated by the radar is directly proportional to the radar's azimuth beamwidth and its pulse repetition rate, and inversely proportional to the radar's azimuth scan rate, we must compromise among these three factors. Reduction of the azimuth scan rate decreases the radar's data rate, but increases the number of hits per scan for a fixed PRF. A similar result can be obtained by increasing the radar's azimuth beamwidth. The penalty paid in the latter case, however, is reduced gain and azimuth accuracy.

Most MTIs in use today are coherent, that is, they employ complex addition and subtraction using I and Q video or digital words to represent the echo amplitude and phase. In these MTIs, either a pulse-to-pulse phase or amplitude change in an echo will produce a finite residue at the output of the MTI. Prior to the availability of long, accurate, and economical delays, noncoherent MTIs were used. These radars were called clutter-gated MTIs [17–19, 21]. An example of a two-pulse noncoherent MTI is illustrated in Fig. 4–13.

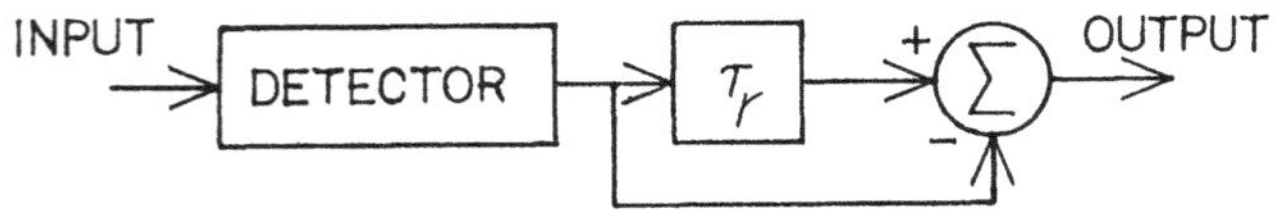

Fig. 4–13 Two-Pulse Noncoherent MTI.

Noncoherent MTIs depend on distributed clutter echoes adding to target echoes having phases which vary from pulse to pulse due to the target motion between pulses. This addition prior to detection causes the target-plus-clutter echo amplitude to fluctuate from pulse to pulse after detection. This fluctuation produces a residue at the output of the MTI, which indicates the presence of a moving target. In the absence of background clutter, both moving and nonmoving reflector echoes cancel in a noncoherent MTI, unless the target scintillates from pulse to pulse.

Because noncoherent MTI's are not commonly used, they will not be discussed further in the main body of this text. However, they are discussed in detail by Kretschmer *et al.* in the readings at the end of this chapter.

4.2.1 Requirements For Good Clutter Cancellation

Ideal clutter cancellation requires the radar's transmitted frequency and phase to be the same as that of the radar's local oscillator signal. This requirement ensures that coherently detected echo signals from nonmoving reflectors will have the same phase from pulse to pulse. Therefore, if each successive pulse has the same amplitude, phase relative to a reference, and wave shape, then one pulse subtracted from the next will completely cancel. The requirement for successive echo signals from nonmoving reflectors to have the same amplitude and wave shape, in turn, requires that the radar's transmitted pulses have the same amplitude and wave shape from pulse to pulse. Magnetron transmitters, used in some radars, have no coherent phase relation from pulse to pulse, and so a sample of the transmitted phase is used to provide a reference phase for the return signal in order to determine the phase difference between the transmitted and received pulses.

Ideal cancellation also requires that the interpulse delay in the MTI be exactly equal to the radar transmitter's interpulse period. This is ensured, in MTIs employing digital shift-register delays, by synchronizing the radar's transmitted pulse to a subharmonic of the clock signal, which is used to shift the signals in the digital shift registers. This synchronization ensures that the pulses being subtracted from each other in the MTI come from the same radar reflector. The accuracy with which these requirements must be met can be appreciated as we work out the following error calculations.

A phase error of $\theta < 1$ radian from pulse to pulse will leave an MTI amplitude residue nearly equal to $\sin\theta$. An error of one milliradian will limit the cancellation ratio to 60 dB. The amplitude of θ can be found to be

$$\theta = (4\pi R\delta f)/c \tag{4–3}$$

for a reflector at a range R with a frequency error δf. Also, a time delay error in the MTI's interpulse period delay lines of δT will produce a phase error:

$$\theta = 2\pi f_i \delta T \quad (4\text{–}4)$$

where f_i is the frequency passing through the delay line.

From (4–3), with $R = 10^4$ m and $c = 3 \times 10^8$ m per second, $\theta = 1$ milliradian limits δf to

$$\delta f = \theta c/(4\pi R) = 3 \times 10^5/(4\pi \times 10^4) = 2.387 \text{ Hz} \quad (4\text{–}5)$$

From (4–4), with $f_i = 3 \times 10^7$ Hz and $\theta = 10^{-3}$ radians, we have

$$\delta T = 10^{-3}/(2\pi \times 3 \times 10^3) = 5.3 \times 10^{-8} \text{ s} \quad (4\text{–}6)$$

Some MTIs today are designed for a clutter cancellation ratio on the order of 60 to 65 dB. With carrier frequencies on the order of 10^9 Hz, this requires a frequency accuracy on the order of one part in 10^{11} and time accuracies on the order of one part in 10^{14}.

Phenomena other than timing errors and antenna scanning modulation that can limit clutter cancellation are clutter decorrelation from pulse to pulse, clutter average velocity, second-time-around echoes from clutter in the absence of filler pulses, and noise modulation of the radar's transmitted signal [1, 5]. Radars using magnetrons are unable to correct for second-time-around echoes because of the random phase from pulse to pulse on transmission. Clutter decorrelation from pulse to pulse is caused by time-changing clutter characteristics, which are produced, for example, by relative scatterer motion, platform motion, wind, and wind shear. Raindrops filling a radar resolution cell change in number and location with time due to their variable production rate, gravity, and wind-induced motions. Similarly, disturbances on a water surface induced by wind vary in structure and location with time.

Clutter having a nonzero average velocity causes the clutter Doppler frequency to move away from zero and out of the zero response of the MTI. To cancel clutter having nonzero Doppler, the radar's local oscillator can be offset by the Doppler frequency of the clutter, or an appropriate phase shift can be inserted into one of the legs of one or more of the MTI canceler stages, as illustrated in Fig. 4–14.

For a coherently related transmission signal, second-time-around pulse echoes from ranges greater than the radar's unambiguous range will not cancel in an MTI

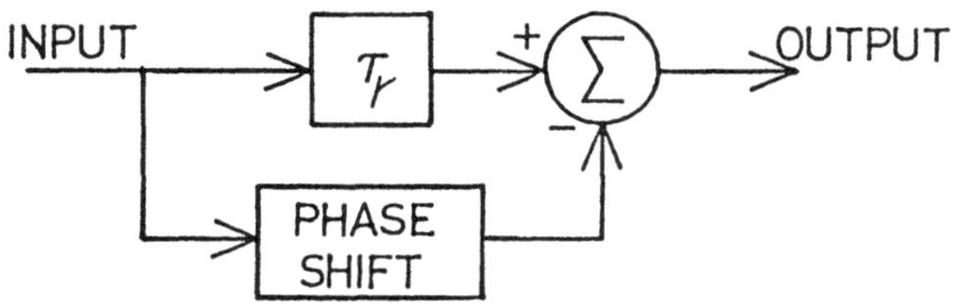

Fig. 4–14 MTI Zero-Frequency Response Control via Phase Shift.

until the clutter stabilizes. In general, for *n*th-time-around clutter and an *N*-pulse MTI, the MTI's output is not useful until $(N+n)$ pulses are received and processed. However, echoes from *n*th-time-around clutter cannot be cancelled when using a magnetron transmitter because of the random phase relationship from pulse to pulse.

Noise modulation of the radar's transmitted signal causes clutter to vary from pulse to pulse. This variation from pulse to pulse leaves residues in the MTI output.

4.3 ADAPTIVE MTI

Radar clutter characteristics (amplitude, distribution, Doppler, and Doppler spread) vary with range, angle, and time. As a consequence, conventional fixed-weight cancelers are not optimum in all conditions. The clutter variation requires the use of *adaptive cancelers* that sense the clutter characteristics and adjust their weights accordingly. These weights are generally complex, rather than real valued, and thereby allow the nulls to be steered in Doppler frequency to cancel clutter as appropriate. The adaptive cancelers discussed in Chapter 3 can be used to obtain nearly optimum MTI performance under any spatial or spectral clutter conditions.

Conventional adaptive MTI was invented by Paul Howells and analyzed by Sidney Applebaum while they were working on coherent sidelobe cancellation. They used Howells' analog closed-adaptive loop, which was discussed in Chapter 3. Figure 4–15 illustrates a two-pulse adaptive MTI.

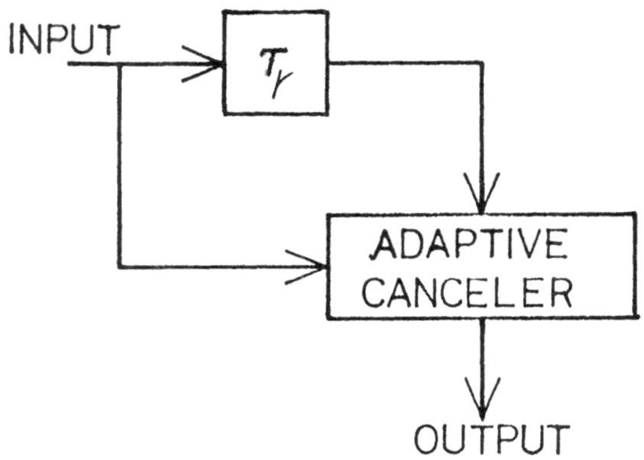

Fig. 4–15 Two-Pulse Adaptive Canceler (According to Howells).

The operation of such an analog MTI at an intermediate frequency and use of analog interpulse delay lines resulted in tuning and delay-line stability problems. The adaptive MTI with closed-loop cancelers avoided cancellation of the desired target echoes by use of heavy averaging to form the weight.

Figure 4–16 illustrates a conventional analog three-pulse adaptive MTI. Here, the output of the MTI is fed back to each adaptive loop. The loop measures the correlation between the MTI output and its auxiliary signal input, and uses this measurement to form a weight which multiplies the auxiliary signal. The loop then subtracts the weighted auxiliary signal from the main input to the loop (the line entering the top of the box in the figure represents the main input). This subtraction decorrelates the loop's output from its auxiliary signal by removing correlated components.

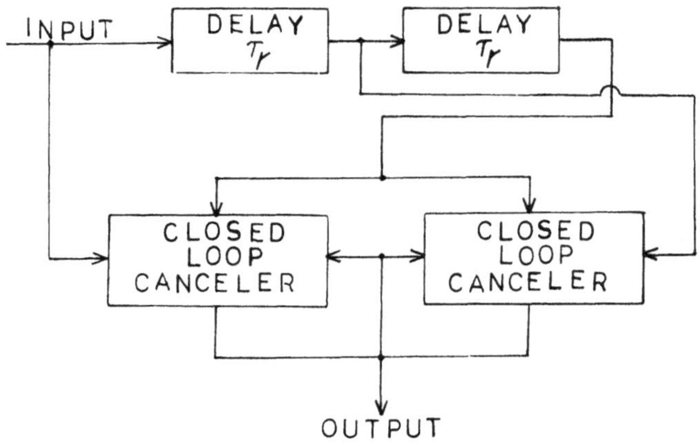

Fig. 4–16 Three-Pulse Adaptive MTI (After Howells).

The output that is fed back to each loop provides a means for data to recirculate among the loops. This type of feedback effectively places every loop in the feedback path of all other loops. The loop-to-loop coupling can result in long settling times, system instability, or both. The long settling times or system instabilities caused by the conventional analog parallel-loop configuration, illustrated in Fig. 4–16, can be eliminated by using the Gram-Schmidt configuration of digital adaptive open-loop cancelers (described in Chapter 3) illustrated in Fig. 4–17.

The open-loop adaptive processor is unconditionally stable because of its open-loop configuration. However, it cancels so well that there are problems in preserving

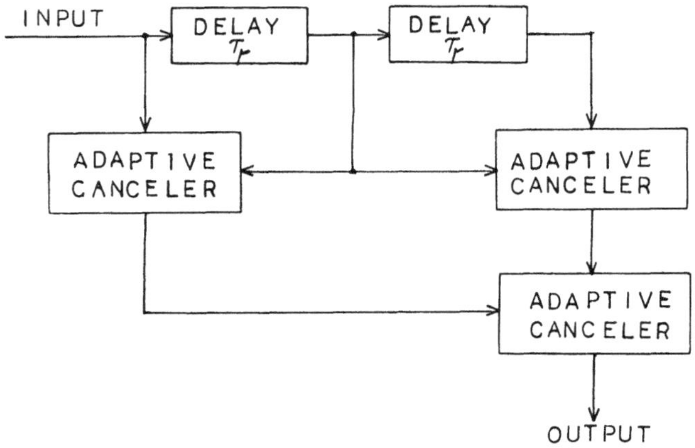

Fig. 4–17 Three-Pulse Gram-Schmidt Configuration of an Adaptive MTI.

the desired targets because targets are present with equal magnitude on both the main and auxiliary signal inputs to the processor. In this way, the adaptive MTI differs from coherent sidelobe cancelers. In sidelobe cancelers, target signals in the main input are always larger than those in the auxiliary input.

Targets can be preserved in the digital open-loop processor in several ways. One way is to offset the samples that are being used to form the sliding-window or batch-processed weight. The amount of offset required, however, may become large if MTI is done prior to pulse compression with a large pulse-compression ratio. Another way to preserve targets is to monitor the I and Q values of the weight and invert them prior to use if I is negative, or if Q/I has a magnitude greater than some preset value. The latter two characteristics are those associated with a moving target.

4.3.1 Comparison Of Adaptive MTI With Fixed Weight MTI

In conventional fixed-weight MTI systems, the weighting is real, so the clutter-rejection notch of the transfer function is placed at zero frequency. Any deviation of the average clutter spectrum from this frequency results in severe degradation of MTI performance. A nonzero-average Doppler spectrum is usually sensed by a *phase-rate estimation* method, which is used to change the coherent oscillator frequency. This new frequency, in turn, translates the average clutter spectrum frequency to zero. However, this method does not perform properly when separate clutter spectra, such as from sea clutter and weather, occur simultaneously in the same range cells.

In order to find the fixed optimal weights that maximize the MTI improvement factor [18, 19, 22], we must have *a priori* knowledge of the clutter spectra or covariance matrix. The optimal weights may be found by maximizing the ratio of the output signal-to-clutter ratio S_o/C_o to the input signal-to-clutter ratio S_i/C_i. The output average clutter power can be written as

$$C_o = C_i \, W^{t*} M_c W \tag{4–7}$$

where W is an n-element weight vector, t denotes a transposition, asterisk denotes conjugation, and M_c is an n-by-n clutter covariance matrix. Let us denote the most recent clutter return by x_1, the clutter return from the previous pulse occuring one repetition period earlier by x_2, and so on up to return x_n, then the ijth element of M_c is given by

$$M_{ij} = E(x_i^* x_j) \tag{4–8}$$

where E is the expectation operator. We have assumed above, without loss in generality, that $E(x_i) = 0$. In a similar manner, the average output signal power can be written as

$$S_o = S_i\, W^{t*} M_c W \tag{4–9}$$

where M_c is the signal covariance matrix. For the case in which the Doppler frequency of the signal is not known *a priori* and is assumed to be equally likely to occur over the PRF interval, the signal covariance matrix M_c becomes the identity matrix. Under these conditions, the MTI improvement factor I may be written as

$$I = W^{t*}W/W^{t*}M_cW \tag{4–10}$$

Were we to maximize I with respect to W, this would be equivalent to minimizing the output clutter power subject to the constraint that $W^{t*}W$ is a constant. This may be formulated by minimizing the quantity J, with respect to W, given by

$$J = W^{t*}M_cW + \lambda(W^{t*}W) \tag{4–11}$$

where the constant λ is a Lagrangian multiplier. The solution found by setting the derivative of J with respect to W equal to zero is given by

$$M_cW = \lambda\, W \tag{4–12}$$

which is recognized as an eigenvector equation. The desired optimal weight W is then the eigenvector corresponding to the minimum eigenvalue of M_c. For the optimal weighting, we have, from (4–10) and (4–12),

$$I_{\text{opt}} = 1/\lambda_{\text{min}}. \tag{4–13}$$

The practical problem is that M_c is generally unknown for all conditions, so that a worst-case situation, or some compromise thereto, is assumed. Any departure of the actual clutter spectra from this assumption results in degraded MTI performance. Also, having arrived at an optimal weighting in the fixed-weight MTI system, an additional problem is that the MTI improvement factor is very sensitive to the implementation accuracy of the eigenvector weighting. Thus, any error incurred in the implementation of this weighting results in degraded MTI performance, even if the assumed clutter spectrum is correct. For zero-mean Gaussian clutter spectra, the binomial weighting provides an improvement factor that is within several dB of the optimum. However, for nonzero-mean, non-Gaussian, or different simultaneous clutter spectra occuring at the same range or for a staggered-PRF MTI, the binomial weighting can result in poor performance. Adaptive MTI attempts to overcome these difficulties by estimating the correlation among the various signals and developing a proper weighting to minimize the output clutter residue for any clutter spectra. Also, the adaptive MTI eliminates the degradation due to scanning modulation in fixed-weight MTI systems.

4.3.2 Theoretical Performance of an Adaptive Three-Pulse MTI as a Function of Clutter Spectral Characteristics

A computer program has been developed to simulate the MTI illustrated in Fig. 4–17, which makes use of digital open-loop cancelers with long-term averaging and multiple noise-free clutter signals. This program calculates the *clutter-cancellation ratio,* and the MTI *improvement factor,* and plots the MTI *transfer function*. The MTI improvement factor is defined as the signal-to-clutter improvement averaged over all signal Doppler frequencies, and the cancellation ratio is defined as the ratio of the clutter power at the input of the first delay element to the clutter power from the MTI. In the following plots, the transfer function is normalized to make 0 dB equal to the MTI improvement factor, which is the sum, in decibels, of the cancellation ratio and the average target gain, referred to as the *enhancement factor*. In this analysis, target returns are assumed to have much smaller duty factors than clutter, and the averaging time to form the weight is large enough so that targets do not influence the weights to cause self-cancellation.

In the case of the three-pulse adaptive MTI, there are two zeros to be deployed in the available Doppler space. The first illustration of the adaptive MTI is shown in the plot of the steady-state transfer function (Fig. 4–18) for two input clutter frequencies C_1 and C_2. The phase angles associated with C_1 in Fig. 4–18, reading from left to right, indicate the phases of the most recent signal, the phase one repetition period earlier, and so on. Thus, C_1 and C_2 are seen to have phase increments of $-25°$ and $-170°$ in the interpulse period τ_r. The abscissa of the plot is Doppler frequency, normalized to the PRF. With the negative phase increments chosen, corresponding to a negative Doppler shift, the two normalized clutter frequencies $-25°$/

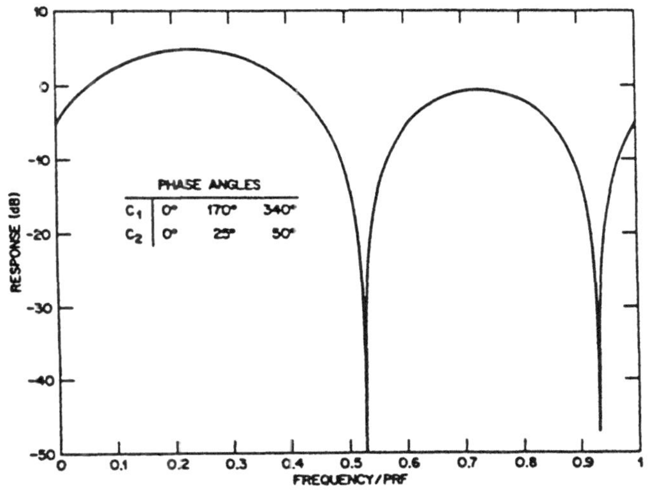

Fig. 4–18 Three-Pulse Adaptive MTI Response (Two Clutter Signals).

360° ≈ −0.07 and −170°/360° ≈ −0.47 for C_1 and C_2, respectively. Because of the periodic nature of the transfer function, these frequencies appear in Fig. 4–18 at (1−0.07) and (1−0.47).

Note that the adaptive MTI placed one of its two zeros on each of the two clutter frequencies, which, because the transfer function is periodic, are shown to occur at 0.93 and 0.53.

In Fig. 4–19, three clutter signals were introduced at negative Doppler phase angles of 170°, 30°, and 20° per interpulse period. The resulting plot of the transfer function indicates that one zero is placed approximately midway between the frequencies of the clutter signals, which were phase shifted −20° and −30° in one interpulse period, and the second zero was placed on the −170° phase-shifted signal.

Fig. 4–20 is identical to Fig. 4–19, except that the −170° clutter signal was changed to −215°, and once again the zero tracks this signal.

Figure 4–21 is identical to Fig. 4–20, except that the clutter phase shift of −215° was changed to −50°, in effect, a normalized Doppler frequency of −50°/360° or about −0.14 or 0.86. Here again the canceler zero followed the changing frequency.

Figure 4–22 is the same as 4–21, except that the clutter phases increase with time, corresponding to positive Doppler frequency shifts. Note that the canceler zeros appear between 0 and 0.5 instead of between 0.5 and 1.0.

To assess the behavior of the adaptive canceler for clutter signals having a continuous Doppler spectrum, a Gaussian spectrum with a standard deviation σ was approximated by 19 discrete-frequency clutter signals. The results are shown in Fig. 4–23, which portrays a family of transfer functions for different values of σT, where T corresponds to τ_r in our present usage. This product of $\sigma\tau_r$ is the same as σ normalized to the PRF.

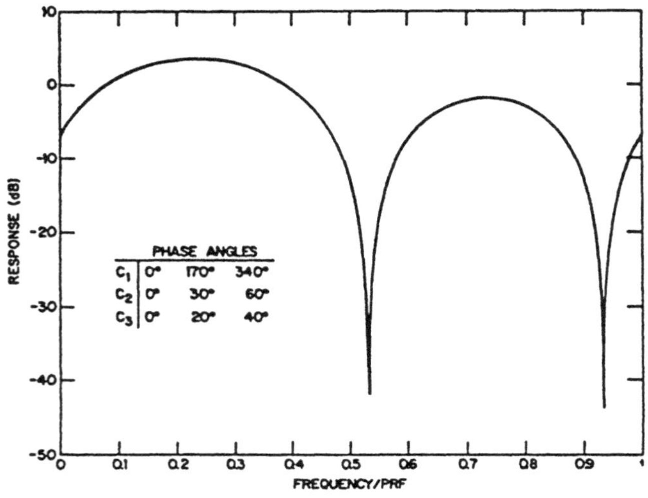

Fig. 4–19 Three-Pulse Adaptive MTI Response (Three Clutter Signals, Velocity Distribution 1).

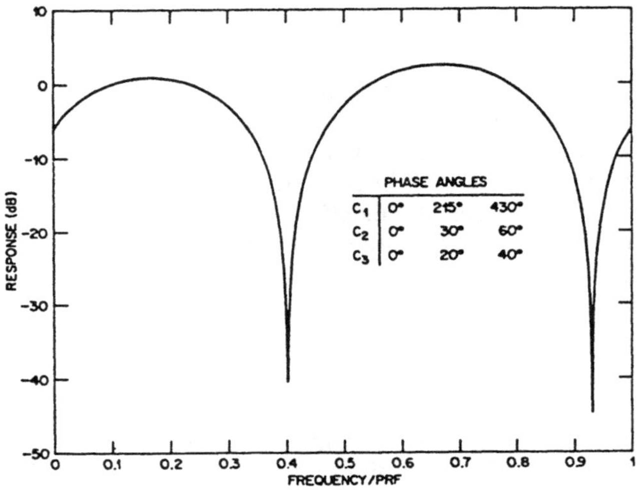

Fig. 4–20 Three-Pulse Adaptive MTI Response (Three Clutter Signals, Velocity Distribution 2).

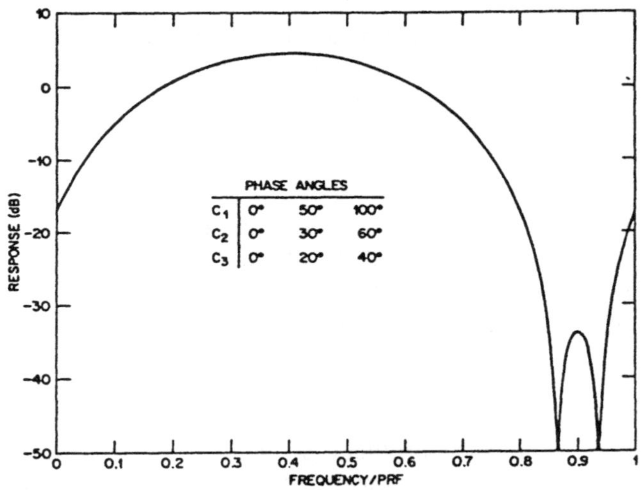

Fig. 4–21 Three-Pulse Adaptive MTI Response (Three Clutter Signals, Velocity Distribution 3).

The center frequency of the Doppler spectrum was arbitrarily selected to be at 0.6 on the abscissa of Fig. 4–23. The improvement factor corresponding to the adaptive MTI is indicated in Fig. 4–23 along with the optimal improvement factor computed for the eigenvector weighting. Comparison of the improvement factors indicates that the adaptive MTI is theoretically nearly optimum in the region of practical interest.

The four-pulse transfer functions and improvement factors are shown in Fig. 4–24 for an approximated Gaussian clutter spectrum, where the assumptions and approximations are the same as peviously discussed in regard to the three-pulse adaptive MTI.

The effect of scanning modulation was simulated with a three-pulse adaptive MTI by using three clutter signal frequencies as depicted in Fig. 4–21, but having a parabolic amplitude variation at the different delayed values of the input. The weights developed by the adap-

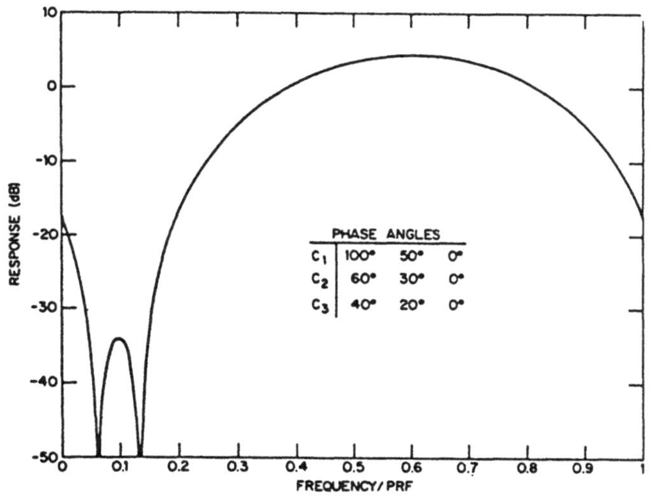

Fig. 4–22 Three-Pulse Adaptive MTI Response (Three Clutter Signals, Velocity Distribution 4).

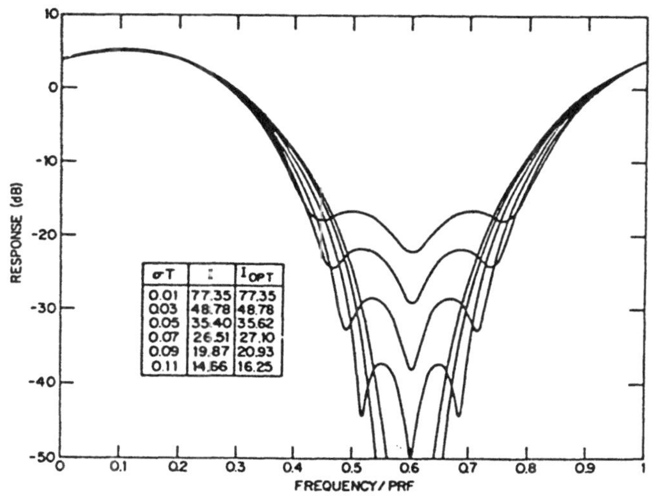

Fig. 4–24 Four-Pulse Adaptive MTI Response (Gaussian Clutter Spectra).

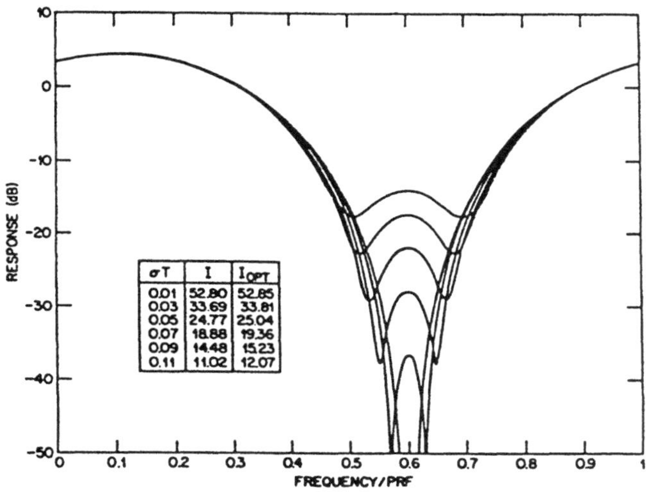

Fig. 4–23 Three-Pulse Adaptive MTI Response (Gaussian Clutter Spectra).

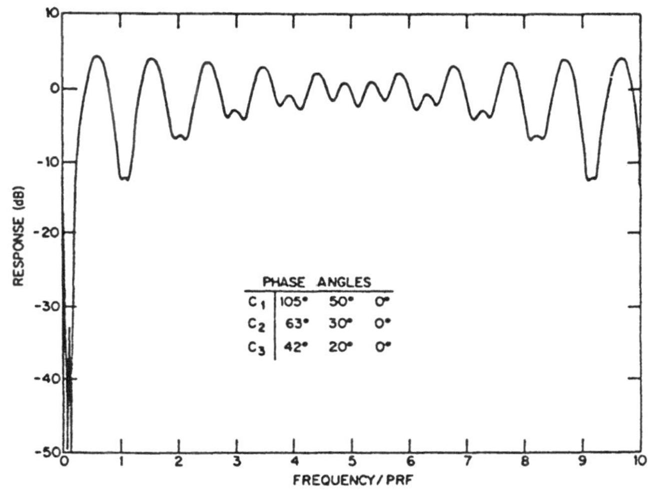

Fig. 4–25 Three-Pulse Adaptive MTI Response For A Staggered-Destaggered Waveform (Three Clutter Signals).

tive MTI canceler loops resulted in the same cancellation that was achieved for the unmodulated case. This suggests that the improvement factor limitation due to scanning modulation, which is normally associated with fixed-weight MTI systems, is not a limitation for the adaptive MTI. This is particularly important for scanning radar systems that have a small number of hits per beamwidth.

4.4 MTI BLIND-SPEED COMPENSATION

Blind speeds in MTI are those target velocities that produce Doppler frequencies which occur at cancellation frequencies in the periodic MTI transfer function. These frequencies are separated by phase shifts which are multiples of 2π radians in an MTI interpulse delay and they are canceled in the MTI.

Blind-speed compensation is necessary to permit targets at all expected Doppler frequencies to be detected by a radar using an MTI. With an N-pulse MTI, blind-speed compensation can be achieved, for example, by changing the radar carrier frequency or PRF every N transmitted pulses. In this process, we must ensure that expected target Doppler frequencies do not have simultaneous blind speeds on each transmitted PRF. For example, if the radar alternates between two PRFs with ratios of 1.1 to 1, there will be simultaneous blind speeds at a Doppler frequency 10 times the higher PRF and 11 times the lower PRF for an MTI cancelling clutter at zero frequency. Blind speeds can also be compensated using a *staggered pulse repetition frequency* [1, 5, 18, 23–27]. However, pulse-staggered MTIs cannot be used in the presence of nth-time-around echoes from clutter because they will not cancel properly. Also, a loss is incurred when using fixed binomial weighting rather than optimal improvement weighting [28].

Figure 4–25 shows, for an adaptive MTI, the effect of *pulse repetition interval staggering* when the interval between the first pulse and the second is 10/11 times the interval between the second and third pulses. The cancellation notch of a staggered MTI is very similar to the notch of an unstaggered MTI for the same clutter configuration. The primary effect of staggering is to extend the Doppler space available for the first Doppler blind speed.

4.5 MTI NOISE INTEGRATION LOSS

MTI signal processing correlates the receiver noise from pulse to pulse, and this results in degraded detection performance when the MTI output pulses are integrated, which is called *noise integration loss*. This has been expressed in terms of the number of effective independent pulses at the output of an MTI [29, 30] and also in terms of *S/N* detection losses [31, 32]. The losses for two- to five-pulse MTIs are approximately 1.0, 1.8, 2.2, and 2.5 dB, respectively. More detail on this phenomenon is given by ([31]; *see* Readings, Chapter 4). The noise integration loss also takes place in the absence of clutter if the MTI is employed in the clear, or the clutter-free region, using fixed weights. However, it does not take place in the clear with an adaptive MTI because there is no clutter to generate weights in this case.

4.6 DOPPLER FILTER BANKS

Doppler filter banks are employed in radar to suppress clutter, to improve the signal-to-noise ratio, and to provide quantitative information about target velocities [1, 18, 33]. They are sometimes implemented as fast Fourier transform (FFT) circuits. These circuits provide a bank of identical, contiguous bandpass filters, which subdivide the FFT sampling rate F_s into N subbands of width F_s/N. Here, N is the number of time samples, or points, used in the FFT. Note that, in Fig. 2–9, four time samples in I and Q at a rate equal to the signal bandwidth can be used to steer each of four frequencies unambiguously to different output terminals, called *frequency ports*. Doppler filter banks offer a distinct advantage over simple MTI in their ability to control the passband and stopbands in the available Doppler space F_s. Useful Doppler filter banks require that the input time samples to an FFT be amplitude-weighted to control the frequency sidelobes on each subband filter. Blind speeds still exist, though they can be made much narrower in frequency than with a simple MTI by having N sufficiently large.

An important point to observe in the design and use of MTI or pulse Doppler filter banks is that they must be completely charged with data before they are stabilized so that their output is meaningful. For example, a four-pulse MTI charged with only three pulses will not be properly weighted to ensure that the sum of the weighted clutter signal equals zero. Also, for MTI and Doppler filter bank processing, if *n*th-time-around clutter is expected, *n* filler pulses are needed so that each tap in the filter has a sampling of the ambiguous range clutter, which can then be properly weighted. The effects of ambiguous range clutter on Doppler filter bank processing has been discussed by Ward [34].

In radar, the output of MTI or Doppler filter banks are usually read after every completely new charge. More frequent reading of the outputs would not provide much more useful information because the processor correlates the time samples.

In addition, several observations should be made about Doppler-steered pulse compressors with thumbtack ambiguity diagrams. The Doppler steering prevents loss of the target signal and indicates the target velocity in the absence of clutter. However, clutter having a different velocity or appearing at different ranges than the target, is folded into the target filter, and also the other filters, and thus can mask the target [35].

In recent years, clutter-suppression systems have been analyzed and built that employ MTI and FFT Doppler filter banks in series [1, 36–41]. The MTI is usually placed ahead of the FFT to cancel ground clutter and to reduce the dynamic range required in the FFT. Systems using multiple pulses have been developed, with two or three devoted to MTI and the remainder devoted to a Doppler filter bank.

Recently, the combination of a digital adaptive MTI and a digital Doppler filter bank has been investigated and found to have much to offer. An example of such a study is illustrated by Kretschmer *et al.* ([42]; *see* Readings, Chapter 4). In this study, it was found that for a clutter-limited environment and a fixed number of pulses processed, improved clutter rejection was attainable by using an adaptive MTI prior to Doppler filtering. The adaptive MTI would also reduce the dynamic range requirements of the Doppler filter bank. A particular form of an adaptive MTI was described, based on Gram-Schmidt orthogonalization. This adaptive MTI is capable of very fast convergence and was shown to have optimal properties. A problem area with adaptive MTI is the inadvertent cancellation of desired targets. However, there are several techniques available to prevent this, such as offsetting the processor weight by one range cell so that it is applied to a range cell which is not included in determining the weight. In this study, it was also shown for bimodal clutter spectra that an adaptive MTI is capable of much better performance than a fixed-weight binomial MTI.

Finally, we should note that in some systems a bank of customized digital transversal filters is used. The individual filters are designed to reject the expected clutter, thereby obviating the need for an MTI. This contrasts with the FFT filter bank approach, where each of the filters is identical. A comparison between the different types of MTI and FFT Doppler filter processing has been made by various authors [43–46].

4.7 RANGE-DOPPLER COUPLED MTI

An MTI based on the *range-Doppler coupling* property of the chirp and step-chirp waveforms as well as the palindromic P2 and P4 polyphase codes is described by Kretschmer *et al.* ([47]; *see* Readings, Chapter 4). The principle of operation is based on the fact that for zero Doppler shift, the compressed pulses of successively transmitted up and down chirp signals (separated by τ_r seconds) are the same, and after subtraction the resultant equals zero. In the presence of Doppler shift, however, the compressed waveforms move in opposite directions in range, or equivalently in time, due to range-Doppler coupling, and the resultant is nonzero after subtraction. This forms the basis for an MTI without blind speeds, which is described in detail at the end of this chapter in the readings [47].

4.8 OTHER CLUTTER PROCESSING TECHNIQUES

Other clutter processing techniques consist of *maximum entropy* methods [48–51] and the application of *modern spectrum analysis* or *lattice filtering* techniques [51–53] to the cancellation and classification of radar clutter [53, 54]. Gabriel [55] has pointed out that *linear prediction* is equivalent to use of the maximum entropy criterion. Adaptive MTI, as previously described in this chapter, may be regarded as linear prediction [56]. Many of the clutter-cancellation and interference-cancellation techniques are similar.

PROBLEMS

Problem 4–1 W. B. Gordon described S-band rain clutter experiments in [10]. Part of one experiment involves the following parameters: frequency of operation was 3 GHz, slant range was 33.34 km, elevation angle was 3.2°, and the elevation 3 dB beamwidth was 2°. Assuming a vertically stratified atmospheric region with a horizontally operating wind shear and given a single wind-velocity null occurring between the upper and lower 3 dB points of the radar beam, estimate the maximum and minimum values of the wind shear (in meters per second per kilometer) if the surface wind speed was 12.4 m/s.

Problem 4–2 The table below provides wind velocity profile data for the geographic location 30° N, 100° W [9]. By using a velocity equal to the mean value plus one standard deviation (SD), which is provided in the tabulated data, compute the wind-shear values for the 10,000-ft height intervals.

Height (ft)	*Velocity* (AVE + SD) (knots)
10 000	26
20 000	39
30 000	59
40 000	74
50 000	44
60 000	20
70 000	20
80 000	20
90 000	30
100 000	30

Problem 4–3 Suppose that a stationary target produces unit-amplitude echoes on successive transmissions of a radar. Graphically analyze the operation performed on the succession of these echoes by the two-pulse canceler of Fig. 4–6. Use (4–1) with zero Doppler to describe this process quantitatively.

Problem 4–4 Consider a hypothetical radar's linear gain curve given by $G(\theta) = 1 - |\theta|$, $-1 \leq \theta \leq 1$. Suppose that, because of antenna scanning motion, successive returns from a stationary reflector, such as described in problem 4–3, are received at times for which $\theta = 0.25$, 0.50, and 0.75, respectively. Describe the output of the two-pulse MTI of Fig. 4–1 to these three inputs, taking into account the changing antenna gain.

Problem 4–5 Repeat problem 4–4 by using the three-pulse MTI of Fig. 4–8. Use (4–1) to analyze the operation quantitatively.

Problem 4–6 Consider a hypothetical radar's gain curve given by $G(\theta) = 1 - \theta^2$ for $-1 \leq \theta \leq 1$ and repeat problems 4–4 and 4–5, in the case of echoes received at times for which $\theta = 0.25$, 0.5, and 0.75.

Problem 4–7 Plot (4–2) as amplitude $|S_o|$ *versus* Doppler frequency from zero to $6\pi/\tau_r$. Use $a(t) = 1$ for computational simplicity.

Problem 4–8 Suppose that the first two pulse echoes of problem 4–4 are produced by a target having a Doppler frequency $\omega = 0.5\ \pi/\tau_r$. Use equation 4–2 to compute the amplitude of the output of a two-pulse MTI.

Problem 4–9 Compute the number of echoes returning to a radar antenna per scan from a reflector if the antenna beamwidth is 2°, the PRF is 200 Hz, and the antenna scan rate is 6 revolutions per minute. Repeat the computation for a beamwidth of 1°. Compute the maximum unambiguous range.

Problem 4–10 For problem 4–9, determine the PRF required for the 1° beamwidth to produce the same number of returns per scan as found for the 2° beamwidth operating with a 200 Hz PRF. Compute the associated maximum unambiguous range.

Problem 4–11 A radar using a PRF of 150 Hz receives echoes from a target located 10^7 m away. Find the range that the radar would determine for this target.

REFERENCES

1. M. I. Skolnik, *Introduction to Radar Systems,* (2nd ed.) McGraw-Hill, New York, 1980.
2. D. E. Kerr, *Propagation of Short Waves,* Technical Publishers, Boston, 1964.
3. D. K. Barton, *Radar System Analysis,* Artech House, Dedham, MA, 1977.
4. F. E. Nathanson, *Radar Design Principles,* McGraw-Hill, New York, 1969.
5. W. W. Shrader, "MTI Radar," Ch. 17 in *Radar Handbook,* M. I. Skolnik (ed.), McGraw-Hill, New York, 1970.
6. E. J. Barlow, "Doppler Radar," *Proc. IRE,* Vol. 37, April 1949, pp. 340–355.
7. H. Goldstein, "The Effect of Clutter Fluctuations on MIT," MIT Rad. Lab. Report 700, Dec. 27, 1945.
8.* D. B. Trizna and R. O. Pilon, "An Empirical Model for Wind Shear Over the Ocean for Chaff Application," NRL Report 8832, June 14, 1983.
9. S. L. Valley (ed.), *Handbook of Geophysics and Space Environments,* McGraw-Hill, New York, 1965.
10.* W. B. Gordon, "Analysis of Rain Clutter Data from a Frequency Agile Radar," *Radio Science,* Vol. 17, No. 4, July-Aug. 1982, pp. 801–816.
11. W. B. Gordon and J. Wilson, "Rain Clutter Statistics," NRL Report 8639, Sept. 30, 1982.
12.* B. L. Lewis, J. P. Hansen, I. D. Olin, and V. Cavaleri, "High Resolution Radar Scattering Characteristics of a Disturbed Sea Surface and Floating Debris," NRL Report 8131, July 29, 1977.
13.* B. L. Lewis and I. D. Olin, "Some Recent Observations of Sea Spikes," *Radar 77,* IEE International Radar Conference, London, UK, IEE Conf. Pub. 155, 1977, pp. 115–119.
14.* B. L. Lewis and I. D. Olin, "Experimental Study and Theoretical Model of High-Resolution Radar Backscatter from the Sea," *Radio Science,* Vol. 15, No. 4, July-Aug. 1980, pp. 815–828
15.* J. P. Hansen and V. F. Cavaleri, "High-Resolution Radar Sea Scatter, Experimental Observations and Discriminants," NRL Report 8557, March 5, 1982.
16.* I. D. Olin, "Characteristics of Spiky Sea Clutter for Target detection," IEEE National Radar Conference, Atlanta, GA, 1984.
17. R. S. Berkowitz (ed.), *Modern Radar Analysis, Evaluation and System Design,* John Wiley and Sons, New York, 1965.
18. D. C. Schleher (ed.), *MTI Radar,* Artech House, Dedham, MA, 1978.
19. R. C. Emerson, "Some Pulsed Doppler MTI and AMTI Techniques," RAND Corp. Report No. R-274, March, 1954 (reprinted in Schleher [18]).
20. J. Kroszczynski, "The Two-Frequency MTI System," *Radio and Electronic Engineer,* Vol. 39, March 1970, pp. 172–176.
21.* F. F. Kretschmer, Jr., F. C. Lin, and B. L. Lewis, "A Comparison of Noncoherent and Coherent MTIs," NRL Report 8591, June 22, 1982.
22. T. Murakami and R. S. Johnson, "Clutter Suppression by Use of Weighted Pulse Trains," *RCA Review*, Vol. 32, Sept. 1971, pp. 402–428.
23. O. J. Jacomini, "Weighting Factor and Transmission Time Optimization in Video MTI Radar," *IEEE Trans. on Aerospace and Electronic Systems*, Vol. AES-8, July 1972, pp. 517–527.
24. P. J. A. Prinsen, "Elimination of Blind Velocities of MTI Radar by Modulating the Interpulse Period," *IEEE Trans. on Aerospace and Electronic Systems,* Vol. AES-9, Sept. 1973, pp. 714–724 (reprinted in Schleher [18]).
25. G. W. Ewell and A. M. Bush, "Constrained Improvement MTI Radar Processors," *IEEE Trans. on Aerospace and Electronic Systems,* Vol. AES-11, Sept. 1975, pp. 768–780.
26. J. K. Hsiao and F. F. Kretschmer, Jr., "Design of a Staggered-PRF Moving Target Indication Filter," *Radio and Electronic Engineer,* Vol. 43, pp. 689–693, Nov. 1973.
27. L. E. Brennan and I. S. Reed, "Optimum Processing of Unequally Spaced Radar Pulse Trains for Clutter Rejection," *IEEE Trans. on Aerospace and Electronic Systems,* Vol. AES-4, May 1968, pp. 474–477.
28.* F. F. Kretschmer, Jr., "MTI Weightings," *IEEE Trans. on Aerospace and Electronic Systems,* Vol. AES-10, No. 1, Jan. 1974, pp. 153–155.
29. W. M. Hall and H. R. Ward, "Signal to Noise Loss in Moving Target Indicators," *Proc. IEEE* (letters), Vol. 56, no. 2, Feb. 1968, pp. 233–234.
30.* F. F. Kretschmer, Jr., "Correlation Effects of MTI Filters," *IEEE Trans. on Aerospace and Electronic Systems,* Vol. AES-13, May 1977, pp. 321–322.

31.* G. V. Trunk, "MTI Noise Integration Loss," NRL Report 8132, July 15, 1977.
32. G. M. Dillard and J. T. Rickard, "Performance of an MTI Followed by Incoherent Integration for Nonfluctuating Signals," *Record of the IEEE International Radar Conference,* 1980, pp. 194–199.
33. J. K. Hsiao, "Comb Filter Design," Naval Research Laboratory, Memorandum Report 2433, Washington, DC, May 1972.
34. H. R. Ward, "Doppler Processor Rejection of Range Ambiguous Clutter," *IEEE Trans. on Aerospace and Electronic Systems,* Vol. AES-11, No. 4, July 1975, pp. 519–522.
35. F. C. Lin, B. L. Lewis, and F. F. Kretschmer, Jr., "Parameter Estimation and Target Detection in a Distributed-Clutter Environment," NRL Report 8681, March 23, 1983. (*See* Readings, Chapter 2.)
36.* G. A. Andrews, Jr., "Performance of Cascaded MTI and Coherent Integration Filters in a Clutter Environment," NRL Report 7533, March 27, 1973.
37. G. A. Andrews, Jr., "Optimal Radar Doppler Processors," NRL Report 7727, May 29, 1974.
38.* G. A. Andrews, Jr., "An Airborne Radar Doppler Processing Philosophy," NRL Report 8073, Jan. 31, 1977
39. G. A. Andrews, Jr., and S. L. Sheller, "A Matched Filter Doppler Processor for Airborne Radar," NRL Report 8700, July 13, 1983.
40. C. E. Muehe, "Digital Signal Processor for Air Traffic Control Radars," *IEEE NEREM 74 Record, Part 4: Radar Systems and Components,* Oct. 28–31, 1974, pp. 73–82. (IEEE Cat. No. 74 CHO 934–0 NEREM).
41. C. E. Muehe, "Advances in Radar Signal Processing," *IEEE Electro 76 Record,* Boston, MA, May 11–14, 1976.
42.* F. F. Kretschmer, Jr., B. L. Lewis, and F. C. Lin, "Adaptive MTI and Doppler Filter Bank Clutter Processing," *Proc. IEEE National Radar Conference,* Atlanta, GA, 1984, pp. 69–73.
43. V. G. Hansen and D. Michelson, "A Comparison of the Performance Against Clutter of Optimum, Pulsed Doppler and MTI Processors," *Record of the IEEE International Radar Conference,* 1980, pp. 211–218.
44. V. G. Hansen, "Optimum Pulse Doppler Search Radar Processing and Practical Approximations," *Radar 82,* London, UK, IEE Conf. Pub 216, Oct. 1982, pp. 138–141.
45. J. W. Taylor, Jr., "Sacrifices in Radar Clutter Suppression due to Compromises in Implementation of Digital Doppler Filters," *Radar 82,* London, UK, IEE Conf. Pub. 216, October 1982, pp. 46–50.
46. D. C. Schleher, "Performance Comparison of MTI and Coherent Doppler Processors," *Radar 82,* London, UK, IEE Conf. Pub. 216, Oct. 1982, pp. 154–158.
47.* F. F. Kretschmer, Jr., B. L. Lewis, and F. C. Lin, "Range-Doppler Coupled Moving Target Indicator (MTI) Analysis and Assessment," NRL Report 8789, March 20, 1984.
48. J. H. Saywers, "The Maximum Entropy Method Applied to Radar Adaptive Filtering," *Proc. 1977 RADC Spectrum Estimation Workshop,* Griffiss AFB, Rome, NY, Oct. 1977.
49. J. H. Saywers, "Adaptive Pulse-Doppler Radar Signal Processing Using the Maximum Entropy Method," *EASCON 80 Record,* Sept. 1980, p. 545.
50. S. Haykin, B. W. Currie, and S. B. Kessler, "Maximun Entropy Spectral Analysis of Radar Clutter," *Proc. IEEE,* Vol. 70, No. 9, Sept, 1982, pp. 953–962.
51. D. G. Childers, (ed.), "Modern Spectral Analysis," IEEE Press, New York, 1978.
52. B. Freidlander, "Lattice Filters for Adaptive Signal Processing," *Proc. IEEE,* Vol. 70, No. 8, Aug. 1982, pp. 829–868.
53. C. Gibson and S. Haykin, "Radar Performance Studies of Adaptive Lattice Clutter-Suppression Filters," *IEE Proc.,* Vol. 130, Pt. F, No. 5, 1983, pp. 357–367.
54. W. Stehwien and S. Haykin, "Statistical Classification of Radar Clutter," *Proc. IEEE National Radar Conference,* Los Angeles, CA, 1986, pp. 101–106.
55.* W. F. Gabriel, "Nonlinear Spectral Analysis and Adaptive Array Superresolution Techniques," NRL Report 8345, Feb. 1, 1980.
56.* F. F. Kretschmer, Jr., and F. C. Lin, "Effects of the Main Tap Position in Adaptive Clutter Processing," *Record of the IEEE International Radar Conference,* Washington, DC, 1985, pp. 303–307.

*Reference reprinted herein. *See* Readings, Chapter 4.

AN EMPIRICAL MODEL FOR WIND SHEAR OVER THE OCEAN FOR CHAFF APPLICATIONS

INTRODUCTION

Broadband chaff is created by cutting variable length metal strips which act as resonant quarter-length dipoles over a continuous range of radar frequencies appropriate to the range of cut lengths used. The total cross section is then dependent upon the number density of chaff distributed and its spread in volume, which we will not consider here. (We shall assume that enough chaff is present to saturate the radar receiver for a given radar range, as well as doppler frequency.) The extent of the doppler frequencies, or equivalent aircraft radial velocities, which is affected by chaff is determined by the velocities which the chaff attains due to local winds, or more precisely, the range of velocities, or the wind shear within a given range bin. The greater distribution capability essentially creates a greater volumetric cross section for a given doppler frequency—the doppler bandwidth being controlled by local winds as a function of altitude. *The spectral distribution of the chaff, cross section vs doppler frequency, is thus determined by the number density distribution of scatterers vs altitude, and wind shear with altitude.* If a very dense chaff distribution capability exists, then saturation of the radar signal can be assumed to exist for nearly all doppler bins appropriate to the wind shear present, so the chaff capability becomes simply a function of local wind shear only. For example, for no wind shear, all chaff returns would fill a single doppler bin, that associated with the local wind, constant with altitude. Thus, the problem presented to the radar designer becomes one of being able to predict wind shear accurately, preferably its characteristics vs season, location on the earth's surface, etc.

The model currently used by the Navy and other radar interests is one formulated by Nathanson [1]. Figure 1 is taken from Nathanson's environmental diagram, a plot of doppler frequency vs range, with the third dimension being signal amplitude. Chaff is shown at the highest doppler, assuming it is distributed at higher altitudes than rain clouds with assumed higher wind speeds. He states that the vertical extent of a chaff corridor is usually less than that of a rainstorm, so the standard deviation, and thus spectral Doppler width, will have a lower value than the rainstorm. Our results in this work indicate that this is not necessarily true.

The model for the typical chaff cloud is 10,000 ft in vertical extent, with wind shear of 4 m/s/km continuous over this altitude. This value of shear was apparently based upon measurements over land, and may not be the same as that encountered over the sea. In addition, a fixed vertical extent is not warranted for fall rates which are a function of atmospheric density, but will vary with range based upon the source aircraft altitude and speed. For these reasons, a better model was needed for search radar applications.

In this report we formulate a model for chaff based upon the distribution aircraft altitude and speed and chaff fall rate measurements vs ambient pressure, and a model for wind shear based upon over-water measurements as a function of season for three different regions—the North Atlantic, the Eastern Pacific, and the Mediteranean/Eastern Atlantic. Averages over season with aspect angle effects taken into account yield shear values of 2.3, 2.1, and 1.7 m/s/km for the three regions. Finally, some examples of a realistic environmental diagram are presented for the cloud models calculated and using real radiosonde data.

Manuscript approved December 19, 1983.

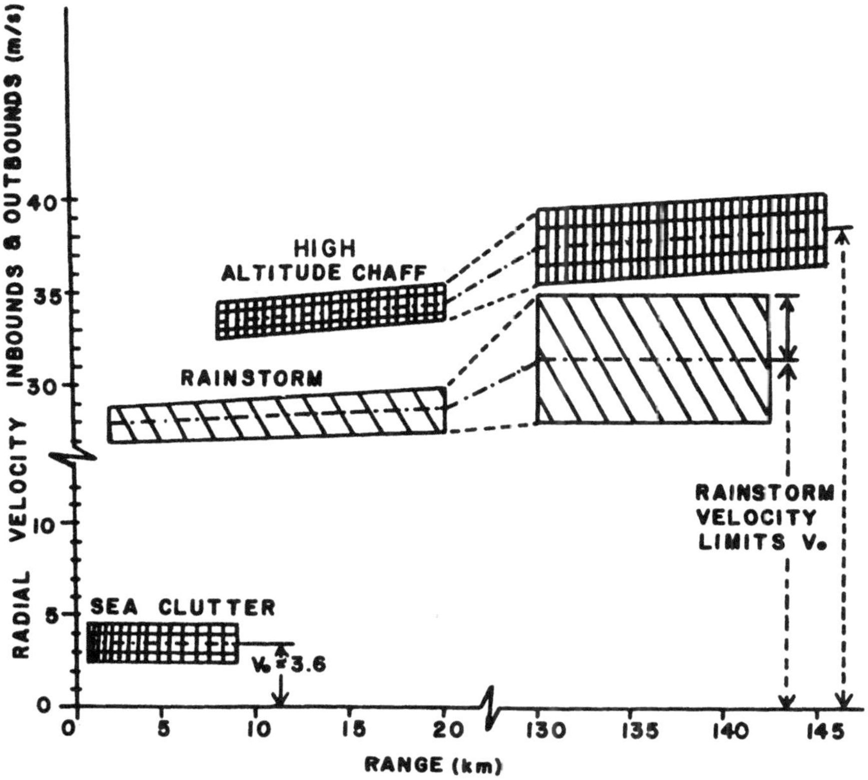

Fig. 1 — A doppler-range radar display is shown, with the regions of the plane affected by various types of clutter shown. These include land and sea clutter, rain clutter, and returns from chaff.

UNPUBLISHED REFERENCES ON WIND-SHEAR MEASUREMENTS

A discussion was held with personnel from the Atmospheric Sciences branch at NRL, regarding radiosonde data collected over water in general and whether the standard radiosonde has the height resolution to identify wind shears accurate enough for our purposes. Although specific references do not exist dealing with this point specifically, it has been their experence that the standard radiosonde does provide sufficient resolution in altitude to satisfy our needs, and that the next generation high-resolution minisondes are aimed at identifying ducting layers, which can be located over a very narrow range of heights. Wind shear, on the other hand, is typically distributed over a relatively wide range of altitudes in general. An inquiry was also made as to the possible existence of wind shear at the top of the atmospheric boundary layer, the mixed region between the water surface and altitudes anywhere from 50 meters to a few kilometers, depending on the time of day and latitude. They indicated that unpublished measurements made looking for this phenomenon demonstrate that it does not exist. As the instrumentation penetrates the top of the boundary layer, and a rapid change in humidity to dry air occurs, no noticeably strong wind shear is found. Our conclusion from this discussion was that the current radiosondes in use are sufficient for our needs of wind shear as a data source, with one caveat on wind direction which shall be discussed later.

CHAFF CLOUD MODEL

Review of the literature on the current model for a chaff corridor cloud indicated a very general model: distribution over 10 km in altitude, with an assumed maximum altitude of the order of 10 km.

The thought was that for altitudes much higher than this the chaff falls too rapidly to maintain an effective cloud.

Our cloud model is based upon two inputs, one theoretical and the other a laboratory measurement of chaff elements falling in a chamber under different atmospheric pressures. The theoretical paper [2], derived a Gaussian shaped cross section vs altitude, which slowly spread in altitude with time, from which we abstracted a 2σ spread in altitude with time. Figures 2, 3, 4, and 5 show the results of this theoretical estimate. The only experimental results available for our interests were fall-rate measurements of resonant dipoles cut for both X- and S-band, made in an evacuated chamber which could be used to simulate high altitude pressures (Juisto, et al. [3]). Juisto made measurements for both cylindrical dipoles and foil, with the foil dipole results having a greater distribution between the minimum and maximum values (Figs. 6a,b). He states that little difference was observed for foil at X- and S-bands. A rough comparison of the two indicates that the minimum foil fall rate is of the order of 60% that of the cylinder, and the maximum fall rate is 250% that of the cylinder. The minimum is smaller because individual foil elements are held aloft longer because of their higher surface-to-volume ratio, whereas the maximum is greater because of *bird-nesting* effects or clumping, during which the individual clumps continue to give off individual elements. Hence, the foil chaff presents a worst-case scenario for vertical spreading, which is the one we used. As can be seen, these experimental results bracket the theoretical values, although not symmetically in altitude. This would appear to confirm the theoretical results as a valid estimate, while the lack of symmetry indicates the cloud is probably not truly Gaussian, but skewed about a maximum value which falls with time.

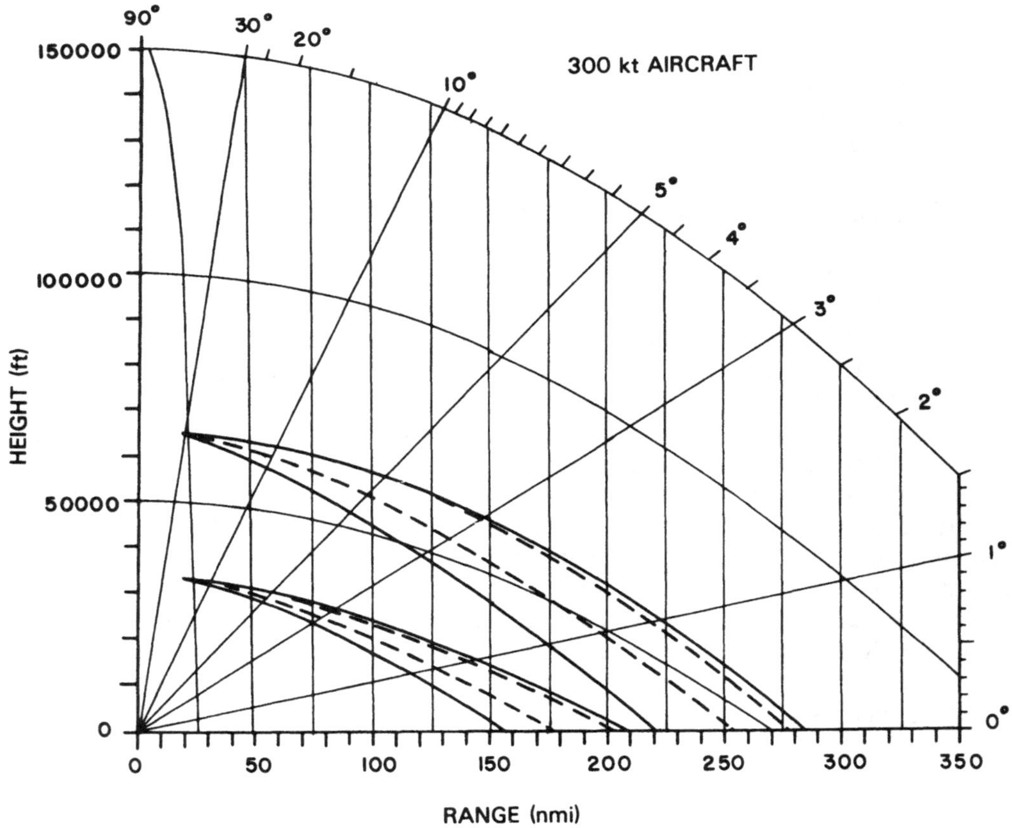

Fig. 2 — A model for a chaff cloud altitude dispersion history is shown. An aircraft flying at 300 knots, at 10- and 20-km altitude, with a terminal range of 20 nmi is shown. The cloud will continue to disperse in time of course.

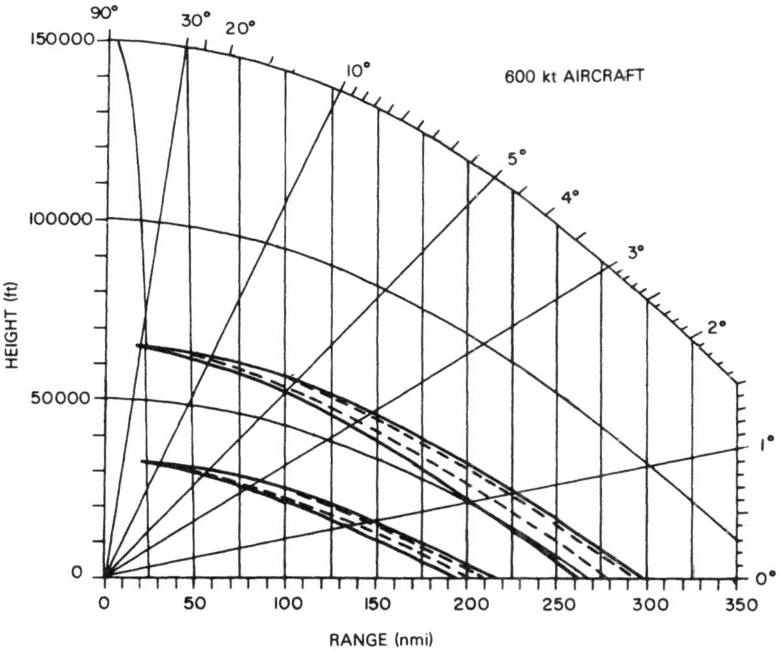

Fig. 3 — A model for a chaff cloud altitude dispersion history is shown. An aircraft flying at 600 knots, at 10- and 20-km altitude, with a terminal range of 20 nmi is shown. The cloud will continue to disperse in time of course.

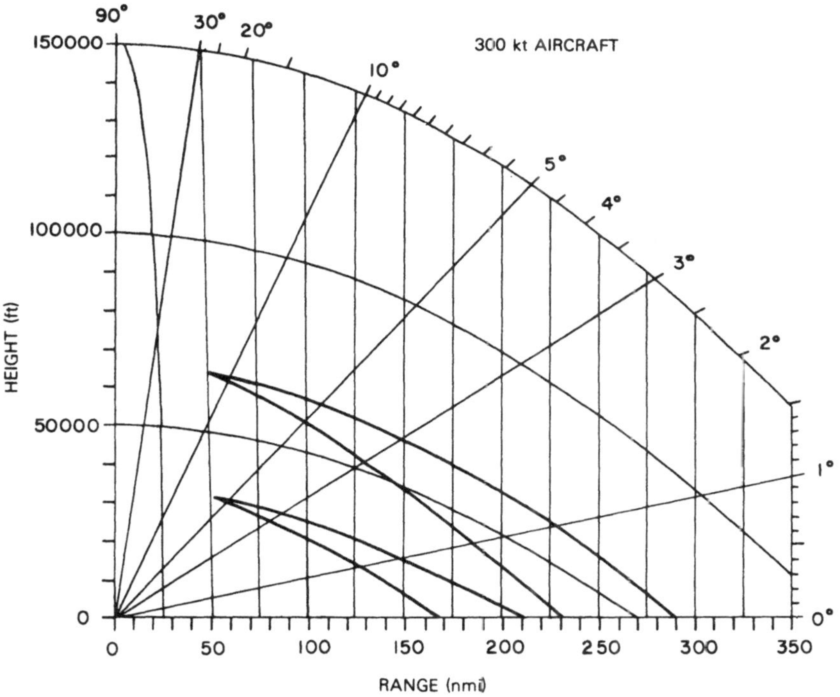

Fig. 4 — A model for a chaff cloud altitude dispersion history is shown. An aircraft flying at 300 knots, at 10- and 20-km altitude, with a terminal range of 50 nmi is shown. The cloud will continue to disperse in time of course.

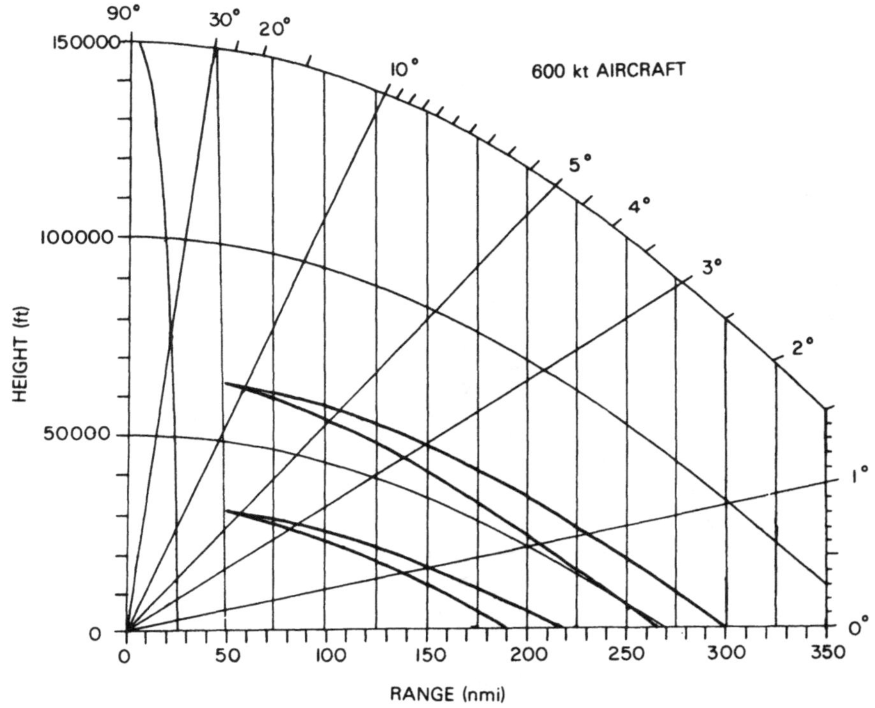

Fig. 5 — A model for a chaff cloud altitude dispersion history. An aircraft flying at 600 knots, at 10- and 20-km altitude, with a terminal range of 20 nmi is shown. The cloud will continue to disperse in time of course.

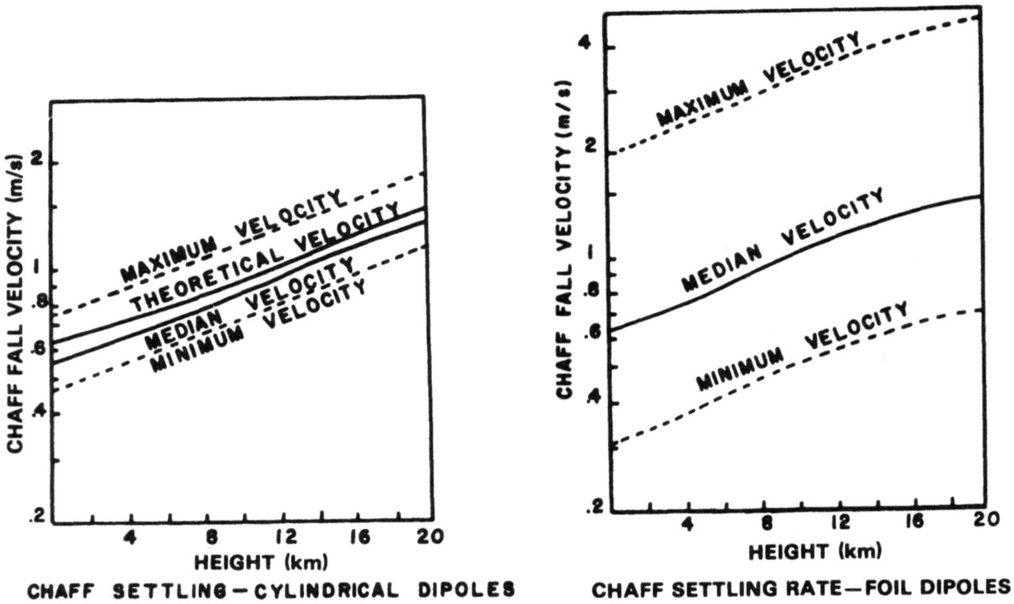

Fig. 6 — Dipole fall rates [3]. These are for both cylindrical and foil dipoles, and represent both X- and S-band results which were identical. The foil dipole results are used in the chaff cloud modeling done in this work.

We then modeled an aircraft flying at speeds of either 300 or 600 knots radially toward the radar and assumed chaff spread at the minimum and maximum values measured by Juisto as a function of atmospheric pressure (or density), and used a standard atmosphere to relate pressure to altitude. The picture that emerges is a cloud distributed in altitude, with a maximum and minimum spreading apart with distance from the aircraft. Two different altitudes were used for the aircraft flight, 10 km and 20 km. Figures 2, 3, 4, and 5 show the results for terminal distances of the aircraft of 20 and 50 nmi distance from the aircraft, on range-height-angle charts for radar application.

The spread of the cloud with altitude is greater for the higher altitude aircraft for a fixed speed since the dispersion is greater at the lower atmospheric pressures. For the lower altitude aircraft, much of the chaff at longer ranges has already fallen below the horizon and is not effective unless ducting were present. Hence, the higher altitude flights appear on two counts to be more effective for chaff distribution.

Where we compare the two different aircraft speeds for a given altitude, it appears that the slower speed is more effective since it allows the chaff a longer time to disperse, and thus creates a cloud which is distributed over a wider range of altitudes. This would appear to be the most effective way of distributing the chaff for aircraft following in time behind the chaff distribution. For the 300-kt aircraft, the time it takes to distribute the cloud, assuming a 200-nmi-long cloud (20 to 220 nmi coverage), would thus be about 40 min.

WIND-SHEAR MODEL

With a model selected for the spread of a chaff corridor cloud in altitude and range, we must consider the third radar dimension, doppler frequency. For this work we are assuming a worst-case situation, so that the amplitude for a given doppler frequency will be assumed to be saturated due to an very efficient volume distribution capability of the aircraft. (This point should be considered in greater detail for a more realistic study of amplitude-spectral distribution of chaff returns.) The problem of chaff distribution is considered from the same point of view as that of rain clutter, regarding the velocity distributions involved [1]. That is, a doppler spectrum of a rain or chaff cloud will be centered at a velocity corresponding to the mean line of sight wind velocity component, and will have a half-width determined by several assumed Gaussian processes present within the scattering volume. These include effects due to wind shear, turbulence, fall velocity distribution, and beam broadening. For chaff in particular, only the first two are of any significance. Because all of the processes are assumed Gaussian, the total variance will simply be the sum of variances of the four processes. The spectral width is then taken as twice the standard deviation of the process, the square root of the variance. This relationship has been verified in a series of experiments conducted with a 5.2-cm radar with a 1.4° beamwidth [1].

Wind-Turbulence Effects

Figure 7 shows three spectral distributions of turbulence for different atmospheric conditions, on a log-log scale. The total variance is simply the integrated area under each of the curves. Assuming that the last data point to the left is the maximum of each curve, and that an increment of wavenumber is roughly 0.01 near the peak, then, because the plots have such a steep variation with wavenumber, the area is estimated to be the leftmost point times the wavenumber increment as an order of magnitude estimate. This yields values of 10, 2, and 1 m/s squared for a severe storm, cumulus cloud, and clear air. The standard deviation or spectral widths for the three will be of the order of 3.16, 1.4, and 1 m/s. This is much less than typical wind-shear values, and turbulence is henceforth neglected in the model, as is done in Nathanson's model. Actually, for severe storms, the turbulence effects can be of the same order of magnitude in some cases, but we shall limit ourselves to nonstorm conditions. One reason for this is that all of the data which we shall use for the model are typically fair-weather data. This is because radiosondes typically have a low survival rate if launched in the midst of storms or high

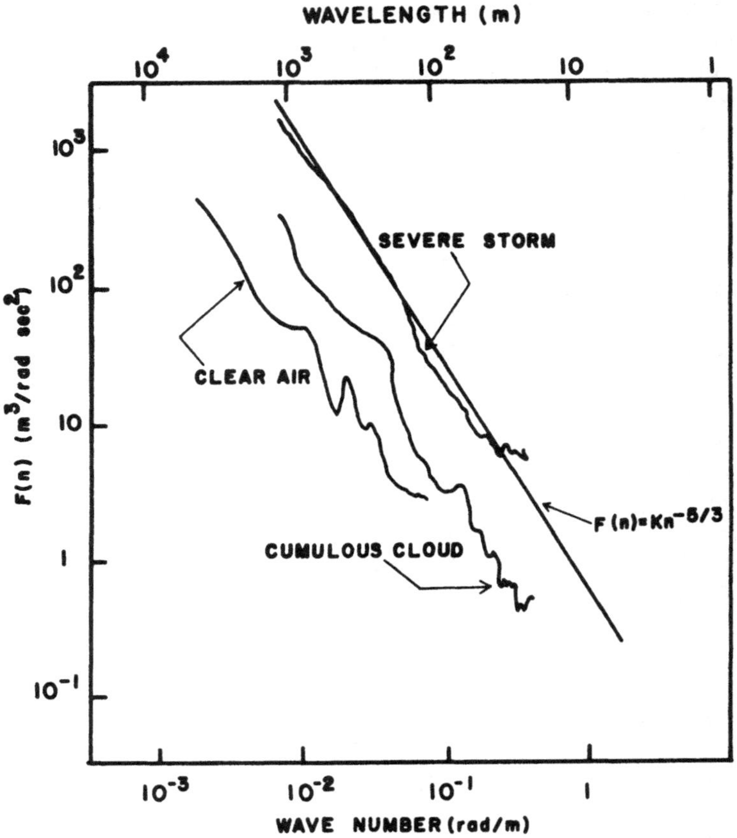

Fig. 7 — Turbulence spectra are shown for three different atmospheric conditions: clear air, cumulous cloud, and a severe storm. The total variance is the integration of the area under each curve.

winds. Because the radiosondes which are used for worldwide synoptic measurements of winds aloft are not sophisticated enough to allow measurements of storm-conditions wind shear, the model which we shall derive will be deficient by virtue of the data source limitations.

Wind-Direction Effects

The next point to consider regarding shear is the contributions from changes in velocity magnitude with height vs changes in direction. One can write the equation for the line-of-sight component of velocity observed by a radar, $\overline{v}$, due to both causes as follows:

$$d\overline{v}/dh = \delta\overline{v}/\delta h + \overline{v}\,\delta\theta/\delta h \tag{1}$$

The second term represents the wind direction change contribution to the shear, and is maximized for a radar look angle parallel to the mean wind between two levels under consideration. However, the maximum value of the total quantity does not necessarily occur along the direction of the wind, and will be different for each set of shear conditions considered. A review of several data sets, as well as a recent paper on the subject, indicate that when severe wind direction shifts occur (e.g., 180°), they are for relatively weak winds or winds constant in speed vs height. That is, severe changes in direction do not generally accompany severe wind magnitude changes with altitude.

Figure 8 shows a plot of wind magnitude and direction for a rather severe wind direction shift [4]. There occurs a shift of nearly 180° in direction over a 3-km altitude, but with a minor change in wind speed. This pattern characterizes warm air advection below with cold air advection above [4], a condition which can trigger thunderstorms. In the set of radiosonde data which we used as a data source, we virtually never found major directional changes associated with large wind-speed changes over the same altitude range. When the few large direction changes did occur, they were very similar to those shown in Fig. 8, generally with low wind speeds, but sometimes as high as 20 knots in this worst case. Assuming a 180° change in wind direction, a shear was observed by radar to extend from -20 to +20 knots with the worst-case radar aspect. Since the data analysis was done manually with input into a programmable calculator, only a single feature was considered for the level of effort desired for this work. The combination of both contributions to Eq. (1) should be included in any additional work in which computer analysis of the data set is used. Additional comments are discussed later regarding direction contributions due to radiosonde direction errors.

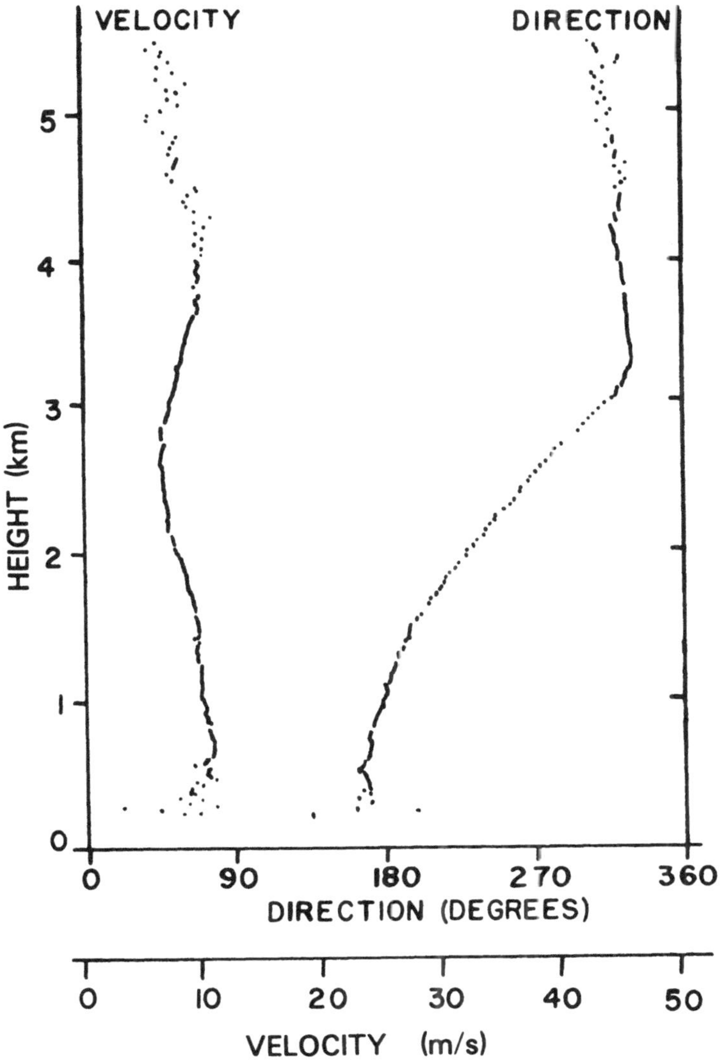

Fig. 8 — A mean wind profile collected using a tracking radar (Rabin, 1981). This technique is far more accurate than the standard radiosonde method of angle tracking to determine wind velocity.

Data Source

The source of data used for this model is a set of radiosonde data obtained from Fleet Numerical Ocean Center, Monterey. These data were obtained by radiosonde launches aboard weather ships on station at sea and U.S. Navy ships on patrol. Figure 9 shows the positions of the data sources available, with those used for this work annotated. For the calculation of cumulative distributions, one generally requires a minimum of 50 data points, with 100 points preferred. Since a month was chosen as the distribution period so that seasonal variations would not be averaged, not all of the locations provided enough data to meet this criterion. Those positions marked with a diamond indicate a scarcity of data and were not used for this reason.

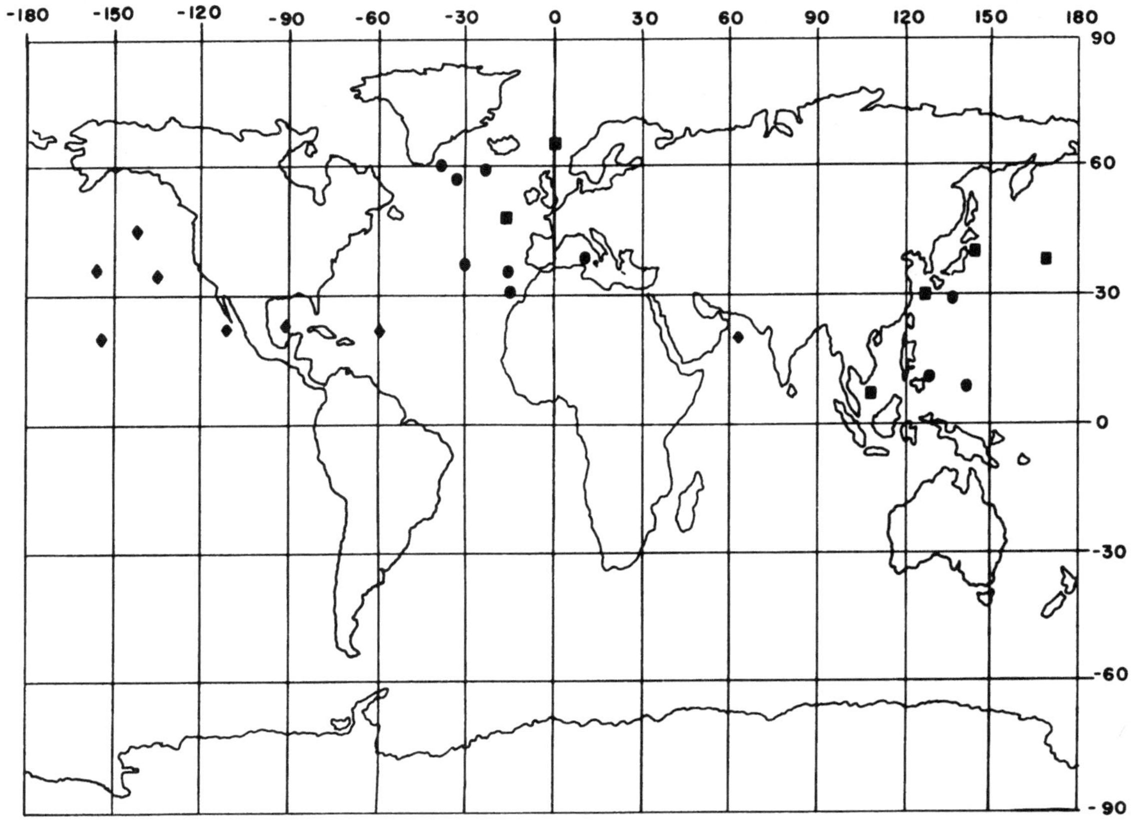

Fig. 9 — Sources of radiosonde data over the ocean available from Fleet Numerical Ocean Center, Monterey. The data key is: circles, data used for this report; diamonds, a scarcity of data on a monthly basis, so as not to be useful; squares, other data sources.

Three different regions were used as representative of Naval applications: the North Atlantic, the Eastern Pacific, and, in addition to the Mediterranean for which there were not sufficient data, the Eastern Atlantic pair of data sources just west of Gibraltar. The locations which were used for these areas are indicated by circles. All other locations, those which were not used for the analysis or those which did not contain a sufficiently large number of data points to be useful, are identified by squares.

Data representative of a one-month period were assembled to produce the statistics. The first 100 data points of a month were used, or the total within that month period if there were less than 100.

For the Pacific area, periods covering two months were used due to scarcity of the data, with the number of points indicated. On some occasions data were judged as unusable because of an error in readings at a small (greater than one) number of altitudes for a given radiosonde, so the entire radiosonde set was eliminated.

A program was written to plot representative data as a function of altitude, and an example of the output is shown in Figure 10a and b. The raw radiosonde outputs of wind speed, direction, air temperature, and dew point are plotted in the lower left corner. Since the data are collected as a function of atmospheric pressure, a translation to height is required. The standard exponential varying atmosphere was used of the form:

$$P(h) = P(0) \exp(-h/H), \tag{2}$$

where H is taken as 7 km [5]. The first reading of the radiosonde is taken as the surface value, and Eq. (2) was inverted to find h as a function of the atmospheric pressure reading. On the right-hand side are shown the change in wind speed and direction, calculated from the raw data at left. Finally, Eq. (1) was used to find the total wind shear, the magnitude of which is plotted in the upper left corner. Although such plots are useful for a small number of radiosondes, they are too costly to produce for the level of effort expended here, providing more information than is manageable.

One other point should be made regarding the measurement methods of radiosondes, since this affects the processing technique used here, namely, the omission of the wind direction contribution to the shear. The standard radiosonde measures pressure, temperature, and dew point as it rises in the atmosphere. Wind velocity is estimated by passively tracking the signal transmitted from the balloon with time, which is done with a small communications antenna. Wind vectors are inferred from changes in elevation and azimuthal angle measurements, using pressure telemetry data to define altitude. Hence, the wind speed and direction will both be affected by tracking errors of the antenna used, and these can be quite high for the wind direction in relation to the contribution it makes to the total wind shear given by the magnitude of Eq. (1). Figure 10a is representative of relatively stable wind direction data, with wind direction shear typically 10°/3 km, and a very minor contribution to the total wind shear at upper left. However, Fig. 10b is representative of somewhat noisy direction data, with the true profile probably being quite constant with altitude. It should be compared with the radiosonde shown in Fig. 8 regarding the noisiness of the data. The contribution of the direction shear to the total shear is quite large for this case, but probably unrealistic. For this reason, and also because the analysis would have been made far more complicated than was warranted for this effort, the contribution of the directional shear was not included in the calculations used to determine shear statistics.

Analysis Methodology

The bulk of the data was processed using a programmable calculator, to determine the total shear over a 3-km altitude region, which was chosen to include the largest wind shear. The 3-km region was a representative height over 1 km because, according to the chaff cloud model, it is a more realistic altitude regime over which a radar return from chaff will be summed. The total shear over such a region gives a less pessimistic estimate of integrated shear than using the value of 4 m/s/km and multiplying by 3 to get 12 m/s for a 3-km altitude regime. This is because when severe wind shear does occur, it is generally over a distance less than 1 Km. Expressing this shear on a per kilometer basis generates a pessimistic value, and extending the same value to 3 km is even more unrealistic.

The way the data were used is described herein. An eyeball estimate of the largest wind-speed difference between consecutive altitudes generally coincided with the largest wind shear as well. (i.e., the largest change in wind speed generally occurred between the same two altitudes as the largest ratio of wind-speed change divided by altitude difference.) The analyst then continued to work in adjacent data points on either side of this first pair selected to a total height increment of 3 km, or the first value just greater than 3 km was reached. Then the total shear over the fixed region (e.g., 12.8 m/s over 3.2 km) was normalized to a 3-km altitude increment (e.g., 12.0 m/s/3 km).

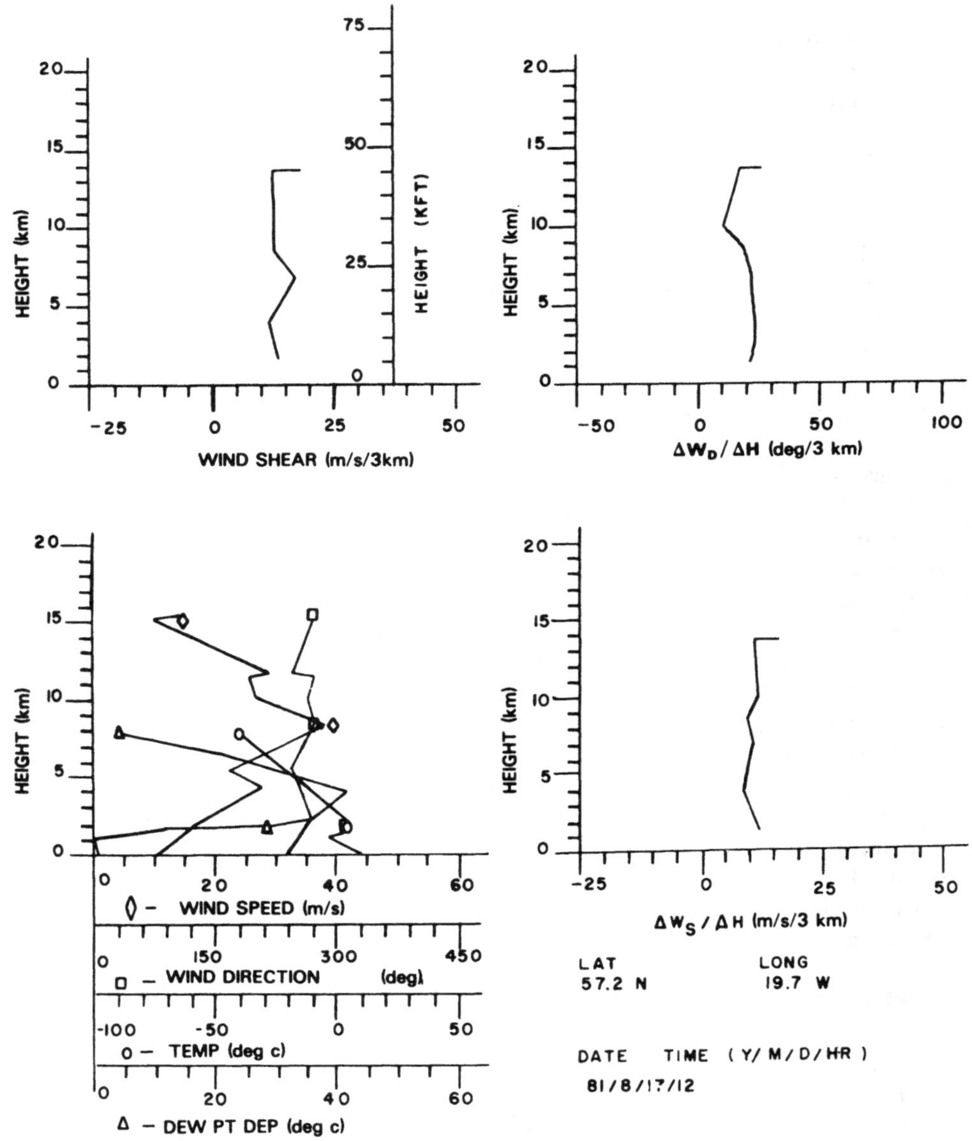

Fig. 10a — Examples of raw data are plotted in the lower left corner. Data include temperature and dew point, which are measured in situ, along with pressure, from which the altitude is determined. Wind speed and direction are determined from tracking the balloon, so noise may be inherent for these values. Wind speed shear is plotted at lower right, averaged over 3 km in a sliding sense, the first point representing the total shear from ground to 3 km, and so on up.

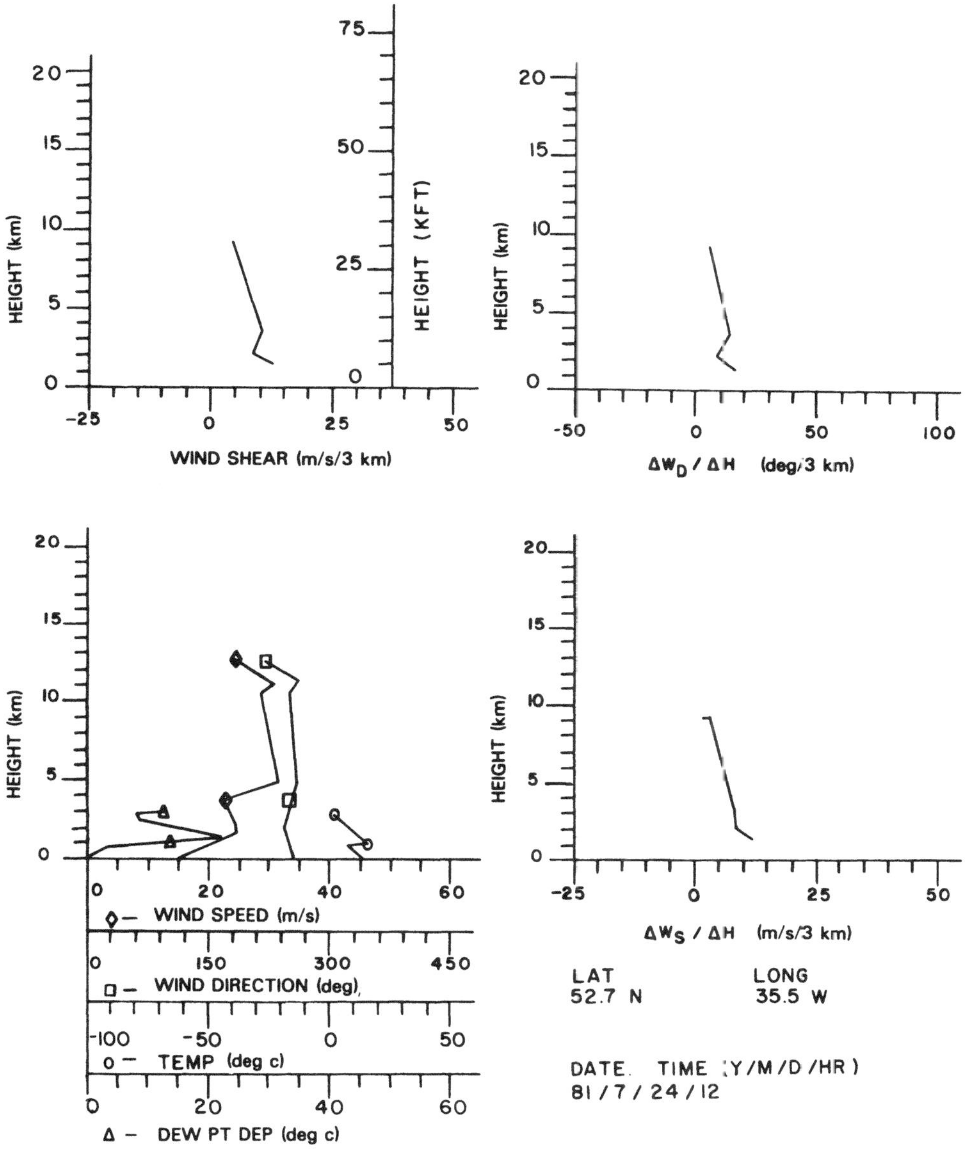

Fig. 10b — Wind shear data as in Fig. 10a, but for a less severe wind shear condition

In some cases the wind-speed increment with altitude changed sign as one continued up in altitude, i.e., a maximum value of the wind speed had been reached in less than 3 km and no further increase was going to occur with further altitude increase or decrease (e.g., 12.5 m/s over 2.5-km increment). For this case, one does not want to normalize the value to 3 km, as this would also give a more pessimistic value for the shear than is warranted. For this case, normalization would give 15.0 m/s/3 km. Since the radar would observe only a total change of 12.5 m/s in illuminating the 3-km altitude which included this 2.5-km shear, a value of 12.5 m/s/3 km is the realistic value. The data were processed according to this set of criteria, representing seasonal variations, for the three representative areas described earlier.

Each radiosonde chosen for inclusion thus yielded a maximum value for wind shear over a 3-km increment in altitude. A distribution of shear values was created, using bins between 0-0.5, 1.5-2.5, 2.5-3.5 m/s/3 km, etc., until a sufficient number of points were accumulated as previously described.

Choice of a 3-km Chaff Cloud Height for Shear Determination

As discussed earlier, 3 km was chosen as a typical height range over which to determine shear based upon the model for the spread of the chaff cloud. Since 3 km was typically the greatest range of altitude over which the chaff was distributed for the most effective cloud distribution scenario, it was felt that the chaff should actually be measured over this region for each radiosonde, rather than simply over one kilometer, and then multiplied by three. To justify this technique, we reviewed a set of 100 radiosonde records, and passed sliding windows in altitude for each set to determine the maximum shear for each radiosonde data set; these results were entered as one point on the histogram for each window that was used. Windows of 1, 2, 3, and 5 km were used and shear statistics established for each. Figures 11-14 show the histograms with a synopsis of these results in Fig. 15. The synopsis is a plot of the 10th, 50th, and 90th percentile for each of the four windows used. For each of the three percentile values, we have also drawn a straight line, scaling the result for the 1-km window with altitude [1]. In each case the actual data points fall below the scaling line, the higher the percentile. The interpretation of this is that for weak shear (low percentile values), the shear does almost scale with altitude increment in a linear fashion. However, the stronger the shear, the less likely it is to scale with the altitude linearly, indicating that strong shear is confined to layers less than 2 km thick. Hence, if one is interested in shear across 2 km or more, then one should actually calculate the shear over that height range, rather than scale the results from a very narrow window, such as one kilometer.

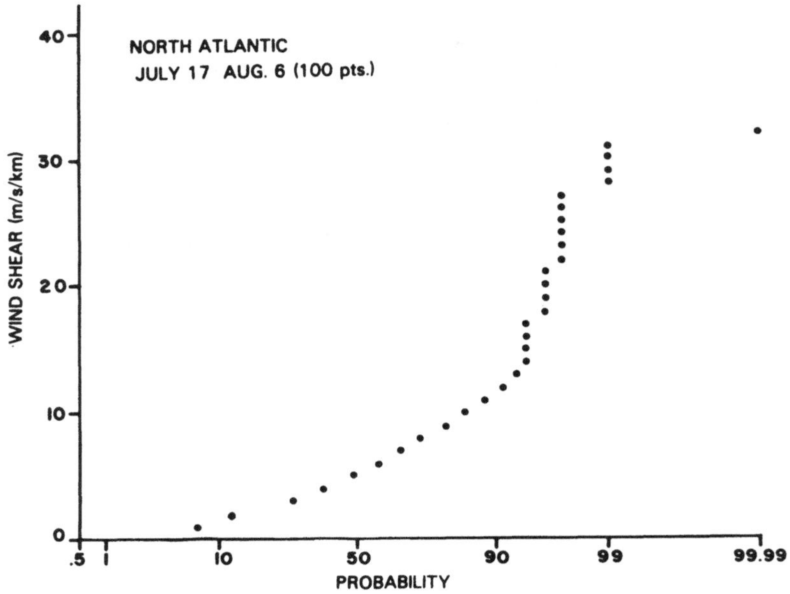

Fig. 11 — Cumulative probability of wind shear is plotted using a 1-km window to calculate the shear, as described in the text.

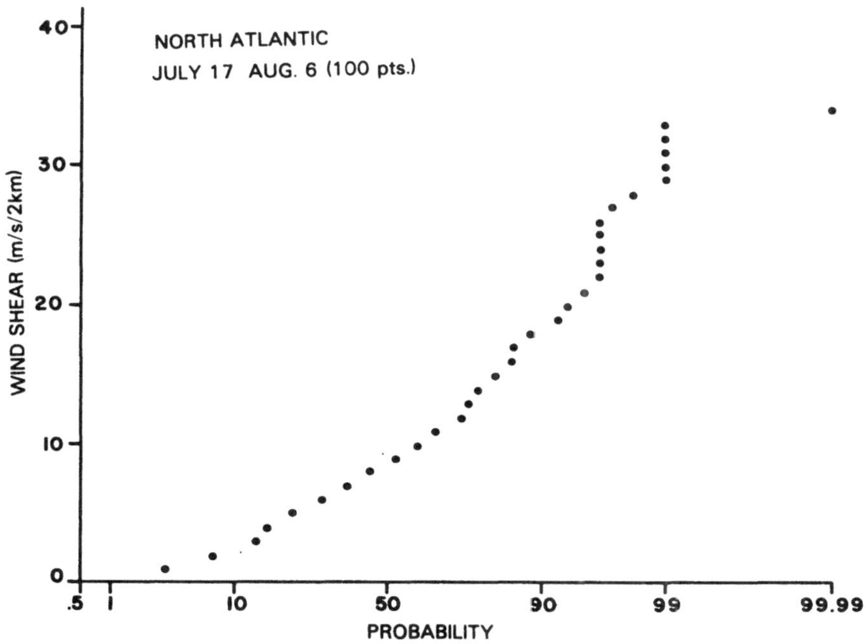

Fig. 12 — The cumulative probability plot is calculated for wind shear as in Fig. 11, but for a 2-km sliding window. It is seen that the median value, 8.7 m/s, is less than twice the median for a 1 km window, indicating wind speed shear does not scale linearly with altitude.

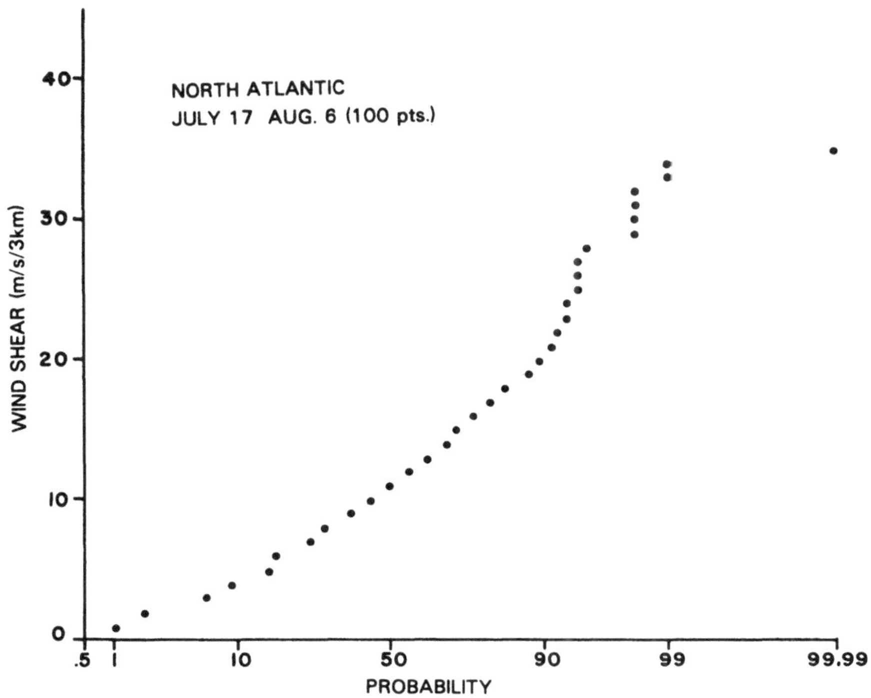

Fig. 13 — The cumulative probability plot is again calculated, but for a 3-km window, as in Fig. 11. The median value, 11.0 m/s, is again less three times the median for a 1-km window.

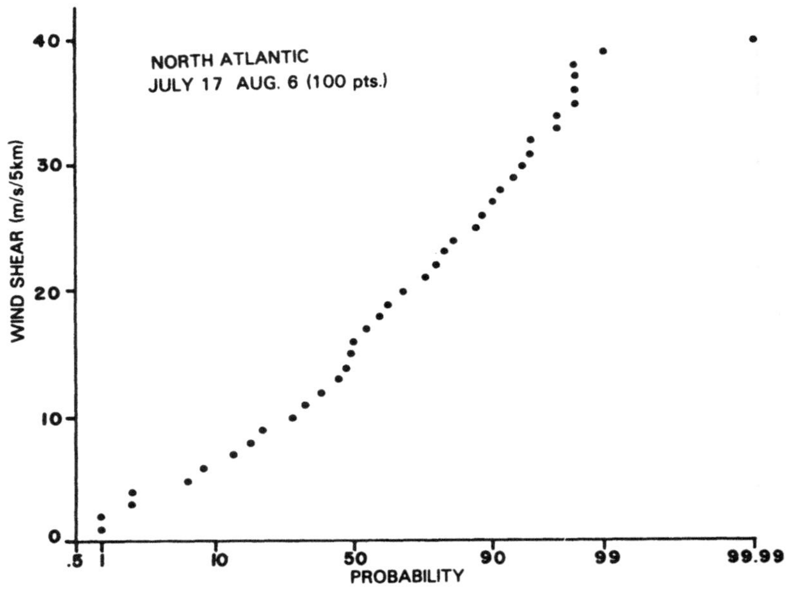

Fig. 14 — The same calculation is again performed, but for a 5-km window, with a median value of 16.0 m/s resulting, again far less than five times the median for a 1-km window.

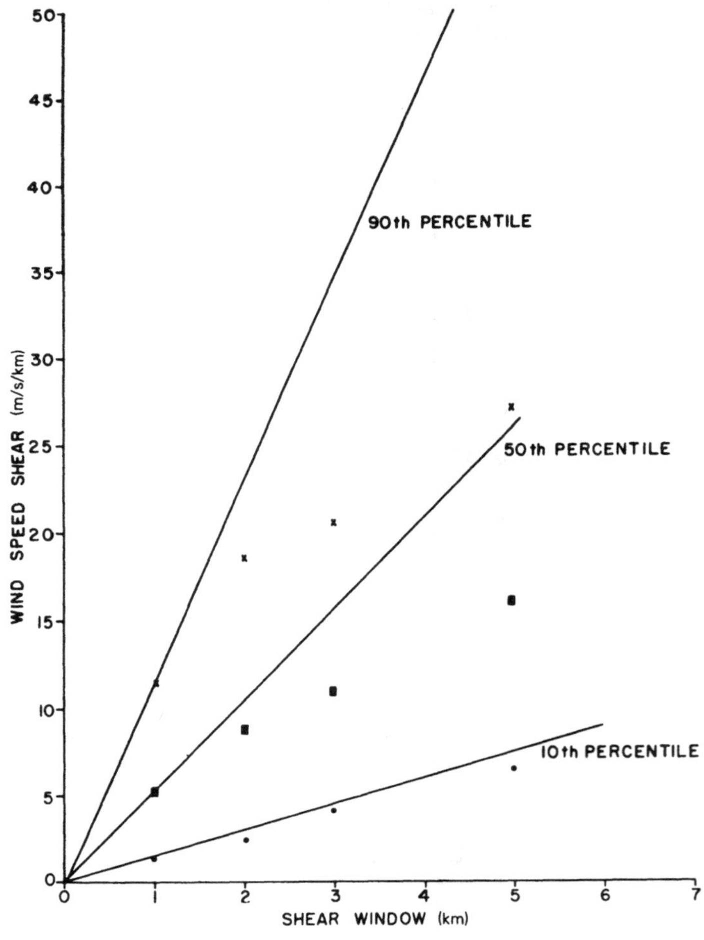

Fig. 15 — The results of the previous four figures are summarized, as plots of 10th, 50th, and 90th percentiles. The straight lines plotted are fits to the one kilometer window results, along which the results for the other windows should fall if the shear scales linearly with window size, as is usually assumed with Nathanson's results. The fact that data points fall farther from the lines the larger the windows, and the higher percentiles, indicates that very high wind shear values typically lie within altitude regimes certainly less than two kilometers thick.

STATISTICAL ANALYSIS RESULTS

Figures 16 - 29 present the results of this type of analysis for the three regions for the remainder of the data. These results are then encapsulated in the form of 10th, 50th, and 90th percentiles plotted vs time of year for the three regions in Figs. 30 - 32, which are easier to compare than the cumulative distributions themselves.

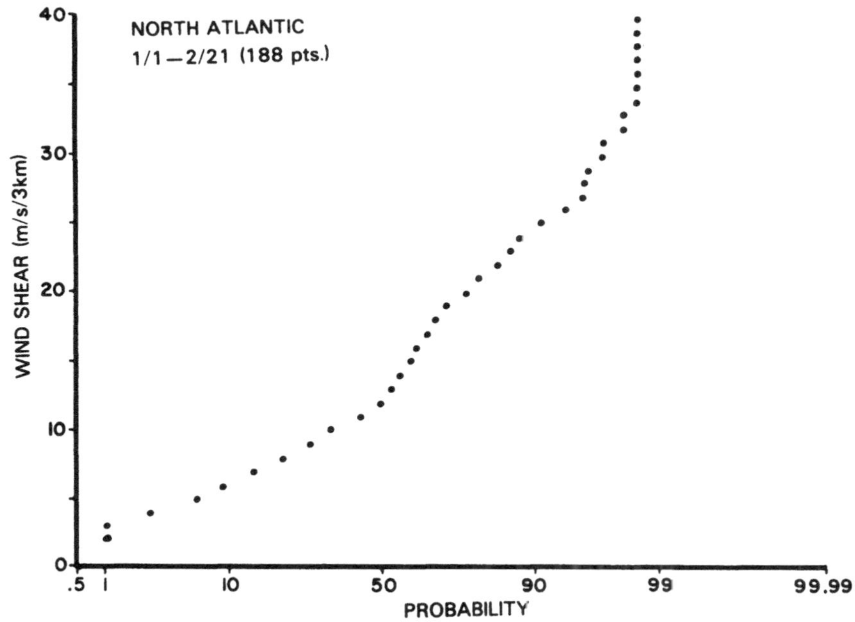

Fig. 16 — Cumulative distributions of the occurrence of wind shear calculated over a 3-km interval, as in Fig. 13, for January and February data in the North Atlantic.

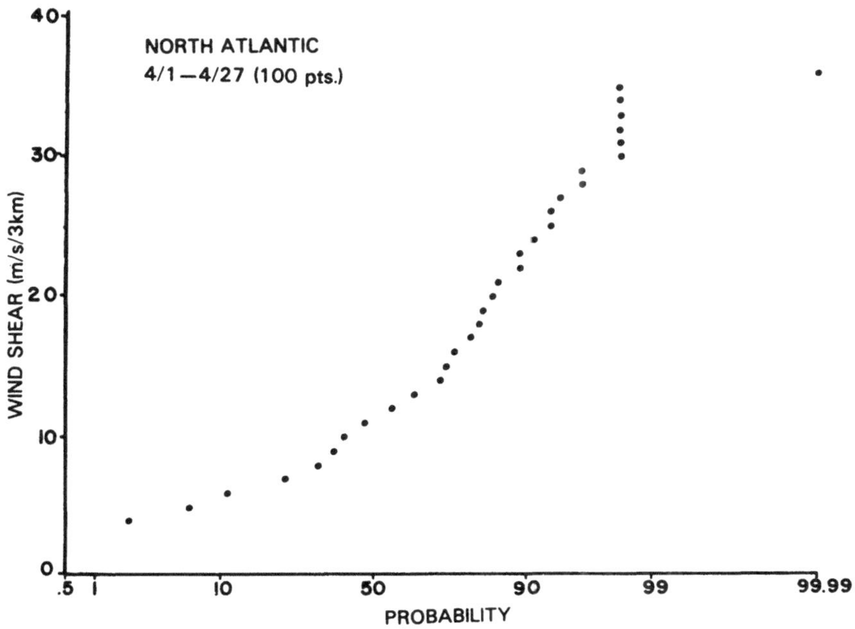

Fig. 17 — Cumulative distributions of the occurrence of wind shear calculated over a 3-km interval, as in Fig. 13, for April data in the North Atlantic.

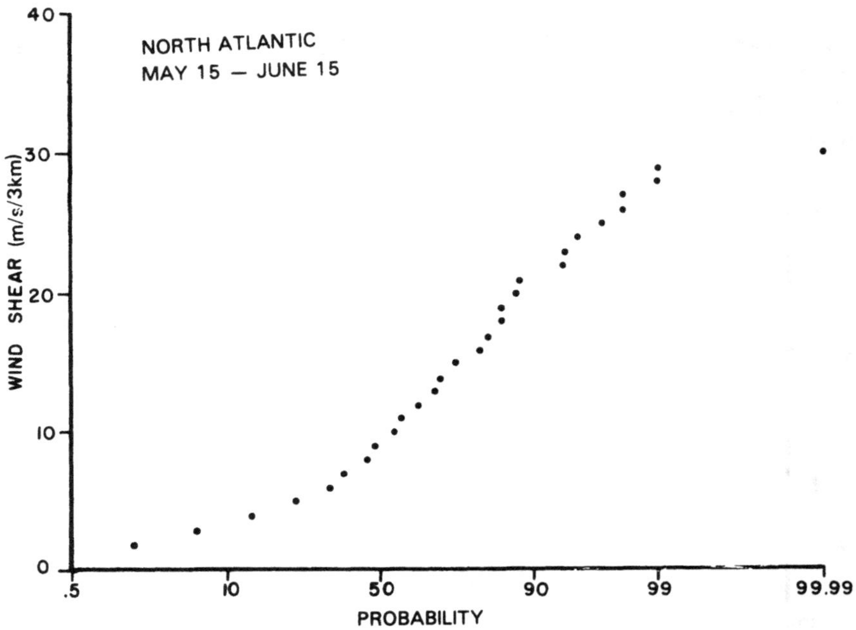

Fig. 18 — Cumulative distributions of the occurrence of wind shear calculated over a 3-km interval, as in Fig. 13, for May and June data in the North Atlantic.

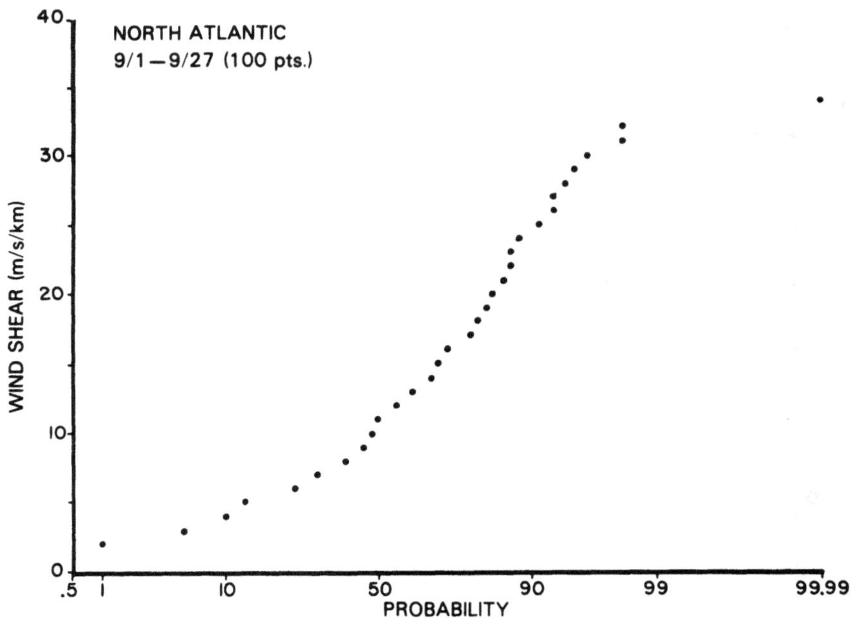

Fig. 19 — Cumulative distributions of the occurrence of wind shear calculated over a 3-km interval, as in Fig. 13, for September data in the North Atlantic.

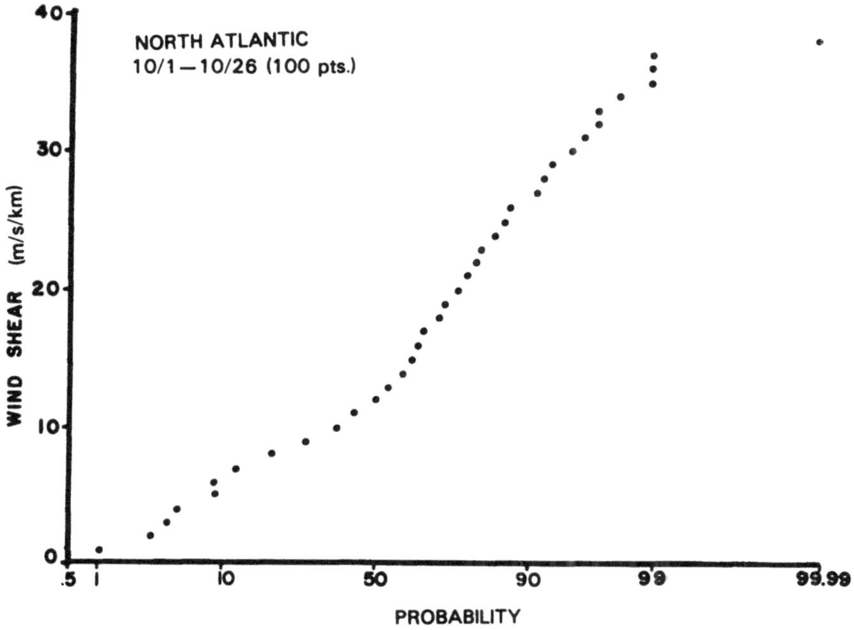

Fig. 20 — Cumulative distributions of the occurrence of wind shear calculated over a 3-km interval, as in Fig. 13, for October data in the North Atlantic.

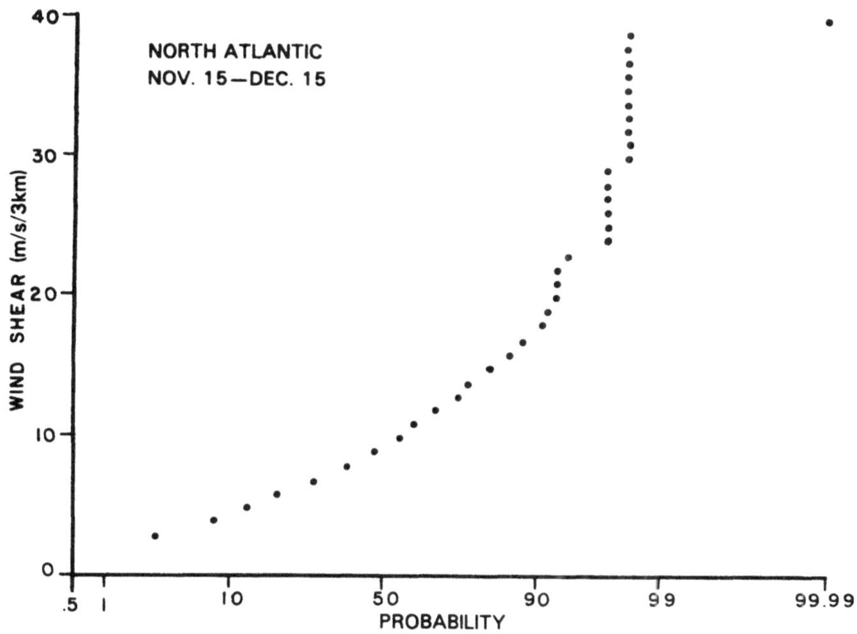

Fig. 21 — Cumulative distributions of the occurrence of wind shear calculated over a 3-km interval, as in Fig. 13, for November-December data in the North Atlantic.

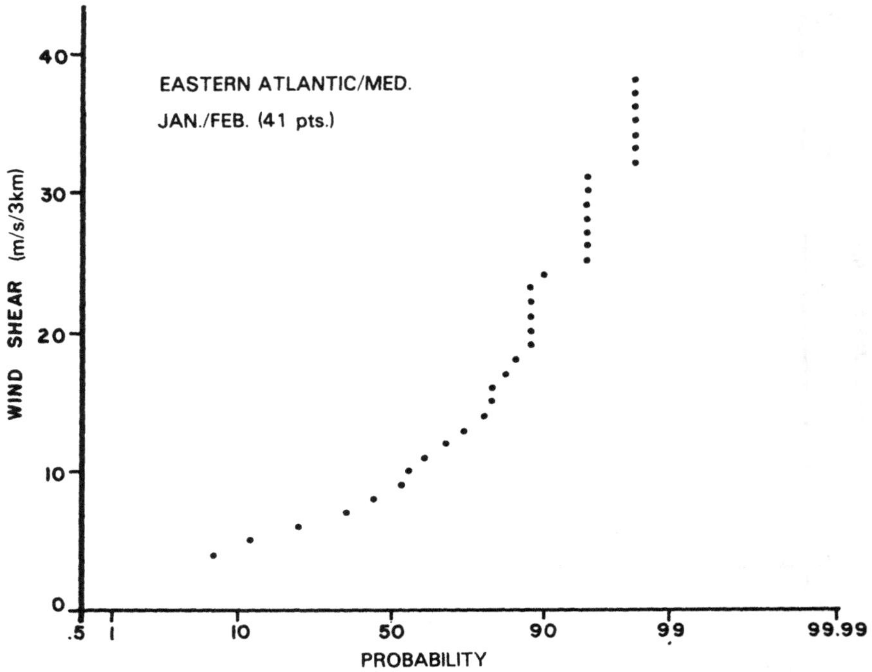

Fig. 22 — Cumulative distributions of the occurrence of wind shear calculated over a 3-km interval, as in Fig. 13, for January-February data in the Mediterranean/Eastern Atlantic.

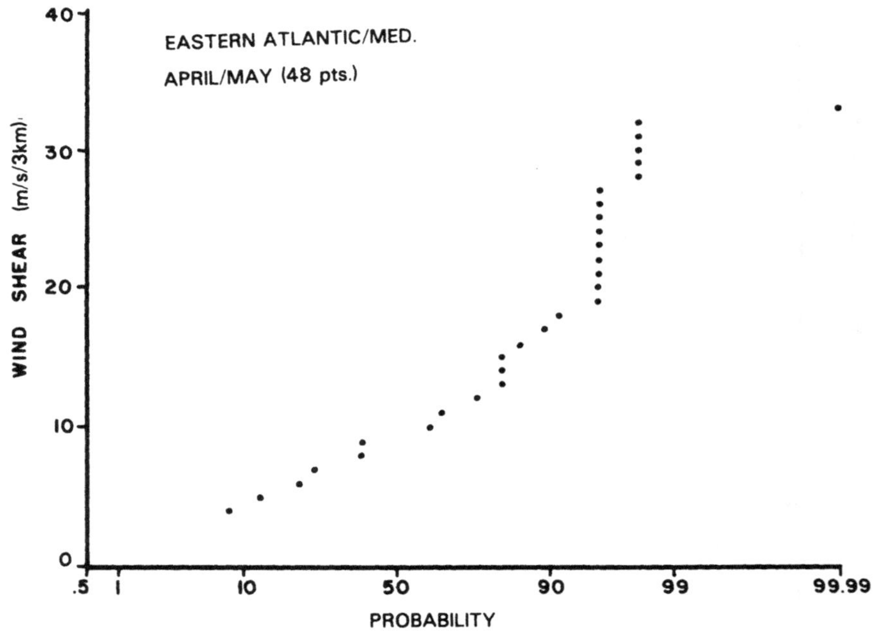

Fig. 23 — Cumulative distributions of the occurrence of wind shear calculated over a 3-km interval, as in Fig. 13, for April-May data in the Mediterranean/Eastern Atlantic.

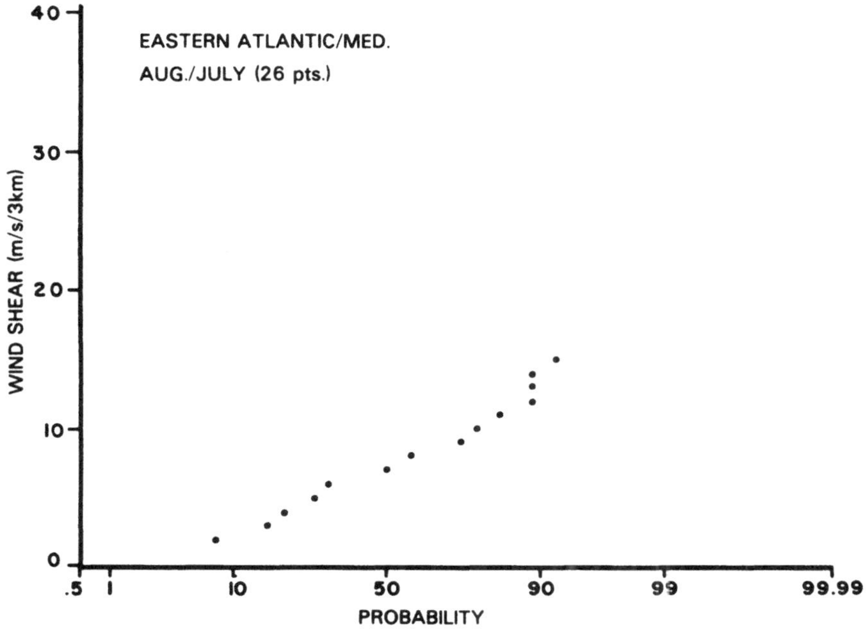

Fig. 24 — Cumulative distributions of the occurrence of wind shear calculated over a 3-km interval, as in Fig. 13, for July-August data in the Mediterranean/Eastern Atlantic.

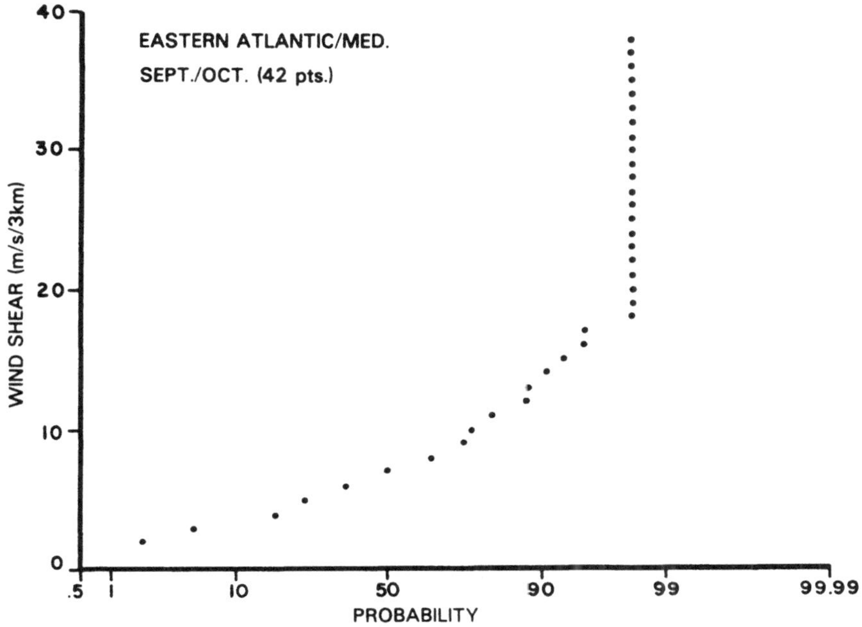

Fig. 25 — Cumulative distributions of the occurrence of wind shear calculated over a 3-km interval, as in Fig. 13, for September-October data in the Mediterranean/Eastern Atlantic.

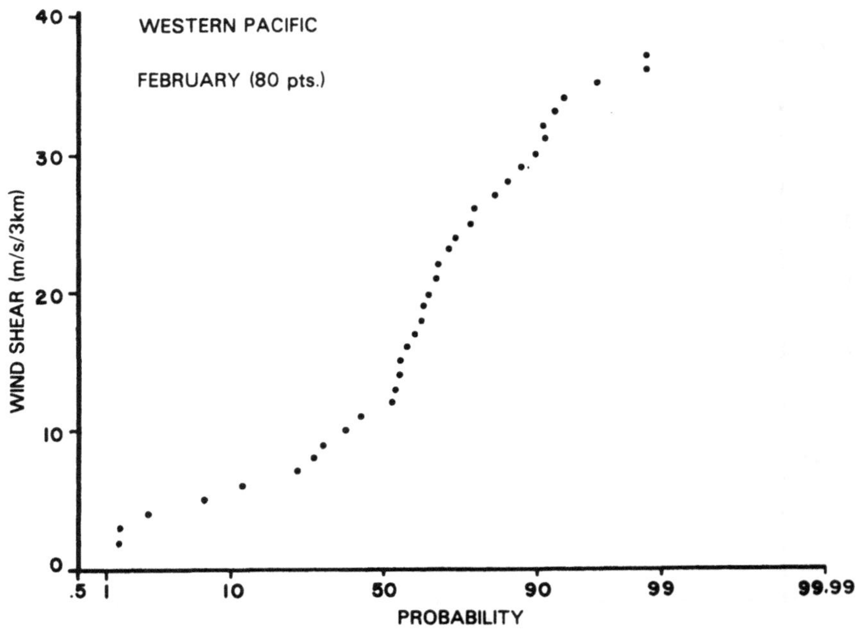

Fig. 26 — Cumulative distributions of the occurrence of wind shear calculated over a 3-km interval, as in Fig. 13, for February data in the Western Pacific.

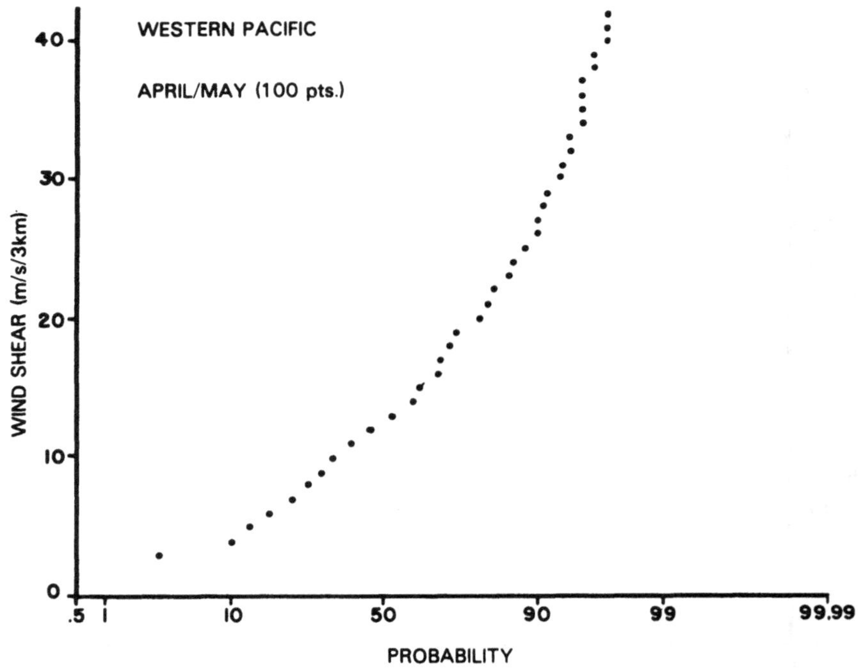

Fig. 27 — Cumulative distributions of the occurrence of wind shear calculated over a 3-km interval, as in Fig. 13, for April-May data in the Western Pacific.

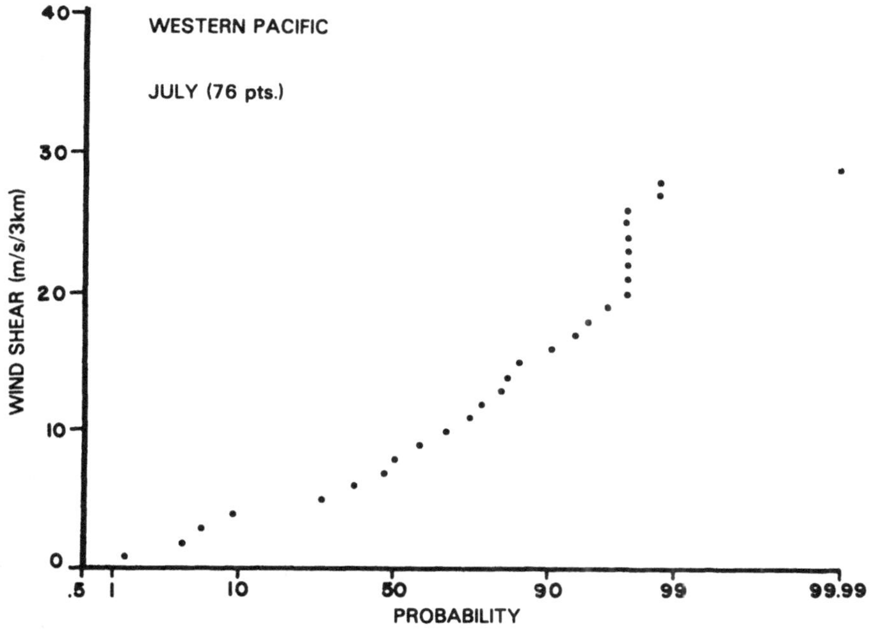

Fig. 28 — Cumulative distributions of the occurrence of wind shear calculated over a 3-km interval, as in Fig. 13, for July data in the Western Pacific.

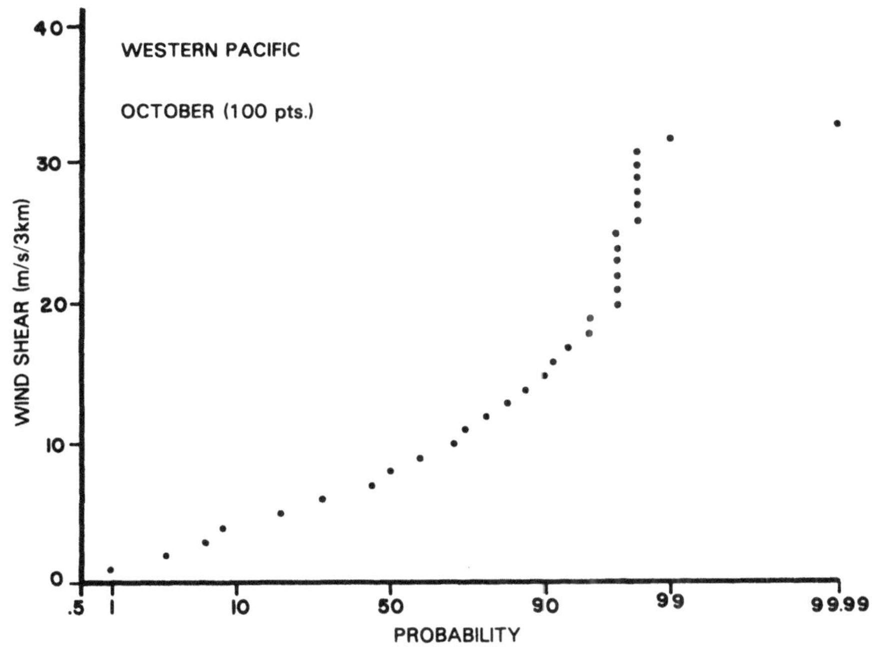

Fig. 29 — Cumulative distributions of the occurrence of wind shear calculated over a 3-km interval, as in Fig. 13, for October data in the Western Pacific.

Cumulative Distributions

The cumulative distributions in general do not show straight line slopes, which would imply Gaussian statistics. If one considers the data for January-February in the North Atlantic (Fig. 16), one sees a hump in the distribution near the 70th percentile. This is an indication of a bimodal distribution, as one might expect for unusually severe storms with high shear conditions superimposed upon the standard distribution describing the rest of the data. In this case, this secondary hump probably affects the median value of the cumulative somewhat, but its effect is felt particularly for the 70th through 90th percentiles. The data for October show a slight hump in the lower percentile region, and similar effects occur for the February Pacific data. The only conclusion to draw from this kind of effect is that the median alone is not a sufficient descriptor of the statistics of the wind shear, and that other percentile marks are necessary if one wishes to use the results for radar design considerations.

Data Summary

As previously mentioned, the data were encapsulated so that trends could be identified for the cumulative distributions. The length of the lines indicates the time over which the data were collected. For the Mediterranean and Pacific data, a period longer than a month was required to achieve meaningful statistics. Figure 30 shows seven different time periods. The first and the fourth time period also show dotted lines, which are the results for the calculation of the maximum shear over 1 km, multiplied by three, the comparison discussed in the previous section. The differences become very dramatic when displayed in this manner, particularly the 90th percentile data. The medians for the entire year lie between 10 and 12.5 m/s/3 km, slightly less than the Nathansons result of 12 m/s/3 km. The 90th percentiles show a little more variation, with values as 27 m/s/3 km reached in the late fall months.

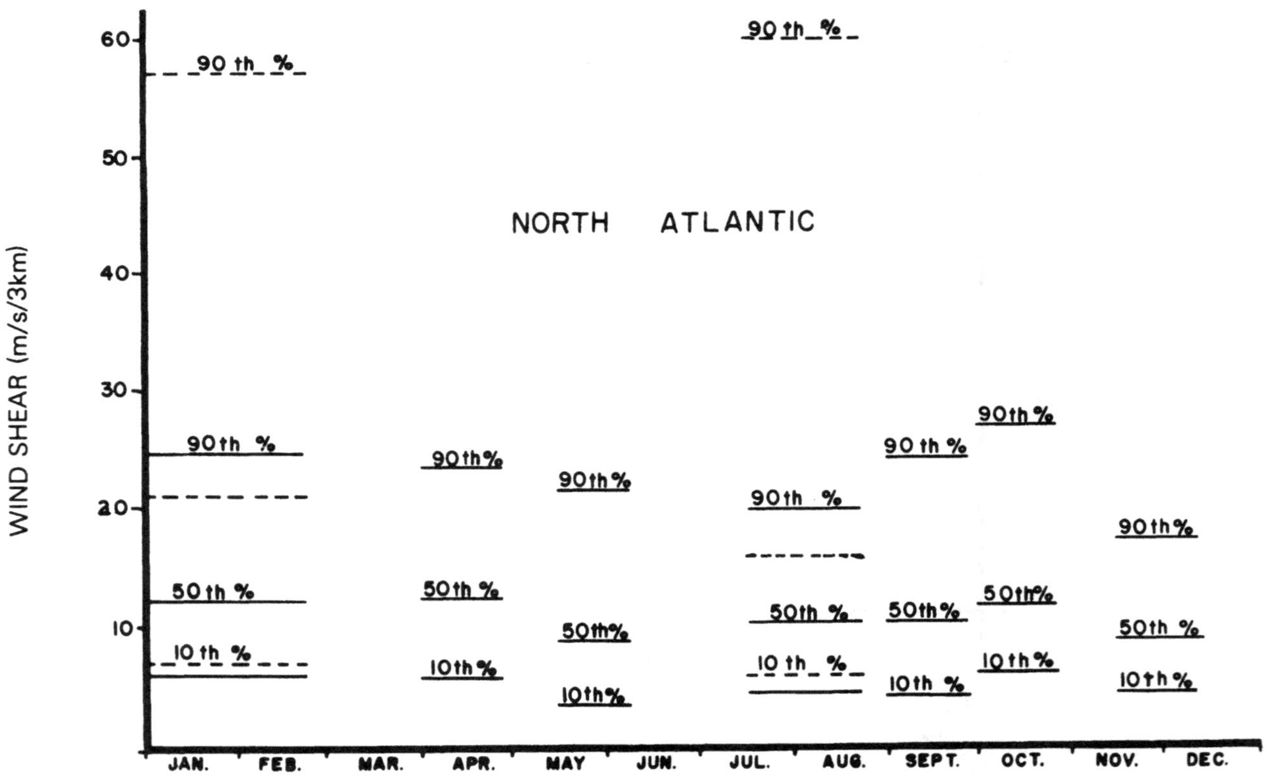

Fig. 30 — The results of Figs. 13 and 16-21 are summarized as percentiles as a function of time of year. The dotted lines represent three times the results of the calculations made for shear over a 1-km increment times three, and are seen to be much larger than the results for the same radiosonde data, but for calculations done over a 3-km interval. The latter are therefore more meaningful in the context of a radar illuminating a 3-km high chaff cloud.

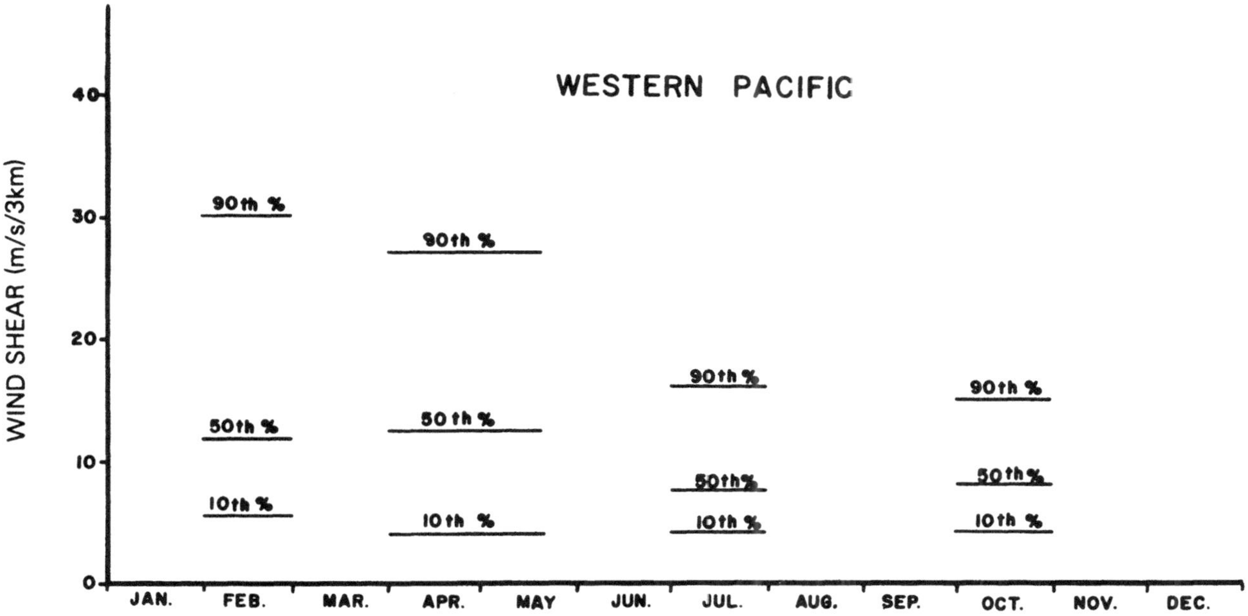

Fig. 31 — The results of Figs. 13 and 16-21 are summarized as percentiles as a function of time of year. The dotted lines represent three times the results of the calculations made for shear over a 1-km increment times three, and are seen to be much larger than the results for the same radiosonde data, but for Western Pacific radiosonde data. The latter are therefore more meaningful in the context of a radar illuminating a 3-km high chaff cloud.

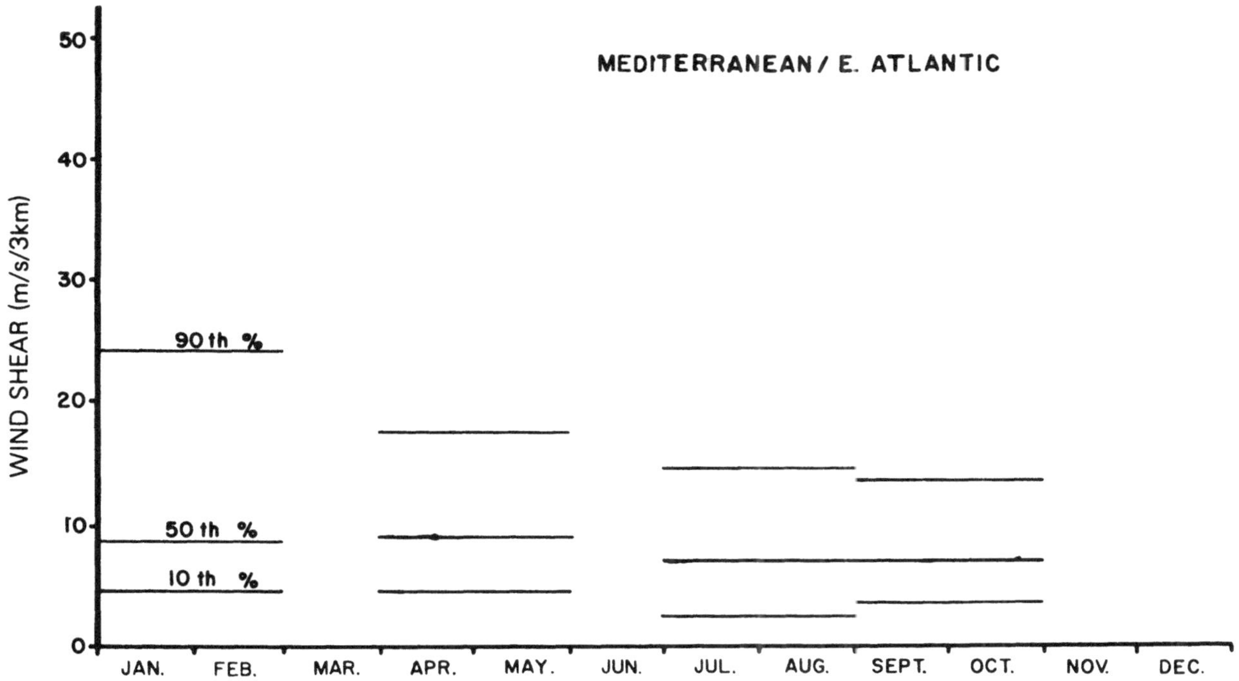

Fig. 32 — The results of Figs. 13 and 16-21 are summarized as percentiles as a function of time of year. The dotted lines represent three times the results of the calculations made for shear over a 1-km increment times three, and are seen to be much larger than the results for the same radiosonde data, but for Mediterranean/Eastern Atlantic radiosonde data. The latter are therefore more meaningful in the context of a radar illuminating a 3-km high chaff cloud.

The results for the Western Pacific in Figure 31 are similar, but with medians different than the North Atlantic, ranging from 7.5 to 12.5. The largest value for the 90th percentile of all the data is reached in February, 30 m/s/3 km. The seasonal variation is different from the North Atlantic data as well, as one might expect. The Eastern Atlantic/Mediterranean data in Fig. 32 show the weakest shear, with medians between 7 and 8.5 m/s/3 km through the year, with a seasonal variation similar to the Pacific data. Evidently the winter storms which are created over the North American land mass and run through the North Atlantic do not affect the Eastern Atlantic, which has weather variation independent of these systems.

Radar-Wind Aspect and a Yearly Average

The results so far are worst-case, in the sense that they assume the radar is looking into the wind for all cases considered, which would not be true for a typical search radar scenario with full 360° rotation. Since the wind shear due to wind-speed difference alone is modified by a term which is the sine of the difference angle (radar aspect minus wind direction), it is appropriate to average the sine function over one-half a period, normalize to that period, and multiply the median shear values by the result, which is 0.637 or roughly 2/3 of the worst case value.

If one further averages over periods of the year the values of Figures 30 - 32, one can determine a single yearly median value, typical for each of the three areas considered; they are:

North Atlantic:	6.9 m/s/3 km	2.3 m/s/ km
Western Pacific	6.3 m/s/3 km	2.1 m/s/ km
Med/E. Atlantic:	5.0 m/s/3 km	1.7 m/s/ km

These values are much less than those given in Ref. 1 using a similar analysis technique, but using data sources not as typical of conditions which are encountered at sea. They are of the same order, in fact, as the shear contribution by turbulence for conditions between cumulus clouds and severe rainstorms. Of course, the shear encountered for rainstorm conditions is not included in this work because of the nature of the data collection scenario which precludes radiosonde launch during rainstorms or heavy weather with rain present.

MODIFIED ENVIRONMENTAL DIAGRAM

Although the statistical representation of the data discussed so far is useful for general design criteria, it is useful to consider the effects of shear for particular radiosonde readings in greater detail, particularly in so far as how a given shear condition will affect the radar operation via the environmental diagram of Fig. 1. For a given radiosonde reading, and using the cloud models discussed earlier, one can estimate worst-case effects of chaff vs radar range for different cloud-spreading scenarios. We limit ourselves to the case of the final chaff aircraft range terminating at 20 nmi. Of course the cloud will continue to spread and fall with time, so that the results we present represent only one instant in the evolution of the cloud, that given by the chaff aircraft at its shortest typical distance to the radar. In our modeling of the environmental diagram, we shall assume the chaff cross section is of significant amplitude for all doppler bins which it affects, i.e., that the backscatter amplitude from the chaff at a given wind speed is larger than any target at that radial velocity and range. Hence, for a cloud extending between 5 and 10 km which experiences winds between 20 and 40 m/s, all doppler bins from 20 to 40 m/s will present a larger chaff return in the doppler spectrum than any target present. This is obviously a worst-case scenario; cross section vs altitude has been studied to some degree and should be included in a more refined model.

We have calculated eight different environmental diagrams corresponding to the two sets of radiosonde data in Figures 10a and b, and the four models of the chaff cloud (2 aircraft speeds at two altitudes), and these are presented in Figs. 33 to 40. The theoretical and experimental results of Figures 2 and 3 were used to generate the dotted and solid plots of these figures, respectively. We have plotted the bins occupied by chaff for all ranges illuminated by the radar for the four different speed/height configurations, indicating which doppler bins are unavailable for target detection due to chaff. The plots run from 0 to 300 nmi, with the dotted areas representing folded ambiguous ranges for a maximum unambiguous radar range of 200 nmi. Target doppler radial speeds are plotted on the ordinate, with no regard to ambiguous doppler at this time, assuming that the radar PRF and operating frequency are such to allow 50 m/s to be displayed unambiguously. The same radiosonde is assumed valid over all ranges. This assumption of horizontal homogeneity is used for duct propagation applications, and is sufficient for our needs.

These results indicate, first of all, that the effects of spreading chaff vary depending on height and aircraft velocity one uses. Figure 33 represents the highest flying, slowest aircraft, which appears to be the most effective in filling the doppler-range area. Because the aircraft is above the highest altitude indicated in the radiosonde of Fig. 10a, the shortest ranges show 14 m/s chaff velocities, but with no spread. (Actually, the spread due to turbulence discussed earlier would be observed at these shortest ranges, a few m/s). As the chaff falls to lower altitudes, such that the top and bottom of the chaff cloud intercept a range of velocities, the chaff return spreads in doppler in the environmental diagram. The greatest spread in doppler is observed for this first case because of two factors when compared to the following cases: the higher altitude allows the chaff to interact with the higher shear experienced at these higher altitudes, and the slower velocity allows the chaff cloud to fall a longer distance while the aircraft moves horizontally.

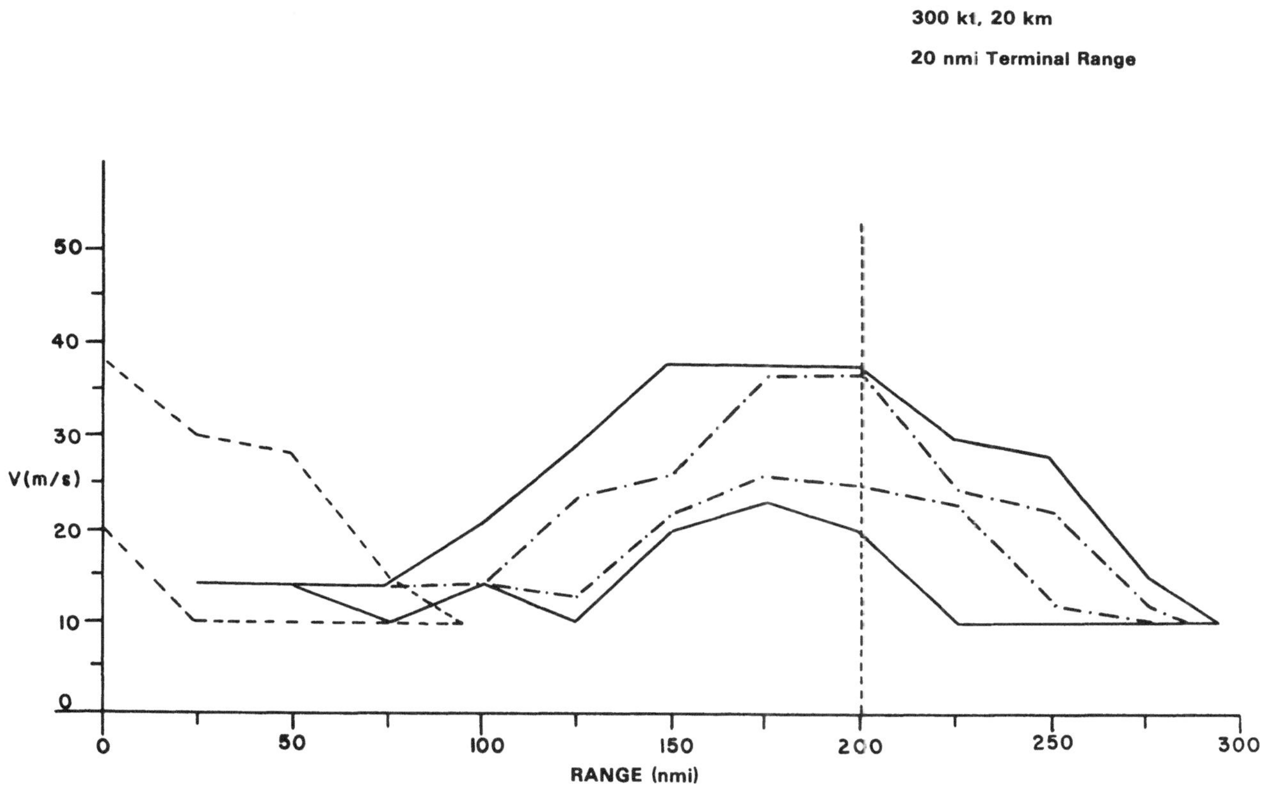

Fig. 33 — Doppler-range display, or Nathanson's environmental diagram of Fig. 1, with a realistic estimate of doppler bins affected by chaff returns which was generated by combining the radiosonde of Fig. 10a with the chaff cloud model of Fig. 2, for an altitude of 20 km. This cloud is expected to produce the greatest area affected on the doppler-range display because it results in the greatest altitude dispersion of the four chaff cloud models. For a given range on the cloud model, the actual maximum and minimum altitudes of the cloud were determined and the range of doppler frequencies (or velocities) encompassed by this altitude range was determined from the radiosonde wind speed data.

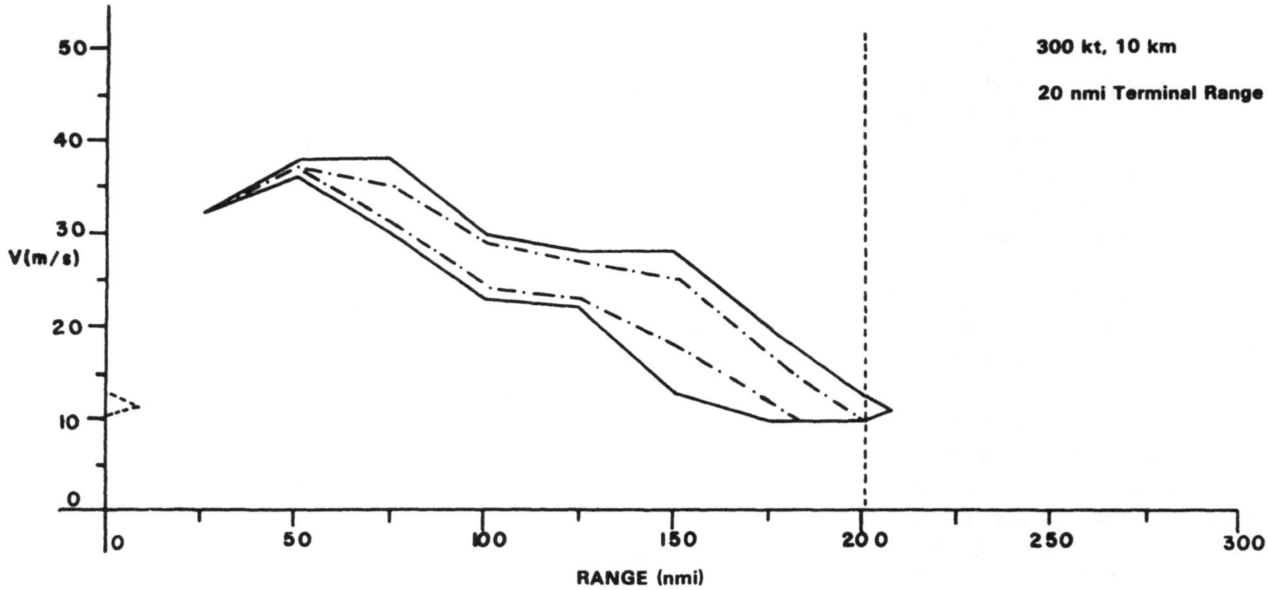

Fig. 34 — Doppler-range display, or Nathanson's environmental diagram of Fig. 1, with a realistic estimate of doppler bins affected by chaff returns which was generated by combining the radiosonde of Fig. 10a with the chaff cloud model of Fig. 2, for an altitude of 10 km. This cloud is expected to produce the greatest area affected on the doppler-range display because it results in the greatest altitude dispersion of the four chaff cloud models. For a given range on the cloud model, the actual maximum and minimum altitudes of the cloud were determined and the range of doppler frequencies (or velocities) encompassed by this altitude range was determined from the radiosonde wind speed data.

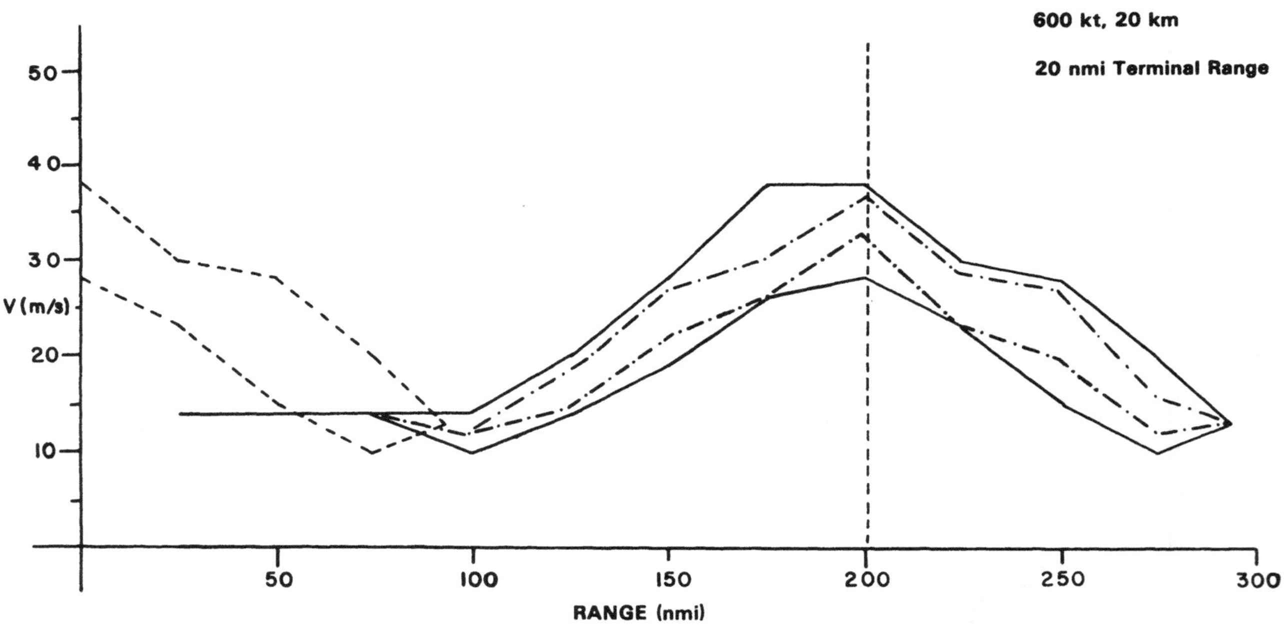

Fig. 35 — Doppler-range display, or Nathanson's environmental diagram of Fig. 1, with a realistic estimate of doppler bins affected by chaff returns which was generated by combining the radiosonde of Fig. 10a with the chaff cloud model of Fig. 4, for an altitude of 20 km. This cloud is expected to produce the greatest area affected on the doppler-range display because it results in the greatest altitude dispersion of the four chaff cloud models. For a given range on the cloud model, the actual maximum and minimum altitudes of the cloud were determined and the range of doppler frequencies (or velocities) encompassed by this altitude range was determined from the radiosonde wind speed data.

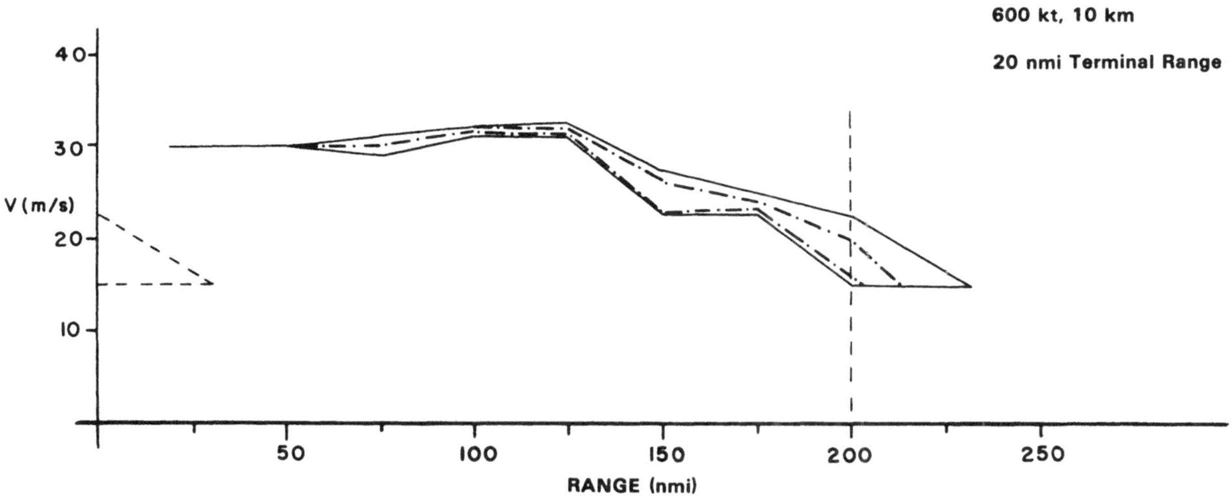

Fig. 40 — The chaff cloud model of Fig. 3, for 10 km, is again used to calculate and environmental diagram, but for the radiosonde data of Fig. 10b. These data show an example of shear near the mean value for the season considered, and the environmental diagram shows far less doppler-range area covered than did the result in Fig. 36, derived for a much more severe wind shear condition.

Figure 34 shows the case of the same 300-knot aircraft speed, but flying at an altitude of just 10 km. The spread in doppler is much less for this radiosonde data. Figure 35 shows the case of the aircraft at 20 km again, but with a velocity of 600 knots, so that the cloud is not falling and spreading as far as the aircraft advances rapidly toward the radar. Figure 36 shows the case for flying at just 10 km, and at a ground speed of 600 knots, again giving a relatively narrow spread in doppler frequency.

Figures 37-40 give similar results, using the shear data of Figure 10b.

The most important point to note from these diagrams is that the results plotted in histogram form earlier are worst-case wind shears determined for a 3-km altitude range for a given radiosonde, but that this worst-case statistic will apply only for one or a very few range bins for a given environmental diagram. For the radiosonde considered here, which was a relatively high shear case, the maximum shear indicated in the calculation used to determine the histogram statistics, 20 m/s/3 km, occurred only at 250 nmi range, and only for the cloud associated with the slowest aircraft speed and highest velocity. From this one can conclude that worst case wind shear will be observed in only a fraction of the range bins processed by the radar, if any, even for worst case aspect considered here.

SUMMARY

We have reconsidered the problem of wind shear over the ocean as it relates to radar chaff applications. Considering radiosonde data from three different operationally interesting ocean areas, we calculated seasonal histograms for wind-shear effects due to wind-speed contributions only, since the wind-direction contributions were estimated as being primarily due to measurement error in the data, with true major directional effects being important only for relatively mild wind conditions. A rigorous analysis should include these effects if the data quality allows it. Taking radar aspect effects into account, median yearly values of wind shear averaged over 3-km altitudes, but expressed as 1-km values for comparison with previous works, are 2.3, 2.1, and 1.7 m/s/km for sites in the North Atlantic, Western Pacific, and combined Mediterranean/Eastern Atlantic. Furthermore, detailed considerations of the environmental diagram, or doppler-range radar display, indicate that when severe shear does occur over a given altitude regime with a worst-case aspect angle, it may in fact affect the radar doppler at only a small selection of ranges, and perhaps not as severely if the chaff distribution is not conducted in an optimum manner. In summary, it appears that the Ref. 1 model is probably too severe for radar design applications and these results present a more realistic assessment of wind-shear shear effects on chaff distribution.

REFERENCES

1. Nathanson, F., and J.P. Reilly, "Clutter Statistics which Affect Radar Performance Analysis," *IEEE Trans. Aerospace & Electronic Systems*, **AES-3**, 386-398, 1967.

2. G.A. Silverman, "A Theory of Chaff Dispersion and an Estimate of the Chaff Threat to the SAM-D System," (U) Tech. Rpt. ECOM-ST14, U.S. Army Electronics Com., Ft. Monmouth, NJ, 1975. (Secret)

3. J.E. Juisto, and W.J. Eadie, Terminal Fall Velocity of Radar Chaff, *Journal of Geophysical Research*, **9**, 2858-2861, 1963.

4. R. Rabin, "Observations of the Optically Clear Boundary Layer Using Doppler Radar," *Proc. 19th Conf. Radio Meterology, American Meteorological Society*, Boston, 1981.

5. B.R. Bean and E.J. Dutton, *Radio Meteorology*, Dover Publications, N.Y., 1968.

Analysis of rain clutter data from a frequency agile radar

William B. Gordon

Radar Division, Code 5311, Naval Research Laboratory, Washington, D. C. 20375

(Received November 4, 1981; revised March 15, 1982; accepted March 15, 1982.)

Measurements of rain clutter were taken with a frequency agile *S* band radar for the purpose of studying the Doppler power spectra and the variation of backscattered power with space, time, and frequency. The clutter was found to be nonhomogeneously distributed in both space and time, but the mean backscattered power versus frequency relation seemed to involve only a single parameter having a spatial and temporal variation. Although the radar was looking into the surface wind, a significant portion of the backscattered signal energy was always found to lie in the negative frequency range, and the Doppler power spectra were sometimes found to have multiple peaks at both positive and negative Doppler frequencies. The simplest explanation for such spectra is that of a wind shear operating on a vertically stratified distribution of clutter, and assuming this to be the case, it can be estimated that the wind shear was of the order of 8 m s^{-1} km^{-1} and that the peaks were produced by horizontal bands or clumps of clutter whose vertical extent was of the order of 200 m.

1. INTRODUCTION

The rain clutter data presented in this report were taken with a frequency agile *S* band radar whose parameters are shown in Table 1. Five equispaced frequencies from 2.1 to 3.9 GHz were transmitted and varied from pulse to pulse in such a manner that every fifth pulse had the same frequency. A single antenna is used for both transmission and reception over the entire range of frequencies. The data was taken on September 5, 1979, at the Randle Cliff facility of the Naval Research Laboratory from the remnants of hurricane David. The wind at the surface was from the NE at approximately 12 m s^{-1}. The facility is located on the western shore of the Chesapeake bay, and the radar coverage lay in the northeast quadrant, looking over the Bay and into the wind. The rain was heavy and widespread, but the 'clutter' might also have included the return from suspended water cloud droplets.

Our main concerns are the functional relationship between the mean backscattered power and frequency, the spatial and temporal variation of the backscattered signal, and the Doppler power spectra. Among our results, those concerned with the clutter Doppler spectra have the greatest potential importance for the radar engineer, since it is the properties of these spectra which determine the effectiveness of clutter rejection filters. A Gaussian model for rain clutter spectra is generally accepted by the radar community (see e.g., *Nathanson*, [1969]), and the early experimental works of *Barlow* [1949] and *Kerr* [1951] are still cited as primary sources on this subject. However, we found that the Doppler spectra sometimes had multiple peaks, and we would like to learn something about the nature of the processes which can produce such multipeaked spectra and how frequently they can occur. The simplest explanation for the occurrence of such spectra is that of a wind shear operating on a vertically stratified distribution of clutter, but there are other possibilities. One such possibility frequently discussed in the literature is that of Bragg-type scattering from turbulent eddies, but the observed variation of backscattered power with space, time, and frequency seems to indicate that although Bragg scattering effects may have been present, they were not a major factor in the production of multipeaked clutter spectra.

From the spatial and temporal variation of certain statistics we found that the mean backscattered power versus frequency relation conformed to the model $P(f) = AQ(f)$, where $Q = Q(f)$ is some function which depends only on frequency and A is a frequency independent parameter which varies with time and position. For the Bragg-type scattering ef-

Paper number 2S0412.

TABLE 1. Radar Parameters

Frequencies, GHz*	Pulse Repetition Frequency	Beamwidth at 3.0 GHz	Pulse Width	Antenna Rotation, rpm	Polarization
2.1	1600/s	1° azimuth	compressed to 0.7 μs	0	vertical
2.55		2° elevation		2	
3.0				6	
3.45					
3.9					

* Varied from pulse to pulse.

fects discussed in the literature we have $Q(f) = f^n$, where the exponent n varies between 0 and 2. But from the data we conclude that the exponential rate of growth of power with frequency lies in the range $3 < n < 7$, with the preferred value being approximately $n = 5$, which happens to be the value of n which corresponds to incoherent scattering from narrow horizontal bands. This analysis is complicated by the possible existence of calibration errors whose effects cannot be 'averaged out' because they are constant and multiplicative. For this reason (as well as others) we do not estimate the value of n by 'fitting a curve' to the data in the usual sense of least squares regression. Instead, we calculate the smallest rms calibration error that would be required to produce the departure of measured values of n from an assumed 'true' value. We judge that the rms calibaration error was no larger than 2 dB (with the exception of a certain 'glitch' at the highest frequency, described in section 3), and values of n lying in the range $0 \leq n \leq 2$ are rejected as reasonable possibilities because they would imply an rms calibaration error greater than 3 dB.

A final remark concerning methodology: we are not very interested in establishing trends in data which are heavily averaged over space and time, and stationarity is assumed only for data originating from a single range cell over a short duration of time.

2. ORGANIZATION OF THE DATA

As previously mentioned, our analysis involves a close examination of the spatial and temporal variation of certain statistics, and it is therefore necessary to go into some detail concerning the organization of the data. The basic data consist of I and Q channel voltages, which are squared and summed to form values of instantaneous power. The data are organized into 'files,' each of which consists of data taken from a number M_c of contiguous range resolution cells within a given azimuthal sector, with different files consisting of data taken from different regions of space. Each file consists of a number M_r of 'records,' each of which is a block of data taken over a fixed time interval. In effect, the files consist of data from a selected clutter cell appearing on the PPI scope, and the records consist of as much data (28,130 words) as can be stored into a minicomputer before being dumped onto tape.

More specifically, each radar pulse gives rise to a block of data consisting of M_c data words (one for each range cell), and each word consists of two 8-bit bytes recording the I and Q channel voltages. There are also two words devoted to frequency and azimuth information, and hence the total number N_p of pulses transmitted per record is given by

$$N_p = \frac{28{,}130}{2 + M_c}$$

The time duration of each record is equal to $N_p/1600$ s, since the pulse repetition frequency is always 1600 Hz, as indicated in Table 1.

The basic blocks of data on which we operate consist of single records from single range cells. The radar pulse width of 0.7 μs converts to a (spatial) range cell width of 105 m, and the time duration (length) of each record is rather short, being either 0.27 s or 1.76 s. For reasons to be explained, the time interval between consecutive records is usually much larger than the record length. Our working assumption is that the backscattered voltages from each single record from each single range cell consist of samples drawn from a stationary process characterized by its second order statistics (mean power and autocorrelations), but from the nature of

TABLE 2. File Descriptions

File Number	General Description	Antenna Rate, rpm	Range km	Azimuth	Elevation	Number of Records	Time Duration Record, s	Number of Range Cells
1	Multifrequency	6	18.95	359°	3.6°	10	0.27	64
2	Multifrequency	6	29.20	57°	3.6°	10	0.27	64
3	Single-frequency	6	29.20	57°	3.6°	5	0.27	64
5	Multifrequency	2	31.78	354°	3.2°	5	1.76	8
6	Multifrequency	0	33.34	40°	3.2°	10	1.76	8
8	Multifrequency noise file	...	...	...	...	5	0.52	32

the data we have to allow that these statistics vary from record to record and range cell to range cell.

There are some other details that have to be mentioned:

For reasons not connected with this experiment, the data are sampled at a rate of 3 MHz, which is twice the reciprocal of the compressed pulse width of 0.7 μs. Hence there is a 50% overlap between successive range cells.

When the antenna is rotating, the recording device is triggered to take data centered at a given azimuth. Hence, for example, there is a 10 s delay between records when the antenna is rotating at 6 rpm. When the antenna is not rotating, there is approximately a 2 s delay between records.

The data files are either 'single-frequency' or 'multifrequency'. The single-frequency files always use the carrier frequency of 3.0 GHz, whereas the multifrequency files use each of the five frequencies shown in Table 1. In the multifrequency case the frequencies are varied in order from pulse to pulse, and hence the pulse repetition frequency (PRF) for pulses at each given frequency is reduced from 1600 to 320 in the multifrequency case.

Table 2 gives the file descriptions for the files referred to in this report. Of special interest is the file numbered 8, which was obtained by turning off the transmitter. This file will be referred to as the 'noise' file and is used to measure the receiver noise statistics. The file numbered 3 is the only single-frequency file used in this report. Figure 1 shows a schematic PPI display of the regions covered by the four multifrequency files 1, 2, 5, and 6. Owing to the effects of antenna rotation, the azimuthal coverage of files 1, 2, and 5 is much larger than the antenna beamwidth.

3. SYSTEM CALIBRATION

Small systematic biases were detected in the voltages, which appeared to be the same in the noise file as in those files with signal present. These biases were thereafter removed numerically, and the bias corrected voltages thus obtained were found to conform to the usual Gaussian model. In particular, the cross correlation between the two channels was found to be negligible [*Gordon and Wilson*, 1982].

Previous measurements had shown that the transmitter signal-to-noise power ratios were 38 dB or better over the entire frequency range. Hence the transmitter noise was not an important factor in our data analysis, since the corresponding receiver noise levels were orders of magnitude larger.

For each of the five frequencies used, measurements were made of the antenna gain, receiver gain, transmitted power, and cable losses. These measurements were used to obtain 'calibration factors' $k_i (i = 1, 2, 3, 4, 5)$ which are proportional to the reciprocals of the total system response at each frequency. The calibrated values of power shown in the text are obtained by multiplying the 'raw' values by the factors k_i. These 'calibrated' values are given in terms of

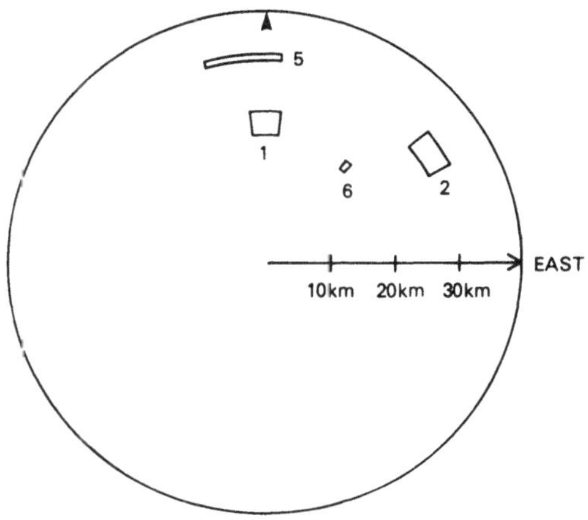

Fig. 1. Coverage areas for files 1, 2, 5, and 6.

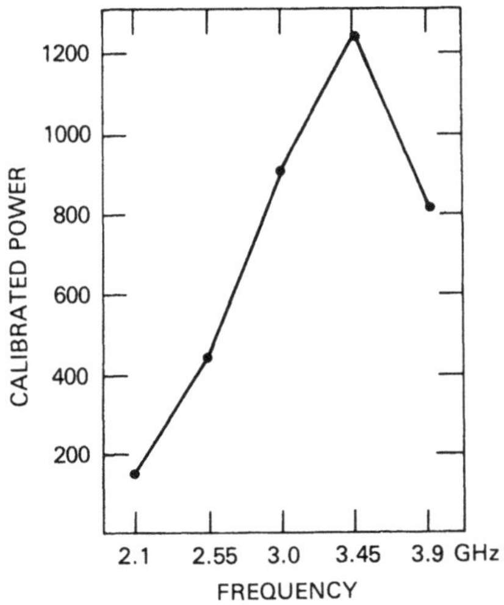

Fig. 2. Calibrated power versus frequency.

unspecified units of power, since no attempt was made to measure the power in watts. Unfortunately, we were not able to use a test sphere in the calibration.

The power versus frequency curve shown in Figure 2 was obtained by averaging all the data in file 6; however, the curves obtained from data in any individual record from any individual range cell all have the same general appearance. In particular, there is always a sharp drop-off in power at the highest frequency, 3.9 GHz. A spectrum analyzer was used in an attempt to discover malfunctions in the signal synthesizer or transmitter at this frequency, but none was detected, and we therefore provisionally ascribe this anomaly to an 'unidentified calibration error.' In the sequel it will be seen that the existence of this anomaly has no effect on the nature of our conclusions; in particular, the returns from the first four frequencies were sufficient to exclude Bragg scattering as a significant factor in the production of the Doppler spectral peaks.

4. SPATIAL, TEMPORAL, AND FREQUENCY VARIATION

Notation

There are three types of averages which appear in our subsequent discussion, and which are indicated by the use of boldface type, single bars, and double bars. To illustrate, we consider the three types of 'mean power'.

For each of the five frequencies f_i, $\mathbf{P}_i = \mathbf{P}_i\ (r, c)$ denotes the theoretically 'true' or ensemble mean power which is characteristic of the return at the ith frequency from the rth record from the cth range cell. The sample mean $\bar{P}_i = \bar{P}_i(r, c)$ is the numerical average of the N values of instantaneous power contained in the rth record from the cth range cell, where $N = 85$ for files 1 and 2 and $N = 562$ for files 5 and 6. Similarly, $\bar{\bar{P}}_i$ denotes the numerical average of all the values of instantaneous power (all records from all range cells) contained in a given file. Thus each file gives rise to $M_r \cdot M_c$ values $\bar{P}_i(r, c)$, $(1 \leq r \leq M_r,\ 1 \leq c \leq M_c)$, but to only one value of $\bar{\bar{P}}_i$, which is equal to

$$\bar{\bar{P}}_i = \frac{1}{M_r \cdot M_c} \sum_r \sum_c \bar{P}_i(r, c)$$

In theory one attempts to relate the quantities $\mathbf{P}_i$ to the physical properties of the clutter. In practice, the true means $\mathbf{P}_i$ are approximated by the sample means $\bar{P}_i$, and the experimental verification of any physical model is complicated by random statistical fluctuations in the quantities $\bar{P}_i$, which are due to the Gaussian nature of the voltages, and would exist even in the absence of receiver noise.

In general, data from different files are never collated, and the above notation will be followed consistently. These general rules are violated only in the case of receiver noise. The quantities $\mathbf{N}_i$ are approximated by the file averages $\bar{\bar{N}}_i$ obtained from the noise file, and we shall use the same symbol $\mathbf{N}_i$ for both the true value and its estimate $\bar{\bar{N}}_i$. Thus, for example, we shall write $\bar{P}_i - \mathbf{N}_i$ instead of $\bar{P}_i - \bar{\bar{N}}_i$ for the 'noise-corrected' value of the indicated sample mean. In practice, the difference $|\mathbf{N}_i - \bar{\bar{N}}_i|$ is very small, since each $\bar{\bar{N}}_i$ is the numerical average of 26,400 values of instantaneous noise power.

The variation of power with position, time, and frequency

From a straightforward analysis of the effects of receive noise and calibration errors, one obtains the relation

$$\mathbf{P}_i = Z_i \mathbf{F}(f_i) + \mathbf{N}_i \tag{1}$$

where $\mathbf{F} = \mathbf{F}(f)$ is the 'ideal' power versus frequency relation that would be observed in the absence of calibration errors and noise and Z_i is the calibration error at the ith frequency. Hence $Z_i \equiv 1$ if the calibaration is perfect, and $\mathbf{N}_i \equiv 0$ in the absence of noise. Our problem is to use the data $\bar{P}_i$ to obtain information about the function $\mathbf{F} = \mathbf{F}(f)$, which is

TABLE 3. File-Averaged Values of Calibrated Power $\bar{\bar{P}}_i$

Frequency, GHz	File 1	File 2	File 5	File 6	File 8
2.1	141.5	77.5	14.9	39.7	6.3
2.55	451.2	244.9	42.5	127.9	12.9
3.0	889.5	480.6	64.1	268.5	12.6
3.45	1241.1	705.1	95.5	389.4	12.9
3.9	814.1	434.8	79.8	246.3	24.3

dependent on the radar range R as well as other parameters which are descriptive of the clutter.

Table 3 shows the calibrated values of instantaneous power averaged over all the data in each file, i.e., $\bar{\bar{P}}_i$. These data are tabulated only for the multifrequency files, including the noise file 8. The noise power is not flat across the frequency range because the calibration involved certain factors (such as antenna gain) which do not affect the receiver noise. Another point to be noted is the file-to-file variation of the signal-to-noise ratios $\bar{\bar{P}}_i/\mathbf{N}_i$. File 5 is by far the noisiest file, the signal-to-noise ratios in this file all being less than 9 dB.

We now consider the 'relative values of noise-corrected power' defined by

$$\bar{\bar{Q}}_i = (\bar{\bar{P}}_i - \mathbf{N}_i)/\bar{\bar{P}}_1 - \mathbf{N}_i)$$
$$\bar{Q}_i = (\bar{P}_i - \mathbf{N}_i)/(\bar{P}_1 - \mathbf{N}_1) \tag{2}$$

The quantities $\bar{\bar{Q}}_i$ are tabulated in Table 4, and, as can be seen, the corresponding values of $\bar{\bar{Q}}_i$ are fairly constant from file to file. This suggests the model

$$P(f_i) = AQ(f_i) + \mathbf{N}_i$$

where A is a frequency independent parameter which varies from file to file and $Q = Q(f)$ is a function which depends on frequency only and no other parameters. However, for reasons mentioned in the introduction, we are not very interested in the behavior of heavily averaged data, and we therefore restrict our attention to data contained in a single file (or portion thereof) and consider whether the data within a given file conforms to model

$$\mathbf{P}_i = A\mathbf{Q}_i + \mathbf{N}_i \tag{3}$$

where the parameter A varies with time and position but not with frequency, and the function

$$\mathbf{Q}_i = \mathbf{Q}(f_i) = (\mathbf{P}_i - \mathbf{N}_i)/(\mathbf{P}_1 - \mathbf{N}_1) \tag{4}$$

is only dependent on frequency. Comparing (3) to (1), we see that the model (3) asserts that the 'ideal' mean power versus frequency relation is given by

$$\mathrm{F}(f_i) = A \cdot (\mathbf{Q}_i/Z_i)$$

where the first factor contains all the spatial and temporal dependence and the second factor contains all the frequency dependence.

We recall that $\mathbf{P}_i = \mathbf{P}_i(r, c)$ varies with record r and range cell c, and the hypothesis now to be tested is that, for data within a given file, $\mathbf{Q}_i$ does not vary with r and c. The quantities $\mathbf{Q}_i$ are approximated by the quantities $\bar{Q}_i$, and to test the hypothesis the standard deviations of the $\bar{Q}_i$ are first calculated theoretically, on the assumption that $\mathbf{Q}_i$ are constant, and then these theoretically calculated standard deviations are compared with the actual sample standard deviations computed directly from the $M_r \cdot M_c$ samples of $\bar{Q}_i$ contained in the given file. The results are shown in Table 5. In our theoretical calculations we neglect the effects of receiver noise, and this explains why the agreement is not as good in the noisy files 5 and 6 as in the files 1 and 2. Keeping in mind the nature and effects of these simplifying assumptions, we judge that the results presented in Table 5 represent good fit between the data and the model (3). Details of the statistical analyses are given in the appendix.

Although the relative power values $\bar{Q}_i$ were thus found to be as constant as the Gaussian nature of the voltages permits, the absolute values $\bar{P}_i = \bar{P}_i(r, c)$ were found to be highly variable with time and position. Tabulated values of $\bar{P}_3(r, c) - \mathbf{N}_3$ are shown in Table 6 for $r = 4$, 5 and $1 \leq c \leq 64$ in files 1 and 2. The expected standard deviation standard in these quantities due to purely random fluctuations in the signal voltages is approximately $(0.1)\bar{P}_3$, but the observed variation is much larger than this, and one can sometimes discern well-defined gradients in the clutter distribution.

In the work of *Gordon and Wilson* [1982] we present data on the spatial and temporal autocorrelation

TABLE 4. Relative Values of File-Averaged, Noise-Corrected Power $\bar{\bar{Q}}_i$

Frequency, GHz	File 1	File 2	File 5	File 6
2.1	1.00	1.00	1.00	1.00
2.55	3.24	3.26	3.44	3.44
3.0	6.49	6.57	5.99	7.66
3.45	9.08	9.72	9.61	11.27
3.9	5.34	5.76	6.45	6.64

TABLE 5. Theoretically Calculated Versus Sample Standard Deviations of $\bar{Q}_i$

Frequency, GHz	File 1			File 2			File 5			File 6		
	Sample	Theory	Ratio	Sample	Theory	Ratio	Sample	Theory	Ratio	Sample	Theory	Ratio
2.1	...	...	...	...	...	...	...	...	...	...	...	...
2.55	1.00	0.90	1.11	0.86	0.87	0.99	0.75	0.44	1.70	0.31	0.26	1.21
3.0	1.86	1.82	1.02	1.94	1.92	1.01	0.96	0.78	1.22	0.71	0.58	1.22
3.45	2.49	2.52	0.99	2.71	2.78	0.98	1.65	1.29	1.28	0.95	0.82	1.16
3.9	1.70	1.64	1.04	1.57	1.67	0.94	1.02	0.83	1.23	0.62	0.48	1.29

of the clutter, but here we shall only note that the temporal decorrelation times (= first zero crossing) vary with frequency, being about 10 ms at 2.1 GHz and 6 ms at 3.9 GHz, and that, similarly, the spatial decorrelation distances vary between 2 and (3/2) of a range resolution cell width. This information is used in theoretical calculations given in the appendix.

5. POLYNOMIAL LAWS

Generalities

In the literature one finds discussions of various scattering effects, each of which gives rise to an ideal power versus frequency relation of the type

$$\mathbf{F}(f) = Af^n \tag{5}$$

If several such effects are operating simultaneously, then

$$\mathbf{F}(f) = \sum A_j (f/f_1)^{n_j} \tag{6}$$

with one term for each effect. In (6) we have normalized the frequency f by f_1 to make the coefficients A_j have the same physical dimension (i.e., power). In the previous section we showed that the data are consistent with the operation of an ideal power versus frequency relation F involving only a single variable parameter A, of which (5) is the simplest type. However, the more general type (6) is also consistent with the data provided that the ratios A_j/A_k are constant, as can be seen by setting

$$A = \sum A_j \qquad F(f) = A \sum (A_j/A)(f/f_1)^{n_j}$$

We shall refer to (5) as the monomial law to distinguish it from the more general 'polynomial' type (6). For the classisical case of incoherent scattering from drops uniformly distributed throughout the radar resolution cell we have a monomial law with $n = 4$, and there are other incoherent scattering effects from nonuniform distributions (described below) for which $n = 5$ or 6. In addition, there are certain coherent effects (discussed by *Kerr* [1951]) and Bragg-type scattering effects caused by turbulence-induced inhomogenieties in drop distribution (discussed by *Naito and Atlas* [1966], *Gossard* [1979], and *Gossard and Strauch* [1980]) which produce values of n which fall in the range of $0 \leq n \leq 2$. In all these theoretical formulations it is assumed that the radar cross section (rcs) σ of each drop follows the Rayleigh scattering law $\sigma \propto f^4$.

At any instant of time the total backscattered power is merely the squared sum of the fields scattered by each of the individual drops, and the scattering is 'incoherent' at the extent to which the 'cross terms' average to zero. [cf. *Feynman et al.*, 1964]. In this case the total rcs of an assemblage of N drops is $N\sigma$, and the mean backscattered power from a radar resolution cell containing N drops is therefore [*Skolnik*, 1962]

$$\mathbf{F} = \frac{P_t A}{(4\pi R^2)^2} GN\sigma$$

where P_t is the transmitted power, A is the capture area of the antenna, and G is the antenna gain. We take into account the dependence of G and the beam area Ω (in steradians) on frequency, given by

$$G = 4\pi A/\lambda^2 = 4\pi/\Omega$$

and we now consider the variation of the factor $GN\sigma$ with frequency f. When the drop distribution is uniform, N will vary directly as the cell volume V, which in turn varies as the beam area Ω. However, the quantity GN will be frequency independent due to the relation $G\Omega = 4\pi$. Therefore, in the case of a uniform drop distribution, the variation $\mathbf{F}$ with f comes about entirely through the variation of σ with f, and hence $n = 4$. As another example, we consider incoherent scattering from a thin layer aligned parallel to

TABLE 6. Record-Averaged, Noise-Corrected Power $\bar{Q}_3$ Versus Range Cell

File 1			File 2		
Range Cell	Record 4	Record 5	Range Cell	Record 4	Record 5
1	163.93	110.24	1	173.28	178.76
2	158.16	105.21	2	222.55	199.23
3	215.29	121.16	3	270.30	246.79
4	181.95	140.36	4	254.44	233.01
5	168.23	112.73	5	301.91	278.24
6	210.08	108.14	6	381.97	333.75
7	232.01	164.65	7	384.43	351.03
8	259.35	192.03	8	341.37	346.16
9	261.11	246.94	9	299.08	390.74
10	253.45	261.67	10	353.78	455.12
11	200.16	321.06	11	526.65	430.40
12	223.20	319.85	12	538.09	418.65
13	278.63	291.59	13	523.79	551.30
14	296.53	297.09	14	602.55	586.94
15	317.79	232.75	15	621.46	651.76
16	428.91	211.11	16	672.00	815.69
17	455.43	311.56	17	832.99	840.75
18	602.76	378.09	18	866.55	735.04
19	695.36	467.98	19	639.72	659.79
20	682.16	453.22	20	591.72	861.64
21	466.94	540.24	21	778.02	891.25
22	521.89	713.58	22	831.37	781.01
23	803.76	809.74	23	702.62	799.22
24	973.81	778.86	24	685.38	897.97
25	1068.90	989.52	25	792.05	838.75
26	1152.64	962.83	26	926.96	655.86
27	1100.74	796.81	27	746.35	492.04
28	1079.18	727.39	28	706.39	531.99
29	1017.65	795.35	29	831.69	498.71
30	1112.04	1245.36	30	822.87	467.36
31	1489.19	1367.94	31	695.41	446.44
32	1769.84	1193.32	32	555.47	460.21
33	1620.96	1052.68	33	416.80	466.17
34	1488.48	1140.50	34	391.64	438.13
35	1334.06	1526.23	35	371.10	362.48
36	1631.59	1629.15	36	344.59	351.86
37	1646.35	1275.93	37	367.83	389.25
38	1445.51	1189.55	38	348.08	369.44
39	1273.94	1122.02	39	346.94	390.45
40	1173.75	1302.71	40	443.90	349.97
41	1091.18	1554.43	41	504.90	383.08
42	1135.10	1538.92	42	470.01	402.79
43	1135.81	1378.86	43	341.11	410.01
44	1128.04	1393.89	44	382.57	353.60
45	1289.82	1583.34	45	384.74	289.66
46	1289.44	1632.78	46	325.75	300.09
47	1732.34	1445.54	47	414.50	347.80
48	2118.45	1314.08	48	411.51	332.99
49	1660.51	1263.01	49	348.02	379.02
50	1354.62	1188.23	50	365.67	361.28
51	1531.50	1234.49	51	344.20	312.34
52	1765.66	1438.20	52	297.53	322.77
53	1693.00	1626.90	53	299.35	358.48
54	1545.73	1459.48	54	243.01	443.04
55	1411.74	1487.97	55	296.14	446.76
56	1386.58	1617.51	56	430.98	400.71
57	1384.94	1645.16	57	420.86	401.76
58	1389.48	1402.46	58	481.95	317.12
59	1465.61	1300.81	59	481.75	283.85
60	1552.63	1389.47	60	373.71	250.32
61	1447.44	1237.25	61	349.18	315.91
62	1420.55	1167.37	62	306.37	324.04
63	1507.13	1072.47	63	365.45	330.07
64	1611.69	979.02	64	427.05	399.82

the radar line of sight, e.g., a horizontal layer when the radar is pointed at the horizon. In this case N varies directly as the horizontal beamwidth, which in turn varies as f^{-1}. Then $GN\sigma \propto f^2 f^{-1} f^4$, and $n = 5$. As a final example, we consider the case of a uniform smear of clutter in which there is embedded a sparce collection of very dense 'clumps' whose sizes are small with respect to the radar beam but whose spacing is larger than the beam. For simplicity, we assume the use of only two frequencies which give rise to two beams, the smaller of which (produced by the higher frequency) lies within the larger. Because of wind and/or antenna rotation, a clump will variously appear within both, one, or none of the beams. Due to the uniform smear, a fourth power law will be observed over a period of time during which no clump appears in either beam. During a time period in which a clump appears in the inner beam, the number N of drops in the clump which are illuminated by the beams will be independent of f, and hence the clump by itself would give rise to a sixth power law. And finally, during a time in which a clump appears inside the larger beam but outside the inner one, the exponential growth of power with frequency will be less than 4, since in this case the relative increase of power with frequency will be less than if no clump existed in either beam. Hence, if such clumps existed, they might give rise to a polynomial law (6) with at least three terms, with exponents n_j equal to 6, 4, and something less than 4.

TABLE 7. Values of Exponent Estimates $\bar{\bar{n}}_M$

	$\bar{\bar{n}}_2$	$\bar{\bar{n}}_3$	$\bar{\bar{n}}_4$	$\bar{\bar{n}}_5$
File 1	6.0	5.2	4.5	3.1
File 2	6.2	5.4	4.6	3.2
File 5	5.8	4.6	4.2	3.0
File 6	6.3	5.7	4.9	3.4

Testing the monomial law

We now assume the operation of a monomial law (5) and estimate the exponent n by means of the weighted logarithmic sums.

$$\hat{n}_M = \sum_{i=1}^{M} \xi_i^{(M)} \log (\bar{P}_i - \mathbf{N}_i) \tag{7}$$

i.e., each $\hat{n}_M$ is an estimate of n formed by using only the first M frequencies ($M = 2, 3, 4, 5$). The weights $\xi_i^{(M)}$ are chosen according to certain criteria discussed below, and once they are chosen, each file gives rise to $M_r \cdot M_c$ values of $\hat{n}_M$ (one for each record from each range cell).

By assumption, $\mathbf{F}(f) = Af^n$, and assuming large signal-to-noise ratios, and using the approximation $\log (1 + x) \simeq x$ for small $|x|$, from (1) we get the approximation

$$\hat{n}_M = (\log A) \sum \xi_i^{(M)} + n \sum \xi_i^{(M)} \log f_i + \sum \xi_i^{(M)} \log Z_i + \sum \xi_i^{(M)} (D_i/\mathbf{P}_i)$$

where for convenience, we have set

$$D_i = \bar{P}_i - \mathbf{P}_i$$

It is therefore natural to require the weights $\xi_i^{(M)}$ to satisfy the two constraints

$$\sum_{i=1}^{M} \xi_i^{(M)} = 0 \quad \sum_{i=1}^{M} \quad \log f_i = 1 \tag{8}$$

so that

$$\hat{n}_M = n + \sum_{i=1}^{M} \xi_i^{(M)} \log Z_i + \sum_{i=1}^{M} \xi_i^{(M)} (D_i/\mathbf{P}_i) \tag{9}$$

For each M the collection of samples $\hat{n}_M$ in each file is averaged to form an estimate denoted by $\bar{\bar{n}}_M$, and we now consider the effects of the second and third term in (9) on the estimates $\bar{\bar{n}}_M$. The third term is a zero mean process, and the variance of each factor (D_i/P_i) is on the order of $(1/N)$, where $N(=85$ or $562)$ is the number of pulses per frequency in each record. Therefore, the effects of the third term are very much reduced in forming the average $\bar{\bar{n}}_M$. On the other hand, the second term gives rise to a constant bias which is not averaged out, and we therefore choose the weights $\xi_i^{(M)}$ to minimize the effects of this bias. In the absence of any knowledge about the calibration errors Z_i, it would appear that the best that one can do to minimize this bias is to choose the weight $\xi_i^{(M)}$ to have minimum length, subject to the constraints (8). By using the method of Lagrange multipliers, the 'minimum length' weights are found to be given by

$$\xi_i^{(M)} = \left[\log f_i - \frac{1}{M} \sum_{j=1}^{M} \log f_j \right] \cdot \left[\sum_{j=1}^{M} (\log f_j)^2 - \frac{1}{M} \left(\sum_{j=1}^{M} \log f_j \right)^2 \right]^{-1} \tag{10}$$

The results obtained using these weights are shown in Table 7. Because of the 50% overlap between successive range cells, we only use the return from every other range cell in these calculations in order to insure the statistical independence of the $(\frac{1}{2})M_r \cdot M_c$ samples of $\hat{n}_M$ which are thus obtained from each file. Figure 3 is a histogram of the 320 samples of $\hat{n}_4$ contained in file 1, and the distribution is apparently Gaussian. Table 8 shows the sample standard deviation of the $\hat{n}_M$. It should be kept in mind that the expected standard deviation (SD) in $\bar{\bar{n}}_M$ equals $\mathrm{SD}(\hat{n}_M)/(N)^{1/2}$, where $N = (\frac{1}{2})M_r \cdot M_c$.

Using standard statistical techniques it can be shown that filt-to-file variation in the $\bar{\bar{n}}_M$ is statistically significant. It follows that the file-to-file variation in the distribution of values of relative power $\bar{Q}_i$ (defined by (2)) is also statistically significant, since the quantities $\hat{n}_M$ are linear combinations of $\log \bar{Q}_i$. However, although these small file-to-file variations are significant in a purely statistical sense, their practical or physical significance is not at all clear. In particular, we would have liked to establish some correlation between variations in observed features of

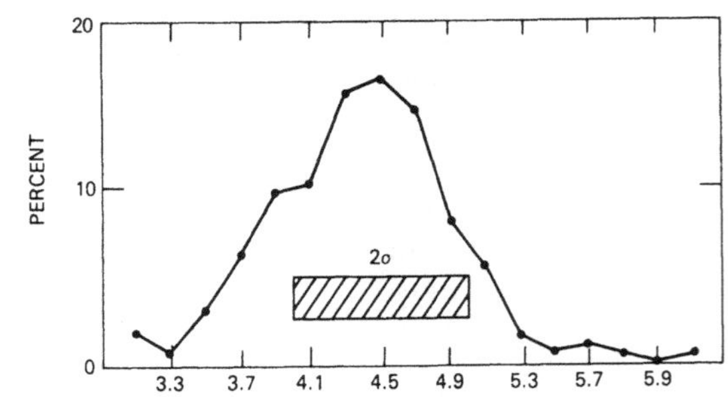

Fig. 3. Histogram of $\hat{n}_4$. Data from file 1: 320 samples. Mean = 4.5, standard deviation = 0.49.

TABLE 8. Sample Standard Deviations of $\hat{n}_M$

	SD($\hat{n}_2$)	SD($\hat{n}_3$)	SD($\hat{n}_4$)	SD($\hat{n}_5$)
File 1	1.33	0.74	0.49	0.37
File 2	1.15	0.68	0.47	0.34
File 5	0.79	0.45	0.41	0.26
File 6	0.38	0.25	0.18	0.13

the power versus frequency relation and properties of the Doppler spectra. For example, one might have expected to find that certain values of $\hat{n}_M$ are associated with multipeaked spectra, while other values are associated with single-peaked spectra. However, no such correlation was observed.

We refer the reader to the appendix for details concerning the statistical analysis of the file-to-file variation in $\bar{\hat{n}}_M$.

We now consider two possibilities for the observed systematic variation of $\bar{\hat{n}}_M$ with M, viz., (1) the operation of a polynomial law (6) with two or more terms, and (2) the effects of calibration errors.

Assuming the operation of a polynomial law (6), the expansion (9) is modified by replacing the first term ($= n$) with the expression

$$\sum_{i=1}^{M} \xi_i^{(M)} \log \mathbf{F}(f_i)$$

whose value varies with M and always lies between the smallest and largest exponent n_j appearing in (6). Therefore, the variation of $\bar{\hat{n}}_M$ with M could be explained by the operation of a polynomial law, but we also have to consider the effects of the calibration errors, i.e., the second term in (9).

As explained above, the effects of the third term in (9) are greatly reduced when forming the numerical average $\bar{\hat{n}}_M$, and therefore, assuming the operation of a monomial law, we have

$$\bar{\hat{n}}_M \simeq n + \sum_{i=1}^{M} \xi_i^{(M)} \log Z_i \tag{11}$$

Let L denote the 5 vector whose components are log Z_i, and for $M = 2, 3, 4$ let $\{\xi_i^{(M)}\}$ be made a 5 vector by setting $\xi_i^{(M)} = 0$ for $M < i \leq 5$. Then (11) can be written

$$\xi^{(M)} \cdot L = \bar{\hat{n}}_M - n \tag{12}$$

One can then apply the Schwarz inequality to obtain a lower bound on the length $\| L \|$ of the vector L in terms of the quantities $\bar{\hat{n}}_M - n$. More generally, letting α vary over a subset of the set $\{2, 3, 4, 5\}$, one can show that

$$\| L \|^2 \geq \sum_{\alpha} \sum_{\beta} M_{\alpha,\beta}^{-1}(\bar{\hat{n}}_\alpha - n)(\bar{\hat{n}}_\beta - n) \tag{13}$$

where $M_{\alpha,\beta} = \xi^{(\alpha)} \cdot \xi^{(\beta)}$. By using various subsets of the set 2, 3, 4, 5 we get various lower bounds using (13) with the observed values of $\bar{\hat{n}}_\alpha$ and various assumed values for n. The length of L is a constant multiple of the rms calibration error (decibels) defined by

$$\text{rms} = \left[\frac{1}{M} \sum_{i=1} (10 \log_{10} Z_i)^2\right]^{1/2}$$

where M is the largest integer in the subset.

The lower bounds obtained in this way are shown in Table 9, where the data $\bar{\hat{n}}_\alpha$ is from file 6. The head of each column displays an assumed value of n, in integral steps from 0 to 10. It should be kept in mind that these lower bounds to the rms calibration errors are not necessarily equal to the actual values. Each of the four rows in this table gives the lower bound for among the first M frequencies, $M = 2, 3, 4, 5$. Note that the lower bounds in the row $M = 5$ never fall below 2.0 dB, and it is tempting to exclude this row, which is the only one involving the anomaly at 3.9 GHz. On the other hand, we judge that an rms calibration error of 2.0 dB has to be allowed for in the remaining (i.e., first four) frequencies, and we therefore conclude that the data are consistent with the operation of a simple nth power law with n lying in

TABLE 9. Lower Bounds to Root Mean Square Calibration Error $\left(= \left[(1/M) \sum_{1}^{M} (10 \log_{10} Z_i)^2\right]^{1/2}\right)$ Versus Assumed Exponent n

M	$n = 0$	$n = 1$	$n = 2$	$n = 3$	$n = 4$	$n = 5$	$n = 6$	$n = 7$	$n = 8$	$n = 9$	$n = 10$
2	2.7	2.2	1.8	1.4	1.0	0.6	0.1	0.3	0.7	1.1	1.6
3	4.4	3.7	2.9	2.1	1.4	0.6	0.4	1.1	1.8	2.6	3.4
4	5.6	4.5	3.4	2.3	1.2	0.7	1.4	2.5	3.6	4.7	5.8
5	5.3	3.8	2.5	2.0	2.1	2.7	4.1	5.6	7.0	8.5	10.0

the range $3 < n < 7$. In particular, the operation of the classical fourth power law appears to be a real possibility, although the results shown in Table 9 (for $M \leq 4$) appear to favor values of n greater than 4.

Our results on the clutter Doppler spectra will be presented in the next section; however, at this point it may be useful to summarize the main results of this and the previous sections and indicate what bearing they might have on the possible causes for multipeaked spectra. We shall only consider two competing models for the production of multiple peaks, viz., wind shear operating on a vertically stratified clutter distribution and Bragg-type scattering from turbulence-induced inhomogeneities. First, it has been shown that the data are consistent with the operation of an ideal power versus frequency law $F = F(f)$ involving only a single space and time varying parameter A, which is multiplicative. The simplest law of this type is the monomial law (5) (although the more general polynomial type (6) is also possible provided that the ratios A_j/A_k are constant); second, our measurements of the exponent n (shown in Table 7) yield values of $n \geq 3$; and third, as shown in Table 9, values of $n \leq 2$ imply rms calibration errors significantly larger than 3.0 dB, which is beyond the bounds of credibility. The evidence therefore favors the vertical stratification model over that of Bragg-type scattering, since the latter would produce values of n lying in the range $0 \leq n \leq 2$. In fact, even if we allow the possibility of a polynomial law (6) with two or more terms, an analysis along the lines given above favors exponential values n_j which are significantly greater than 2.

The case for vertical stratification is especially compelling if we exclude the data at the anomalous frequency 3.9 GHz, for if we only use the results from the first four frequencies, then we get measured values of $n \geq 4$, and the results of Table 9 also favor values of $n \geq 4$. In fact, the smallest lower bound for the rms calibration error in the first four frequencies is obtained at $n = 5$, and we recall that the exponent $n = 5$ corresponds to the case of scattering by narrow horizontal bands.

On the other hand, there is some negative evidence against the simple statification model that should be noted. Namely, if this model were correct, one would expect to find higher measured values of n associated with multipeaked spectra. However, no such correlation was observed.

Finally, we mention that one of the reviewers has suggested the possibility that some of the spectral peaks might be due, in part, to birds. Although birds have sometimes been observed by other, more powerful, radars located at the same site, such detections have been made at extremely short ranges (a few kilometers), and it is extremely doubtful that birds could have produced a detectable signal at the ranges of 20–30 km used in this experiment. Moreover, owing to the general nastiness of the weather, it is doubtful that many birds were active at the time, and none were observed visually.

6. DOPPLER POWER SPECTRA

Figures 4–13 show Doppler spectra from the first five consecutive records from the third range cell in the multifrequency file 6, which is the only file with the antenna not rotating. The PRF for pulses at a given frequency is 320 ($=1600/5$), and hence the Nyquist interval is ± 160 Hz and (for reasons to be explained) the Doppler resolution is 10 Hz. The spectra shown in Figures 4–8 and 9–13 are taken from the backscattered return at 2.1 and 3.0 GHz, respectively. The relation between Doppler frequency and velocity v is given by $\delta = 2v/\lambda$. At 2.1 GHz, 1 Hz = 0.1388 knots, and at 3.0 GHz, 1 Hz = 0.0972 knots. In round numbers, to convert from hertz to knots, we divide by 7 or 10 when the carrier frequency is 2.1 or 3.0 GHz. Hence the Doppler resolution at these two frequencies is approximately 1.4 or 1.0 knots.

The Doppler power spectra are computed according to

$$I(\nu) = \frac{1}{N}\left|\sum_{n=0}^{N-1} z(nt) \exp\left(-2\pi i \nu n t\right)\right|^2 \tag{14}$$

where N is the number of pulses used in the calculation, $z(t) = x(t) + iy(t)$ is the complex data formed from the I and Q channel voltages $x(t)$, $y(t)$, and t is the interpulse period ($=3.1$ ms). Figures 4–13 show smoothed spectra for each of five records from the third range cell. The smoothing is accomplished by decomposing each record into a number M of smaller subrecords containing N pulses each, where N is a power of 2, and averaging the M 'periodograms' (14) to obtain the spectra shown in the figures. In Figures 4–13, $M = 17$ and $N = 32$, and the Doppler resolution is therefore 10 Hz. In each case the $I(\nu)$ are computed at increments of ν equal to the resolution. For purposes of graphical display, the maximum value of $I(\nu)$ is normalized to unity.

When $x(t)$ and $y(t)$ are two stationary gaussian

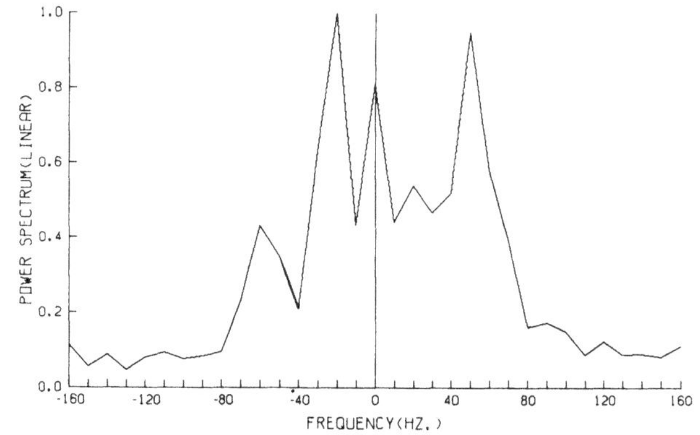

Fig. 4. Doppler power spectrum at 2.1 GHz. Record 1 from range cell 3 of file 6.

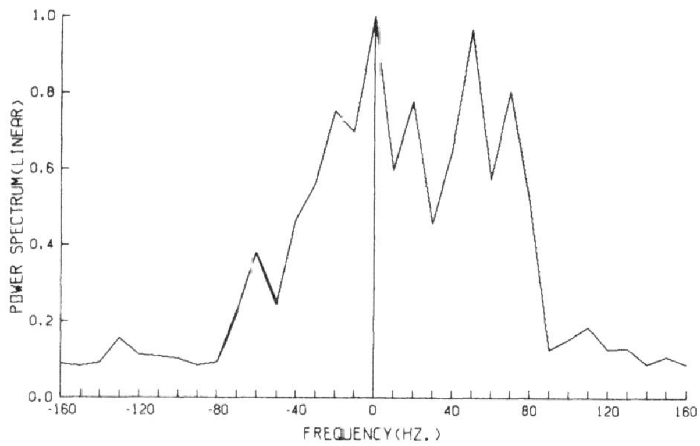

Fig. 6. Doppler power spectrum at 2.1 GHz. Record 3 from range cell 3 of file 6.

zero mean processes having the same autocorrelation function and zero cross correlation, then expanding the right-hand side of (14), one can easily establish that $I(v)$ has the form $U^2 + V^2$, where U and V are two zero mean gaussian variables which satisfy

$$E[U^2] = E[V^2] \qquad E[UV] = 0$$

Hence, for each value of v the variable $w = I(v)$ has an exponential probability density function, and in particular, its mean and standard deviation are equal. When the spectra are averaged as described above, the mean remains the same but the standard deviation is reduced by the factor $1/(M)^{1/2}$, where (as before) M is the number of subrecords into which the original record has been divided. The averaging process therefore reduces the levels and numbers of spurious spectra peaks but at the cost of reduced resolution. With these consideration in mind, it appears that some of the spectra shown in Figures 4–8 are truly multimodal; i.e., that at least some of the peaks off the main lobe are physically real and not the spurious effects of random fluctuations. The width of these peaks, when they occur, is generally on the order of about 20 Hz, i.e., 3 knots. As previously mentioned, the appearance of multipeaked spectra is in sharp distinction from the usual Gaussian model discussed in the radar literature [*Nathanson*, 1969].

The existence of multiple peaks is less apparent in the return at 3.0 GHz, and all of the 'spikyness' in Figures 9–13 can be reasonably attributed to purely random fluctuations. The 'most typical' spectrum at 3.0 GHz is shown schematically in Figure 14 (which was obtained by averaging the spectra from the first five records from the sixth range cell). The hump appearing in the negative frequency range is significant; for although the radar was looking into the

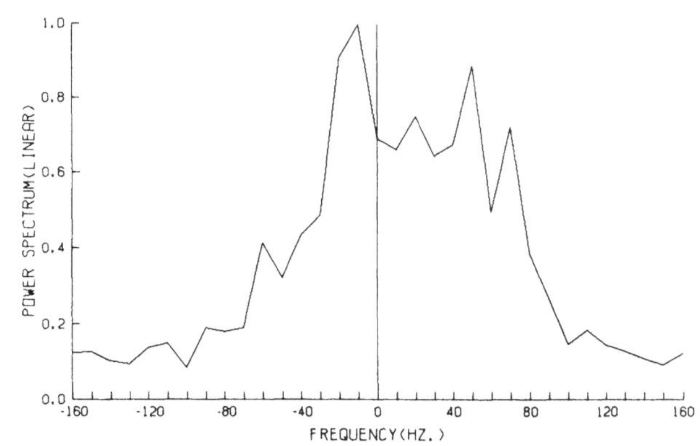

Fig. 5. Doppler power spectrum at 2.1 GHz. Record 2 from range cell 3 of file 6.

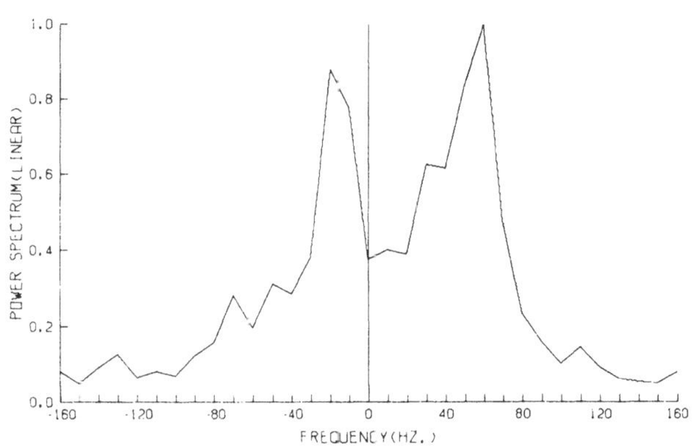

Fig. 7. Doppler power spectrum at 2.1 GHz. Record 4 from range cell 3 of file 6.

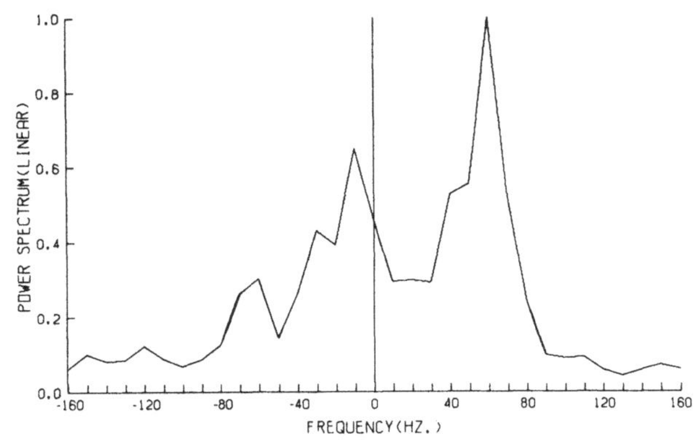

Fig. 8. Doppler power spectrum at 2.1 GHz. Record 5 from range cell 3 of file 6.

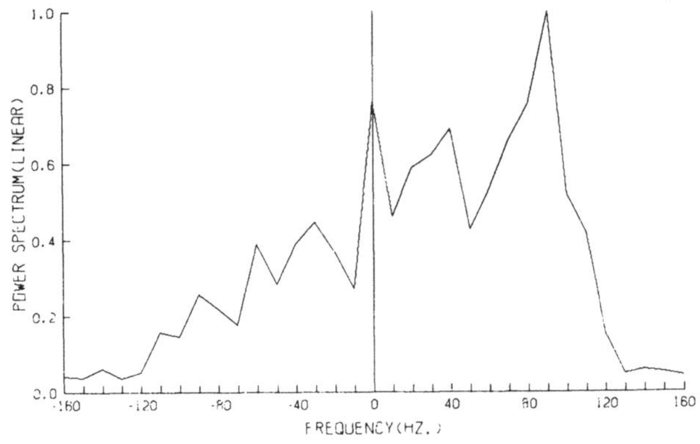

Fig. 10. Doppler power spectrum at 3.0 GHz. Record 2 from range cell 3 of file 6.

wind, in every single spectrum examined (at either frequency) we always found that a significant portion of the backscattered signal energy lies in the negative frequency range.

Assuming that these spectra are caused by a wind shear operating on a vertically stratified medium, the wind shear can be crudely estimated by supposing the wind velocity null to occur between the upper and lower 3-dB points of the radar beam. For the 3.0 GHz beam at the range and elevation angle indicated for file 6 in Table 2, these two points occur at the heights of 1.28 and 2.44 km. And given the fact that the surface wind speed was 23 knots = 12.4 m/s, we conclude that the wind shear was between 5 and 10 m s^{-1} km^{-1}. The extent of the radar beam between the first upper and lower nulls is approximately double that between the two 3-dB points, and in particular, the lower null of the 2.1 GHz beam lay just above the surface. It therefore appears that the vertical coverage of the beams contained rain falling near the surface and extended well into the cloud above.

The estimated value of wind shear can also be used to gauge the vertical extent Δh of the clutter, producing a spectral peak whose velocity spread is Δv: $\Delta h = \Delta v/\text{shear}$. For the narrow 3-knot peaks discussed above, we get $\Delta h = 160$ m or 320 m for the wind shear values of 5 or 10 m s^{-1} km^{-1}.

Figures 15 and 16 show two spectra taken from the single frequency file 3, in which the antenna was rotating. Both spectra are computed from the same set of data (i.e., from the same record and range cell), but the first is the average of six 64-point periodograms, whereas the second is the average of thirteen 32-point periodograms. The Doppler resolutions are, respectively, 25 Hz (=2.5 knots) and 50 Hz (=5 knots), and the Nyquist interval has been expanded to ± 800 Hz, since the single frequency is transmitted at a PRF of 1600. In all of the spectra from this file we find that almost all the signal energy is located in

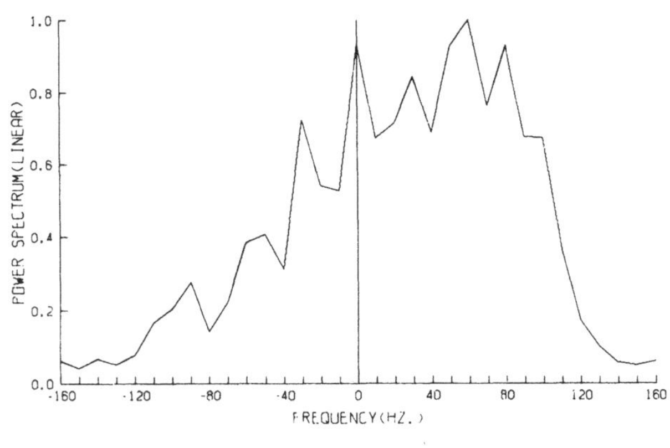

Fig. 9. Doppler power spectrum at 3.0 GHz. Record 1 from range cell 3 of file 6.

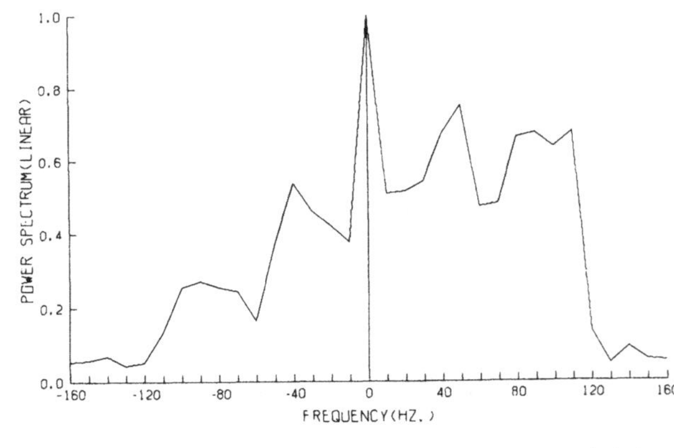

Fig. 11. Doppler power spectrum at 3.0 GHz. Record 3 from range cell 3 of file 6.

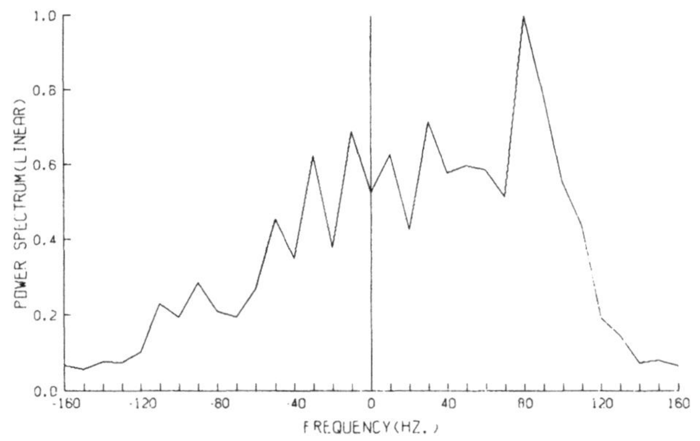

Fig. 12. Doppler power spectrum at 3.0 GHz. Record 4 from range cell 3 of file 6.

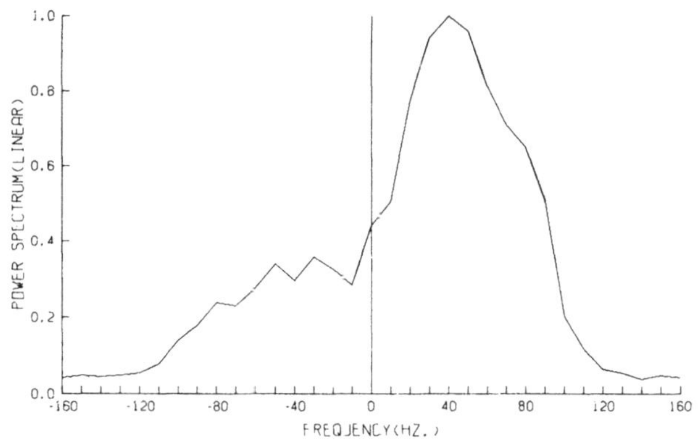

Fig. 14. 'Most typical case' at 3.0 GHz. Average of five 17 × 32 point spectra from range cell 6 of file 6.

the frequency interval between ± 200 Hz, which increases our confidence that there is no appreciable spectral foldover in the spectra shown in Figures 4–8, which is at a lower carrier frequency. It should be kept in mind that, owing to antenna rotation, each of the M subrecords into which the total record is divided originates from a different circular arc in space. (Counting the beam width, each of these arcs is about 2° wide, which at a range of 30 km amounts to 1 km.) Therefore, the averaging process will tend to smooth out features which vary with azimuth. The narrowness of the central peak is a striking feature that appears in many of the spectra from this file, and (again) indicates that a large fraction of the total clutter return originates from clumps or horizontal bands whose vertical extent is rather small.

Finally, we again note that an attempt was made to correlate the appearance of multiple peaks with features in the observed power versus frequency relation, i.e., the data $\bar{Q}_i$; but we could discern no difference between those data records in which multiple peaks occurred and those records in which multiple peaks did not occur.

7. SUMMARY

Before summarizing our results we recall that the data are organized into files, each of which consists of data from a selected clutter cell appearing on the radar PPI scope; that the files are composed of records, each of which consists of a block of data whose time duration is either 0.27 s or 1.76 s; and that the data from different files are never collated. We refer the reader to section 2 for definitions of the 'true' or 'ensemble' means $\mathbf{P}_i$, the record averaged quantities $\bar{P}_i$, and the file averaged quantities $\bar{\bar{P}}_i$.

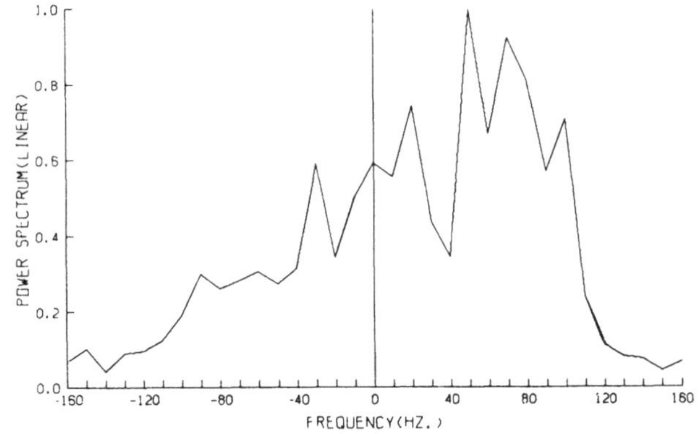

Fig. 13. Doppler power spectrum at 3.0 GHz. Record 5 from range cell 3 of file 6.

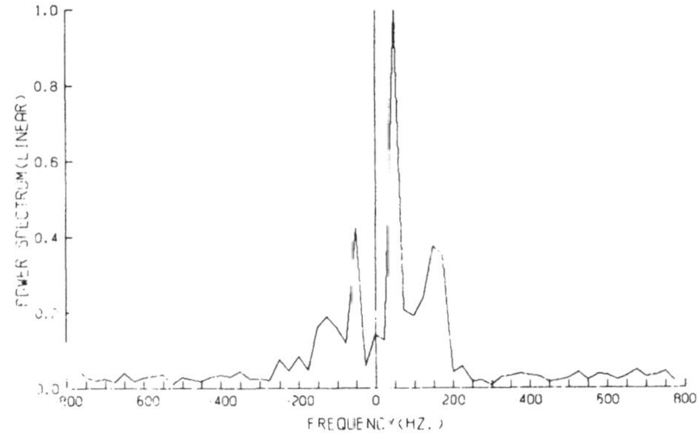

Fig. 15. A 6 × 64 point spectrum from the single-frequency file 3. The antenna is rotating at 6 rpm.

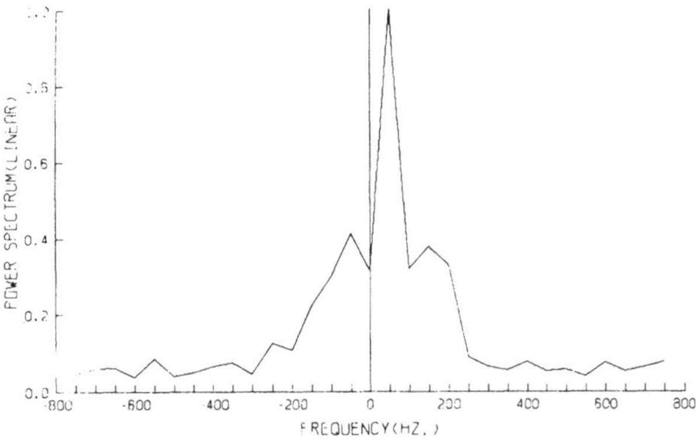

Fig. 16. The 13 × 32 point spectrum calculated from the same data as Figure 15. The antenna is rotating at 6 rpm.

Summary of results concerning the power versus frequency relation

The data within each separate file conforms to the model

$$\mathbf{P}_i = A\mathbf{Q}(f_i) + \mathbf{N}_i$$

where A is a frequency independent parameter which varies from record to record and range cell to range cell, and the function $\mathbf{Q} = \mathbf{Q}(f)$ is dependent on frequency only. In other words, apart from the effects of the receiver noise $\mathbf{N}_i$, the mean backscattered power $\mathbf{P}_i$ is dependent on only a single variable parameter A, which is multiplicative. By convention, we can always set $\mathbf{Q}(f_1) = 1$, so that

$$\mathbf{Q}(f_i) = (\mathbf{P}_i - \mathbf{N}_i)/(\mathbf{P}_1 - \mathbf{N}_1)$$

i.e., the $\mathbf{Q}(f_i)$ are the relative values of the noise-corrected mean power at each frequency, and the parameter A measures the intensity of the clutter. The test of this model consists in determining whether the variation of the ratios $\bar{Q}_i = \{(\bar{P}_i - \mathbf{N}_i)/(\bar{P}_1 - \mathbf{N}_1)\}$ falls within statistical expectations, on the assumption that the function $\mathbf{Q} = \mathbf{Q}(f)$ contains no random or variable elements (other than the frequency f).

Although the corresponding values of $\bar{\bar{Q}}_i \equiv \bar{\bar{Q}}(f_i)$ are remarkably constant from file to file, the file-to-file variation of these quantities (which is usually of the order of 10% or less) is significant in a purely statistical sense. However, the practical or physical significance of this variation is not clear.

In the literature one finds discussions of various scattering effects, each of which give rise to a power versus frequency law of the type $\mathbf{Q}(f) = f^n$. Assuming the operation of such a law, the exponent n is estimated by the weighted logarithmic sums

$$\hat{n}_M = \sum_{i=1}^{M} \xi_i^{(M)} \log(\bar{P}_i - \mathbf{N}_i)$$

i.e., each $\hat{n}_M$ is an estimate of n formed by using only the first M frequencies ($M = 2, 3, 4, 5$). Each file gives rise to $M_r \cdot M_c$ values of $\hat{n}_M$ (one for each record from each range cell), whose average over the file is denoted by $\bar{\hat{n}}_M$. The file-to-file variation of corresponding values of $\bar{\hat{n}}_M$ is usually quite small, but within any given file the variation $\bar{\hat{n}}_M$ with M is large and systematic. More specifically, for $M = 2$, 3, 4, and 5 the corresponding values obtained for $\bar{\hat{n}}_M$ are approximately 6.1, 5.2, 4.6, and 3.2.

The variation of $\bar{\hat{n}}_M$ with M might be partly explained by the existence of calibration errors or by the simultaneous action of two or more of the scattering effects mentioned above, which would give rise to a power versus frequency relation of the type

$$\mathbf{P}_i - \mathbf{N}_i = Af_i^{\,n} + Bf_i^{\,k} + \cdots$$

If one assumes the operation of a simple nth power law

$$\mathbf{P}_i - \mathbf{N}_i = Af_i^{\,n}$$

with the exponent n being specified, then from the data one can infer something about the sizes of the calibration errors; and conversely, any set of a priori bounds assumed for the calibration errors imply restrictions of the range of possible values for the exponent n. Using this kind of analysis, it appears that the data is consistent with the operation of some nth power law, with n lying somewhere in the range $3 < n < 7$.

Summary of results concerning the spatial and temporal variation of the clutter

One can often discern large systematic variations in the sample means $\bar{P}_i$ from record to record and range cell to range cell which are much larger than could have been produced by purely random fluctuations due to the Gaussian nature of the signal voltages.

The temporal decorrelation times (= first zero crossing) vary with frequency, being about 10 ms at 2.1 GHz and 6 ms at 3.9 GHz.

Similarly, the spatial decorrelation distances vary between 2 and (3/2) of a range resolution cell width.

Summary of results concerning the Doppler power spectra

Each of the Doppler spectra discussed below were computed from single data records from single range cells and were smoothed to reduce the level of spurious spectral peaks.

Although the radar was looking into the surface wind, a significant amount of the backscattered signal energy (a third or more) was always contained in the negative frequency range, and some of the spectra contained two or more narrow spectral peaks located at both positive and negative Doppler frequencies.

The simplest explanation for the appearance of such spectra is that of a wind shear operating on a vertically stratified medium. Assuming this model to be correct, the wind shear can be estimated to be on the order of 8 m s^{-1} km^{-1}. The width of the multiple peaks, when they occurred, corresponded to a clutter velocity spread of about 3 knots or less. Combining this result with the value of wind shear given above, it would appear that such peaks were produced by clumps or horizontal bands of clutter whose vertical extent was on the order of 200 m.

The observed values of the exponential rate of growth n of power with frequency tend to favor the vertical stratification model over that of Bragg-type scattering from turbulence induced inhomogeneities in the clutter spatial distribution. In the former case one would expect the values of n to hover around $n = 5$ (which was observed), whereas much smaller values of n would be produced by Bragg scattering.

On the other hand, no correlation was observed between the occurrence of multiple spectral peaks and large values of n.

APPENDIX: DETAILS OF THE STATISTICAL ANALYSES

This appendix is concerned with details of the statistical analyses involved in testing the model (3) and in establishing the significance in the file-to-file variation in the exponent estimates $\bar{n}_M$.

Testing the model (3). We recall that the model is tested by comparing the actual sample variances of the quantities $\bar{Q}_i = (\bar{P}_i - \mathbf{N}_i)/(\bar{\mathbf{P}}_1 - \mathbf{N}_1)$ to their theoretical values, which are calculated on the hypothesis that the quantities $\mathbf{Q}_i = (\mathbf{P}_i - \mathbf{N}_1)/(\mathbf{P}_1 - \mathbf{N}_1)$ are constant on each file. Our working assumption is that the backscattered voltages from each single record from each single range cell consist of samples drawn from a zero mean stationary gaussian process characterized by the signal power $\mathbf{P}_i$ and the autocorrelation coefficients of the voltages. Let $\rho_i(k)$ be the autocorrelation coefficient at the kth lag for the voltages at the ith frequency, and let $R_i(k)$ be the corresponding autocorrelation coefficient for the instaneous power. Then $R_i(k) = [\rho_i(k)]^2$, and the variance of the sample mean $\bar{P}_i$ is given by

$$\text{Var}\,[\bar{P}_i] = C_i\,\mathbf{P}_i^2/N \tag{A1}$$

where N is the number of pulses per frequency in each record (which equals 85 or 562 for our multifrequency files) and

$$C_i = 1 + \frac{2}{N}\sum_{k=1}^{N}(N-k)\,R_i(k) \tag{A2}$$

We now consider the variation of $\bar{Q}_i$. Again setting $D_i = \bar{P}_i - \mathbf{P}_i$, we have

$$\bar{Q}_i = \frac{\mathbf{P}_i - \mathbf{N}_i + D_i}{\mathbf{P}_1 - \mathbf{N}_1 + D_1} = \mathbf{Q}_i\left\{\frac{1 + D_i/(\mathbf{P}_i - \mathbf{N}_i)}{1 + D_1/(\mathbf{P}_1 - N_1)}\right\}$$

and assuming large signal-to-noise ratios $\mathbf{P}_j/\mathbf{N}_j$, we have the approximation

$$\bar{Q}_i \simeq \mathbf{Q}_i\,\frac{1 + D_i/\mathbf{P}_i}{1 + D_1/\mathbf{P}_1}$$

We also have

$$\text{Var}\,[D_j/\mathbf{P}_j] = \text{Var}\,[\bar{P}_j/\mathbf{P}_j] = C_j/N \ll 1 \tag{A3}$$

and therefore

$$\bar{Q}_i \simeq \mathbf{Q}_i\left\{1 - \frac{D_i}{P_i} - \frac{D_1}{\mathbf{P}_1}\right\} \tag{A4}$$

Let α be a two-dimensional index which indexes the record and range cell number pair. Then, from (A4) for each i, the $M_r \cdot M_c$ samples $Q_i(\alpha)$ have the form

$$\bar{Q}_i(\alpha) = \mathbf{Q}_i + X(\alpha)$$

where for each α the process $X(\alpha)$ has zero mean. Note that $\mathbf{Q}_i$ is not indexed by α, since, by hypothesis, $\mathbf{Q}_i$ is constant on the file. Setting $M = M_r \cdot M_c$, the sample variance of $\bar{Q}_i$ is therefore equal to

$$\frac{1}{M}\sum_{\alpha}\left\{\bar{Q}_i(\alpha) - \frac{1}{M}\sum_{\beta}\bar{Q}_i(\beta)\right\}^2 = \frac{1}{M}\sum_{\alpha}\left\{X(\alpha) - \frac{1}{M}\sum_{\beta}X(\beta)\right\}^2$$

and the expected value of this quantity is easily shown to be equal to

$$\left(\frac{1}{M} - \frac{1}{M^2}\right)\sum_{\alpha}\text{Var}\,[X(\alpha)] \simeq \frac{1}{M}\sum\text{Var}\,[X(\alpha)]$$

Hence from (A3) and (A4) the expected value of the

TABLE A1. Values of t Used to Test Significance of File-to-File Variation of $\bar{\bar{n}}_M$

File	1	2	5
	$M = 2$		
2	1.3		
5	1.3	2.0	
6	3.1	1.9	2.9
	$M = 3$		
2	2.1		
5	5.6	6.8	
6	7.8	6.0	9.8
	$M = 4$		
2	4.8		
5	3.0	3.0	
6	11.2	6.8	7.6
	$M = 5$		
2	2.9		
5	0.8	2.2	
6	9.6	7.1	5.4

sample variance of $\bar{Q}_i$ is equal to

$$\text{Var}(\bar{Q}_i) = \mathbf{Q}_i^2 (\bar{\bar{C}}_i + \bar{\bar{C}}_1)/N \quad \text{(A5)}$$

where $\bar{\bar{C}}_j$ is the file-averaged value of C_j.

Formula (A5) is used to calculate the expected values of the variance of $\bar{Q}_i$ shown in Table 7. The quantities $\mathbf{Q}_i$ were approximated by $\bar{\bar{Q}}_i$, and it only remains to describe how the quantities $\bar{\bar{C}}_i$ are computed from the data. For each record from each range cell we calculate C_i according to (A2), but we only use the first 15 lagged values of $R_i(k)$, the quantities $R_i(k)$ being calculated directly from the data. Then, for reasons of economy, each $\bar{\bar{C}}_i$ is taken to be the numerical average of only the 40 values of C_i obtained from the first five records from the first eight range cells. Within each file we found only about a 10% variation in $\bar{\bar{C}}_i$ with i, but there is an appreciable file-to-file variation in $\bar{\bar{C}}_i$ for each fixed i. More specifically, the values of $(\frac{1}{5})$ $(\bar{\bar{C}}_1 + \cdots + \bar{\bar{C}}_5)$ obtained in files 1, 2, 5, and 6 are, respectively, 3.2, 3.4, 4.9, and 1.5. Note that although the value for file 5 is extremely large, the calculated variances of $\bar{Q}_i$ in this file do not fall out of line with the others.

Testing the significance of the file-to-file variation in n. We recall the definitions of $\hat{n}_M$ and $\bar{\bar{n}}_M$ given in section 5. Let $\bar{\bar{n}}_M(j)$ be the value of $\bar{\bar{n}}_M$ obtained from file j, $j = 1, 2, 5, 6$. From Figure 3, it appears that the $\hat{n}_M$ have a gaussian distribution, and it is a standard technique to test the equality of $\bar{\bar{n}}_M(j)$ and $\bar{\bar{n}}_M(k)$ by means of the statistic

$$t = \frac{|\bar{\bar{n}}_M(j) - \bar{\bar{n}}_M(k)|}{[(S_j^2/M_j) + (S_k^2/M_k)]^{1/2}}$$

where S_j^2 is the sample variance of $\hat{n}_M(j)$ and M_j is the number of samples.

The values of t thus obtained are shown in Table A1. The 5% critical region is given by $t \geq 2$. From the table we see that for $M = 3$, 4, and 5 we have $t \geq 2$, with only one exceptional case (at $M = 5$). We therefore conclude that the file-to-file variation in the standard exponents is statistically significant.

Acknowledgments. The author gratefully acknowledges the help of George J. Linde of the Search Radar Branch who besides collecting the data and calibrating the system was very patient in explaining its operation.

REFERENCES

Barlow, E. J. (1949), Doppler radar, *Proc. IRE*, *37*, 340–355.

Feynman, R. P., R. B. Leighton, and M. Sands (1964), *Lectures on Physics*, vol. 1, sect. 32, Addison-Wesley, Reading, Mass.

Gordon, W. B., and J. D. Wilson (1982), Rain clutter statistics, report, Naval Res. Lab., Washington, D. C., in press.

Gossard, E. E. (1979), A fresh look at the radar reflectivity of clouds, *Radio Sci.*, *14*(5), 1089–1097.

Gossard, E. E., and R. G. Strauch (1980), The internal radio refractive index spectra of clouds from UHF forward scatter radar observations, paper presented at 19th Conference on Radar Meteorology, Am. Meteorol. Soc., Miami, Fla.

Kerr, D. E. (1951), *Propagation of Short Radio Waves*, *MIT Radiat. Lab. Ser.*, vol. 13, 478 pp., McGraw-Hill, New York.

Naito, K., and D. Atlas (1966), On microwave scatter by partially coherent clouds, paper presented at 12th Conference on Radar Meteorology, Am. Meteorol. Soc., Boston, Mass.

Nathanson, F. E. (1969), *Radar Design Principles*, 626 pp., McGraw-Hill, New York.

Skolnik, M. I. (1962), *Introduction to Radar Systems*, 648 pp., McGraw-Hill, New York.

HIGH-RESOLUTION RADAR SCATTERING CHARACTERISTICS OF A DISTURBED SEA SURFACE AND FLOATING DEBRIS

INTRODUCTION

This report will describe work done in developing a concept for an effective debris-avoidance radar for high-speed ships. The work included both experimental and theoretical efforts and resulted in a better understanding of sea scatter phenomena and a new concept for debris detection radar design.

The goal of this study was to find a technique for designing a radar that would detect debris floating on a disturbed sea surface. The problem was to differentiate between echoes from debris and those from the disturbed sea surface.

The approach to the problem was to measure the characteristics of sea return as a function of polarization, carrier frequency, range resolution, angle resolution, and time. The same measurements were then made on rigid debris floating on the water, and comparisons were made to determine whether there were any characteristics that would distinguish one from the other.

INSTRUMENTATION

A schematic of the radar developed to study debris and sea return is shown in Fig. 1. Two continuous-wave (CW) signal generators are used as carrier sources. One generator is tuned to 8.6 GHz, and the other to 9.2 GHz. The outputs of these two generators are added in a hybrid and amplified in a traveling wave tube (TWT) chain; they then excite the input of a normally open diode switch. This switch is closed for the duration of a pulse from a timing generator. The minimum usable pulse length is 20 ns, and the maximum pulse length is limited by the pulse repetition period of the timing generator, which is variable from 0.02-1 ms. The output of the switch passes through a circulator, then through a polarization selector switch, and is radiated from an antenna as either a vertically or horizontally polarized wave. The antenna is a parabolic reflector either 1 or 2.67 m in diameter with orthogonally linearly polarized feeds separately connected to the polarization switch.

Radar echoes that enter the antenna pass through the polarization switch and enter the radar receiver through the circulator and a diode switch that is closed by a receiver gate pulse from the timing generator. The output from this switch passes through a calibrated variable attenuator and is amplified by a low-noise TWT chain. The output of the TWT chain enters port A of a four-port circulator and couples to a bandpass filter that is 200-MHz-wide, tuned to 8.6 GHz, and attached to port B of the circulator.

Manuscript submitted May 18, 1977.

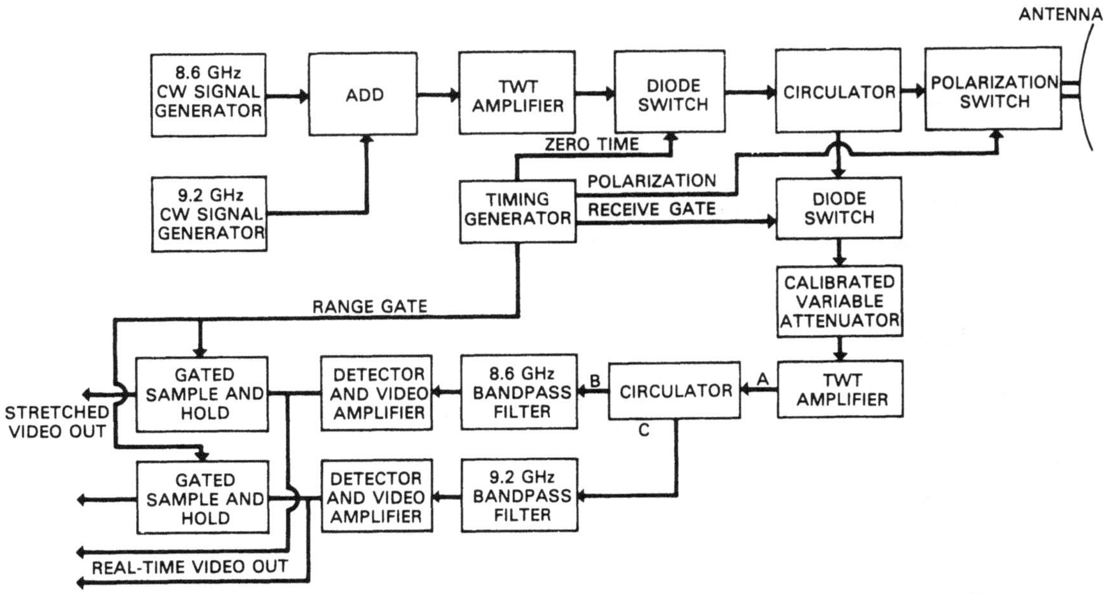

Fig. 1 — Instrumentation radar

The 8.6-GHz signal passes through this filter, and the 9.2-GHz signal reflects back into port B of the circulator. The 9.2-GHz signal exits from port C of the circulator and passes through a bandpass filter 200-MHz-wide and tuned to 9.2 GHz. The outputs of these two filters are rectified in separate detectors and amplified in separate wide-band video amplifiers. The outputs of these amplifiers are displayed as real-time video on a two-channel oscilloscope and are also used as inputs to two range-gated sample-and-hold circuits. The gated stretched outputs of the sample-and-hold circuits are displayed as time functions on a two-channel oscilloscope and are also recorded on a two-channel magnetic tape recorder.

Figure 2 is a schematic of the data recording complex. A television camera with a zoom lens was mounted on the radar antenna with its field of view centered on the radar field of view. Two other television cameras were used to record the real-time video and the range-gated time functions displayed on two-channel oscilloscopes. Split-screen techniques were used to record the output of all three television cameras on the same television recorder. This ensured time synchronization of the radar data and the optical picture of the sea surface that produced the radar echoes.

CALIBRATION

The system was calibrated using a corner reflector with a known equivalent scattering cross section of $\sigma = 24\ \text{m}^2$ (Fig. 3). The variable attenuator at the input to the receiver was adjusted so that both the real-time video and the gated stretched video were at the upper end of the linear range of the system.

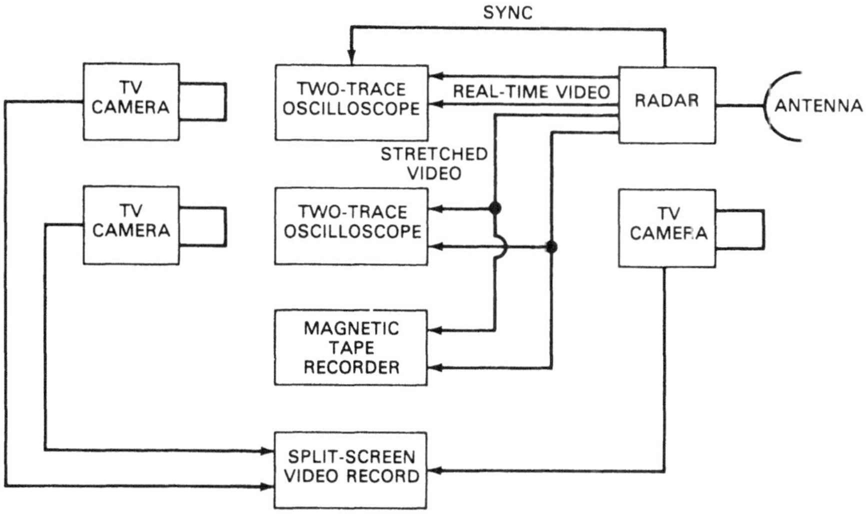

Fig. 2 — Recording instrumentation

In Fig. 3 the right-hand section of the picture is the view from the boresight camera. The inset at the bottom left is the real-time video and the range gage. The lower trace is 8.6 GHz, the middle trace is 9.2 GHz, and the upper trace is the approximate range gate position. The sweep speed for the real-time video was 50 ns/cm, or 500 ns per sweep. The inset at the top left is the gated stretched video with 9.2-GHz top and 8.6-GHz bottom trace. The sweep speed of these traces was 20 ms/cm (0.2 s per sweep). The circular dark spot on the boresight picture is the region filled by the radar beam at ranges long enough to minimize the parallax between the TV camera and the antenna. The dotted lines in this dark spot are the region covered by the 20-ns range gate when the beam strikes the water at a 4.6° grazing angle viewing the disturbed sea surface.

In these measurements, the radar pulse repetition frequency was 50 kHz.

EXPERIMENTAL RESULTS

The first measurements of sea return were made at the Chesapeake Bay with the 1-m diameter radar antenna, 4 m above the mean water level. The radar beamwidth with this antenna was 2.6°, and the center of the beam impinged on the water at a distance of 50 m. These values yielded a 2-m cross-range resolution cell at the center of the beam and a grazing angle of 4.6°. The radar pulse length and range gate were 20 ns long, yielding a range resolution of 3.33 m.

Figure 4 typifies data obtained with vertical polarization and with 1- to 1.3-m waves. The individual pictures were taken sequentially in time with approximately 0.3 s between pictures. This time interval corresponded to the time between sweeps of the oscilloscope

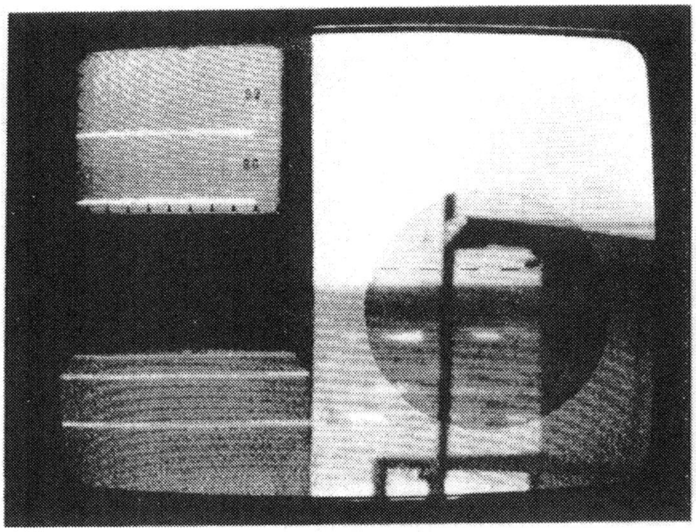

(a) 50-dB attenuation in receiver

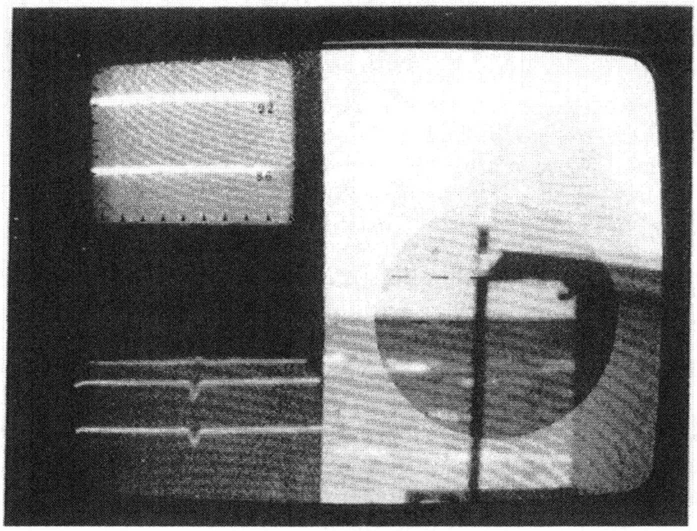

(b) 15-dB attenuation in receiver

Fig. 3 — Calibration via corner reflector (σ = 24 m^2)

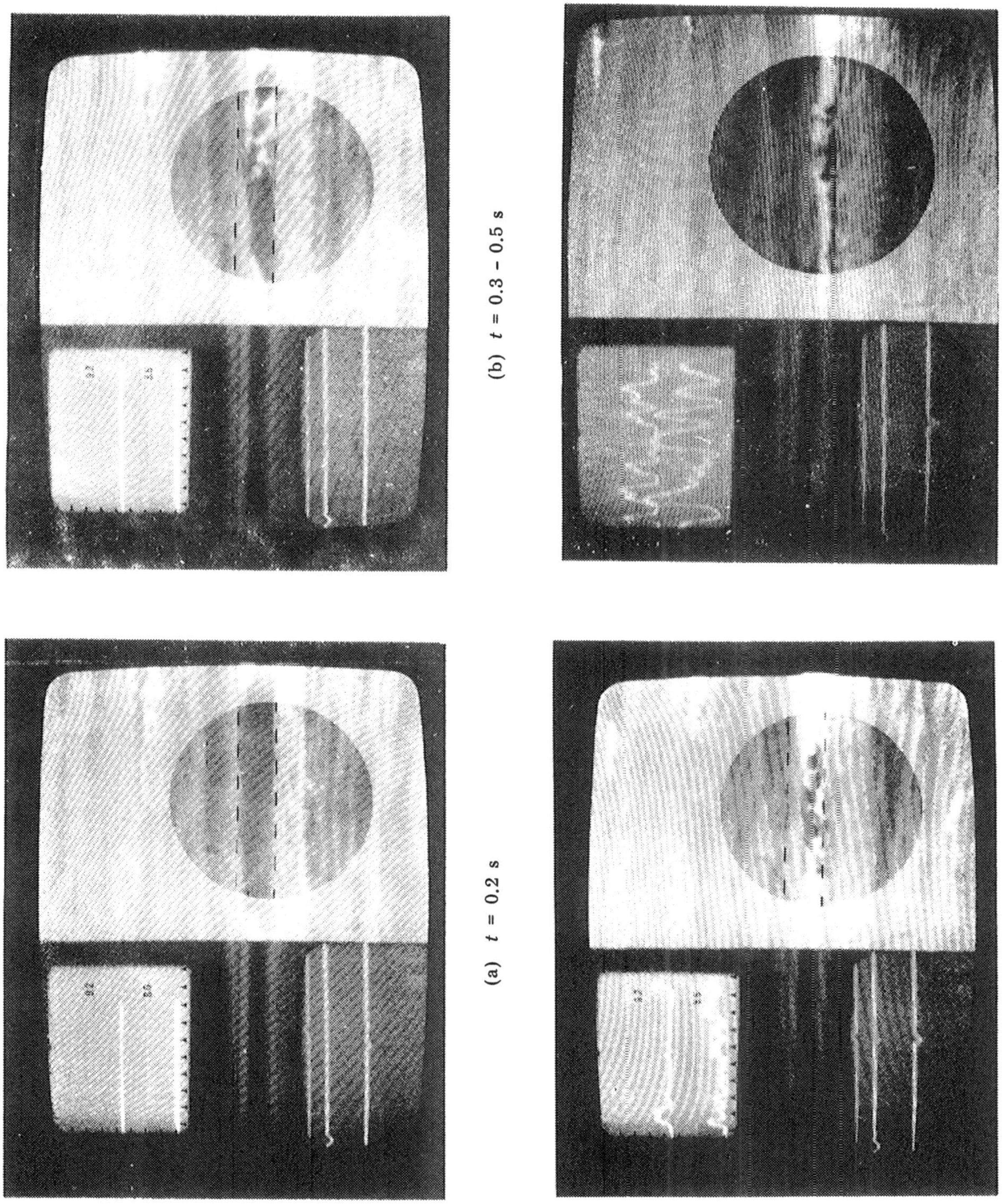

(a) $t = 0.2$ s

(b) $t = 0.3 - 0.5$ s

(c) $t = 0.6$ to 0.8 s

(d) $t = 0.9$ to 1.1 s

Fig. 4 — Sea return data obtained with vertical polarization (6-dB attenuation in receiver

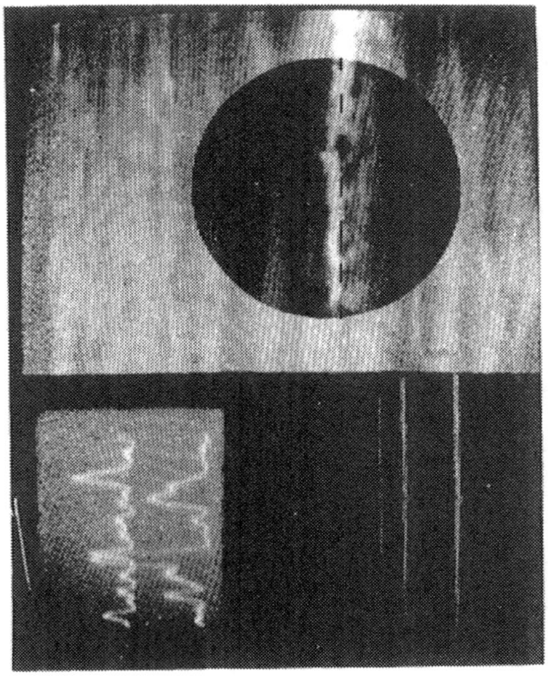

(e) t = 1.2 to 1.4 s

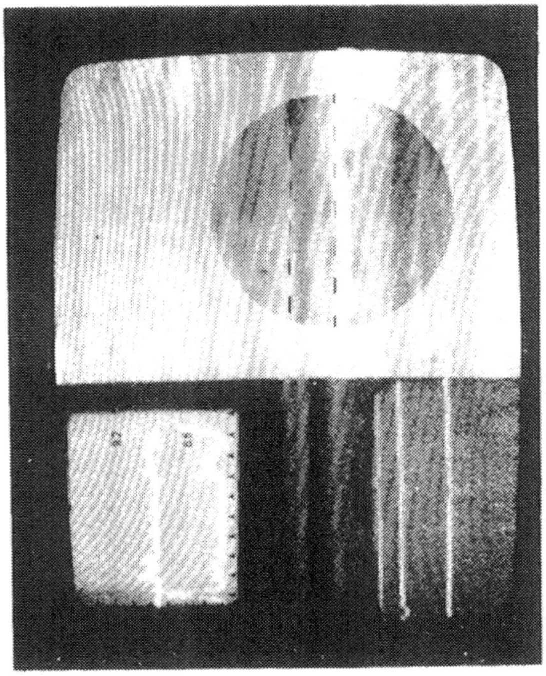

(f) t = 1.5 to 1.7 s

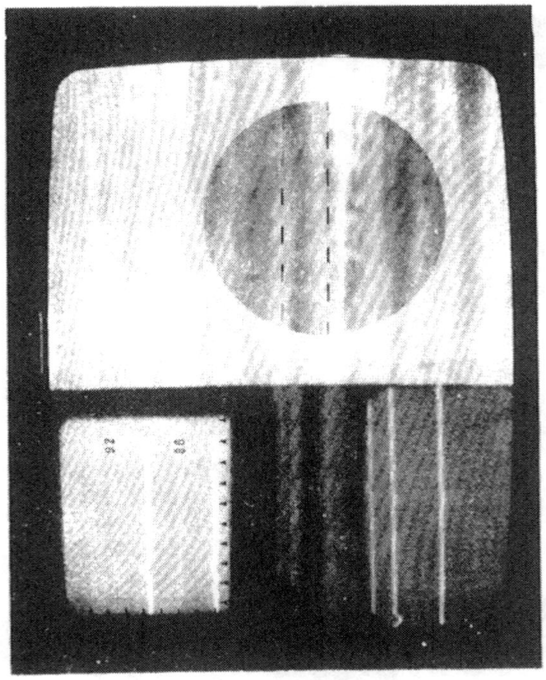

(g) t = 1.8 to 2.0 s

Fig. 4 (Continued)

portraying the gated stretched video. It should be noted that the real-time video displayed in these pictures corresponded to the last displayed stretched video only when the photos were taken before the stretched video sweep ended. In all data, the oscilloscopes sweep from left to right in time.

Figure 4a is the start of the sequence and reveals 8.6-GHz returns (real-time video) from two waves, one in front of and the other behind the range gate. The latter appears to originate from the sharp peak or crest of the second wave, and the former corresponds to the whitecap on top of the first wave. No returns are evident on 9.2 GHz.

Figure 4b portrays the sea surface 0.3 s later, with a large wave in the range gate but no real time video returns evident on either frequency. The stretched video, however, shows evidence of some small returns on both frequencies between pictures.

Figure 4c was taken 0.3 s after 4b and shows whitecaps at the two sides of the wave in the range gate, a real time video return at 8.6 GHz (note the pulse fill-in), and data on both frequencies in the stretched video display. The largest peaks in the stretched video correspond to an equivalent scattering cross section σ on the order of 1.6 m^2 in the resolution cell. The stretched video also shows that the return is heavily amplitude modulated, with nearly 100% amplitude changes in times on the order of 10 ms.

Figure 4d, taken 0.3 s after Fig. 4c, shows a whitecap filling the azimuth resolution cell as the whole wave breaks. Returns are evident on both frequencies in the real-time data, and large returns are evident in the stretched video traces. The width of the visible whitecap in range is on the order of 1 m, and the peak equivalent scattering cross sections evident from the stretched video are in excess of 4 m^2 at 9.2 GHz and 7 m^2 at 8.6 GHz. Note that the two frequency-stretched video time functions appear to be uncorrelated, as one would expect, due to the large frequency difference (600 MHz).

Figure 4e, taken 0.3 s after Fig. 4d, shows the whitecap moved toward the short range side of the range resolution cell. The stretched video shows an extremely large return on 8.6 GHz that actually went off scale and saturated. This required a σ in excess of 10 m^2.

The next picture in the sequence (4f) shows large real-time video almost out of the range gate, and Fig. 4g completes the sequence of the wave moving through the range gate.

Figure 5 is another typical vertical polarization sequence of a whitecap forming and moving out of the range gate. Note that the wave traveled about 1.33 m in 1.2 s and that the size and shape of the broken water region were constantly changing. The largest σ values in this sequence were on the order of 7 m^2. Figure 6 is a wave sequence taken

Figure 6 is a wave sequence taken with the radar polarized horizontally. In these data, the calibrated attenuator was varied to increase the receiver sensitivity by 5 dB. The time intervals between pictures in the sequence were 0.3 s, as in the vertically polarized case, and all trace speeds were maintained.

Figure 6a shows a whitecap forming on a wave at the outer end of the range resolution cell. *Note the optical image of the whitecap in the water in front of the whitecap.*

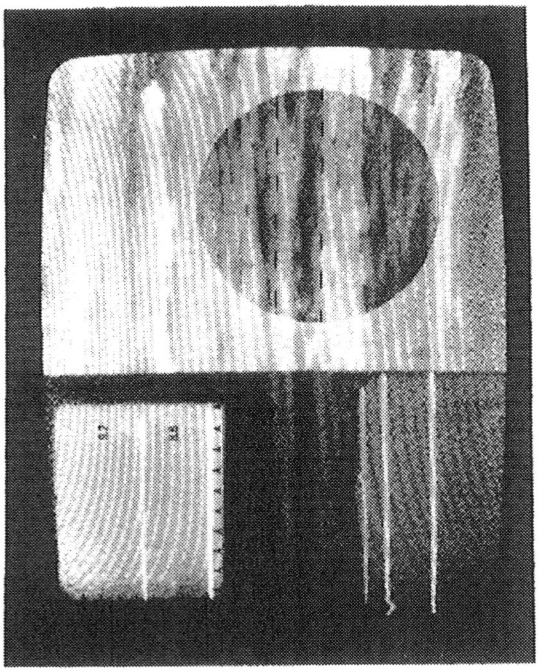

(a) $t = 0$ to 0.2 s

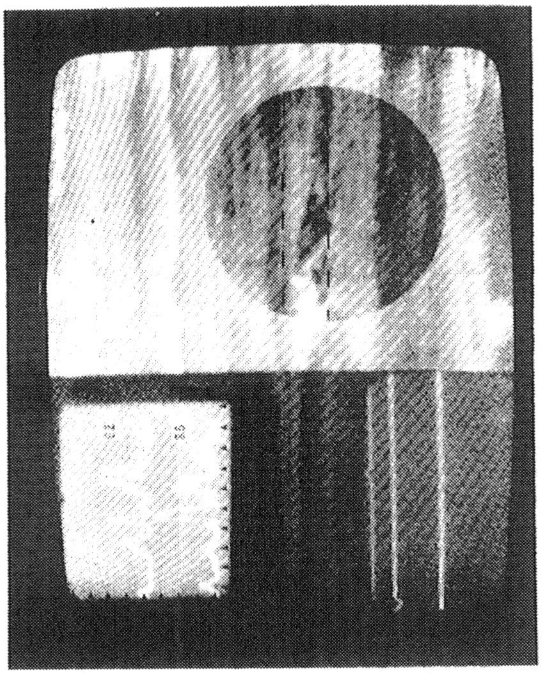

(b) $t = 0.3$ to 0.5 s

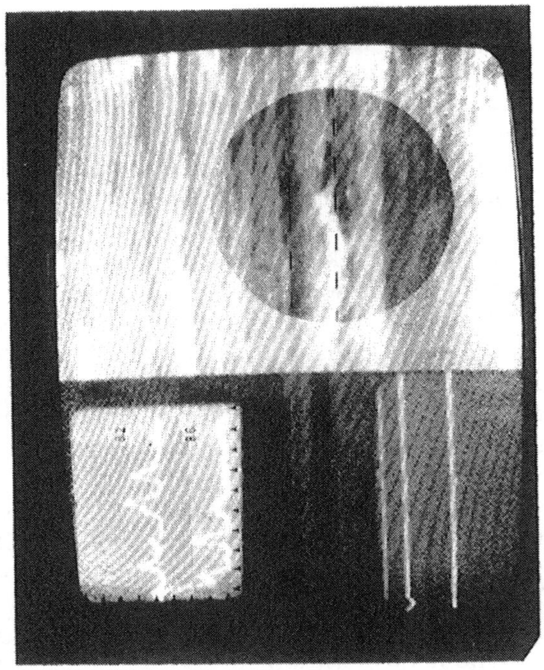

(c) $t = 0.6$ to 0.8 s

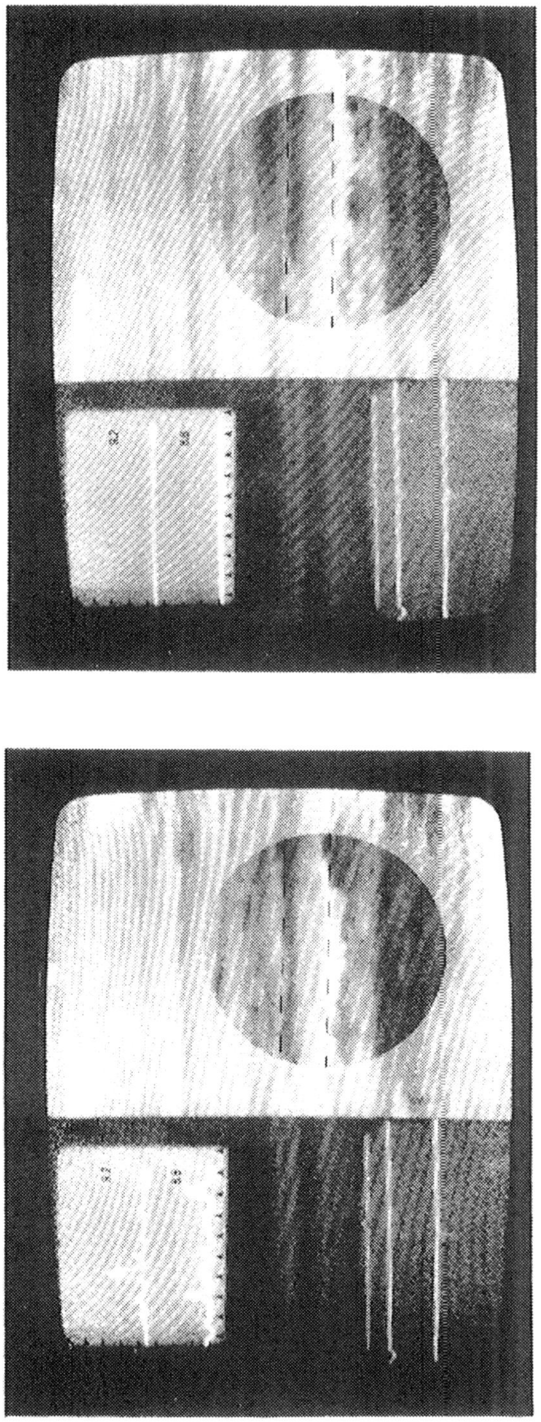

(d) t = 0.9 to 1.1 s

(e) t = 1.2 to 1.4 s

Fig. 5 — Sea return data obtained with vertical polarization (6-dB attenuation in receiver)

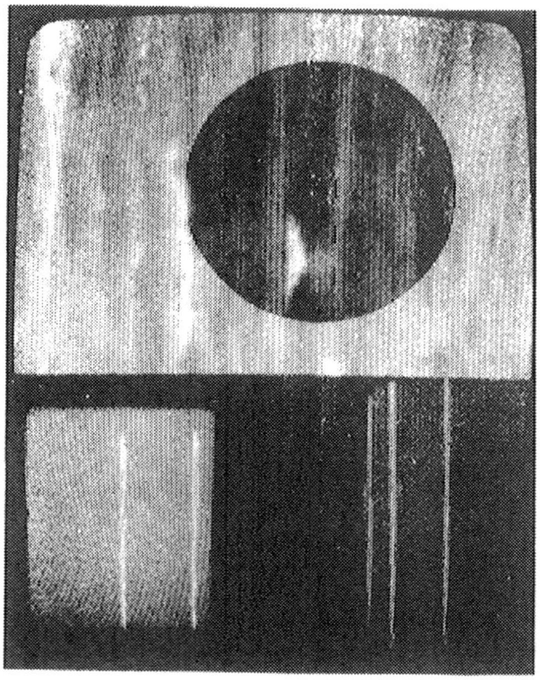

(a) $t = 0$ to 0.2 s

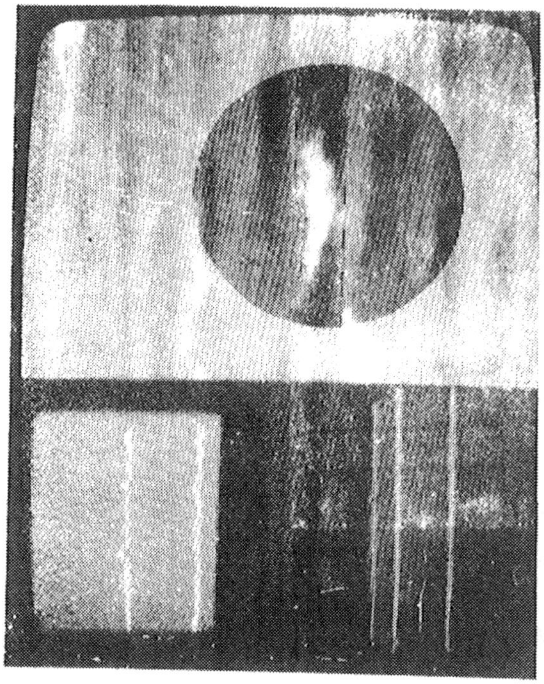

(b) $t = 0.3$ to 0.5 s

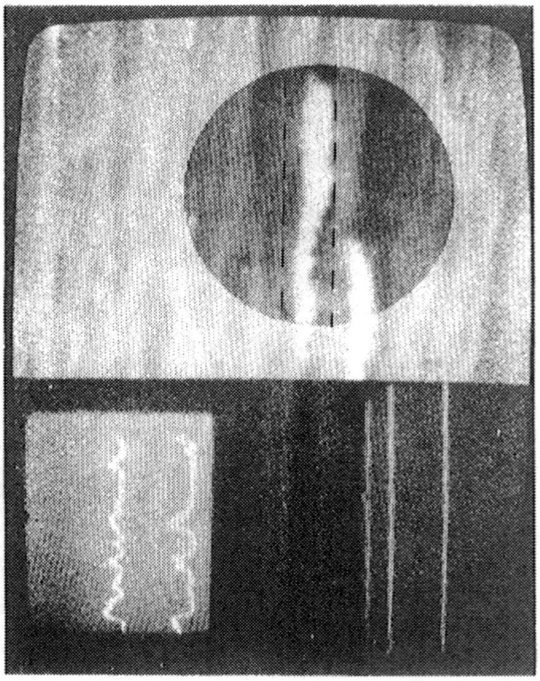

(c) $t = 0.6$ to 0.8 s

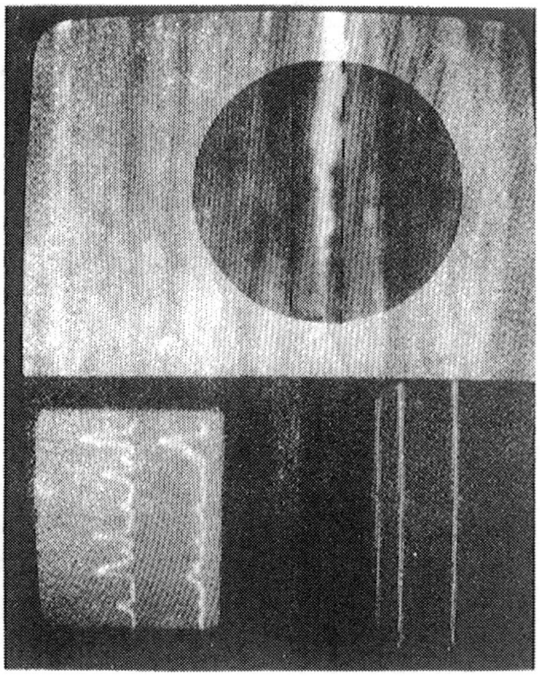

(d) $t = 0.9$ to 1.1 s

(e) $t = 1.2$ to 1.4 s

(f) $t = 1.5$ to 1.7 s

Fig. 6 — Sea return data obtained with horizontal polarization (1-dB attenuation in receiver)

This optical imaging was noted many times in the measurement program and will be shown to be significant in later sections of this report. Real-time video is evident, but the stretched video trace had ended prior to the video occurrence and shows no data.

Figure 6b shows the whitecap greatly increased in size and indications of returns on the gated stretched video traces. The peaks in those traces correspond to σ values of 0.2 m^2.

Figure 6c shows larger echoes on the stretched video traces, with peak σ values of about 1 m^2. Note the characteristic high-frequency, large-percentage amplitude modulation on the echo as a function of time.

Figure 6d shows both real-time video and previous echoes from σ values on the order of 1.5 m^2 peak. Figures 6e and 6f complete the sequence.

Figure 7 is another typical sequence with horizontal polarization. Figure 7a starts the sequence with no visible returns but with a large wave in the resolution cell.

Figure 7b shows the wave starting to break and evidence of radar echoes in the gated stretched video. The peak echoes on 9.2 GHz were from σ values on the order of 1 m^2, and those on 8.6 GHz were on the order of 0.5 m^2.

Figure 7c shows the wave break expanding across the azimuth resolution cell. Real-time video is evident, and rapidly modulated returns are evident in the gated stretched video. The σ values in this picture were in excess of 1.5 m^2 at 9.2 GHz and 0.5 m^2 at 8.6 GHz.

Figure 7d shows evidence of a very large return on 8.6 GHz ($\sigma \approx 2.5$ m^2), and the wave is broken over most of the azimuth resolution cell.

Figure 8 shows the radar return from a semisubmerged platform on both horizontal (Fig. 8a) and vertical (Fig. 8b) polarization. Note that the echo is modulated as the water rises and falls. Note also that this modulation is very low in frequency compared to that characteristic of echoes from breaking waves.

DEEP WATER DATA

Sea return measurements were also made at Boca Raton, Florida, in deep water. In these measurements, the 2.67-m-diameter antenna was employed with a 1° beamwidth. The antenna was mounted on top of a building 13 m above mean sea level, and the center of the beam intercepted the water at a distance of 530 m. This produced a grazing angle of about 1.4°. The boresight television camera field of view was narrowed to permit it to cover the radar beam in azimuth (1°) on the split-screen display, and it provided a 2° vertical field of view.

The 530-m range and 1° beamwidth produced an azimuth resolution of about 10 m, and the 20-ns pulse yielded a range resolution of 3.33 m.

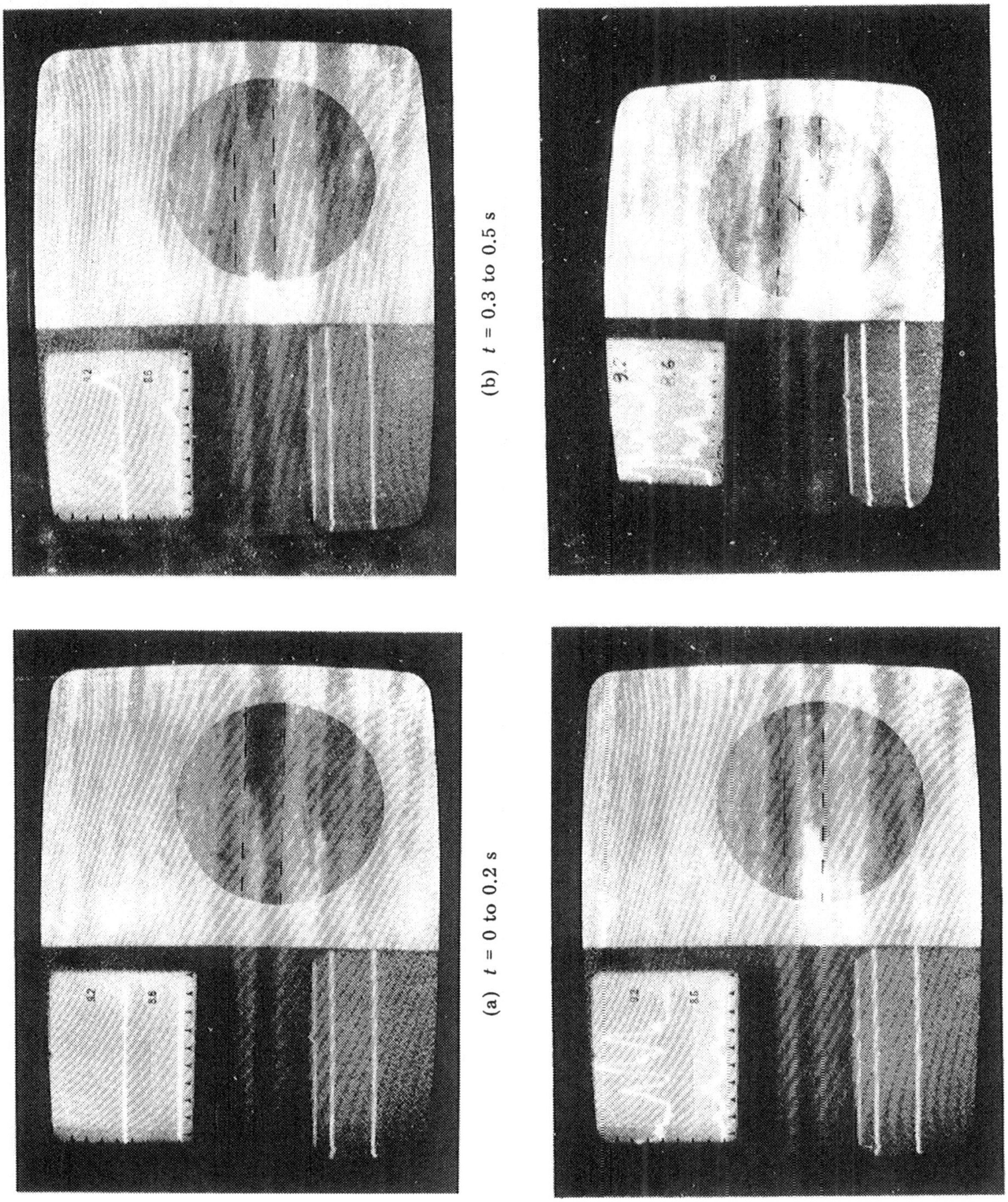

(a) t = 0 to 0.2 s

(b) t = 0.3 to 0.5 s

(c) t = 0.6 to 0.8 s

(d) t = 0.9 to 1.1 s

Fig. 7 — Sea return data obtained with horizontal polarization (1-dB attenuation in receiver)

(a) horizontal polarization

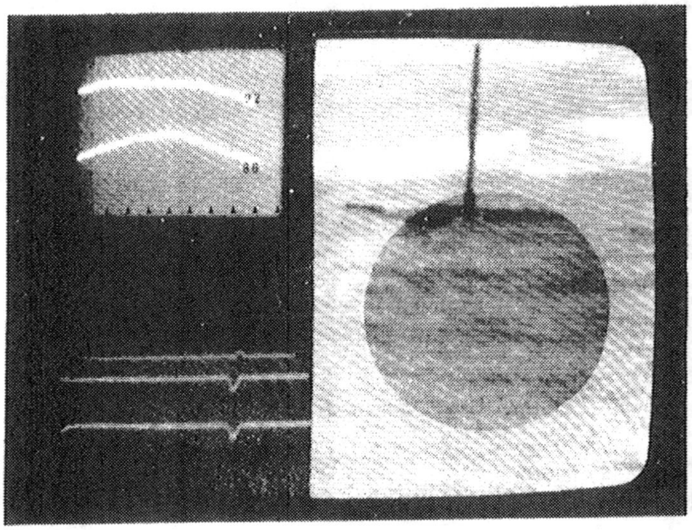

(b) vertical polarization

Fig. 8 — Return from partially submerged platform
(1-dB attenuation in receiver)

Whitecap data obtained from this site with 1- to 1.3-m waves were very similar to data obtained at Chesapeake Beach. However, an opportunity occurred to observe whitecaps from waves traveling across the field of view. Data from these waves revealed much higher frequency amplitude modulation than those from waves breaking toward the radar. Autocorrelation times on the order of 3 ms were observed. Figure 9 is characteristic of the returns for vertical and horizontal polarization. (The triangular marker on the side of the boresight picture marks the range gate position.)

The increased frequency of the amplitude modulation can be explained by the increased rate of growth of the whitecap in range due to the angle between the wavefront and the line of sight. Note that the growth rate of the Chesapeake Beach whitecaps (Fig. 6) was much higher along the water wavefront than normal to the wavefront. Differences of 4 to 1 were common.

Data were also obtained on small unbreaking waves at Boca Raton, Florida. These waves ranged from 0.15 to 0.3 m in height. Analysis of this data has not been completed at this time. However, several obvious facts were noted. These were as follows:

1. The echoes (Fig. 10) had the same temporal and spatial characteristics as those from whitecaps but were much smaller in magnitude. They had autocorrelation times of 10 ms or less and were widely separated in range (relatively improbable in any given range resolution cell).

2. Echoes on horizontal polarization were less probable than on vertical polarization but were higher in amplitude when received. The latter effect was attributed to the fact that the angle of incidence on the waves was near Brewster's angle for vertically polarized radiation.

3. The lifetime of any return was on the order of 1 s or less.

MEDIAN EQUIVALENT SCATTERING CROSS SECTION OF WHITECAPS

The large equivalent scattering cross section of whitecaps at small radar depression angles can be explained theoretically by assuming them to be very rough surfaces at the wavelength of interest (3 cm). Such rough surfaces scatter incident radiation in all directions in the upper half space.

The grazing angle of the radiation on the whitecap can be approximated by the sum of the radar depression angle and half the slope of the wave when it breaks (30°). Thus, for small depression angles, the power intercept area of the whitecap can be taken to be

$$A_S = A \sin 15^\circ = A/4, \qquad (1)$$

where A is the actual area of the visible whitecap.

With this power intercept area, the equivalent scattering cross section is

$$\sigma = A_S\, G_S = A G_S/4, \tag{2}$$

where G_S is the gain of the scatterer in the direction of the radar receiver. G_S in this case will be determined by the roughness of the whitecap, by the fact that energy can scatter only into the upper half space, and by the fact that an image of the whitecap will form in the surrounding unbroken water, i.e.,

$$G_S = G_R\, G_{HS}\, G_{TI} \tag{3}$$

where G_R is the roughness factor, G_{HS} is the half-space factor, and G_{TI} is the target-image factor.

For very rough surfaces $G_R \approx 1$, $G_{HS} = 2$, and $G_{TI} = 2$, since the image of the whitecap can have the same power intercept area as the whitecap. With these values, Eq. (2) becomes

$$\sigma = A_S. \tag{4}$$

It should be noted at this point that the σ, given by Eq. (4), is the mean value. Since the target and its image are illuminated by coherent radiation, they can interfere constructively and destructively. In addition, the coherent scatter from the different portions of the rough surface can interfere constructively or destructively. In fact, the whitecap's height above mean sea level and its size and structure will change with time. This will cause severe amplitude modulation of σ with time, as observed in the data obtained in sea return measurements.

The measurement program revealed that Eq. (4) is useful and reasonably accurate. Dividing this equation by A to obtain σ_0 yields $\sigma_0 = 1\ m^2/m^2$. This is on the order of 35 dB greater than the σ_0 quoted in standard tables for sea scatter in low-resolution radars for sea states 3 and 4 and 3° grazing angles [1]. The difference can be justified by the large ratio of sea surface not covered by whitecaps to that covered by white caps.

Applying Eq. (4) to the data in Fig. 4d, where A_S was noted to be 1 m deep in range by 2 m in azimuth, would yield an average $\sigma \approx 2\ m^2$. This is in reasonable agreement with the data, considering the statistics involved.

MULTIPATH EFFECTS

Interference between the echo from the whitecap and its image can produce a relatively slow modulation on the composite return as the object and image spacing vary with time. As previously noted, images have been observed at optical frequencies and

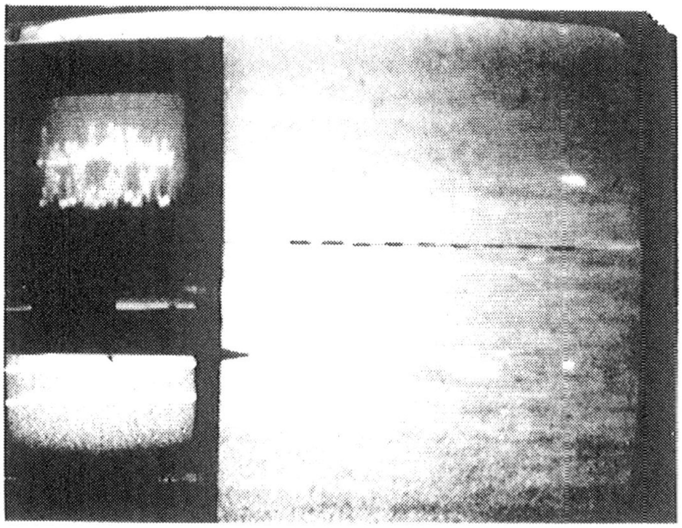

(a) vertical polarization

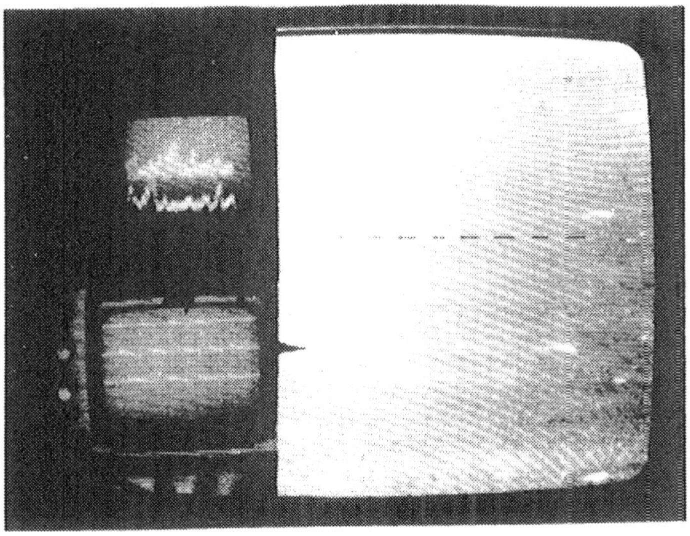

(b) horizontal polarization

Fig. 9 — Returns obtained in deep water, 20 ms/cm sweep; stretched video

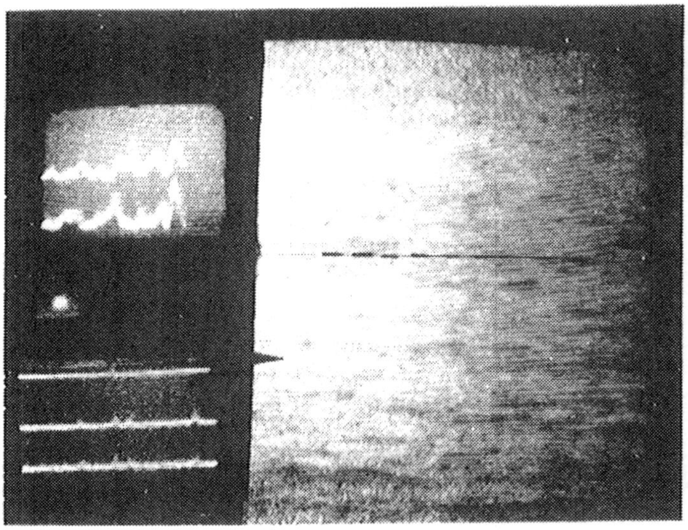

(a) vertical polarization

(b) horizontal polarization

Fig. 10 — Returns from small waves in deep water

would be even more likely at microwave frequencies. Whitecaps appear frequently at sea state 3, for which the wave height and period are about 1.25 m and 4 s, respectively. At a wavelength of 3.3 cm (corresponding to 9.1 GHz), the vertical lobing pattern width ($\theta = \sin^{-1} \lambda/2H$) is about 0.8°. For the CBD tests the grazing angle was 4.6°, so that the radar antenna was located in the sixth lobe of the whitecap interference pattern. For the 4-s period, this suggests a maximum detected modulation frequency of 3 Hz. At most radar operating ranges, the grazing angle will be much less. If it is less than 0.8°, the signal variation due to multipath will be at the wave period, corresponding to a frequency of 1/4 Hz. The elevation lobing phenomenon is suggested as one reason for observed time delays that often occur between the formation of a whitecap and the reception of the sea spikes. It can also explain the relatively long time in which either the 8.6- or 9.2-GHz return is detected in the absence of the other.

POLARIZATION DEPENDENCE OF WHITECAP ECHOES

All the data taken in this study revealed that the equivalent scattering cross section of broken water (whitecaps) was on the order of 5 dB higher for vertical polarization than for horizontal polarization. This agrees with Nathanson [1]. However, this difference is not predictable on the basis of scattering from a randomly rough surface. Such a surface should scatter all polarizations equally.

A possible clue to the cause of this polarization sensitivity was obtained from echoes from sharp-topped, unbroken waves. Such echoes are much higher for vertical polarization than for horizontal polarization. The sharp top acts as an impedance discontinuity to surface currents induced by incident radiation. Such currents partially reflect from this discontinuity and radiate backward like long-wire standing wave antennas [2,3] with significant directivity and gain. An example of such augmentation of equivalent scattering cross section is evident in the variation of σ with incidence angles of radiation on a long wire.

Skolnik [4] shows such a plot for a rod 39 wavelengths long and 1/4 wavelength in diameter. This plot shows σ falling off rapidly with the departure of the grazing angle from 90°. However, σ begins to rise again at small grazing angles and peaks at about 8° to a value several orders of magnitude greater than that at larger grazing angles. This 8° angle is the position of the largest radiation lobe of such a rod when excited by a current, at the frequency of interest, inserted at one end.

This mechanism could explain the polarization sensitivity of sea scatter since the high-gain lobes make small angles with the reflected surface current and since no reflected surface currents are produced by incident radiation with the E field parallel to the sharp top of the wave. Vertically polarized radiation will induce such reflected currents on wave tops with the maximum scatter gain back in the direction of the radar (in monostatic systems), as shown in Fig. 11. Horizontally polarized radiation can induce such reflected signals, but the reflected surface currents will make big angles with the line of sight from the radar (Fig. 12) and the large high-gain long-wire lobes will miss the radar.

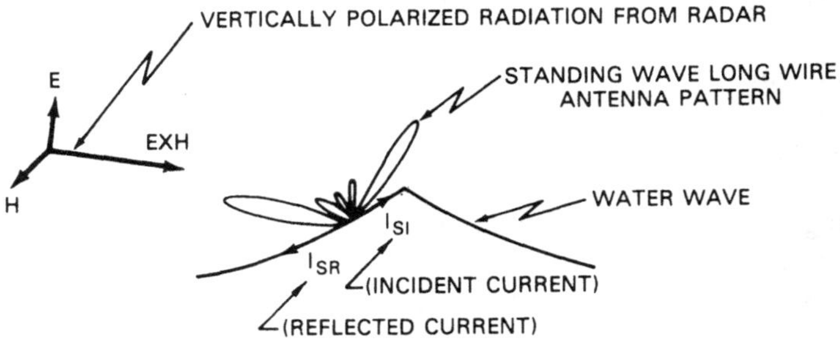

Fig. 11 — Vertically polarized scatter from water wave

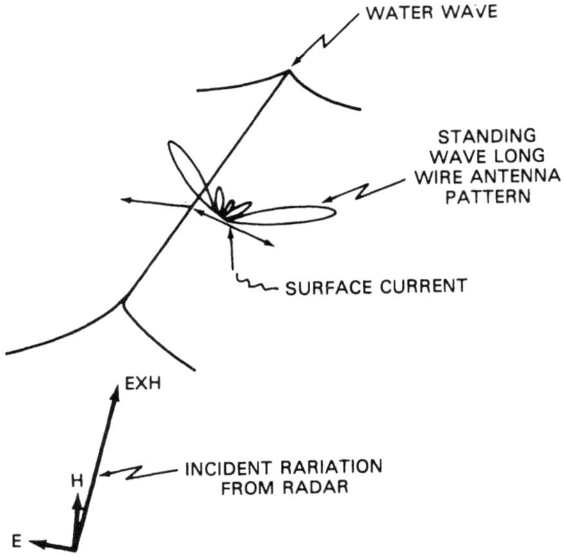

Fig. 12 — Horizontally polarized scatter
from water wave

This phenomenon can be very important in the case of whitecaps, where there is an abrupt discontinuity between the unbroken water and the broken water falling down the face of the wave (Fig. 13). In this case, echoes from horizontally polarized radiation would originate mainly in the rough broken water, while those from vertically polarized radiation would originate both in front of and within the broken water.

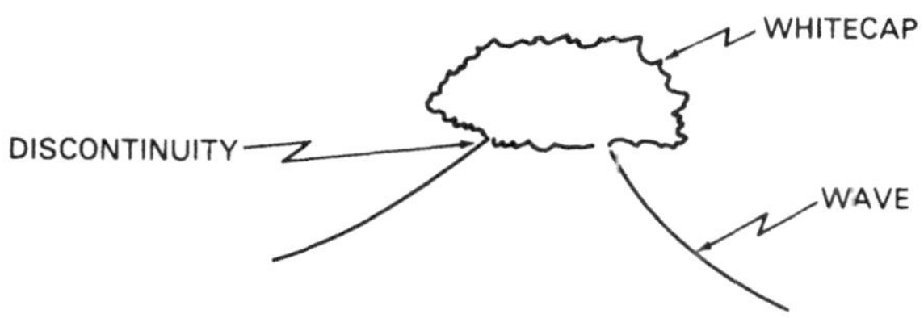

Fig. 13 — Discontinuity at whitecap-wave interface

Future work is planned to test this hypothesis using bistatic radar measurements of sea scatter.

THEORETICAL SCATTERING MODELS

Many hours of high-resolution radar sea return recordings were made in the Chesapeake Bay and in deep water off Boca Raton, Florida. In all cases, large returns were found to be associated with broken water in whitecaps; returns from unbroken water were much smaller. In all cases the returns were found to be amplitude modulated, with relatively high frequencies and high modulation indexes. If these findings are generally applicable, they provide a basis for discriminating between sea return and echoes from rigid debris floating on the water, because the latter would display much lower frequency-amplitude modulations. The general applicability of these observations would be more acceptable if theoretical analysis explained and predicted the modulation characteristics of sea return. As a consequence, a theoretical analysis was undertaken.

The approach to this analysis was to construct a series of possible models of the surface and to compare calculated scattering from these surfaces with measured data from the sea surface.

In modeling the surface, it was noted that coherent illumination at low grazing angles at a wavelength λ would define the autocorrelation length in range to be $\lambda/4$; i.e., echoes from scatterers separated in range by less than $\lambda/4$ would be correlated and would interfere constructively. It was also noted that the autocorrelation time of the X-band echoes from whitecaps was on the order of 10 ms, and that this corresponded very closely to the observed time necessary for the whitecap to grow or shrink in range length by $\lambda/4$. This growth rate was also measured by B. L. Hicks et al. [5]. As a consequence, all models employed one scatterer per $\lambda/4$ range increment, rather than a continuum of scatterers.

All models of whitecaps were based on the assumption that the whitecap started from zero size, grew in range to a finite value, then shrank back to zero. This growth or shrink rate was assumed to be on the order of $\lambda/4$ in 10 ms. These assumptions were based on hours of visual study of the sea surface via television cameras with known frame rates and thus were not completely arbitrary.

The various models constructed differed in the degree of randomness of the surface and in the ratio of internal to external autocorrelation times. The latter is the time of addition of an extra $\lambda/4$ zone, and the former is the decorrelation time of scatterers in zones already formed.

The first model of the surface to be constructed assumed that the amplitude of the reflected signal from the nth $\lambda/4$ zone added would be represented by

$$e_n = (1 - kV_n)(-1)^{n-1} \tag{5}$$

where V_n was picked from a table of random numbers uniformly distributed between 0 and 1, and k was the assumed randomness, having a value from 0 to +1. The equivalent scattering cross section of a surface composed of n zones was then defined to be

$$\sigma_n = \left(\sum_{i=1}^{n} e_i \right)^2 . \tag{6}$$

Differences in internal and external autocorrelation times were incorporated by updating (drawing new random numbers for) already formed zones when such zones were m zone additions old. For example, an $m = 1$ would imply equal internal and external autocorrelation times and was instrumented by summing new random numbers for each zone every time a new zone was added. Thus, σ_n involves different random numbers from those forming σ_{n-1} or σ_{n+1}. An $m = 5$ would imply an internal autocorrelation time five times longer than the time required for the surface to grow by one zone. Thus, new random numbers would be drawn for zones that are five zone additions old.

For all models, σ was plotted against n (the number of zones formed) for five different realizations of the surface (five different sets of random numbers) and for k values of 1, 0.50, and 0.25. In all cases n_{max} was taken to be 128, which at X-band would correspond to a whitecap 1.05 m long. In addition, the spectrum of the resultant time function was plotted by passing the function through a 128-point fast Fourier transform (FFT) program. Since only real numbers were used, the plot was extended only to 65 frequencies, because frequencies from 65 to 129 would be the mirror images of those plotted.

Figure 14 is a plot of σ vs n for $k = 0$. Note the σ variation from 0 to 1 to 0, etc., as zones are added (as would be expected). The spectrum in Fig. 14 is the result of passing the σ values through the FFT, which produced a DC term and a term at half the sampling rate.

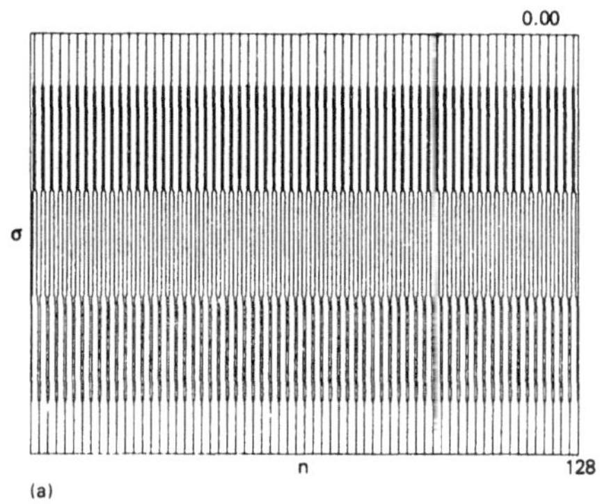

(a) σ vs n

(b) power P vs frequency f

Fig. 14 — Random-amplitude model, k = 0

Figures 15a through 15e are σ and spectra plots of different realizations (different random numbers) with $k = 0.25$ and $m = 1$. In these plots, the vertical scale is fixed by the largest value encountered, and the spectra are the result of passing the amplitudes associated with the σ values through the FFT. In these spectra the amplitudes were equally likely to be positive or negative. Note that the largest peaks are on the order of eight times the value for one zone, which in this case is dominated by the coherent component and is evident at $n = 1$, 3, and 5.

Figure 15f is the average spectrum of the five realizations with $k = 0.25$ and $m = 1$.

Figures 16a through 16e are the σ and spectra plots of five different realizations of $k = 0.5$ and $m = 1$, with the vertical scale normalized to the largest peak. Note that the largest peaks have increased with respect to the average σ for one zone due to the increase in k. Again, the spectra resulted from passing the amplitude values from which the σ plot was derived through the FFT. As a consequence, the DC component is relatively small compared to the total power in the spectrum. Figure 16f is the average spectrum of the five realizations.

Figures 17a through 17e are the result of $k = 1$, i.e., a completely random surface. Note that the largest σ values are many times that for one zone and the spectrum contains many high-frequency components plus a DC term. Figure 17f is the average spectrum.

Comparison of these data with the experimental results shows good agreement, especially with $k = 1$.

Figures 18 through 20 are the results of k values of 0.25, 0.5, and 1 and an m value of 5; i.e., zone updates every five zones added. These results give longer spikes (longer autocorrelation times) than were evident in the experimental results.

The agreement of the model used for Fig. 17 with the measured data indicates that whitecaps are very rough surfaces with nearly equal internal and external autocorrelation times. The agreement also adds confidence that the water dynamics will always produce a sea return with high-percentage high-frequency amplitude modulations.

The previously discussed theoretical models employed only random amplitude. The return from each zone was assumed to come from the center of the zone. Another type of model was also developed that allowed both the amplitude and phase of zone returns to vary.

In this model, the return from the nth zone was defined as

$$e_n = (-1)^{n-1}(1-kV_{n1})\left\{\cos\left[k\left(\frac{1}{2} - V_{n2}\right)\pi\right] + j\sin\left[k\left(\frac{1}{2} - V_{n2}\right)\pi\right]\right\}, \tag{7}$$

where k is the randomness factor $0 \leqslant k \leqslant 1$ and V_{n1} and V_{n2} are two random numbers, each uniformly distributed between 0 and 1.

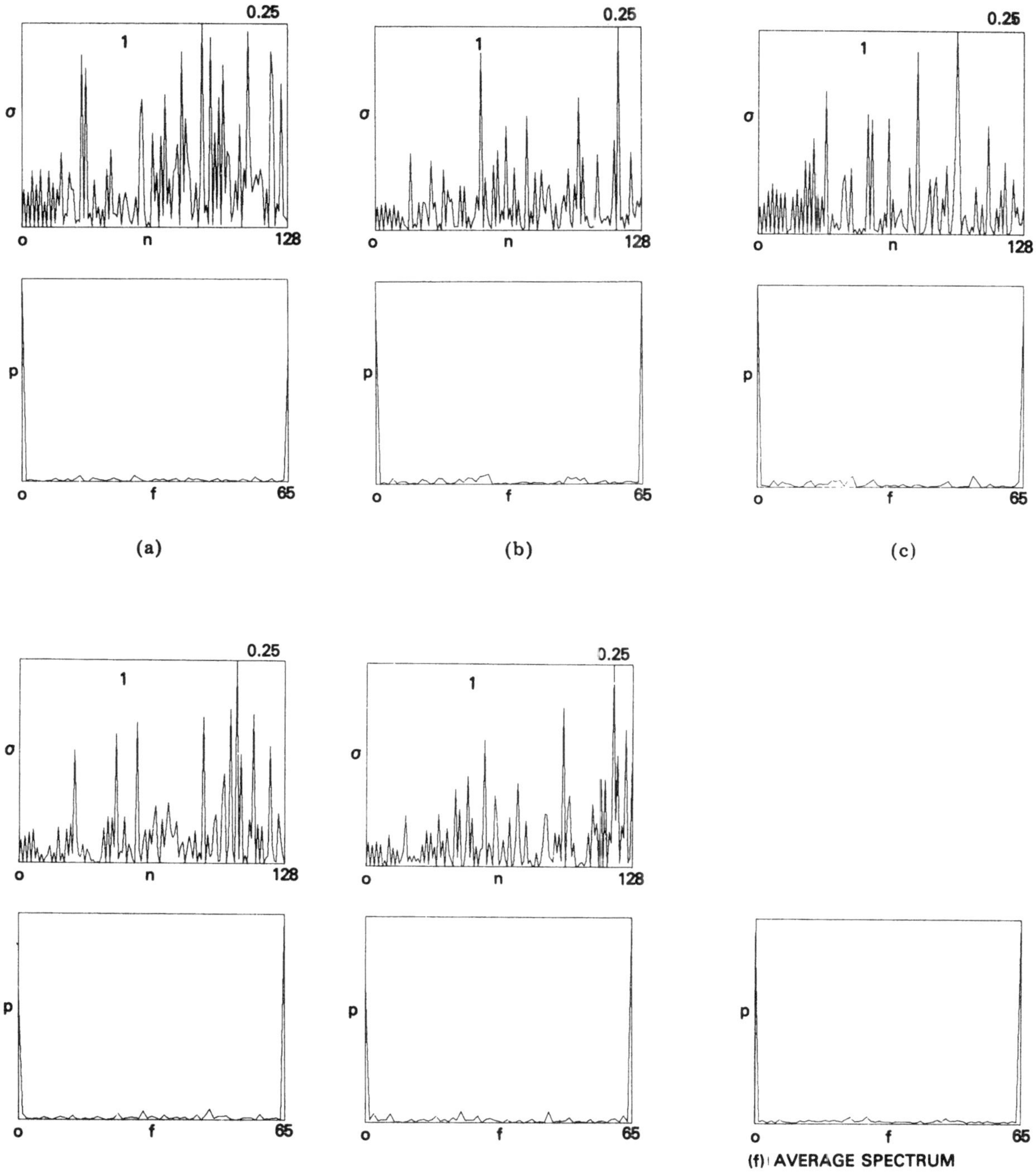

(a) (b) (c)

(d) (e) (f)

Fig. 15 — Random-amplitude model, five realizations (k = 0.25 and m = 1), σ vs n and power P vs frequency f

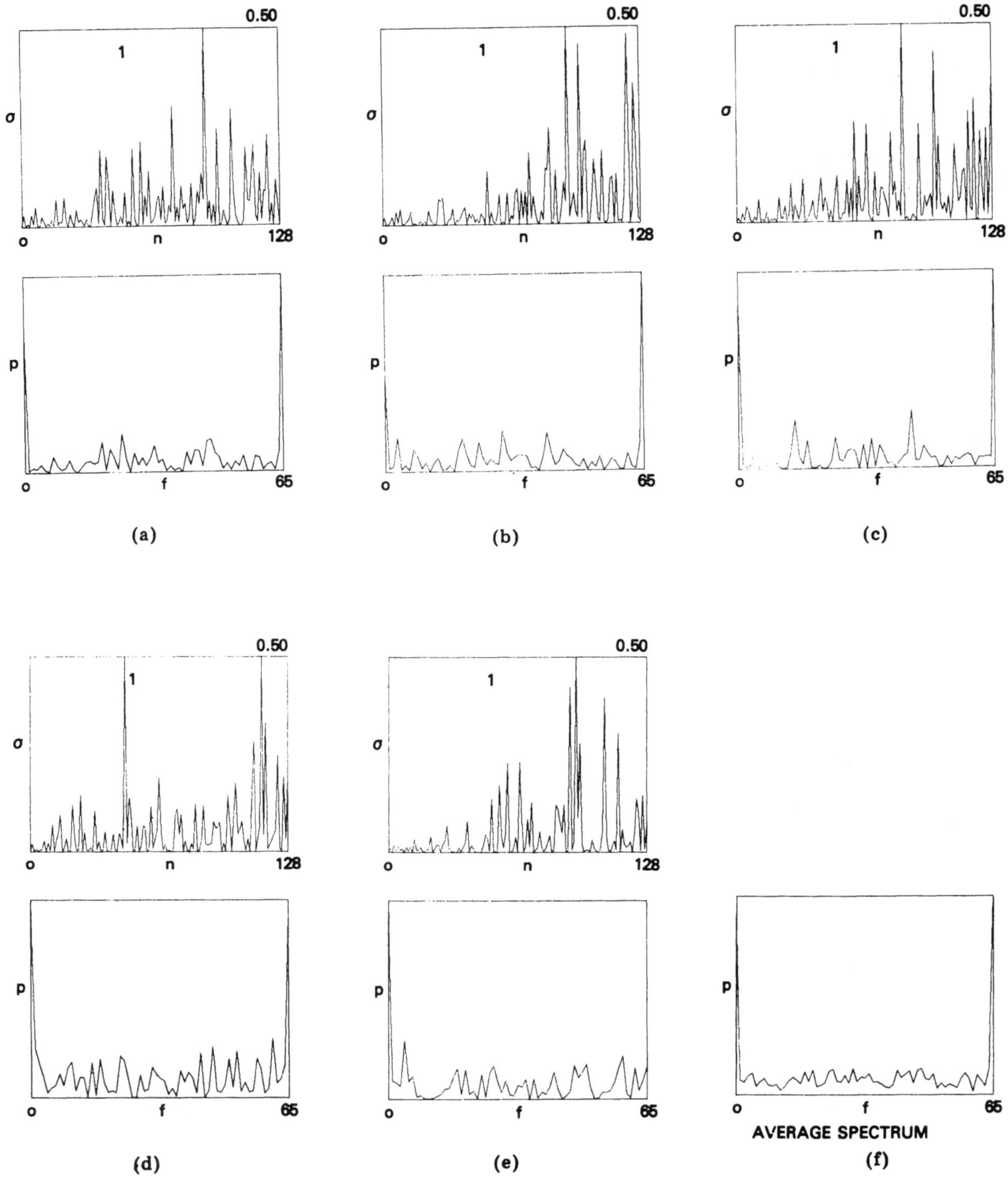

Fig. 16 — Random-amplitude model, five realizations (k = 0.5 and m = 1), σ vs n and power P vs frequency f

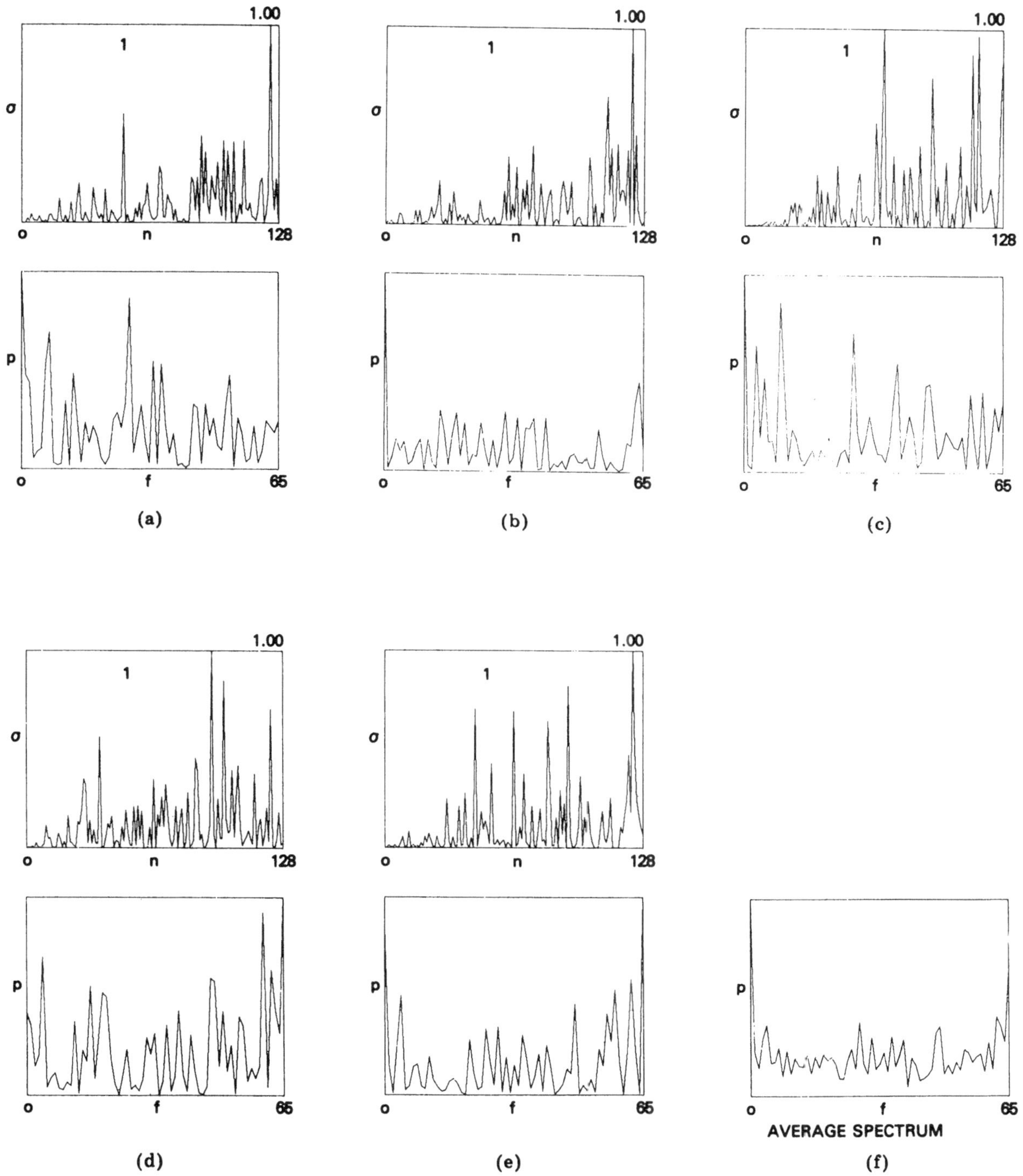

Fig. 17 — Random-amplitude model, five realizations (k = 1 and m = 1), σ vs n and power P vs frequency f

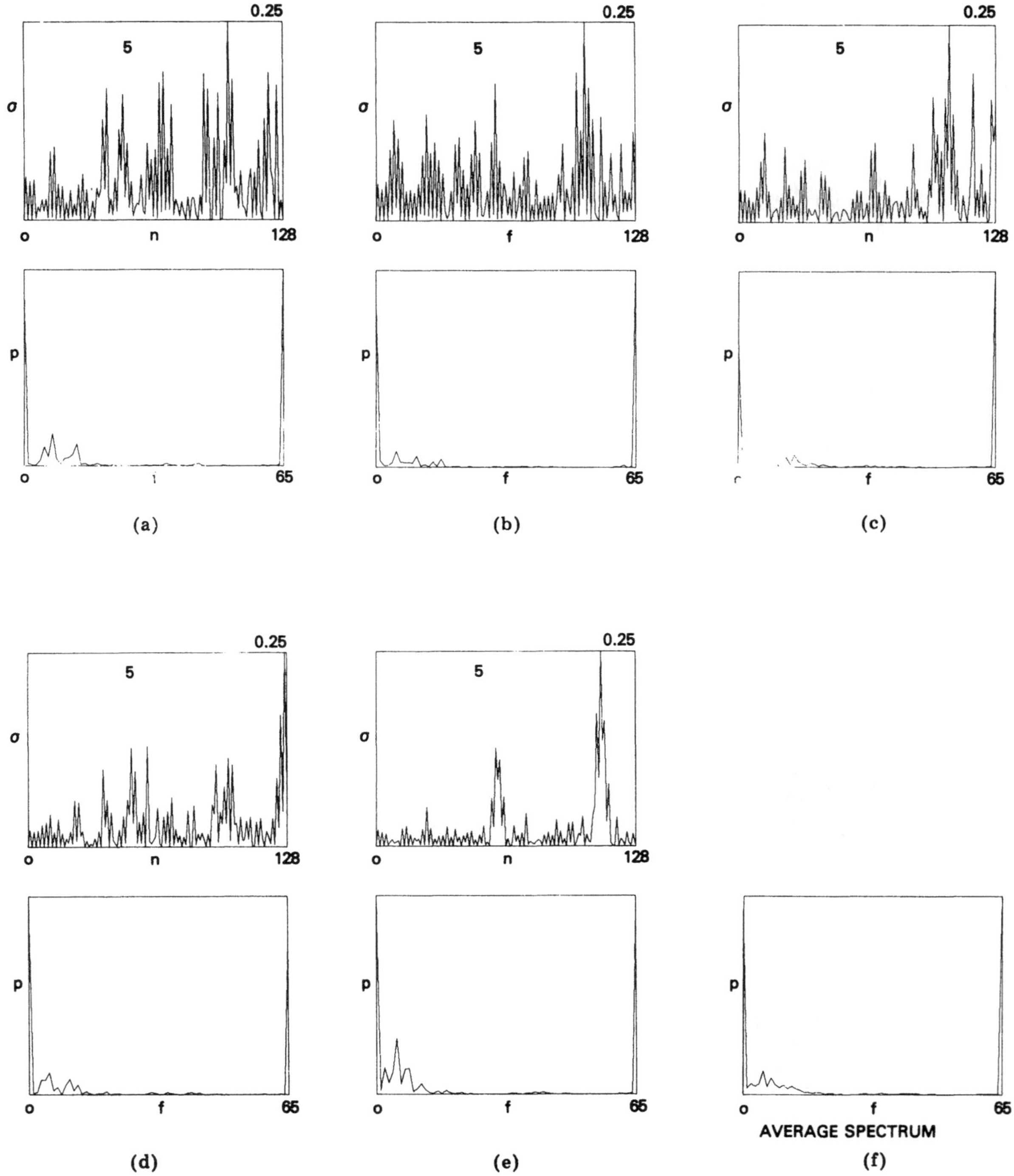

Fig. 18 — Random-amplitude model, five realizations (k = 0.25 and m = 5), σ vs n and power P vs frequency f

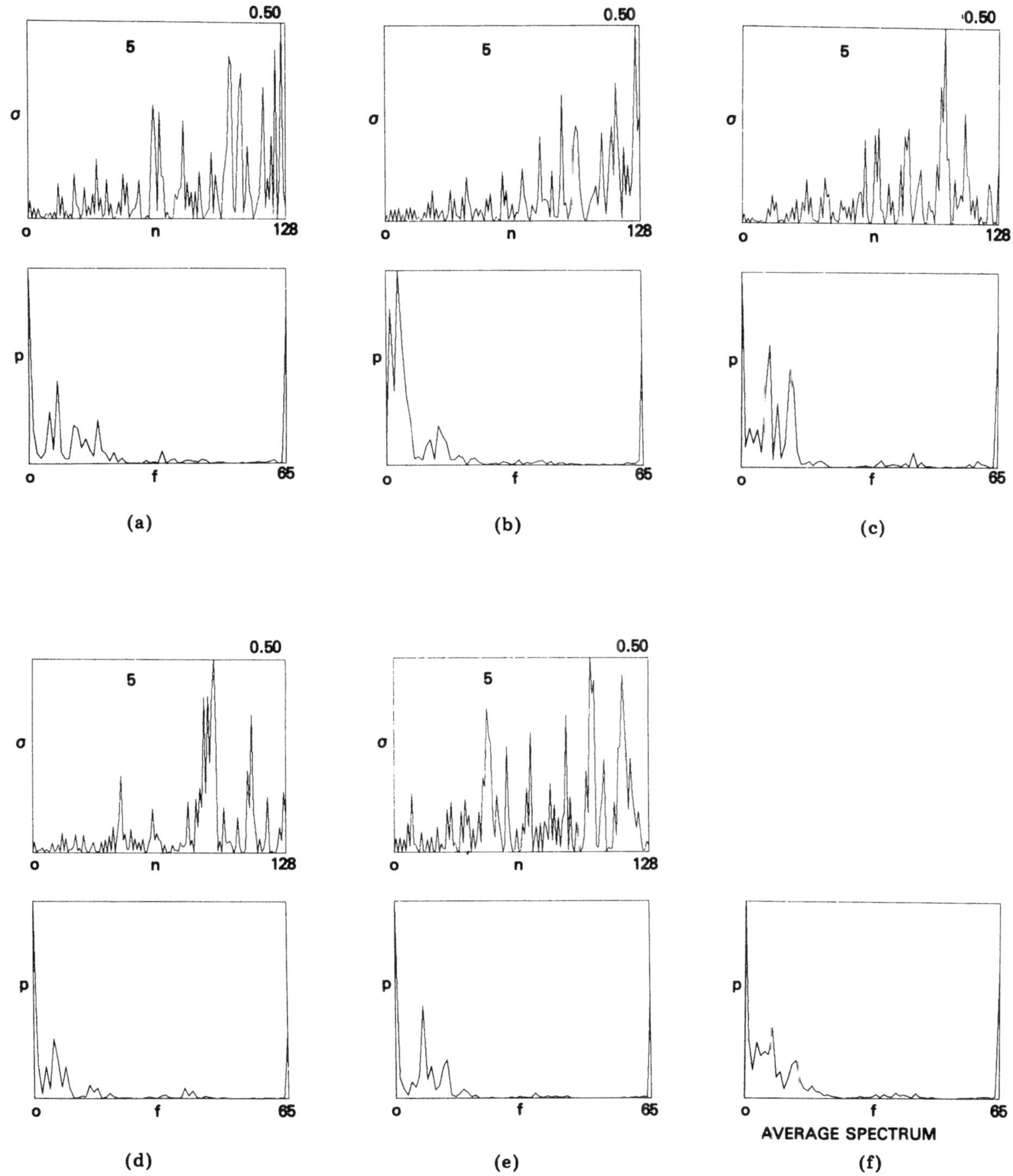

Fig. 19 — Random-amplitude model, five realizations (k = 0.5 and m = 5), σ vs n and power P vs frequency f

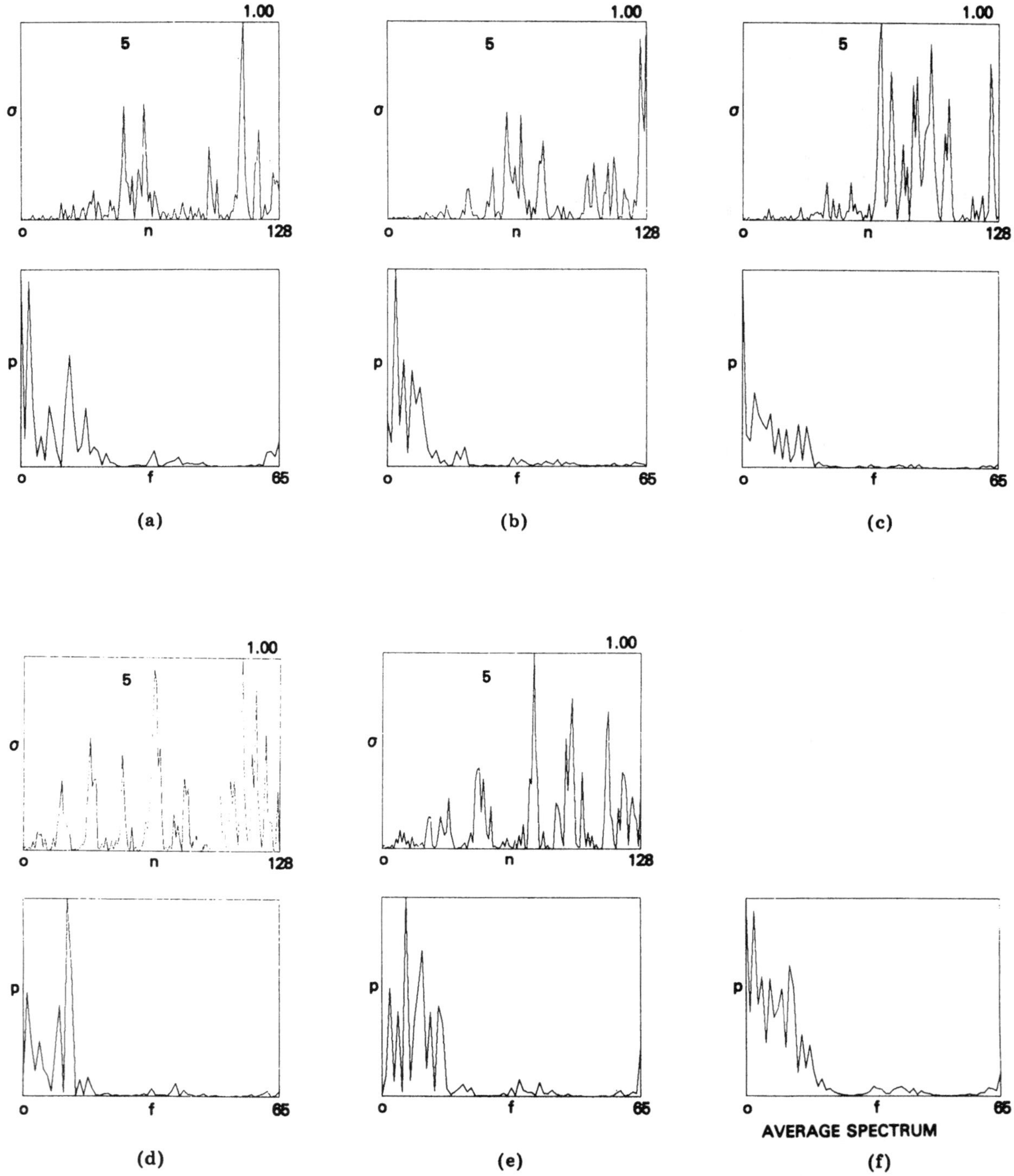

Fig. 20 — Random-amplitude model, five realizations (k = 1 and m = 5), σ vs n and power P vs frequency f

Figures 21 through 23 and 24 through 26 illustrate the results of this model with $k = 0.25$ and 0.5, and 1, and with $m = 1$ and 5. A comparison of these results with those from the random-amplitude-only model reveals very little difference in character. However, the DC component of the spectra of the $m = 1$ data appears to be lower in the random-amplitude and random-phase data.

It should be noted that scatterer separations of greater than $\lambda/4$ could have been used in all these models with no significant changes. However, separations of less than $\lambda/4$ would have correlated the surface and made every zone nearly equal in magnitude with a phase center in the center of the zone.

It should also be noted that these models can be used to explain the small-wave Boca Raton data obtained in the absence of whitecaps.

DEBRIS CHARACTERISTICS

The radar echoing characteristics of rigid debris floating on the sea surface were studied both theoretically and experimentally. Oil drums, logs, and aluminum-covered plastic gallon milk bottles were considered and measured.

A comparison of the echoing characteristics of solid debris and the sea surface revealed that returns from debris floating on the sea had much lower frequency-amplitude modulation than sea return. Figure 27 is typical of the data obtained. This figure is the measured (vertically polarized) echo signal from an anchored plastic gallon milk bottle covered with aluminum foil. The gated stretched video was recorded on a visicorder. The low-frequency modulation was produced by 2-ft (0.6 m) waves swinging the bottle on its tether and by interference between the bottle and its image in the sea surface. This piece of data was taken in the absence of a whitecap in the resolution cell. Note that large returns at 8.6 and 9.2 GHz were received at different times. This indicated the aforementioned target image interference phenomena.

Figure 28 is equivalent data taken with the radar polarized horizontally. A small whitecap was in the resolution cell in the left hand portion of this recording and added relatively high frequency modulation components to the bottle return on 8.6 GHz.

Figures 29 and 30 show sea return data from a breaking wave on the two different polarizations for comparison purposes. Note the large percentage of high-frequency amplitude modulation on the sea return.

Calculation revealed that a 3.33-m-long log would have to be seen near broadside rotating at 0.5 rad/s about a vertical axis normal to its long axis in order to produce modulation frequencies near those characteristic of sea return. It is obvious that such a condition could not exist for any extended period of time. This, coupled with the observed fact that whitecaps are very improbable in any given resolution cell, provided the concept necessary for design of an effective debris-avoidance radar.

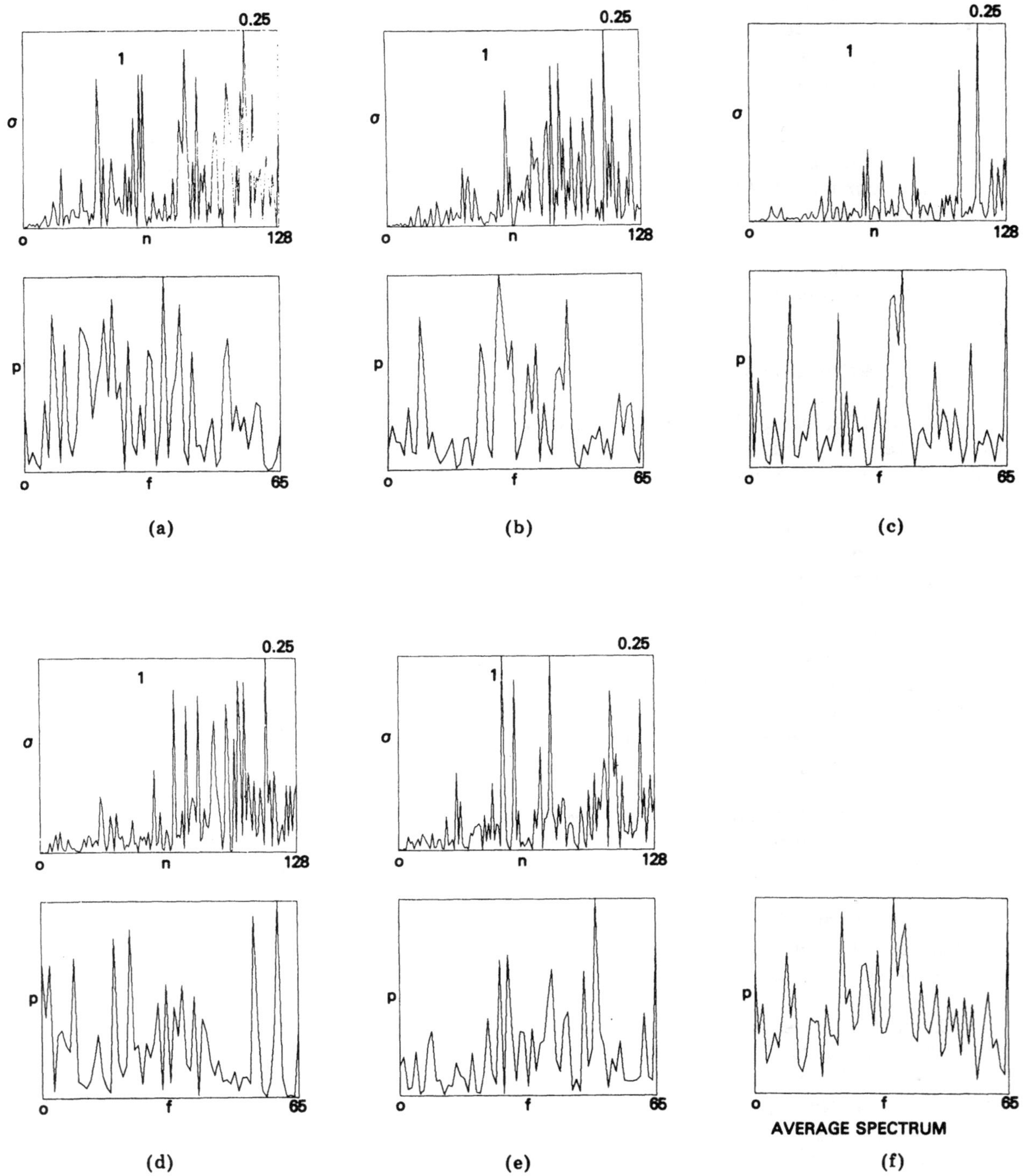

Fig. 21 — Random-amplitude and phase model (k = 0.25 and m = 1), σ vs n and power P vs frequency f

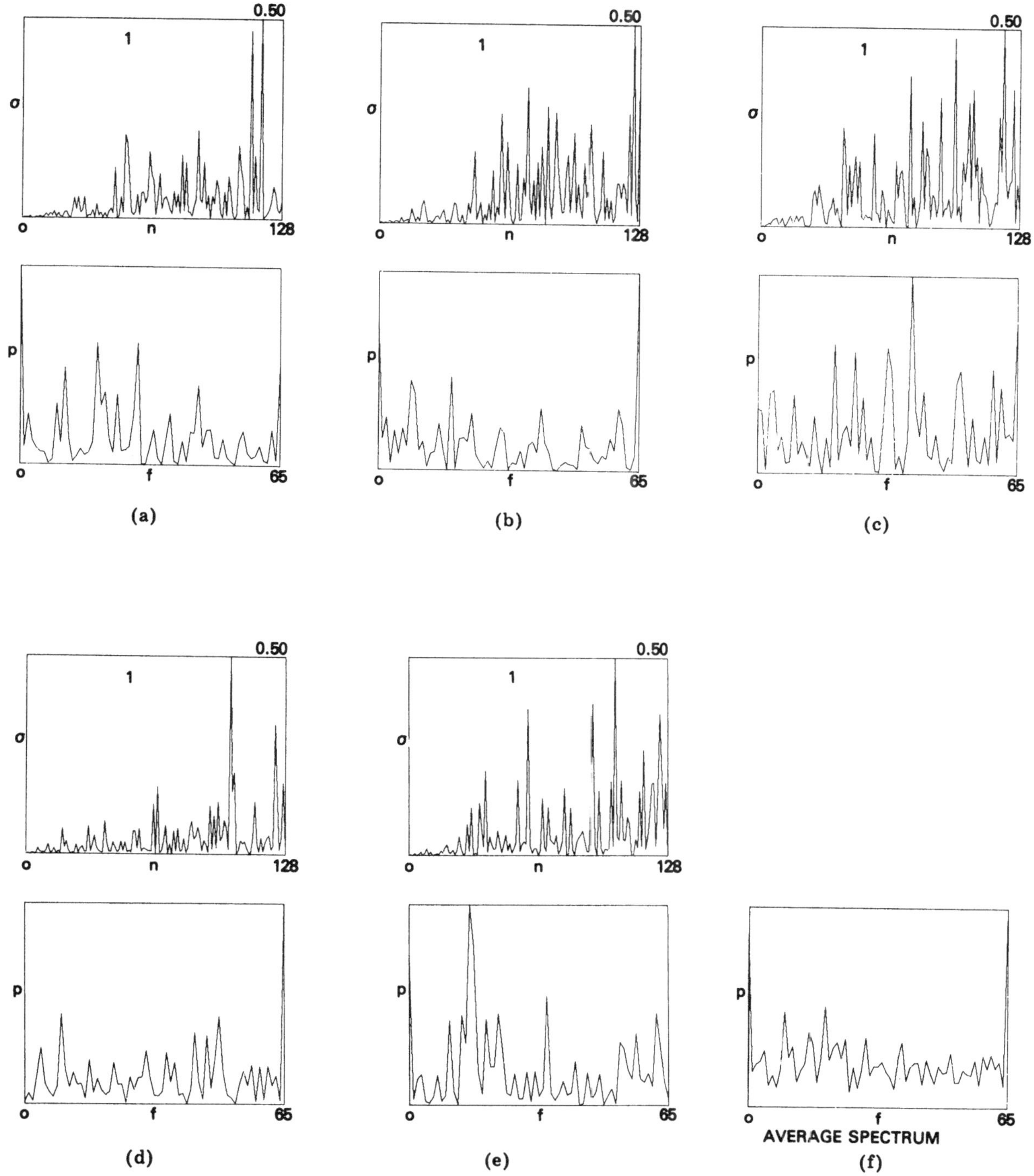

Fig. 22 — Random-amplitude and phase model (k = 0.5 and m = 1), σ vs n and power P vs frequency f

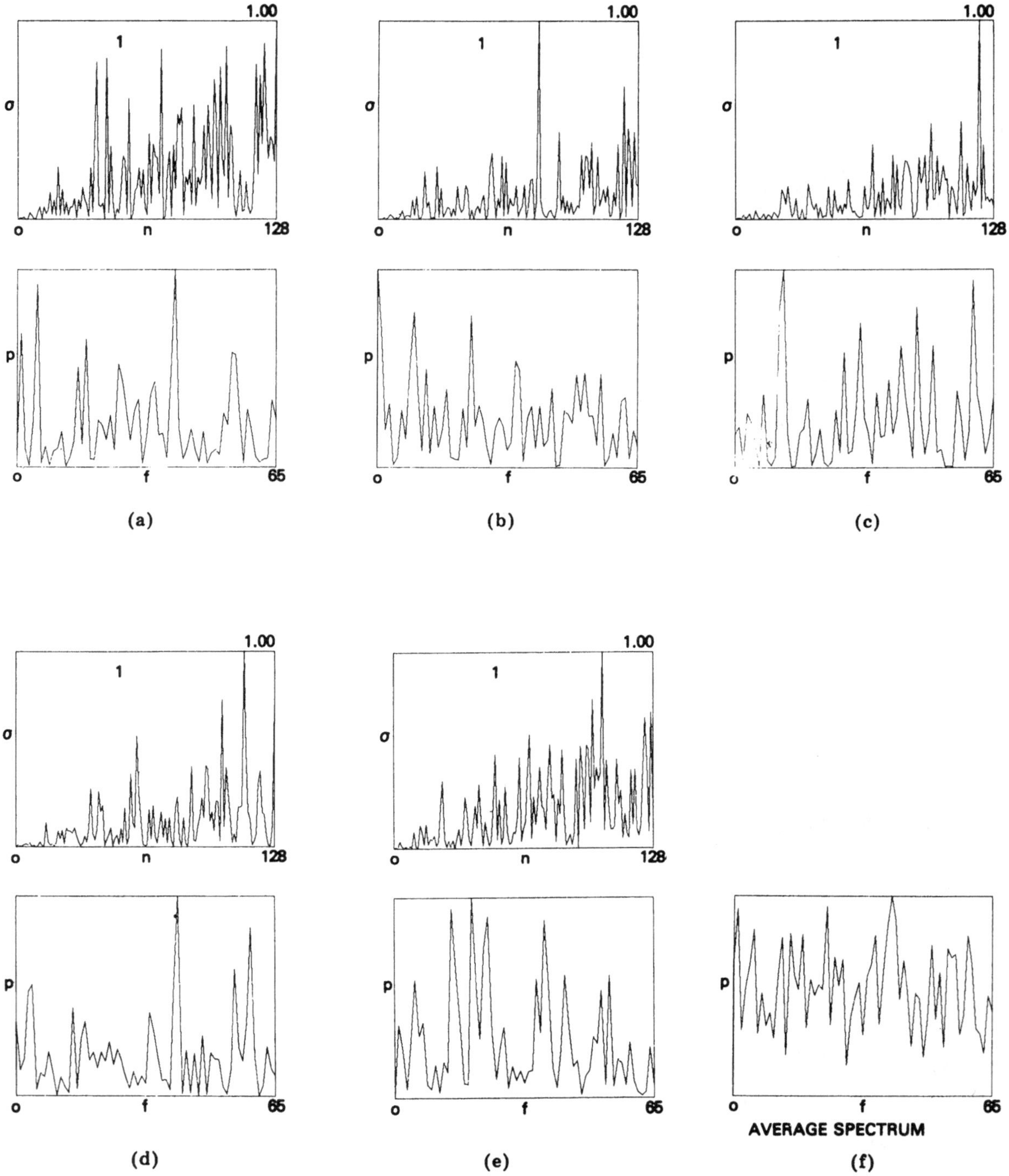

Fig. 23 — Random-amplitude and phase model (k = 1 and m = 1), σ vs n and power P vs frequency f

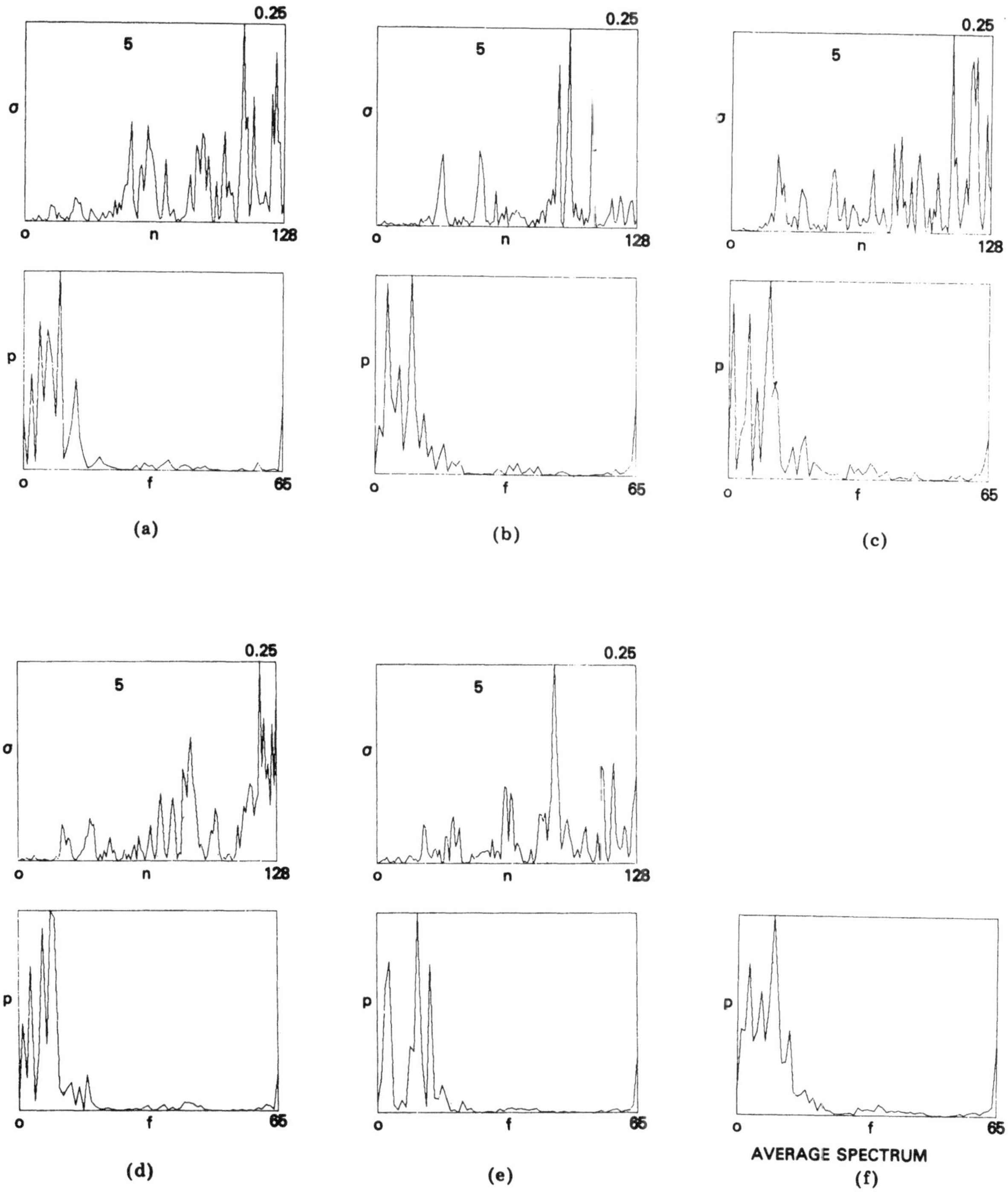

Fig. 24 — Random-amplitude and phase model (k = 0.25 and m = 5), σ vs n and power P vs frequency f

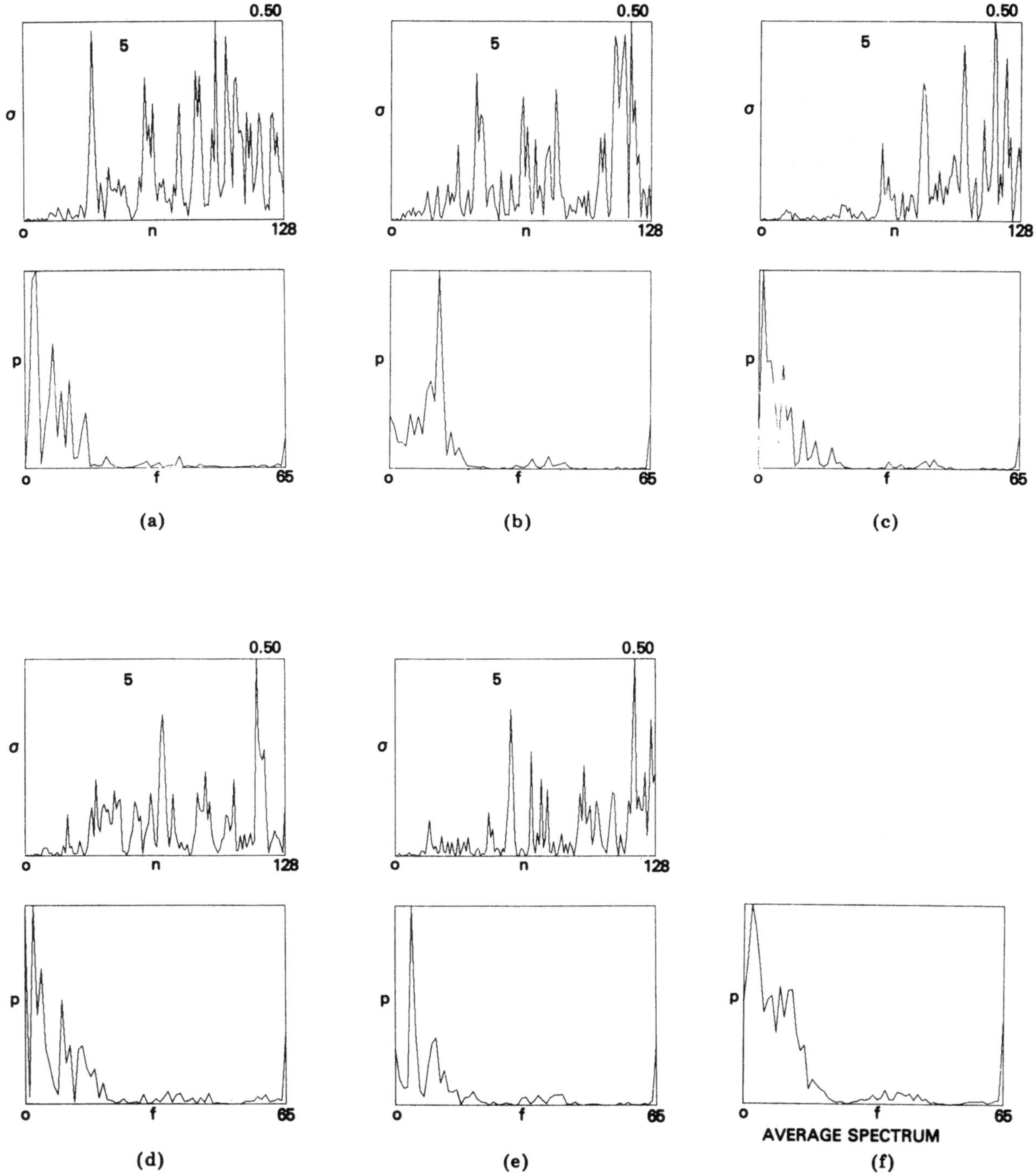

Fig. 25 — Random-amplitude and phase model (k = 0.5 and m = 5), σ vs n and power P vs frequency f

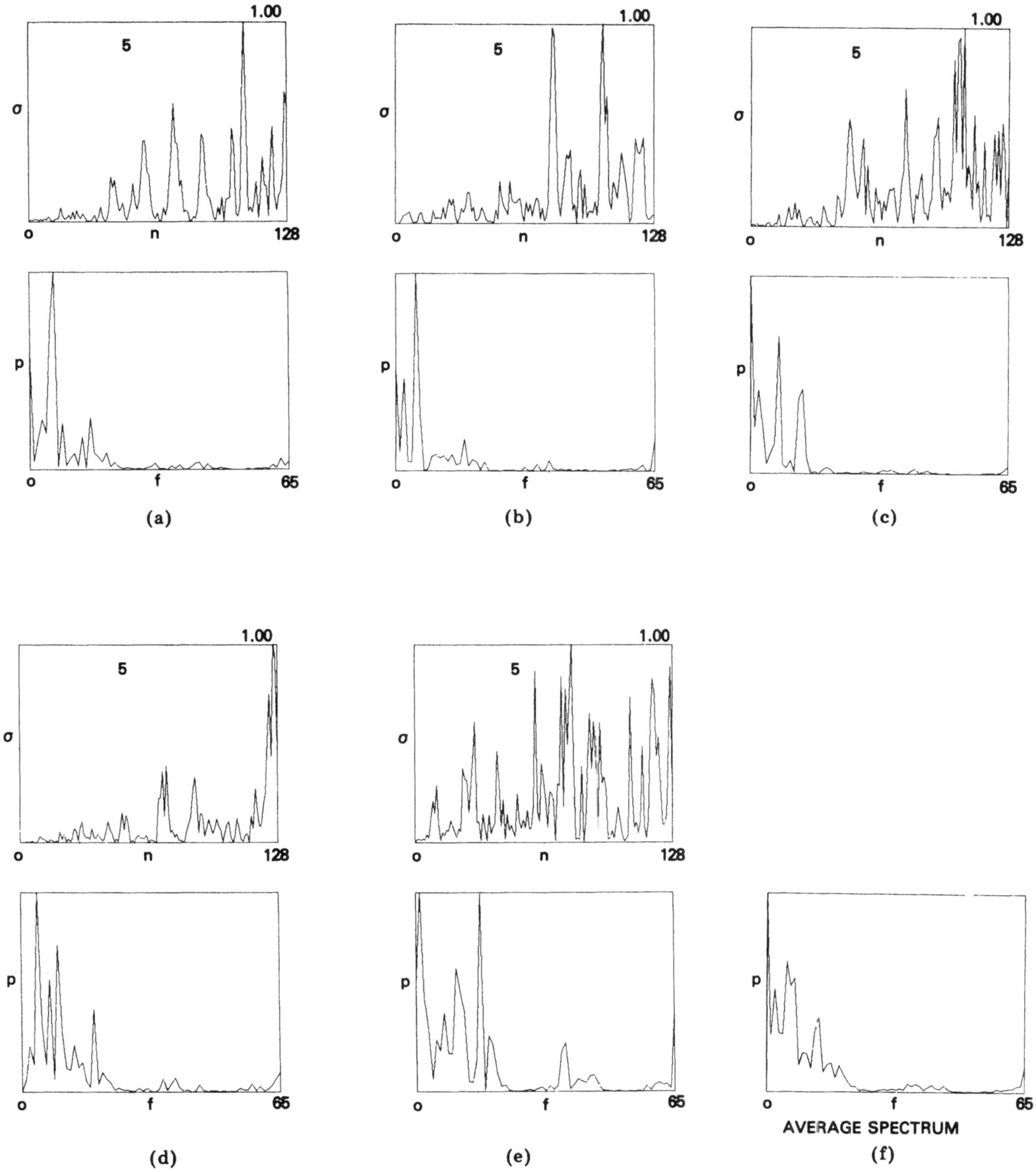

Fig. 26 — Random-amplitude and phase model (k = 1 and m = 5), σ vs n and power P vs frequency f

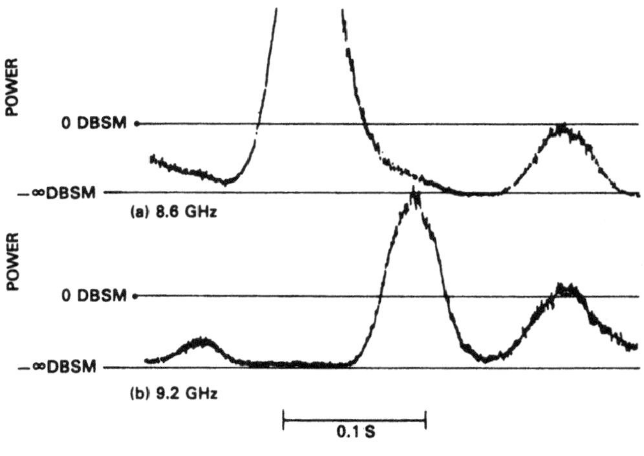

Fig. 27 — Vertically polarized return from foil-covered plastic bottle

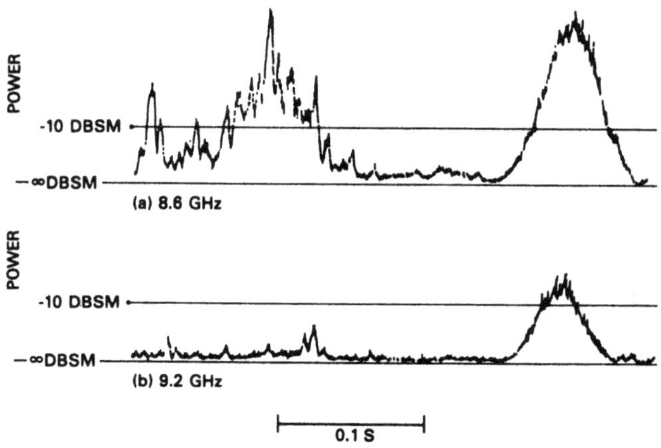

Fig. 28 — Horizontally polarized return from foil-covered plastic bottle

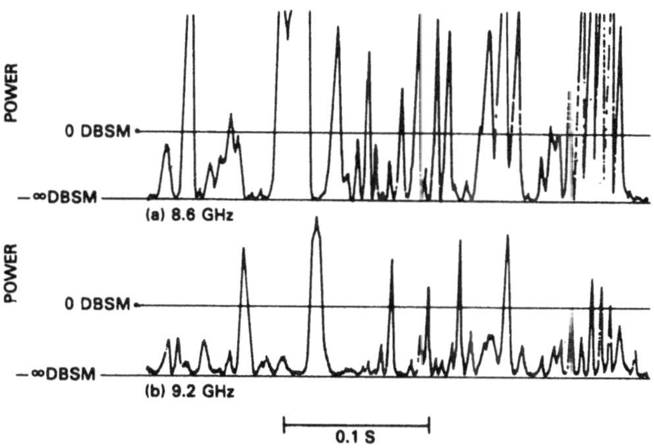

Fig. 29 — Vertically polarized return from wave

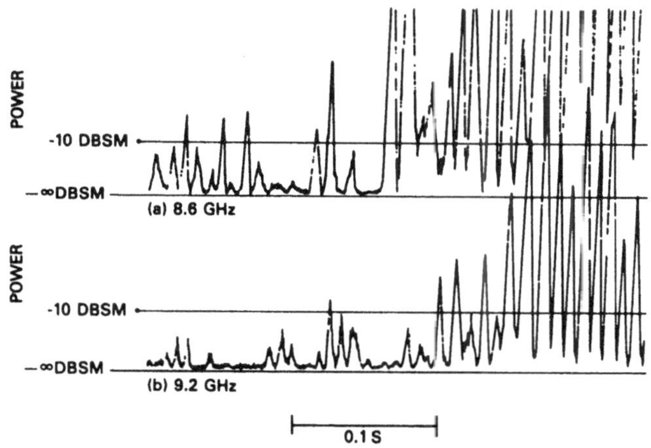

Fig. 30 — Horizontally polarized return from wave

DEBRIS-AVOIDANCE RADAR CONCEPT

The difference in the amplitude modulation characteristic of rigid floating debris and sea return suggests that percentage modulation vs time might be used as a discriminant to suppress sea return. Figure 31 illustrates one way in which such a discriminant might be instrumented. The radar return signals are delayed by integral multiples of the transmitted pulse repetition interval in a tapped delay line similar to those used in moving target indicators (MTI). The outputs of the taps are subtracted from each other as indicated to detect high-frequency, large-percentage amplitude modulation on echoes in any given range resolution cell. If such modulation is detected, the output from one or more of the subtractors will exceed a preset threshold and inhibit a gate to block passage of that video to a display. Since sea return fluctuates by more than 6 dB in less than 20 ms and debris echoes remain constant for much longer intervals, this instrumentation will block all sea return video and pass debris video.

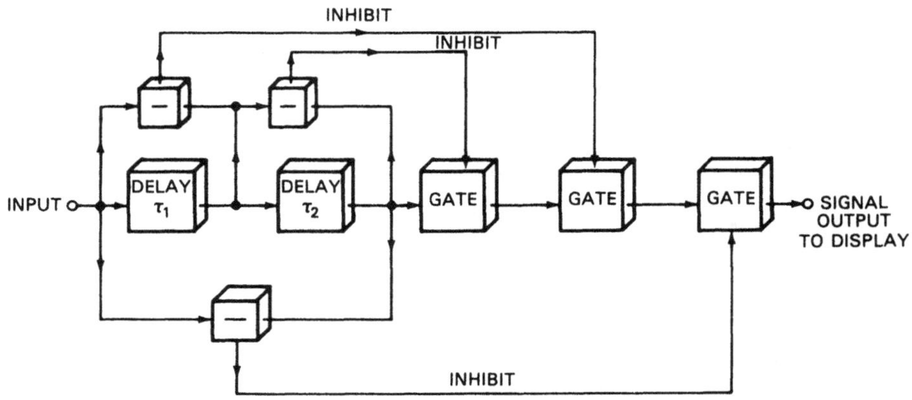

Fig. 31 — Proposed sea return suppressor

This instrumentation should not materially reduce the probability of detection of debris, since sea return is expected only about 1% of the time in any given range cell. Thus, when sea return and debris return come from the same cell and that cell is blocked, the blockage will correspond to a mere 1% reduction in viewing time.

To employ the percentage-modulation-vs-time discriminant, the radar must dwell on the same region long enough to suppress sea return and ensure debris detection. Such long dwells can be provided by employing a roll- and pitch-stabilized antenna with multiple azimuth beams filling a 10° or 20° sector looking forward in the direction of the ship motion. The beams can be spoiled downward to permit observation of the sea surface near the ship but should be limited in width above the horizon to reduce weather clutter.

The azimuth beamwidth and range resolution should be designed to ensure low probability of including a whitecap in any given resolution cell. A study of the measurements made to date reveals that 1° beamwidths and 20-ns pulse widths should be adequate for ranges between 1.6 and 4.8 km.

Frequency diversity in the form of simultaneous short pulses on different carrier frequencies would improve the probability of detecting debris by filling in the elevation lobing nulls produced by target and image interference.

PLANS FOR THE FUTURE

Plans for the future include the following:

1. Bistatic measurements of sea scatter as a function of polarization and frequency will be made, to test the long-wire scattering model.

2. Monostatic sea return measurements using short pulses on a 3-GHz carrier will be made, to verify the linkage between echo autocorrelation time and the time required to add or subtract one $\lambda/4$ zone to a whitecap.

3. Three extra sample-and-hold circuits will be added on each X-band frequency, and four contiguous range cells will be gated. A target will be anchored in one of the gated range cells, and target and/or sea returns will be recorded on magnetic tape from each stretcher. These data will then be processed to determine the optimum thresholds for suppressing sea return and detecting debris.

4. A multibeam radar will be instrumented and fitted with the proposed data processor and will be tested at sea.

SUMMARY AND CONCLUSIONS

The experimental measurement program revealed that high-resolution X-band radar echoes from a disturbed sea surface are very improbable in any given resolution cell and that when they occur they are always amplitude modulated with a high modulation index and have a lifetime on the order of 1 s. The modulation function was observed to have autocorrelation times of 10 ms or less. Echoes from rigid objects floating on the disturbed sea surface were found to have much longer autocorrelation times than echoes from the sea. This difference was used in designing a signal processor that would suppress sea echoes and permit detection of debris echoes.

Theoretical models of the sea scattering phenomena were developed to explain the temporal and spatial characteristics of sea return. In addition, the effect of radar polarization on sea returns was considered and an explanation for its effect was suggested. The theoretical work aided in justifying the proposed debris-avoidance radar concept that was derived from the experimental data.

The principal conclusion that can be drawn on the basis of this effort is that it should be possible to design an effective debris-avoidance radar for high-speed ships. On the basis of this conclusion, continuation of the program is recommended.

REFERENCES

1. Nathanson, F.E., *Radar Design Principles*, p. 228-238, McGraw-Hill, New York, 1969.

2. Jasik, H., ed., *Antenna Engineering Handbook*, p. 4-1 to 4-7, McGraw-Hill, New York, 1961.

3. Wolf, E.A., *Antenna Analysis*, p. 372-379, John Wiley and Sons, Inc., New York, 1966.

4. Skolnik, M.I., *Introduction to Radar Systems*, p. 42, McGraw-Hill, New York, 1962.

5. Hicks, B.L., Knable, N., Kovaly, J.J., Newell, G.S., Ruina, J.P., and Sherwin, C.W., "The Spectrum of X-Band Radiation Backscattered from the Sea Surface," *J. Geophys. Res.* **65** (3), 825 (Mar. 1960).

SOME RECENT OBSERVATIONS OF SEA SPIKES

B. L. Lewis and I. D. Olin

Radar Division, Naval Research Laboratory, Washington, D.C. 20375

INTRODUCTION

Radar echoes from a disturbed sea surface are generally objectionable because they form a clutter background against which desired targets must compete. An estimate of the radar cross section (RCS) background for various sea states, grazing angles, polarizations and frequencies can usually be made from tables such as those given by Nathanson (1). These tables list the normalized RCS, σ_0, which results from dividing the measured RCS by the comparatively large area illuminated by the measurement radar. However, when the radar pulse length is reduced from microseconds to nanoseconds or the beamwidth is made very narrow, the character of the sea clutter changes significantly and the tables no longer apply. The relatively constant average value of the sea echo in any range resolution cell is replaced by a very low background with occasional large amplitude signals called "sea spikes." These signals, often lasting for seconds, produce false alarms for low detection thresholds and thus set a limit on the detection of low cross section targets unless long integration of the target and clutter signals is achievable. To learn more about sea spikes, a dual frequency X-band radar using 20 nsec pulses and operating with horizontal and vertical polarization was used to observe the backscatter from a small patch of disturbed sea while simultaneously displaying a television view of the illuminated area.

MEASUREMENT TECHNIQUES

The measurement system was selectably dual polarized and transmitted 20 nsec pulses at a 50 kHz rate simultaneously at two carrier frequencies (8.6 GHz and 9.2 GHz) through a single antenna. Separate receivers with a common range gate were used for the two frequencies and video in the range gates of the receivers was stretched pulse-to-pulse prior to display. Calibrations were furnished by a six inch metal sphere or by a corner reflector. Generally it was possible to calibrate and record data at the same radar range. In order to provide an accurate comparison of the radar detected video with the water surface, the collimated TV video was mixed with the video from two other cameras observing gated and ungated scope displays, and presented on a single TV monitor. Video recordings of this monitor display and gated stretched video constituted the data recording. Fig. 1 clearly shows the often periodic nature of the sea spike detail revealed by high time resolution of the sea spike signal. Although many sea spikes contain pronounced periodicity, others show additional modulation components and still others, although highly modulated, appear random. Moreover, differences in the 8.6 GHz and 9.2 GHz backscatter suggest a multipath effect between the "target" and its image in the surrounding water. Similar modulation and occurrence characteristics have been noted by Long (2).

EXPERIMENTAL OBSERVATIONS

The first measurements were made at the Chesapeake Bay Division (CBD) of NRL with a 0.9 m diameter antenna. The range was very short (about 45 m) to take advantage of the water-shore interaction which produced high amplitude spikes with moderate wind. Subsequent measurements in Boca Raton, Florida, using a 2.4 m diameter antenna at ranges in excess of 480 m and grazing angles of about 1.4° resulted in similar signal characteristics, but often with higher frequency fluctuations. Fig. 2 shows a sequence of TV monitor frames for horizontal polarization which show the development of the whitecap and the corresponding signals. The circular dark spot on the right is the region filled by the radar beam. The dotted lines delineate the region covered by a 20 nsec range gate. TV camera recordings on the left of the frames show the real time video (below) and the range gated and stretched video (above) for the two operating frequencies. By using this split screen recording technique, time synchronization of the radar data and the optical view was assured. It should be noted that the gated video was recorded for 0.2 sec while sweeping from left to right. The sea surface was photographed at the end of this period. Fig. 3 shows two frames of data using vertical polarization.

Several significant observations resulted from study of the CBD recordings:

1. Very large backscatter amplitudes (sea spikes) were associated with the whitecaps. During this time the whitecaps first grew in range extent, then decayed.

2. The foam following the whitecap decay did not produce high amplitude spikes.

3. The backscatter was amplitude modulated with relatively high frequencies (approximately 50 Hz) and a high modulation index.

4. The presence or absence of return from the two transmitted frequencies suggests a multipath effect. When both components were present they appeared uncorrelated.

Another observation pertains to the timing between the whitecap occurrence and the reception of sea spikes. Studies of the video frames indicate that they generally occur together, but that exceptions depended on the transmitted polarization. With vertical polarization, sea spikes could occur before the formation of whitecaps, however, with horizontal polarization the opposite was observed. It is suggested that for vertical polarization, currents induced in the front surface of the sharply cresting wave (prior to whitecap formation) result in scattered lobes with the largest amplitude in the direction of the radar. Under the same conditions, horizontal polarization results in the largest lobes scattered nearly normal to the radar line of sight. Then, however, two other conditions can result in a short delay of the radar backscatter which would be observed with horizontal polarization. First, if the backscatter is considered as resulting from the random buildup of constituent scatterers, it can take some time before the surface grows sufficiently to produce significant amplitudes. Second, depending on the height of the whitecap, multipath can result in a null until the geometry changes. As whitecap images have been observed optically, (see Fig. 2-top) they are even more likely at microwave frequencies.

It is readily shown that fluctuations from multipath (which modulate the components due to surface area changes) occur at a relatively slow rate. Whitecaps appear frequently at sea state 3, for which the wave height and period are about 1.25 meters and 4 seconds, respectively. With a whitecap on the crest, the vertical lobing pattern width (= $\sin^{-1} \lambda/2H$) for λ = 3.3 cm is about 0.8°. For the CBD tests the grazing angle was 4-1/2° so that the radar antenna was in the sixth lobe of the interference pattern. With a 4 second period, the whitecap falls to zero height in two seconds, resulting in a maximum detected modulating frequency of 3 Hz. Of course the frequency can be less, for example, when the whitecap is not on the wave crest. For most operating conditions the grazing angle will be smaller with a corresponding reduction in frequency due to multipath.

The physical characteristics of the whitecaps and therefore the observed backscatter are affected by the mechanism producing the whitecaps. For the CBD tests interaction with the shore caused the waves to break along the line of sight (LOS). This produced whitecaps which grew slowly along and rapidly across the LOS. However, during tests at longer ranges (deeper water), whitecaps were produced by interacting waves. For particular conditions observed, the line of rapid growth was along the LOS. Data from these waves revealed amplitude modulation at a higher frequency than observed during the CBD tests; decorrelation times as low as 3 msec were measured. Peak amplitudes of the spikes observed, however, were about the same as those from near shore.

Table I lists peak and average values from the vertical and horizontal polarized measurements made at CBD. The peak values are the result of recording the largest amplitude reached during each of thirty-five sea spike occurrences and averaging the results. The average values were obtained by determining the average amplitude during each sea spike occurrence and again averaging over thirty-five occurrences. Temporal statistics of the occurrence frequency revealed that the average probability of occurrence for a large sea spike in the 6.7 m² patch observed at CBD was about .06. In addition to the very large amplitude spikes represented in Table I which always accompany whitecaps, the deep water data also shows some infrequent spikes at least an order of magnitude lower in amplitude, but with similar fluctuations. These occur when the local water surface is very rough, yet unbroken. Nonetheless, most of the time the high resolution data shows long periods of very low residual backscatter.

TABLE I - RCS values (m²) of sea spikes

	Vertical Pol. (8.6/9.2 GHz)	Horizontal Pol. (8.6/9.2 GHz)
Peak RCS (m²)	5.9/4.8	1.7/1.9
Avg. RCS (m²)	0.87/0.90	0.31/0.36

CHARACTERISTICS MODELING

The origin of the large amplitude sea spikes has been observed as the backscatter from the whitecap of a breaking wave which varies in size, expanding during formation and later contracting. No attempt is given here to characterize the actual scatter elements, however, both droplet spray and the rough surface beneath appear likely. For analysis purposes the whitecap will be represented by some arbitrary distribution of scatterers. During illumination by the pulse (always assumed wider than the whitecap extent), the scatterers in a narrow range interval or zone can be considered fixed and backscatter is the result of summing the component scatterers using the appropriate round-trip phase delays. For convenience consider a zone $\lambda/4$ wide, so that all scatterers within contribute constructively to the signal. The total backscatter from such a zone can be represented as coming from a single resultant scatterer. Assuming homogeniety, its location is at the zone center and its amplitude constant from zone to zone. Without this assumption its magnitude and location will generally be random. The selection of $\lambda/4$ zones also has a physical basis, since the observed decorrelation time of the spike components was on the order of 10 msec at X-band, with the wave breaking in the direction of the line of sight. In addition, the growth rate of the whitecap was estimated to be about 100 cm/sec. This implies an approximate decorrelation distance consistent with the $\lambda/4$ zone assumption. Hicks, et al. (3), indicates the width of the whitecap velocity distribution to be about 3 knots. This is about 50% greater than the estimate from the current observations; however, in view of the approximate nature of the measurements, they are probably not inconsistent.

Fig. 4 depicts the whitecap region of width W divided into zones each containing an equivalent scatterer of arbitrary magnitude and location. The resulting backscattered signal for $N = 4W/\lambda$, referred to the initial boundary is given by

$$\Gamma = \sum_{k=1}^{N} S_k e^{-j\frac{4\pi}{\lambda}\left[(k-1)\,\lambda/4 + d_k\right]} . \qquad (1)$$

If X is a random variable (R.V.) in the interval $0 \le X \le 1$, then $d_k = X_k \lambda/4$ so that Eqn. (1) becomes

$$\Gamma = \sum_{k=1}^{N} (-1)^{k-1} S_k e^{-j\pi X_k} . \qquad (2)$$

For a homogeneous surface, $S_k = S$, $X_k = 1/2$ and the backscatter is given by

$$\Gamma_{homog.} = Se^{-j\pi/2} \sum_{k=1}^{N} (-1)^{k-1} \quad . \qquad (3)$$

Thus as W increases, $\Gamma_{homog.}$ varies from zero to S in a periodic manner. For the more general case considered here, S_k and X_k are considered random so that a periodic backscatter variation will only be observed occasionally.

Realizations of the relative radar cross section (RCS) as a function of N were computed using Eqn. (2), where RCS $\propto |\Gamma|^2$. Since many of the whitecaps grow to about 1 m, N = 128 was selected as being consistent with X-band illumination and also facilitating computation of the Fast Fourier Transform (FFT). During the approximate two second "lifetime" of the whitecap, the scattering properties of each of the zones will not remain static. To account for this "internal" decorrelation time, selection of random numbers (S_k, X_k) representing the kth zone were changed using different updating alternatives. Representative results are shown in Figs. 5,6. Fig. 5 is one realization assuming S_k has only a 25% random component (uniformly distributed), while X_k is constant (= 1/2). In addition, an update interval of 5 signifies that a new value of S_k is selected for every 5-zone additions. Thus the internal decorrelation time is 5 times longer than the time required for the surface to grow by one zone. With only a small random component and a long update interval the strong periodic component evident from Eqn. (3) is observed. The envelope of the fine structure, however, does not correspond to the experimental data. Fig. 6 is one realization with both S_k and X_k completely random and all previous (S_k, X_k) changed each time a new zone is added. Note that the larger RCS values are many times that for one zone. Generally this case compares favorably with the experimental results. Figs. 7,8 show the averaged FFT's for the conditions corresponding to Figs. 5,6. For the 128 zone sample and a 10 msec zone addition period, the actual frequency varies from 0 to 50 Hz.

An estimate of the average RCS can be made by considering the intercept area and equivalent gain on reflection of the whitecap and multipath effects. The illuminating grazing angle can be approximated by the sum of the grazing angle and one-half the slope of the wave when it breaks (30°). If A is the actual whitecap area, the power intercept area for low grazing angles is A sin 15° ≈ A/4. Since the whitecap image will have the same intercept area, the total intercept area is A/2. The average gain of the whitecap and image in the direction of the radar, assuming isotropic scatter in the upper half space, is G = 2, so that the average RCS is $\sigma = G \cdot A/2 = A$. Thus the area normalized value is $\sigma_0 = 1$, which is in reasonable agreement with the observations. This is about 35 dB greater than σ_0 quoted in tables for low resolution radars for sea states 3 and 4 and grazing angles of 3° (1). The difference can be justified by the large ratio of sea surface not covered by whitecaps to that covered by whitecaps.

CONCLUSIONS

The very large amplitude sea spikes actually consist of a comparatively long random sequence of short high amplitude signals. These are the result of reflection from the whitecap region following the breaking of waves. This region continually changes in area from buildup through decay, often producing components of several square meters in RCS with those from vertical polarization 4 to 5 dB greater than those from horizontal polarization. When the observed sea patch is small, sea spikes occur with relatively low probability; between these events a very low clutter background is observed. The frequency components of the sea spikes were in the range of 50-170 Hz, depending on the direction of whitecap growth. With the wave breaking in the LOS direction, the growth rate of the whitecap is estimated to be 100 cm/sec. Sea spike signals appear approximately simultaneous with whitecap growth and decay, with variations depending on polarization. Foam following decay, however, did not produce high amplitude signals. Multipath appears to affect the backscattered signals, so that sea spike components observed at one frequency are not always observed using a different frequency at the same time.

ACKNOWLEDGMENTS

The authors express their appreciation to J. P. Hansen and V. F. Cavaleri for their part in acquiring the data and to T. H. Gauss for his part during analysis.

REFERENCES

1. Nathanson, F. E., 1969, "Radar Design Principles," McGraw-Hill Book Company, New York, N.Y.

2. Long, M. W., 1975, "Radar Reflectivity of Land and Sea," D. C. Heath and Company, Lexington, Mass.

3. Hicks, B. L., Knable, N., Kovoly, J. J., Newell, G. S., Ruina, J. P., and Sherwin, C. W., 1960, Journal of Geophysical Research 65, No. 5, 825-837.

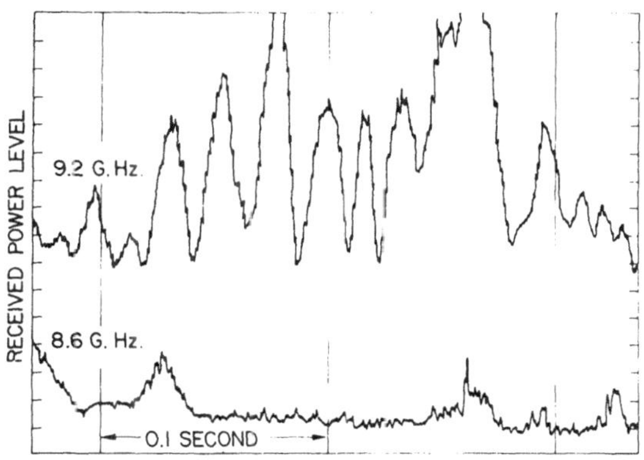

Figure 1 Expanded plot of typical sea spike

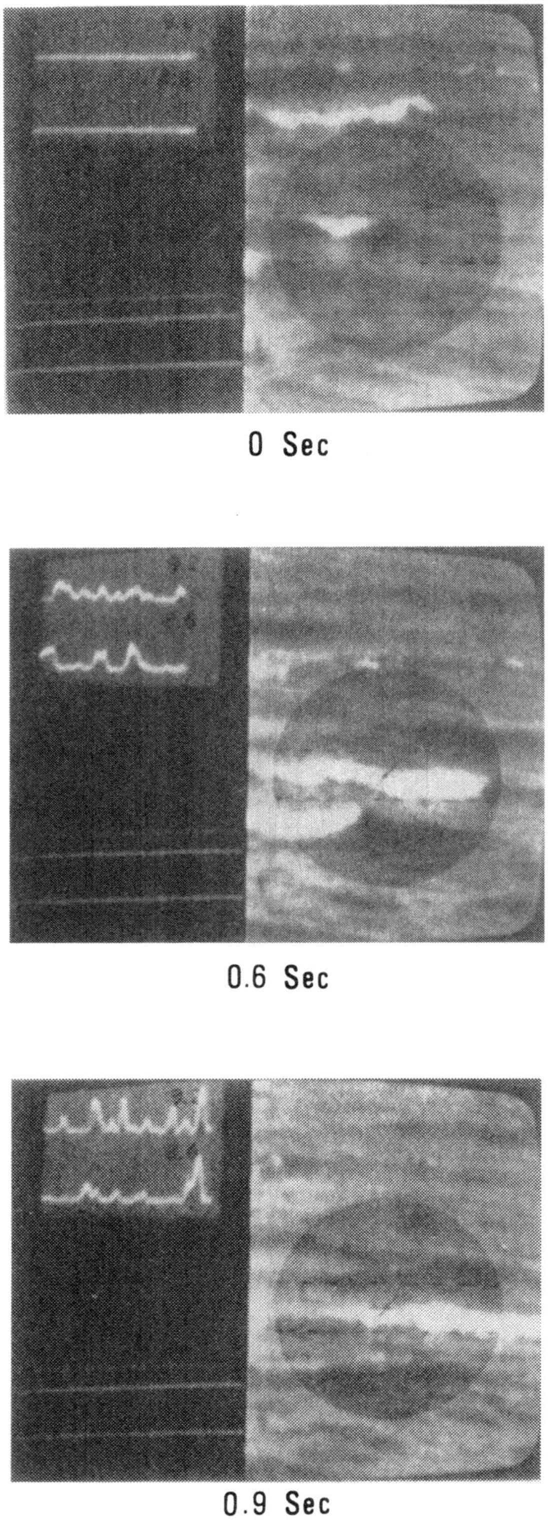

Figure 2 Whitecap region development and corresponding gated (top) and ungated (bottom) signals for horizontal polarization

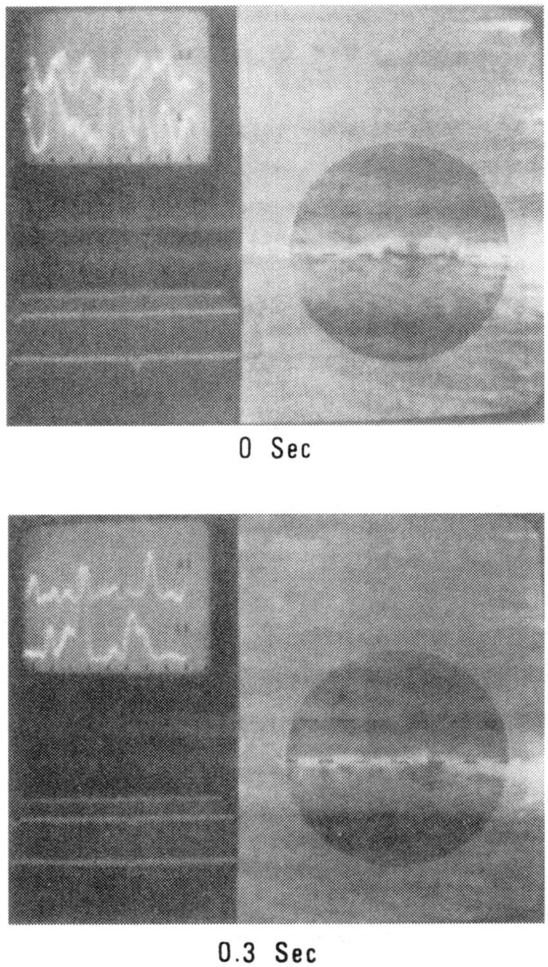

Figure 3 Whitecap region development for vertical polarization

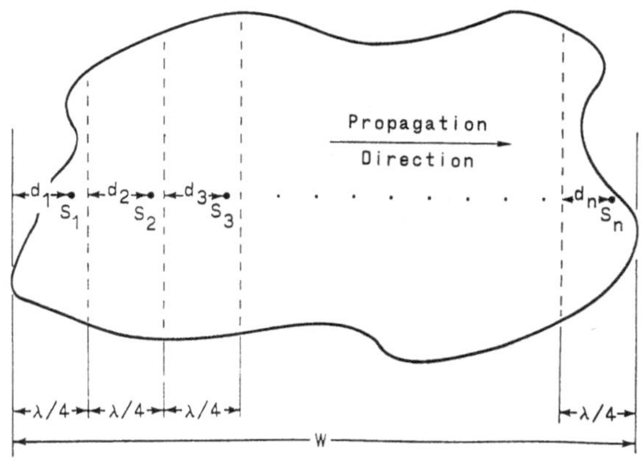

Figure 4 Representation of whitecap divided into λ/4 wide zones

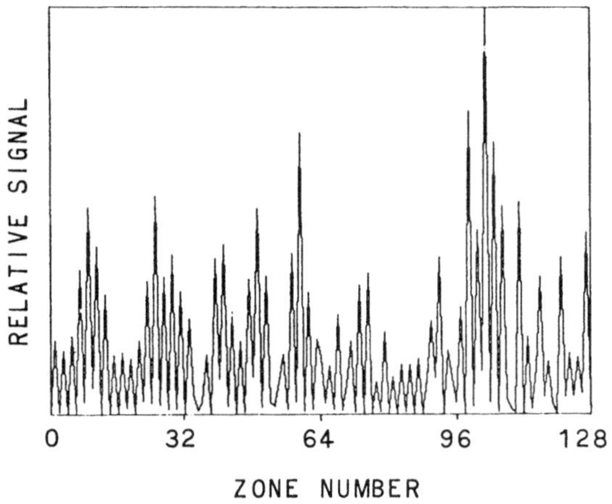

Figure 5 Sea spike realization for 25% random amplitudes and constant phase per $\lambda/4$ zone. Zones updated after every 5 zone additions

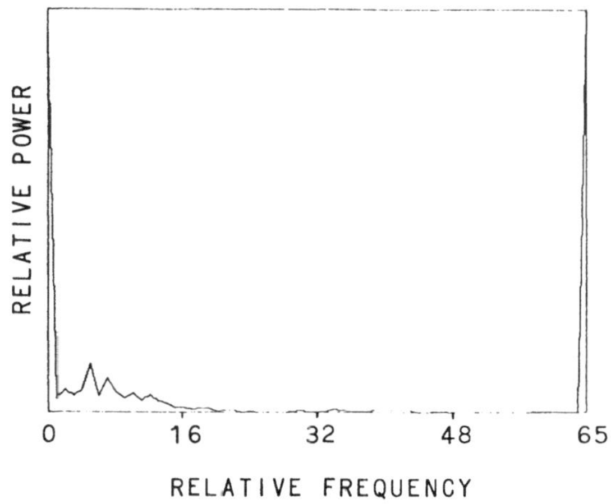

Figure 7 Averaged FFT for five realizations each with the conditions of Figure 5. For 10 msec zone addition period and 128 zone sample, a relative frequency of 65 corresponds to 50 Hz

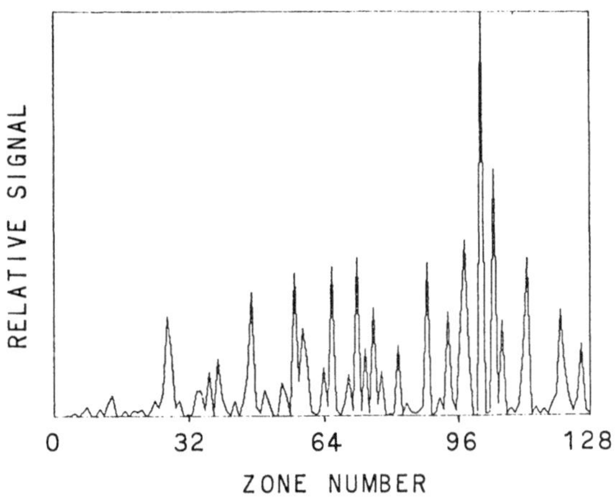

Figure 6 Sea spike realization for 100% random zone amplitudes and phases. Zones updated after every zone addition

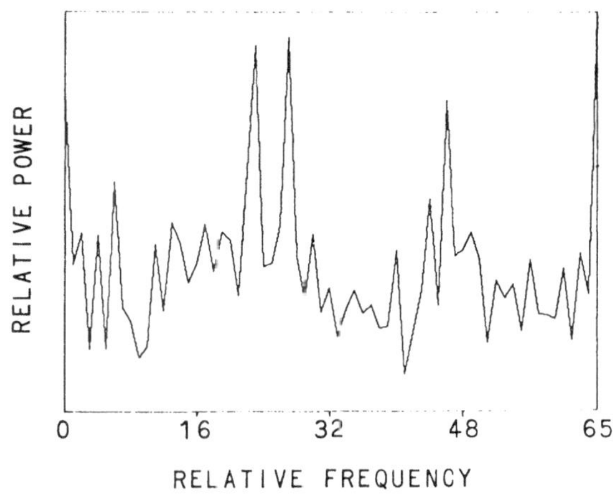

Figure 8 Averaged FFT for five realizations each with the conditions of Figure 6

Experimental study and theoretical model of high-resolution radar backscatter from the sea

Bernard L. Lewis and Irwin D. Olin

Naval Research Laboratory, Washington, D.C. 20375

(Received July 5, 1978; revised December 5, 1978.)

The temporal characteristics of radar backscatter from the sea depend on the pulse width for surface resolutions less than about 20 m. While low-resolution backscatter appears continuously noiselike, as the resolution is increased, the backscatter is punctuated by substantially quieter periods. This results in a noticeably 'spikey' characteristic. Measurements at 3 cm using several different pulse widths have been made near grazing incidence together with synchronized television views of the radar-illuminated surface. These observations confirm that the largest backscatter is due to whitecaps. However, even the quieter periods disclose spikes, although they are about 3 orders of magnitude lower in power. A simple model of the salient scattering regions consists of a random collection of effective scattering centers which change in number with whitecap growth or decay. By comparing modeled surface realization with radar data it is concluded that the decorrelation time of the surface and scatterers within the surface are each about 10^{-2} s at X band.

INTRODUCTION

A description of the radar backscatter from the sea surface is important in predicting radar detection of targets on or near the surface, since the probability of a detection or of a false alarm depends upon the observed signal-to-clutter ratio. Early work by *Goldstein* [1951] characterized sea clutter as noiselike and used a single dimensionless parameter σ_0 and an associated distribution function to describe its amplitude characteristics. The parameter σ_0 is defined as the average radar cross section (RCS) per unit area of illuminated sea surface; values from numerous sources have been tabulated by *Nathenson* [1969] for a variety of radar wavelengths, sea states, polarizations, and incident angles. For example, with X-band horizontally polarized illumination of a moderate (state 3) sea at a grazing angle of 1°, σ_0 of about 40 dB below 1 m^2/m^2 is given. Following the usual assumption that a large number of independent scatterers are illuminated by the radar, the probability density function of RCS is given by the exponential form $p(\sigma) = (1/\bar{\sigma})e^{-\sigma/\bar{\sigma}}$, where $\bar{\sigma}$ is the average RCS. Since the corresponding distribution density of the backscattered signal voltages is Rayleigh, this function is often referred to as the Rayleigh form. Then, given σ_0 and assuming an acceptable false alarm probability, the probability of detecting a target can be calculated. To achieve detection of lower RCS targets against the sea clutter background the resolution cell (illuminated area) must be decreased. The usual approach is to reduce the radar pulse width; however, measurements of the sea clutter for very narrow widths have demonstrated that the RCS is no longer Rayleigh distributed. Instead there is an increased tendency toward high-amplitude spikes (commonly known as 'sea spikes') with the result that the 'tail' of the distribution function is higher than that given by the Rayleigh function. Several different distributions have been used [*Trunk and George,* 1970; *Fay et al.,* 1977] to describe this characteristic. Nonetheless, in order to maintain a satisfactory false alarm probability, detection thresholds must be kept high, thereby limiting the benefits of the higher radar resolution unless a more optimum detector is used.

In the work described here the temporal and spectral characteristics of the sea surface using high-resolution radar observing a very small sea patch are examined experimentally, and a model fitting the observations is proposed. Results indicate that the largest sea spikes have a definite relationship to whitecap formation and have characteristic signatures which should be useful in substantially improving target detection over that realized by raising detection thresholds.

EXPERIMENTAL OBSERVATIONS

To study the characteristics of sea spikes, a dual frequency, dual polarization radar operating at about 3 cm and capable of generating 20-, 40-, 100-, and 400-ns pulses at a 50-kHz rate was constructed. Pulses were simultaneously radiated at 8.6 and 9.2 GHz with either horizontal or vertical polarization. Two different parabolic antennas were used, depending on the observation ranges. Initially a 1-m-diameter antenna yielding a 2.6° beam was used at a shore site at Chesapeake Beach, Maryland, about 4 m above mean water level. At a range of only 45 m the effective illuminated patch with 20-ns pulses was about 2 × 3 m. The short range enabled clear observation of the backscattering surface, and the water-shore interaction produced 1- to 1.3-m-high breaking waves with moderate wind. Later measurements were made from a shore site at Boca Raton, Florida, using a 2.7-m-diameter antenna with a 1° beam. The antenna was 13 m above mean sea level, and the observation range was 530 m, so that for 40-ns pulses the effective illuminated patch was 9 × 6 m. The water depth was about 13 m, and the water-shore interaction had little effect on the direction of breaking waves; in fact, most of them traveled across the field of view.

Data were recorded by two means. A video tape recording using split screen techniques simultaneously combined visual images of the radar field of view, the real-time video corresponding to radar backscatter from the water surface over the entire range sweep, and the modulation envelope of backscatter from a single range cell obtained by stretching gated video. In addition, the gated and stretched video was recorded on magnetic tape for further detailed analysis. Figure 1 contains a single TV frame of just the range sweep portion of the composite video tape recording made at Boca Raton.

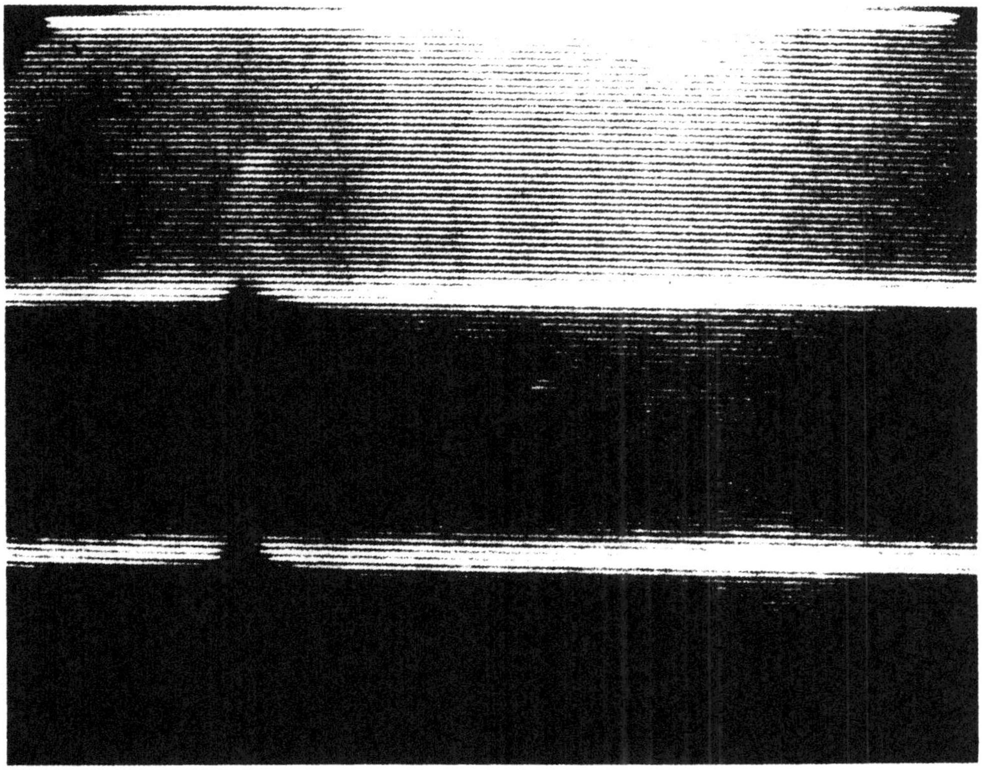

Fig. 1. Range sweeps showing radar video return for an illuminated range of 300 m with about 1/30 s signal integration. Gating signal, used for observing echoes in a fixed range cell, is shown above. Lower traces show ungated echoes from 9.2 GHz (above) and 8.6 GHz (below) for a 40-ns transmitted pulse width and horizontal polarization (Boca Raton data).

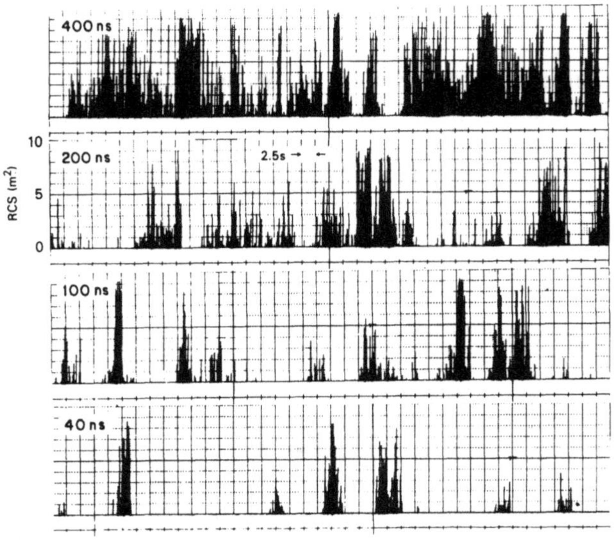

Fig. 2. Backscatter from whitecaps as a function of time using vertically polarized 9.2-GHz pulses 40, 100, and 400 ns wide. Data at each pulse width were taken at different times. Ordinate scales are identical for each plot (Boca Raton data).

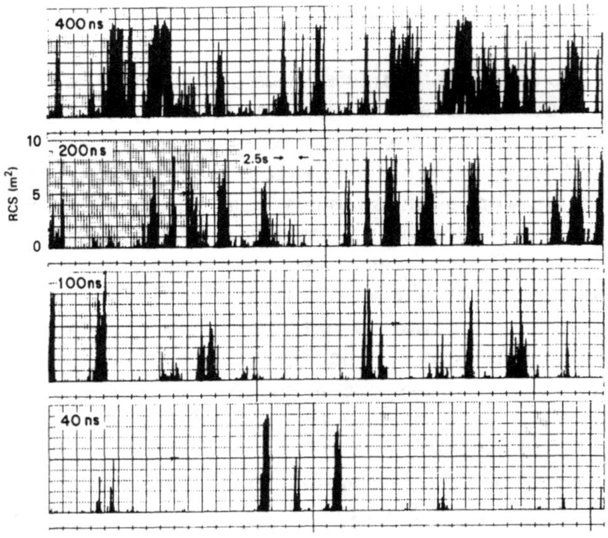

Fig. 3. Backscatter from whitecaps using horizontal polarization for the same conditions as in Figure 2.

The sparseness of high-amplitude return at this radar resolution is typical of observations in the presence of whitecaps. Radar backscatter from the illuminated water surface over a range of about 300 m are shown for 40-ns transmitted pulses. The top trace shows the range gate width (equal to the transmittal pulse width) and its location. The lower traces show range sweeps from 9.2-GHz (above) and 8.6-GHz (below) backscatter. The dominant backscatter pulse width is observed to be comparable with the range gate width, and this proved to be the case regardless of transmitted pulse width as long as whitecaps were sufficiently far apart. It was concluded therefore that backscatter was similar to that from radar targets with range extents comparable with or smaller than the range resolution cell defined by the transmitted pulse width.

Figure 2 shows recordings of the gated and stretched video made at Boca Raton of the 9.2-GHz vertically polarized backscatter from windblown sea with many whitecaps using several pulse widths. These represent backscatter at a fixed range and with a gate width equal to the transmitted pulse width. As the radar pulse width decreased, the character of the backscatter clearly changed; at 40 ns, backscatter consists of sparce peaked intervals (sea spikes) with amplitudes comparable with the largest backscatter from the 400-ns pulses (about 9 m^2). As was noted by *Long* [1975], the sea spikes persist for significant periods, often over 1 s. However, in between there are substantial periods of very low backscatter. Similar recordings using horizontal polarization are shown in Figure 3. A comparison of the results from the two polarizations

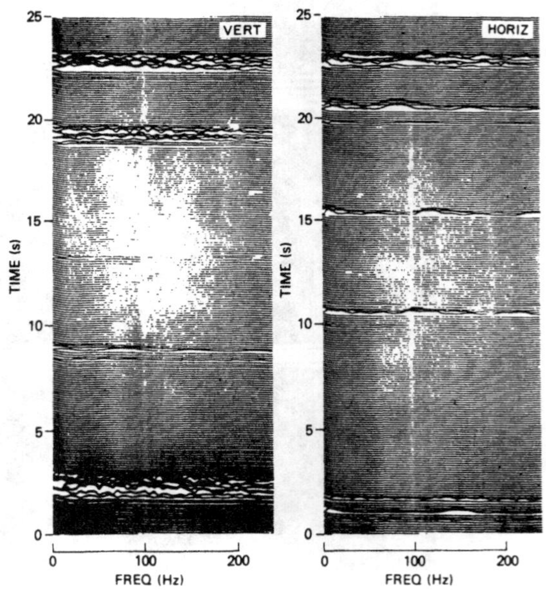

Fig. 4. Video spectra versus time for whitecap backscatter using 40-ns vertically and horizontally polarized pulses at 9.2 GHz (Boca Raton data).

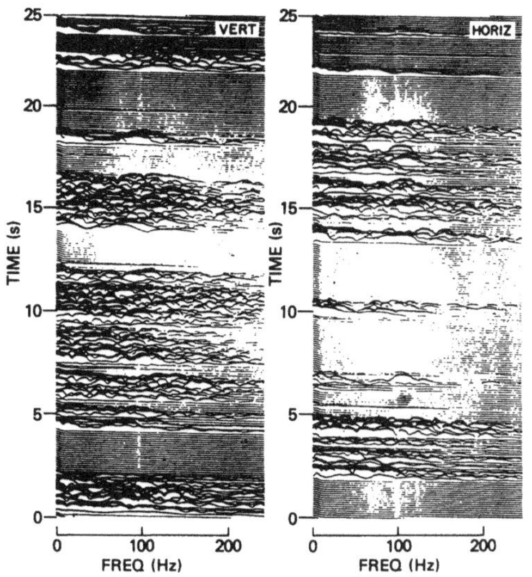

Fig. 5. Video spectra versus time for whitecap backscatter using 400-ns vertically and horizontally polarized pulses at 9.2 GHz (Boca Raton data).

discloses less frequent occurrence and increased spikey character when horizontal polarization is used. The spectral content of the backscatter, regardless of the pulse width, contains significant high-frequency components. Figure 4 shows spectral-time plots of the backscatter amplitudes from a gated range cell using 40-ns pulses for both polarizations. The occurrence of the relatively infrequent and wide spectral backscatter corresponds to the sea spike events. Figure 5 shows corresponding plots using 400-ns pulses. Although the spectral characteristics are similar, the sea spike events occur almost constantly. The amplitude characteristics of the sea spikes in a gated range cell are better illustrated by the expanded plot shown in Figure 6. This represents earlier data by *Lewis and Olin* [1977] from Chesapeake Beach using 20-ns pulses and operating simultaneously at the two frequencies. The backscatter occasionally discloses an almost periodic power variation. (The small steplike variations on the signal are due to the signal recording technique and should be ignored.) Comparing signal envelopes near the beginnings and ends of the records, it is observed that diminished levels at one frequency are accompanied by increased levels at the other frequency. This effect is attributed to multipath between the signal reflected directly from the scattering surface and its image in the unbroken water of the surrounding wave crest.

In order to better understand the nature of the scattering surface producing the sea spikes and its relationship to the radar pulse width, the video tapes containing the visual images together with the corresponding received signals were reviewed. Figure 7 shows a sequence from the Chesapeake Beach measurements using horizontal polarization and 20-ns pulses. The portion of sea surface corresponding to the gated range cell is delineated by the dotted lines within the circular spot which, in turn, defines the 3-dB points of the radar beam. The signals received from that region in the 9.2- and 8.6-GHz channels are shown (top left) in each television frame at a sweep speed of 20 ms per division (0.2 s full sweep). The image of the sea surface, which takes 1/30 s to scan, corresponds in time to the right end of these sweeps. As in Figure 1, the remaining three traces at the bottom left display the gate signal (above) and the entire radar range sweep (below) for the two channels, although the backscattered signals on this scale are obscured owing to reproduction limitations. As a result of studying considerable amounts of similar data, both from Chesapeake Beach and Boca Raton, it has been confirmed that very high amplitude sea spikes are associated with the development and decay of the whitecaps. (A similar association has also been described by *Long* [1975].) Typically, a very low signal background exists within the gate prior to the whitecap occurrence in the corresponding illuminated area. With the development of the whitecap a large highly modulated signal is observed, followed by a return to low background signal with either decay of the whitecap within the gate or movement of the wave out of the gate. The frequency range of the modulation was comparatively low for waves breaking in a direction toward the radar, as is shown in these figures. For example, the decorrelation time of the spike components was estimated to be about 10^{-2} s. At Boca Raton, however, with waves breaking in a direction across the field of view, the frequency range of the modulation was much higher. This result should be expected when the relatively slow range development of the whitecap for the waves breaking toward the radar are compared with the very rapid range development when the wave breaks across the field of view. In addition, the growth rate for these whitecaps as measured from the photos was

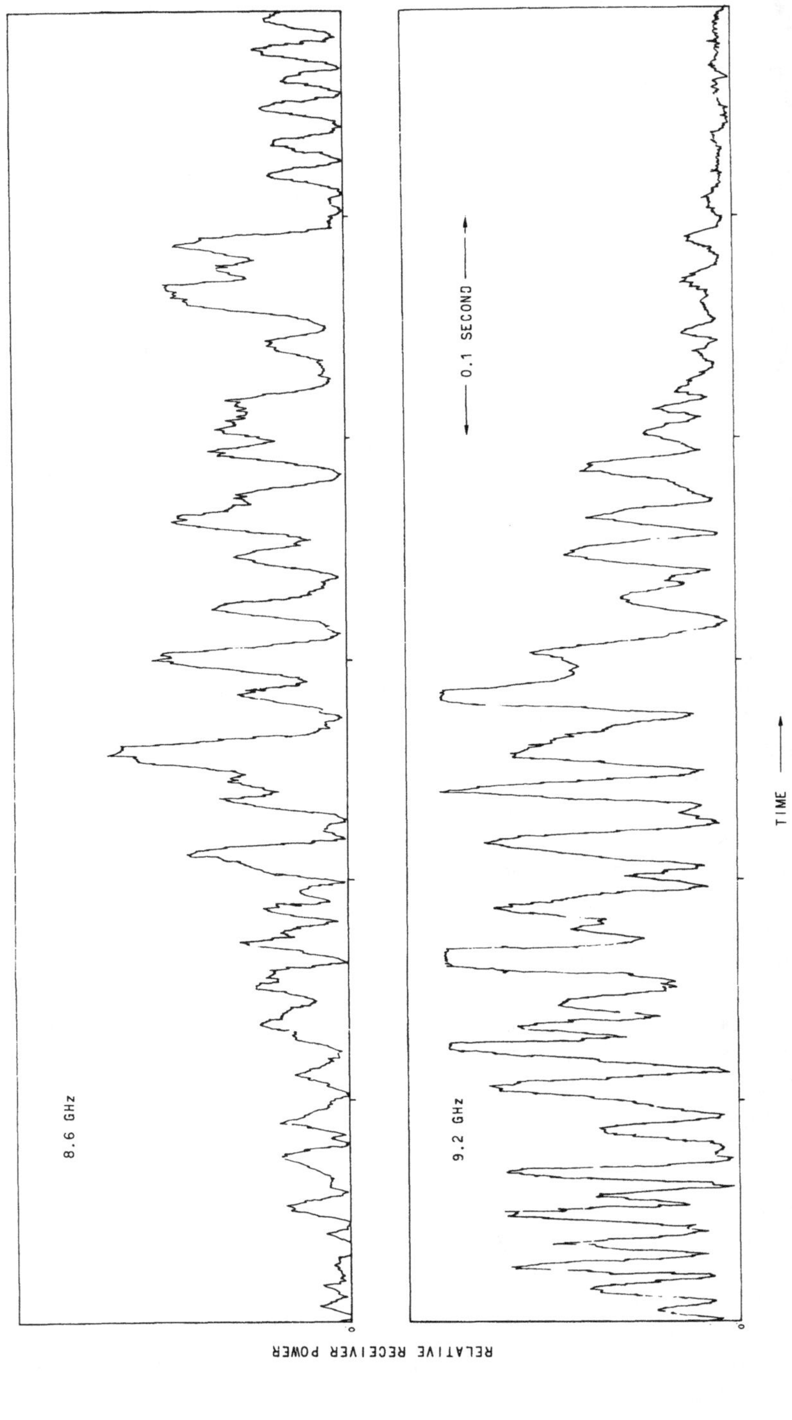

Fig. 6. Expanded amplitude versus time recordings of sea spike backscatter at two frequencies. The recordings are time coincident (Chesapeake Beach data).

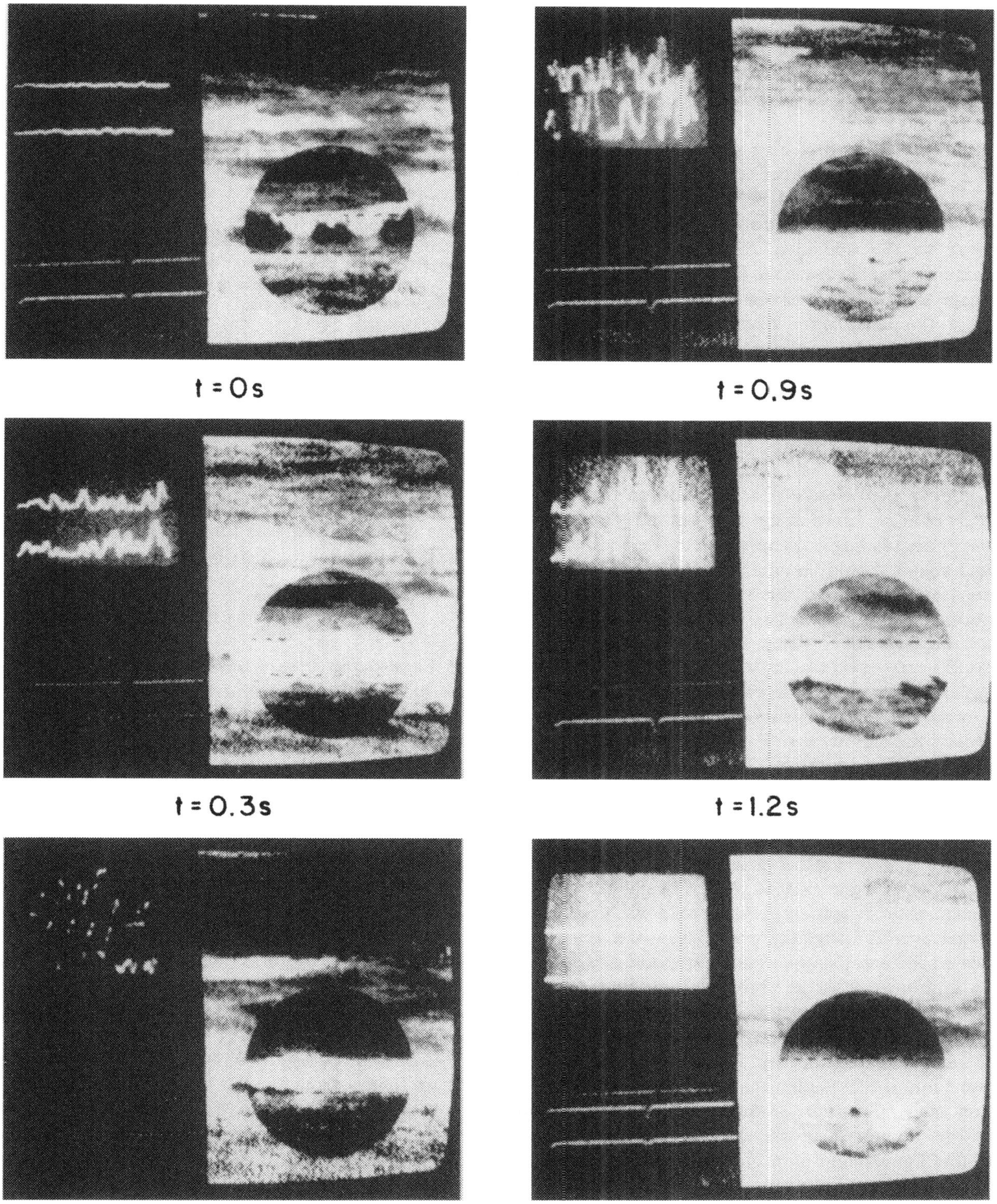

Fig. 7. Display of sea surface and accompanying backscattered signals using horizontally polarized 20-ns pulses (Chesapeake Beach data).

about 100 cm/s. This agrees approximately with the results of *Hicks et al.* [1960], who indicated a whitecap velocity distribution of about 3 knots (150 cm/s).

Figure 7 ($t = 0$ s) also shows the whitecap and an optical image as reflected in the surrounding water wave. It should be noted that the point of reflection is on the wave surface and therefore very close to the whitecap. Since the radar and TV cameras observe the same field, a similar microwave image should be observed by the radar. Moreover, since the radar wave length is much greater than the optical, the image should be better and can be expected to give rise to multipath effects in the radar signal, as is illustrated in Figure 8. R and R′, for example, define two paths which can interfere depending on the point of reflection Q. With the rise and fall of the wave and its associated whitecap the relative phasing between these paths will change, producing the familiar frequency-dependent fading. Furthermore, the periods of fades and signal should be comparatively long, since the separation between the two paths is small. This geometry contrasts with that usually considered for a radar target over water. For this case the whitecap would only produce an image as observed by the radar if the water in the vicinity of point P in Figure 8 was sufficiently smooth. For the Chesapeake Beach measurements, P would be about 10 m in front of the whitecap. With sufficiently smooth water unlikely at this point, such multipath geometry must be ruled out.

Another observation concerns the effect of polarization. Using vertical polarization, the sea spikes persisted longer and in the Chesapeake Beach measurements tended to follow the wave in range, whereas with horizontal polarization the sea spikes developed and disappeared at about the same range. In addition, although the peak amplitudes of the sea spikes for the Boca Raton data were essentially independent of polarization, data from Chesapeake Beach indicated that for vertical polarization they were about 5 dB higher than for horizontal polarization. An explanation for those observations in terms of the alignment of the wave crests with respect to the line of sight is proposed later in the paper.

In between whitecaps, or under calmer conditions, recorded signals were several orders of magnitude lower. But even here the signals were spikey in appearance as shown in Figures 9 and 10, the data for which were taken in the absence of whitecaps. Corresponding spectral-time plots are shown in Figures 11 and 12. Observation of range sweeps similar to that in Figure 1 still disclosed targetlike backscatter. Although within the 300-m range sweep as many as three or four spikes were observed, none were distributed in range.

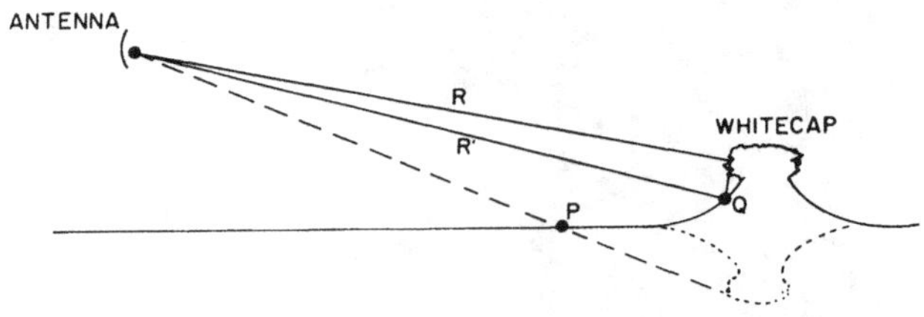

Fig. 8. Multipath effect produced by reflection of the whitecap in the water wave at Q. Shown dotted are the path and radar image which would be produced if the water in the vicinity of P were sufficiently smooth.

SURFACE MODELING

A backscatter theory based on a composite surface model consisting of a long water wavelength (compared with radar wavelength) surface due to gravity waves, which tilted a slightly rough scattering surface due to capillary waves, was developed by *Wright* [1968]. (A slightly rough surface is characterized by height variations small in comparison with the radar wavelength and with slopes much less than unity.) Such a surface exhibits resonance effects between radar and water wavelength. For the usual case of illuminated areas large in comparison with the square of water wavelength, Wright's theory, written in terms of the energy density spectrum of the surface height variations, models the average radar backscatter of large ocean areas fairly well.

Trunk and George [1970] considered the case of backscatter from smaller illuminated areas. Data from an airborne radar flying at 90 m/s and illuminating a surface area about 3 m in range by 33 m in azimuth were used. Non-Rayleigh models were proposed and fitted to the data, although little consideration was given to the physical significance implied in the model selection. Since data runs spanned minutes of flight, they obtained long sam-

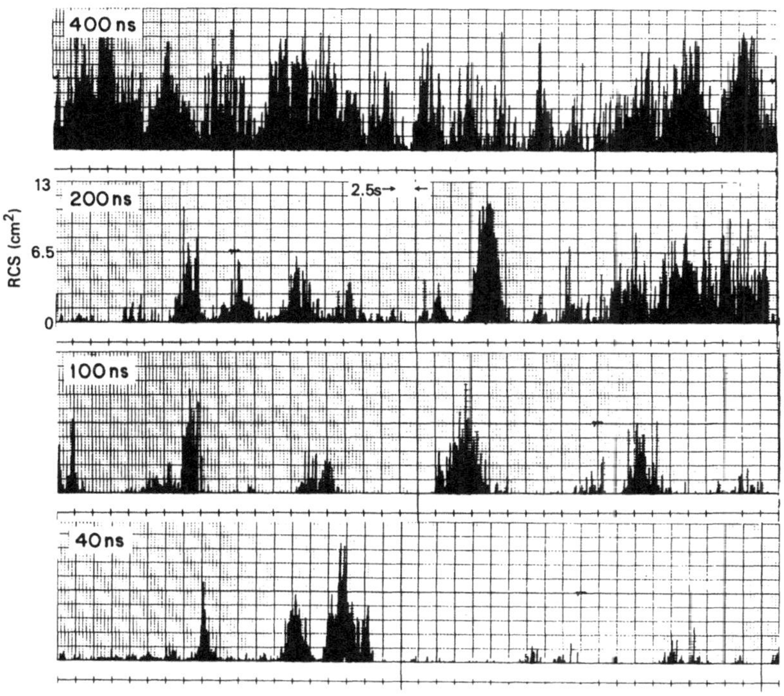

Fig. 9. Backscatter from calm water as a function of time using vertically polarized 9.2-GHz pulses 40, 100, 200, and 400 ns wide. Data at each pulse width were taken at different times. Ordinate scales are identical for each plot (Boca Raton data).

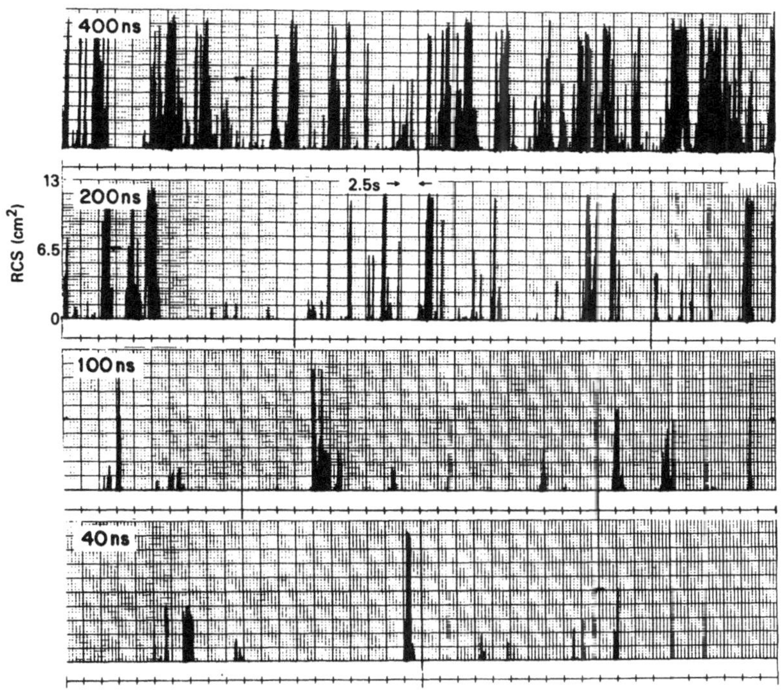

Fig. 10. Backscatter from calm water using horizontal polarization for the same conditions as in Figure 7.

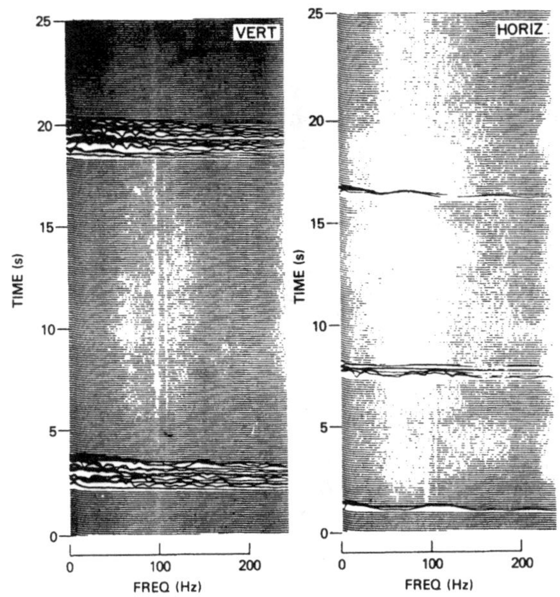

Fig. 11. Video spectra versus time for calm water backscatter using 40-ns vertically and horizontally polarized pulses at 9.2 GHz (Boca Raton data).

ples from which statistically significant results could be obtained. However, *Trunk* [1972] in further examining the earlier data pointed out that a spatial varying Ricean density function is particularly appropriate to high-resolution data. He noted that it fits the experimental data and that it can be physically generated by including a dominant fixed scatterer within a Gaussian scattering background. Current work suggests that whitecaps can provide the dominant scatter component which leads to the Ricean distribution. Considering the aircraft speed, the range resolution used by Trunk allowed very few independent samples in each range cell. (Assuming 10^{-2} s sea decorrelation time, only about three independent samples were measured.) Thus measurements were of an essentially 'frozen' surface, and the whitecaps, which have been observed to exhibit high targetlike backscatter, can appear both as fixed and dominant.

Prior modeling work therefore has been divided between electromagnetic wave solutions of an area extensive surface and investigations of the fit of experimental data to assumed statistical distributions. In both approaches the result has been to characterize the backscatter by a statistical measure. However, since temporal characteristics were not included, these approaches cannot model the sea spike backscatter structure noted during the current measurements. Within the small radar resolution cell used in these investigations the whitecap can grow and decay, so that the dominant backscattered signal is a time dependent function of its size. As was noted in the experimental results, the pulse-to-pulse envelope of the backscatter signals sometimes exhibits an almost periodic amplitude change with growth and decay and often contains short-duration, high-amplitude spikey components. As will be described later, temporal changes in scatterers already in a cell will be characterized by an 'internal' decorrelation time, whereas changes in the extent of the principal scattering region within the cell, as in the buildup and decay of a whitecap, will be characterized by an 'external' decorrelation time. However, even in the absence of a whitecap, temporal changes in the scatterer distribution throughout a fixed cell can be represented by an internal decorrelation time.

Backscatter from a whitecap can arise from both the droplet spray and the roughened surface beneath, which can also be tilted with respect to the horizontal. But rather than make any assumptions

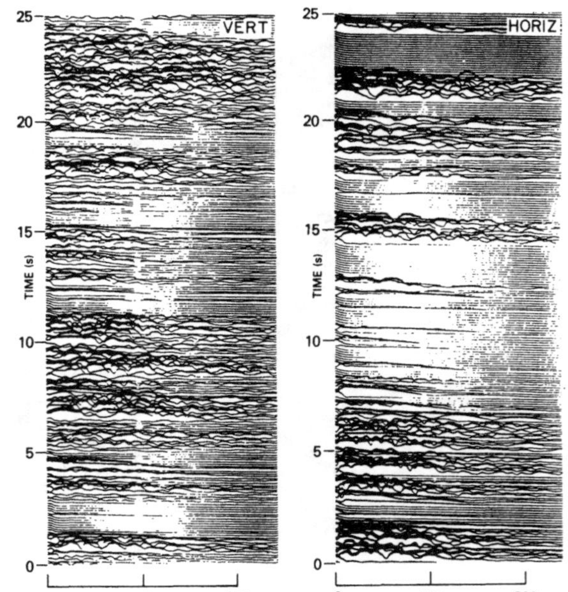

Fig. 12. Video spectra versus time for calm water backscatter using 400-ns vertically and horizontally polarized pulses at 9.2 GHz (Boca Raton data).

regarding the actual origin of the backscatter, a simple model which can be regarded as generally applicable to any collection of scatterers serves to describe the temporal behavior of the whitecap. In addition, by comparing simulations using the model with the measurements the scale of roughness of the whitecap and the relationship between internal and external decorrelation times can be established. Since very rough surface scatter is assumed, polarization dependence is not included.

At low radar grazing angles, backscatter from scatterers separated by less than $\lambda/4$, where λ is the radar wavelength, interfere constructively. Conversely, scatterers separated by more than $\lambda/4$ (but less than $\lambda/2$) interfere destructively and are decorrelated. Therefore the surface can be represented by a single resultant scatterer, specified by amplitude and location, within each $\lambda/4$ range increment. The number of such increments will generally be a time dependent function. This selection is also reasonably consistent with observation of the whitecap growth rate (10^2 cm/s) and 3-cm sea spike decorrelation time (10^{-2} s), since the whitecap expands by $\lambda/4$ every 0.75×10^{-2} s. It is also assumed that whitecaps start from zero size, grow in range to a finite extent, then shrink to zero. On the basis of these few assumptions it is easily demonstrated that simple $\lambda/4$ zone additions to the scattering surface, even when their equivalent amplitudes and reflectivity centers are random, result in the characteristic rapid modulations observed during backscatter measurements. Considering the mth $\lambda/4$ zone, the normalized backscattered signal referred to the initial surface boundary of the whitecap is given by

$$e_m = S_m \exp - j\frac{4\pi}{\lambda}\left[(m-1)\frac{\lambda}{4} + d_m\right] \qquad (1)$$

where $S_m \leq 1$ is the equivalent zonal reflection coefficient and $0 \leq d_m \leq \lambda/4$ represents its location within the zone. Given independent random numbers X_m and Y_m ($0 \leq X_m \leq 1$, $0 \leq Y_m \leq 1$) and letting k represent the fraction of random component in the equivalent zonal amplitude and location, S_m and d_m can be written as

$$S_m = (1 - kX_m) \qquad (2)$$

$$d_m = \frac{\lambda}{8} + \frac{\lambda}{4}k(Y_m - \tfrac{1}{2}) \qquad (3)$$

By using these expressions and shifting the initial reference surface by $\pi/2$, (1) can be written:

$$e'_m = (-1)^{m-1}(1 - kX_m)\exp - j\pi k(Y_m - \tfrac{1}{2}) \qquad (4)$$

If the range extent of the whitecap is W, then the number of zones is $N = 4W/\lambda$, and the total backscattered signal Γ_N is given by

$$\Gamma_N = \sum_{m=1}^{N} (-1)^{m-1}(1 - kX_m)\exp -j\pi k(Y_m - \tfrac{1}{2}) \qquad (5)$$

where k is assumed constant from zone to zone, since it is a measure of the scale of roughness within the limited area of the whitecap. By using (5) the temporal variations can be modeled with the growth (or decay) of the whitecap simulated by changes in N. Since this adds (or subtracts) $\lambda/4$ zones to the total scattering surface, it represents the external decorrelation time of the whitecap, given the appropriate velocity. But the interval required for development and decay of a whitecap is comparatively long, so that during observations, scatterers within zones already formed do not remain static. This can be accounted for by defining an internal decorrelation time. In the model it is represented by an integer D which expresses the ratio of internal to external decorrelation time.

In general, the characteristic of Γ_N will depend on the values of k, X_m, and Y_m. For example, with identical scatterers centered within each zone, $k = 0$, $Y = 1/2$. If N varies linearly with time, simulating the whitecap growth, a periodic change in Γ_N results. Figure 13 shows the amplitude characteristics plotted in terms of radar cross section, $\sigma_N = |\Gamma_N|^2$, for four different surface realizations using uniformly distributed random numbers. Each curve has been scaled so that the largest spike is plotted full scale. In all cases, N was taken to be 128, so that with $\lambda = 3$ cm the maximum length of the corresponding whitecap is about 1 m. Assuming a 10^{-2} s zone addition period, the corresponding growth interval is about 1-1/4 s. Except for the case of $k = 0.25$, $D = 5$, values of $k = 0.25$ and 1.0 correspond to selecting a 25% and 100% random component to represent both scatterer amplitudes and location within zones. $D = 1$ implies that all previous zonal scatterer amplitudes and locations are changed whenever a new zone is added, whereas with $D = 5$ no changes occur in prior zones until

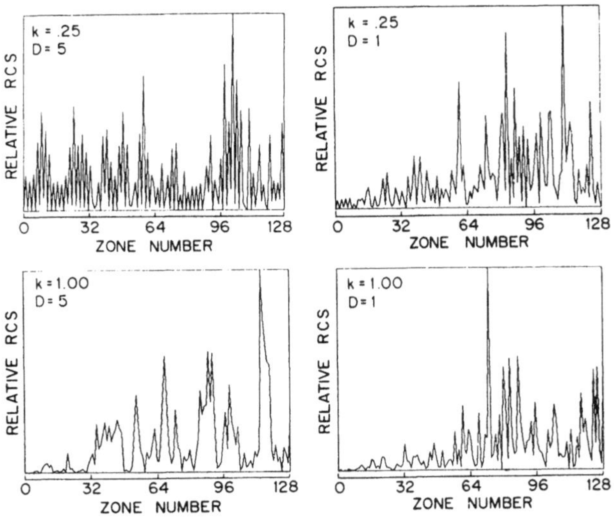

Fig. 13. Sea spike realizations using equivalent scatterers of random amplitude and location. The degree of randomness is represented by k, and D represents the interval in which zone amplitudes and locations are updated (case of $k = 0.25$, $D = 5$ for random amplitudes only).

five new zones are added. The case of $k = 0.25$, $D = 5$ is for a random amplitude component only, i.e., the location of the amplitude representing each zone is at its center. This case shows the fine periodic structure characteristic of resonance from $\lambda/4$ zones. Varying degrees of periodicity, spike buildup and width are illustrated in the other cases. As a result of comparing realizations with measured data, it has been concluded that random amplitudes with or without random locations but with short internal decorrelation times are most nearly in agreement. This is especially true for $k = 1$, $D = 1$ and indicates that whitecaps are very rough surfaces with nearly equal internal and external decorrelation times. The agreement between simulation and experimental data also adds confidence that the water dynamics will always produce backscatter with substantial high-frequency amplitude modulation.

Although the modeling results support observations of the very large sea spikes associated with whitecaps, experimental data also show similarly structured backscatter in the absence of whitecaps. However, peak amplitudes are at least an order of magnitude lower. Physically, the surface is smoother than that of the whitecap; in addition, differences in the relationship between the illuminated area and the water surface exist, as illustrated in Figure 14. In the case of whitecap development the radar-illuminated area is depicted as being larger than the whitecap. Therefore the large backscatter is a function of the time-varying whitecap area, and both internal and external decorrelation descriptors are appropriate. For unbroken water the radar-illuminated area can be assumed smaller than the backscattering surface. Consequently, the radar backscatter is a function of the number of zones represented by the pulse width, and only an internal surface decorrelation descriptor applies.

The randomness factor k has a marked effect on the magnitude of the sea spikes. This is simply shown for the random amplitude (fixed phase) case. For large N and assuming $Y = 1/2$, the normalized backscatter from (5) can be written:

$$\Gamma_N = \sum_{m=1}^{N-1} k(X_m - X_{m+1}) \tag{6}$$

The nonrandom term has been suppressed, since its maximum contribution summed over the surface is small. Since the radar cross section $\sigma \propto \Gamma^2$, it can be shown that the expected value of RCS is

$$\sigma_N \propto Nk^2 \tag{7}$$

N is a function of the pulse width (in the case of unbroken water) or of the whitecap and the radar wavelength, while k depends on the characteristics of the surface. For slightly rough and unbroken water the random component from one quarter-wavelength zone to the next will be small, so that the value of k will be low. Conversely, rough water and whitecaps exhibit much more random backscatter from zone to zone, and values of k will approach unity. Thus given two surfaces equal in extent but

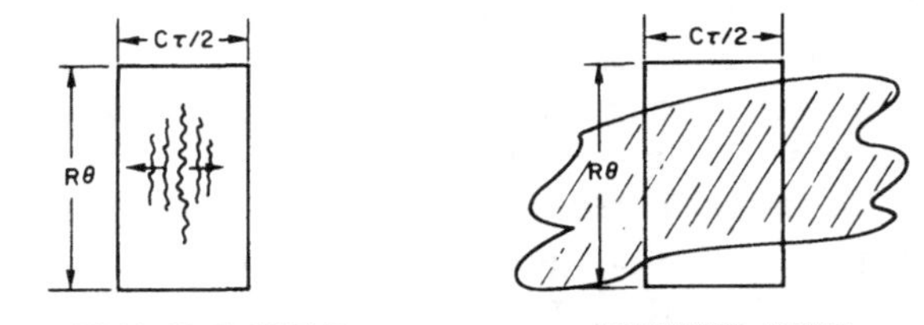

Fig. 14. Comparison of effective radar reflective patch with surface characteristics. R, θ, and τ represent the radar range, beamwidth, and pulse width, respectively; C is the speed of light.

having values of $k = 0.25$ and $k = 1$, their average radar cross section can be expected to differ by 12 dB. The fine structure in the backscatter, however, will be similar.

AMPLITUDE SCALING

The previous analysis modeled the structure of the short pulse radar backscatter. As long as a random component existed from one zone to the next, sea spikes could occur, although their actual amplitudes are a function of the magnitude of the random component. However, an estimate of the maximum amplitude of these spikes can be obtained by assuming, as was indicated in the experimental work, that the surface is very rough, so that the backscatter is diffuse. By definition the equivalent radar cross section of a scatterer is given by $\sigma = G_s A_s$. A_s is the power intercept area of the scatterer, and G_s is its gain in the direction of the receiver, with respect to an isotropic radiator. Since the whitecap is assumed to be a very rough surface, it will scatter radiation in all directions within the upper half space (2π sr), so that its gain with respect to an isotropic radiator is $G_s = 4\pi/2\pi = 2$. Therefore the equivalent radar scattering cross section of a whitecap per unit illuminated area normal to the beam is $\sigma_0 = \sigma/A_s = 2$. This value is in reasonable agreement with the experimental measurements, where the resolution cell was small. For example, the 6-m^2 resolution cell of the Chesapeake Beach measurements was sufficiently small that during development a whitecap would approximately fill the cell. The largest of the peaks were about 9 m^2, implying a value of $\sigma_0 = 1.5$.

While the value of σ_0 for the whitecap alone is large, it is nonetheless consistent with the much smaller values measured using low-resolution radars, since only a small portion of the ocean surface is covered by whitecaps. Thus a $\sigma_0 = -37$ dB would imply a 1-m^2 whitecap in a region 100 m square.

POLARIZATION CHARACTERISTICS

For slightly rough unbroken water (sea states 1 and 2), backscattered peaks using vertical polarization were generally smaller than those from horizontal polarization. Inasmuch as this observation was at low grazing angle, the result is consistent with measurements in the vicinity of Brewster's angle. But with the roughened water at which whitecaps form, measurements at Chesapeake Beach indicated that the backscatter peaks for vertical polarization were about 5 dB higher than for horizontal polarization. This difference is not predictable on the basis of a uniformly random very rough surface; such a surface should scatter all polarizations equally. In fact, for the Boca Raton measurements, as indicated in Figures 2 and 3, the results using the two polarizations are nearly equal.

Although no rigorous explanation for this polarization sensitivity is offered, some insight regarding the mechanism producing polarization differences resulted from observing the backscatter from sharp crested unbroken waves at Chesapeake Beach. These waves, with crests aligned normal to the line of sight, also backscatter higher peak signals with vertical polarization than with horizontal. The significant factor appeared to be the sharp edge of the wave crest. Electromagnetically this edge acts as an impedance discontinuity to currents induced in a direction normal to the edge. The result is a reradiated component with an intensity pattern dependent upon the surface geometry. Figure 15 illustrates such an effect. The pattern is similar to that from a long wire antenna with current fed at one end. The discontinuity at the opposite end acts to produce a radiated pattern with principal lobes near the wire surface. The same characteristic occurs for long wire or rod targets when the incident energy makes a small angle with the target axis [*Peters,* 1958]. The angle between the target and the largest lobe is a function of target length; for example, with a rod 39 wavelengths long this angle is about 8°. For shorter lengths this angle increases. For the geometry illustrated in Figure 15 the relatively high gain radiation lobes will sweep through the vertically polarized transmitting/receiving aperture producing significant backscatter amplitudes. However, with horizontal polarization the surface-induced currents will be parallel to the sharp crest,

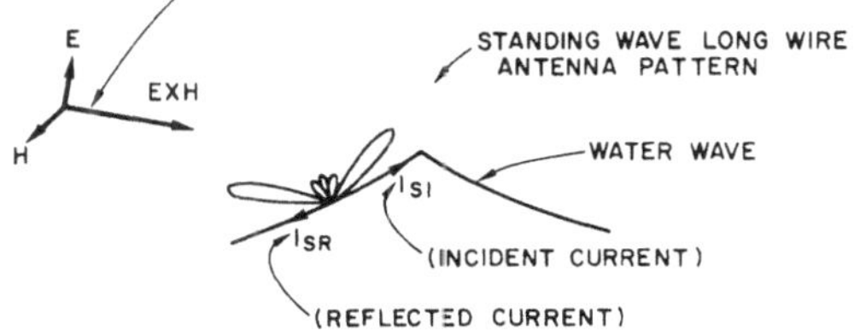

Fig. 15. Vertically polarized scatter from water wave.

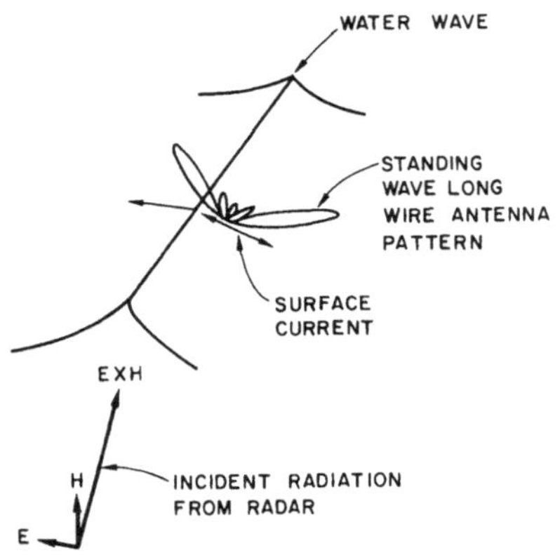

Fig. 16. Horizontally polarized scatter from water wave.

and no edge effect will be detected. Figure 16 shows a horizontally polarized case in which the direction of the incident radiation does induce a current component in the wave surface which results in a pattern similar to that shown in Figure 15. However, the direction of these lobes is such that most of the energy is scattered at large angles to the incident direction, and very little monostatic backscatter from this effect is observed.

For whitecaps the same phenomenon should occur, since there is an abrupt discontinuity to surface-induced currents produced by the whitecap itself. In the case of the Chesapeake Beach measurements, backscatter from horizontal polarization originates mainly from the rough broken water of the whitecap. Vertical polarization, however, results in backscatter from both the whitecap and the frontal wave surface in the manner described. This is also consistent with the more frequent expectation of spikey backscatter when using vertical polarization [*Long,* 1975].

CONCLUSION

Radar echoes from whitecaps (sea spikes) appear to originate from a surface comparable with or less than the range extent defined by the radar pulse width. Maximum amplitudes of about 9 m^2 have been observed, and their development and decay are associated with the development and decay of the whitecap. By using high-resolution radar, sea spikes were observed to occur relatively infrequently. The image of the whitecap in the surrounding water resulted in radar frequency sensitivity characteristic of multipath propagation. In between whitecaps, or under calmer conditions, the observed signals also showed spikes with similar temporal characteristics, but amplitudes were several orders of magnitude lower.

Generally, isolated sea spike spectra were broad and noiselike, although the frequency range depended on the direction the wave broke with respect to the radar line of sight. Waves breaking along the line-of-sight direction exhibited the narrowest spectral width, whereas those breaking across the line-of-sight direction exhibited the broadest spectral width. Nonetheless, the comparatively high amplitude noiselike spectrum of the backscatter envelope within the radar resolution cell represents a characteristic signature which could be utilized to suppress sea return when it exists in any cell and enable detection of low radar cross-section targets whose echoes do not fluctuate as rapidly.

Polarization sensitivity was observed to be a function of both the water surface roughness and the alignment of wave crests relative to the line of sight. With slightly rough unbroken water (sea states 1 or 2), vertical polarization yielded smaller but more probable peaks than horizontal. When whitecaps appeared, the peak amplitudes were essentially independent of polarization unless the radar observation was nearly perpendicular to the wave crest. Then the backscatter using vertical polarization was higher than that observed using horizontal polarization. Greater polarization differences were observed, however, in connection with the duration of the spikey interval and its fluctuation. Spikes occurred more frequently with vertical polarization and persisted longer. Moreover, when the line of sight was perpendicular to the wave crests, the backscatter using vertical polarization tended to follow the wave in range, whereas with horizontal polarization the sea spikes developed and disappeared at the same range. Since the duration of the horizontal polarized signals was less than the vertical, it appeared more spikey in character.

Modeling results suggest that the addition of random amplitude quarter radar wavelength wide zones, each appropriately updated to account for temporal changes in the scattering surface, approximate the observed backscatter characteristics of the sea spikes. Moreover, by comparing the experimental data with the model it was estimated that the internal and external decorrelation times of the rough whitecap surface are each approximately 10^{-2} s for radar wavelengths of about 3 cm.

Acknowledgments. We wish to acknowledge the efforts of J. P. Hansen in designing the equipment and, together with V. F. Cavaleri, in acquiring the experimental data. In addition, we wish to acknowledge the work of T. H. Gauss, who programed the model and provided the simulation results shown.

REFERENCES

Fay, F. A., J. Clarke, and R. S. Peters (1977), Weibull distribution applied to sea clutter, *Radar-77 Conf. Publ. 155,* pp. 101–104, Inst. of Elec. Eng., London.

Goldstein, H. (1951), Sea Echo, in *Propagation of Short Radio Waves, MIT Radiat. Lab. Ser.,* vol. 13, Sect. 6.6, edited by D. E. Kerr, McGraw-Hill, New York.

Hicks, B. L., N. Knable, J. J. Kovaly, G. S. Newell, J. P. Ruina, and C. W. Sherwin (1960), The spectrum of *X*-band radiation backscattered from the sea surface, *J. Geophys. Res., 35*(3), 825–837, 1960.

Lewis, B. L., and I. D. Olin (1977), Some recent observations of sea spikes, *Radar-77 Conf. Publ. 155,* pp. 115–119, Inst. of Elec. Eng., London.

Long, M. W. (1975), *Radar Reflectivity of Land and Sea,* chap. 5, D. C. Heath, Lexington, Mass.

Nathenson, F. E. (1969), *Radar Design Principles,* pp. 234–237, McGraw-Hill, New York.

Peters, L., Jr. (1958), End-fire echo area of long, thin bodies, *IEEE Trans. Antennas Propagat., AP-6*(1), 133–139.

Trunk, G. V. (1972), Radar properties of non-Rayleigh sea clutter, *IEEE Trans. Aerospace Electron. Syst., 8*(2), 196–204.

Trunk, G. V., and S. F. George (1970), Detection of targets in non-Gaussian sea clutter, *IEEE Trans. Aerospace Electron. Syst., 6*(5), 620–628.

Wright, J. W. (1968), A new model for sea clutter, *IEEE Trans. Antennas Propagat., 16*(2), 217–233.

HIGH-RESOLUTION RADAR SEA SCATTER, EXPERIMENTAL OBSERVATIONS AND DISCRIMINANTS

INTRODUCTION

This report describes work done in connection with small-target detection techniques. The primary objective was to develop concepts for improved radar-detection performance against a sea background.

High-resolution radar detection of small objects lying low in the water is often made difficult by the presence of targetlike clutter echoes from the sea surface. (Figure 1 shows examples of typical disturbed sea surfaces. Note the various facets and contours which are presented to an observing radar system.) Techniques which discriminate against sea clutter yet preserve target signals could be applied to a variety of radar problems, ranging from collision avoidance by high-speed ships to periscope detection.

Two goals of this experimental study were to characterize the radar-echoing qualities of the sea surface and of small floating targets and to devise and demonstrate the feasibility of hardware approaches to satisfy the specialized radar requirements. The experimental approach toward these goals consisted of measuring the echoing characteristics of sea clutter and small targets, as functions of the controllable parameters of pulsewidth, frequency, and polarization, while optically observing physical changes in the scattering surface. Sea-return measurements were carried out with several X-band systems at both shallow-water and deep-water sites and over a varied range of sea and wind conditions. These quantitative data have contributed to an initial theoretical model of sea-scatter phenomena [1-3] and have also been used to demonstrate a new sea-clutter discriminant based on differences in the pulse-to-pulse amplitude modulation characteristics of sea return and of rigid targets.

During the experimental program it was found that high-resolution sea scatter was dependent on many dynamic conditions of wind and sea. The physical environment is so changeable that long-term data averages can tend to obscure some of the more interesting features. This report has therefore been organized around specific samples in an attempt to present to the reader some of the varied character seen in actual high-resolution radar observations of the sea surface.

EXPERIMENTAL MEASUREMENT SYSTEMS

Several systems were developed and utilized to study the characteristics of high-resolution radar sea return. Block diagrams of the measurement systems and data-recording complexes are shown in Figs. 2 to 4.

The dual-frequency, single-range-cell system shown in Fig. 2(a) used two continuous-wave (cw) signal generators as carrier sources (nominally tuned to 8.6 GHz and 9.2 GHz). The outputs of these two generators were added in a hybrid circuit and amplified; the hybrid signal then excited the input of a diode switch for pulse shaping. The minimum usable pulse length was 20 ns, and the maximum pulse length was limited by the pulse repetition period of the timing generator, which was variable from 0.02 to 1 ms. The antenna was a parabolic reflector, either 1 or 2.67 m in diameter, with orthogonal, linearly polarized feeds separately connected to a polarization switch.

Manuscript submitted September 25, 1981.

(a)

(b)

Fig. 1 — A disturbed sea surface: (a) sea state 2 , (b) sea state 4

Radar echoes that entered the antenna passed through the polarization switch and entered the radar receiver through the circulator and a diode switch that was closed by a receiver gate pulse from the timing generator. The output from this switch passed through a calibrated variable attenuator and was amplified by a low-noise traveling-wave tube (TWT) chain. The output of the TWT chain entered port A of a four-port circulator and coupled to a 200-MHz-wide bandpass filter tuned to 8.6 GHz and attached to port B of the circulator. The 8.6-GHz signal passed through this filter, and the 9.2-GHz signal reflected back into port B of the circulator. The 9.2-GHz signal exited from port C of the circulator and passed through another bandpass filter tuned to 9.2 GHz. The outputs of these two filters were rectified in separate detectors and amplified in separate wideband video amplifiers. These amplifier outputs were displayed as real-time video on a two-channel oscilloscope and were also used as inputs to two range-gated sample-and-hold circuits. The gated stretched outputs of the sample-and-hold circuits were displayed as time functions on a two-channel oscilloscope and were also recorded on a two-channel magnetic tape recorder. This presentation showed the amplitude-time history of the power received from a single range cell.

Some initial measurements were also carried out with a dual-frequency system transmitting X-band and S-band signals (9.2 and 3.0 GHz). The general layout of the system was similar to the dual X-band system, except that two TWT chains were used with separate parabolic antennas (1 m for X-band and 1.6 m for S-band).

Figure 2(b) is a schematic of the data-recording complex for the dual-frequency systems. A television camera with a zoom lens was mounted on the radar antenna with its field of view centered on the radar field of view. Two other television cameras were used to record the real-time video and the range-gated time functions displayed on two-channel oscilloscopes. Split-screen techniques were used to record the output of all three television cameras on the same television recorder. This ensured time synchronization of the radar data and the optical picture of the sea surface that produced the radar echoes.

Figure 2(c) shows the typical TV video display produced by this data-recording system. The right-hand section of this picture is the view from the boresight camera. The inset at the bottom left is the real-time video and the range gate. The lower trace is 8.6 (or 3.0) GHz, the middle trace is 9.2 GHz, and the upper trace is the approximate range-gate position. The sweep speed for the real-time video was 50 ns/cm, or 500 ns per sweep. The inset at the top left is the gated stretched video, with 9.2-GHz top trace and 8.6-GHz bottom trace. The sweep speed of these traces was 20 ms/cm (0.2 s per sweep). The circular dark spot on the boresight picture is the region filled by the radar beam at ranges long enough to minimize the parallax between the TV camera and the antenna. The dotted lines in this dark spot are the region covered by the 20-ns range gate when the beam strikes the water at a 4.6° grazing angle.

Figure 3 is a block diagram of the single-frequency, contiguous-range-cell measurement system. This system utilized a pulsed-magnetron signal source (centered at 9.3 GHz) as its basic transmitter. Diode switches were used to trim the nominal 300-ns transmit pulse down to 40 ns. The maximum repetition rate was 2.5 kHz. The real-time video for this system was fed into eight paralleled sample-hold circuits, which were sequentially gated every 40 ns over a time span of 320 ns. This time period provided a contiguous-range-cell coverage of approximately 50 m. The first four gated and stretched outputs were displayed as time functions on a four-channel oscilloscope. Seven of the outputs plus a time code were also recorded on an eight-channel magnetic tape recorder.

In the single-frequency system an additional TV video input was added to provide a synthetic range display of the pulse-to-pulse received-power levels of the eight contiguous range cells and also to record visually a time code for synchronization with the audio tape data.

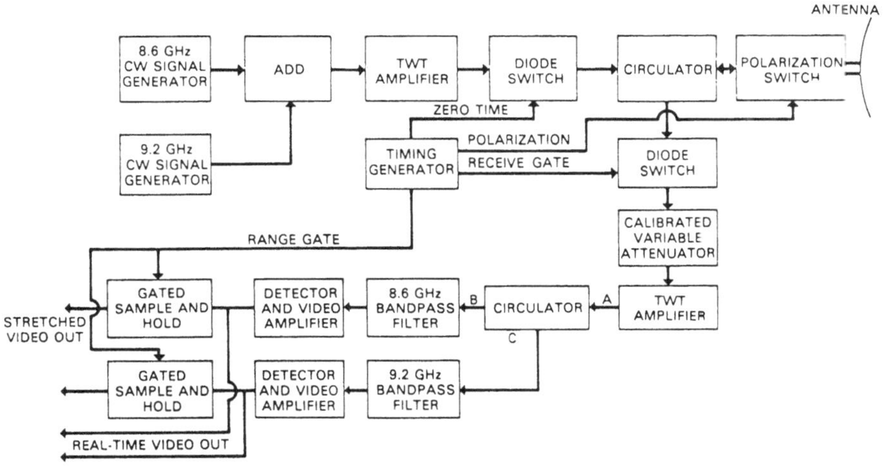

Fig. 2(a) — The dual-frequency, single-range-cell measurement system

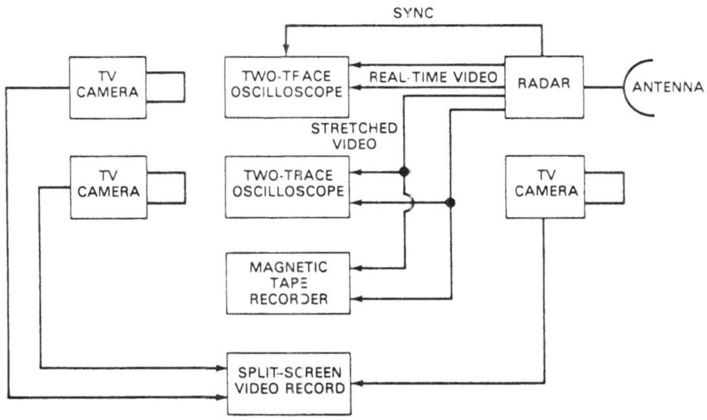

Fig. 2(b) — The dual-frequency, single-range-cell data-recording complex

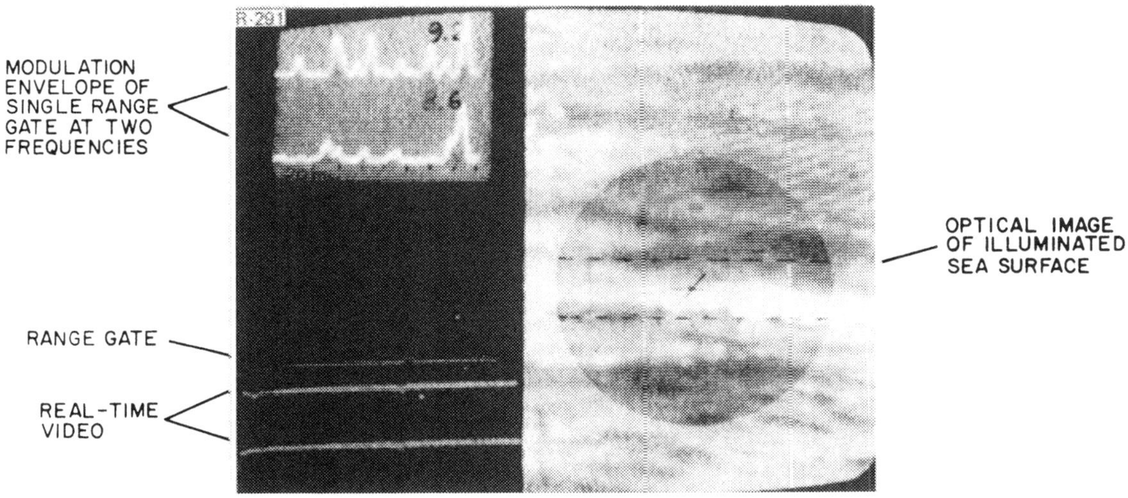

SEA-CLUTTER DATA FORMAT
(TWO-FREQUENCY SYSTEM)

Fig. 2(c) — The TV video display for the dual-frequency, single-range-cell measurement system

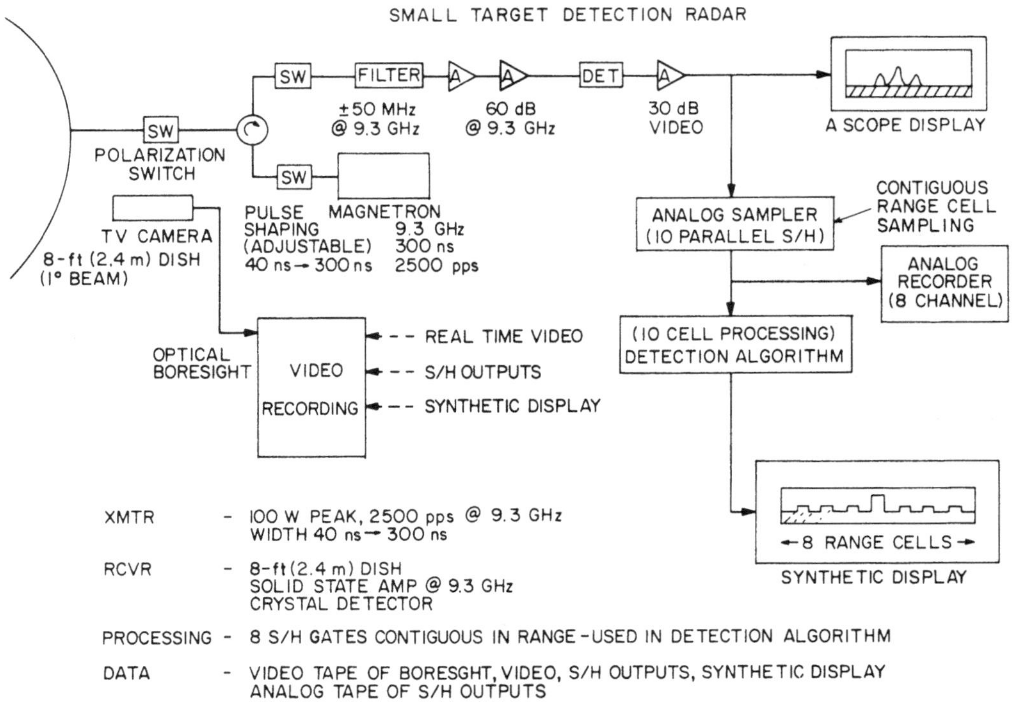

Fig. 3 — The single-frequency, contiguous-range-cell measurement system

Figure 4 shows the TV video display produced with this recording complex. The upper right-hand section of the picture is the view from the boresight camera. The inset at the bottom right shows the real-time video (top trace) and a pulse representing the coverage of the eight contiguous range gates (bottom trace). The video inset at the top left shows the gated stretched return signals for the first four range cells. Sweep speed for the real-time video was nominally 200 ns per division. Sweep speed for the stretched video was 20 ms per small division. The inset in the lower left shows the real-time video levels being received by the eight contiguous range cells, displayed in a synthetic eight-cell contiguous range format, and also the time code.

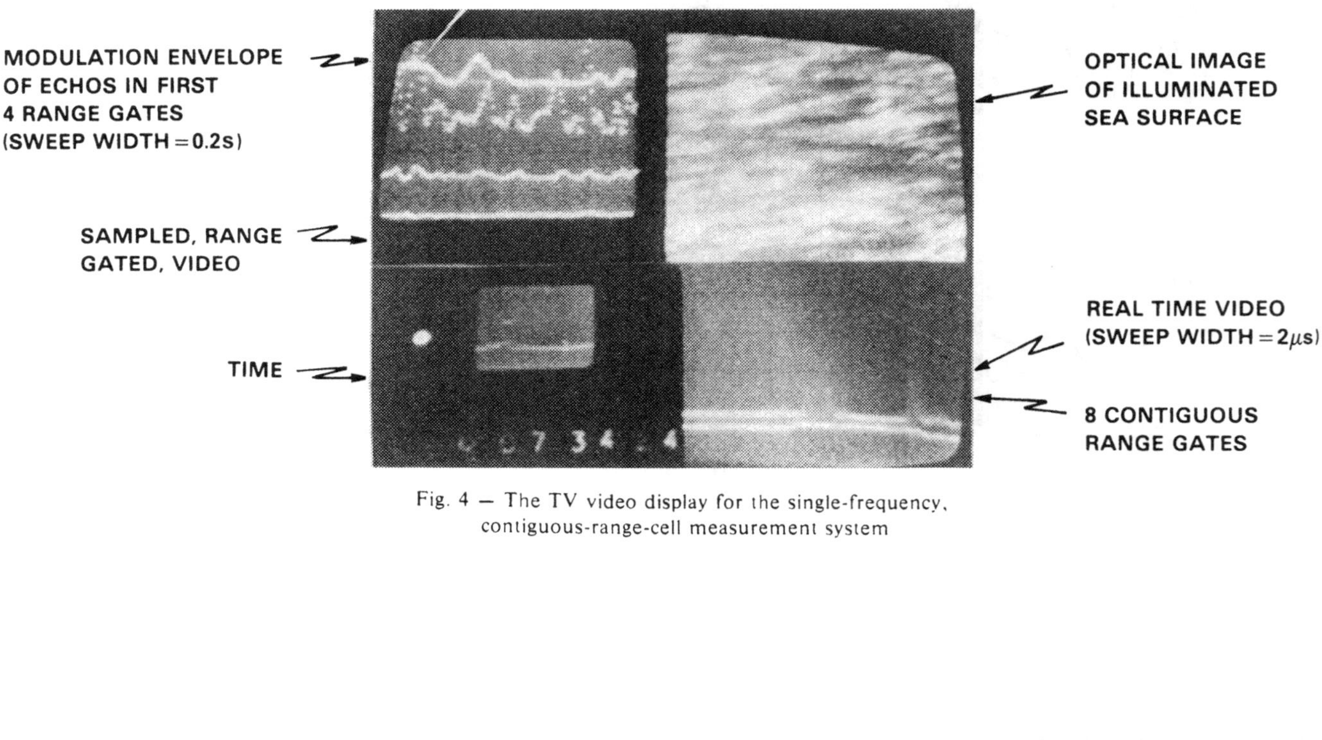

Fig. 4 — The TV video display for the single-frequency, contiguous-range-cell measurement system

CALIBRATION

The systems were calibrated and boresighted using corner reflectors (or spheres) with known equivalent scattering cross sections. Variable attenuators at the input to the receiver were adjusted so that both the real-time video and the gated stretched video were at the upper end of the linear range of the system. Calibration was made in the far field at the two shore sites (Chesapeake Bay, Maryland, and Boca Raton, Florida) and in the near field at the platform site (Stage I, Panama City, Florida). Figures 5 through 7 show pictures of the transmit-receive antennas and calibration setups for the various measurement systems.

Fig. 5(a) — Measurement site, shallow-water data, Chesapeake Bay, Md., showing dual-frequency X-band system antenna

Fig. 5(b) — Measurement site, deep-water data, Boca Raton, Fla., showing dual-frequency X-band system antenna

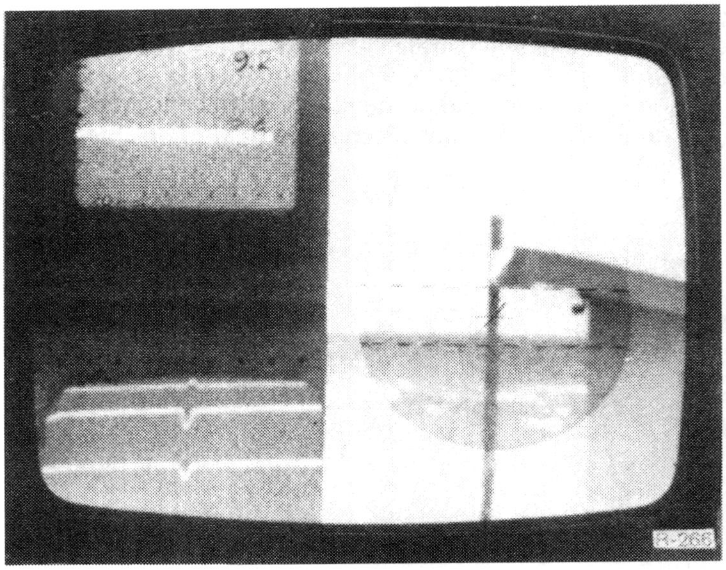

Fig. 5(c) — Calibration setup (corner reflector), dual-frequency X-band system, Chesapeake Bay, Md.

(a)

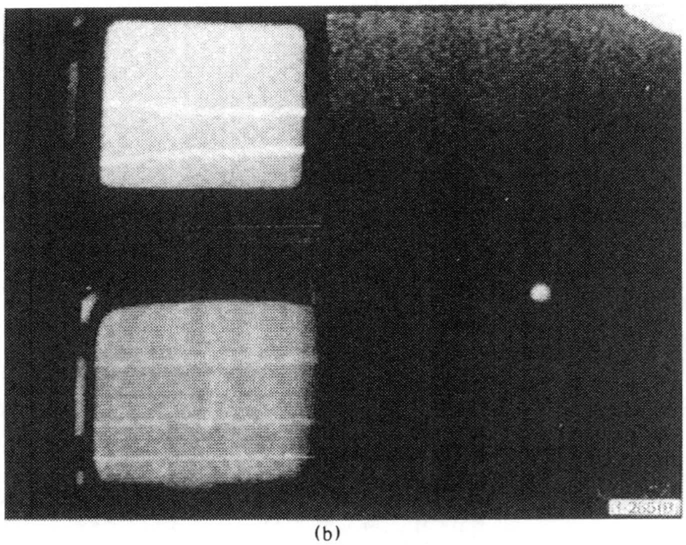

(b)

Fig. 6 — Antenna (a) and calibration setup (sphere) (b), S- and X-band system, Chesapeake Bay, Md.

(a)

(b)

Fig. 7 — Antenna (a) and calibration setup (corner reflector) (b), single-frequency, contiguous-range-cell system, Stage I, Panama City, Fla.

EXPERIMENTAL OBSERVATIONS

Shallow-Water Data (2-m Depth)

The first measurements of sea return were made at the Chesapeake Bay site [see Fig. 5(a)] with the dual-frequency, single-range-cell system and with a 1-m-diameter radar antenna, 4 m above the mean water level. The radar beamwidth with this antenna was 2.6°, and the center of the beam impinged on the water at a distance of 50 m. These values yielded a 2-m cross-range resolution cell at the center of the beam and a grazing angle of 4.6°. The radar pulse length and range gate were 20-ns long, yielding a range resolution of 3.33 m.

Figures 8(a) and 8(b) typify data obtained with vertical and horizontal polarization and with 1- to 1.3-m waves. The individual pictures were taken sequentially with approximately 0.3 s between pictures. This time interval corresponded to the time between sweeps of the oscilloscope which shows the gated stretched video. All data were taken in a generally upwind direction.

These dual-frequency, single-range-cell, shallow-water measurements allowed close correlation between the physical changes in the radar-illuminated sea surface and the received backscatter. The following conclusions were reached from these data [1,2]:

- The backscatter was heavily amplitude modulated with relatively high frequencies, of the order of 20 to 100 Hz. (See Figs. 9(a) and 9(b) for examples of pulse-to-pulse stretched data under two different surface conditions.)

- Large backscatter amplitudes were associated with the turbulent breaking action of the water (whitecaps) but not with the residual foam left after the breaking action. Slightly disturbed surfaces (by wind or rain) also produced a characteristic low-level but spikelike backscatter, as in Fig. 9(b). Figures 10 and 11 show examples of backscatter from wind- and rain-disturbed surfaces, as seen on an A-scope.

- The short-term fluctuations between the the envelopes of the spikelike returns from two transmitted frequencies suggested a multipath effect involving the breaking water and its image in the crest. Simultaneous observation of an optical reflection of the whitecap in the crest supported this conclusion.

- With vertical polarization and normal wave-crest alignments to the radar, high-amplitude spike returns would sometimes occur slightly prior to the visually observed formation of whitecaps.

- Observation of the sea surface with a dual X- and S-band system suggested that the characteristic amplitude modulation tends to scale down in frequency with increased transmitted wavelength. Figure 12 shows examples of these data. Note that the modulation frequencies for the X-band return (top trace) are generally a factor of two to three higher than the S-band return (second trace) for the same radar-illuminated surface.

Deep-Water Data (12.3-m Depth)

Sea-return measurements were also made at Boca Raton, Florida, in deep water with the dual-frequency, single-range-cell system [see Fig. 5(b)]. In these measurements, the 2.67-m-diameter (1° beamwidth) antenna was employed. The antenna was mounted on top of a building 13 m above mean sea level, and the center of the beam intercepted the water at a distance of 530 m. This produced a grazing angle of about 1.4°. The boresight-camera field of view was narrowed to permit it to cover the radar beam in azimuth (1°) on the split-screen display, and it provided a 2° vertical field of view.

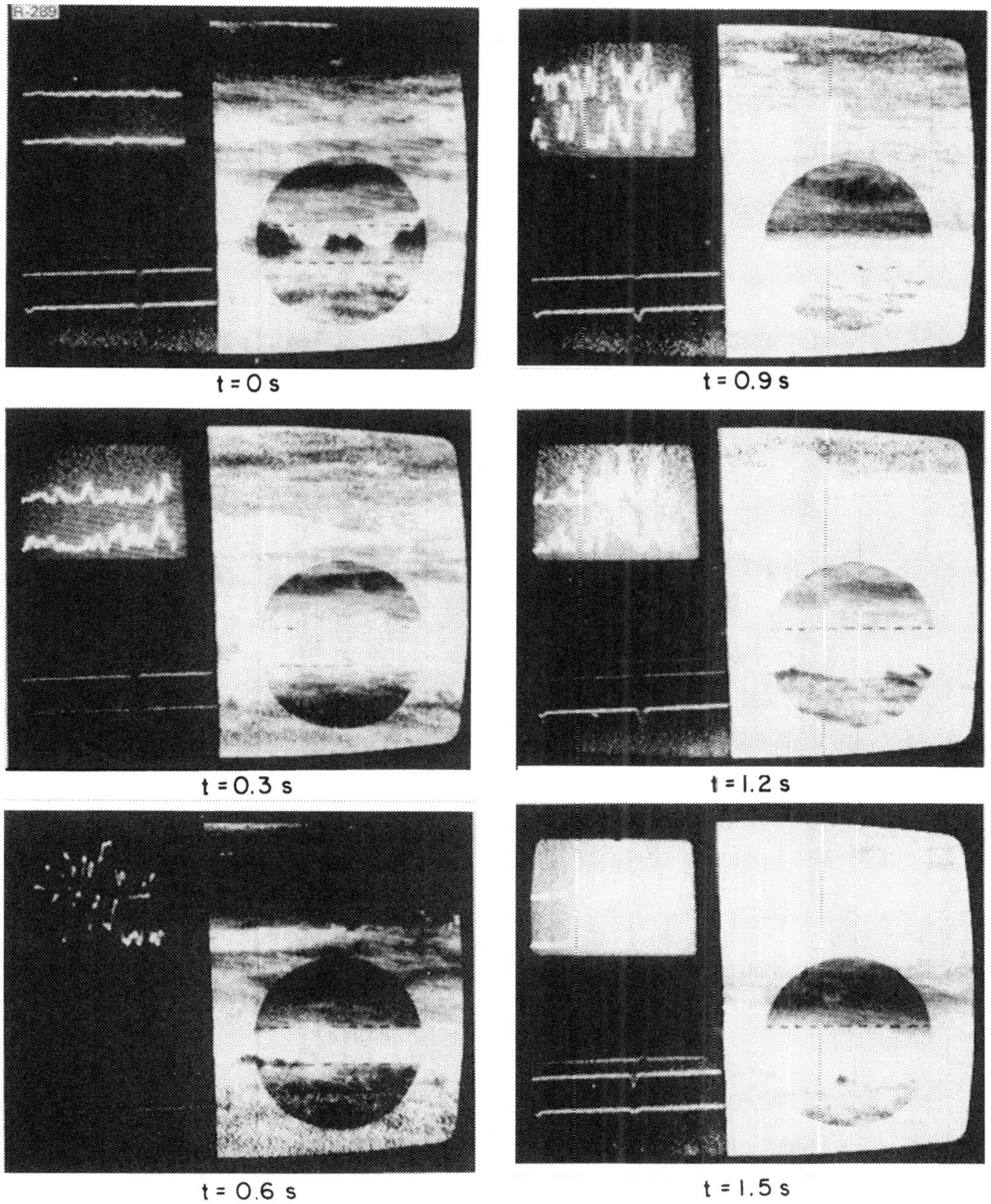

Fig. 8(a) — Typical TV video data, sequential photos, shallow water, dual-frequency X-band system: vertical polarization

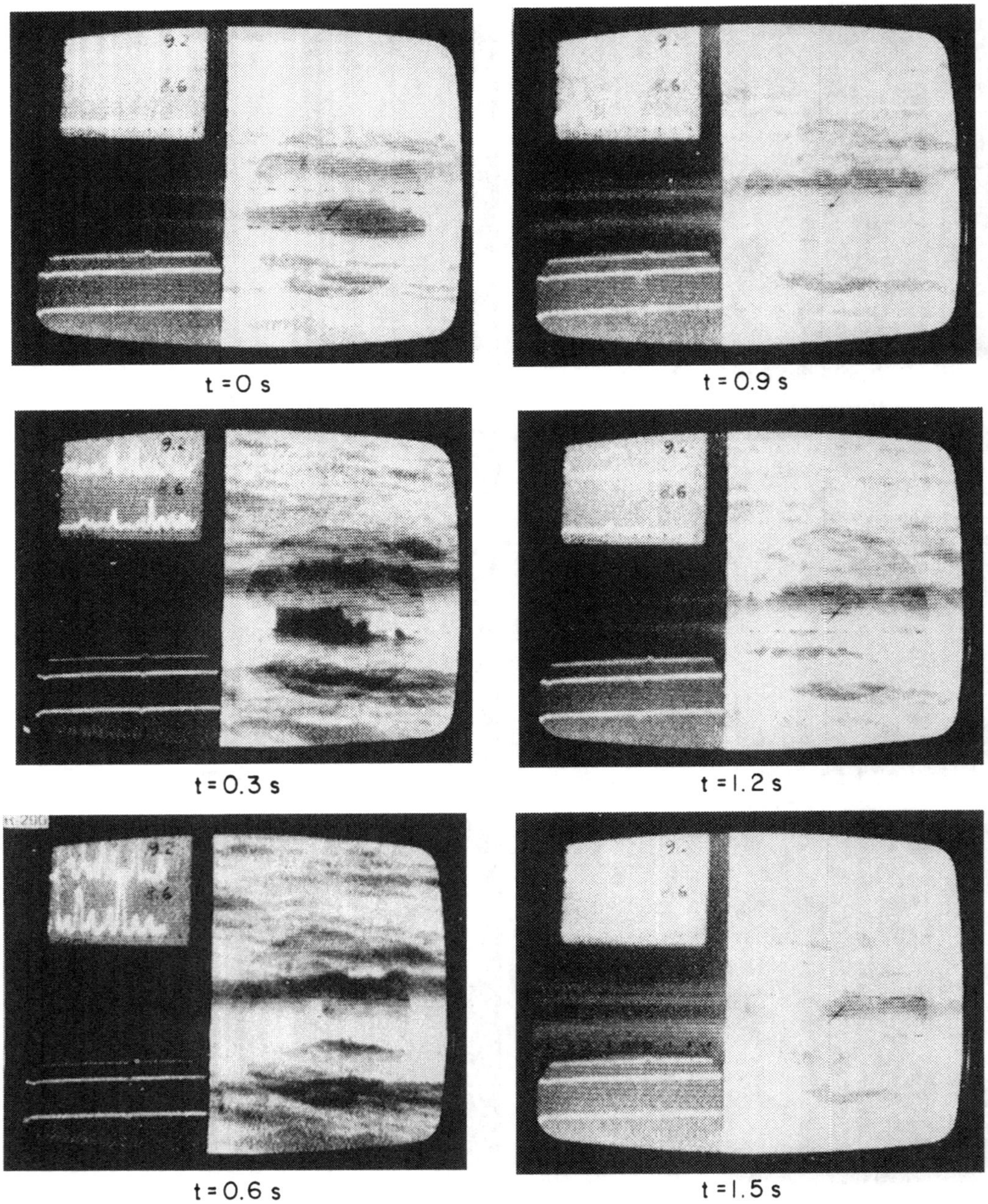

Fig. 8(b) — Typical TV video data, sequential photos, shallow water, dual-frequency X-band system: horizontal polarization

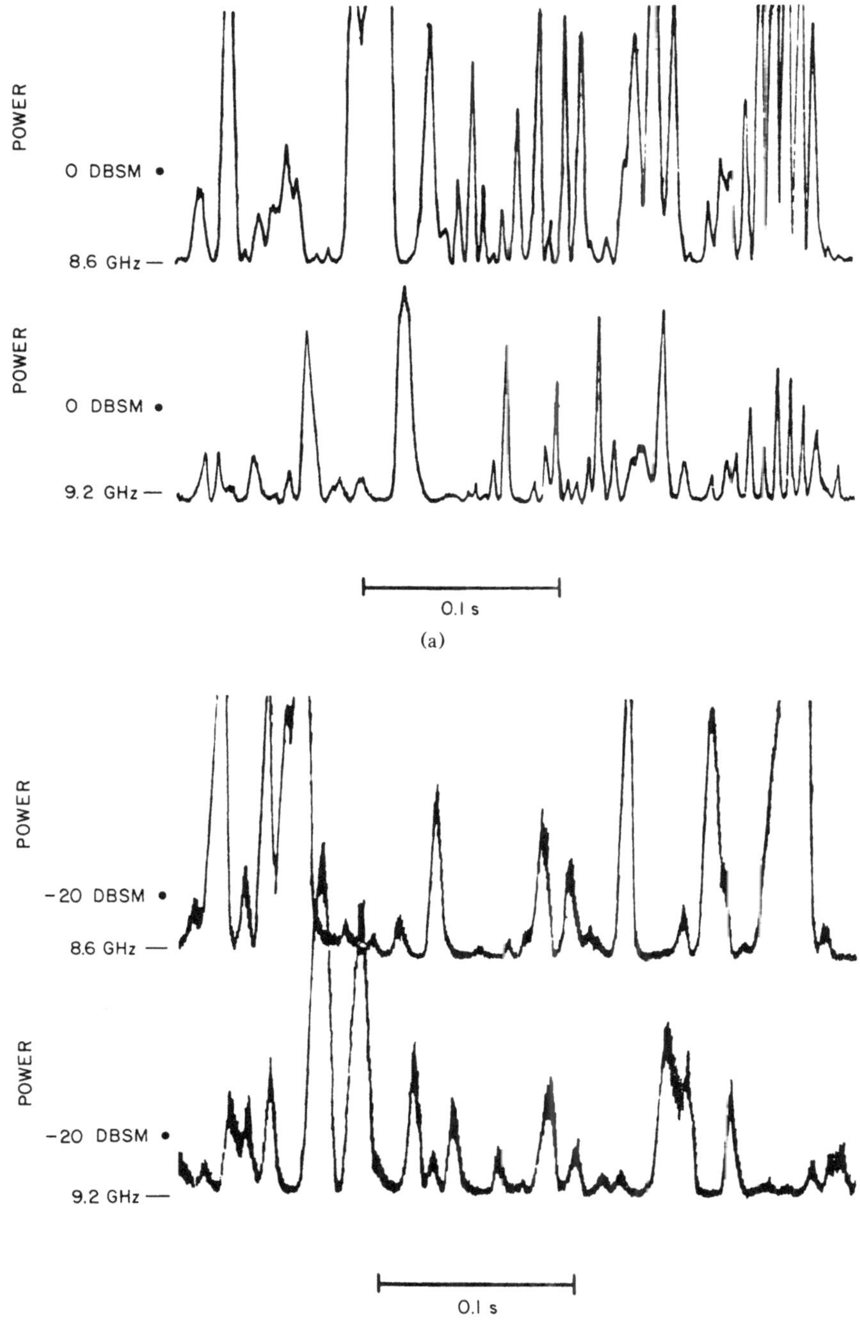

Fig. 9 — Pulse-to-pulse amplitude of scatter, vertical polarization: (a) from breaking wave (whitecap); (b) from relatively undisturbed sea (no whitecaps)

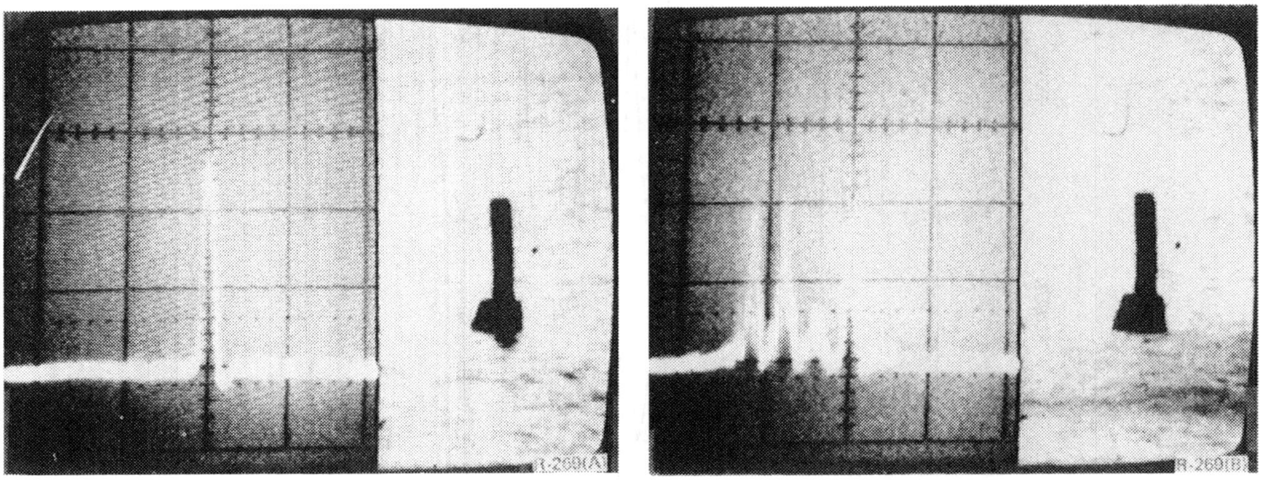

(a) (b)

Fig. 10 — A-scope photos of high-resolution scatter from a wind-disturbed sea surface:
(a) return from wooden post at edge of illuminated sector;
(b) scatter returns from the wind-distributed sea surface surrounding the post

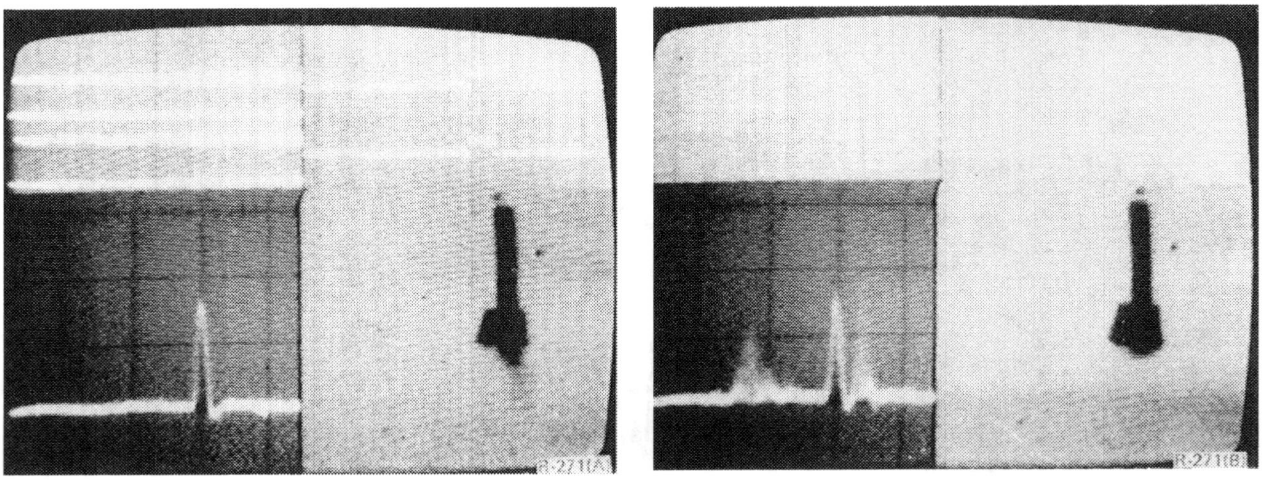

(a) (b)

Fig. 11 — A-scope photos of high-resolution scatter from a calm, rain-disturbed sea surface:
(a) return from wooden post just prior to a rain squall;
(b) scatter returns from the rain-disturbed sea surface surrounding the post

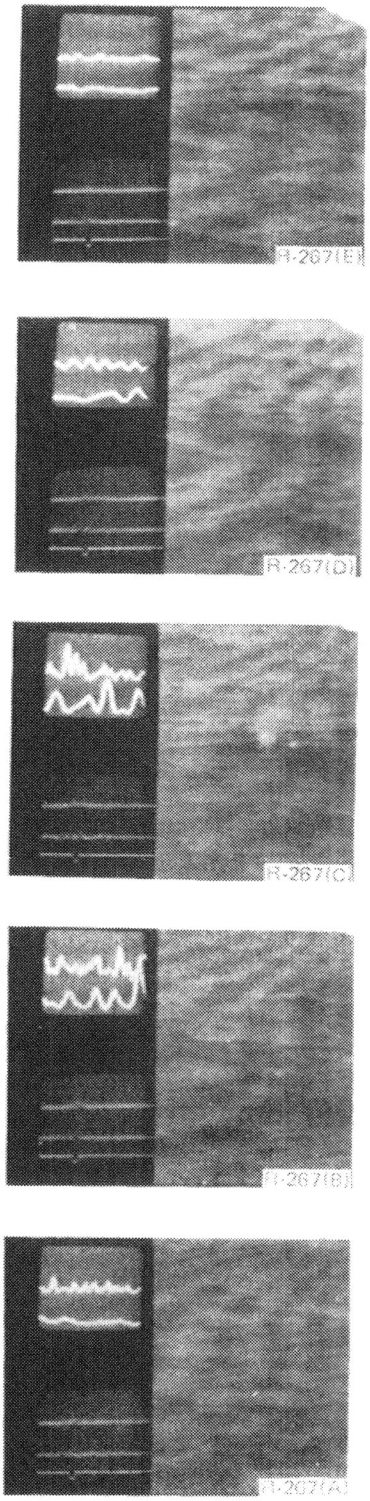

Fig. 12 — Typical TV video data, vertical polarization, shallow water, dual-frequency X- and S-band system. Top trace in each photo is X-band return; second trace is S-band return.

The 530-m range and 1° beamwidth produced an azimuth resolution of about 9 m, and the 40-ns pulse yielded a range resolution of 6.2 m (56 m^2 cell area).

Figure 13 typifies the TV video data obtained at this site in approximately sea states 1 and 3 (sea state 3, relative 30-dB attenuation). These single-range-cell, dual-frequency data were taken with several transmitted-pulse widths for both horizontal and vertical polarization. The data reconfirmed some of the initial shallow-water test results and also demonstrated a backscatter dependence on transmitted-pulse width and polarization [3]:

1. The temporal characteristics of radar backscatter from the sea strongly depend on pulse width for surface resolutions less than about 20 m. Low-resolution backscatter appears continuously noise-like, but as resolution is increased, the backscatter is punctuated by substantially quieter periods. This results in a noticeably spikelike characteristic on an amplitude-time record. The observations confirmed that large backscatter is often associated with whitecaps. However, even the quieter periods disclosed spikelike returns, although they were several orders of magnitude lower in power. The stretched, single-cell data, as displayed in the chart recordings shown in Figs. 14 and 15, illustrate the changes in the frequency of occurrence of sea return as a function of transmitted-pulse width for both calm and rough-sea conditions (RCS = radar cross section). Figures 16 and 17 show examples of video returns on an A-scope display for 1/30-s exposure times at various transmitted-pulse widths.

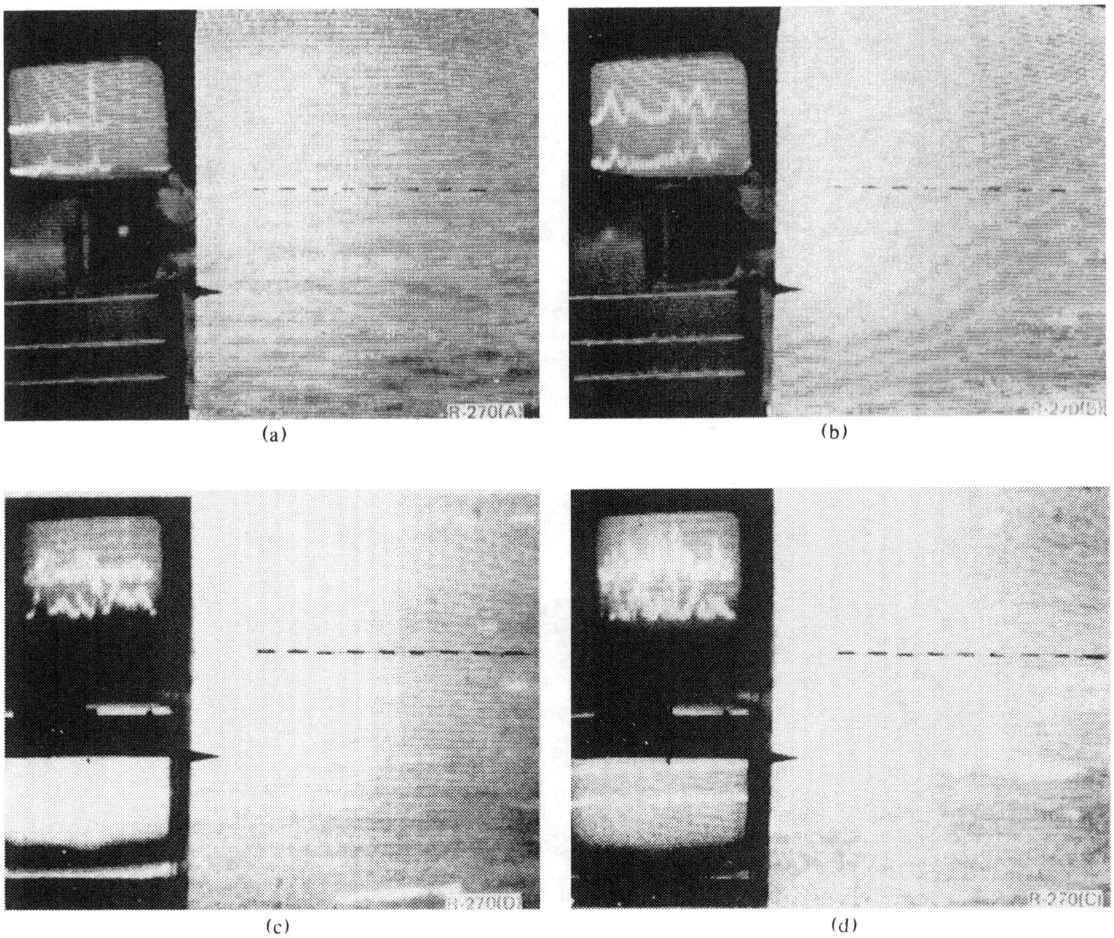

Fig. 13 — Typical TV video data, deep water, dual-frequency X-band system: (a) vertical polarization, sea state 1; (b) horizontal polarization, sea state 1; (c) vertical polarization, sea state 3; (d) horizontal polarization, sea state 3

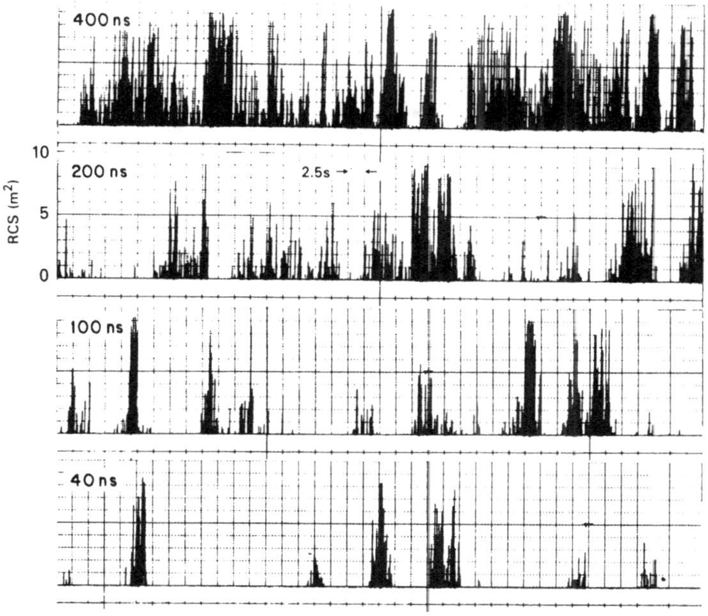

(a)

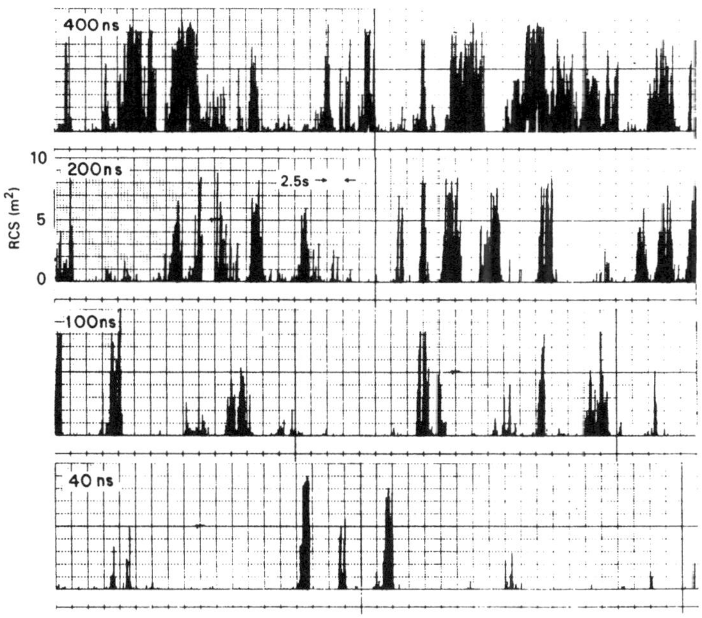

(b)

Fig. 14 — Pulse-to-pulse amplitude of scatter for sea state 3, with pulse widths from 400 to 40 ns (scales: 1 m^2 per div. and 2.5 s per div.): (a) vertical polarization; (b) horizontal polarization

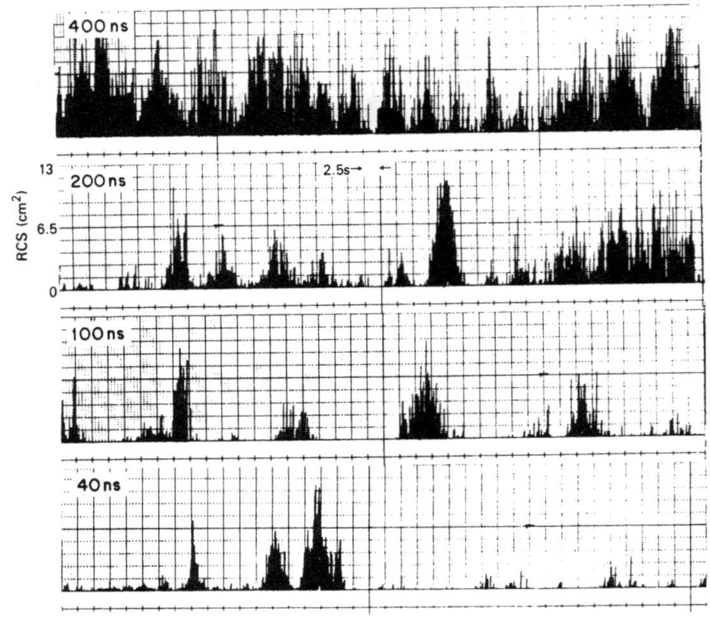

(a)

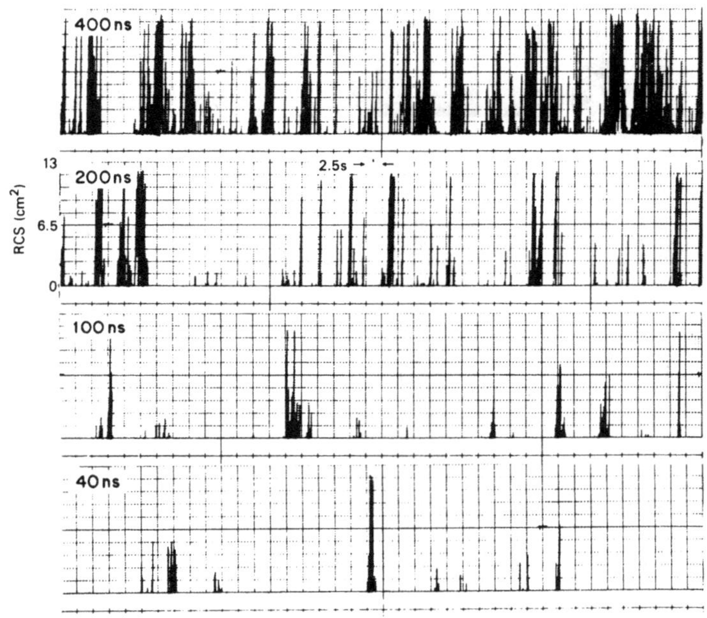

(b)

Fig. 15 — Pulse-to-pulse amplitude of scatter for sea state 1, with pulse widths from 400 to 40 ns (scales: 1.3 cm^2 per div. and 2.5 s per div.): (a) vertical polarization; (b) horizontal polarization

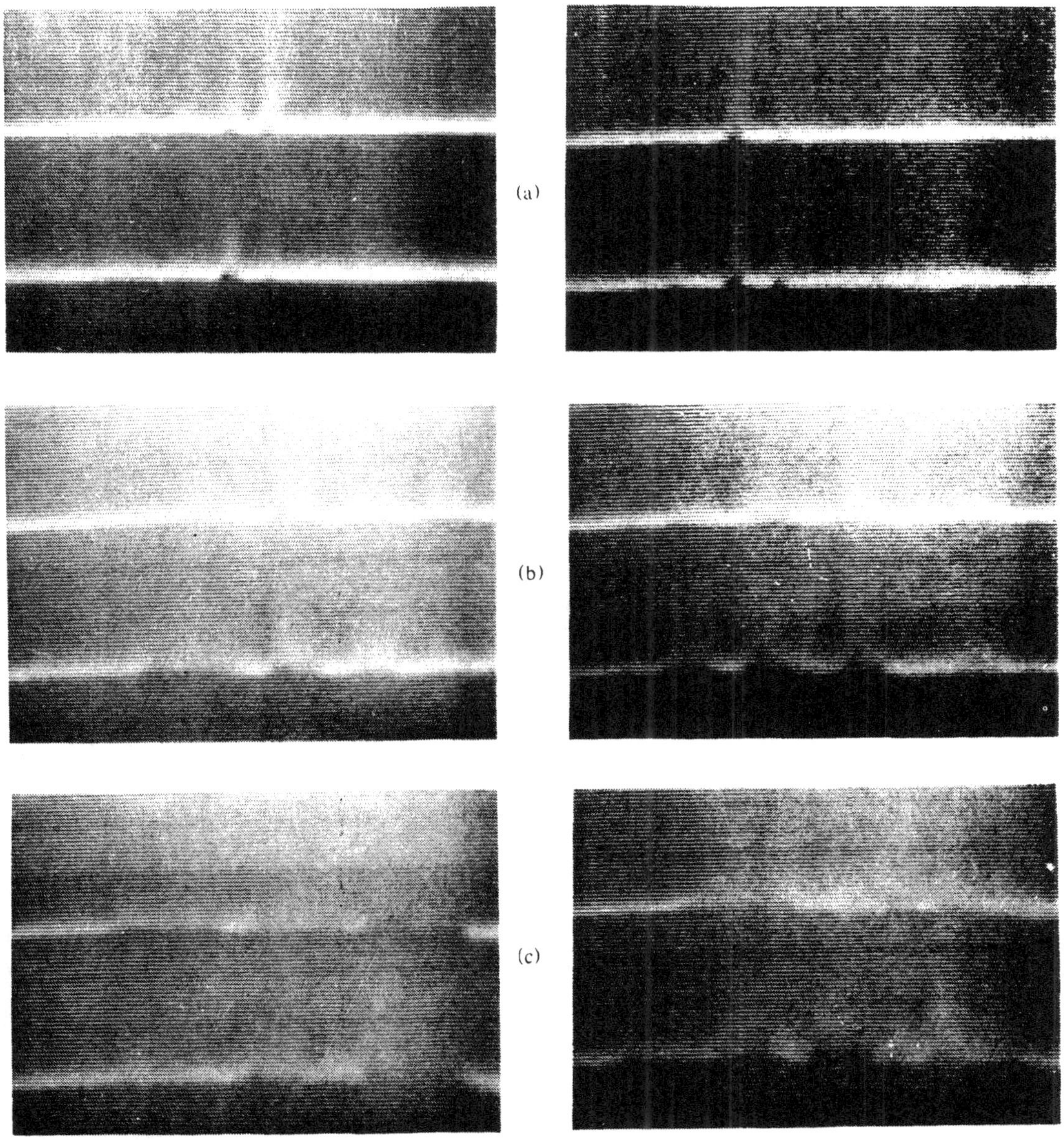

Fig. 16 — Typical A-scope displays in sea state 2 for pulse widths of (a) 40, (b) 200, and (c) 400 ns: vertical polarization, time scale 200 ns per horizontal division

2. As in the shallow-water data, the returns were highly amplitude modulated, with modulation frequencies of the order of 20 Hz to more than 200 Hz. Figures 18 and 19 show examples of video spectral analysis vs time for approximately sea states 3 and 1 with vertical and horizontal polarization. It was noted during these tests that the pulse-modulation frequencies produced by breaking-wave crests were somewhat higher than the frequencies observed during the shallow-water Chesapeake Bay measurements. This difference is attributed to a difference in observation direction: the Chesapeake Bay

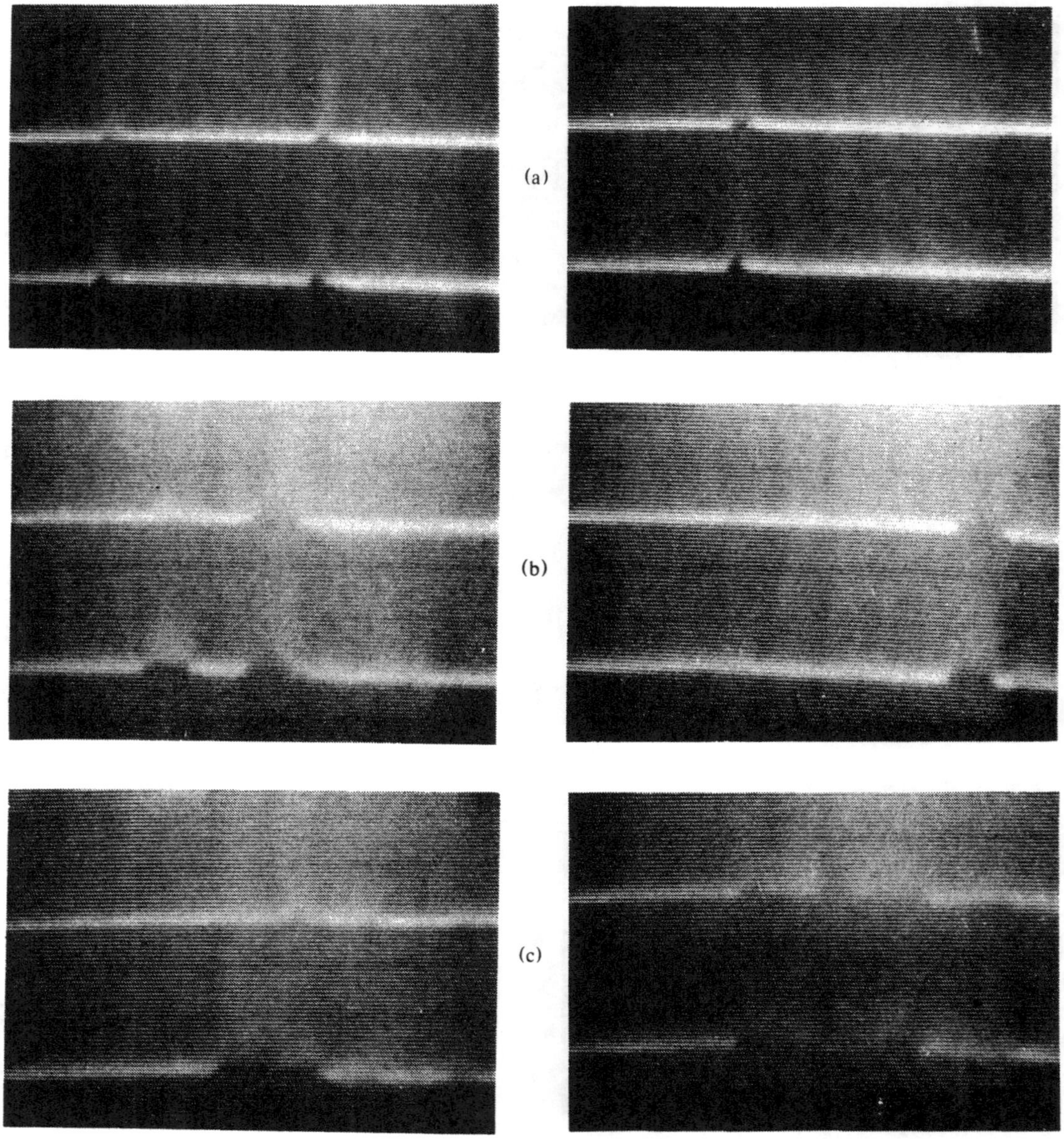

Fig. 17 — Typical A-scope displays in sea state 2 for pulse widths of (a) 40, (b) 200, and (c) 400 ns: horizontal polarization, time scale 200 ns per horizontal division

measurements were primarily upwind along a normal to the breaking wave crests, whereas the Boca Raton measurements were partially crosswind at an angle of 45° to the normal. The higher modulation frequencies may have simply been a result of an apparent higher range growth rate of the breaking wave crests when they were observed from directions away from the normal to the crest.

3. Polarization sensitivity was observed to be a function of both the water surface roughness and the alignment of wave crests relative to the line of sight. With slightly rough, unbroken water (sea states 1 or 2), vertical polarization yielded smaller but more frequent peaks than horizontal. When whitecaps appeared, the peak scatter amplitudes tended to be essentially independent of polarization. (This was true unless the radar observation was nearly perpendicular to the wave crest. Then, as

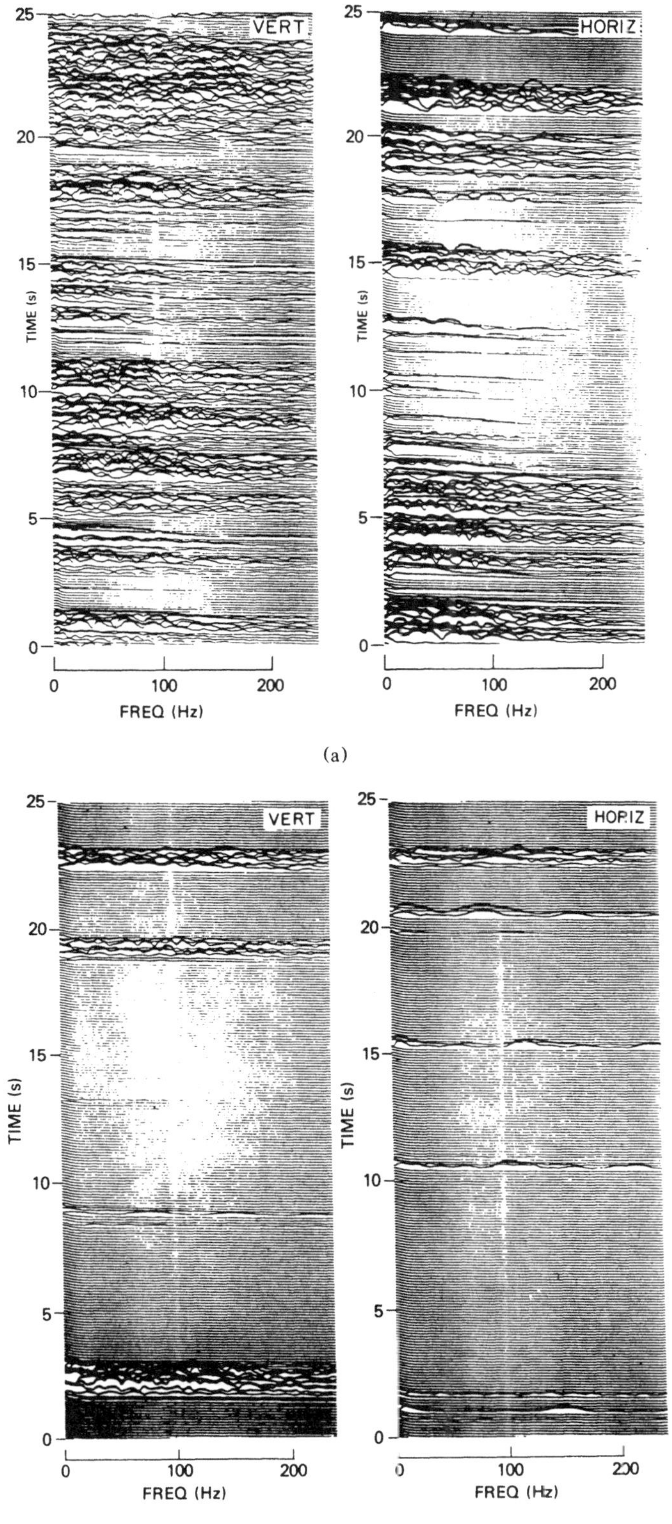

Fig. 18 — Video spectral analysis vs time for sea state 3: (a) 400-ns pulsewidth; (b) 40-ns pulsewidth

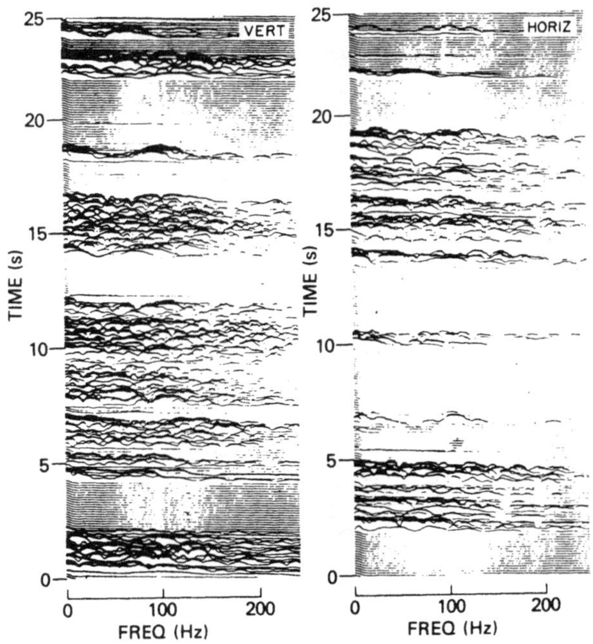

(a)

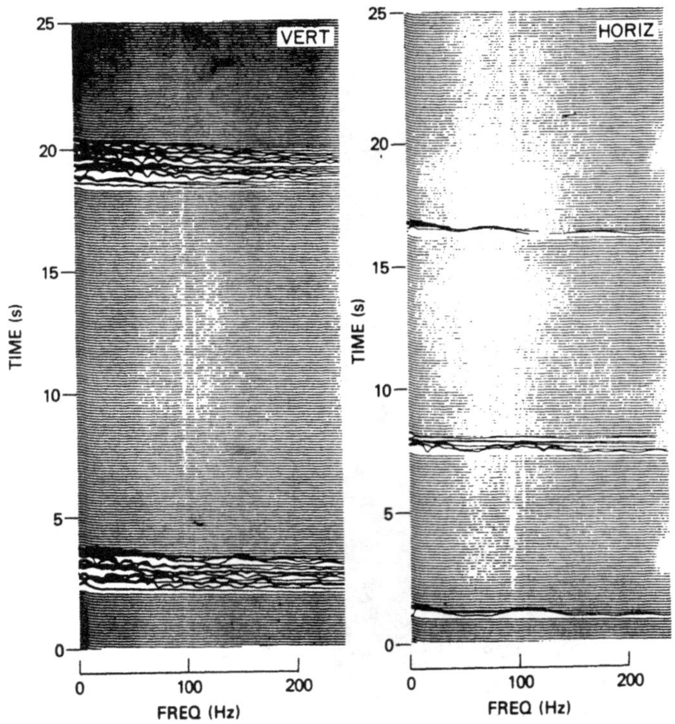

(b)

Fig. 19 — Video spectral analysis for sea state 1:
(a) 400-ns pulsewidth; (b) 40-ns pulsewidth

observed in the shallow-water measurements, the backscatter with vertical polarization was sometimes greater than the backscatter with horizontal polarization.) Greater polarization differences were observed in connection with the duration of the spikelike intervals and their fluctuation. As seen in the data of Figs. 13 through 19, scatter returns occurred more frequently with vertical polarization and persisted longer. Moreover, when the line of sight was perpendicular to the wave crests, the backscatter with vertical polarization tended to follow the wave in range, whereas with horizontal polarization the scatter often tended to develop and disappear at the same range. Since the duration of the horizontally polarized scatter signals was generally shorter than that of the vertically polarized scatter signals, the horizontal scatter appeared more spikelike in character.

Deep-Water Data (30.7-m Depth)

The single-frequency, contiguous-range-cell measurement system was used to make deep-water measurements at Stage I (see Fig. 20), 16 km (10 m) offshore from Panama City, Florida. In these measurements the 2.67-m (1° beamwidth) antenna was employed. The antenna was mounted 15 m above mean sea level, and the center of the beam was aimed to intercept the water at a distance of 300 or 600 m. This produced a grazing angle of 3° or 1.5°. The 300- or 600-m range and 1° beamwidth produced azimuth resolution of 5 or 10 m, and the 40-ns transmitted pulse yielded a range resolution of 6.2 m (cell area was 31 or 62 m^2). It was noted at this open-sea site that for low wind and sea conditions there were often many small crests appearing on the sea surface which were relatively random in alignment (not necessarily aligned perpendicular to the dominant wind and wave direction). In cases of relatively high wind and sea conditions, a dominant crest alignment becomes more apparent, but even in these cases many of the smaller crests are still randomly oriented. Figures 21 and 22 show typical video data obtained at this site in approximately sea states 5 and 2 with vertical and horizontal polarization.

Fig. 20 — Measurement site, deep-water data, Stage I, Panama City, Fla.

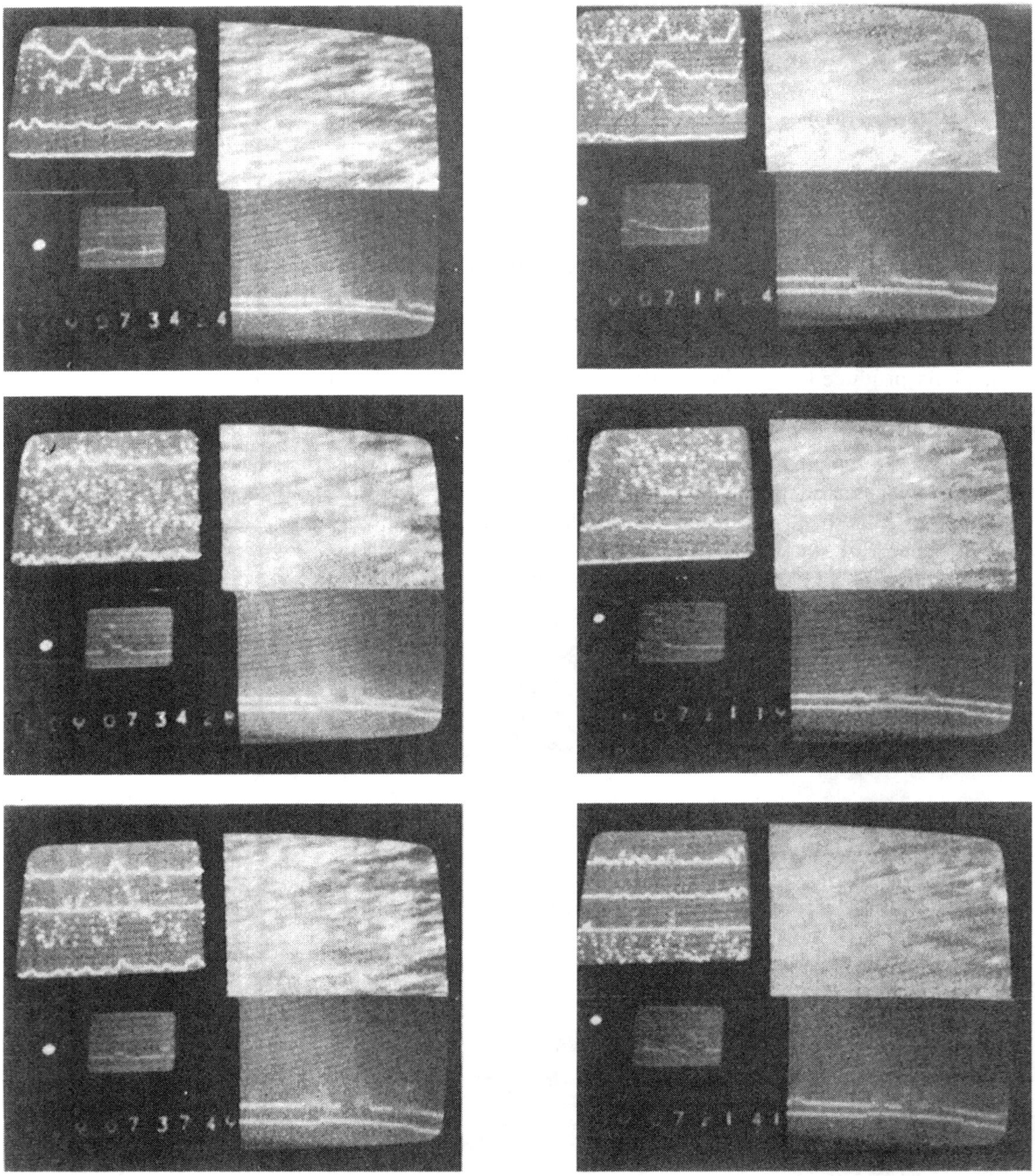

(a) (b)

Fig. 21 — Typical TV video data, deep water, sea state 5:
(a) vertical polarization; (b) horizontal polarization

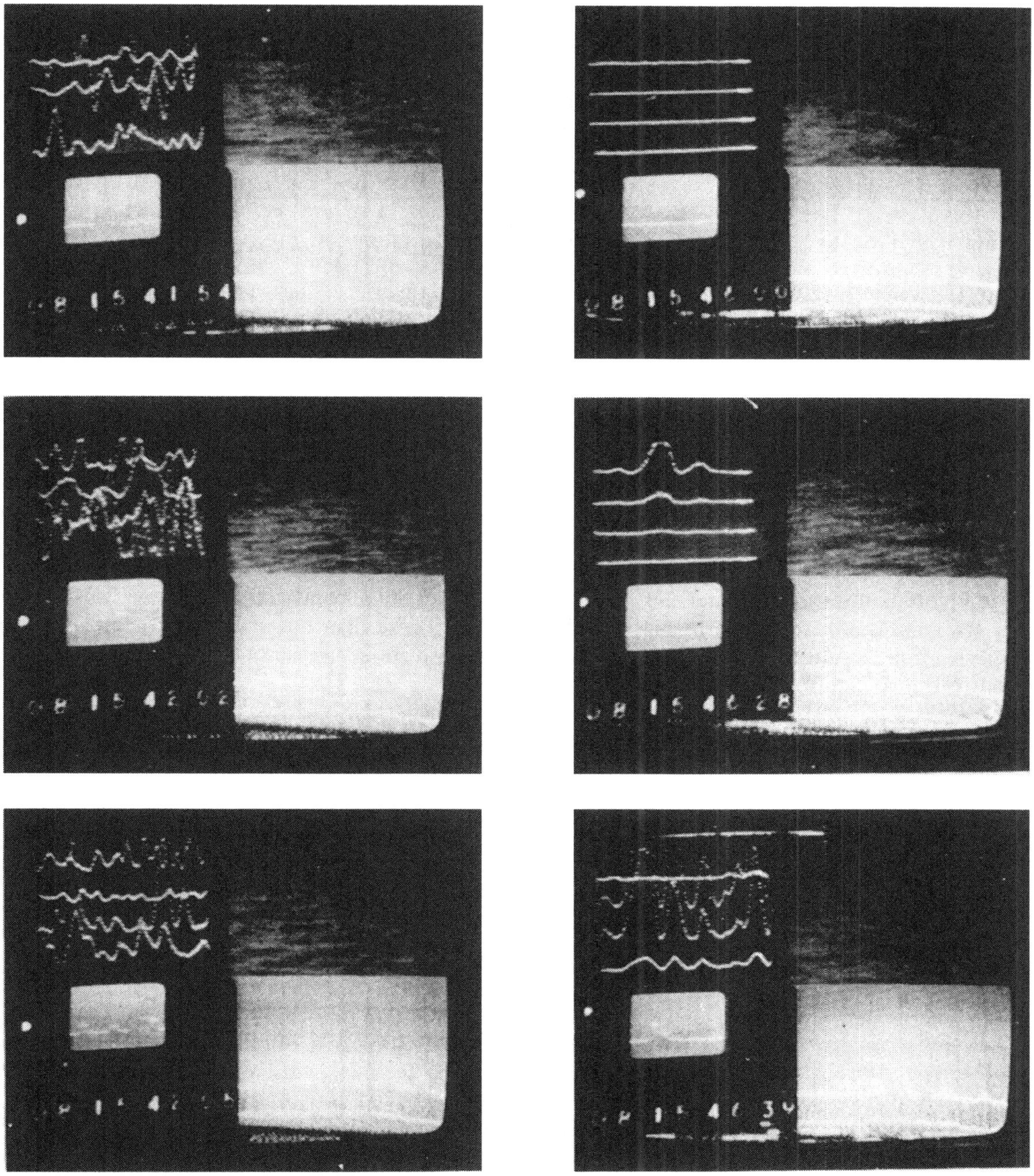

(a) (b)

Fig. 22 — Typical TV video data, deep water, sea state 2: (a) vertical polarization; (b) horizontal polar zation

The measurements at the Stage I site once again reinforced the results from previous experiments. The contiguous-range-cell data also provided some insight into the spatial occurrence of the backscatter, especially as a function of radar polarization:

1. As in previous data, when returns were present in any range cell they were highly amplitude modulated pulse to pulse, with frequencies of the order of 20 Hz and sometimes exceeding 500 Hz. Note the modulation envelopes displayed in the video data of Figs. 21 and 22. The rise and fall times of these modulation envelopes are occasionally fast enough to define individual pulse intervals (as dots) in the sample hold display. Figure 23 illustrates a particular example of extremely high frequency pulse-to-pulse modulation. In this example (sea state 5, upwind, 3° incidence, horizontal polarization) the boresight is centered on a churning wave crest. Although it is not evident in the still photos, a review of the real-time TV video reveals an abrupt cross-range physical movement in a part of the observed wave crest. The pulse-to-pulse modulation envelope at several points during this movement is changing with peak to null spacings comparable to a single interpulse interval (0.4 ms). This example represents a pulse-to-pulse modulation frequency well over 500 Hz.

2. The backscatter at this high resolution (40 ns) has a noticeably spikelike appearance, with periods of extremely low level returns interspaced between relatively high level returns. Examples of this behavior for several sea states, incidence angles, and observation directions are shown as contiguous-channel, amplitude-time (24-s span) histories in Figs. 24 and 25. Note the range resolution of apparent scatter centers.

3. In the sea states tested, peak cross-section values greater than 10 m^2 were observed in individual range cells. They were generally associated with higher sea states and rough breaking water (whitecaps). Between breaking wave crests and under calmer sea conditions the backscatter often showed similar spectral content but with amplitudes of the order of 20 dB below 1 m^2. Amplitude distributions (plotted on Rayleigh statistical paper) for several 80-s data samples of sea scatter under different sea conditions are shown in Figs. 26 through 28. These distributions can be compared with a typical Rayleigh linear curve shown on each plot and also with the response of the data sampling system to filtered test noise, shown in Fig. 29.

Median cross-section values $\bar{\sigma}_0$ in decibels below one square meter per square meter (dB m ^{2}) for the nominal upwind sea states 5 and 2 are compared in Table 1 with data published by other researchers. The measured values with vertical polarization are in reasonably good agreement with those compiled by Sittrop [4] and by Long [5] for similar sea conditions but lower resolutions. The measured horizontal-polarization values, however, are somewhat low. The higher resolution of the Stage I system (40 ns vs 200 to 400 ns for the published values) may in itself account for the difference.

Several trends are apparent from the shapes of the distributions. First, the curves tend to be linear on an exponential scale but have a falloff slower than a typical Rayleigh distribution for both polarizations at the higher sea states. Second, the curves become less linear at the lower sea state with horizontal polarization showing a distinctive bow or inflection in the distribution. These trends are similar to ones observed by Bishop [6] for measurements at 70-ns resolution. Long [7] has also cited data (at 6.3 GHz, 190 ns, horizontal polarization) with a break or bow in the distribution.

A polarization sensitivity with sea state and with crest alignment was very evident in the data taken at Stage I. In the low sea states vertical polarization tended to yield smaller but more frequent peaks than horizontal. In higher sea states, with rough broken water, the scatter peaks tended to be similar for both polarizations. For upwind conditions, backscatter with vertical polarization tended to follow the waves in range while scatter with horizontal polarization tended to be sporadic.

Particularly prominent in the 24-s amplitude-time histories of Figs. 24 and 25 are "Kalmykov bursts" produced with horizontal polarization (a few of these bursts are designated in the figures). This

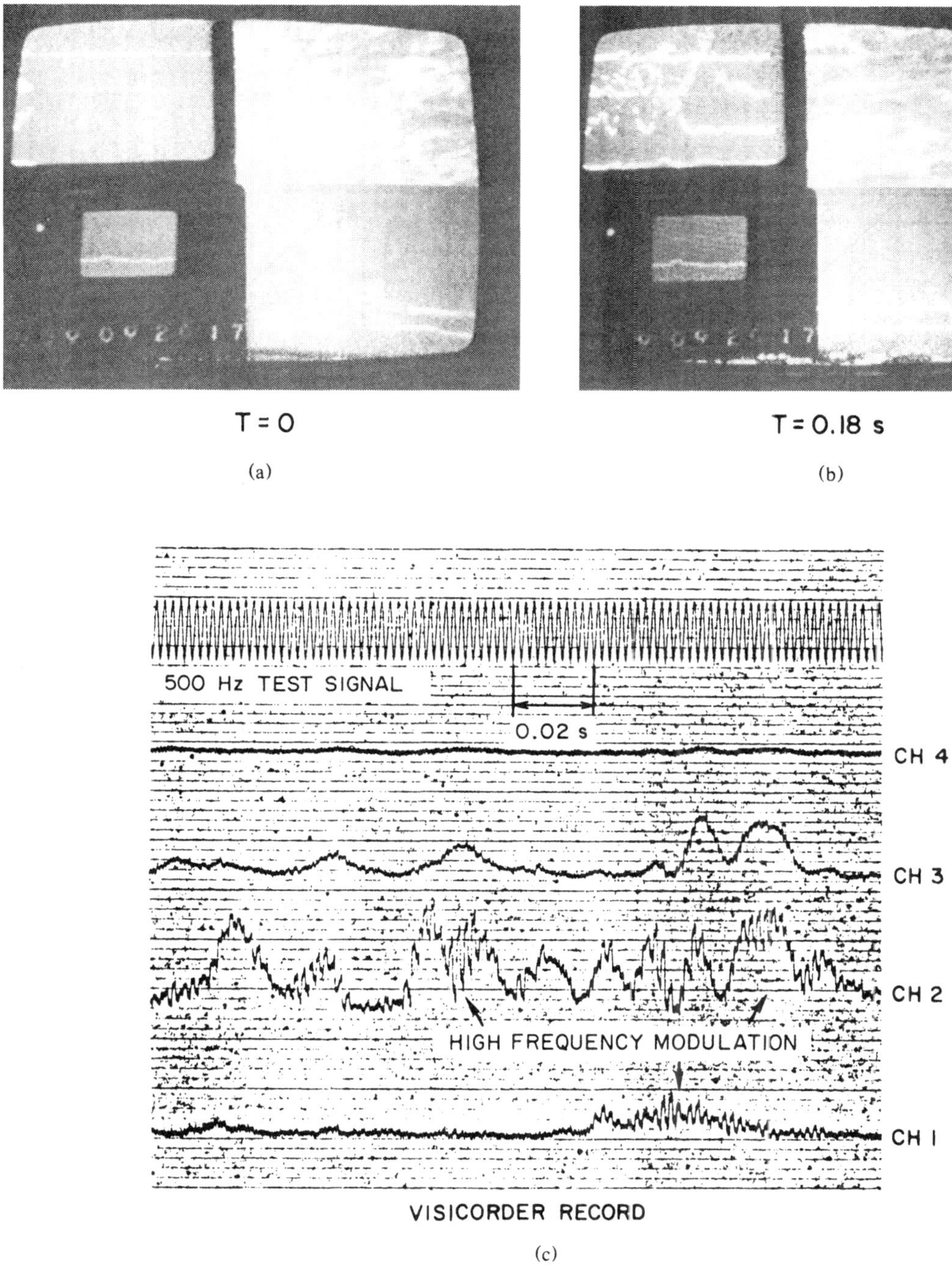

Fig. 23 — Data example showing extremely high frequency pulse-to-pulse modulation: (a) and (b) boresight photos taken 0.18 s apart; (c) visicorder record shows sampled and held scatter return in channels 1 thru 4 plus directly converted 500-Hz test signal. Sea state 5, upwind, 3° incidence with horizontal polarization.

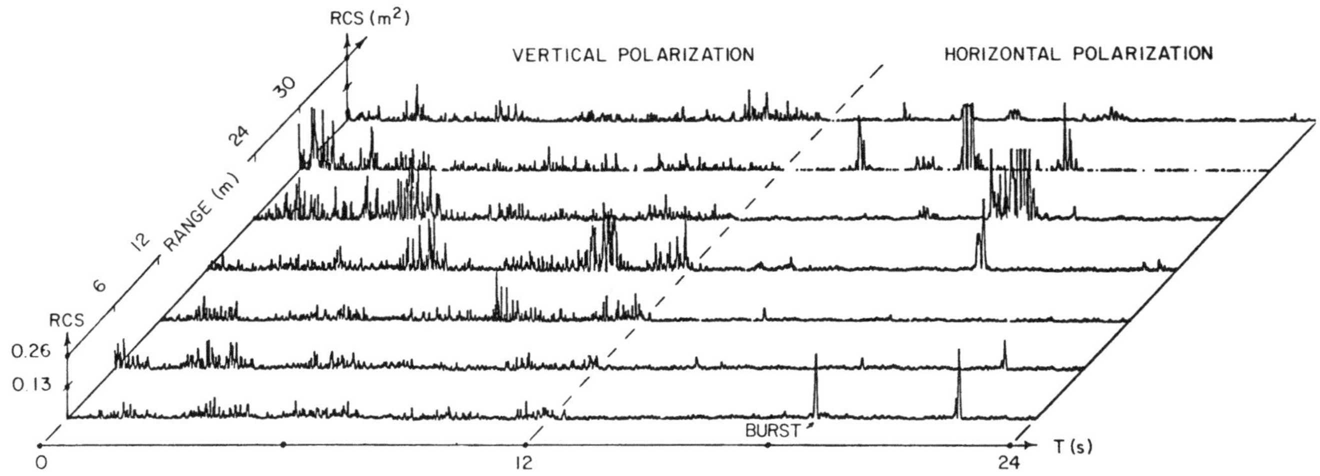

(a)

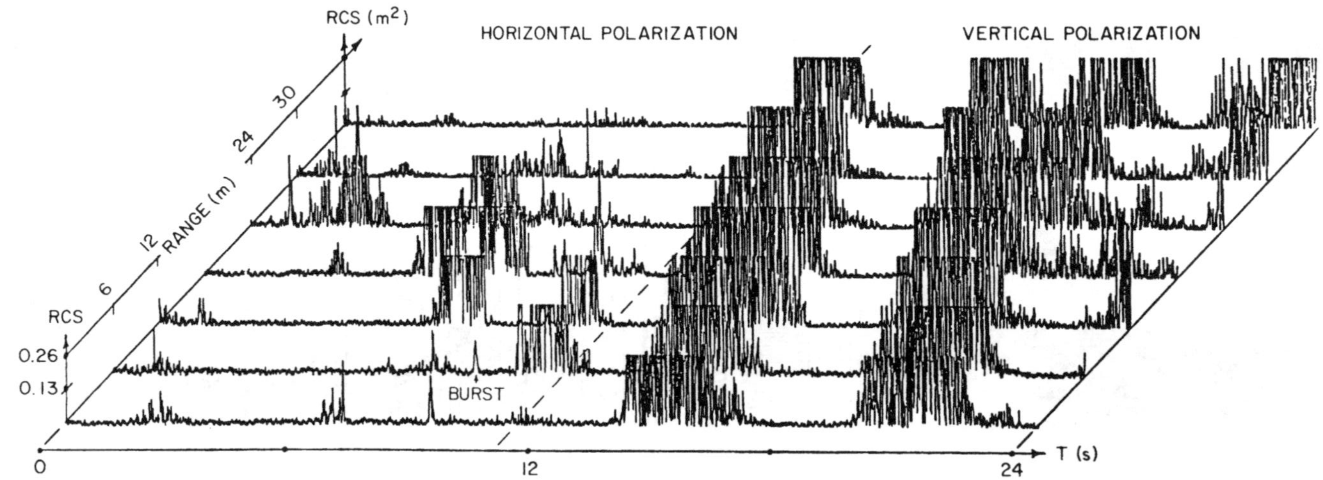

(b)

Fig. 24 — Backscatter amplitude-time history for seven contiguous range cells spaced 6 m apart. Data record length 24 s, 40-ns pulsewidth: (a) sea state 2, winds 8 knots, upwind, 3° incidence angle; (b) sea state 5, winds 22 knots, upwind, 3° incidence angle

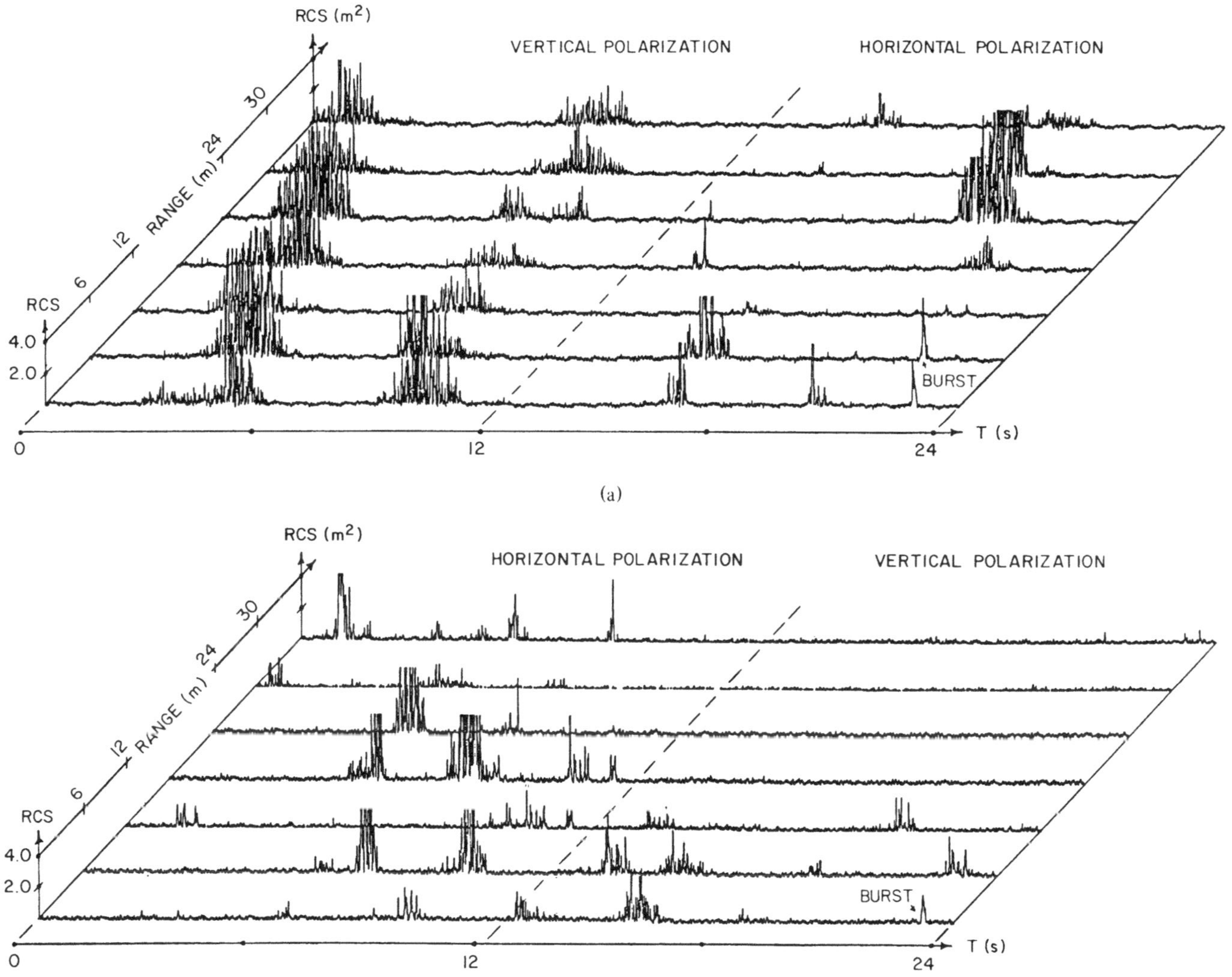

(a)

(b)

Fig. 25 — Backscatter amplitude-time history for seven contiguous range cells spaced 6 m apart. Data record length 24 s, 40-ns pulsewidth: (a) sea state 4 to 5, winds 20 knots, upwind, 1.5° incidence angle; (b) sea state 3, winds 15 knots, crosswind, 1.5° incidence angle

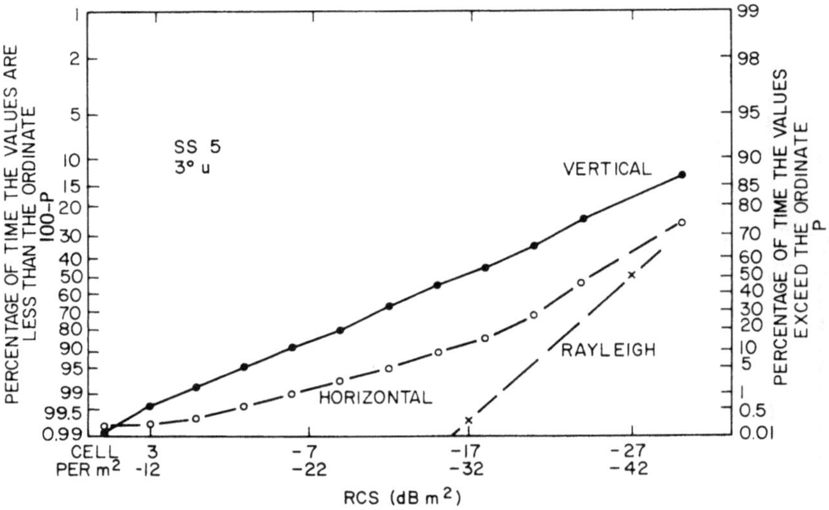

(a)

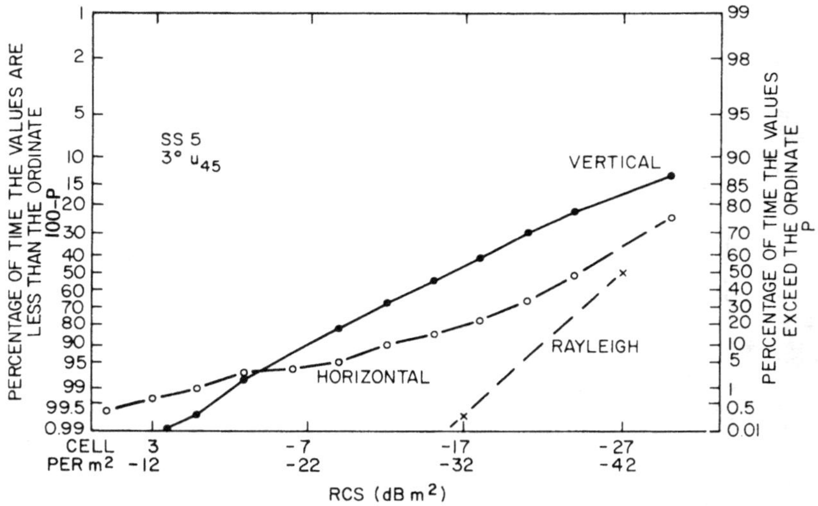

(b)

Fig. 26 — Amplitude distributions, sea state 5, 80-s time record: (a) upwind and (b) 45° from upwind

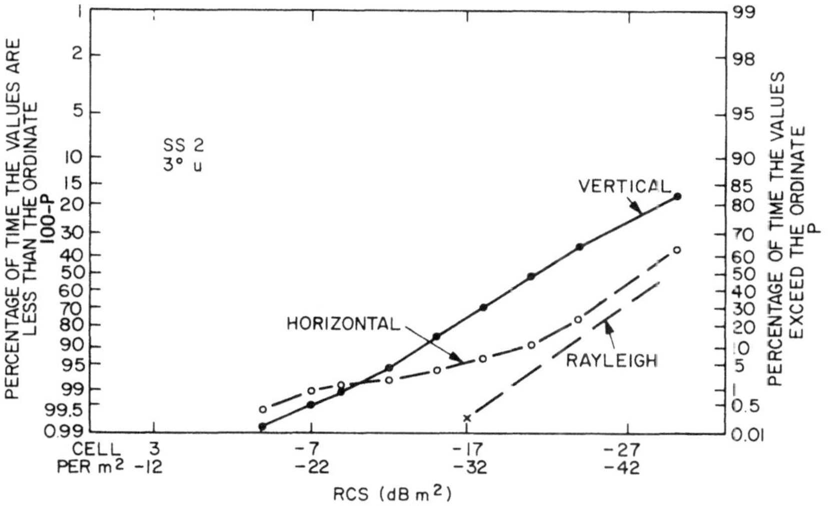

(a)

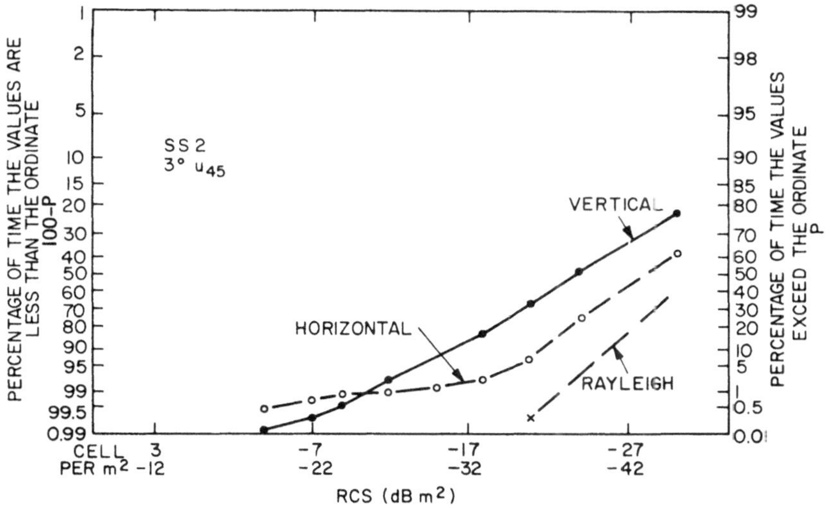

(b)

Fig. 27 — Amplitude distributions, sea state 2, 30-s time record: (a) upwind and (b) 45° from upwind

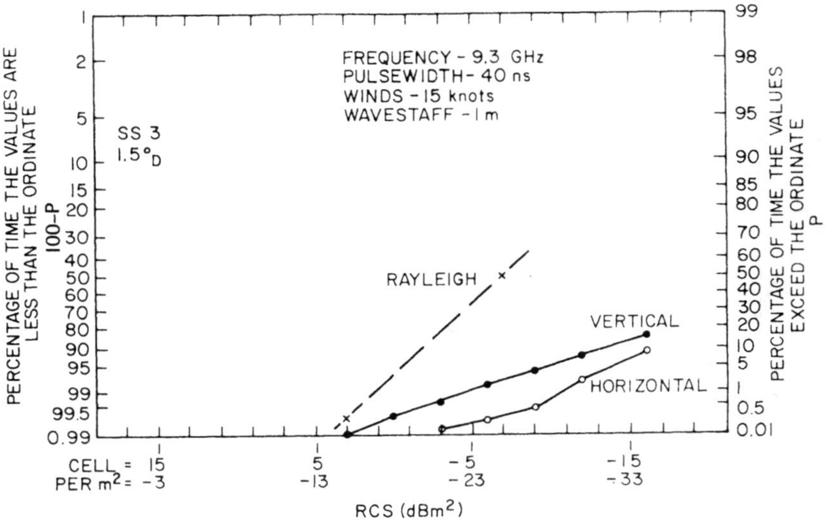

(a)

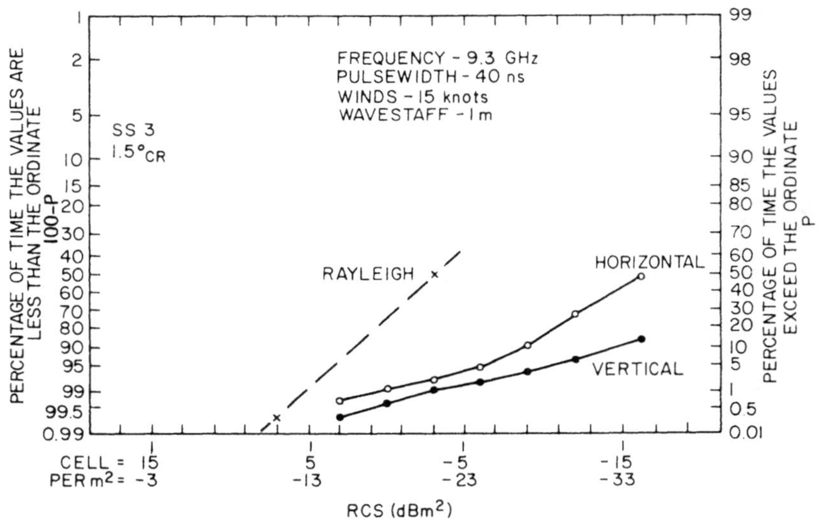

(b)

Fig. 28 — Amplitude distributions, sea state 3, 80-s time record: (a) downwind and (b) cross wind (90°)

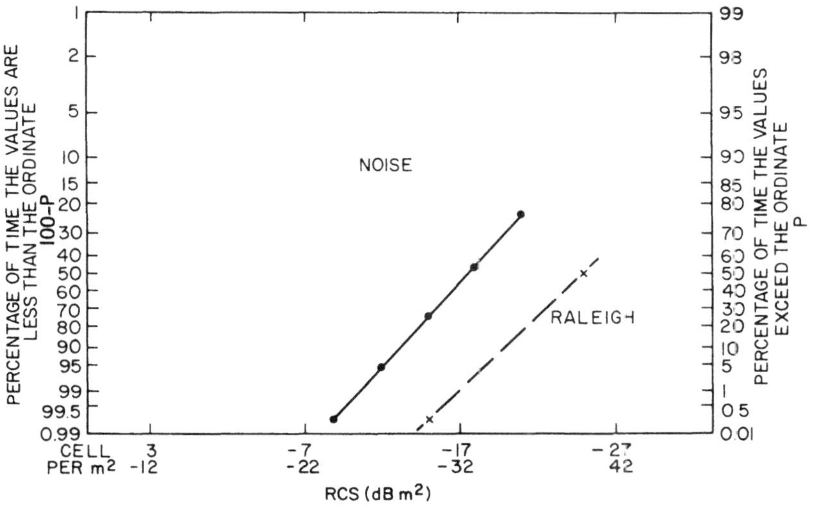

Fig. 29 — Amplitude distribution of filtered and rectified test noise.

Table 1 — Measured Upwind Cross-Section Values at X-Band

Researcher	Resolution (ns)	Wind (knots	Incidence (deg)	Median Cross Section σ_{0_V} (dB m^2)	σ_{0_H} (dB m^2)
Hansen	40	22	3	−31.5	−39
Sittrop	120	22-26	3	−31*	−34*
Nathanson	500	19-24	3	−33.5†	−35†
Daley et al.	500	29	5	−35.5	−37
Guinard et al.	500	17	4	−36	−37
Hansen	40	8	3	−36	−43
Sittrop	500	10	2	−43*	−45*
Nathanson	500	7-12	3	−41†	−45†
Guinard et al	500	7	4	−38	−39
Bishop	270	8	1	−42*	−44*

*Median value obtained by subtracting 3.0 dB from average value as per Long [5]
†Adjusted median values where upwind, downwind, and crosswind results were averaged.

type of modulation was first noted by Kalmykov et al. [8] when observing the modulation envelope of scatter upwind from strongly breaking waves. These bursts were found to be present in virtually all of the Stage I data taken with horizontal polarization. In contrast, a burst formation was rarely observed with vertical polarization. An exception appears in the crosswind vertical polarization example shown in Fig. 25(b), where a few isolated burst modulations are present. In higher sea states, the scatter for horizontal polarization becomes a combination of relatively short duration but low frequency modulated bursts and relatively long duration but high frequency modulated spike returns (similar to the ones observed at Chesapeake Bay and Boca Raton from turbulent breaking water).

Consider the upwind 5-min data runs shown in Fig. 30. A comparison of a wave staff recording and the scatter data indicates that scatter with vertical polarization profiles every approaching wave while many of the waves are missed with horizontal polarization. Also note that the envelope formed

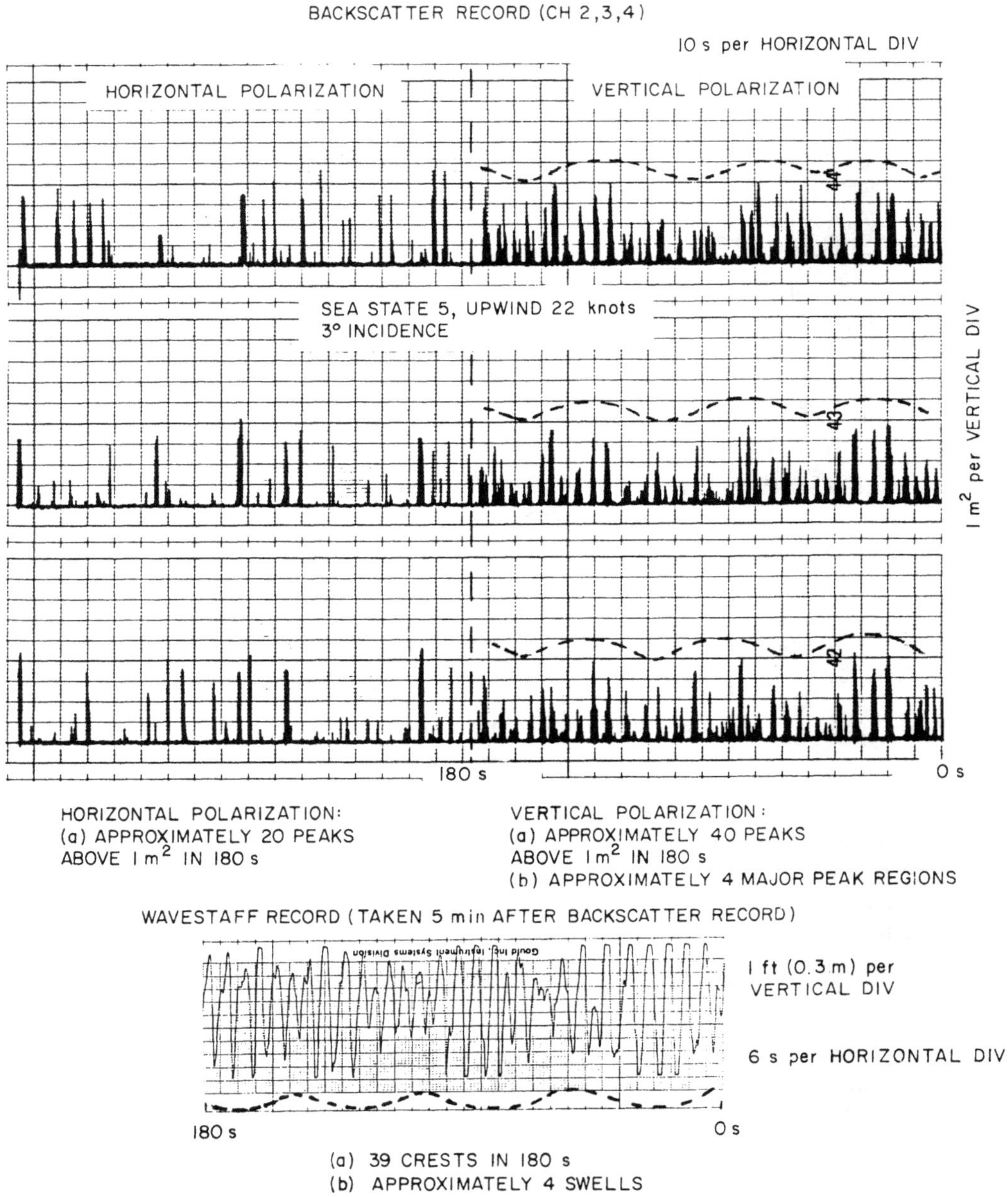

Fig. 30 — Comparison of the scatter amplitudes in three contiguous range cells with a record of wave height and frequency of occurrence. (Wavelength record was taken 5 min after the backscatter record.)

by the peak scatter amplitudes for vertical polarization shows a slow periodicity, which is similar to the swell period present in the wave staff recording. It appears that for upwind conditions and vertical polarization the radar is receiving scatter from contributing features which are present on all of the unshadowed wave surfaces, while with horizontal polarization only very specific and range-resolvable features are producing scatter.

Kalmykov and Pustovoytenko [9], Long [10], and others have suggested that one of the dominant scatter sources for vertical polarization is small waves or ripples, while a dominant scatter source for horizontal polarization may be facets or sharp wave crests (or "wedges"). These suggestions would seem to explain the observed profiling action of the return for vertical polarization, since small ripples would be present on virtually all of the nonbreaking wave surfaces. Occasionally occuring sharp crests could also be the source of the burst modulations seen in scatter data taken with horizontal polarization. However, the fact that the burst modulations do not appear for every wave formation would also indicate that there might be preferential alignment or shape needed to produce the scatter. Lewis and Olin [3] have suggested that some of the scatter is caused by sharp wave crests acting as impedance discontinuities. They speculate that the scatter from the crests should be preferential, with vertical polarization scatter being optimized for crest alignments normal to the radar and with horizontal polarization scatter being more probable for crests aligned away from the normal. These suggestions may have some confirmation in the observed differences in the amplitude distributions shown in Figs. 26 and 27. The distributions show a definite decrease in the peak vertical return and an increase in the horizontal return when the observation direction is changed from normal to 45°. There have also been occasional instances of sea states and observation directions where the scatter for horizontal polarization is substantially stronger than that for vertical polarization. (Note the data shown in Fig. 28.) Finally, there has been general agreement by previous researchers (Long, Kalmykov, and Bishop), as confirmed by the data in this report, that turbulent breaking water is associated with a characteristic high-level scatter which appears to be similar in magnitude and in modulation for both polarizations. A net observation is therefore reached that radar scatter at high resolution may, in fact, be produced by many types of resolvable surface formations, with ripples, sharp crests, and regions of turbulent breaking water being at least three that are evident.

THEORETICAL MODELING OF THE SCATTER SURFACE

As outlined by Lewis et al. [1-3], modeling results have suggested that the temporal characteristics of the sea backscatter might be approximated by a surface containing random amplitude and phase zones spaced at quarter-radar-wavelength intervals (Fig. 31).

Simple addition or subtraction of scattering zones (as in the formation and decay of a whitecap) results in a characteristically rapid modulation. Several sea-spike realizations with different relative internal and external correlation times are shown in Fig. 32. Comparisons of this computer modeling with experimental data have suggested that the internal and external correlation times of typical sea backscatter are of the order of 0.01 s at X-band. The modeling results were also extended to the cases when a slightly disturbed surface (as produced by wind or rain) fills the entire resolution cell.

As a further check on this type of model, a simple experimental simulation was made. This experiment utilized an oval-shaped roughened surface sliding on a tilted plate (see Fig. 33). The rough surface was steadily moved through the beam of the dual-frequency radar system in order to approximate the growth (and decay) of a turbulent sea surface. Figure 34 shows some examples of the backscatter pulse-to-pulse modulation characteristic produced by this experimental model. Note the similarity in the amplitude characteristic to actual scatter from the sea surface.

EXPERIMENTALLY OBSERVED SCATTERING CHARACTERISTICS OF RIGID TARGETS

The radar-echoing characteristics of some rigid targets were also observed in an attempt to establish possible discriminants which could be used against sea scatter. Examples of the scattering character of a submerged platform and a small floating target are shown in Fig. 35. The dominant difference between the backscatter from the sea surface and that from rigid targets was noted to be in the depth and modulation frequency of the pulse-to-pulse amplitude-modulation characteristic. In general, returns from rigid targets are found to have much lower modulation frequencies and percentage of modulation when compared to typical sea return.

As a contrast, the statistical distributions of the scatter from a floating 6-in. (15-cm) metal sphere (in approximately sea state 1) and from the sea surface (at sea state 2) are compared in Fig. 36. Note that the amplitude distributions are very similar. A marked difference is observed, however, in a comparison of the pulse-to-pulse modulation characteristics shown in Fig. 37.

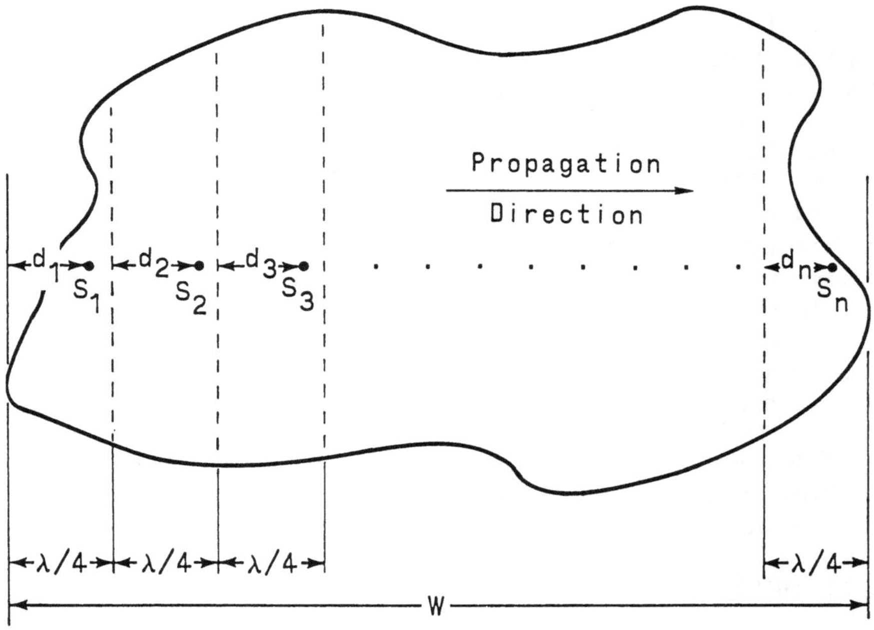

Fig. 31 — Theoretical model of a disturbed sea surface

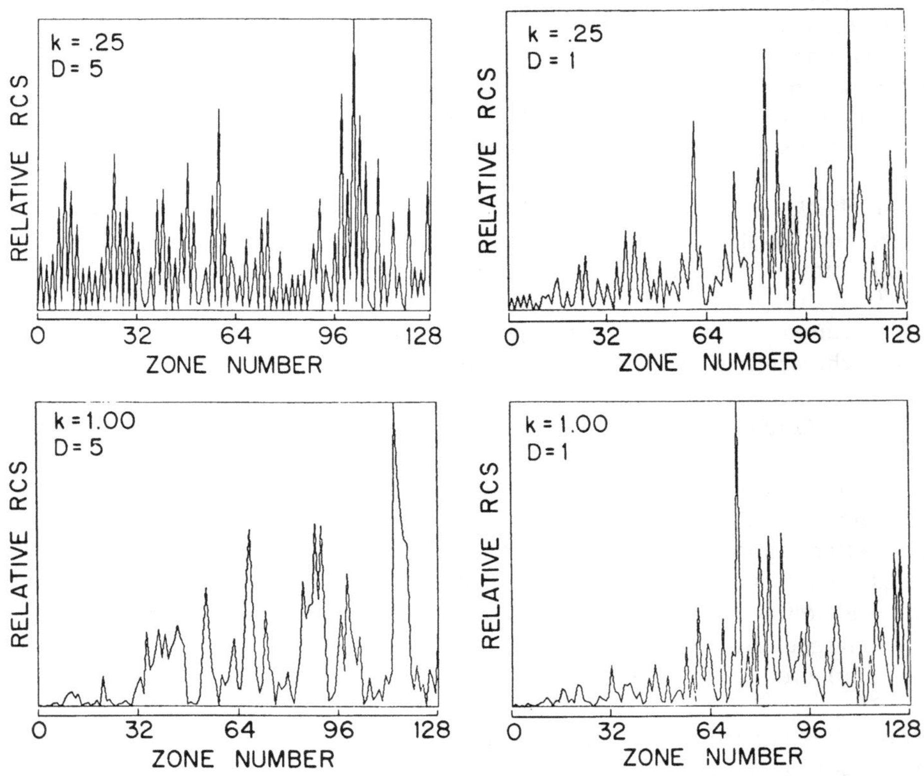

Fig. 32 — Computer simulations of sea-spike realizations for several internal and external growth and decay rates

Fig. 33 — Experimental apparatus for a simulation of sea-scatter model, roughened surface sliding on a tilted conducting plate

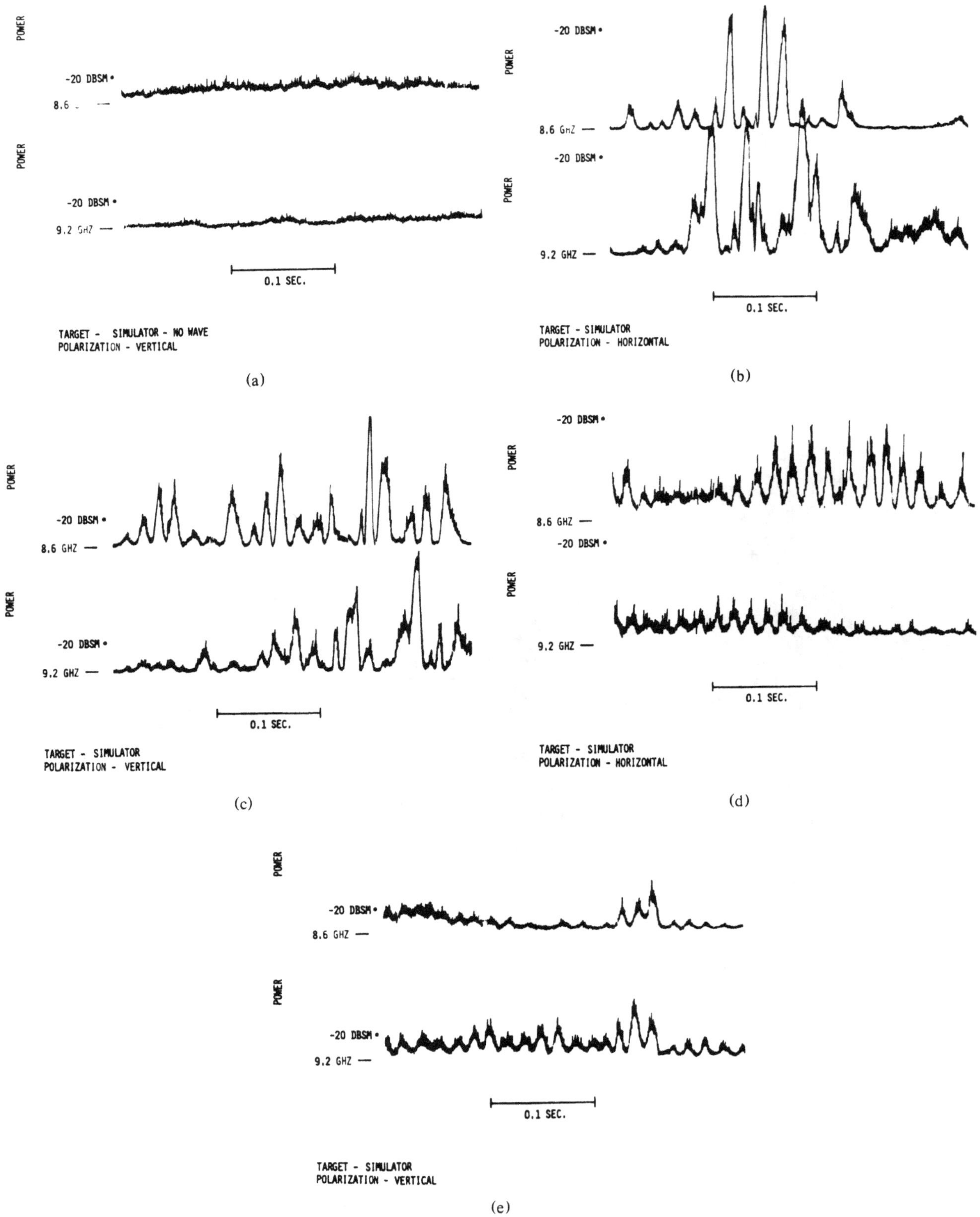

Fig. 34 — Pulse-to-pulse amplitude-time histories produced by observing the experimental simulation apparatus of Fig. 33 with a dual-frequency, high-resolution measurement system.

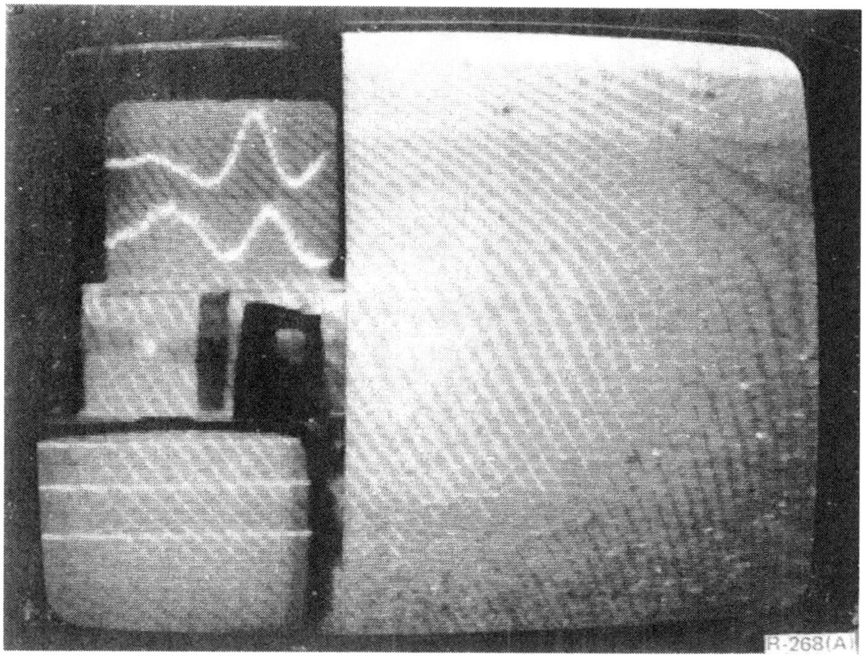

(a)

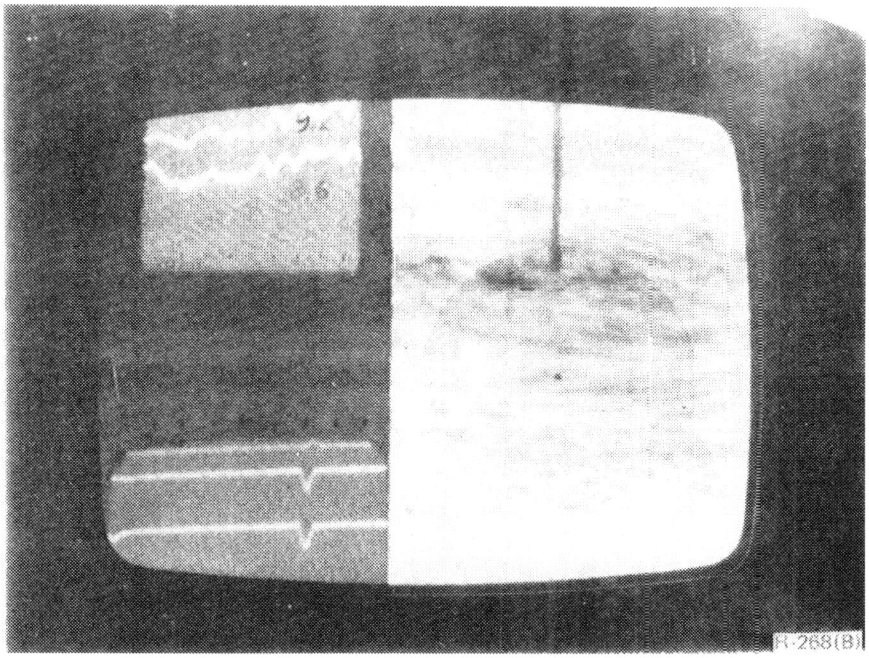

(b)

Fig. 35 — Typical TV video data, dual-frequency X-band system: scatter from a floating corner reflector (a) and from a partially submerged wooden platform (b)

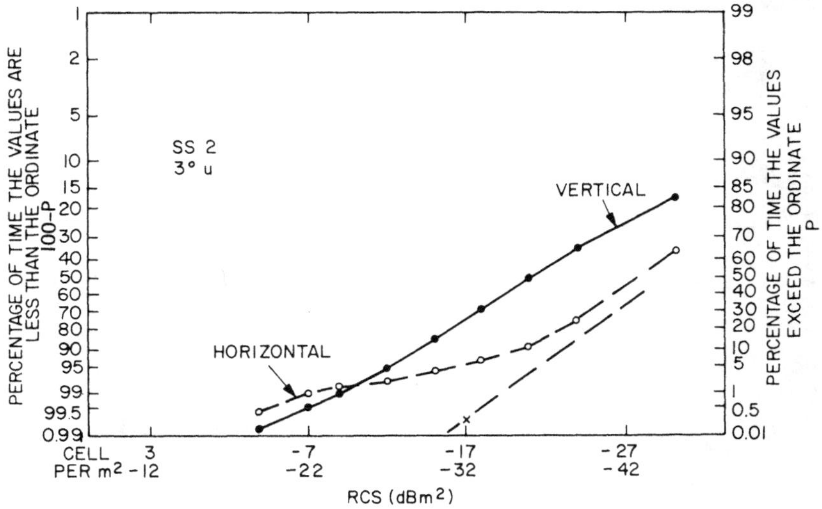

(a)

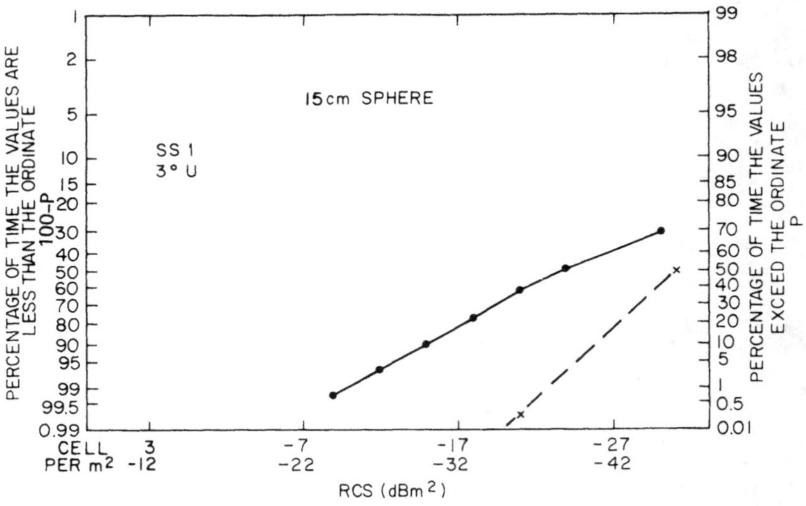

(b)

Fig. 36 — Comparison of the amplitude distributions of sea state 2 (observed from upwind) (a) and of a floating sphere in sea state 1 (b)

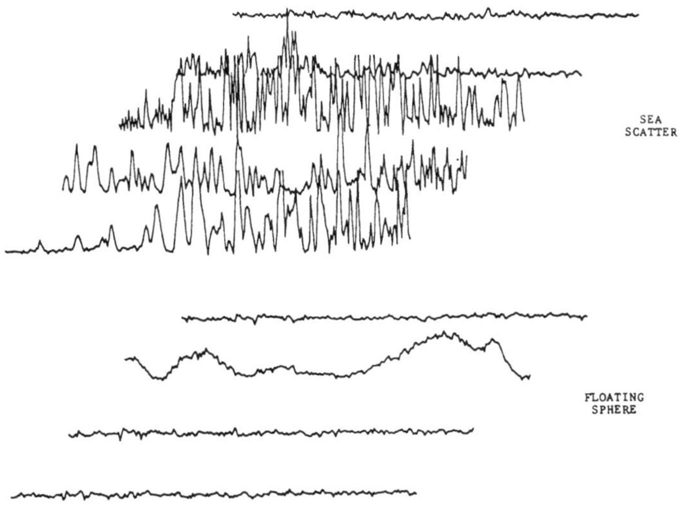

Fig. 37 — Examples of pulse-to-pulse amplitudes of scatter RCS from typical sea surface (a) and from a floating sphere (b) (data record 1 s)

SEA-SCATTER DISCRIMINANTS AND TARGET DETECTION HIGH-RESOLUTION RADAR

The difference in the amplitude-modulation characteristics of rigid floating debris and sea return suggests that percentage modulation vs time might be used as a discriminant to suppress sea return. Figure 38 illustrates one way in which such a discriminant can be instrumented. The radar return signals from individual range cells are delayed by selected intervals via sample/hold circuits. The outputs are subtracted from each other as indicated to detect high-frequency, large-percentage amplitude modulation on echoes in a given range-resolution cell. If such modulation is detected, the output from one or more of the subtractors will exceed a preset threshold and inhibit a gate to block passage of that video to a display. Since sea return generally fluctuates by more than 6 dB in less than 20 ms, and rigid target echoes often remain constant for much longer intervals, this type of instrumentation can block sea-return video and pass typical target video.

This blockage should not materially reduce the probability of detection of rigid targets as long as the high-amplitude sea returns are occurring for only a small amount of viewing time. Loss of target will occur when sea return and target return occur in the same cell (with that cell being blocked). For a reasonable detection threshold, however, this blockage should correspond to relatively small reductions in viewing time.

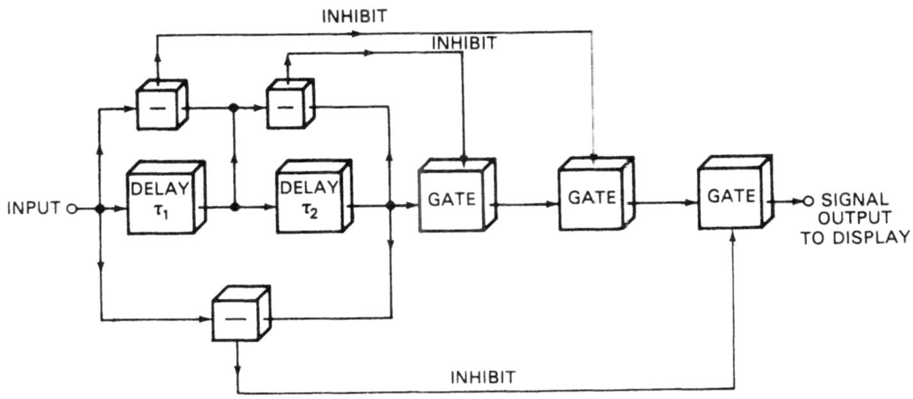

Fig. 38 — A sea-scatter rejection circuit based on amplitude-modulation characteristics of pulse-to-pulse scatter

As a means of testing the amplitude-modulation discriminant, recorded data from the contiguous-range-cell measurements of Stage I were utilized in several different experiments to demonstrate target detection and sea-scatter rejection.

In one test, an analog two-delay circuit was instrumented to operate on the recorded contiguous-range-cell data and display the results in unprocessed and processed range formats. In these tests target signals (simulated) were inserted in one range cell to observe target maintenance. Figures 39 and 40 illustrate typical results. Note that sea clutter is rejected and simulated target signal maintained with few false alarms or target rejections.

The Stage I data were also digitized and were processed by a five-delay rejection discriminant. The results are tabulated in Table 2. These tests indicated that for simulated target modulations below 5 Hz, a multiple-delay discriminator can effectively reject the sea scatter and maintain a target signal. As tested in this simulation, the five-delay discriminant performed best for individual delays of the order of 12 ms (60-ms total delay). A sea-spike false alarm (in the discriminator output) occurred only when the amplitude-modulation characteristic was obscured by system saturation.

SUMMARY OF EXPERIMENTAL RESULTS

The experimental measurement program has provided data which have led to the following observations concerning high-resolution-radar scatter from a disturbed sea surface and from rigid targets:

1. High-resolution X-band sea scatter observed from low incidence angles (less than 5°) is *always* heavily amplitude modulated with relatively high modulation frequencies. (Modulation frequencies on the order of 20 Hz to more than 500 Hz have been detected.) The frequency of modulation appears to be influenced by such parameters as the type of physical surface (breaking water, sharp wave crests, ripples, etc.), the relative wind and wave direction, and the radar polarization. Specifically, it has been observed that, for turbulent breaking water, higher modulating frequencies are observed (with both polarizations) in directions away from a normal to the dominant wave crest. This may simply be a result of the relative range growth rate of the turbulent area. In calmer seas, returns for horizontal polarization are dominated by lower frequency modulated Kalmykov bursts.

2. The time-varying characteristic of sea scatter is also dependent on transmitted pulse width. Low-resolution backscatter tends to appear continuously noiselike. As radar resolution is increased, the backscatter is punctuated by quiet periods and presents a spikelike characteristic with time (especially with horizontal polarization).

3. With high resolution, high-amplitude sea-spike returns often occur during the turbulent breaking action indicated by whitecap development. A spikelike characteristic can still be present during periods of relatively undisturbed sea, but the return is generally two to three orders of magnitude lower in power.

4. A polarization sensitivity is evident which is a function of both water roughness and alignment of wave crests to the radar system. In most (but not all) sea states and crest alignments vertical polarization tends to yield smaller but more frequent spikes. Also, when the scatter occurs, it tends to persist longer for vertical polarization.

For upwind conditions, in particular, return scatter with vertical polarization is evident from virtually all portions of the nonshadowed sea surface, while horizontal polarization produces sporadic returns, often with the burst character. These scatter modes may be an indication of some distinct physical scattering formations: small waves or ripples for vertical polarization and preferentially aligned sharp wave crests for horizontal polarization.

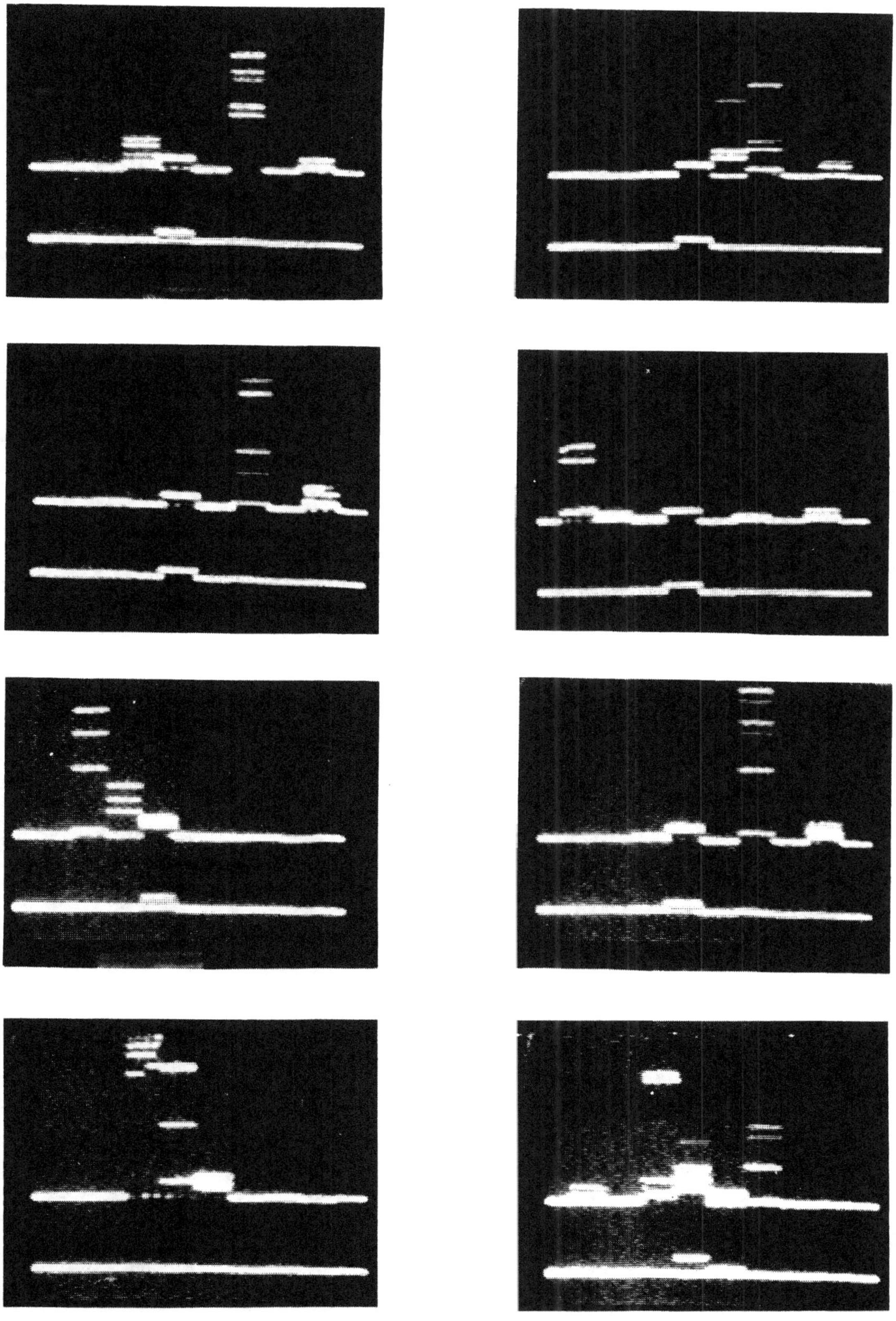

Fig. 39 — Performance of 2-delay sea-scatter rejection circuit (10-ms delays) on contiguous-range-cell data (constant-level simulated-target signal added to range gate 4). Top trace in each unprocessed, bottom trace processed.

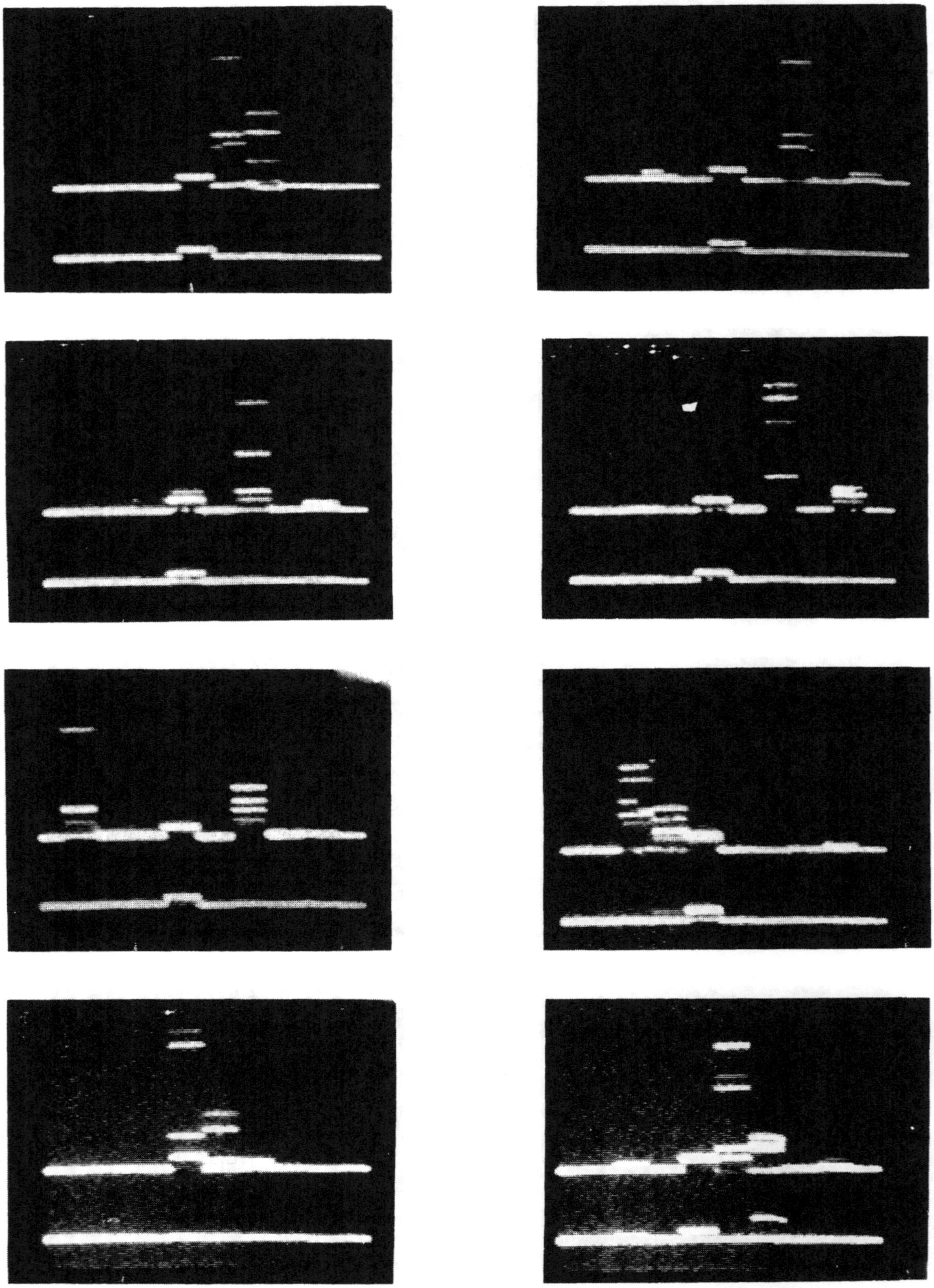

Fig. 39 (Continued)

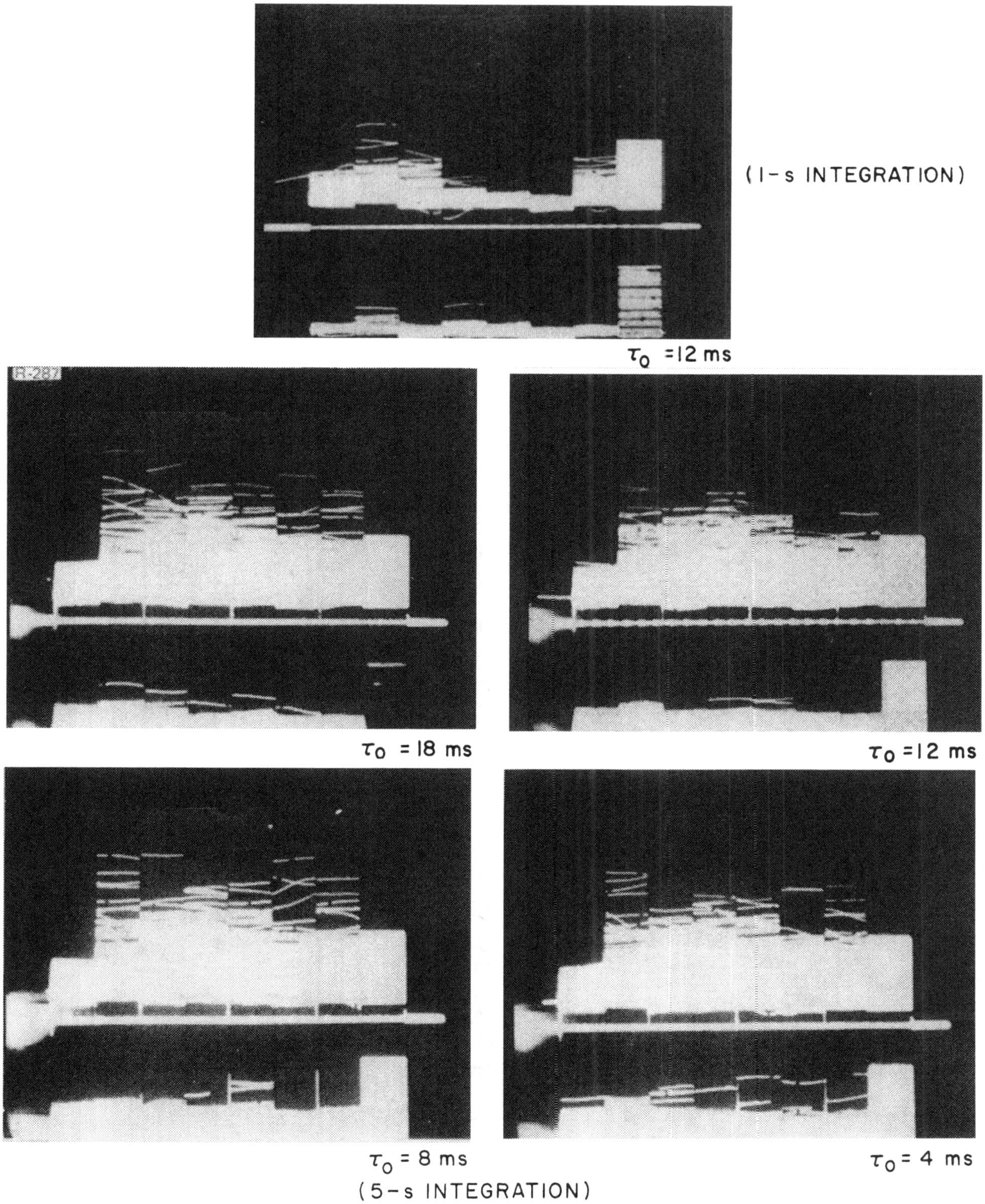

Fig. 40(a) — Performance of 2-delay sea-scatter rejection circuit for several different delay settings (5-Hz modulated target signal in range gate 8). Photos are for 5-s integration time except for top photo, which has 1-s integration time. Top trace in each unprocessed, bottom trace processed; for sea state 4 and vertical polarization.

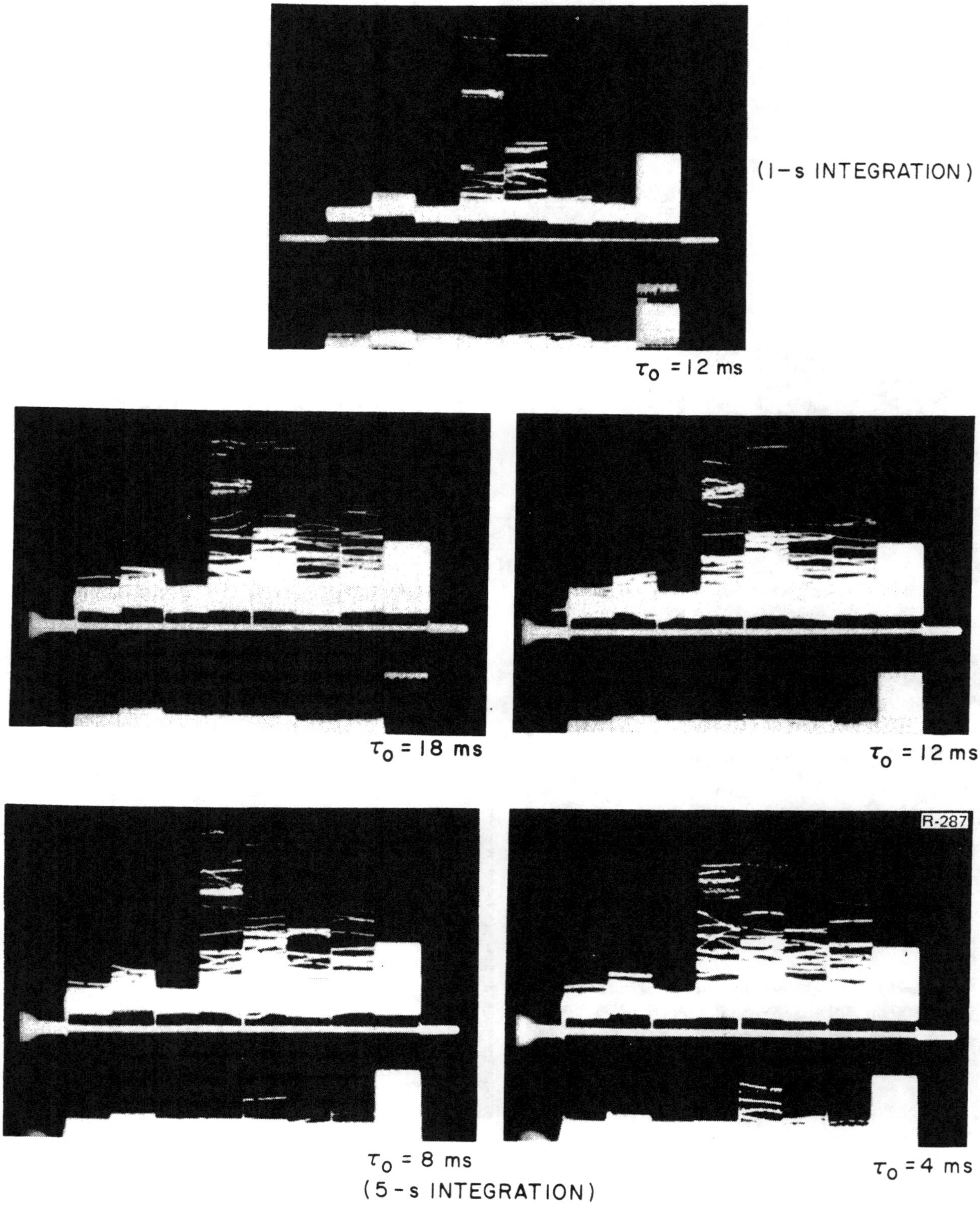

Fig. 40(b) — Performance as in Fig. 40(a), here for sea state 4 and horizontal polarization

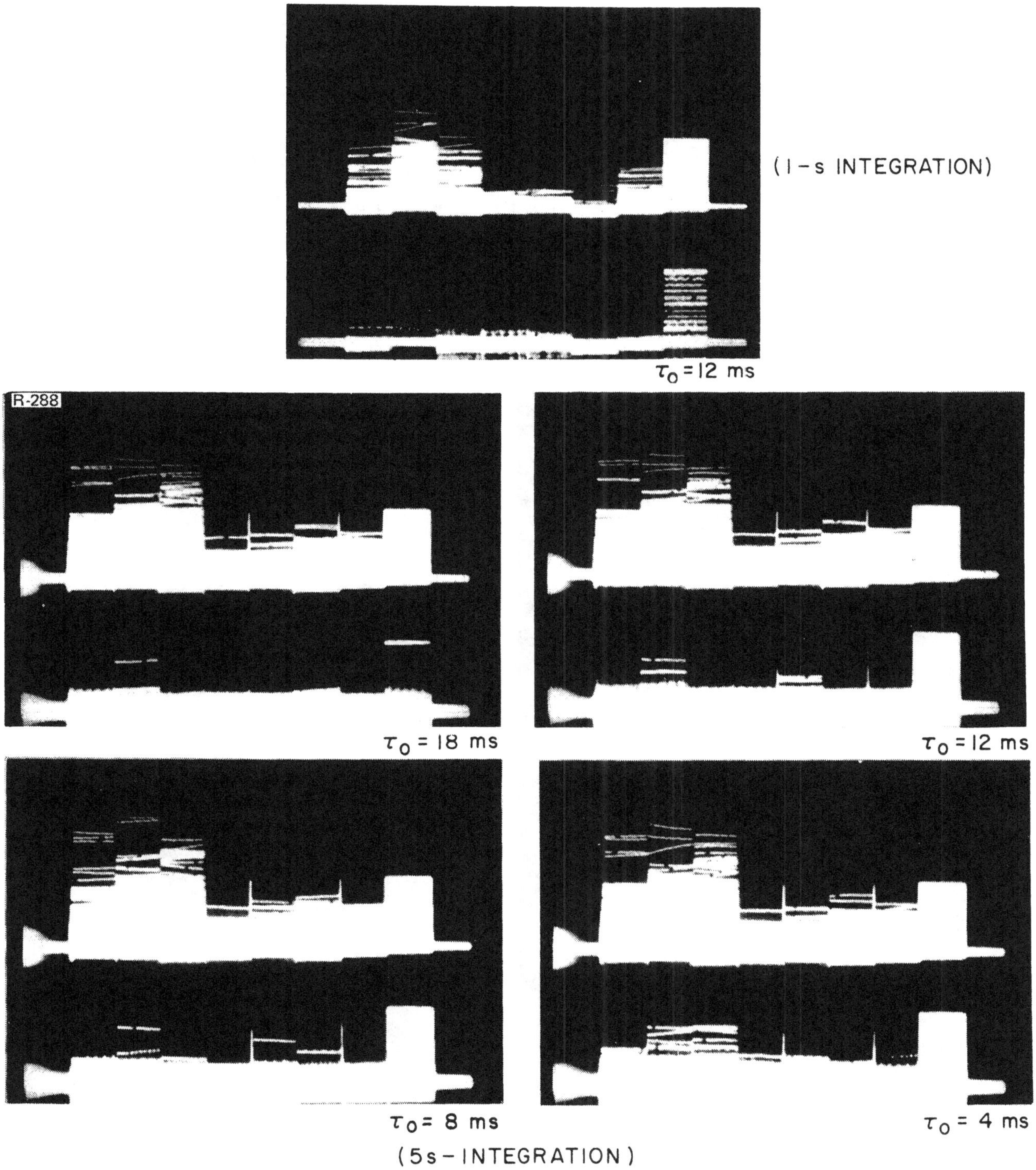

Fig. 40(c) — Performance as in Fig. 40(a), here for sea state 2 and vertical polarization

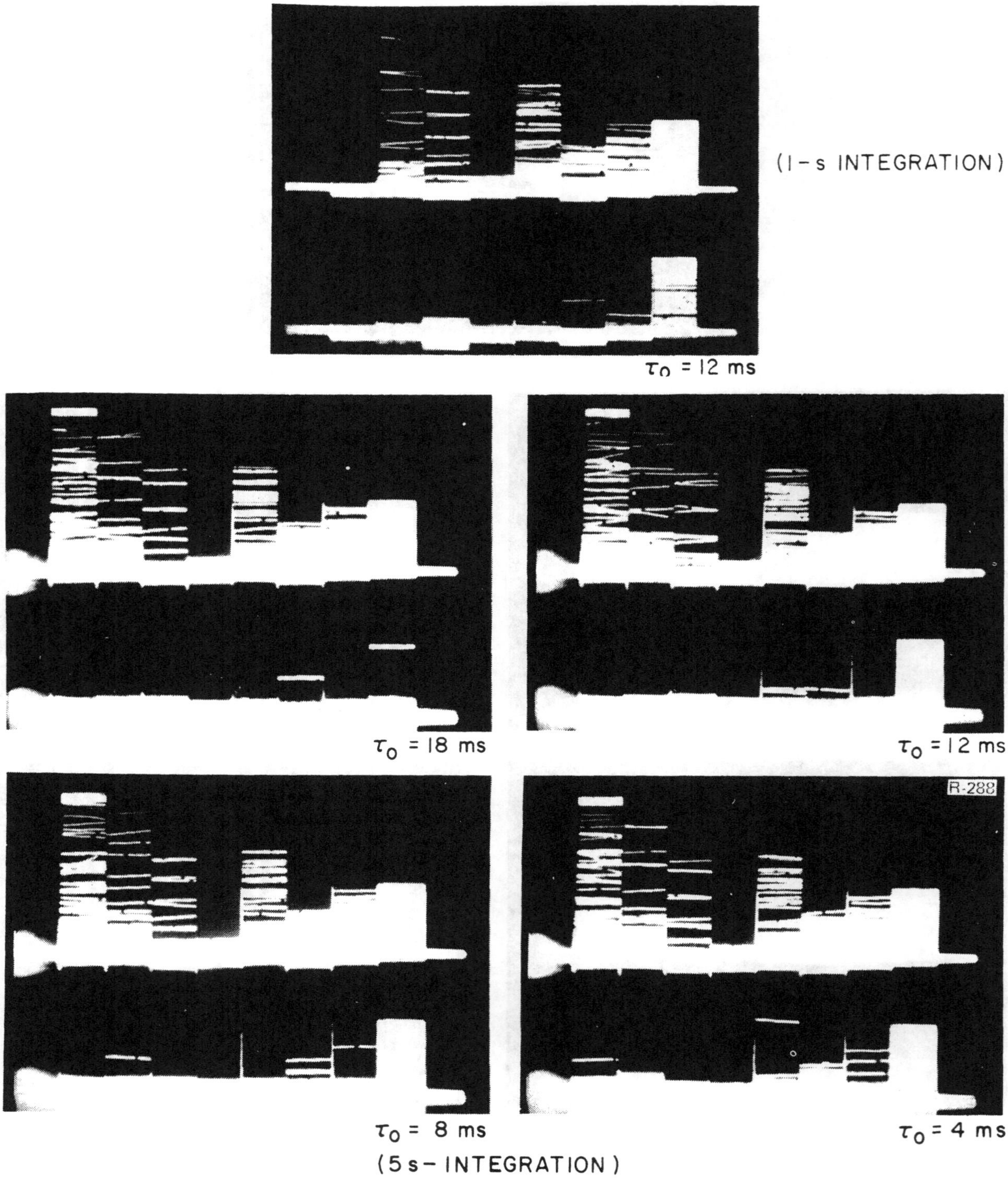

Fig. 40(d) — Performance as in Fig. 40(a), here for sea state 2 and horizontal polarization

Table 2 — Performance of a Five-Delay Sea Scatter Discriminant for Various Delay Times

Data Parameters	Delay Increment (ms)	Sea-Scatter Detections 0.6 m^2 Det. Level (% time)						5-Hz Target Retention (% visible time)
		CH 1	CH 2	CH 3	CH 4	CH 5	CH 6	
Vert. Pol.	0	10.7	14.6	11.4	15.8	13.4	8.8	100
Sea State 5	4	0.26	0.72	0.18	0.46	0.36	0.14	99.666
0° obs. angle	(20 total)							
3° inc. angle	12	0.005	0.01	0.005	0.015	0.01	0.000	99.625
	(60 total)							
	20	0.015	0.03	0.000	0.015	0.005	0.000	98.864
	(100 total)							
Hor. Pol.	0	4.1	4.9	4.0	6.7	7.9*	6.3*	100
Sea State 5	4	0.11	0.22	0.08	0.31	0.49	0.29	99.796
0° obs. angle	(20 total)							
3° inc. angle	12	0.005	0.01	0.005	0.03	0.39	0.12	99.320
	(60 total)							
	20	0.000	0.02	0.000	0.025	0.20	0.05	98.845
	(100 total)							

*Data partially saturated on CHs 5 and 6.

When examined on an amplitude-density basis, the distributions of scatter for vertical polarization tend to be more exponential (linear slope on Rayleigh statistical coordinates) while the distributions for horizontal polarization exhibit a bow or inflection in the low-level portion which becomes more apparent at lower sea states. This inflection is believed to be one result (or indicator) of the sporadic burst character of the scatter observed with horizontal polarization.

5. Multipath effects occur which appear to involve the disturbed sea (whitecaps) and its image on the tilted wave surface.

6. High-resolution-radar return from floating rigid objects generally exhibits lower amplitude-modulation frequencies than does sea return. Amplitude distributions, on the other hand, may be quite similar.

7. A multiple-delay rejection circuit can detect and block most high-frequency-modulated sea-return signals and pass lower-frequency-modulated target-return signals.

CONCLUSIONS

The experimental measurement program has determined that one dominant characteristic of low-incidence-angle, high-resolution radar sea scatter is a relatively rapid and high-percentage pulse-to-pulse amplitude modulation. The high-resolution sea scatter has also been found to be nonuniform in time with quiet periods of relatively low level returns interspersed with high-level, targetlike spikes. High-level returns for both horizontal and vertical radar polarizations have been observed with the formation of turbulent breaking water (as indicated by the visual observation of whitecaps). Lower level returns for vertical radar polarization have been associated with the small-wave or ripple structure, while a distinctive burstlike scatter for horizontal radar polarization has been associated with preferentially aligned sharp wave crests.

The measured differences in the modulation characteristic of a disturbed sea surface and of rigid floating targets suggests that a discriminant based on percentage modulation vs time could be used to reject sea scatter. Tests with recorded sea-scatter data have shown that this type of discriminant will block typical sea-scatter signals and pass a target signal. Specifically, a five-delay rejection circuit with internal delays of 12 ms and a total delay of 60 ms effectively rejected X-band sea-scatter spikes and passed simulated target signals which were sawtooth modulated at 5 Hz.

A subject radar using this type of processing would be required to dwell on or revisit the same region over a sufficiently long time period (60 ms) to detect and suppress the sea return. Within these restrictions, a radar processor could employ the amplitude-modulation discriminant and markedly reduce the probability of sea-scatter false alarms yet still maintain detection of rigid targets.

Although the high-resolution measurements cited in this report were somewhat limited in scope, the observed variability with radar resolution, polarization, and observation direction suggests that much more intensive scatter research is required if a realistic radar sea-scatter model is to be formulated. It is felt that the amplitude-modulation discriminant has shown promise but that a complete evaluation of this and other discrimination techniques will also be dependent on further measurements. Specific areas of interest include the effects of still higher resolution and frequency, the impact of wave shadowing on the detection of small floating targets, and the modulation characteristic of complex-shaped rigid targets.

REFERENCES

1. B.L. Lewis and I.D. Olin, "Some Recent Observations of Sea Spikes," *IEE Conference Publication No. 144*, p. 115, 1975.

2. B.L. Lewis, J.P. Hansen, I.D. Olin, and V. Cavaleri, "High-Resolution Radar Scattering Characteristics of a Disturbed Sea Surface and Floating Debris," NRL Report 8131, July 29, 1977.

3. B.L. Lewis and I.D. Olin, "Experimental Study and Theoretical Model of High-Resolution Radar Backscatter from the Sea," Radio Science **15** (4), 815-828 (July-Aug. 1980).

4. H. Sittrop, "Radar Reflection Characteristics of Sea Clutter," presented at AGARD Conference on New Devices, 1976.

5. M.W. Long, *Radar Reflectivity of Land and Sea*, D. C. Heath and Co., Lexington, Mass., 1975, Chapter 6.

6. G. Bishop and E.H. Boyenval, "Amplitude Distribution Characteristics of X-Band Radar Sea Clutter and Small Surface Targets," Memo No. 2348, Royal Radar Establishment, Fort Halstead, Sevenoaks, Kent, Great Britain, 1978.

7. M.W. Long, *Radar Reflectivity of Land and Sea*, D. C. Heath and Co., Lexington, Mass., 1975, Chapter 5, p. 180.

8. A.I. Kalmykov, A.S. Kurekin, Yu.A. Lamenta, I.Ye. Ostrovskiy, and V.V. Pustovoytenko, "Scattering of Microwave Radiation by Breaking Sea Waves," translated from Gor'kiy Radiofizika **19** (9), 1315-1321 (1976).

9. A.I. Kalmykov and V.V. Pustovoytenko, "On Polarization Features of Radio Signals Scattered from the Sea Surface at Small Grazing Angles," J. Geophys. Res. **81**, 1960-1964 (1976).

10. M.W. Long, "On a Two Scatterer Theory of Sea Echo," IEEE Trans. Antennas Propag. **AP-22** (5), 667-672, (Sept. 1974).

CHARACTERIZATION OF SPIKY SEA CLUTTER FOR TARGET DETECTION

IRWIN D. OLIN

Radar Division, Naval Research Laboratory, Washington, DC

ABSTRACT

Existence of comparatively high amplitude target-like clutter echoes ("sea spikes") when illuminating the sea surface with high resolution radar is recognized as an obstacle in detecting small surface targets. Approaches based on selecting a detection threshold from assumed characteristics or measurements of local clutter can result in unacceptably high false alarm rates or missed detections. An alternative is to exploit the relatively low amplitude backscatter between the sea spikes for target detection, while rejecting the sea spikes based on their rapid fluctuations. In this paper distributions of sea spike width (duration) and inter-spike period from high resolution X-band sea backscatter measurements are presented. A technique for rejecting the sea spikes is described and the change in the resulting clutter statistics discussed. The technique is illustrated using a sea spike rejection circuit with recorded sea clutter data to which a simulated target signal has been added.

INTRODUCTION

Radar detection of targets on the sea surface is generally facilitated by decreasing the size of the range-azimuth cells within which detection tests are made. The decrease lowers the average radar backscatter, thereby increasing the received target-to-clutter signal ratio. But in addition to a reduction in the average clutter backscatter level, changes in the amplitude statistics are observed. With large radar resolution cells, large numbers of random scatterers within each cell result in a Rayleigh density distribution for the backscattered signal voltages. However, smaller resolution cells containing fewer scatterers result in amplitude statistics which depart from the Rayleigh form. This is especially observed in the case when only a small number of breaking waves occur within a resolution cell. These waves produce large target-like backscattered signals which have been called sea spikes. As a result, amplitude statistics disclose substantially longer distribution "tails," i.e., there are many more observations of high radar backscatter within a sample than would be predicted from a Rayleigh distribution with the same median value. Therefore, a detection threshold selected to maintain a low false alarm rate due to the sea spikes will result in a lower overall detection sensitivity to the desired targets.

SEA SPIKE WAVEFORMS

The effect of resolution cell size on radar backscatter signals is shown in Figure 1. Decreasing the radar pulse width results in discernable "spikes" which persist for significant periods and with peak amplitudes comparable to that using lower resolution. Polarization too has a marked effect on high resolution sea backscatter, as illustrated in Figure 2. With vertical polarization every unshadowed approaching wave crest tended to be profiled, whereas with horizontal polarization this did not occur. It has been suggested in Ref. 1 that the presence of breaking waves is critical to the occurrence of the large radar backscatter spikes when using horizontal polarization.

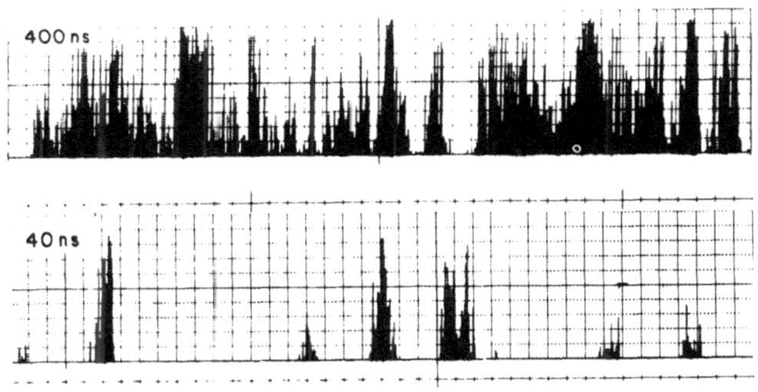

Figure 1. Low angle X-band sea backscatter for 40 ns and 400 ns transmitted pulses. (9 m cross range resolution, vertical polarization, windblown sea, time scale: 2 1/2 s per major division.)

Figure 3 illustrates the details of high resolution sea backscatter considered in this paper. Based on a selected threshold, T, signals exceeding this threshold are termed sea spikes, providing they persist longer than a time t_{min}. In addition, above threshold crossings occurring less than a time t_0 apart are considered part of the same sea spike event. Using these constraints, sea spikes with durations d_1 and d_2 are illustrated. Specifying t_{min} enables short noise impulses to be ignored.

Specifying t_o allows the inclusion of the oscillatory characteristic of sea spikes observed in connection with breaking waves. (Ref. 1)

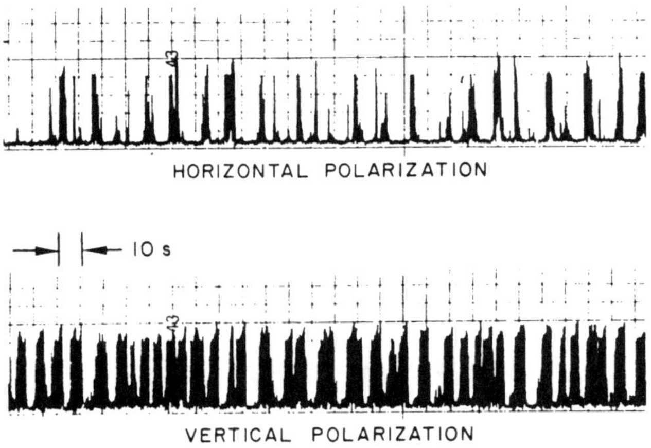

Figure 2. Comparison of high resolution X-band sea backscatter using vertical and horizontal polarization. Experimental conditions similar to those in Fig. 1.

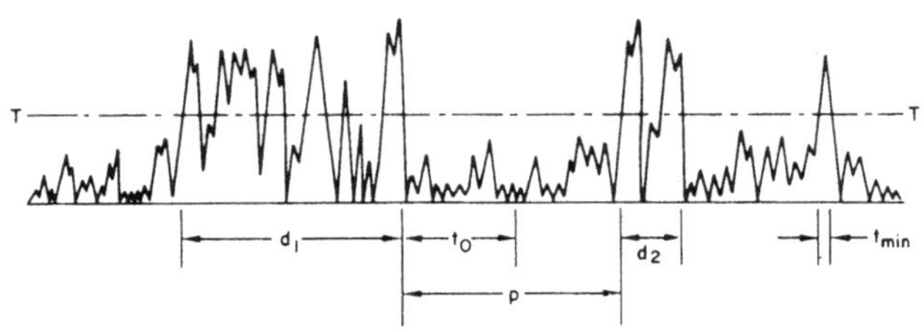

Figure 3. Illustration of high resolution sea backscatter defining sea spikes based on selecting threshold T and parameters d, t_o, t_{min}.

A series of observations at X-band were made using 40 ns pulses from an offshore platform at Panama City, Florida, at a water depth of about 130 m. (Ref. 2) A 1° pencil beamwidth was used with a grazing angle of 2.9°. Most measurements were made at 300 m range, so that the illuminated patch was only 31.6 m^2. Long analog data records (80 s - 480 s) were digitized and analyzed. Sea spikes typically observed had amplitudes as high as 10 m^2; however, between sea spike events a backscatter level often two or more orders of magnitude lower was observed.

TEMPORAL DISTRIBUTIONS

Since the sea spike amplitudes observed with high resolution radar are distinctive, improvement in detecting surface targets could be achieved if the sea spikes are rejected and target detection confined to the inter-spike periods. Rejection of the sea spikes, however, reduces the opportunities for target detection. Two temporal observables relating to this limitation are the sea spike width (d_1 and d_2 in Figure 3) and the inter-spike period (p in Figure 3). Average values of these as a function of threshold for sea state 5 are shown in Figures 4 and 5. In these and subsequent curves values of t_{min} and t_o were 0.01 s and 0.1 s, respectively. With horizontal polarization the sea spike rises and decays rapidly within the radar resolution cell. As a result little variation in width as a function of threshold is observed. Inter-

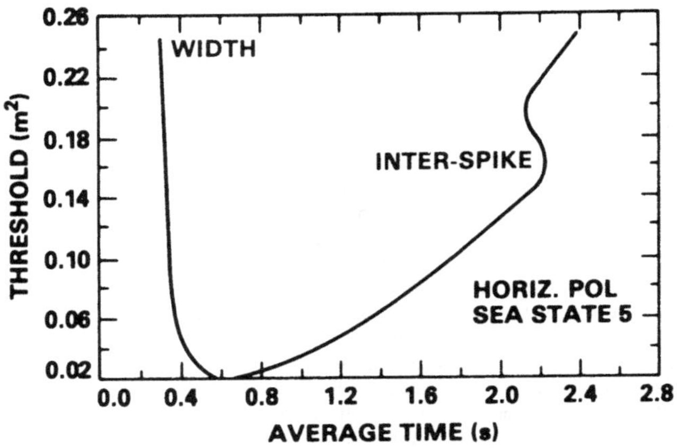

Figure 4. Average values of sea spike width and inter-spike period as a function of selected threshold using horizontal polarization.

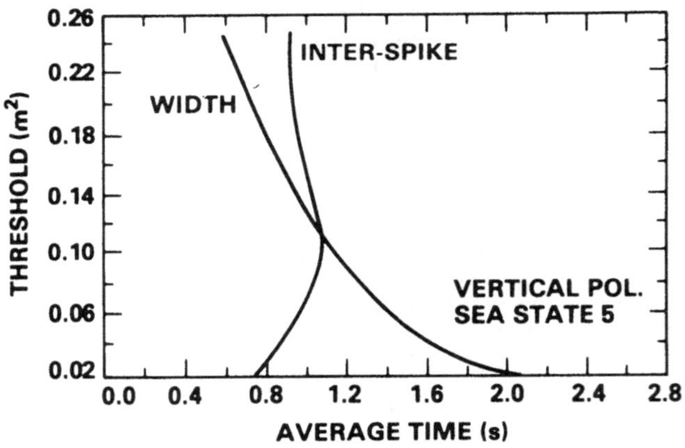

Figure 5. Average values of sea spike width and inter-spike period as a function of selected threshold using vertical polarization.

spike period, however, increases with increasing threshold. The short region of reverse slope is due to the selection of t_o which results in "splitting" a single longer inter-spike period into two shorter periods as the threshold is increased. Vertically polarized backscatter is significantly different. The initiation and decay of the sea spikes within a resolution cell are more gradual so that the average width varies with the threshold. Also, as observed in Figure 2, the somewhat regular occurrence results in comparatively little variation of inter-spike period with threshold. Cumulative distributions of these observables are shown in Figures 6-9, based on heavy smoothing of long records. For this purpose thresholds of 0.1 m^2, 0.175 m^2, and 0.25 m^2 were selected. As in the previous analyses, t_{min} = 0.01 s and t_o = 0.1 s. Clearly, substantial "quiet" periods occur in high resolution sea backscatter, especially using horizontal polarization. The impact of considering only these periods to represent the clutter in which surface targets are to be detected can be assessed by deleting the sea spikes and examining the resulting amplitude characteristics.

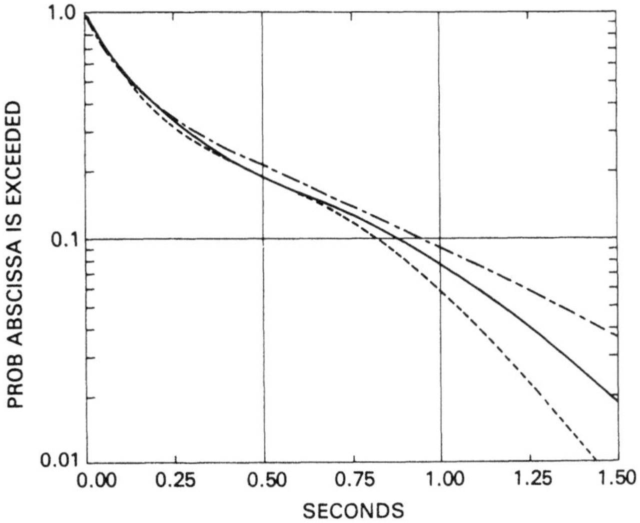

Figure 6. Distributions of sea spike widths using horizontal polarization. T = 0.1 m^2 (——— - ———), T = 0.175 m^2 (———), T = 0.25 m^2 (- - - -).

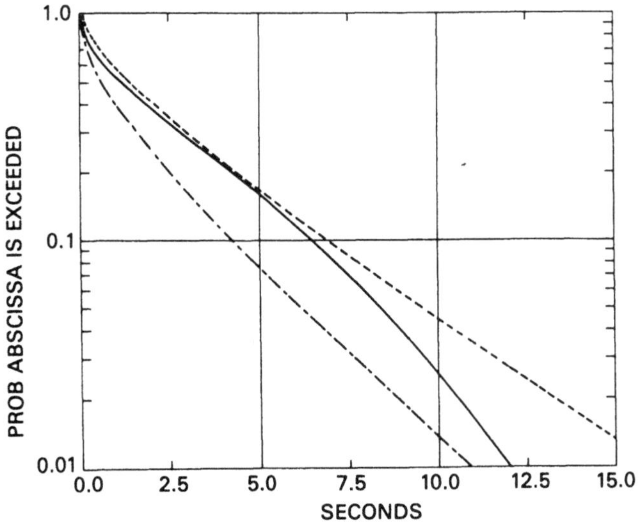

Figure 7. Distributions of sea spike inter-spike periods using horizontal polarization.

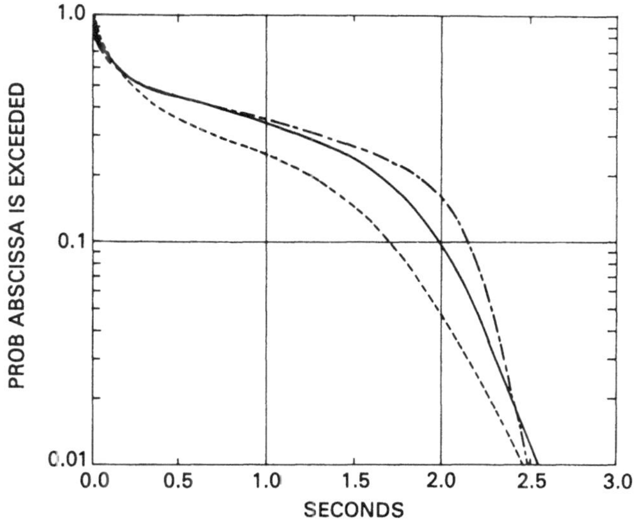

Figure 8. Distributions of sea spike widths using vertical polarization. T = 0.1 m^2 (——— - ———), T = 0.175 m^2 (———), T = 0.25 m^2 (- - - -).

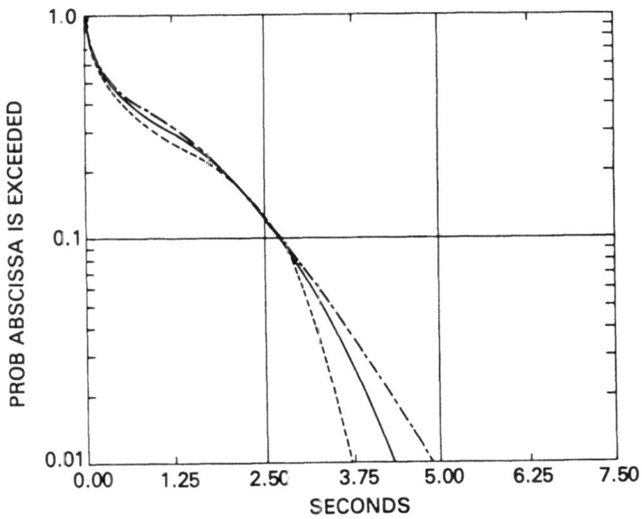

Figure 9. Distributions of sea spike inter-spike periods using vertical polarization.

AMPLITUDE DISTRIBUTIONS

Figure 10 shows the amplitude distributions of unmodified vertically and horizontally polarized sea backscatter along with distributions using the same data following deletion of the sea spikes based on a threshold of 0.1 m^2. For comparison a plot of Rayleigh backscattered signal voltages (which are exponentially distributed in power or radar cross section) is also shown. With vertical polarization deleting the spiky clutter clearly decreases the resulting backscatter. Since the unmodified backscatter as shown in Figure 2 consists of numerous closely spaced sea spike events, this result is expected. Some amplitudes above the 0.1 m^2 threshold occur due to t_{min} used in defining the sea spike. For horizontal polarization, which originally consisted of sparse sea spike events, a marked decrease occurs in the distribution tails of the modified data. For example, the probability of exceeding 0.2 m^2 using the modified data is two orders of magnitude less than if unmodified data were used. For the figure the axes scales were chosen so that the Weibull distributions, for which $P[R > R_t] = \exp[-R^B/A]$, all plot as straight lines. Since the Rayleigh form is represented by B = 1, it is clear that the distributions of the unmodified vertical and the modified horizontal backscatter components fit the Weibull distribution within the the limits plotted. Amplitude distribution fits to these data have been examined in Ref. 2.

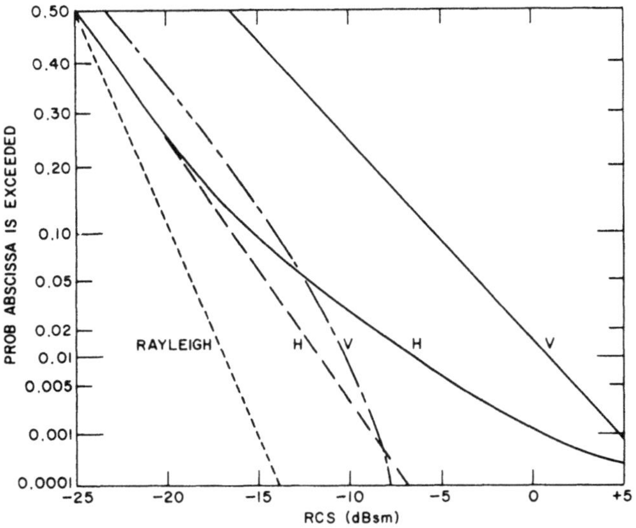

Figure 10. Horizontal and vertical polarization amplitude distributions for unmodified (solid curves) and modified (broken curves) high resolution sea state 5 data. (Theoretical Rayleigh distributed clutter shown for comparison.)

DATA PROCESSING

The result of deleting the sea spikes from the data is a significant decrease in false alarms when a low threshold is used for detecting surface targets. An appropriate means to accomplish this is to recognize the occurrence of the sea spikes based on their distinctive spectral characteristics and to inhibit all video from subsequent processing. Measurements of sea spikes, as noted in Ref. 1, disclosed spectral spreads greater than 200 Hz. A technique which exploits this characteristic as described in Ref. 3, is shown in Figure 11. Delays τ_1 and τ_2 are selected equal to a fraction of the modulation periods observed in the sea spikes. For example, values in the range 8 ms - 20 ms have been used. In operation large amplitude differences in time intervals τ_1, τ_2 and $(\tau_1 + \tau_2)$ are detected and inhibit signal output. Backscatter from surface targets is not expected to vary as rapidly, so that no inhibit signal will be generated. Demonstrated results of this technique, as described in Ref. 3, are illustrated in Figures 12 and 13. For this purpose $\tau_1 = \tau_2 = 12$ ms. A 5 Hz "target" signal was added to one range cell of recorded sea state 4 horizontally polarized 40 ns backscatter. Each of the scope photos show the radar backscatter from eight contiguous range cells for a one or five second integration period. Unprocessed data contains both "target" and clutter, whereas the high amplitude spiky clutter is rejected and the "target" signal in the end range cell is passed.

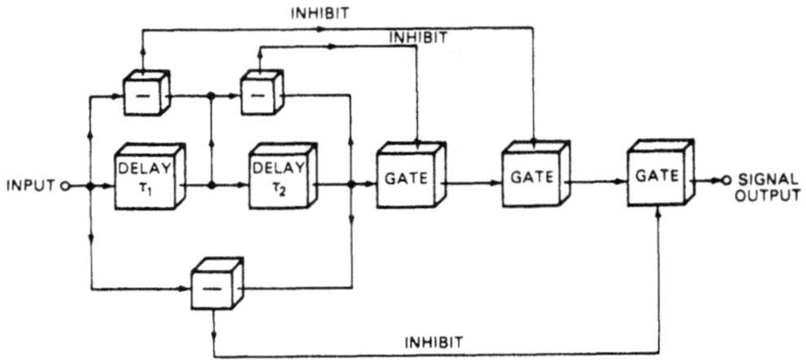

Figure 11. Circuit for rejecting sea spikes based on observed amplitude modulation characteristics.

Basic to the operation of circuitry using this technique are observation times within each antenna azimuth beamwidth of at least $\tau_1 + \tau_2$. For high resolution radar scanning 360°, this could lead to very long revisit times. An alternative described in Ref. 4 is the rotation of N squinted beams as depicted in Figure 14. Beam squint angles and scanning time are proportioned to accommodate the requisite delays while realizing reasonable update times. A pulse burst transmission, as shown in the figure, can be used. One burst is transmitted through the correspondingly numbered beams for each beamwidth of antenna rotation. Successive looks t seconds apart at the same azimuth position are then provided by selecting ω, given m and θ, ($t = m\theta/\omega$; also $f_r = m/t$). Then in order to compare data from the same range-azimuth cells, appropriate delays must be used in each of the receiver beams. ($D_i = (N - i)t$; i = 1, 2, N). The burst frequency, f_b, can be arbitrarily high, since each of the beams looks in a different direction. Alternatively, f_b can be based on a maximum operating range, thereby preventing returns from one beam transmission being coupled into the receiver of another beam. However, this introduces an error in successive beams covering exactly the same azimuth cells. The maximum error, in beamwidths is given by: $\varepsilon = (N - 1)\, f_r/f_b$. In effect this technique enables an equivalent "scan back" during antenna rotation. Other variants are possible, for example, unequal beam squint angles can be used to develop unequal delays between samples as denoted by τ_1 and τ_2 in Figure 11. A disadvantage of this technique is the long delays, D_i, which must each accommodate the large number of range cells to be searched.

CONCLUSIONS

Analysis of the temporal characteristics of high resolution sea clutter suggests that the use of horizontal polarization offers advantages over vertical polarization in detecting small targets on the sea surface. Using horizontal polarization, longer periods of low background clutter are observed between sea spikes, so that a lower detection threshold may be used. Sea spikes can then be rejected in favor of targets based on differences in their spectral characteristics.

ACKNOWLEDGEMENTS

The author wishes to acknowledge the considerable contributions of Mr. B.L. Lewis to this subject in general and to an interpretation of the polarization characteristics in particular. Also to be acknowledged are the efforts of Mr. J.P. Hansen for developing and executing the experimental program which provided the data.

REFERENCES

1. B.L. Lewis, I.D. Olin, "Experimental Study and Theoretical Model of High-Resolution Radar Backscatter from the Sea," Radio Science, 15, Number 4, pp 815-828

2. I.D. Olin, "Amplitude and Temporal Statistics of Sea Spike Clutter," Radar-82, IEE Conference Publication Number 216, pp 198-202

3. J.P. Hansen, V.C. Cavaleri, "High-Resolution Radar Sea Scatter, Experimental Observations and Discriminants," NRL Report 8557, March 5, 1982

4. B.L. Lewis, "Scanning Radar System," U.S. Patent 4357,708, Nov 2, 1982

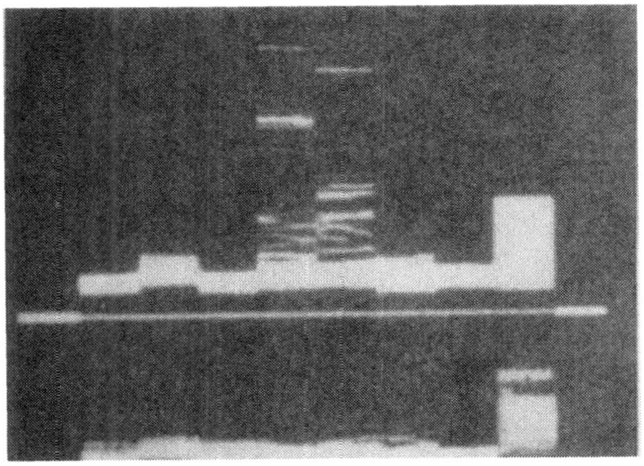

Figure 12. Demonstration of sea spike rejection circuit ($\tau_1 = \tau_2 = 12$ ms) using sea state 4 clutter data with 5 Hz "target" signal added to end range cell, (8 contiguous range cells shown) 1 s integration. Unprocessed data above, processed data below.

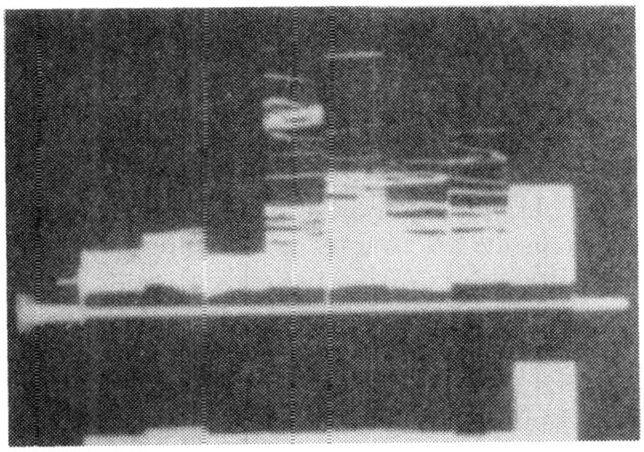

Figure 13. Demonstration of sea spike rejection circuit ($\tau_1 = \tau_2 = 12$ ms) using sea state 4 clutter data with 5 Hz "target" signal added to end range cell, (8 contiguous range cells shown) 5 s integration. Unprocessed data above, processed data below.

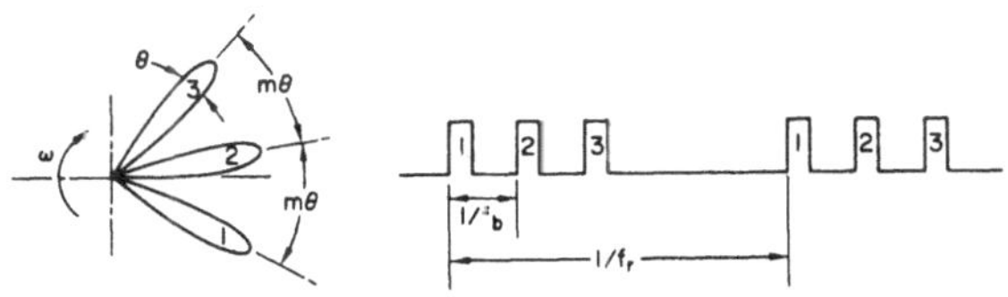

Figure 14. Use of rotating multiple squinted beams and pulse burst waveform to realize signal requirements of Fig. 11.

A COMPARISON OF NONCOHERENT AND COHERENT MTI's

INTRODUCTION

Noncoherent moving target indication (MTI) differs from coherent MTI in that the MTI processing is performed at video after an envelope or square law detector [1-5]. In the coherent MTI, processing is done at IF or more commonly in the complex video inphase I and quadrature Q channels.

The noncoherent MTI is similar in many ways to the coherent MTI but differs in some important respects. With the noncoherent MTI, if the pulse-to-pulse changes in the amplitude of the clutter return are small, the clutter will be heavily attenuated by the MTI filter following the envelope detector. Since the phase of the clutter is not used, the noncoherent MTI is capable of centering the cancellation notch on clutter having a nonzero average radial velocity. In the presence of clutter, a target having a different radial velocity from that of the clutter will cause pulse-to-pulse amplitude variations and the target plus clutter signal will not be as heavily attenuated by the MTI filter. Since the target signal is clutter referenced it will be canceled in the absence of clutter unless the target scintillates.

An examination of the existing literature indicated that little analytical work has been performed on the noncoherent MTI, particularly based on the envelope detector which is of interest in this report. Prior work [3,4] considered the envelope detector to behave approximately as a square law detector and based the results on arguments relating to the spectral characteristics of the clutter signal after a square law detector.

In this report, a statistical approach is taken, in both the analysis and the Monte Carlo simulations, which is not based on any spectral approximations. Clutter attenuation and improvement factors based on generalized definitions are computed and compared to the coherent MTI.

It is shown that the common assumption that the clutter standard deviation is increased by a factor of $\sqrt{2}$ is not valid for the noncoherent MTI using an envelope detector for values of the correlation coefficient which are required for good MTI performance. The factor of $\sqrt{2}$ spectral spread is attributed by other investigators to the alleged square law behavior of the envelope detector which results in a self-convolution of the clutter power spectral density which exists prior to the envelope detector.

COMPARISON OF COHERENT AND NONCOHERENT MTI

Coherent MTI

First we consider the clutter attenuation for a two-pulse coherent MTI. The successive clutter returns are designated by the complex video vectors consisting of the I and Q values, or equivalently the amplitude and phase which are computed from $(I^2 + Q^2)^{1/2}$ and $\tan^{-1}(Q/I)$ respectively.

For a two-pulse canceler the complex residue signal is

$$R = C_1 - C_2 \tag{1}$$

where C_1 and C_2 are successive complex clutter return signals from a given range cell separated in time by the interpulse period T.

Manuscript submitted February 17, 1982.

The average residue power C_0 is taken to be

$$C_0 = \overline{|R|^2} = \overline{|C_1 - C_2|^2} \qquad (2)$$
$$= \overline{(C_1 - C_2)(C_1 - C_2)^*}$$
$$= \overline{|C_1|^2} + \overline{|C_2|^2} - 2\mathrm{Re}\, \overline{C_1 C_2^*}.$$

The cross correlation term Re $\overline{C_1 C_2^*}$ may be determined from the I and Q components

$$\mathrm{Re}\, \overline{C_1 C_2^*} = \mathrm{Re}\, \overline{(C_{1I} + jC_{1Q})(C_{2I} - jC_{2Q})} \qquad (3)$$
$$= \overline{C_{1I} C_{2I}} + \overline{C_{1Q} C_{2Q}} = 2\rho\sigma^2$$

where it is assumed that the I and Q components of each clutter return have the same variance σ^2 and each have the same correlation coefficient ρ. Substituting Eq. (3) in Eq. (2) results in

$$C_0 = \overline{|R|^2} = 4\sigma^2 - 4\rho\sigma^2 = 4\sigma^2(1 - \rho). \qquad (4)$$

Letting the input clutter power be

$$C_{in} = \overline{|C_1|^2} = \overline{|C_2|^2} = 2\sigma^2, \qquad (5)$$

the clutter attenuation or cancellation ratio is

$$CR = \frac{C_{in}}{C_0} = \frac{1}{2(1 - \rho)}. \qquad (6)$$

This result may be generalized in terms of matrix notation as follows. Let W denote the column matrix of weights applied to the successive clutter samples C_i and let M_c denote the covariance matrix of the clutter which is given by

$$M_c = \overline{CC'} \qquad (7)$$

where C' denotes the complex conjugate of the transposed matrix C. Letting T denote the transpose operation, the residue power is

$$C_0 = \overline{(W^T C)(W^T C)'}$$
$$= \overline{W^T C\, C' W^*} = W^T M_c W^*. \qquad (8)$$

Applying Eq. (8) to the two-pulse canceler example described above, we have

$$C_0 = (1\ {-1}) \begin{pmatrix} 2\sigma^2 & 2\sigma^2\rho \\ 2\sigma^2\rho & 2\sigma^2 \end{pmatrix} \begin{pmatrix} 1 \\ -1 \end{pmatrix} = 4\sigma^2(1 - \rho)$$

which agrees with Eq. (4).

The clutter cancellation ratio may be expressed in general form as

$$CR = \frac{2\sigma^2}{W^T M_c W^*}. \qquad (9)$$

The MTI improvement factor I is determined by the ratio of the output target-to-clutter ratio divided by the input target-to-clutter ratio where the input target is averaged over all velocities if no a priori knowledge is available. The average target response, or target enhancement factor, is equivalent to normalizing the filter response to white noise so that the output noise power is equal to the input noise power. It follows that

$$I = \frac{\overline{T_0/C_0}}{T_{in}/C_{in}} = \overline{\left(\frac{T_0}{T_{in}}\right)}\left(\frac{C_{in}}{C_0}\right)$$

$$= (W^T W^*) \cdot (CR) = \frac{2\sigma^2 (W^T W^*)}{W^T M_c W^*} \tag{10}$$

where we have made use of Eq. (9) and the fact that the white noise response of the filter is given by $W^T W^*$.

Noncoherent MTI

For the two-pulse noncoherent MTI, the output clutter residue is

$$R = |C_1| - |C_2| \tag{11}$$

and the output clutter power is

$$C_0 = \overline{R^2} = \overline{(|C_1| - |C_2|)^2} \tag{12}$$

$$= \overline{|C_1|^2} + \overline{|C_2|^2} - 2\overline{|C_1||C_2|}.$$

From Refs. 6 to 8 the cross-correlation term is given by

$$\overline{|C_1||C_2|} = 2\sigma^2 \frac{\pi}{4} \, {}_2F_1 \, (-1/2, -1/2, 1, \rho^2) \tag{13}$$

where ${}_2F_1 \, (a, b, c, x)$ is the Gaussian hypergeometric function [6] given by

$${}_2F_1(a, b, c, x) = 1 + \frac{ab}{c1!}x + \frac{a(a+1)b(b+1)}{c(c+1)2!}x^2 \dots .$$

In particular, we have

$${}_2F_1 \, (-1/2, -1/2, 1, \rho^2) = 1 + \frac{1}{4}\rho^2 + \frac{1}{64}\rho^4 + \frac{1}{256}\rho^6 + \dots . \tag{14}$$

References 6 to 8 use an identity given in Ref. 9 as

$${}_2F_1 \, (-1/2, -1/2, 1, \rho^2) = \frac{4}{\pi} E(\rho) - \frac{2}{\pi}(1-\rho^2) K(\rho) \tag{15}$$

where K and E are complete elliptic integrals of the first and second kinds respectively. These functions are tabulated in Ref. 10 and are also available as a subroutine on the NRL ASC digital computer where they are defined in terms of ρ^2 and are designated as $E(\rho^2)$ and $K(\rho^2)$. Incorrect usage of these tabulated values leads to erroneous results. To be consistent with the literature we retain the notation of $E(\rho)$ and $K(\rho)$.

From Eqs. (12), (13), and (15), the resultant output clutter is

$$C_0 = 4\sigma^2 \, [1 - \pi/4 \;\; {}_2F_1 \, (-1/2, -1/2, 1, \rho^2)]$$

$$= 4\sigma^2 \left[1 - \left\{E(\rho) - \frac{(1-\rho^2)}{2} K(\rho)\right\}\right] \tag{16}$$

which may be written more compactly as

$$C_0 = 4\sigma^2 [1 - F(\rho)] \tag{17}$$

where

$$F(\rho) = \frac{\pi}{4} \, {}_2F_1 \, (-1/2, -1/2, 1, \rho^2)$$

$$= E(\rho) - \left(\frac{1-\rho^2}{2}\right) K(\rho). \tag{18}$$

Comparison of Eq. (17) with the expression for the coherent MTI in Eq. (4) shows the expressions are the same if one interchanges ρ and $F(\rho)$. This result generalizes for higher order cancelers so that for the noncoherent canceler the output clutter may be expressed in a similar way to Eq. (8) as

$$C_0 = W^T M_N W^* \tag{19}$$

where M_N is the covariance matrix of the noncoherent MTI which is the same as M_c in Eq. (8) except that ρ is replaced by $F(\rho)$. In either case, it is noted that ρ is a function of time separation which is equal to the Fourier transform of the clutter power spectral density and is dependent on the interpulse intervals.

From the previous results, the cancellation ratio for the noncoherent MTI is given by

$$CR = \frac{2\sigma^2}{W^T M_N W^*}. \tag{20}$$

A discussion of the improvement factor for the noncoherent MTI is deferred to later in this section.

Comparison of Cancellation Ratios

At this juncture, we compare the cancellation ratios of the coherent and noncoherent MTI for a clutter spectrum which is assumed to be Gaussian and given by

$$G(f) = \frac{1}{\sqrt{2\pi}\sigma_s} e^{-f^2/2\sigma_s^2} \tag{21}$$

where σ_s denotes the standard deviation of the clutter spectrum. The correlation function $\rho(\tau)$ is the Fourier transform of $G(f)$ and is given by

$$\rho(\tau) = e^{-2\pi^2(\sigma_s\tau)^2} \tag{22}$$

where τ is the time delay variable.

A useful relationship for the Gaussian spectrum is that

$$\rho(k\tau) = [\rho(\tau)]^{k^2}. \tag{23}$$

By use of Eq. (8) and Eqs. (20) through (23), the cancellation ratio was computed for the coherent and noncoherent MTI and the results are shown in Fig. 1 for various order cancelers where the standard binomial weighting was used. In Fig. 1 the abscissa denotes the spectral width σ_s which is normalized to the pulse repetition frequency (prf). Figure 1(a) shows the comparison for the two- and three-pulse cancelers, and Fig. 1(b) shows the comparison for the four- and five-pulse cancelers.

For the two-pulse canceler, it is shown that the noncoherent MTI is approximately 3 dB better than the coherent MTI in contrast to prior beliefs. Skolnik [1] noted this result by use of a simple vector relationship. A more detailed derivation using vectors is given in the appendix. For higher order cancelers the noncoherent clutter attenuation is seen in Fig. 1 to be generally worse than the coherent MTI except for the three-pulse canceler whose curves cross over for σ_s/prf approximately equal to 0.07.

The clutter attenuation of the coherent and noncoherent MTI's was also simulated on a digital computer using Monte Carlo techniques, which are described later, and excellent agreement was obtained.

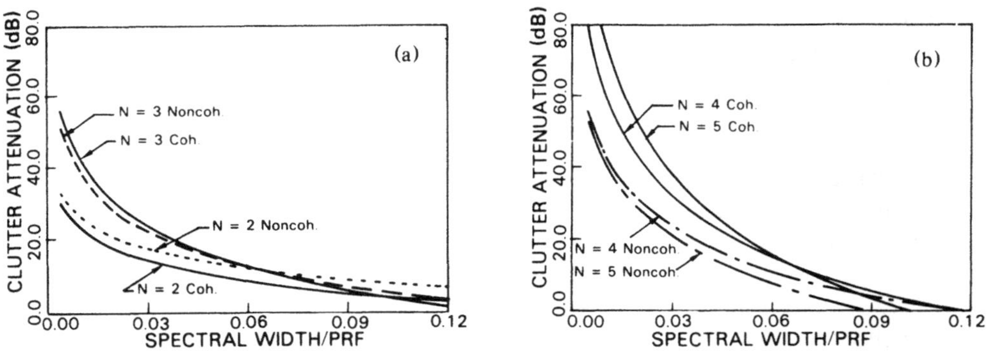

Fig. 1 — Clutter attenuation of coherent and noncoherent (envelope detector) MTI's. (a) 2- and 3-pulse canceler and (b) 4- and 5-pulse canceler.

At first glance, it may appear strange that the clutter attenuation curves for the coherent MTI cross over as shown in Fig. 1. However, since the improvement factor given in Ref. 10 asymptotically approaches unity the clutter attenuation cannot become less than the reciprocal of the target enhancement factor. This also follows from the relation

$$\lim_{\rho \to 0} CR = \lim_{\rho \to 0} \frac{2\sigma^2}{W^T M_c W^*} = \frac{2\sigma^2}{2\sigma^2 W^T W^*} = \frac{1}{W^T W^*} \tag{24}$$

where we have made use of the fact that M_c approaches $2\sigma^2$ times the identity matrix as ρ approaches 0.

Spectral Spreading Evaluation

Next, we examine the relationship between the coherent and noncoherent cancelers in terms of a spectral spreading factor. That is, for a given σ_s/prf in Fig. 1 and the associated noncoherent clutter attenuation, we find the multiplicative factor for σ_s/prf which results in the corresponding coherent canceler having the same cancellation ratio as the noncoherent canceler.

This is plotted in Fig. 2 where it is seen that there is no simple relationship which may be stated in regard to the spectral spreading factor.

It was previously shown that the clutter output power of the coherent or noncoherent MTI filters can be computed from Eqs. (8) or (19), where the difference lies in the covariance matrices M_c and M_N. As previously described, the difference between the matrices is that ρ in the coherent covariance matrix is replaced by $F(\rho)$ given by Eqs. (18) and (14) as

$$F(\rho) = \frac{\pi}{4}\, {}_2F_1\,(-1/2,\, -1/2,\, 1,\, \rho^2).$$

$$= \frac{\pi}{4}\left\{1 + \frac{\rho^2}{4} + \frac{\rho^4}{64} + \frac{\rho^6}{256} + \ldots\right\}.$$

From this expansion several observations may be made. First, the envelope detector corresponds to an expansion in even powers of ρ. Also, the only term which can be identified with a square law detector is the ρ^2 term, which corresponds to a convolution of the power spectral densities in the frequency domain. It is found that in the region of interest where good MTI performance is obtained, ρ is nearly unity and even using as many as the first three terms of the expansion yields inaccurate results for the envelope detector case. Note that the spectrum corresponding to $F(\rho)$ consists of a summation

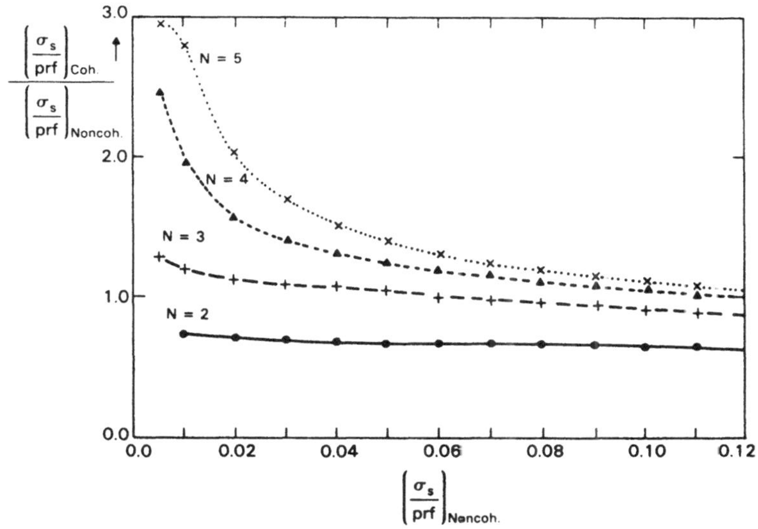

Fig. 2 — Spectral spreading factor for noncoherent MTI using an envelope detector

of Gaussian terms having unequal variances and that the resultant spectrum is non-Gaussian, in contrast to the output spectrum of the square law detector. Hence, the spectral spreading factor is an equivalence only in terms of the resultant value of the cancellation ratio.

Improvement Factor Comparison

Because of the nonlinearity of the envelope detector in a noncoherent MTI system it is not appropriate to separately determine the doppler-averaged target response and the clutter attenuation and combine these factors as was discussed for the coherent MTI whose improvement factor was given by Eq. (10). To circumvent this problem we define the output target-to-clutter power ratio as

$$\frac{T_0}{C_0} = \frac{(T + C)_0 - C_0}{C_0}. \tag{25}$$

The improvement factor using the generalized definition becomes

$$I = \frac{\overline{(T + C)_0 - C_0}}{T_{in}} \frac{C_{in}}{C_0} \tag{26}$$

where the averaging is over all target velocities.

Due to the nonlinearity of the noncoherent MTI, the improvement factor will in general be dependent on the input target-to-clutter ratio. In Ref. 7 an approximate relationship is derived which gives the value of $\overline{|C_1||C_2|}$, needed for the noncoherent MTI, for the small target-to-clutter ratio. However, this relation was found to be unsatisfactory since for a zero target-to-clutter ratio the expression reduces to the expression containing the first three terms of the Gaussian hypergeometric function which is not sufficiently accurate for values of ρ near unity. Therefore, computer simulations were performed using Monte Carlo techniques. For the different order cancelers successive clutter samples, corresponding to the returns from a given range cell on successive sweeps, were generated on the computer. The correlation between successive sweeps was specified, and the MTI residue was computed for each trial consisting of N sweeps for an N-pulse MTI. Three thousand independent trials were run, and the residues were averaged for each case.

On each trial the first return was taken to have a Rayleigh distributed amplitude and a uniformly distributed phase, and successive returns were correlated with the first return. This was achieved as follows. The first clutter sample was generated from two independent Gaussian distributions associated with the I and Q components of C_1, with each distribution having a mean of 0 and a variance equal to σ^2. The next clutter sample was obtained from a distribution C_2' which had a correlation of $\rho(T)$ with C_1. For a three-pulse canceler, for example, a third clutter sample was taken from the distribution C_3' which had a correlation $\rho(T)$ with C_2' and $\rho(2T)$ with C_1 which we renamed C_1'. Thus in the primed coordinates the distributions are mutually correlated complex Gaussian distributions that are normalized to have zero mean and a variance of σ^2 in both the I and the Q components.

The above remarks may be summarized in mathematical terms for the three-pulse example as

$$\begin{aligned} C_1' &= a_{11}\, C_1 \\ C_2' &= a_{21}\, C_1 + a_{22}\, C_2 \\ C_3' &= a_{31}\, C_1 + a_{32}\, C_2 + a_{33}\, C_3 \end{aligned} \tag{27}$$

where

$$a_{11} = 1, \qquad a_{22} = [1 - \rho^2(T)]^{1/2},$$

$$a_{21} = \rho(T), \qquad a_{32} = \rho(T)\frac{[1 - \rho\ (2T)]}{a_{22}},$$

$$a_{31} = \rho(2T), \qquad a_{33} = [1 - \rho^2(2T) - a_{32}^2]^{1/2}.$$

The MTI was then simulated by applying the binomial weighting to the complex C_i' samples, or to the amplitude of the C_i' samples, for the coherent and noncoherent cases respectively. For different values of $\rho(T)$, 3000 independent trials were run and the output residue powers were averaged. The cancellation ratios were computed for the two cases, where the input clutter power for each was taken as the average of the input $(I^2 + Q^2)$ value which is $2\sigma^2$. The output clutter power for the coherent case was taken as the average residue, again computed from $(I^2 + Q^2)$, averaged over the 3000 independent trials. The noncoherent residue power was computed by averaging the square of the residue. Thus, both systems are computed with common input and output terminals.

The target signal was added to the clutter signal with a random initial phase angle and a specified phase shift corresponding to the target's doppler. The average target output power as determined from the generalized definition (Eq. (25)) was determined for 10 different target velocities uniformly spaced across the prf interval. Excellent agreement was obtained with the known improvement factor for the coherent MTI case. The results for the coherent and noncoherent MTI are plotted in Fig. 3 for an input clutter-to-target ratio of 20 dB. The results were nearly the same for all clutter-to-target ratios above 10 dB. It is seen that the improvement factors for the two-pulse canceler are nearly the same while the noncoherent MTI improvement factor degrades relative to the coherent MTI as the number of pulses increases. A comparison with the improvement factor computed from the relation in Steinberg [3], for a square law device, is shown in Fig. 4 where it is seen that the improvement factor for the square law detector is equal to or greater than that for the envelope detector. In Fig. 5, the target enhancement factor of the simulated noncoherent MTI is compared with the coherent MTI for a clutter-to-target ratio of 20 dB. It is seen that the noncoherent MTI target enhancement factor is approximately 3 dB less than that for the coherent MTI.

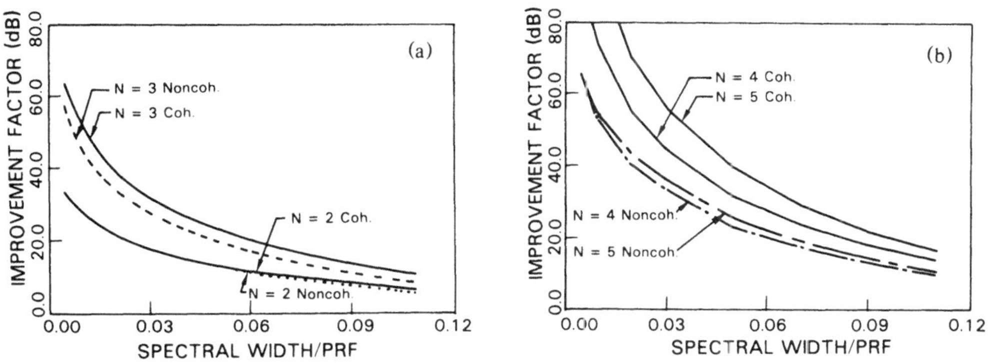

Fig. 3 — Improvement factors of coherent and noncoherent (envelop detector) MTI's. (a) 2- and 3-pulse canceler and (b) 4- and 5-pulse canceler.

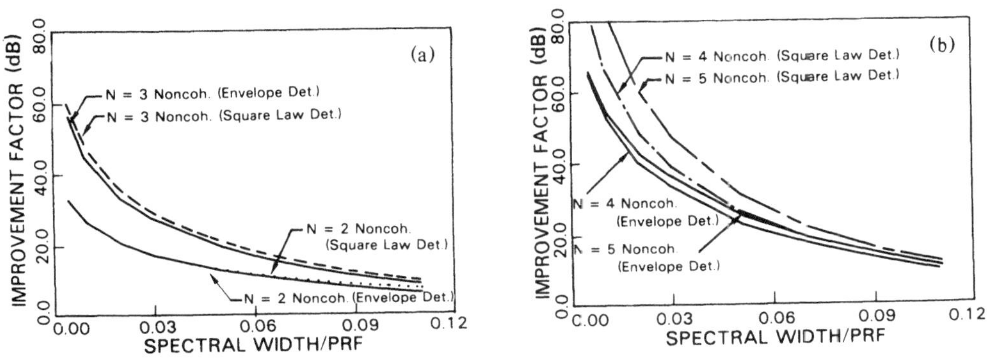

Fig. 4 — Improvement factors for noncoherent MTI's using an envelope and a square law detector. (a) 2- and 3-pulse canceler and (b) 4- and 5-pulse canceler.

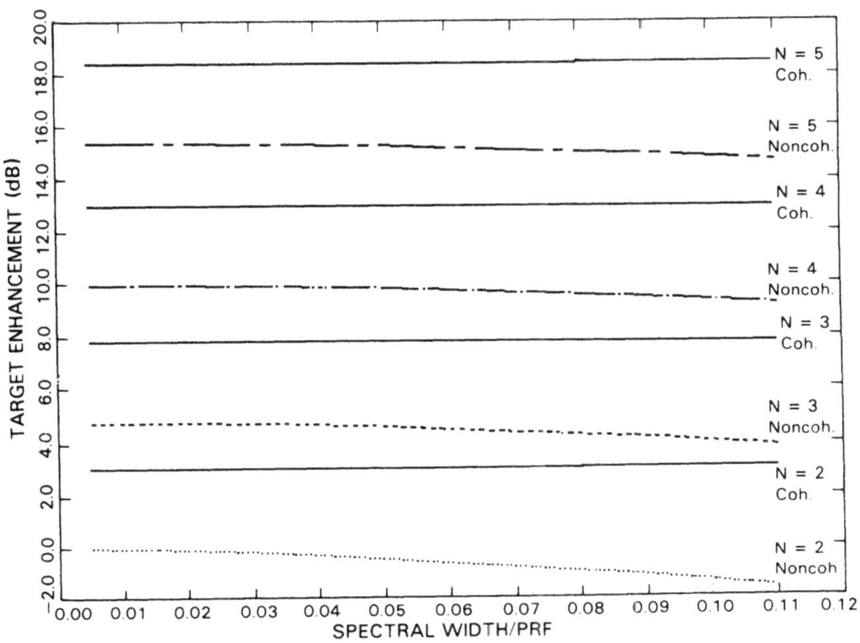

Fig. 5 — Target enhancement factor vs normalized spectral width for coherent MTI and noncoherent MTI using an envelope detector

SUMMARY AND CONCLUSIONS

A comparison between a coherent MTI and a noncoherent MTI using an envelope detector has been made based on statistical analysis and on computer simulations using Monte Carlo techniques. This approach differs from prior investigations which base the results on the envelope detector behaving approximately as a square law detector from which it is argued that the standard deviation of the power spectral density of the clutter is increased by $\sqrt{2}$ due to the self-convolution of the clutter spectra.

The results of the analysis in this report indicate that in general, the envelope detector cannot be regarded as a square law detector which increases the standard deviation of the clutter spectrum by $\sqrt{2}$. The equivalent spectral spread was found to depend on the number of pulses used in the MTI as well as on the correlation of the returned clutter signals. It was noted that the spectrum after envelope detection is not Gaussian since the correlation function consists of higher order terms than ρ^2 which cannot be ignored. Thus, the equivalent spectral spreading of the input Gaussian clutter spectrum prior to envelope detection is an equivalency only in terms of the value of the cancellation ratio. It was also found that the clutter attenuation for the two-pulse noncoherent canceler using an envelope detector is 3 dB better than the clutter attenuation for the coherent canceler in contrast to some prior conceptions.

In terms of improvement factors, it was found that the two noncoherent MTI's and the coherent MTI are nearly the same for the two-pulse canceler. As the number of pulses increases, the improvement factors become more unequal. The coherent MTI has the largest improvement factor, followed in order by the square-law and envelope detector MTI's.

REFERENCES

1. M.I. Skolnik, *Introduction to Radar Systems,* McGraw-Hill, New York, 1962.
2. F.E. Nathanson, *Radar Design Principles,* McGraw-Hill, New York, 1969.
3. B.D. Steinberg, in *Modern Radar Analysis, Evaluation and System Design,* R.S. Berkowitz, ed., John Wiley and Sons, New York, 1965, Chaps. 1 and 2, Part IV.
4. D.C. Schleher, ed., *MTI Radar,* ARTECH House, 1978.
5. R.C. Emerson, "Some Pulsed Doppler MTI and AMTI Techniques." (This paper appears in Ref. 4.)
6. D. Middleton, *An Introduction to Statistical Communication Theory,* McGraw-Hill, New York, 1960.
7. J.L. Lawson and G.E. Uhlenbeck, eds., *Threshold Signals,* MIT Radiation Lab. Series 24, McGraw-Hill, New York, 1950.
8. W.B. Davenport, Jr., and W.L. Root, *An Introduction to the Theory of Random Signals and Noise,* McGraw-Hill, New York, 1958.
9. W. Magnus and F. Oberhettinger, *Special Functions of Mathematical Physics,* Chelsea, New York, 1949.
10. M. Abramowitz and I.A. Stegun, eds., *Handbook of Mathematical Functions,* National Bureau of Standards, 1964.

Appendix

APPROXIMATE CLUTTER RESIDUES FOR A TWO-PULSE CANCELER USING VECTOR RELATIONS

Letting C_1 and C_2 denote independent complex Gaussian variables having 0 mean and a variance of σ^2 for the I and Q components, we make the linear transformation

$$C_1' = C_1$$

$$C_2' = \rho C_1 + (1 - \rho^2)^{1/2} C_2 = \rho C_1 + d C_2 \tag{A1}$$

where $d = (1 - \rho^2)^{1/2}$. Then C_1' and C_2' are also Gaussian random variables whose correlation is ρ and whose mean and variance are the same as for C_1 and C_2.

We next consider the residue vector of a two-pulse coherent MTI given by

$$R = C_1' - C_2'. \tag{A2}$$

These relations are shown in Fig. A1. The average residue power using Eqs. (A1) and (A2) is then

$$\overline{|R|^2} = (1 - \rho)^2 \overline{|C_1|^2} + d^2 \overline{|C_2|^2}.$$

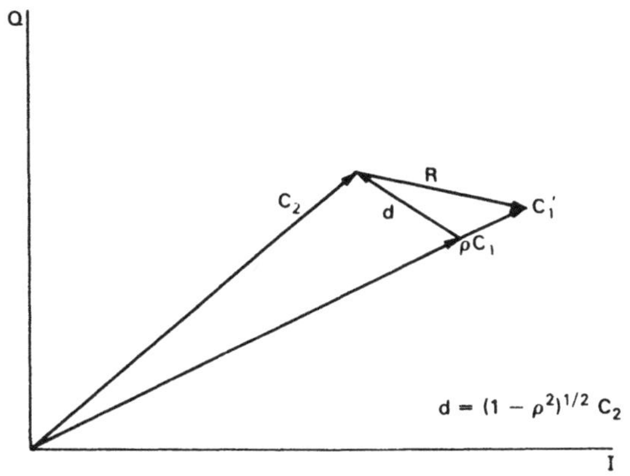

Fig. A1 — 2-pulse MTI vector relations

Letting $\rho = 1 - \epsilon$, it is seen that the term

$$(1 - \rho)^2 = [1 - (1 - \epsilon)]^2 = \epsilon^2$$

which is small compared to the term

$$d^2 = (1 - \rho^2) \approx 2\epsilon$$

so that

$$\overline{|R|^2} = 2\epsilon(2\sigma^2) = 4\sigma^2(1 - \rho) \tag{A3}$$

in agreement with Eq. (4).

For the noncoherent MTI the term dC_2 may be regarded as being composed of inphase and quadrature components relative to C_1' and since the orthogonal component contributes little to the amplitude of C_2', we approximate the vector length of C_2' as

$$\overline{|C_2'|} = \rho|C_1| + \frac{d}{\sqrt{2}}|C_2|$$

so that

$$\overline{R^2} = \overline{(|C_1'| - |C_2'|)^2} = \left[(1-\rho)^2 + \frac{d}{2}\right]^2 2\sigma^2$$

which, using the previous approximations may be written as

$$\overline{R^2} \approx \frac{d^2}{2}(2\sigma^2) \approx \epsilon 2\sigma^2. \tag{A4}$$

Comparison of Eqs. (A3) and (A4) shows that the residue clutter power of the two-pulse noncoherent MTI is 3 dB less than that for the coherent MTI.

MTI Weightings

Abstract

A comparison is made between optimum weighting, which maximizes P_d for a given α, and several MTI weightings. It is shown that a special case of the optimal weighting is approximately equal to the MTI weighting which maximizes the MTI improvement factor.

Introduction

For a Gaussian clutter and noise model, the weighting function which is most efficient in the Neyman-Pearson sense is given by the statistic [1]

Manuscript received September 7, 1973.

$$cK^{-1}V^* \tag{1}$$

where

c = constant
K = covariance matrix of interference consisting of clutter and noise
V^* = the conjugate of the known signal waveform vector.

This weighting, which corresponds to the matched filter in a colored noise situation, was also investigated by Ares [2] For the usual case where the Doppler frequency is unknown a priori, a bank of filters may be implemented using the weightings of (1) for various Doppler frequencies.

If one seeks the optimum linear filter which maximizes the ratio of the average signal-to-interference power at the end of the processing interval, the optimum weights for the known target are again given by (1) [1]. On the other hand, the optimum linear weighting for the averaged-Doppler frequency linear filter is provided by the eigenvector corresponding to the smallest eigenvalue of K. It is pointed out in [1] that this weighting optimizes the average S/C improvement factor, which is denoted by I, but it does not necessarily optimize the performance for a single filter in terms of detection and false-alarm probabilities. However, computation of the optimum weighting is difficult and no results are available for the staggered prf case.

In the following discussion it is shown that a particular case of the weighting given by (1) approximates the weights which optimize I. A comparison is made between these weightings and the binomial weightings, which are the coefficients of the expansion $(1 - x)^n$ for the uniform sampling case.

Uniform prf

It is assumed hereafter that the interference is clutter dominated, and that in (1),

$$V = [\exp(j2\pi f_d t_1), \exp(j2\pi f_d t_2), \cdots, \exp(j2\pi f_d t_n)] \tag{2}$$

where f_d is the Doppler frequency and t_i is the sampling time.

We first consider the uniform sampling case. For a target having a Doppler frequency equal to one half the repetition frequency, the consecutive values of V are alternately plus and minus 1, which is denoted by the vector S. The MTI response is a maximum and symmetric about this point for the binomial and optimum I weightings, which consist of real weights. For this Doppler frequency, the weighting of (1) is also real, and it is found that $K^{-1}S$ provides an excellent approximation to the weights which maximize I. That is, the improvement factor is approximately maximized for the weighting that maximizes the detection probability for a target assumed to be at the midband Doppler frequency or its harmonics.

For the simple two-pulse canceller, the binomial, maximum I, and $K^{-1}S$ weights are exactly the same, (1, −1), irrespective of the variance of the assumed Gaussian clutter spectrum. This is because there is only one corresponding zero in the Z plane and the optimum location is at $Z = 1$, which corresponds to zero Doppler frequency. Since the weightings are the same, the values of I are also the same for the two-pulse canceller. The results for the three-pulse canceller are summarized in Table I. The quantity σT represents the normalized clutter-width parameter, with σ being the standard deviation of the assumed Gaussian clutter spectrum and T being the interpulse interval. I_{bin} refers to the improvement factor obtained by using the binominal weights.

TABLE I

I in Decibels for the Three-Pulse Canceller

	σT					
	0.01	0.03	0.05	0.07	0.09	0.11
I_{max}	52.85	33.81	25.04	19.36	15.23	12.07
$I_{K^{-1}S}$	52.85	33.81	25.04	19.35	15.23	12.06
I_{bin}	51.1	32.13	23.48	17.96	14.03	11.07

Similar results are obtained for higher-order, uniform prf cancellers, where the agreement is excellent for the cases of interest. For the three-pulse canceller, there are two corresponding Z-plane zeroes. The zeroes corresponding to the binomial weights (1, −2, 1) are both at $Z = 1$. For small σT, the other two weightings are very nearly the same as the binomial weights, so that the corresponding zeroes are also nearly at $Z = 1$. For uniform sampling, the zeroes corresponding to the I_{max} weights lie on the unit circle [3], and the $K^{-1}S$ zeroes do also. As σT is increased, the zeroes corresponding to I_{max} and $K^{-1}S$ move symmetrically around the unit circle. As σT increases, the zeroes of the $K^{-1}S$ weighting begin to lead, or have a greater angle, than the I_{max} zeroes. For large σT, the $K^{-1}S$ zeroes approach 60 degrees, and those of I_{max} approach 45 degrees. At 60 degrees the $K^{-1}S$ weights are ju[illegible] ti[illegible]es of S, which corresponds to the matched filt[illegible] a white noise situation.

Staggered prf

For the staggered prf case, the results from an arbitrarily selected 5: 6: 7 stagger pattern are presented. The zeroes for a staggered prf correspond to the roots of a polynomial whose degree is equal to the sum of the stagger values. For this example, the zeroes are the roots of an 18th degree polynominal which are, in general, complex. The Doppler, f_b, corresponding to the first blind speed is determined from

$$f_b T_i = k_i \tag{3}$$

where T_i is the ith interpulse interval and k_i is the ith number in the stagger pattern.

The midband Doppler frequency is located at one-half of f_b and the response is symmetric about this midband Doppler frequency with a periodicity of f_b Hz. However, it is next shown that the maximum response does not, in general, occur at the midband Doppler, but rather at one half the Doppler frequency associated with the average interpulse period $\bar{T}$.

The S/C gain, $G(f_d)$, is given by [4]

$$G(f_d) = \frac{\sum_i w_i^2}{\sum_i \sum_j w_i w_j \rho_{ij}} \cdot \frac{\sum_i \sum_j w_i w_j \cos 2\pi f_d (t_i - t_j)}{\sum_i w_i^2} \tag{4}$$

where w_i is the ith weight, ρ_{ij} is the clutter covariance function evaluated at $(t_i - t_j)$, and all indices range from 1 to n. The first term in (4) is the same as I, while the second term is the normalized frequency response function, which is denoted $E(f_d)$. In abbreviated form, we have

$$G(f_d) = IE(f_d). \tag{5}$$

$E(f_d)$ may be expressed as

$$E(f_d) = \frac{1}{\sum w_i^2} \cdot \sum_i \sum_j w_i w_j \cos [2\pi f_d \bar{T}(t_i - t_j)/\bar{T}] \tag{6}$$

and, for small stagger, (6) may be approximated by

$$E(f_d) = \frac{1}{\sum w_i^2} \sum_i \sum_j w_i w_j \cos 2\pi f_d \bar{T}(i - j). \tag{7}$$

At $f_d = (1/2\bar{T})$, (7) may be written as

$$E\left(\frac{1}{2\bar{T}}\right) = \frac{1}{\sum w_i^2} \sum_i \sum_j w_i w_j (-1)^{i-j}. \tag{8}$$

But, since the sign of w_i alternates, we have that

$$E\left(\frac{1}{2\bar{T}}\right) = \frac{\left(\sum |w_i|\right)^2}{\sum w_i^2} \tag{9}$$

which is also a maximum. The approximation utilized in the determination of (9) is based on a small degree of staggering, but (9) is exact for the unstaggered case.

TABLE II

I in Decibels for a Staggered Four-Pulse Canceller

	$\sigma\bar{T}$					
	0.01	0.03	0.05	0.07	0.09	0.11
I_{max}	77.61	49.05	35.90	27.40	21.25	16.60
$I_{K^{-1}S}$	77.61	49.05	35.90	27.39	21.25	16.58
$I_{K^{-1}V}$	77.42	47.65	32.93	23.10	16.11	11.15
I_{bin}	73.65	45.26	32.43	24.35	18.68	14.50

The response at midband is obtained by use of (3) in the expression for $E(f_d)$ given by the second term in (4). In particular, we have for this cosine part of the expression

$$\cos 2\pi f_d (t_i - t_j) = \cos 2\pi f_d \sum_{n=i}^{j} T_n$$

$$= \cos \pi f_b \sum_{n=i}^{j} T_n$$

$$= \cos \pi \sum_{n=i}^{j} k_n$$

$$\begin{aligned} &1 \text{ for } \sum k_n \text{ even} \\ &= -1 \text{ for } \sum k_n \text{ odd.} \end{aligned} \tag{10}$$

Since this term does not always have the correct polarity to correct the alternating sign of the weights, the midband gain is not generally a maximum.

At $f_d = (1/2\bar{T})$, the successive phase shifts are nearly 180 degrees for a staggered prf. Using the same weighting function $K^{-1}S$ again provided excellent agreement with the I_{max} weighting. On the other hand, at the midband Doppler frequency, V of (1) is (1, −1, −1, 1) for the assumed stagger. Evaluation and utilization of $K^{-1}V$ weights for this case did not agree well with the I_{max} weightings, except for small values of the clutter width parameter $\sigma\bar{T}$. Some of the results are tabulated in Table II to provide a comparison. The entry I_{bin} refers to the improvement factor obtained by using weighting which is a generalization of the binomial weights. These weights have the property of being able to perfectly cancel samples from an $(n-2)$-degree polynomial (also see [2]).

Summary

It is shown how, for interference dominated by clutter, optimization of P_d versus α for a target assumed to be at the center of the average prf frequency results in a weighting which is approximately equal to the weighting required for maximizing the MTI improvement factor.

FRANK F. KRETSCHMER
Naval Research Lab.
Washington, D.C. 20375

References

[1] L.E. Brennan and I.S. Reed, "Optimum processing of unequally spaced radar pulse trains for clutter rejection," *IEEE Trans. Aerospace and Electronic Systems*, vol. AES-4, pp. 474-477, May 1968.

[2] M. Ares, "Some anticlutter waveforms suitable for phased array radars," General Electric Heavy Military Electronics Dept., Syracuse, N.Y., Rept. R66EMH8, November 1965.

[3] E.A. Robinson, "Optimum weighting functions for the detection of sampled signals in noise," *IEEE Trans. Information Theory*, vol. IT-11, p. 452, July 1965.

[4] R.C. Emerson, "Some pulsed Doppler MTI and AMTI techniques," RAND Corp., Rept. R-274, March 1954; ASTIA Doc. 65881.

Correlation Effects of MTI Filters

Abstract

A frequency domain approach for determining the effective number of independent pulses at the output of a square law detector in a moving target indicator (MTI) system leads to a definition which gives the same asymptotic results as given by a previous definition.

Introduction

It has been previously shown [1], [2] that the effect of a moving target indicator (MTI) is to cause a loss in signal-to-noise ratio or in detection probability. In [1] Hall and Ward give an expression for the effective number of independent pulses which result from postdetection integration of pulses which have been subjected to MTI filtering. They define the number of effective independent pulses in terms of the ratio of the reduced normalized noise variance as

$$N_e = [\sigma^2/m^2]_{\text{in}}/[\sigma^2/m^2]_{\text{out}} \qquad (1)$$

where σ and m correspond to the standard deviation and mean of the input or integrated output pulse. In this correspondence another definition is stated in terms of the correlation of the output noise caused by the MTI filter. It is

Manuscript received May 27, 1976; revised October 1, 1976.

observed that this definition is analogous to the definition of the aperture efficiency of an antenna.

Frequency Domain Approach

For an $(n + 1)$ pulse MTI canceller having binomial weights, the power spectrum of the MTI filter is given as [3]

$$H(f_d) = 2^{2n} \sin^{2n} [\pi f_d / f_r] \tag{2}$$

where f_d is the Doppler frequency and f_r is the pulse repetition frequency.

Let us compare the correlation time of this filter with that of unfiltered white noise over the same interval. Since the periodicity does not contribute any information, let us regard only the unrepeated power spectrum over one repetition interval. Moreover, let us redefine the frequency scale so that the MTI power spectrum is defined relative to the center of the band and let us also ignore the unimportant constant associated with (2). In essence, we wish to compare the correlation times associated with the spectral densities given by

$$H(f) = \cos^{2n} [\pi f / f_b] \tag{3}$$

$$G(f) = \text{rect} [f / f_b] \tag{4}$$

where f_b is the frequency bandwidth of interest.

We define the relative number of independent pulses as the ratio of the correlation time of the band-limited white noise spectrum to the correlation time of the MTI filter spectrum. The correlation functions associated with (3) and (4) are determined by taking the Fourier transform of these functions. However, the determination of the effective number of independent pulses as described above may be associated with a similar calculation in antenna theory. The directive gain of an antenna in one dimension is given by [4]

$$G = 2\pi P(\theta)_{max} / \int P(\theta)\, d\theta = 2\pi / B \tag{5}$$

where $P(\theta)$ is the antenna radiated power pattern, the Fourier transform of aperture distribution, and $B = \int P(\theta) d\theta / P(\theta)_{max}$, effective rectangular beamwidth.

The relative gain or aperture efficiency of an antenna is defined as the gain of a given antenna relative to that of an antenna having a uniform aperture distribution. This is the parallel of determining the ratio of the correlation times corresponding to the band-limited white noise spectrum and the MTI filter spectrum. Regarding $H(f)$ in (3) as the aperture distribution and defining the correlation time as the effective rectangular width of the correlation function, the aperture efficiency for such a distribution is given by Silver [5] as

$$L = \{[1 \cdot 3 \cdot 5 \cdots (m-1)] / [2 \cdot 4 \cdot 6 \cdots m]\} \cdot \{[(m+2)(m+4) \cdots 2m] / [(m+1)(m+3) \cdots (2m-1)]\} \tag{6}$$

where $m = 2n$, which is equal to twice the number of MTI delays. This formulation is found to give identical asymptotic results as [1] for S given by

$$S = N_e/N = N / [N + 2 \sum_{k=1}^{N-1} (N-k) R^2(k)] \xrightarrow{\text{large } N} 1/[1 + 2 \sum_{k=1}^{N-1} R^2(k)] \tag{7}$$

where N is the number of pulses integrated after detection and $R(k)$ is the normalized correlation between MTI output pulses, separated by k pulses. For example, for a four-pulse binomial weight MTI, the weights are $(1, -3, 3, -1)$. From this it is found that $R(1) = -15/20$, $R(2) = 6/20$, $R(3) = -1/20$, where the value of 20 in the denominators is the normalization factor obtained by summing the squares of the weights. Substituting in (7) gives $S \to 100/231$. Since the four-pulse canceller has three delays, $m = 6$ in (6) and

$$L = [(1 \cdot 3 \cdot 5)/(2 \cdot 4 \cdot 6)]\,[(8 \cdot 10 \cdot 12)/(7 \cdot 9 \cdot 11)] = 100/231.$$

Summary

A definition for the relative number of independent pulses at the output of an MTI followed by a square law detector is given in terms of the ratio of the correlation times associated with the band limited white-noise power spectrum and the MTI power spectrum. This ratio of the two correlation times gives a measure of the increase in correlation time of the noise due to the MTI. It is noted that this definition is analogous to the aperture efficiency of an antenna and that an MTI with binomial weights may be associated with an aperture distribution of the form $\cos^{2n} (\pi f / f_b)$. The aperture efficiency expression for this distribution [5] is found to give the same asymptotic results as [1].

Acknowledgment

The author thanks W.K. Saunders of the Naval Research Laboratory for stimulating the frequency domain approach to the problem.

FRANK F. KRETSCHMER, JR.
Radar Division
Naval Research Laboratory
Washington, D.C. 20375

References

[1] W.M. Hall and H.R. Ward, "Signal-to-noise loss in moving target indicator," *Proc. IEEE.* pp. 233-234, Feb. 1968.

[2] G.M. Dillard, "Signal-to-noise loss in an MTI cascaded with coherent integration filters," *IEEE Int. Radar Conf.*, pp. 117-119, 1975.

[3] F.E. Nathanson, *Radar Design Principles.* New York: McGraw-Hill, 1969.

[4] M.I. Skolnik, *Introduction to Radar Systems.* New York: McGraw-Hill, 1962.

[5] S. Silver, *Microwave Antenna Theory and Design* (M.I.T. Rad. Lab. Ser.), vol. 12. New York: McGraw-Hill, 1949, table 6.1, p. 187.

MTI NOISE INTEGRATION LOSS

INTRODUCTION

MTI signal processing correlates the receiver noise and thus results in degraded detection performance when the MTI pulses are integrated. Previous investigators [1,2] have described the decreased performance in terms of a reduction in the effective number of independent pulses integrated. However, since the effective number of pulses N_e can be represented by

$$N_e = \frac{(\sigma^2/m^2)_{\text{in}}}{(\sigma^2/m^2)_{\text{out}}},$$

where σ and m are the standard deviation and mean of the input samples, N_e has a precise meaning (in terms of detection performance) only if the output noise distribution is completely specified by N_e. For instance, when the number of pulses integrated (N) is large, the integrated output is approximately Gaussian distributed and integration improvement varies as the square root of the number of pulses integrated. Thus the loss (due to the MTI correlating the receiver noise) in signal-to-noise ratio (S/N) for a large number of integrated pulses is

$$L = 10 \log (N/N_e)^{1/2}.$$

In this report the MTI integration loss is calculated when the number of integrated pulses is small and thus the output is not Gaussian distributed. This calculation is performed using simulation techniques. First, the appropriate thresholds for a given probability of false alarm P_{fa} are calculated using importance-sampling techniques. Next, probability of detection P_D curves are generated by simulation of the pulse-by-pulse video. Finally, the MTI integration loss is found by comparing the generated P_D curves with those for independent samples [3].

FALSE-ALARM THRESHOLDS

Although Monte-Carlo simulations have been used for many years to calculate P_D curves, they have not been used to calculate P_{fa} curves because of the enormous number of repetitions usually required: approximately $10/P_{fa}$. However this difficulty can be overcome by using importance sampling [4]. The fundamental principle of the importance-sampling technique is to modify the probabilities that govern the outcome of the basic experiment of the simulation in such a way that the event of interest (the false alarm) occurs more frequently. This distortion is then compensated for by weighting each event by the ratio of the probability that this specific event would have occurred if the true probabilities had been used in the simulation to the probability that this same event would occur with the distorted probabilities. Consequently by proper choice of the distorted

Manuscript submitted May 18, 1977.

probabilities the number of repetitions can be reduced greatly. For instance, the mean of a function $Q(x)$ is given by

$$E\{Q(x)\} = \int Q(x)\,dP(x),$$

where $P(x)$ is the distribution of x. The mean of $Q(x)$ can be estimated by selecting M independent samples x_i from $P(x)$ and associating the probability $1/M$ with each event. Then $E\{Q(x)\}$ can be estimated by

$$\frac{1}{M}\sum_{i=1}^{M} Q(x_i). \tag{1}$$

The importance-sampling technique uses the Radon-Nikodyn derivative to express the mean value of $Q(x)$ by

$$E\{Q(x)\} = \int Q(x)\frac{dP(x)}{dG(x)}\,dG(x),$$

where $G(x)$ is a distribution function. The mean $E\{Q(x)\}$ can be estimated by selecting M independent samples from $G(x)$ and associating the probability $dP(x_i)/MdG(x_i)$ with each event $Q(x_i)$. Thus $E\{Q(x)\}$ is estimated by

$$\frac{1}{M}\sum_{i=1}^{M} Q(x_i)\frac{dP(x_i)}{dG(x_i)}. \tag{2}$$

Since (1) and (2) are both unbiased estimates of $Q(x)$, it is possible to select $G(x)$ so that the variance of (2) is less than the variance of (1).

In our problem of determining the threshold for a given P_{fa}, when MTI samples are noncoherently integrated, it is necessary to estimate the distribution curve

$$P(Z_j \leqslant T) = 1 - P_{fa}, \tag{3}$$

where

$$Z_j = \sum_{i=1}^{N} Z_{ij}, \tag{4}$$

in which

$$Z_{ij} = \left[(x_{ij}'^2 + y_{ij}'^2)/P(k)\right]^{1/2} \tag{5}$$

where, for a two-pulse MTI,

$$x'_{ij} = x_{ij} - x_{i-1,j} \tag{6}$$

and

$$y'_{ij} = y_{ij} - y_{i-1,j}, \tag{7}$$

with x_{ij} and y_{ij} being independent Gaussian variables with zero mean and a variance of σ and $P(k)$ being the noise power out of a k-pulse MTI: $P(2) = 2$, $P(3) = 6$, $P(4) = 20$, and $P(5) = 70$. The straightforward way of estimating (7) is to generate Gaussian samples by

$$x_{ij} = \sigma(-2 \ln u_{ij})^{1/2} \sin 2\pi v_{ij} \tag{8}$$

and

$$y_{ij} = \sigma(-2 \ln u_{ij})^{1/2} \cos 2\pi v_{ij}, \tag{9}$$

with u_{ij} and v_{ij} being independent random numbers uniformly distributed on the interval (0,1). To estimate (3), M independent sums $\{Z_j, j = 1, M\}$ are formed using (4) through (7), and the estimated distribution is

$$\hat{P}(Z \geqslant T) = \frac{1}{M} \sum_{j=1}^{M} \delta_j,$$

where

$$\delta_j = 1, \quad Z_j \geqslant T,$$
$$= 0, \quad Z_j < 0.$$

Importance sampling differs from the previous procedure by generating samples using

$$x_{ij} = \alpha(-2 \ln u_{ij})^{1/2} \sin 2\pi v_{ij} \tag{10}$$

and

$$y_{ij} = \alpha(-2 \ln u_{ij})^{1/2} \cos 2\pi v_{ij}, \tag{11}$$

where $\alpha > \sigma$, a device which yields more false alarms. Using (10) and (11) and (4) through (7), M sums Z_j are generated. Then the estimated distribution is

$$\hat{P}(Z \geqslant T) = \frac{1}{M} \sum_{j=1}^{M} \delta_j P_j,$$

where

$$\delta_j = 1, \quad Z_j \geqslant T,$$
$$= 0, \quad Z_j < 0,$$

and

$$P_j = \prod_{i=2-k}^{N} \frac{\frac{1}{2\pi\sigma^2} e^{-(x_{ij}^2 + y_{ij}^2)/2\sigma^2}}{\frac{1}{2\pi\alpha^2} e^{-(x_{ij}^2 + y_{ij}^2)/2\alpha^2}}.$$

With use of $\alpha = 2.0$ and $M = 20{,}000$ for $N = 4$, $\alpha = 1.7$ and $M = 10{,}000$ for $N = 8$, $\alpha = 1.5$ and $M = 10{,}000$ for $N = 16$, and $\alpha = 1.3$ and $M = 2500$ for $N = 32$, threshold curves were generated for two-, three-, four-, and five-pulse (binary weighting) MTIs and are shown in Fig. 1. The reference curve for independent samples was generated using detection curves in Robertson [3].

PROBABILITY OF DETECTION

Since the S/N out of the MTI is a function of the target doppler, the doppler frequency where the input and output S/N are equal will be used. The S/N gain (or loss) provided by the k-pulse MTI is

$$\frac{\left(\sum_{i=1}^{k} a_i \cos i\phi_k\right)^2 + \left(\sum_{i=1}^{k} a_i \sin i\phi_k\right)^2}{\sum_{i=1}^{k} a_i^2}, \tag{12}$$

where $\{a_i, i = 1, \ldots, k\}$ are the MTI coefficients and ϕ is the change in target phase between successive PRFs. Setting (12) equal to 1 and solving for ϕ_k yields the solutions $\phi_2 = 90°$, $\phi_3 = 103°$, $\phi_4 = 110.9°$, and $\phi_5 = 116.5°$.

Thus the P_D for a k-pulse MTI and a given P_{fa} can be found by generating sample video using

$$x_{ij} = \sigma(-2 \ln u_{ij})^{1/2} \sin 2\pi v_{ij} + A \sin i\phi_k \tag{13}$$

and

$$y_{ij} = \sigma(-2 \ln u_{ij})^{1/2} \cos 2\pi v_{ij} + A \cos i\phi_k, \tag{14}$$

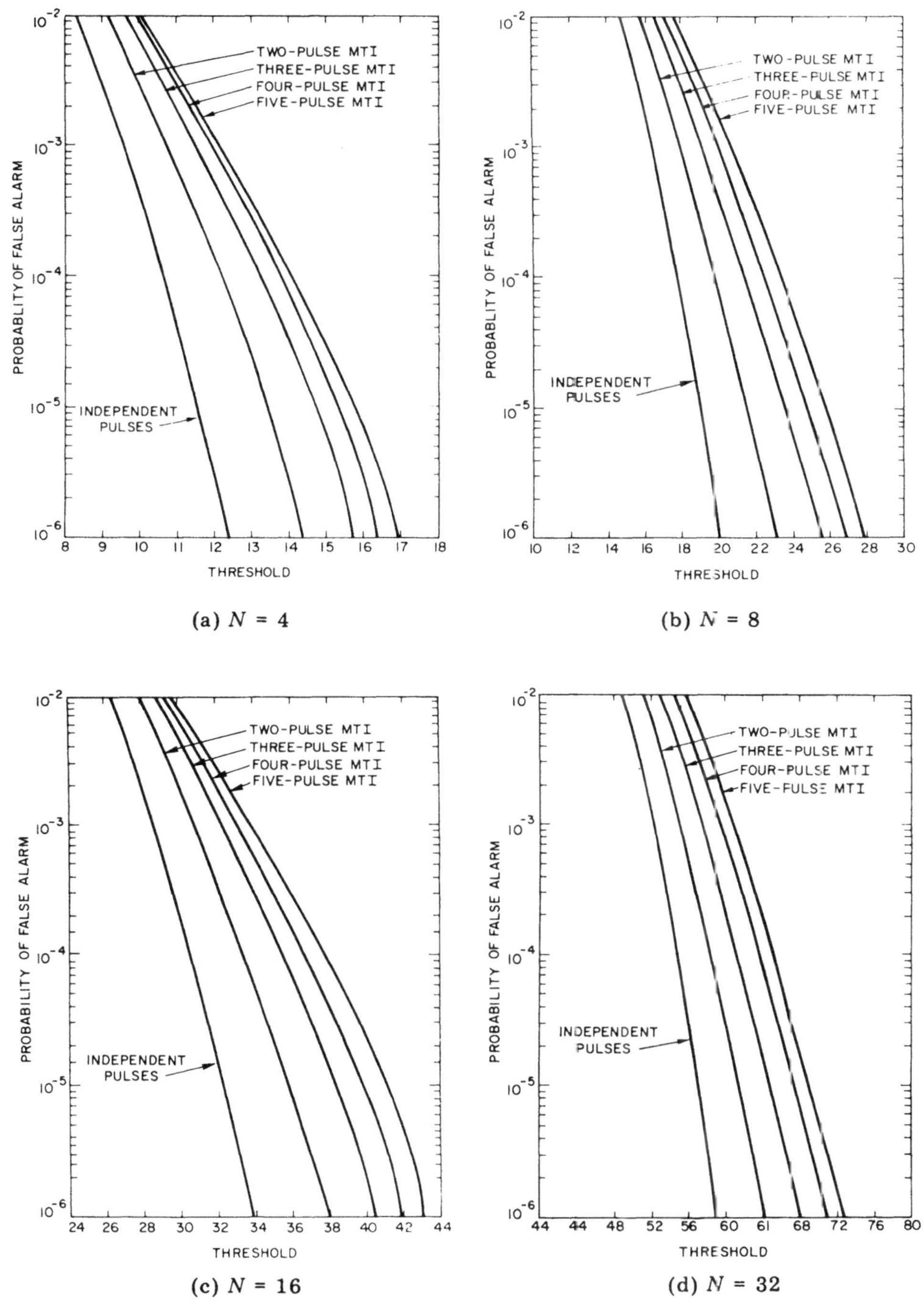

Fig. 1 — Threshold curves for N pulses integrated

where S/N(dB) = 10 log $(A^2/2\sigma^2)$. By use of (13) and (14) and (3) through (7), M = 1024 Z_j values were generated for each S/N and compared to the appropriate threshold. The P_D curves for $P_{fa} = 10^{-6}$ are shown in Fig. 2.

The difference between the P_D curves for the various MTIs and the curve for independent pulses is the MTI noise integration loss. This loss is given in Table 1 for the P_D and P_{fa} values indicated. The loss appears to be fairly independent of both N, the number of pulses integrated, and P_{fa}.

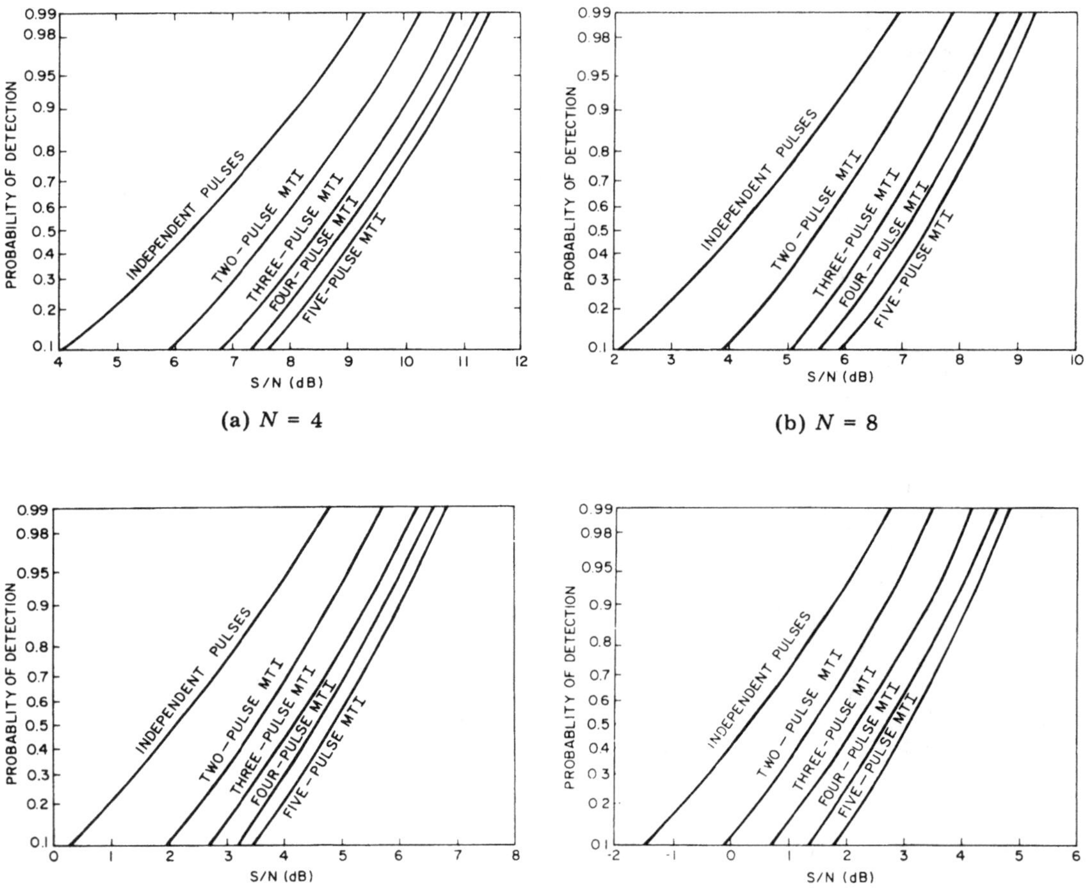

(a) N = 4

(b) N = 8

(c) N = 16

(d) N = 32

Fig. 2 — Probability of detection curves for N pulses integrated with $P_{fa} = 10^{-6}$

Table 1 — MTI Noise Integration Loss for $P_D = 0.9$ and N Noncoherent Pulses Integrated

MTI Pulses	Loss (dB)				Average Difference (dB)
	$N = 4$	$N = 8$	$N = 16$	$N = 32$	
$P_{fa} = 10^{-6}$					
Two	1.1	1.1	1.1	0.9	1.0
Three	1.8	1.9	1.7	1.7	1.8
Four	2.2	2.4	2.1	2.1	2.2
Five	2.5	2.7	2.3	2.4	2.5
$P_{fa} = 10^{-4}$					
Two	1.1	0.9	0.9	0.8	0.9
Three	1.8	1.7	1.6	1.5	1.6
Four	2.1	2.1	1.9	1.9	2.0
Five	2.3	2.5	2.1	2.2	2.3

COMPARISON WITH PREVIOUS RESULTS

The number of effective pulses integrated for a k-pulse MTI is given [1] by

$$N_e(k) = \frac{N^2}{N + 2 \sum_{J=1}^{N-1} (N - j) R_k^2(j)},$$

where $R_k(j)$ is the correlation coefficient

$$R_k(j) = \frac{E\{x_i' \, x_{i+j}'\}}{P(k)}.$$

Thus, to find the MTI noise integration loss, the difference must be found between the required S/N for N_e and N independent pulses. To accomplish this, a curve of S/N versus N for $P_D = 0.9$ and $P_{fa} = 10^{-6}$ was generated using the detection curves in Robertson [3] and is shown in Fig. 3. From this curve the MTI noise integration loss was calculated and is shown in Table 2. These losses are about 0.2 dB higher than the corresponding losses in Table 1.

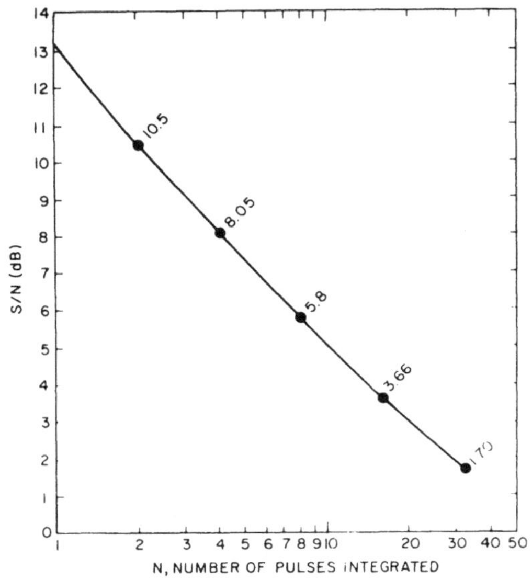

Fig. 3 — S/N for P_D = 0.9 and P_{fa} = 10^{-6} as a function of the number of independent pulses integrated

Table 2 — MTI Noise Integration Loss Using the Effective Number of Pulses N_e Integrated for P_D = 0.9 and P_{fa} = 10^{-6}

MTI Pulses	Loss (dB)				Average Difference (dB)
	$N = 4$	$N = 8$	$N = 16$	$N = 32$	
Two	1.1	1.2	1.2	1.1	1.1
Three	1.8	1.7	1.9	1.8	1.8
Four	2.4	2.6	2.4	2.3	2.4
Five	2.7	2.9	2.9	2.7	2.8

SUMMARY

MTI signal processing correlates the receiver noise, and this results in an MTI noise integration loss. The losses for two-, three-, four-, and five-pulse MTIs are approximately 1.0, 1.8, 2.2, and 2.5 dB respectively. The P_D for a given target can be found using the following procedure:

1. Calculate the input S/N (to the MTI) using the radar range equation;
2. Calculate the output S/N from the MTI using (12)
3. Use Fig. 2 to determine P_D or else assume all N pulses are independent, reduce S/N by the MTI noise integration loss, and find P_D from standard detection curves such as given in Robertson [3].

REFERENCES

1. W.M. Hall and H.R. Ward, "Signal-to-Noise Loss in Moving Target Indicator," IEEE Proceedings **56** (No. 2), 233-234 (Feb. 1968).

2. F.F. Kretschmer, Jr., "Correlation Effects of MTI Filters," IEEE Trans Aerospace and Electronic Systems **AES-13** (No. 3), 321-322 (May 1977).

3. G.H. Robertson, "Operating Characteristics For a Linear Detector of CW Signals in Narrow-Band Gaussian Noise," Bell Sys. Tech. J. **46**, 755-774 (Apr. 1967).

4. V.G. Hansen, "Detection Performance of Some Nonparametric Rank Tests and an Application To Radar," IEEE Trans. Inform. Theory **IT-16** (No. 3), 309-318 (May 1970).

PERFORMANCE OF CASCADED MTI AND COHERENT INTEGRATION FILTERS IN A CLUTTER ENVIRONMENT

I. INTRODUCTION

Recent advances in velocity filtering for radar systems have been made possible mainly through the use of digital signal processing. The detection of moving targets in the presence of strong returns from fixed objects (clutter) has classically been accomplished with moving target indicators (MTI) or with pulse doppler radars (Chap. 17 and 19, Ref. 1). The flexibility of digital signal processing allows improved implementation of both techniques, as well as processing configurations which make use of the advantages of both.

MTI is normally used with low-pulse-repetition-frequency (low-PRF) radars. The MTI filter rejects clutter by means of a notch in its passband centered on the clutter doppler spectrum. A single output provides moving target detection over the remaining doppler spectrum between the PRF harmonics. Pulse doppler is usually associated with high-PRF radars. A contiguous bank of narrow-band filters is used to detect moving targets outside of the clutter spectrum.

A pulse doppler radar can usually achieve a greater improvement in the signal-to-clutter ratio than can an MTI. This improvement is needed to contend with the higher clutter levels caused by clutter foldover at the range ambiguities of the high-PRF pulse doppler radar. The increased clutter rejection results from the high PRF and from the larger number of pulses processed by the pulse doppler radar.

McAulay (2) formulated the MTI problem as a classical detection problem. He determined the optimum process by maximizing the resulting likelihood ratio and showed that the optimum receiver structure could be interpreted as a clutter filter in cascade with a narrow-band doppler filter bank. This arrangement has the advantage that the dynamic range at the input to the narrowband processor is greatly reduced by the clutter rejection of the clutter filter.

A practical approximation to this processor consists of a conventional MTI filter cascaded with a coherent integrator formed by a contiguous bank of narrow-band filters. The coherent integrator has been implemented by using the Fast Fourier Transform (FFT) algorithm (3,4). The processing gain, i.e., the improvement in signal-to-clutter ratio that can be achieved by this configuration, is considered in this report.

The improvement factor that can be achieved using an MTI alone, and then a coherent integrator alone, is presented first. This gives a basis for comparing the advantages of cascading the two filters. A Gaussian clutter spectrum is assumed in computing the improvement factors. Although not completely accurate, this assumption is generally made in the MTI literature and it has been found to give a reasonable prediction of MTI performance.

In general, the number of pulses that can be coherently processed is limited by the number of pulses that are transmitted during the antenna dwell time. This number can range from one or two to several hundred, depending on the particular radar application. In practice, the number of pulses may be constrained by the cost, weight, or size of the processor hardware. In this report, the MTI filters are considered to be single, double, or triple cancellers. The coherent integrators are considered to be 8- or 16-pulse FFTs; the weights used to shape the FFT filter response are either uniform (all weights equal) or Tschebyschev (-25 dB sidelobes). These choices seem to be typical of the processors that might be considered for many low-PRF radars. However, the computer programs used to evaluate these processors are neither restricted by these choices nor by the assumed Gaussian clutter spectrum.

II. GENERALIZED PROCESSOR MODEL

Matrix Equivalent

Both the MTI and the coherent integrator can be modeled as a transversal filter of the type shown in Fig. 1. A tapped delay line (or multiple delay lines) is used to provide $n+1$ time samples of the input $x(t)$. The output $y(t)$ consists of a weighted sum of these samples:

$$y(t) = W_1 xt + W_2 x(t - T) + \ldots + W_{n+1}\, x(t - nT) \tag{1}$$

where T is the time between samples. $x(t)$, $y(t)$, and the weights W_i may be complex.

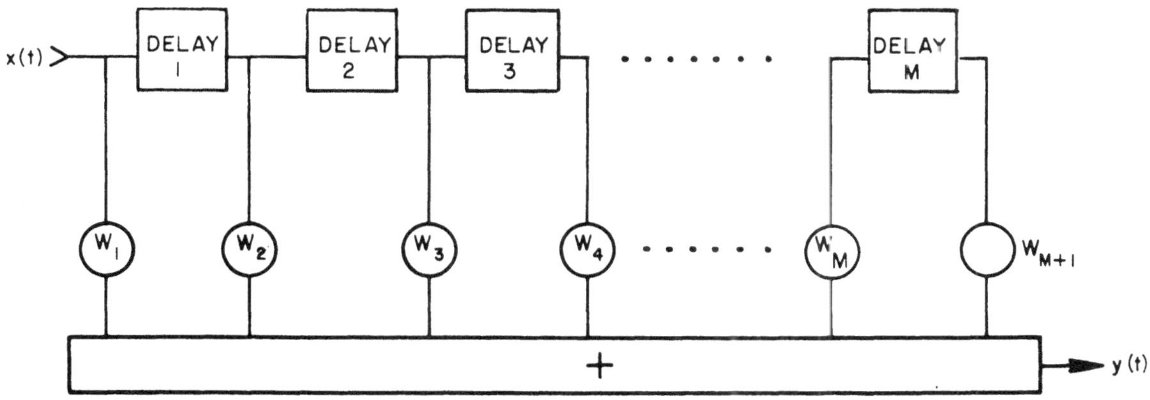

Fig. 1 — Generalized model for an MTI or coherent integrator processor consisting of a single filter and a single output

For purposes of analysis, it is convenient to write Eq. (1) as a matrix operation. Therefore, define W and X as column vectors with elements given by

$$W = \begin{bmatrix} W_1 \\ W_2 \\ W_3 \\ \cdot \\ \cdot \\ \cdot \\ W_{n+1} \end{bmatrix} \quad \text{and} \quad X = \begin{bmatrix} x_1 \\ x_2 \\ x_3 \\ \cdot \\ \cdot \\ \cdot \\ x_{n+1} \end{bmatrix}$$

where W_i is the ith weight shown in Fig. 1, and

$$\begin{aligned} x_1 &= x(t) \\ x_2 &= x(t\text{-}T) \\ x_3 &= x(t - 2T) \\ &\cdot \\ &\cdot \\ &\cdot \\ x_{n+1} &= x(t\text{-}nT). \end{aligned}$$

With these matrices, Eq. (1) can be written as the product of the transpose of W, W_T, times X:

$$y = W_T X.$$

The output power, as shown in Ref. 5, is given by

$$\begin{aligned} P &= |y|^2 = yy^* = yy_T^* \\ &= W_T X (W_T X)_T^* \\ &= W_T X X_T^* W \end{aligned} \tag{2}$$

where use has been made of the fact that

$$y_T = y.$$

The output signal-to-clutter ratio can be computed by taking the ratio of Eq. (2) when $x(t)$ is the input signal to Eq. (2) when $x(t)$ is the input clutter.

Expected Output Clutter Power

The expected output clutter power $\overline{P}_c$ is found from the expected value of Eq. (2):

$$\overline{P}_c = E\left\{W_T X X_T^* W^*\right\} = W_T \overline{X X_T^*} W^*. \tag{3}$$

Letting

$$M_c = \overline{X X_T^*}$$

Then

$$\overline{P}_c = W_T M_c W^* \tag{4}$$

where M_c, the covariance matrix of the clutter, can be derived from the Fourier transform of the normalized clutter power density spectrum. For example, using a Gaussian clutter spectrum yields

$$P_c(f) = \frac{1}{\sqrt{2\pi}\,\sigma_c} e^{-\frac{1}{2}\left(\frac{f-\mu_c}{\sigma_c}\right)^2}. \tag{5}$$

The mean μ_c implies relative motion between the radar platform and the clutter. The standard deviation σ_c of the clutter spectrum is a measure of the bandwidth of the clutter spectrum.

The Fourier transform of Eq. (5) is

$$\psi(\tau) = \exp\left(-2\pi^2\sigma_c^2\,\tau^2 - j\,2\pi\,\mu_c\,\tau\right).$$

Assuming stationarity, the k,l element of M_c is

$$M_c\,(k,l) = \exp[-2\pi^2\sigma_c^2\,(k-l)^2 T^2 - j2\pi\,\mu_c\,(k-l)\,T]. \tag{6}$$

Expected Output Signal Power

Letting the input $x(t)$ correspond to a signal input, the expected output signal power P_s is found in the same way as above for the clutter:

$$P_s = W_T M_s W^* \tag{7}$$

where M_s is the covariance matrix for the signal. The spectrum of a target is affected by such things as the target's characteristics, transmitted waveform, radar stability, antenna scanning, and relative velocity. The received signal waveform $r(t)$ is given by

$$r(t) = s(t)\,e^{j\omega_d t} \tag{8}$$

where $s(t)$ includes all the above effects, except relative velocity, and ω_d is the doppler frequency corresponding to the relative velocity between radar and target. Since no apriori knowledge is assumed about ω_d, it is given a uniform probability distribution ($0 \leqslant \omega_d \leqslant$ PRF).

If the doppler shift is the dominant effect, i.e., the bandwidth of the other effects are small relative to the doppler shifts expected, then Eq. (8) is approximately

$$r(t) = e^{j\omega_d t}. \tag{9}$$

The autocorrelation of Eq. (9) is

$$\psi(\tau) = e^{j\omega_d \tau}. \tag{10}$$

The processor shown in Fig. 1 depicts only one filter and one output. However, in general, multiple independent outputs are possible by selecting appropriate weight vectors. In particular, the FFT weights form n+1 filters with n+1 outputs. The signal gain of a particular filter then will be the expected gain, provided that ω_d is such that a signal exists somewhere between the crossover points of adjacent filters. This is shown in Fig. 2 for an 8-pulse FFT.

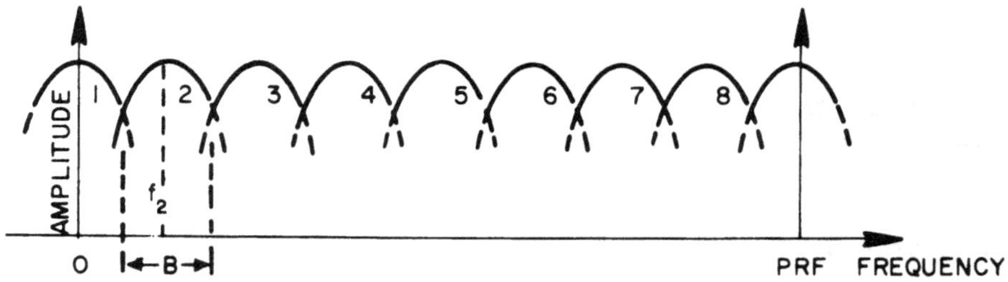

Fig. 2 — Signal amplitude vs PRF for a multiple-filter, multiple-output integrator. The case illustrated here is for an 8-pulse FFT.

From Fig. 2, the gain of the second filter is derived by considering a target doppler f_d that can occur with equal probability in the region

$$f_2 - (B/2) \leqslant f_d \leqslant f_2 + (B/2)$$

where f_2 is the center of the response of the second filter.

The gain of the ith filter is derived in the same way by considering the region

$$f_i - (B/2) \leqslant f_d \leqslant f_i + (B/2).$$

The probability density function for f_d associated with the ith filter is

$$P(f_d) = \begin{cases} 1/B, \text{ for } (f_i - B/2) \leqslant f_d \leqslant (f_i + B/2 \\ \\ 0, \text{ elsewhere.} \end{cases}$$

The expected value of Eq. (10) is

$$E\left\{\psi(\tau)\right\} = \int_{-\infty}^{\infty} e^{j2\pi f_d \tau} P(f_d)\, df_d = e^{j2\pi f_i \tau} \frac{\sin(\pi B\tau)}{\pi B\tau}. \tag{11}$$

Assuming stationarity, the k,l element of the signal covariance matrix M_s is given by

$$m_s(k,l) = \exp\,[j2\pi\, f_i(k - l)\, T]\, \frac{\sin\,[\pi B(k - l)\, T]}{\pi B(k - l)\, T}. \tag{12}$$

Improvement Factor

The processing gain of any arbitrary transversal filter specified by a weight vector W can be computed by taking the ratio of the signal output power [Eq. (7)] to the clutter output power [Eq. (4)]. The processing gain, or improvement factor I, is

$$I = \frac{W_T\, M_s\, W^*}{W_T\, M_c\, W^*}. \tag{13}$$

The matrix M_s is generated from Eq. (12), and M_c is generated from Eq. (6).

III. MTI FILTERS

A simplified diagram of an n-stage MTI is shown in Fig. 3. This diagram is equivalent to the model in Fig. 1 if the weights are given by the binominal coefficients with alternating signs, i.e.,

$$w_i = (-1)^{i-1} \binom{n}{i-1},\ i = 1, 2, \ldots, n{+}1.$$

Using these weights in Eq. (7), the power transfer function can be computed by setting the bandwidth B equal to zero in Eq. (12) and varying f_i from zero to $1/T$, where $1/T$ is the PRF. This corresponds to a sine wave input varying from zero to the PRF. The normalized power transfer function P_s, as defined in Eq. (7), is found for $n = 1$ to $n = 7$, and the results are as shown in Fig. 4.

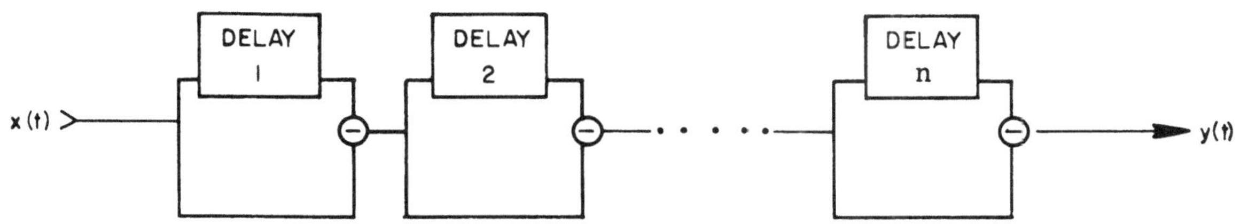

Fig. 3 — An n-stage, or n-canceller, MTI

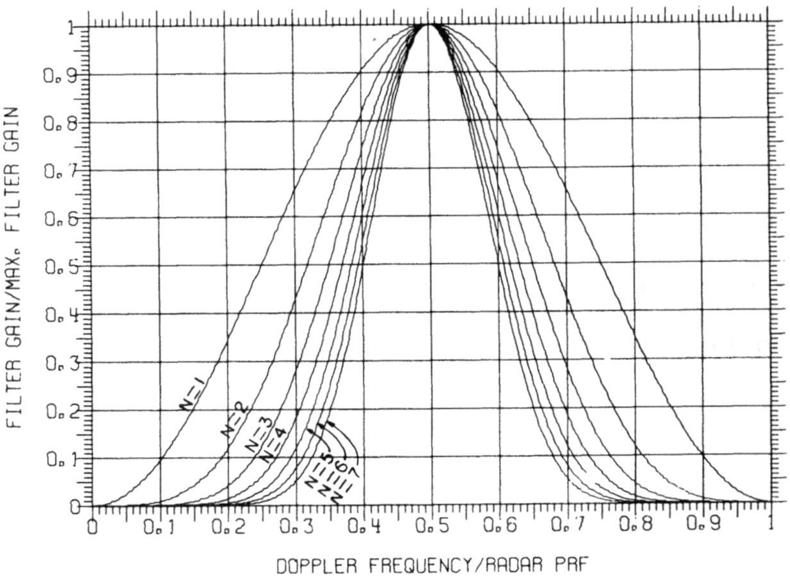

Fig. 4 — Normalized power transfer function P_s for an MTI having the indicated number of cancellers

The power transfer function is normalized by dividing by the maximum gain W_N given by

$$W_N = \sum_{i=1}^{n+1} w_i^2.$$

The maximum gain W_N is listed below for $n = 1$ to $n = 7$:

n	W_N
1	2
2	6
3	20
4	70
5	252
6	923
7	3432

The improvement factor for an n-stage MTI is obtained by using Eq. (13). The matrix M_s is generated by letting $B = 1/T$ in Eq. (12), i.e., the target doppler may occur at any frequency from zero to the PRF (PRF = $1/T$). Also, let $f_i = 1/2T$, i.e., the center of the filter is at one-half the PRF. The covariance matrix M_n is generated using Eq. (6), with the average clutter doppler μ_c equal to zero. The results are plotted in Fig. 5, where the clutter spectral width σ_c ranges from 10^{-3} to 10^{-1} times the PRF. Curves similar to Fig. 5 have been derived using various methods and published in several places (6,7).

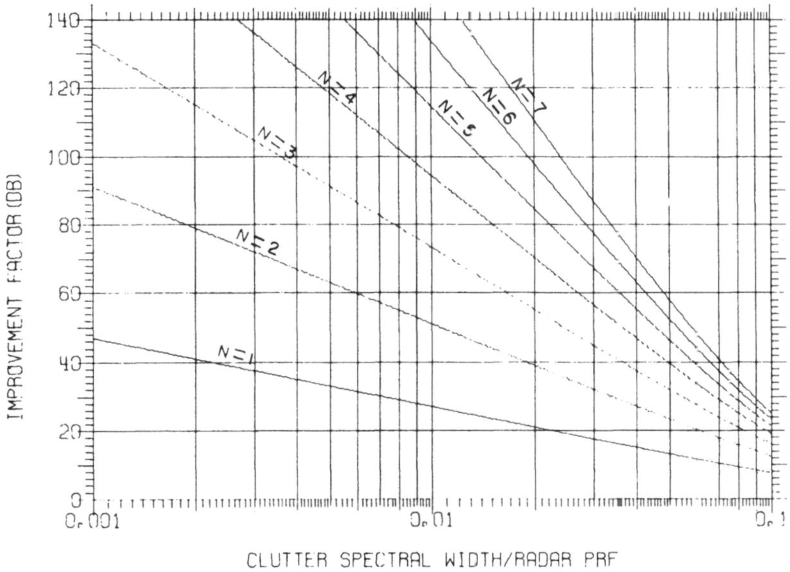

Fig. 5 — Improvement factor I [given by Eq. (13)] for an MTI having the indicated number of cancellers (filters)

IV. COHERENT INTEGRATOR (FFT)

Uniform Weights

The discrete Fourier transform can be characterized by

$$y_k = \sum_{n=1}^{N} x_n \, e^{-j2\pi \frac{(n-1)(k-1)}{N}}, \quad k = 1, 2, \ldots, N \tag{14}$$

where x_n is the input time sample, and the y_k are the output spectral components. Therefore, Eq. (14) describes N filters with outputs $y_1, y_2, \ldots, y_N$, as sketched in Fig. 2. The effective weights of the kth filter are

$$W_{kn} = e^{-j2\pi \frac{(n-1)(k-1)}{N}}, \quad n = 1, 2, \ldots, N. \tag{15}$$

These filters have a $(\sin x)/x$ shape and are translated in frequency by the index k. When $N = 16$, the filter responses for $k = 1$ and $k = 2$ are as shown in Figs. 6(a) and 6(b), respectively. The normalization used for these curves is

$$W_N = \sum_{n=1}^{N} |W_{kn}|^2 = 256$$

for $N = 16$.

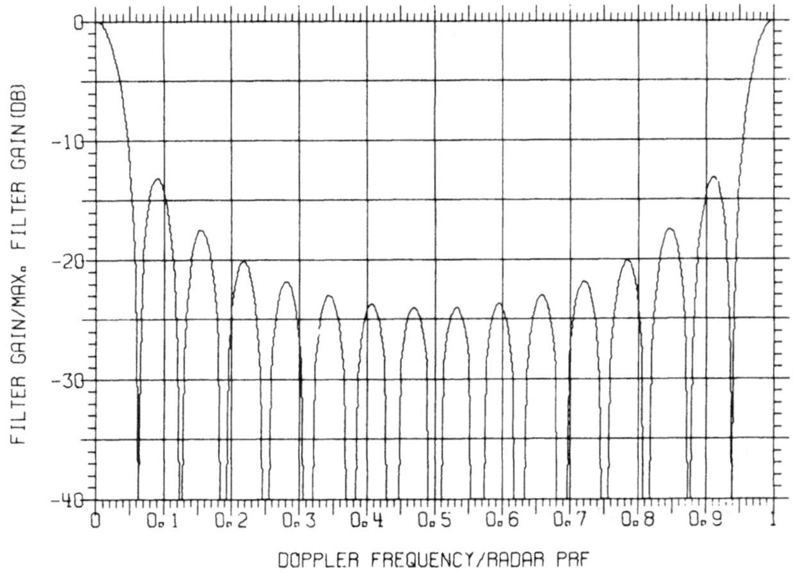

(a)

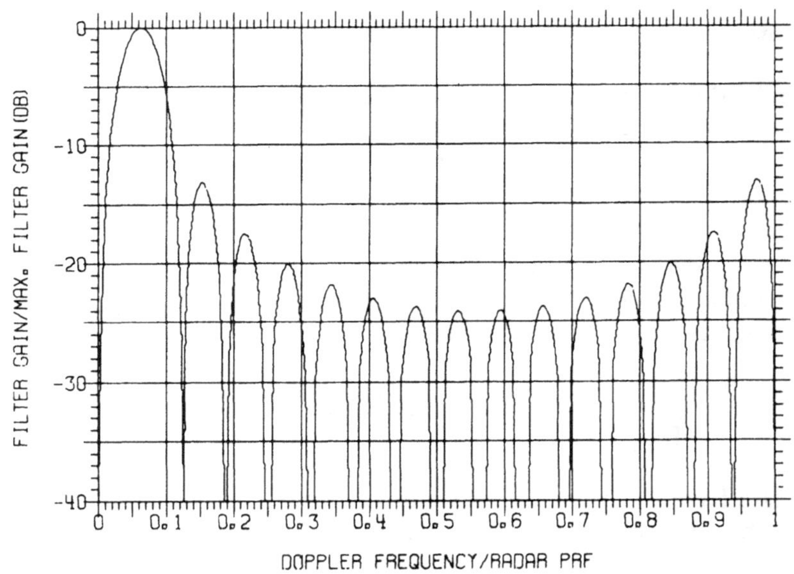

(b)

Fig. 6 — Normalized power transfer function (filter response) for (a) filter no. 1 and (b) filter no. 2 of a 16-pulse (16-filter) MTI integrator

Improvement Factor — The improvement factor for the coherent integration is computed the same way as for the MTI discussed in Section III. Equation (13) is used with the weight vector W generated by Eq. (15). The signal covariance matrix M_s is generated using Eq. (12), with the bandwidth B given by

$$B = \frac{1}{N}$$

where N is the number of pulses integrated. The center of the ith filter, f_i, is given by

$$f_i = \frac{i-1}{N},\ i = 1, 2, ..., N.$$

The improvement factor is shown in Fig. 7(a) for an 8-pulse integrator and in Fig. 7(b) for a 16-pulse integrator. The gains of the filters are averaged and shown by a dotted curve in each figure. The average gain corresponds to the expected improvement factor for a target with equal probability of occurring in any filter. This average improvement factor should be compared with the MTI improvement factor shown in Fig. 5, which was computed for a target with equal probability of occurring at any doppler.

The irregularities in Fig. 7 are caused by the interaction between the clutter spectrum and the filter transfer function. Note that the width of the sidelobes of the filters, as shown in Fig. 6, is given by PRF/N, i.e., by 0.125 × PRF for an 8-pulse integrator and by 0.0625 × PRF for a 16-pulse integrator. At the left side of Fig. 7 the clutter spectrum is narrow as compared to the sidelobe width so that more of the clutter energy coincides with the null between sidelobes. The improvement factor is reduced as the clutter width is increased and more energy coincides with a peak in the sidelobes. This effect continues until the clutter width becomes wide enough for an appreciable amount of clutter energy to coincide with the next sidelobe null. In this region, less clutter energy is passed, and the improvement factor decreases at a slower rate as the clutter spectral width is increased. This effect continues until an appreciable amount of clutter energy coincides with the next peak in the sidelobes.

For narrow clutter spectral widths, the improvement factors for the 8- and 16-pulse integrators are approximately the same, as can be seen in Figs. 7 (a) and (b). Intuition would seem to indicate that the integration of more pulses should result in more gain in the signal-to-clutter ratio. When the interference consists of "white" noise, the improvement I in signal-to-noise ratio is given by

$$I = 10 \log N$$

which, as expected, does give a larger improvement as N is increased.

For correlated (or "colored") noise, the gain that can be realized depends on the shape of the noise spectrum and the shape of the filter transfer function. Except for filter number one, all the filters generated by both the 8- and 16-pulse integrators have sidelobe nulls at zero doppler. The width of the null for an 8-pulse integrator is twice as wide as for a 16-pulse integrator. Apparently, the additional clutter rejection caused by the wider null of the 8-pulse integrator offsets the additional integration gain of the 16-pulse integrator.

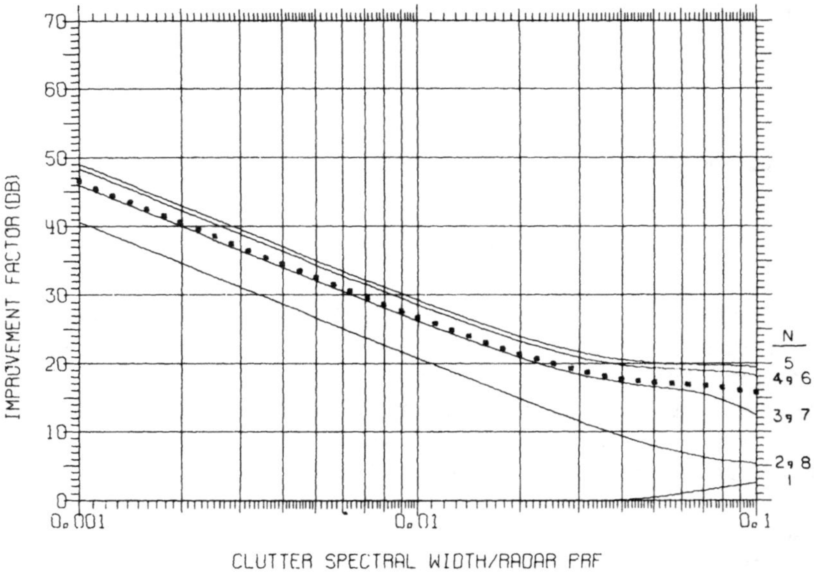

(a)

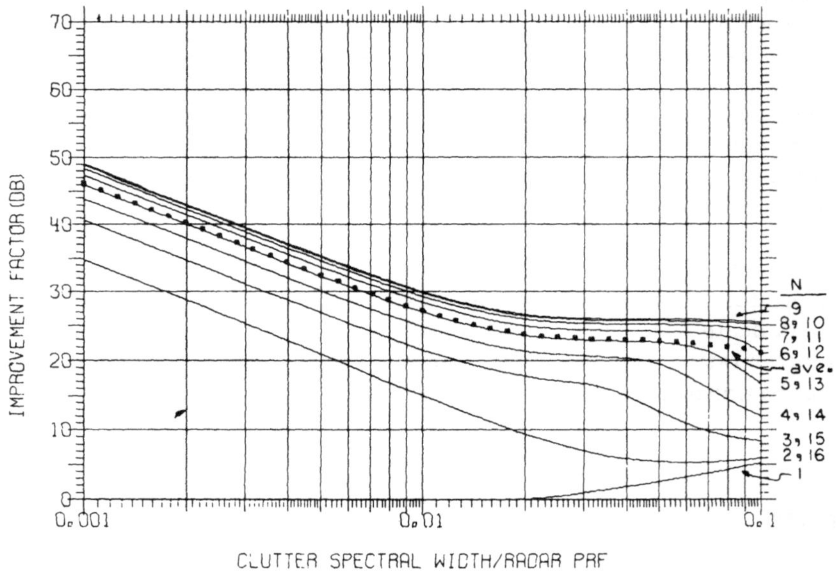

(b)

Fig. 7 — Improvement factor I [given by Eq. (13)] for the indicated filter number N in (a) an 8-pulse and (b) a 16-pulse coherent integrator with uniform weighting. The average improvement for all filters is indicated by the dotted curve.

For wide clutter spectral widths, this effect disappears and the 16-pulse integrator achieves more gain. The wider spectral width is more characteristic of "white" noise.

Notice that filter number one has its mainlobe at zero doppler, as shown in Fig. 6(a). The clutter spectrum is also centered at zero doppler. For these reasons, the improvement factor for filter number one, as shown in Fig. 7, increases as the clutter spectral width increases since this results in more clutter energy outside the filter mainlobe region.

Tschebyschev Weights

The sidelobes levels of the filters shown in Fig. 6 can be reduced by using a data "window" function (Ref. 8). This corresponds to weighting the input time samples x_n of Eq. (14). Let $x_n \to a_n x_n$ in Eq. (14) where a_n is the weight vector for a desired filter shape. Then, from Eq. (15), the total effective weights of the kth filter become

$$W_{kn} = a_n e^{-j2\pi \frac{(n-1)(k-1)}{N}}, \quad n = 1, 2, \ldots, N. \tag{16}$$

When the weights a_n correspond to the familiar Tschebyschev weights for a —25 dB sidelobe level, a 16-pulse coherent integrator results in 16 filters, two of which are shown in Figs. 8(a) and (b).

Improvement Factor — The improvement factors for 8- and 16-pulse integrators with —25 dB sidelobes are shown in Figs. 9(a) and 9(b), respectively. Notice that at narrow clutter spectral widths, the average improvement factor of the 8-pulse integrator is greater than the average improvement of the 16-pulse integrator. Also, notice the irregular ordering of the filters in Fig. 9(b) for narrow clutter spectral widths. Those filters whose center frequencies are farther from the clutter spectrum center (zero doppler) might be expected to have a higher improvement factor than filters nearer to the clutter. However, for a narrow clutter spectrum, this does not happen. These peculiarities are a result of the irregular sidelobe and null widths produced by the Tschebyschev weighting. A filter that is closer to the clutter may have a wider null near zero doppler and, therefore, realize a higher improvement factor.

An examination of Figs. 7 and 9 reveals a certain degree of unpredictability for coherent integration filters in a clutter (or "colored" noise) environment. This unpredictability could be reduced by preceding the integration filters with a "clutter" filter, such as an MTI canceller, to reduce the coherency of the clutter input to the integration filters. This configuration will be examined in the following section.

V. CASCADED MTI AND COHERENT INTEGRATOR

Optimum Cascaded Filters

The maximum improvement in signal-to-clutter ratio could be achieved by following a "prewhitening" filter with a filter matched to the target return. Since the target doppler spectrum considered by this report is a single line component, the matched filter cannot

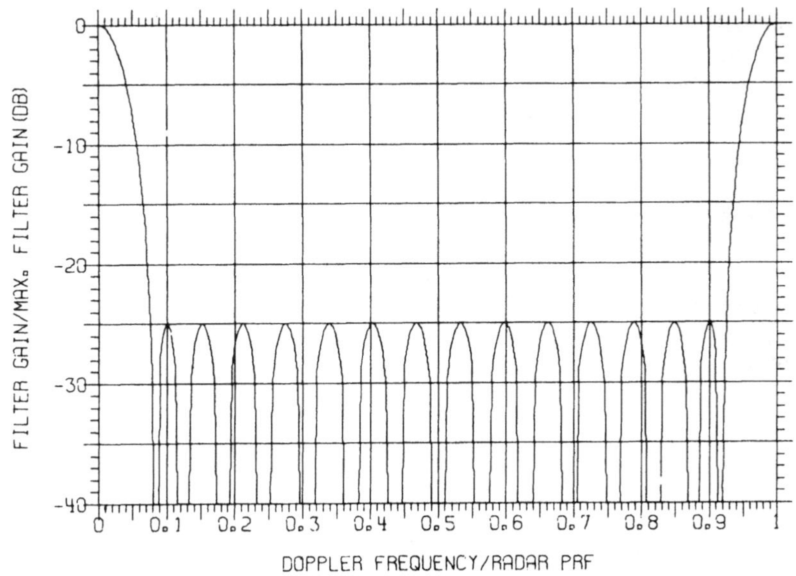

(a)

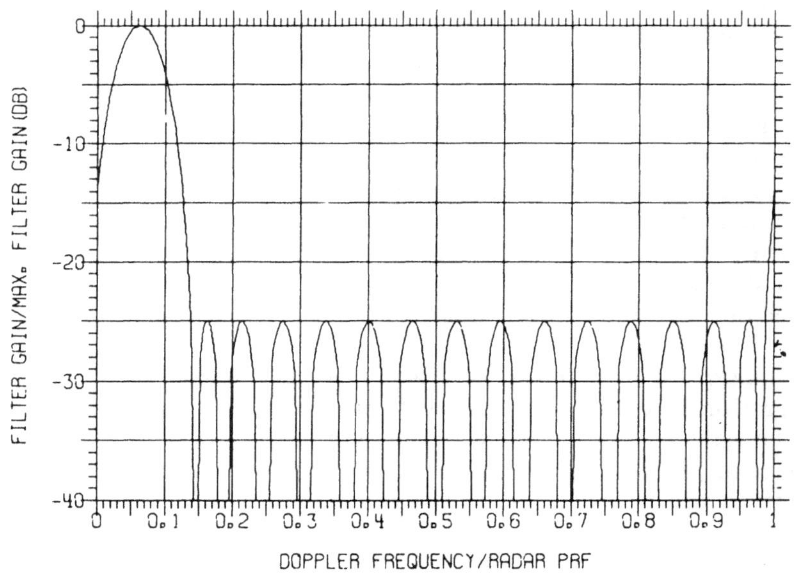

(b)

Fig. 8 — Normalized power transfer function (filter response) for (a) filter no. 1 and (b) filter no. 2 in a 16-pulse coherent integrator. The input time samples x_n are Tschebyschev weighted for a −25 dB side-lobe level. (Compare with Fig. 6.)

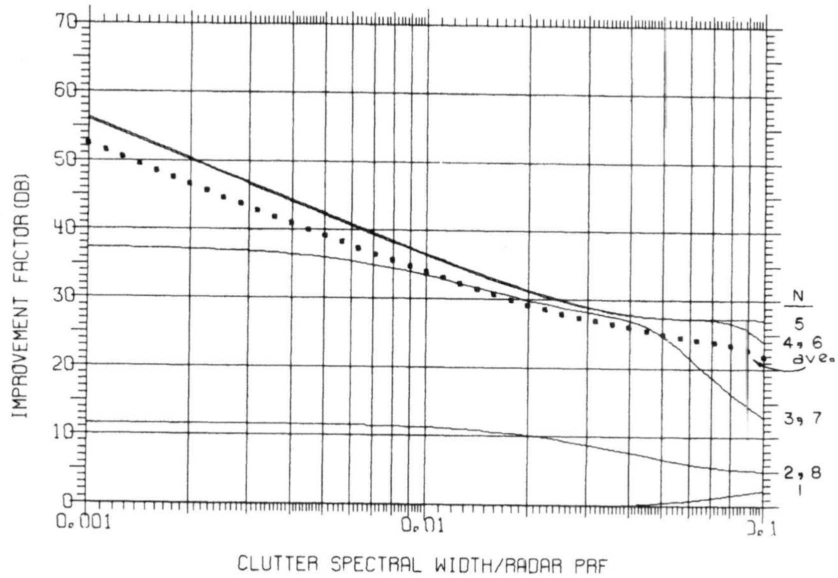

(a)

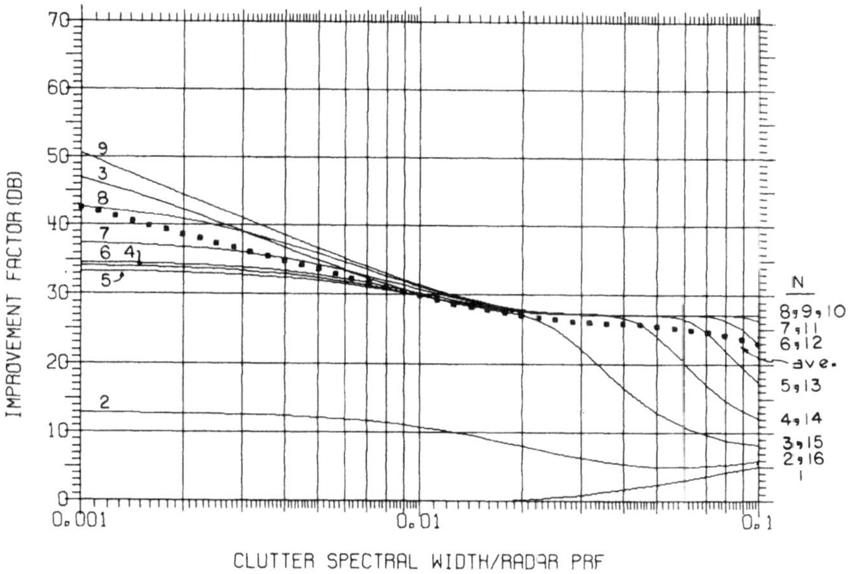

(b)

Fig. 9 — Improvement factor I [given by Eq. (13)] for (a) an 8-pulse and (b) a 16-pulse coherent integrator with Tschebyschev weighting. The average improvement for all filters is indicated by the dotted curve.

be realized by a processor with a finite processing time. However, the coherent integration filters considered in the previous section are an approximation to a matched filter. The approximation is limited only by the processing time (or the number of pulses integrated).

To completely "prewhiten" the clutter prior to coherent integration, the prewhitening filter must have a processing time equal to the processing time of the coherent integrator, i.e. each filter must process the same number of pulses. Although this approach gives the classical matched filter in "colored" noise, and therefore gives the maximum improvement in signal-to-clutter ratio, it is worthwhile to consider substituting for the ideal prewhitening filter a more simple filter with a shorter processing time. Any clutter notch filter that precedes the integration filters reduces the dynamic range of the input to the integration filters by an amount equal to the clutter rejection realized by the clutter notch. This requires less storage for the integration filters.

In this report, the three clutter notch filters considered are MTIs consisting of (a) a single canceller, (b) a double canceller, and (c) a triple canceller. The integration filters considered are an 8- and a 16-pulse FFT. Therefore, the processing time of the clutter filter is less than the processing time of the integration filter.

Some care is required in the analysis of a cascaded combination of two sampled data filters with different processing times. Theoretically, the order of the filters can be interchanged with the same results since both filters are linear. For analysis it is convenient to consider the filter with the shorter processing time first, and then the filter with the longer processing time. Also, for practical reasons, the MTI should precede the integration filter to reduce the dynamic range that the integration filter has to handle.

Matrix Equations For Cascaded Filters

To analyze the cascaded filters in terms of the matrix operation of the preceding sections, consider the filter arrangement shown in Fig. 10. The first filter is characterized by the number M of pulses processed and by the weight vector A. The second filter is characterized by the number N of pulses it processes and by its weight vector B. The number M is always less then N.

The output $z(t)$ of these filters can be expressed in terms of the input $y(t)$ to the second filter by means of the matrix equation

$$z = B_T Y \tag{17}$$

where

$$Y = \begin{bmatrix} y_1 \\ y_2 \\ \cdot \\ \cdot \\ \cdot \\ y_N \end{bmatrix} \text{ and } B = \begin{bmatrix} b_1 \\ b_2 \\ \cdot \\ \cdot \\ \cdot \\ b_N \end{bmatrix}$$

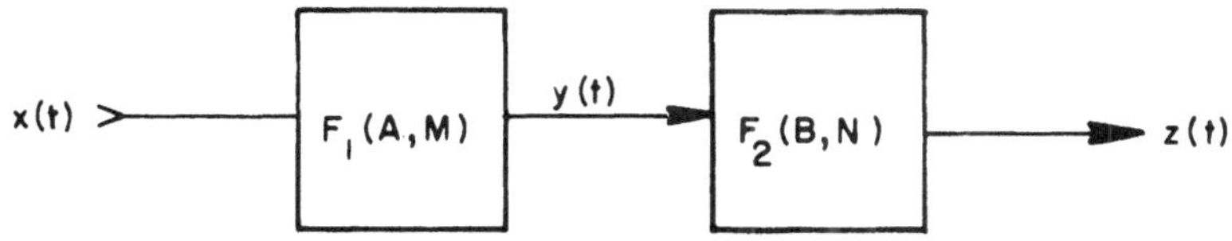

Fig. 10 — Cascaded filters with different processing times. M and N are the number of pulses processed ($M < N$), and A and B are weight vectors.

with B_T being the transpose of the weight vector B. The components of Y can be related to the input $x(t)$ by

$$\begin{aligned} y_1 &= a_1 x_1 + a_2 x_2 + \dots + a_M x_M \\ y_2 &= a_1 x_2 + a_2 x_3 + \dots + a_M x_{M+1} \\ &\vdots \\ y_N &= a_1 x_N + a_2 x_{N+1} + \dots + a_M x_{M+N-1}. \end{aligned}$$

Let

$$X = \begin{bmatrix} x_1 \\ x_2 \\ \cdot \\ \cdot \\ \cdot \\ x_{M+N-1} \end{bmatrix} \quad \text{and} \quad A_1 = \begin{bmatrix} a_1 \\ a_2 \\ \cdot \\ \cdot \\ \cdot \\ a_M \\ 0 \\ 0 \\ \cdot \\ \cdot \\ \cdot \end{bmatrix} \left.\begin{matrix} \\ \\ \\ \\ \\ \end{matrix}\right\} N-1 \text{ zeros}.$$

Then

$$y_1 = (A_1)_T X.$$

In the same way

$$y_2 = (A_2)_T X,$$

where

$$A_2 = \begin{bmatrix} 0 \\ a_1 \\ a_2 \\ \cdot \\ \cdot \\ \cdot \\ a_M \\ 0 \\ 0 \\ \cdot \\ \cdot \\ \cdot \end{bmatrix} \begin{matrix} \}\text{ 1 zero} \\ \\ \\ \\ \\ \\ \\ \left.\begin{matrix} \\ \\ \\ \\ \\ \end{matrix}\right\} N-2 \text{ zeros} \end{matrix},$$

and

$$y_N = (A_N)_T X$$

where

$$A_N = \begin{bmatrix} 0 \\ 0 \\ \cdot \\ \cdot \\ \cdot \\ a_1 \\ a_2 \\ \cdot \\ \cdot \\ \cdot \\ a_M \end{bmatrix} \begin{matrix} \left.\begin{matrix} \\ \\ \\ \\ \\ \end{matrix}\right\} N-1 \text{ zeros} \\ \\ \\ \\ \\ \\ \\ \end{matrix}.$$

Combining all of these matrix relations gives

$$Y = \hat{A}_T X \tag{19}$$

where X is given by Eq. (18) and

$$\hat{A} = [A_1 A_2 \ldots A_N].$$

Substituting Eq. (19) into Eq. (17), the matrix relationship between input and output becomes

$$\begin{aligned} z &= B_T \hat{A}_T X \\ &= W_T X \end{aligned} \tag{20}$$

since the total effective weight vector is given by

$$W = \hat{A}B. \tag{21}$$

Therefore this cascaded combination of filters can be evaluated in the same way that was outlined in section II and used in section III for the MTI filter and in section IV for the integration filter. The only difference is that the weight vector W is defined by Eq. (21) and generated as shown above.

Transfer Function

The power transfer function of the cascaded filters can be computed by means of Eq. (7) and the weights given by Eq. (21). The signal covariance matrix M_s is generated from Eq. (12) by setting the bandwidth B equal to zero and varying the center frequency f_i from zero to $1/T$. The weight vector W is computed from Eq. (21), with the matrix $\hat{A}$ generated from the weight vector A given by

$$A = \begin{bmatrix} a_1 \\ a_2 \\ \cdot \\ \cdot \\ \cdot \\ a_M \end{bmatrix}.$$

If the first filter is an MTI consisting of $M - 1$ binomial cancellers, then

$$a_i = (-1)^{i-1} \binom{M-1}{i-1}, \; i = 1, 2, ..., M \tag{22}$$

as shown in section III. If the second filter is an FFT with uniform weights, then the weight vector B can be generated from Eq. (15).

Using this procedure, the power transfer functions generated by a triple-canceller MTI cascaded with a 16-pulse coherent integrator are shown in Figs. 11(a)-(i) for nine of the sixteen filters. Filter no. 10 is the mirrow image of filter no. 8; filter no. 11 is the mirrow image of filter no. 7, etc. Uniform integrator weights are used. When 25-dB Tschebyschev weights are used, the resulting transfer functions are shown in Figs. 12(a)-(i). The filter patterns are analogous to the antenna patterns obtained with a 16-element linear array with elements equally spaced at one-half wavelength. The element pattern is analogous to the MTI filter transfer function. All filter transfer functions are normalized to the maximum gain of the center filter (filter no. $(N/2) + 1$, where N is the number of pulses integrated).

Improvement Factor

The improvement in signal-to-clutter ratio is shown in Figs. 13(a) and (b) for a single canceller cascaded with an 8-pulse integrator. With 25-dB Tschebyschev weights [Fig. 13(b)] there is a larger variation in gain between filters. However, the average gain is greater by 5 to 6 dB than for uniform weights [Fig. 13(a)]. The integrator with uniform weights has more gain in the filters near the clutter spectrum, i.e., filters no. 2 and 8. The MTI null at zero doppler overrides the gain that was achieved by filter no. 1 without MTI.

Figures 14(a) and (b) show the improvement factors for a single canceller cascaded with a 16-pulse integrator. Again, with 25-dB Tschebyschev weights [Fig. 14(b)] there is a larger variation in gain between filters. The filters near the clutter spectrum again have less gain than the same filters with uniform weights [Fig. 14(a)]. However, the average filter gain is slightly higher (about 3 dB), for narrow clutter spectral widths (0.001 × PRF) using uniform weights, but slightly higher (about 2 dB) for wide clutter widths (0.1 × PRF) using 25-dB weights. Comparing Figs. 13 and 14 shows that the average improvement factor is still less for a 16-pulse integrator than for an 8-pulse integrator for narrow clutter. This effect was noticed for the 8- and 16-pulse integrators without MTI.

Figures 15(a) and (b) show the improvement factors for a double canceller cascaded with an 8-pulse integrator. The characteristics of these curves are similar to Fig. 13 for a single canceller and an 8-pulse integrator. With 25-dB weights [Fig. 15(b)] there is a larger variation in filter gains, but also a larger (again 5 to 6 dB) average filter gain. The additional improvement factor realized by preceding an 8-pulse integrator with a double canceller (Fig. 15) instead of with a single canceller (Fig. 13) varies from 42 dB for narrow clutter widths (0.001 × PRF) to 4 dB for wide clutter widths (0.1 × PRF) using either uniform or 25-dB weights.

Figures 16(a) and (b) show the improvement factor for a double canceller cascaded with a 16-pulse integrator. Comparing uniform weights [Fig. 16(a)] with 25-dB weights [Fig. 16(b)], the difference in the variation in filter gain is reduced, although the filters near the clutter spectrum still have less gain with 25-dB weights for narrow clutter. For narrow clutter widths, uniform weights give about 2 dB more average gain. For wide clutter widths, 25-dB weights give about 2 dB more average gain. The additional improvement factor realized by preceding a 16-pulse integrator with a double canceller (Fig. 16) instead of with a single canceller (Fig. 14) varies from about 42 dB for narrow clutter to about 4 dB for wide clutter using either uniform or 25-dB weights. This is the same improvement that was achieved by preceding the 8-pulse integrator with a double instead of with a single canceller.

Figures 17(a) and (b) show the improvement factor for a triple canceller cascaded with an 8-pulse integrator. Comparing uniform weights [Fig. 17(a)] with 25-dB weights [Fig. 17(b)], the 25-dB weights still cause a wider variation in filter gains, with the filters near the clutter having appreciably less gain than the same filters with uniform weights. The average gain is always greater with 25-dB weights, varying from about 6 dB with narrow clutter to about 4 dB for wide clutter. The additional improvement factor realized by preceding an 8-pulse integrator with a triple canceller (Fig. 17) instead of with a double canceller (Fig. 15) varies from more than 30 dB for narrow clutter to about 3 dB for wide clutter using uniform weights, and varies from more than 30 dB to about 2 dB using 25-dB weights.

Figures 18(a) and (b) show the improvement factor for a triple canceller cascaded with a 16-pulse integrator. Again, the 25-dB weights cause a wider variation in filter gains. The average filter gain is always about 2 dB greater for 25-dB weights. The additional improvement realized by preceding the 16-pulse integrator with a triple canceller (Fig. 18) instead of with a double canceller (Fig. 16) varies from more than 30 dB for narrow clutter to about 3 dB for wide clutter. This is the same increase that was achieved by preceding the 8-pulse integrator with a triple canceller instead of with a double canceller.

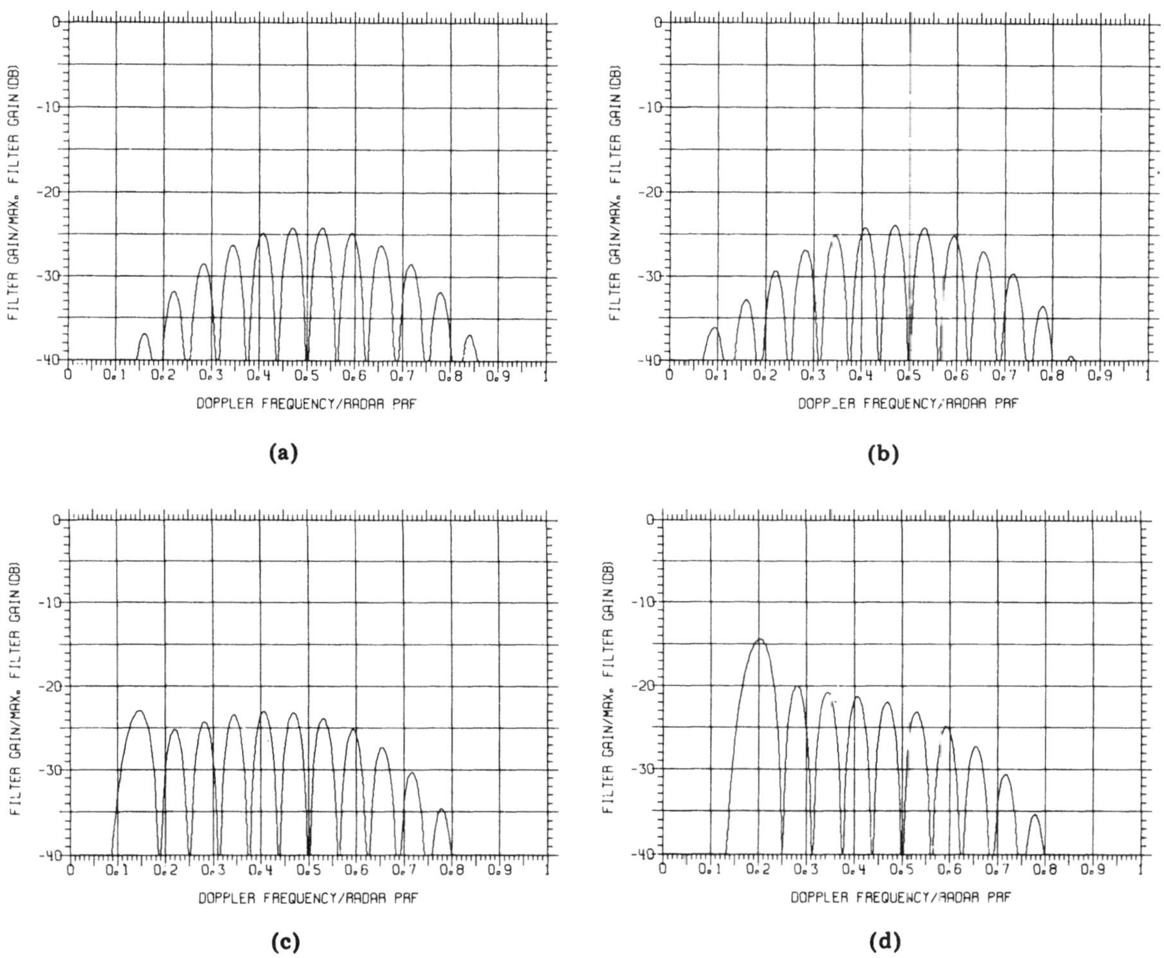

Fig. 11 — Normalized power transfer function for a triple-canceller MTI cascaded with a 16-pulse coherent integrator. The transfer functions for nine of the sixteen filters (filters no. 1 through 9) are shown in the graphs labeled (a) through (i). The integrator filters have been uniformly weighted.

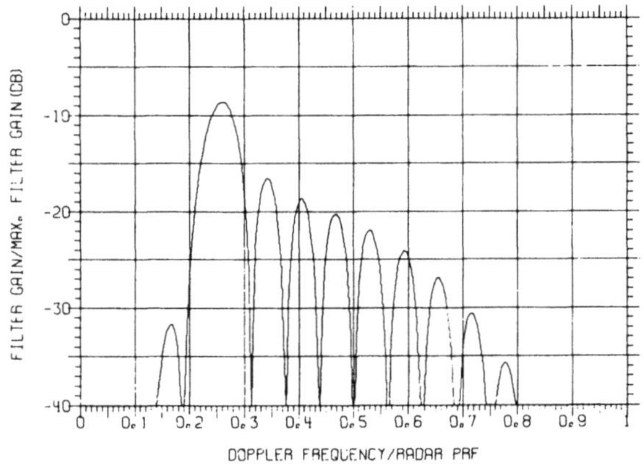
FILTER GAIN/MAX. FILTER GAIN (DB)
0
-10
-20
-30
-40
0
0.1
0.2
0.3
0.4
0.5
0.6
0.7
0.8
0.9
1
DOPPLER FREQUENCY/RADAR PRF

(e)

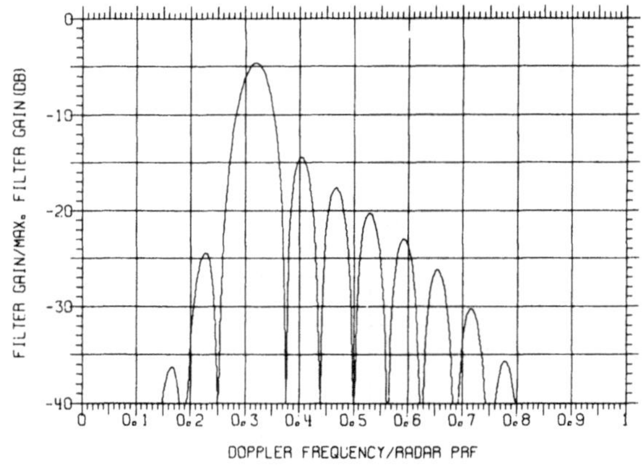
FILTER GAIN/MAX. FILTER GAIN (DB)
0
-10
-20
-30
-40
0
0.1
0.2
0.3
0.4
0.5
0.6
0.7
0.8
0.9
1
DOPPLER FREQUENCY/RADAR PRF

(f)

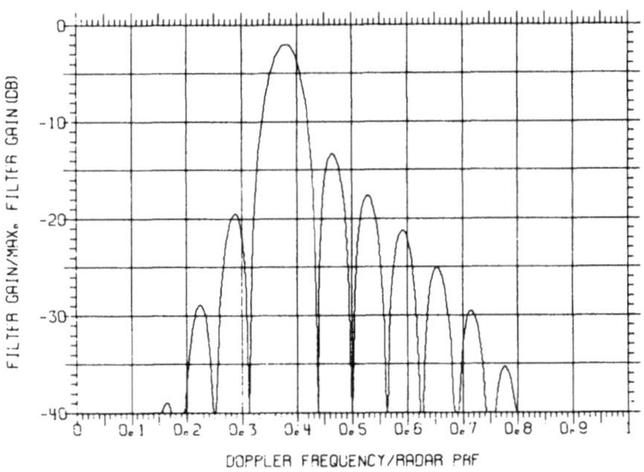
FILTER GAIN/MAX. FILTER GAIN (DB)
0
-10
-20
-30
-40
0
0.1
0.2
0.3
0.4
0.5
0.6
0.7
0.8
0.9
1
DOPPLER FREQUENCY/RADAR PRF

(g)

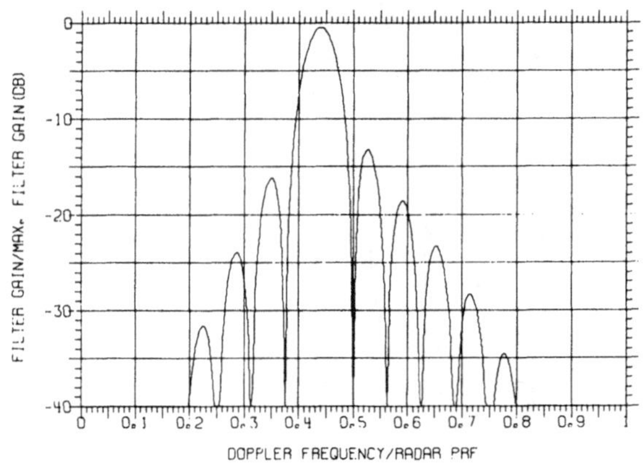
FILTER GAIN/MAX. FILTER GAIN (DB)
0
-10
-20
-30
-40
0
0.1
0.2
0.3
0.4
0.5
0.6
0.7
0.8
0.9
1
DOPPLER FREQUENCY/RADAR PRF

(h)

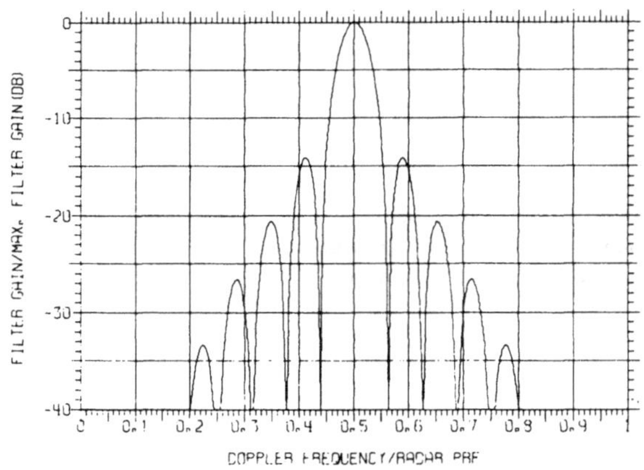
FILTER GAIN/MAX. FILTER GAIN (DB)
0
-10
-20
-30
-40
0
0.1
0.2
0.3
0.4
0.5
0.6
0.7
0.8
0.9
1
DOPPLER FREQUENCY/RADAR PRF

(i)

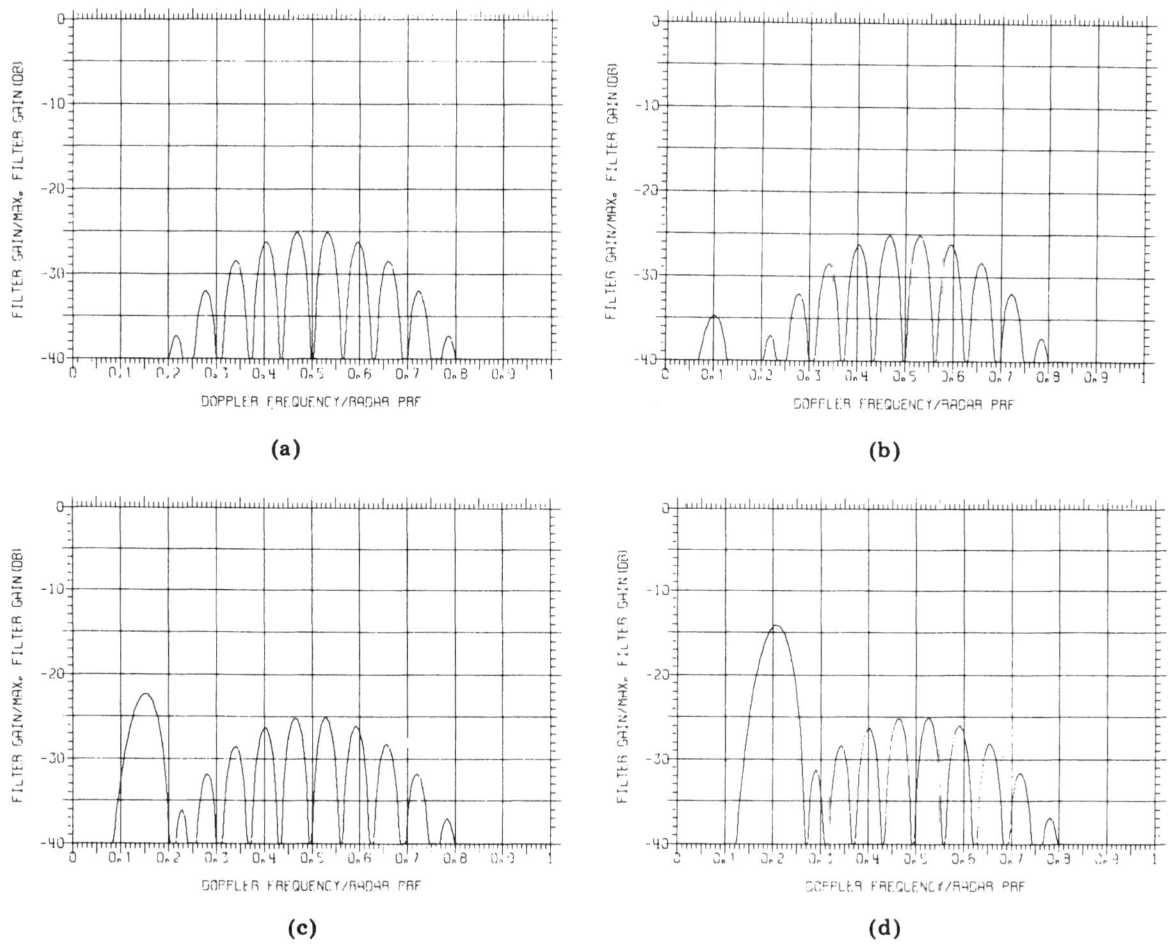

Fig. 12 — Normalized power transfer function for a triple-canceller MTI cascaded with a 16-pulse coherent integrator. The transfer functions for nine of the sixteen filters (filters no. 1 through 9) are shown in the graphs labeled (a) through (i). The integrator filters have been Tschebyschev weighted for a -25 dB sidelobe level (Compare with Fig. 11.)

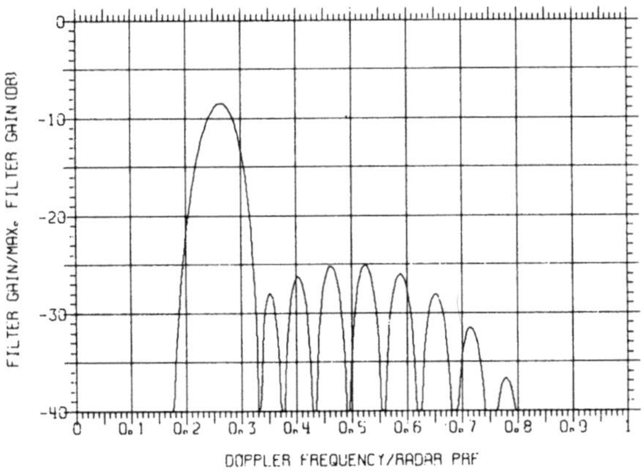
FILTER GAIN/MAX. FILTER GAIN (DB)
DOPPLER FREQUENCY/RADAR PRF

(e)

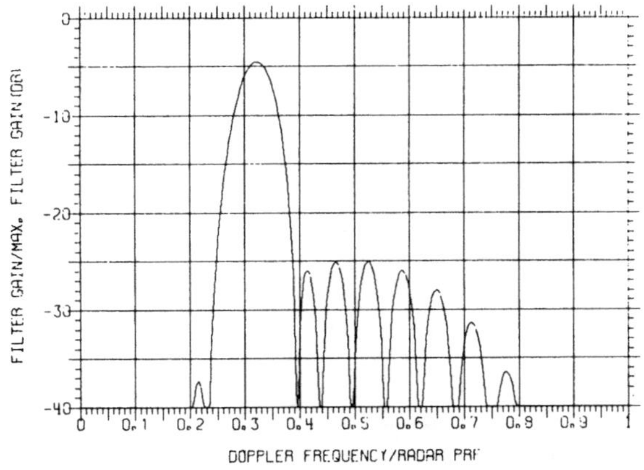
FILTER GAIN/MAX. FILTER GAIN (DB)
DOPPLER FREQUENCY/RADAR PRF

(f)

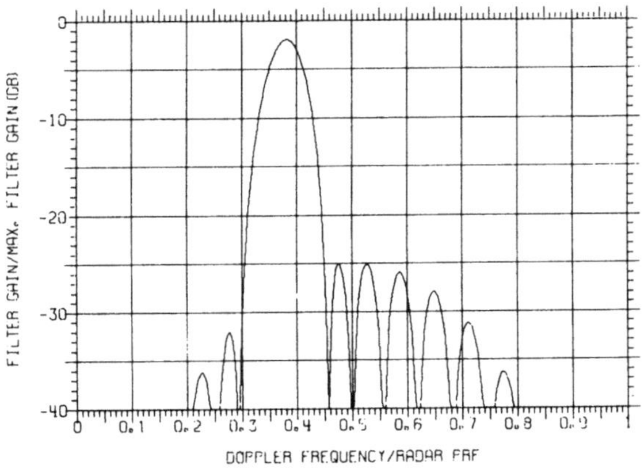
FILTER GAIN/MAX. FILTER GAIN (DB)
DOPPLER FREQUENCY/RADAR PRF

(g)

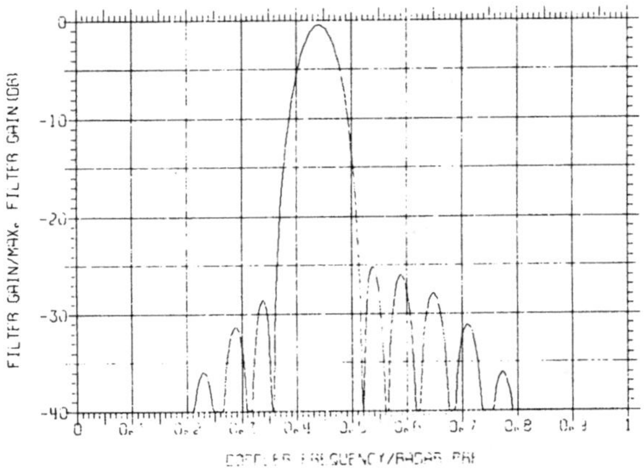
FILTER GAIN/MAX. FILTER GAIN (DB)

(h)

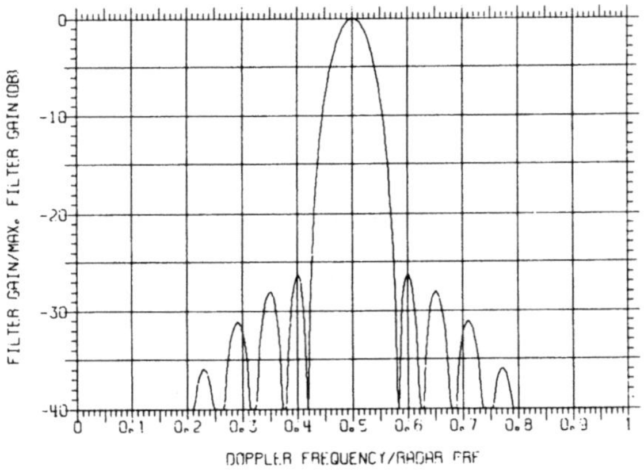
FILTER GAIN/MAX. FILTER GAIN (DB)
DOPPLER FREQUENCY/RADAR PRF

(i)

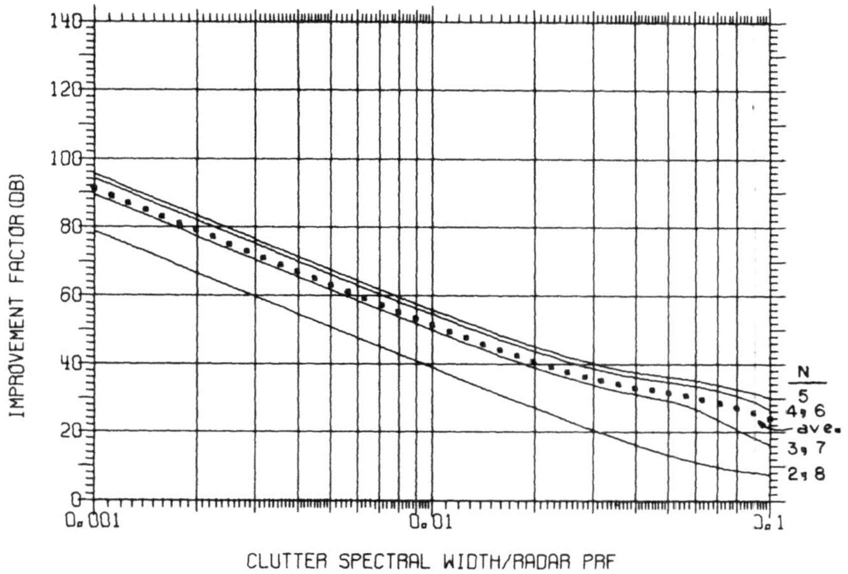

(a)

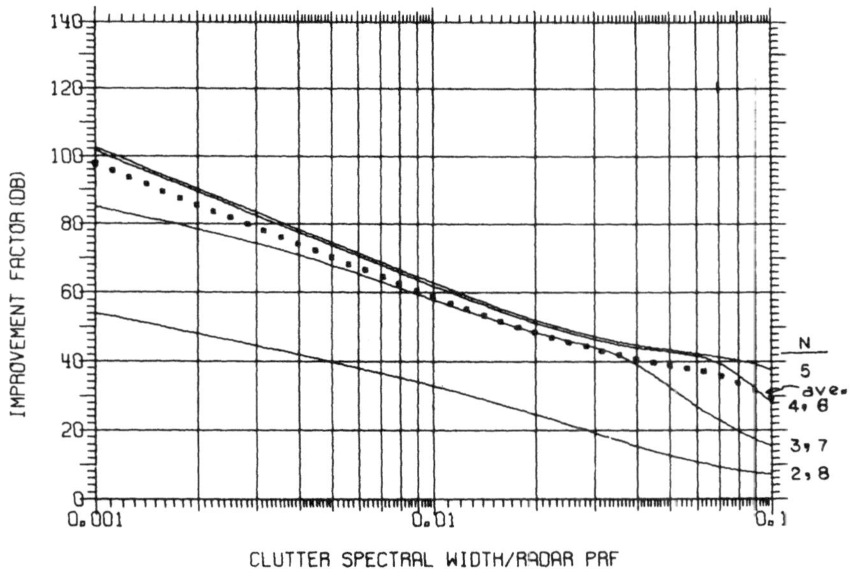

(b)

Fig. 13 — Improvement factor I [given by Eq. (13)] for a single-canceller MTI cascaded with an 8-pulse integrator. The integrator filters have been (a) uniformly weighted and (b) 25-dB Tschebyschev weighted. The average improvement for all filters is indicated by the dotted curve.

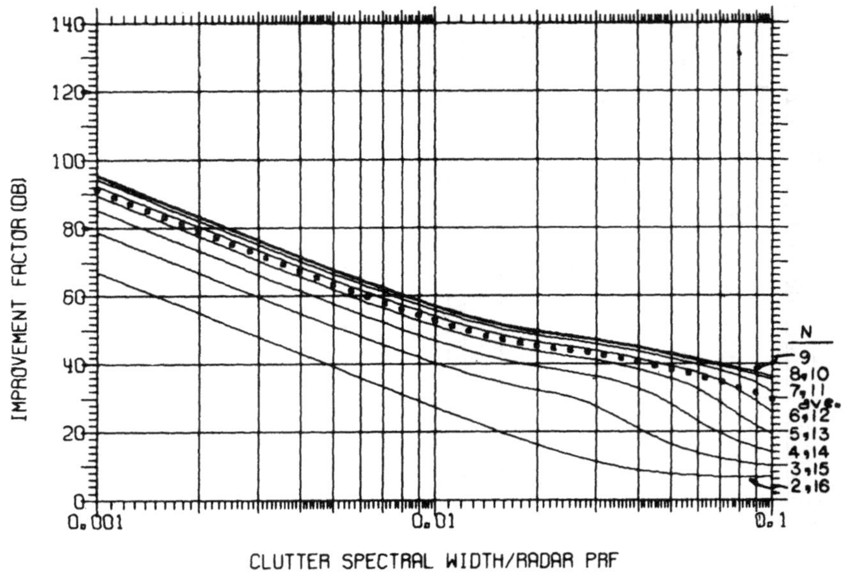

(a)

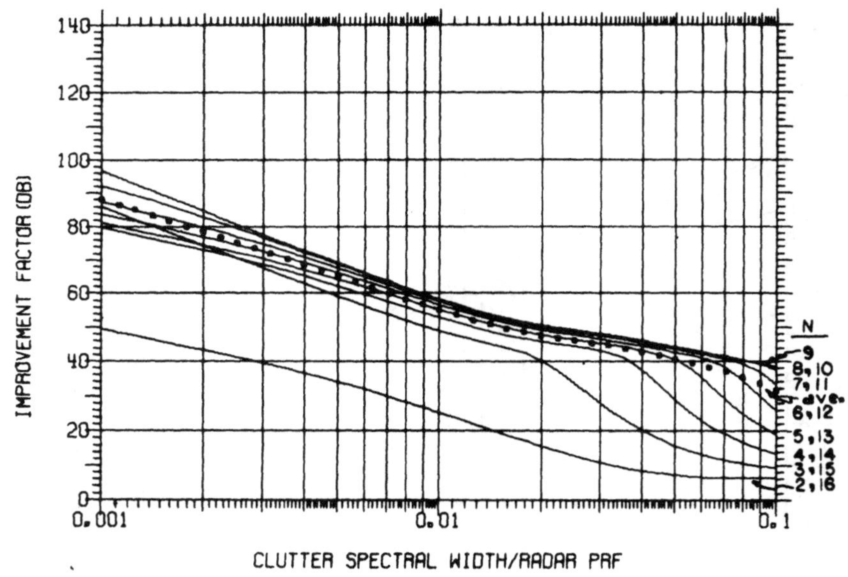

(b)

Fig. 14 — Improvement factor I [given by Eq. (13)] for a single-canceller MTI cascaded with a 16-pulse integrator. The integrator filters have been (a) uniformly weighted and (b) 25-dB Tschebyschev weighted. The average improvement for all filters is indicated by the dotted curve.

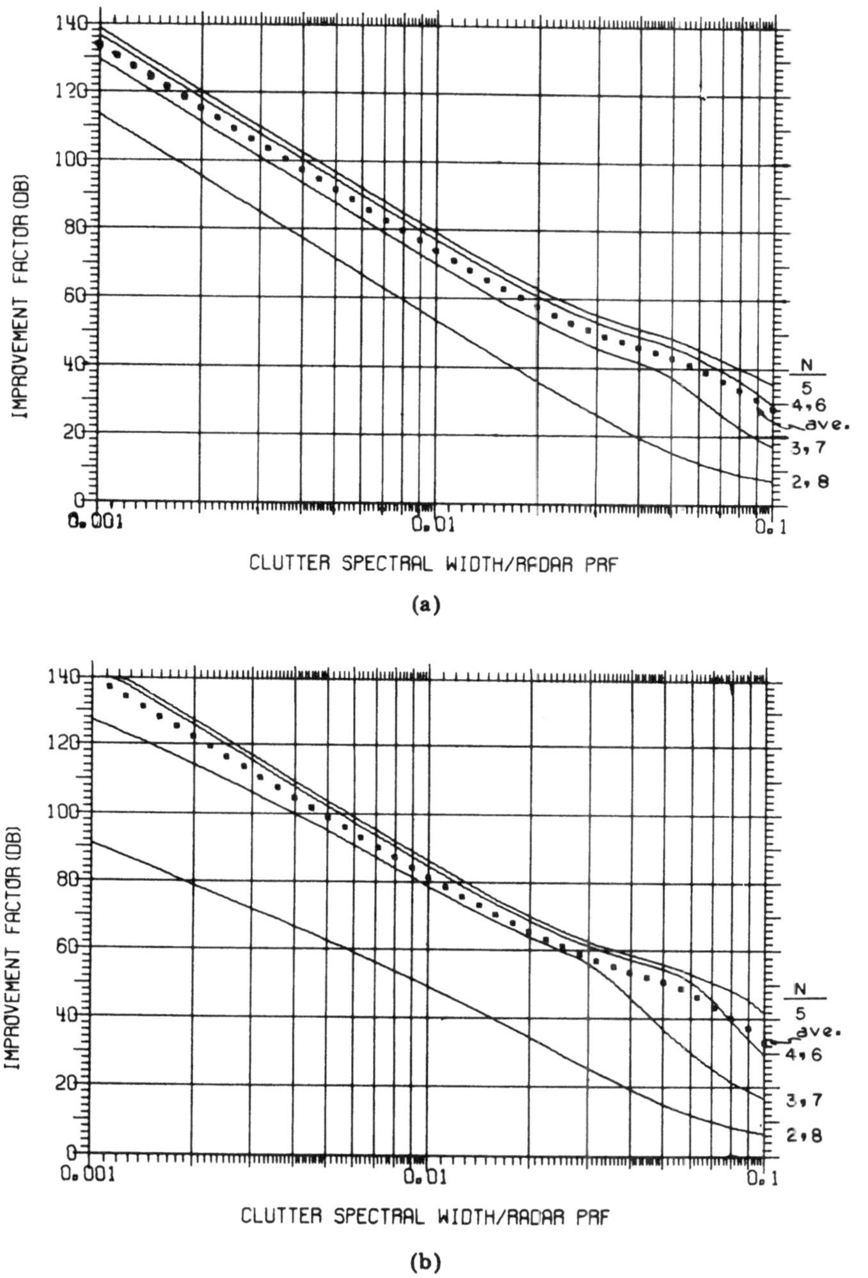

Fig. 15 — Improvement factor I [given by Eq. (13)] for a double-canceller MTI cascaded with an 8-pulse integrator. The integrator filters have been (a) uniformly weighted and (b) 25-dB Tschebyschev weighted. The average improvement for all filters is indicated by the dotted curve. (Compare with Fig. 13.)

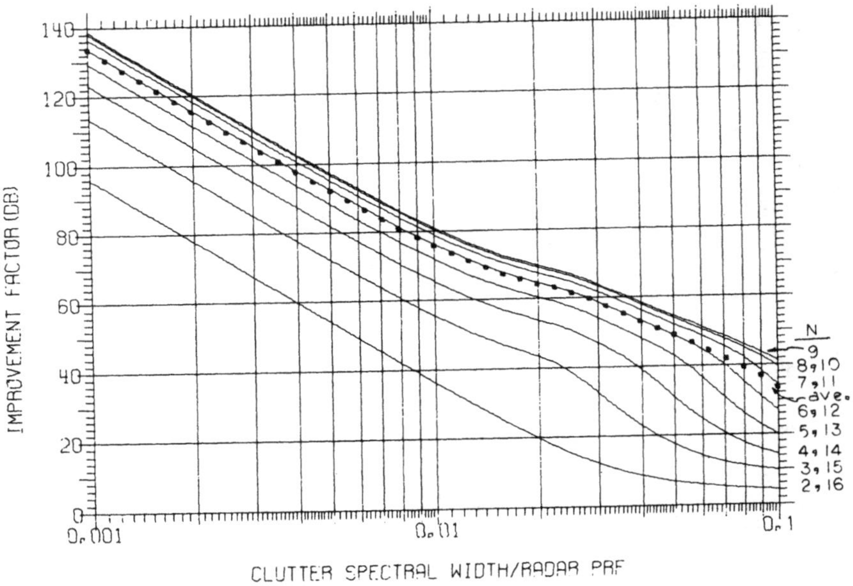

(a)

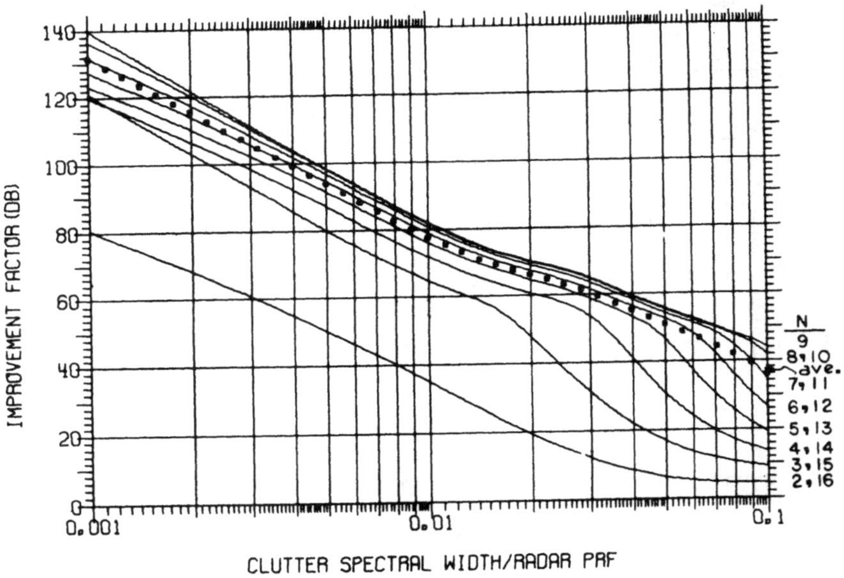

(b)

Fig. 16 — Improvement factor I [given by Eq. (13)] for a double-canceller MTI cascaded with a 16-pulse integrator. The integrator filters have been (a) uniformly weighted and (b) 25-dB Tschebyschev weighted. The average improvement for all filters is indicated by the dotted curve. (Compare with Fig. 14.)

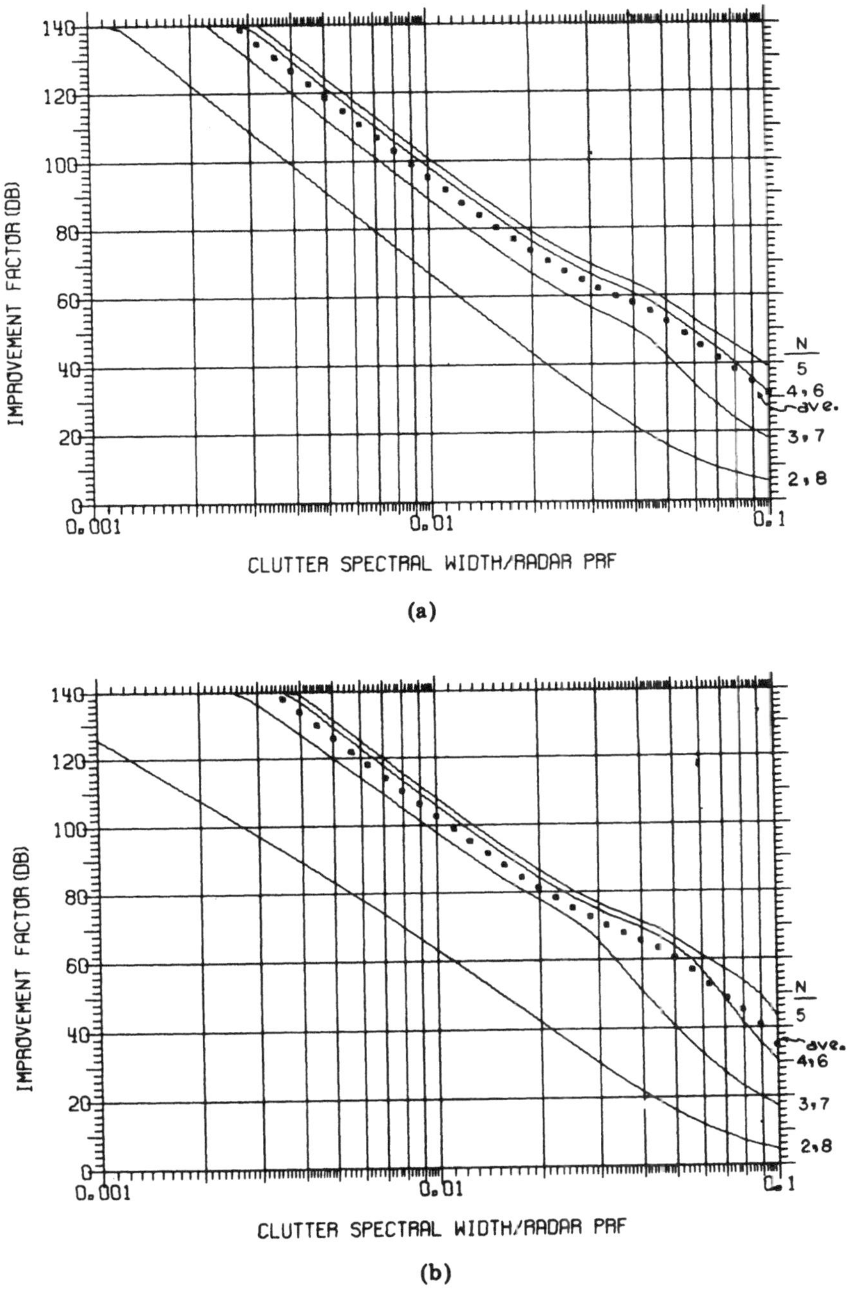

Fig. 17 — Improvement factor I [given by Eq. (13)] for a triple-canceller MTI cascaded with an 8-pulse integrator. The integrator filters have been (a) uniformly weighted and (b) 25-dB Tschebyschev weighted. The average improvement for all filters is indicated by the dotted curve. (Compare with Fig. 15.)

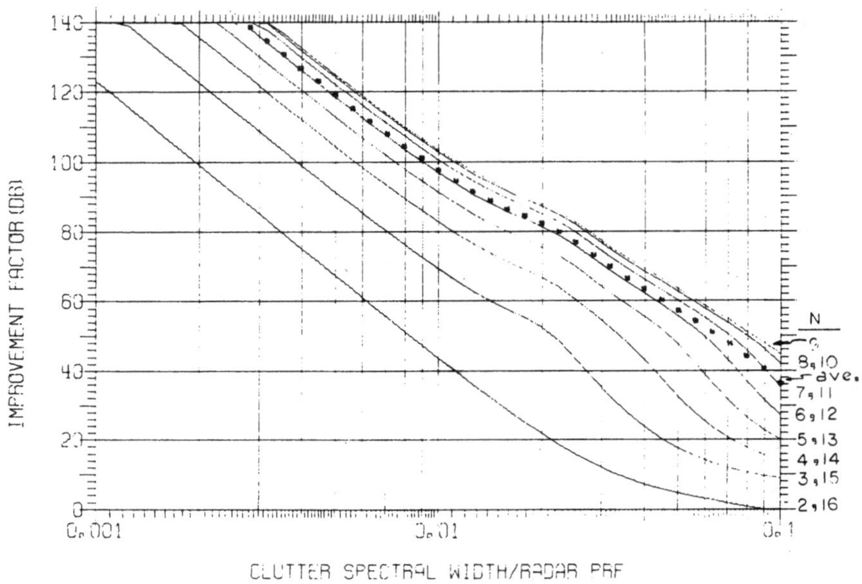

(a)

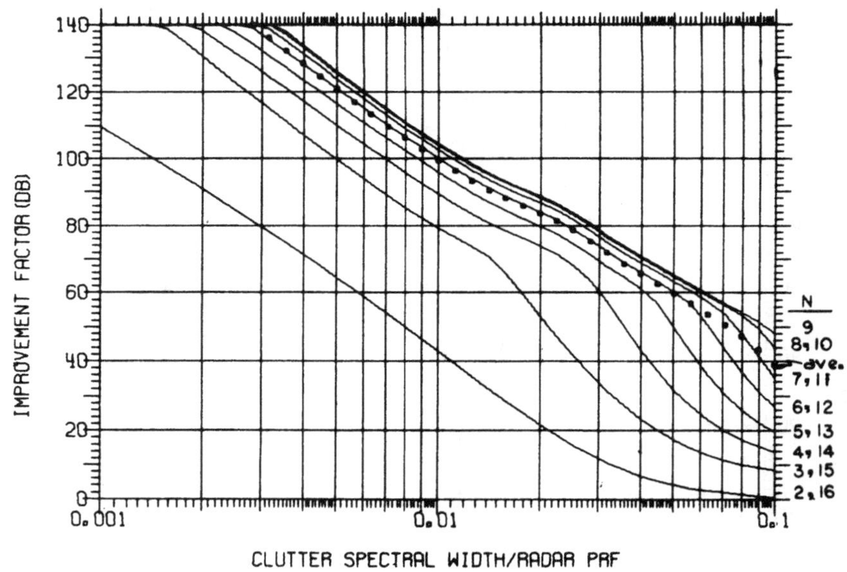

(b)

Fig. 18 — Improvement factor I [given by Eq. (13)] for a triple-canceller MTI cascaded with a 16-pulse integrator. The integrator filters have been (a) uniformly weighted and (b) 25-dB Tschebyschev weighted. The average improvement for all filters is indicated by the dotted curve. (Compare with Fig. 16.)

VI. CONCLUSIONS

Coherent integration is a technique that has been used in radar systems to improve detection in a "white" noise (i.e., receiver noise) environment and also to reject clutter in high-PRF radars. More recently this technique has been considered for clutter rejection in low-PRF radars where the ratio of the clutter spectrum width to the PRF is much larger than for high-PRF systems. Less analysis is available for this application to aid in understanding and evaluating the performance of coherent integration.

This report has considered several particular doppler filters which consist of cascaded MTI and coherent integration filters. These filters are considered typical of the filters that would be applicable to a low-PRF radar which has a relatively small number of hits per dwell time. The computed improvement factors presented in Figs. 7, 9, and 13-18 show that coherent integration filters perform in an irregular, almost impredictable, way then their input is "colored" noise (or clutter). The performance of coherent integrators against "white" noise is well established and easily predicted. Against "clutter", the performance is highly dependent upon the shape of the clutter spectrum and the shape (or weights) of the coherent integrator transfer functions. In particular, it is not necessarily true that more improvement is achieved by integrating more pulses for a given weight vector. On the other hand, it is theoretically true that weights exist which will allow more gain to be achieved when more pulses are processed. The computation of weight vectors that will give the maximum improvement in signal-to-clutter ratio under various optimization criteria will be presented in a later report.

The optimum weights obviously depend on the optimization criteria as well as on the shape of the interference spectrum. Of the integration filter weights discussed in the report, either uniform or 25-dB Tschebyschev weights could be selected as the "better" weights depending on the criterion used. This is pointed out by the following specific results:

a. 25-dB Tschebyschev weights give more average gain than uniform weights.

b. Uniform weights generally give more gain in the filters near the clutter spectrum.

c. Uniform weights result in less variation in gain from filter to filter.

d. For very narrow clutter spectral widths, the integration of more pulses does not necessarily lead to more average gain.

These conclusions apply regardless of whether or not an MTI precedes the integrator. When an MTI is cascaded with the integrator, the following general conclusions can be made:

a. The additional average gain achieved by cascading an MTI with any particular integrator is only very slightly affected by the type of integrator. That is, cascading a single canceller with an 8-pulse integrator results in about the same additional gain above an 8-pulse integrator alone as the additional gain above a 16-pulse integrator achieved by cascading a single canceller with a 16-pulse integrator. [Compare Fig. 7(a) with Fig. 13(a) and Fig. 7(b) with 14(a)]. An equivalent comparison can be made for double and triple cancellers.

b. The average gain achieved by cascading an integrator with any particular MTI is affected to some degree by the type of MTI. This can be seen by comparing the improvement factors for the MTI, shown in Fig. 5, with those in Figs. 13-18 for the MTI cascaded with integrators.

It might be concluded that the chief advantage of the coherent integrator lies not with its average performance in a clutter-dominated environment but with:

a. Increased gain for targets near the clutter spectrum. It narrows the "blind" velocity regions and could increase the detection of low-speed targets.

b. Increased gain in a "white" (i.e., receiver) noise environment. This also applies to other forms of wideband interference such as clutter received through the antenna sidelobes when the radar is on a moving platform. MTI alone has no capability against this type of interference.

c. Improved velocity measurement of targets for tracking or identification.

These factors must be considered in any tradeoff between cancellation and integration for a particular radar application and for a particular interference environment.

VII. ACKNOWLEDGMENT

I would like to acknowledge the very special contributions to this effort by Mr. D. L. Ringwalt and Dr. T. L. ap Rhys through our many interesting discussions. I would also like to thank Mr. Benjamin Koo upon whose mathematical and programming advice this effort depended so heavily. Finally, I would like to thank Mrs. Rosalie Valentine and Mrs. Evelyn Starrett for their determination in typing from a barely legible manuscript.

VIII. REFERENCES

1. Skolnik, M.I., ed., "Radar Handbook," New York:McGraw-Hill, 1970.
2. McAulay, R.J., "A Theory for Optimal MTI Digital Signal Processing," Tech. Note 1972-14, Lincoln Laboratory, Mass. Inst. of Tech., Lexington, Mass., Feb. 1972.
3. "Performance of a Cascaded Digital Canceller and Coherent Integrator," Report SR01-01, General Electric Company, Utica, NY, 13503, Mar. 1971.
4. An AEW AMTI Technique for Obtaining Reliable Detection Against Ground Clutter," Tech. Publ. 156A, TRG Division of Control Data Corporation, Rosemont, Pa., Sept. 1964.
5. Applebaum, S.P., "Adaptive Arrays," Tech. Rept. SPL TR 66-1, Syracuse University Research Corporation, Syracuse, NY, 13210, Aug. 1966.
6. Nathanson, F.E., "Radar Design Principles," New York:McGraw-Hill, 1969.
7. Andrews, G.A., "Airborne Radar Motion Compensation Techniques, Evaluation of TACCAR," NRL Rept. 7407, Apr. 1972.
8. Blackman, R.B., and Tukey, J.W., "The Measurement of Power Spectra," New York: Dover Publ. 1958.

AN AIRBORNE RADAR DOPPLER PROCESSING PHILOSOPHY

INTRODUCTION

Detection of moving targets by an airborne radar system is limited not only by usual clutter rejection problems but also by the effects of platform motion. Returns from stationary objects contain a doppler shift caused by the platform velocity. Therefore, an integral part of any effective airborne radar doppler processing is platform motion compensation [1, 2]. The techniques for this are applied to the design of both antennas and doppler filters.

Rejection of returns from stationary objects, referred to as clutter, is usually limited by such factors as clutter motion, system instabilities, or antenna scanning modulation. The performance of several current doppler filtering configurations, as limited by these factors, is described in Ref. 3. It has been shown [4] that doppler filters can be designed to maximize detection of moving targets if the clutter spectrum is known.

Platform motion both shifts the clutter spectrum and increases its width [5]. The performance of the optimal filters decreases as the width of the clutter spectrum is increased. For this reason, it is necessary to remove the mean doppler shift and any nonstochastic contributions to spectral width, such as the effects of platform motion. If the mean doppler shift is removed by a clutter-locking technique such as Time Averaged Clutter Coherent Airborne Radar (TACCAR), the remaining effects of platform motion on an MTI canceler can be compensated for by the displaced phase center antenna (DPCA) technique. With a properly designed antenna correction pattern [5], the improvement in signal-to-clutter ratio (MTI improvement factor), will be independent of platform motion for most applications.

In this report, an MTI canceler with motion compensation is cascaded first with a fast fourier transform (FFT) doppler filter bank, then with an additional canceler and an FFT doppler filter bank, and finally with a doppler filter bank designed to maximize the signal-to-clutter ratio of the canceled residue.

EFFECTS OF PLATFORM MOTION

The designs of antenna patterns and doppler filter shapes must be considered together to minimize the effects of platform motion. Since compensation techniques are available for MTI cancelers, several configurations of an MTI canceler cascaded with additional doppler filtering will be evaluated. The canceler gives two advantages; it uses the available motion compensation and the additional clutter rejection this allows, and it reduces the dynamic range and in turn the number of bits per word that must be stored for the additional doppler filters.

Manuscript submitted October 4, 1976.

Motion Compensation

It has been shown [5] that realizable antenna correction patterns can be designed to minimize MTI improvement factor loss caused by platform motion. It was further shown that this loss is insignificant for antenna patterns with very low sidelobes; -34 dB relative to the mainlobe peak was illustrated. These examples used clutter with a spectral width which results in a 24-dB cancellation of clutter. This spectral width can be dominated by internal clutter motion, system instabilities, or antenna scanning modulation. One of these will limit the clutter cancellation of a single canceler to less than 24 dB for most practical systems; these conditions therefore pertain to many radar applications.

The other important factor in determining the loss due to platform motion is the velocity of the platform itself. The velocity parameter used in Ref. 5 is

$$\frac{v_p T}{\lambda} \cos \phi_s = 0.7 \tag{1}$$

where v_p is the platform velocity, T is the interpulse period, λ is the transmitted wavelength, and ϕ_s is the vertical angle between the platform velocity vector and the line of direction to the scatterer.

Thus, if the expected clutter cancellation of the MTI canceler is less than 24 dB and if the distance the platform travels between transmitted pulses is less than 0.7 of the transmitted wavelength, then the loss in MTI improvement factor due to platform motion is insignificant. Figure 1 shows the upper limits of platform velocity and transmitted frequency and the lower limit radar pulse repetition frequency necessary to meet the conditions of Eq. 1. With these curves, the limitation of any one of these parameters can be determined if the other two are known.

The following conclusions about motion-compensated MTI cancelers [5] can be made:

1. The loss in improvement factor against mainlobe clutter due to platform motion is insignificant. This conclusion is at least not strongly dependent on mainlobe shape.

2. The loss in antenna gain due to the addition of the correction pattern is insignificant.

3. The sidelobe clutter residue is increased for low-sidelobe antenna patterns, because of the relatively high sidelobes of the optimal correction patterns.

4. The sidelobe clutter residue is flatter relative to azimuth angle than the input sidelobe clutter. Since the frequency of the sidelobe clutter also varies with angle, its spectrum is more nearly white and therefore uncorrelated from pulse to pulse.

From these conclusions, the effect of platform motion on the canceled clutter residue can be characterized by a perfect compensation of mainlobe clutter and uncorrelated sidelobe clutter. The level of the sidelobe clutter is determined by the level of the

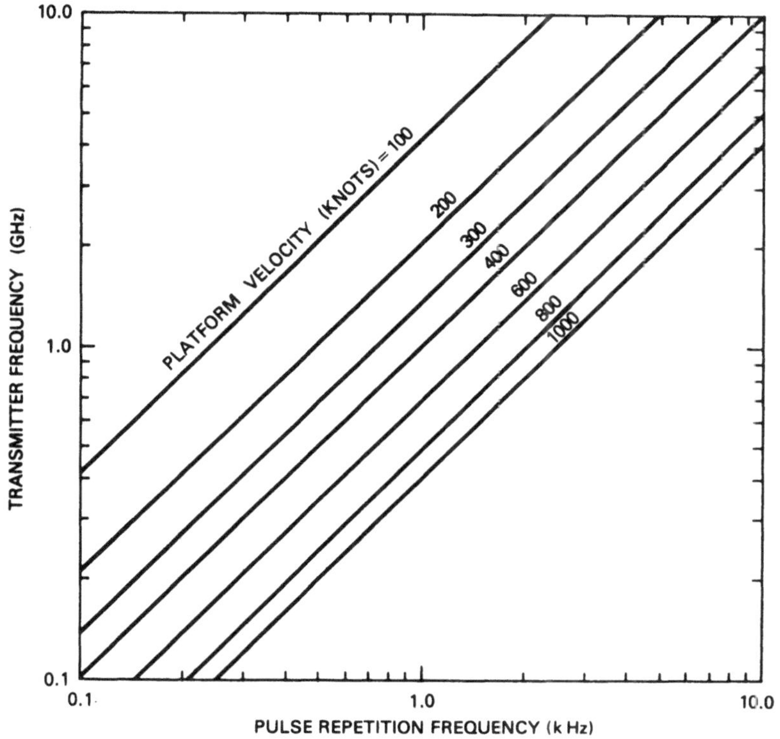

Fig. 1—Relationship among platform velocity, transmitter frequency, and pulse repetition frequency as expressed by Eq. (1) for small grazing angle ϕ_s

sidelobes of the correction pattern relative to the mainlobe of the primary receive pattern. The peaks of this sidelobe clutter seem to be reduced by about 10 dB by the MTI canceler [5].

Correlation Function of a Normalized Clutter Spectrum

It has been stated that platform motion both shifts the mean of the clutter spectrum and increases the spectrum width. This can be illustrated by an incremental patch of clutter with a Gaussian doppler spectrum, so that

$$W(f) = \frac{1}{\sqrt{2\pi}\sigma_c} \exp\left(-f^2/2\sigma_c^2\right). \tag{2}$$

Here, σ_c is the standard deviation of the clutter spectrum. When this clutter energy is received through an antenna pattern at an angle θ relative to the center of the pattern, the clutter spectrum becomes

$$W_1(f, \theta) = |G(\theta)|^4 \frac{1}{\sqrt{2\pi\sigma_c}} \exp(-f^2/2\sigma_c^2).$$

The effect of platform motion is to shift each frequency component by an amount f_d', which after a TACCAR correction is given by [5] becomes

$$f_d'(\theta) = 2\frac{v_p}{\lambda}\cos\phi_s[\cos\theta_a(\cos\theta - 1) - \sin\theta_a \sin\theta] \tag{3}$$

where θ_a is the pointing angle of the antenna pattern relative to the platform velocity vector and ϕ_s is the vertical angle to this clutter patch, measured relative to the horizontal plane.

Including platform motion, the observed doppler spectrum of the incremental clutter patch becomes

$$W_2(f, \theta) = |G(\theta)|^4 \frac{1}{\sqrt{2\pi\sigma_c}} \exp[-(f - f_d')^2/2\sigma_c^2]. \tag{4}$$

The normalized mainlobe clutter is found by integrating between the first nulls of the antenna pattern $(-\theta_0, \theta_0)$:

$$W'(f) = \frac{\displaystyle\int_{-\theta_0}^{\theta_0} |G(\theta)|^4 \frac{1}{\sqrt{2\pi\sigma_0}} \exp[-(f - f_d')^2/2\sigma_c^2]\, d\theta}{\displaystyle\int_{-\theta_0}^{\theta_0} |G(\theta)|^4 d\theta}. \tag{5}$$

The autocorrelation function is formed by taking the Fourier transform of $W'(f)$,

$$\psi_c(\tau) = \int_{-\infty}^{\infty} W'(f) \exp(j2\pi f\tau) df.$$

This leads to

$$\psi_c(\tau) = \frac{\displaystyle\int_{-\theta_0}^{\theta_0} |G(\theta)|^4 \exp[-2\pi^2\sigma_c^2\tau^2 + j2\pi f_d'\tau]\, d\theta}{\displaystyle\int_{-\theta_0}^{\theta_0} |G(\theta)|^4 d\theta}, \tag{6}$$

which can be factored into

$$\psi_c(\tau) = \exp(-2\pi^2\sigma_c^2\tau^2)\,\frac{\displaystyle\int_{-\theta_0}^{\theta_0} |G(\theta)|^4 \exp(j2\pi f_d'\tau)\,d\theta}{\displaystyle\int_{-\theta_0}^{\theta_0} |G(\theta)|^4\,d\theta}\,. \tag{7}$$

Equation (7) gives the autocorrelation function of a Gaussian clutter spectrum with the effects of platform motion. It consists of two factors. The first is simply the autocorrelation function of the clutter spectrum itself, given by Eq. (2). The second shows the effects of platform motion and of the antenna pattern. When platform velocity is zero, then f_d', given by Eq. (3), is zero and the second factor of Eq. (7) is unity.

Correlation Function of Canceled Clutter Residue

Since platform motion correction patterns can be designed to result in an insignificant loss in improvement factor, an MTI canceler can be considered to be perfectly corrected for mainlobe clutter. The resulting power transfer function for the compensated canceler can be written [2]

$$|H(f,\theta)|^2 = [2\sin\pi(f - f_d'(\theta))T]^2, \tag{8}$$

where T is the reciprocal of the pulse repetition frequency (PRF). To show that such a canceler is perfectly compensated, consider an input power spectrum defined by Eq. (4). The output power spectrum is given by

$$W_0(f) = \int_{-\theta_0}^{\theta_0} W_2(f,\theta)\,|H(f,\theta)|^2\,d\theta\,. \tag{9}$$

The autocorrelation function for this clutter residue is defined as

$$\psi_c(\tau) = \int_{-\infty}^{\infty} W_0(f)e^{j2\pi f\tau}\,df\,. \tag{10}$$

If the input clutter power is normalized to unity, the clutter attenuation can be written as

$$CA = \frac{\displaystyle\int_{-\infty}^{\infty} W_0(f)\,df}{\displaystyle\int_{-\theta_0}^{\theta_0} |G(\theta)|^4\,d\theta}\,. \tag{11}$$

From Eqs. (4), (8), and (9), we have

$$CA = \frac{\int_{-\theta_0}^{\theta_0} |G(\theta)|^4 \left[\frac{4}{\sqrt{2\pi}\sigma_c} \int_{-\infty}^{\infty} \sin^2(\pi T x) \exp(-x^2/2\sigma_c^2) dx \right] d\theta}{\int_{-\theta_0}^{\theta_0} |G(\theta)|^4 d\theta}, \tag{12}$$

where $x = f - f'_d$.

Since the bracketed term in Eq. (12) is independent of θ, it can be brought outside the integral. This results in

$$\begin{aligned} CA &= \frac{4}{\sqrt{2\pi}\sigma_c} \int_{-\infty}^{\infty} \sin^2(\pi T x) \exp(-x^2/2\sigma_c^2) dx \\ &= 2(1 - e^{-2\pi^2 T^2 \sigma_c^2}). \end{aligned} \tag{13}$$

Equation (13) is the reciprocal of the usual cancellation ratio of an MTI canceler [6]. It is independent of platform motion, and therefore, the transfer function of Eq. (8) represents a perfectly compensated MTI canceler.

If the input power of Eq. (4) is normalized and Eq. (10) is used, the autocorrelation function of the canceled clutter residue can be determined to be

$$\hat{\psi}_0(\tau) = \frac{\int_{-\infty}^{\infty} W_0(f) e^{j2\pi f\tau} df}{\int_{-\theta_0}^{\theta_0} |G(\theta)|^4 d\theta}.$$

From Eqs. (4) and (9), we have

$$\hat{\psi}_0(\tau) = \frac{\int_{-\theta_0}^{\theta_0} |G(\theta)|^4 \left[\frac{4}{\sqrt{2\pi}\sigma_c} \int_{-\infty}^{\infty} (\sin^2 \pi x T) e^{-\frac{x^2}{2\sigma_c^2}} e^{j2\pi x\tau} dx \right] e^{j2\pi f'_d \tau} d\theta}{\int_{-\theta_0}^{\theta_0} |G(\theta)|^4 d\theta}. \tag{14}$$

Since the bracketed term in Eq. (14) is not a function of θ, it can be brought outside the integral. This leads to

$$\hat{\psi}_0(\tau) = \left[2e^{-2\pi^2\sigma_c^2\tau^2} - e^{-2\pi^2\sigma_c^2(\tau+T)^2} - e^{-2\pi^2\sigma_c^2(\tau-T)^2}\right] \frac{\int_{-\theta_0}^{\theta_0} |G(\theta)|^4 e^{j2\pi f'_d \tau} d\theta}{\int_{-\theta_0}^{\theta_0} |G(\theta)|^4 d\theta}. \quad (15)$$

Equation (15) gives the autocorrelation function of the canceled residue of a Gaussian clutter spectrum, including the effects of platform motion with a perfect motion compensation of the MTI canceler. It consists of two factors. The first is simply the autocorrelation function of the canceled clutter spectrum itself, as given by Eq. (2), and the second shows the effects of platform motion and the antenna pattern.

Comparing Eq. (15) with Eq. (7) reveals that the second factors of each are identical. Therefore, it has been shown that *perfect motion compensation allows the canceler to cancel the original clutter spectrum with no loss, but the effect of platform motion is passed through the canceler unaltered.* Any coherent filters that follow this canceler must deal with the clutter residue whose autocorrelation function is given by Eq. (15).

Antenna Pattern Effects

The effect of the antenna pattern on the canceled clutter residue has two components; mainlobe and the sidelobe effects. The mainlobe effects are determined by evaluating the second factor of Eq. (15) for a particular antenna pattern $G(\theta)$. The sidelobe residues are considered uncorrelated because of their relatively constant level with respect to angle, which is related to frequency [5].

Sidelobe Effects—The optimum correction pattern must be designed and the resulting sidelobe clutter residue of the compensated canceler must be determined [5] to define the level of the uncorrelated sidelobe clutter for a given radar antenna pattern. For the low (-34-dB) sidelobe antenna patterns considered in Ref. 5, the sidelobe level of the composite antenna pattern is dominated by the correction pattern sidelobes (-25 dB relative to the peak of the mainlobe). This results in sidelobe clutter at the -50-dB level. It was seen that this clutter was canceled by about 10 dB, leaving a sidelobe clutter residue at about the -60-dB level. For these patterns, the uncorrelated component can be approximated by doubling the correction pattern sidelobe level and subtracting 10 dB.

Mainlobe Effects—For narrow beamwidth antennas, i.e., for small θ_0, Eq. (3) becomes

$$f'_d(\theta) \approx 2\frac{v_p}{\lambda} \cos\phi_s \sin\vartheta_a \sin\theta$$

$$\approx k_v\theta \quad (16)$$

where

$$k_v = 2\frac{v_p}{\lambda}\cos\phi_s \sin\theta_a . \tag{17}$$

If Eq. (16) is used, the autocorrelation function of the output clutter residue given by Eq. (15) becomes

$$\hat{\psi}_0(\tau) = \left[2e^{-2\pi^2\sigma_c^2\tau^2} - e^{-2\pi^2\sigma_c^2(\tau+T)^2} - e^{-2\pi^2\sigma_c^2(\tau-T)^2}\right] \frac{\int_{-\theta_0}^{\theta_0} |G(\theta)|^4 e^{j2\pi k_v \tau\theta} d\theta}{\int_{-\theta_0}^{\theta_0} |G(\theta)|^4 d\theta} . \tag{18}$$

Since the exact shape of the mainlobe does not significantly affect the clutter rejection, $G(\theta)$ can be approximated by

$$G(\theta) = \cos\left(\frac{\pi}{2}\frac{\theta}{\theta_0}\right) \qquad \text{for } -\theta_0 \leqslant \theta \leqslant \theta_0 \tag{19}$$

where

$$\theta_0 \approx \frac{\lambda}{a}, \tag{20}$$

in which a is the aperture length of the antenna.

From Eqs. (18), (19) and (20), the output autocorrelation function is

$$\begin{aligned}\hat{\psi}_0(\tau) = {} & \left[2e^{-2\pi^2\sigma_c^2\tau^2} - e^{-2\pi^2\sigma_c^2(\tau+T)^2} - e^{-2\pi^2\sigma_c^2(\tau-T)^2}\right] \\ & \times \left[\frac{\sin(k_a\pi/2)}{k_a\pi/2} + \frac{2}{3}\frac{\sin(k_a+2)\pi/2}{(k_a+2)\pi/2} + \frac{2}{3}\frac{\sin(k_a-2)\pi/2}{(k_a-2)\pi/2}\right. \\ & \left. + \frac{1}{6}\frac{\sin(k_a+4)\pi/2}{(k_a+4)\pi/2} + \frac{1}{6}\frac{\sin(k_a-4)\pi/2}{(k_a-4)\pi/2}\right]\end{aligned} \tag{21}$$

where

$$k_a = \frac{8v_p\tau}{a}\cos\phi_s \sin\theta_a \tag{22}$$

Improvement Factor

The processing gain of any arbitrary transversal filter can be computed by taking the ratio of signal output power to clutter output power when the input powers of both signal and clutter are normalized to unity [3]. The processing gain, or improvement factor I, is

$$I = \frac{\mathbf{W}_T^* \mathbf{M}_s \mathbf{W}}{\mathbf{W}_T^* \mathbf{M}_c \mathbf{W}} \tag{23}$$

where $\mathbf{W}$ is the weight vector associated with the transversal filter of interest, $\mathbf{M}_s$ is the covariance matrix of the input signal, and $\mathbf{M}_c$ is the covariance matrix of the input clutter, such that

$$\mathbf{W} = \begin{bmatrix} w_1 \\ w_2 \\ \cdot \\ \cdot \\ \cdot \\ w_N \end{bmatrix}, \quad \mathbf{M}_s = \begin{bmatrix} m_s(1,1) & m_s(1,2) \cdots m_s(1,N) \\ m_s(2,1) & m_s(2,2) \cdots m_s(2,N) \\ \cdot & \\ \cdot & \\ \cdot & \\ m_s(N,1) & m_s(N,2) \cdots m_s(N,N) \end{bmatrix}$$

$$\mathbf{M}_c = \begin{bmatrix} m_c(1,1) & m_c(1,2) \cdots m_c(1,N) \\ m_c(2,1) & m_c(2,2) \cdots m_c(2,N) \\ \cdot & \\ \cdot & \\ \cdot & \\ m_c(N,1) & m_c(N,2) \cdots m_c(N,N) \end{bmatrix}.$$

If the output of the motion-compensated MTI canceller is used as the input, the (i, k) element of clutter covariance matrix $\mathbf{M}_c$ is given by

$$m_c(i, k) = (1 - E_{SL})\hat{\psi}_0((k - i)T) + E_{SL}S(k - i). \tag{24}$$

where $\hat{\psi}_0$ is given by Eq. (21), T is the radar interpulse period, and E_{SL} is the level of the sidelobe clutter at the compensated canceler output relative to the peak of the main-lobe clutter at the canceler input. For the low (-34-dB) sidelobe patterns considered in

Ref. 5, this corresponds in decibels to about twice the level of the peak correction pattern sidelobes relative to the peak of the mainlobe of the primary pattern (2 × (-25 dB) = - 50 dB) minus 10 dB (cancellation by the canceler) equals - 60 dB. Therefore, for this case,

$$E_{SL} = 10^{-6}. \tag{25}$$

At the input to the compensated canceler, "white" system noise has a correlation function

$$\psi_N(\tau) = E_W \delta(\tau)$$

where E_W is the level of the system noise relative to the normalized input clutter level, or approximately the noise-to-clutter ratio.

At the output of the canceler, the noise correlation function is

$$\hat{\psi}_N(\tau) = E_W [2\delta(\tau) - \delta(\tau - T) - \delta(\tau + T)].$$

From Eq. (24), the (i, k) element of the clutter-plus-noise covariance matrix for the canceled residual is given by

$$m_{CN}(i, k) = (1 - E_{SL} - 2E_W)\hat{\psi}_0[(k - i)T] + (E_{SL} + 2E_W)\delta(k - i)$$
$$- E_W \delta(k - i - 1) - E_W(k - i + 1). \tag{26}$$

The signal covariance function for a signal that is on the antenna boresight ($\theta = 0$) and has a doppler shift equally probable of being anywhere within an interval B is given by [3]

$$\psi_s(\tau) = e^{j2\pi f_c \tau} \frac{\sin(\pi B \tau)}{\pi B \tau} \tag{27}$$

where f_c is the center of the interval and B is the length of the interval.

When this signal is passed through the motion-compensated canceler, the output autocovariance function is

$$\hat{\psi}_s(\tau) = 2\psi_s(\tau) - \psi_s(\tau + T) - \psi_s(\tau - T). \tag{28}$$

From Eqs. (27) and (28), the (i, k) element of the signal covariance matrix is

$$m_s(i, k) = \hat{\psi}_s[(k - i)T]. \tag{29}$$

If Eqs. (26) and (29) are used to compute the clutter and signal covariance matrixes, the improvement factor of Eq. (23) can be computed for the transversal filter defined by weight vector **W**. It should be noted that this improvement factor includes the improvement factor of the motion-compensated canceler, since signal and clutter are normalized at the input to this canceler.

CASCADED MTI CANCELERS AND FFT FILTERS

Doppler processing configurations are evaluated in this report using the improvement factor defined by Eq. (23). The signal and clutter-plus-noise covariance matrixes for the output of the motion-compensated canceler are computed from Eqs. (26 and (29). The total improvement factor is then determined for the additional doppler processing defined in terms of the transversal filter weight vector **W**.

For example, consider cascading the motion-compensated canceler with $N - 1$ additional cancelers, as shown in Fig. 2. The transversal filter weights corresponding to these additional cancelers are given by [3]

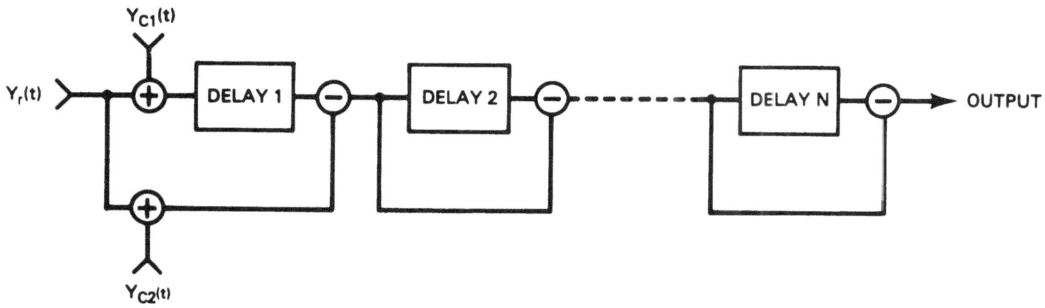

Fig. 2—Motion-compensated MTI canceler followed by $N = 1$ additional cancelers

$$w_i = (-1)^{i-1}\binom{N-1}{i-1}, \qquad i = 1, 2, \cdots, N. \tag{30}$$

Equation (30) for the weight vector, Eq. (29) for the signal covariance matrix, and Eq. (26) for the clutter-plus-noise covariance matrix can be used to derive the improvement factors for N cancelers ($N = 1, 2, \cdots, 7$) as a function of the width of the clutter spectrum $\sigma_c T$. This is shown by the solid curves of Fig. 3 for the following conditions:

$$\left.\begin{aligned} 0.001 \leqslant \sigma_c T &\leqslant 0.1 \\ E_{SL} &= 0 \\ E_W &= 0 \\ \frac{v_p^T}{\lambda}\cos\phi_s &= 0.7 \\ a &= 10\,\lambda \\ \theta_a &= 90^\circ \\ B &= 1/T \\ f_c &= 1/2T \end{aligned}\right\} \tag{31}$$

Antenna aperture a is an important parameter in determining the effects of platform motion. The effects of the fraction of the aperture that the platform moves between pulses are given by Eq. (22). A typical aperture of 10 λ is used throughout this report to evaluate and compare doppler processors. Setting the antenna pointing angle to 90° maximizes the effect of platform motion, as seen by Eq. (22).

In Eq. (27), letting $B = 1/T$ and $f_c = 1/2T$ defines a target whose doppler shift is equally probable of having any value in the ambiguous doppler interval (0, $1/T$).

Only mainlobe clutter is considered in Fig. 3, since sidelobe level E_{SL} and noise level E_W are set equal to zero. The dashed curves of Fig. 3 show the improvement factors when platform velocity is zero. Therefore, the differences between solid and dashed curves represent loss in improvement factor due to platform motion. When the clutter spectral width is small, this loss in cancellation of mainlobe clutter is significant.

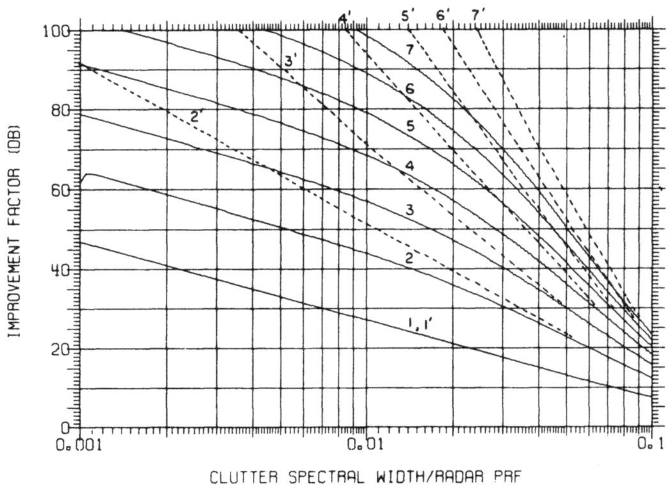

Fig. 3—Improvement factors of cancelers in Fig. 2 for $N = 1$ through 7, plotted against mainlobe clutter. The solid curves illustrate the improvement factor with platform motion as defined by Eq. (1) and an antenna pointing angle of 90°. The dashed curves illustrate the improvement factor with no platform motion.

Figure 4 shows the same cascaded cancelers when sidelobe clutter is added. In this case,

$$E_{SL} = 10^{-6},$$

which corresponds to the -25-dB correction pattern sidelobe level of Ref. 5. With this sidelobe clutter level, the cascaded cancelers are limited by sidelobe clutter to an improvement factor of about 66 dB when the clutter spectral width is very narrow. Comparing Figs. 3 and 4 shows that losses due to platform motion for narrow clutter spectra are

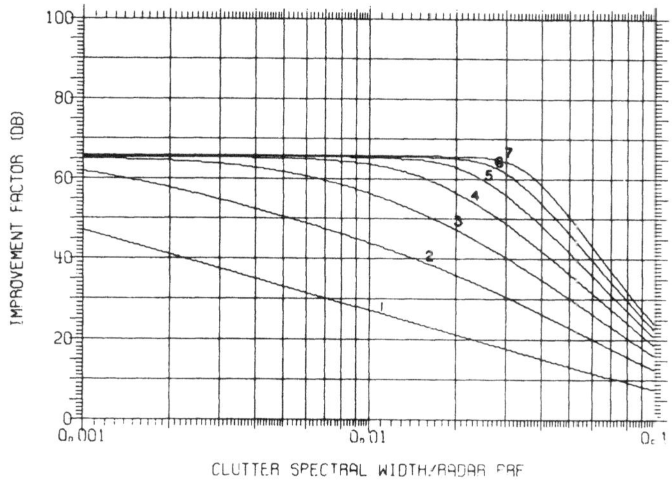

Fig. 4—Improvement factors of cancelers in Fig. 2 for $N = 1$ through 7, plotted against mainlobe clutter plus sidelobe clutter 50 dB below mainlobe clutter

overshadowed by the uncanceled sidelobe clutter residue. For this reason, in most practical systems the sidelobe clutter level is a more important consideration than the effect of platform motion on mainlobe clutter.

Figure 4 shows the *average* improvement factor of a target with unknown doppler as a function of the width of the clutter spectrum. To illustrate the improvement factor for a target with a known doppler shift f_c, let

$$B = 0$$

$$0 \leqslant f_c \leqslant 1/T$$

in Eq. (27). This is shown in Fig. 5 for the conditions of Fig. 4, with the ratio of clutter spectral width to radar PRF of 0.01. Figure 4 shows that the average improvement factor increases as the number of cancelers is increased until the sidelobe clutter limitation is reached. Figure 5 shows that targets with very low doppler shift or a very nearly "blind" velocity ($f_c = 0, 1/T, 2/T, \cdots$) may have larger improvement factors when fewer cancelers are cascaded. Therefore, although the average improvement factor is larger if more cancelers are used, the "blind" velocity regions are also larger. This should be considered in selecting the number of cancelers to be used in a particular clutter environment.

Since the left half of Fig. 5 is the mirror image of the right half, only the left half needs to be shown as in Fig. 6. The advantage of fewer cancelers for low-speed targets or targets very near a "blind" velocity is more clearly seen in this figure.

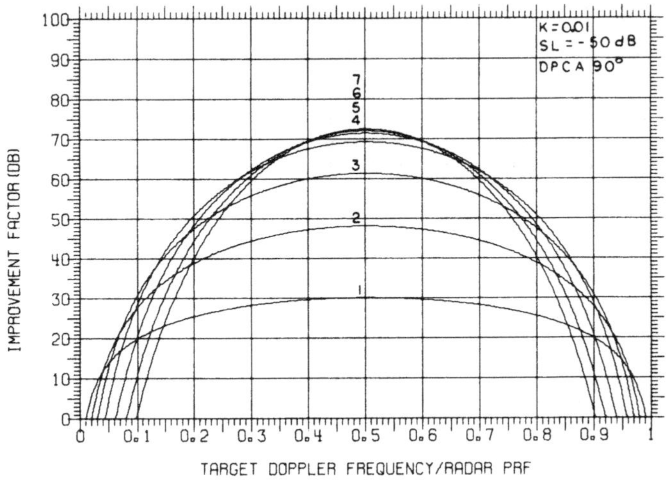

Fig. 5—Improvement factors of cancelers in Fig. 2 for $N = 1$ through 7 plotted against target doppler shift. Sidelobe clutter is 50 dB below mainlobe clutter. The ratio of clutter spectral width to radar PRF is 0.01.

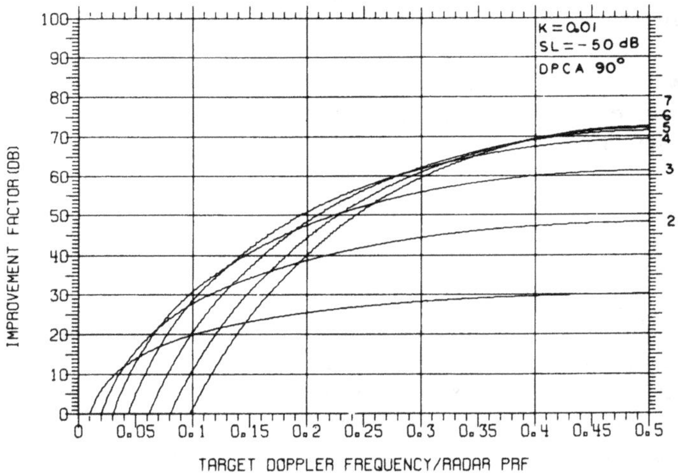

Fig. 6—Improvement factors shown in Fig. 5, with horizontal axis expanded. Target doppler shift from zero to one-half the PRF is shown.

Single Canceler and FFT Filters

Since the advent of the FFT algorithm and continually improving digital hardware technology, an attractive configuration of radar doppler processing consists of MTI cancelers cascaded with FFT filters. To analyze the performance of a single canceler cascaded with an FFT, the signal and clutter-plus-noise covariance matrixes defined by Eqs. (26)

and (29) can be used with Eq. (23) to compute the improvement factor. The weight vector **W** of Eq. (23) is simply the digital Fourier transform equation. Since the N-point FFT is an implementation of N contiguous filters, the weight vector for the nth filter is computed from

$$w_{ni} = u_i \exp\left[-j2\pi(n-1)(i-1)/N\right], \qquad i = 1, 2, \cdots, N \tag{32}$$

where u_i are weights for controlling the filter sidelobes. From Eq. (32) and Eq. (23), the improvement factor of an MTI canceler cascaded with the nth filter of the FFT can be computed. In Fig. 7 the conditions of Eq. (31) are used, except that

$$\begin{aligned} \sigma_c T &= 0.01 \\ 0 \leqslant f_c T &\leqslant 0.5 \\ B &= 0\,. \end{aligned} \tag{33}$$

The improvement factors of the N resulting filters are illustrated for $N = 16$, 32, and 64, and for $u_i = 1$, $i = 1, 2, \cdots, N$. As target doppler shift f_c is varied from zero to half the PRF, the improvement factor of the filter with the maximum response is plotted. The filter numbers shown correspond to the FFT outputs; $n = 1$ is the zero frequency output, $n = 2$ is the $1/NT$ frequency output, $n = 3$ is the $2/NT$ output, and so on.

Comparing Figs. 7a, 7b, and 7c, which illustrate the performance of a 16-pulse FFT, a 32-pulse FFT, and a 64-pulse FFT respectively, shows that the increase in performance is about 5-6 dB each time the number of pulses is doubled. This can be compared to the performance against "white" noise, which increases 3 dB each time the number of pulses is doubled. These conclusions do not apply for the filters very near the clutter (or MTI notch) at zero doppler shift.

In Fig. 8, the sidelobe clutter is increased by 10 dB, which corresponds to a primary antenna sidelobe level of about −29 dB, instead of the −34 dB used in Fig. 7. Comparing Fig. 8a with Fig. 7a and Fig. 8b with Fig. 7b shows that although sidelobe clutter is increased by 10 dB the loss in improvement factor is less than 1 dB. Therefore, for these conditions doppler processing is not limited by the antenna sidelobes.

In Fig. 9, antenna pointing angle θ_a is set to 0°, sidelobe clutter E_{SL} is set to zero, and white noise level E_W is set to 10^{-6}. Comparing Fig. 9 with Fig. 7 shows that filters far away from the clutter (zero doppler) have about the same improvement factors in each figure. Therefore, these doppler processors in Fig. 9 are not limited by white noise under these conditions. Comparing the filters very near the clutter (zero doppler) in Fig. 9 and Fig. 7, shows that filters 2 and 3 of the 16-pulse FFT in Fig. 9a have an improvement factor 3-4 dB higher than the same filters in Fig. 7a. Equivalent conclusions can be made for 32-pulse and 64-pulse FFTs. This increased improvement factor is due to the narrower clutter spectrum at this antenna pointing angle.

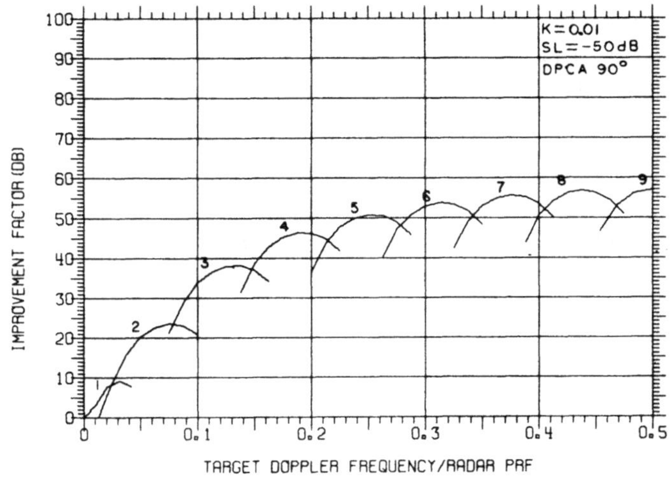

(*a*) MTI canceler and a 16-pulse FFT

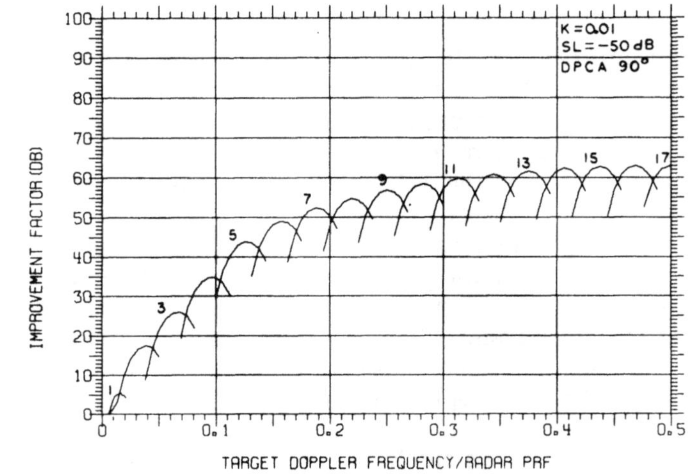

(*b*) MTI canceler and a 32-pulse FFT

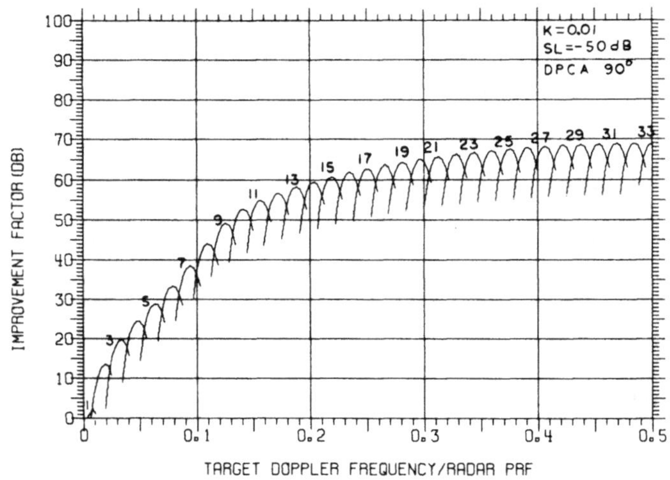

(*c*) MTI canceler and a 64-pulse FFT

Fig. 7—Improvement factor for motion-compensated MTI canceler cascaded with FFT filter bank. Platform motion is as defined by Eq. (1), the antenna pointing angle is 90°, and the ratio *K* of clutter spectral width to radar PRF is 0.01. Sidelobe clutter is at the -50-dB level.

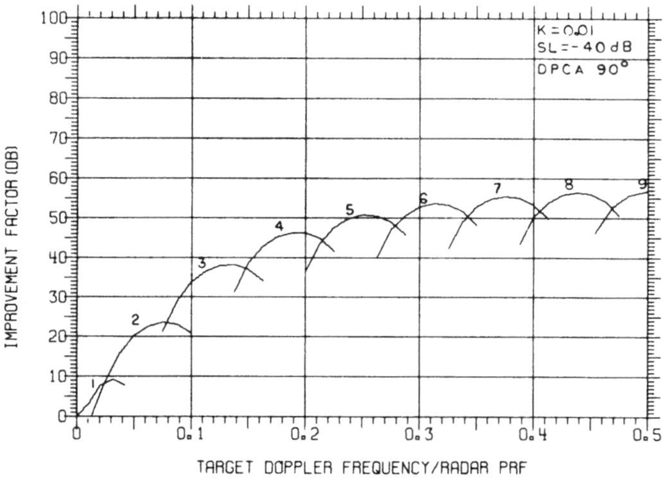

(*a*) MTI canceler and a 16-pulse FFT

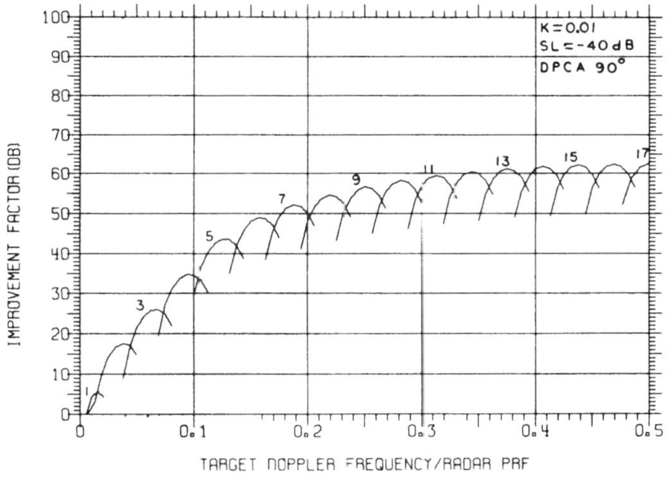

(*b*) MTI canceler and a 32-pulse FFT

Fig. 8—Improvement factor for motion-compensated MTI canceler cascaded with FFT filter bank. Platform motion is as defined by Eq. (1), the antenna pointing angle is 90°, and the ratio K of clutter spectral width to radar PRF is 0.01. Sidelobe clutter is at the -40-dB level.

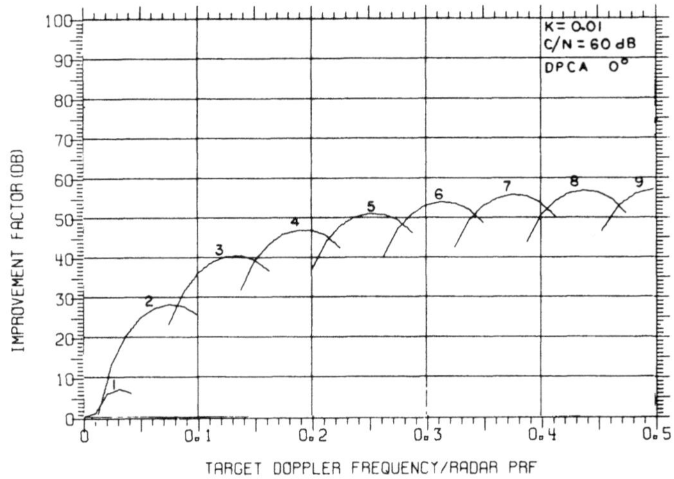

(*a*) MTI canceler and a 16-pulse FFT

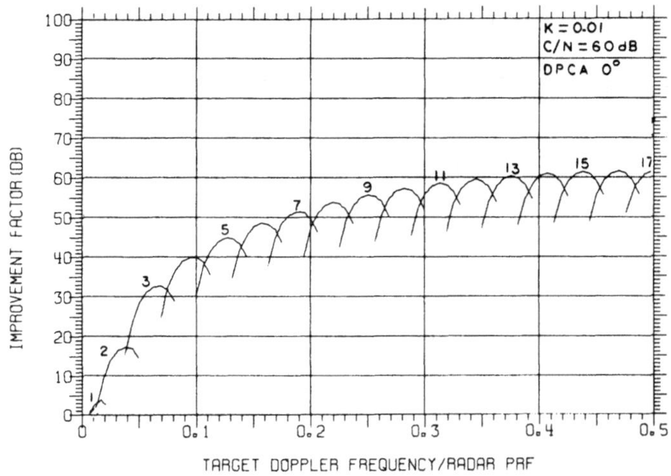

(*b*) MTI canceler and a 32-pulse FFT

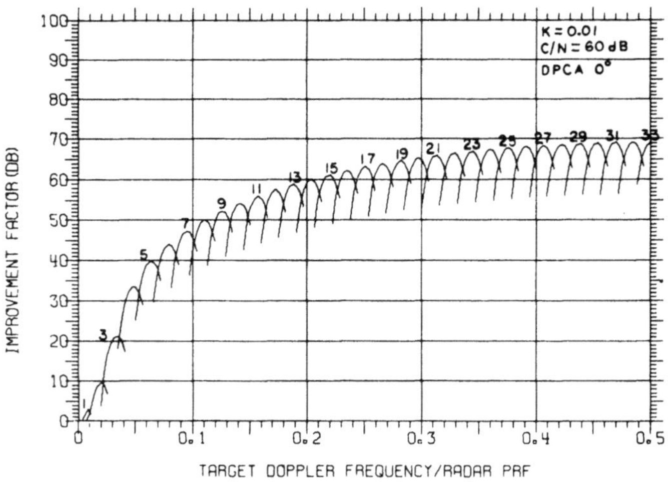

(*c*) MTI canceler and a 64-pulse FFT

Fig. 9—Improvement factor for MTI canceler cascaded with FFT filter bank. There is no platform motion, and the ratio K of clutter spectral width to radar PRF is 0.01. The clutter-to-noise ratio is 60 dB.

Double Canceler and FFT Filters

The covariance matrixes of Eqs. (26) and (29) correspond to the output of a single motion-compensated canceler. Therefore, the performance of a *double* canceler cascaded with FFT filters can be analyzed by applying to these covariance matrixes the equivalent transversal filter weights of a *single* canceler cascaded with FFT filters.

The equivalent transversal filter weights of cascaded digital filters, derived in Ref. 3, are given by

$$\mathbf{W} = \hat{\mathbf{A}}\mathbf{B} \tag{34}$$

where

$$\mathbf{B} = \begin{bmatrix} b_1 \\ b_2 \\ b_3 \\ \cdot \\ \cdot \\ \cdot \\ b_N \end{bmatrix} = \text{the FFT weights of Eq. (32)}$$

and

$$\hat{\mathbf{A}} = \begin{bmatrix} a_1 & 0 & 0 & \cdots & 0 \\ a_2 & a_1 & 0 & & \cdot \\ a_3 & a_2 & a_1 & & \cdot \\ \cdot & a_3 & a_2 & & \cdot \\ \cdot & \cdot & a_3 & & 0 \\ \cdot & \cdot & \cdot & & a_1 \\ a_m & \cdot & \cdot & & a_2 \\ 0 & a_m & \cdot & & a_3 \\ \cdot & \cdot & a_m & & \cdot \\ \cdot & \cdot & \cdot & & \cdot \\ \cdot & \cdot & \cdot & & \cdot \\ 0 & 0 & 0 & \cdots & a_m \end{bmatrix}$$

(the first five entries of the last column, $0, \cdot, \cdot, \cdot, 0$: $N - 1$ zeros; the last five entries of the first column, $0, \cdot, \cdot, \cdot, 0$: $N - 1$ zeros)

M is the number of pulses processed in the MTI canceler. In this case, for a single canceler in addition to the motion-compensated canceler,

$$\left.\begin{aligned} M &= 2 \\ a_1 &= 1 \\ a_2 &= -1 \end{aligned}\right\}. \tag{35}$$

The equivalent weight vector of Eq. (34) is

$$\mathbf{W} = \begin{bmatrix} w_1 \\ w_2 \\ w_3 \\ \cdot \\ \cdot \\ \cdot \\ w_{M+N-1} \end{bmatrix}. \tag{36}$$

The dimension of this vector, $M + N - 1$, is the total number of pulses processed by this filter, and the dimension of the square covariance matrixes of Eqs. (26) and (29) is also $M + N - 1$. From the covariance matrixes and the weight vector described above, the improvement factor of Eq. (23) can be computed for a double canceler cascaded with FFT filters.

Figure 10 shows the improvement factors of a double canceler cascaded with a 16-pulse FFT, a 32-pulse FFT, and a 64-pulse FFT. The conditions of Fig. 10 are the same as those of Fig. 7; that is, the conditions of Eqs. (31) and (33). Comparing Figs. 10a, 10b, and 10c, except for the filters very near the clutter (zero doppler), shows that once again the increase in improvement factor is 5-6 dB as the number of pulses is doubled.

The increased improvement factor due to the second canceler is about 13-16 dB, as shown by comparing Fig. 10 with Fig. 7. This applies to all the filters except the ones very near the clutter of "blind" velocities. Those near the clutter maintain about the same improvement factor for either a single or double canceler. Therefore, the additional canceler does not widen the "blind" velocity notches when cascaded with the FFT filters.

In Fig. 11 sidelobe clutter is increased 10 dB, but the loss in improvement factor is only 6-7 dB. This shows that these doppler processors are being limited by both mainlobe and sidelobe clutter and that adding additional cancelers would only cause the system to be more limited by sidelobe clutter.

In Fig. 12, as in Fig. 9, the following conditions are changed:

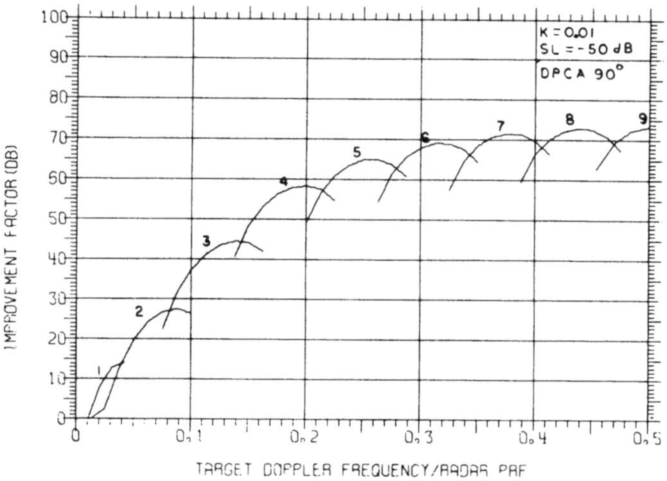

(*a*) Double canceler and a 16-pulse FFT

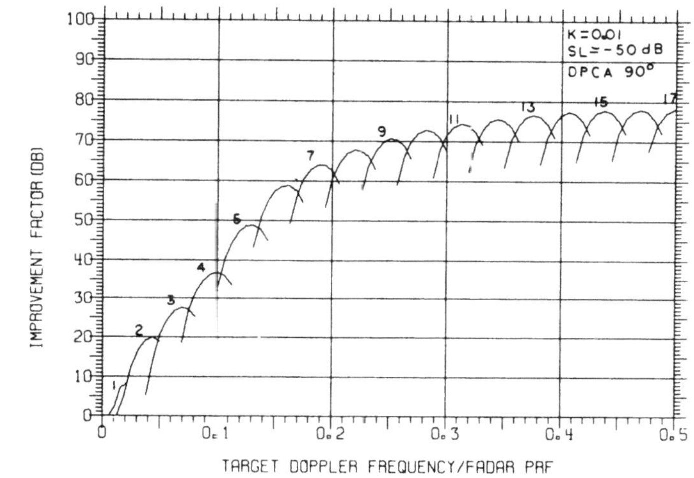

(*b*) Double canceler and a 32-pulse FFT

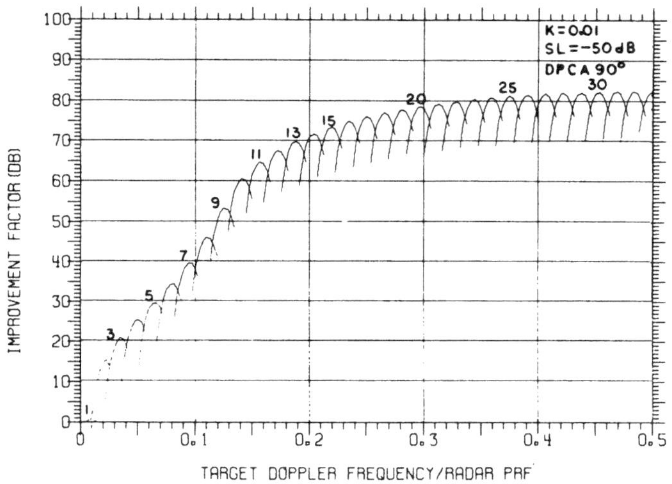

(*c*) Double canceler and a 64-pulse FFT

Fig. 10—Improvement factor for a motion-compensated canceler cascaded with an additional canceler and an FFT filter bank. Platform motion is as defined by Eq. (1), the antenna pointing angle is 90°, and the ratio *K* of clutter spectral width to radar PRF is 0.01. Sidelobe clutter is at the -50-dB level.

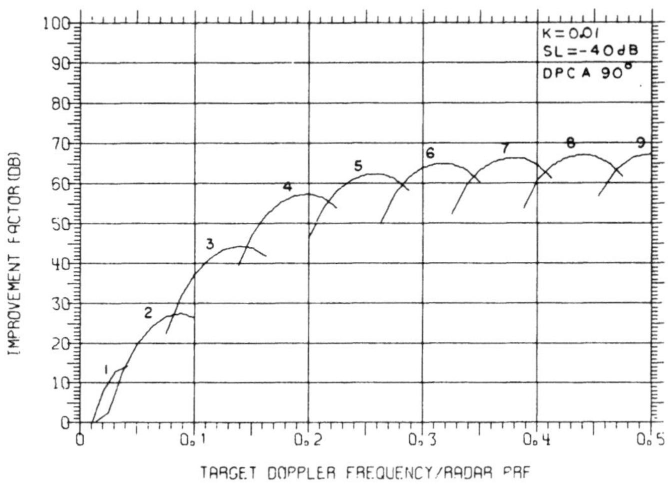

(*a*) Double canceler and a 16-pulse FFT

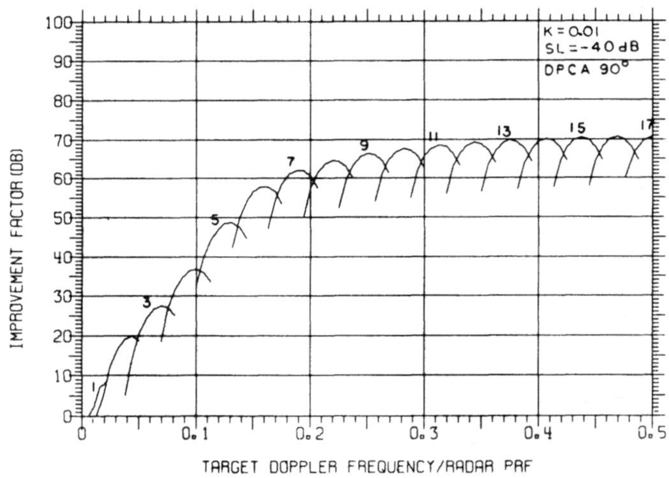

(*b*) Double canceler and a 32-pulse FFT

Fig. 11—Improvement factor for a motion-compensated canceler cascaded with an additional canceler and an FFT filter bank. Platform motion is as defined by Eq. (1), the antenna pointing angle is 90°, and the ratio K of clutter spectral width to radar PRF is 0.01. Sidelobe clutter is at the -40-dB level.

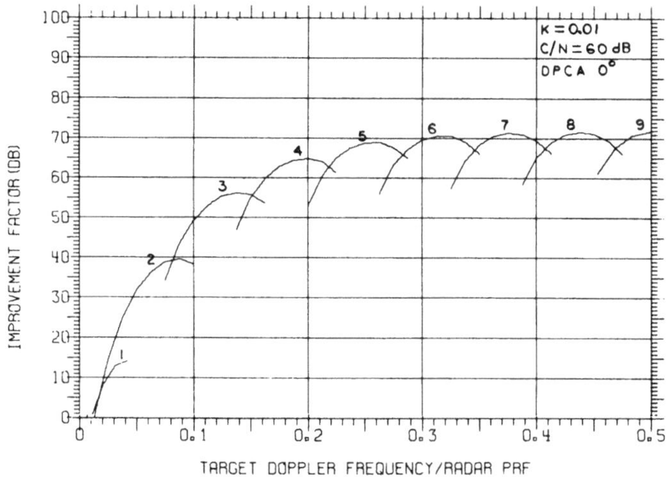

(*a*) Double canceler and a 16-pulse FFT

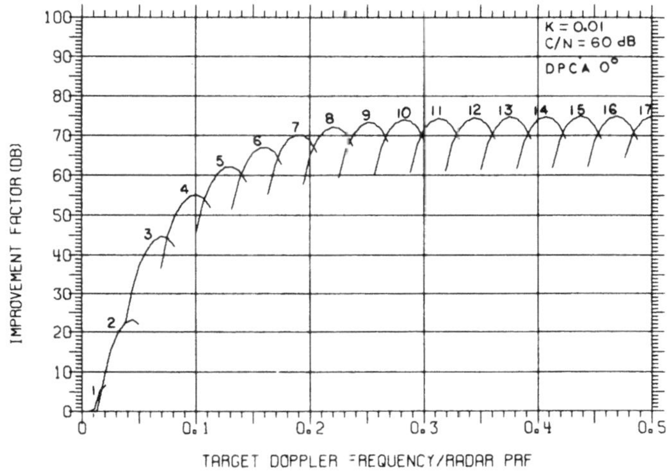

(*b*) Double canceler and a 32-pulse FFT

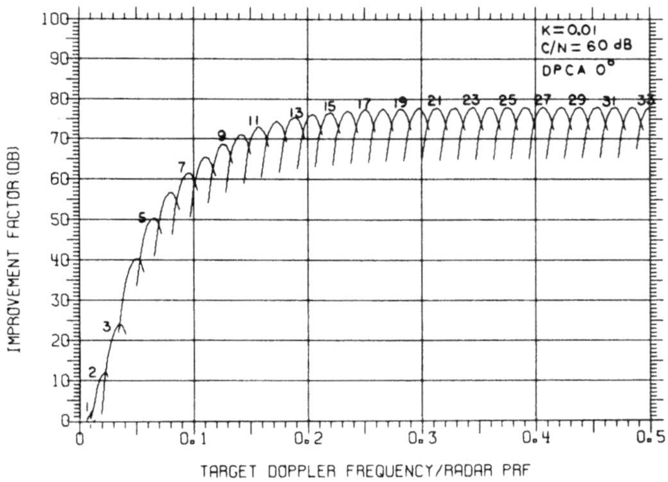

(*c*) Double canceler and a 64-pulse FFT

Fig. 12—Improvement factor for double canceler MTI cascaded with an FFT filter bank. There is no platform motion, and the ratio K of clutter spectral width to radar PRF is 0.01. The clutter-to-noise ratio is 60 dB.

$$\theta_a = 0$$

$$E_{SL} = 0$$

$$E_W = 10^{-6}.$$

It has been shown [4] that the maximum improvement factor for this class of doppler processors is limited to

$$I = (C/N) + N_I \tag{37}$$

where C/N is the clutter-to-noise ratio and N_I is the number of pulses processed by the FFT. For this case,

$$C/N \approx 1/E_w = 10^6,$$

which corresponds to a maximum improvement factor of 72 dB for the 16-pulse FFT, 75 dB for the 32-pulse FFT, and 78 dB for the 64-pulse FFT. These values are achieved by filters farthest from the clutter spectrum. Adding more cancelers would not increase the improvement factor for these filters, but it would increase the improvement of other filters so that more filters would achieve the maximum improvement factor and the response curves of Fig. 12 would become more nearly flat.

Comparing Fig. 12 with Fig. 10 shows that the narrower clutter spectral width with $\theta_a = 0°$ results in higher improvement factors for the filters near the clutter, as was the case for the single-canceler/FFT configuration. This means the "blind" velocity notches are narrower when the antenna is pointing straight ahead or astern.

CASCADED MTI CANCELERS AND OPTIMAL FILTERS

Optimization of radar doppler filters consists of finding the weight vector **W** that maximizes the improvement factor of Eq. (23). The form of the solution depends on a priori knowledge of the signal and clutter covariance matrixes. Emerson [7] showed that when a narrowband signal with a doppler shift equally probable of any value in the ambiguous doppler region $(0, 1/T)$ is assumed, the optimum weight vector is given by the eigenvector associated with the smallest eigenvalue of the clutter covariance matrix. Brennan and Reed [8] showed that for a known target doppler shift the optimum weight vector is

$$\mathbf{W}_{\text{opt}} = \mathbf{M}_c^{-1}\mathbf{S}^*, \tag{38}$$

where $\mathbf{S}^*$ is the complex conjugate of the vector corresponding to the signal with a known doppler shift. The ambiguous doppler region is covered by a bank of such filters optimized for equall,' spaced doppler shifts.

In Ref. 9 it is shown that when the ambiguous doppler region is divided into N equal intervals, the weight vector that maximizes the improvement factor is the eigenvector associated with the largest eigenvalue of the matrix formed by the product of the inverse of the clutter covariance matrix times the signal covariance matrix. This signal covariance matrix can be derived from Eqs. (27) and (28) by letting

$$B = 1/NT. \tag{39}$$

Since this solution is optimized over the entire ambiguous doppler region, it gives the maximum improvement factor that can be achieved by processing N returns.

Optimal Filter Design

The optimal filter design considered in this report is described in Ref. 4, which shows that the eignevector solution for the optimal filter bank described above can be approximated by

$$\mathbf{W}_{\text{opt}} = \mathbf{M}_c^{-1}\overline{\mathbf{S}}^*, \tag{40}$$

where the signal vector $\overline{\mathbf{S}}$ for the nth filter corresponds to a target whose doppler shift has equal probability of occurring anywhere in an interval

$$\left(\frac{2n-1}{2NT}, \frac{2n+1}{2NT}\right) \qquad M = 1, 2, \cdots, N.$$

The ith component of this signal vector for the nth filter is

$$\bar{s}_i = \exp\left[j\frac{2\pi(n-1)}{N}\left(i - \frac{N+1}{2}\right)\right] \frac{\sin\left[\frac{\pi}{N}\left(i - \frac{N+1}{2}\right)\right]}{\frac{\pi}{N}\; i - \frac{N+1}{2}} \tag{41}$$

where

$$i = 1, 2, \cdots, N.$$

Since we are interested in the optimal filters to follow a motion-compensated canceler, the clutter covariance matrix for Eq. (40) is computed from Eq. (26), and the signal vector of Eq. (41) must be extended to include the effects of the MTI canceler. A target at the center of the antenna pattern is not affected by platform motion if there is no error in the TACCAR loop. This can be seen from the second term on the right-hand side of Eq. (18), which shows the effect of platform motion of the clutter residue. This term is unity when $\theta = 0$. The output signal for such a canceler is

$$y(t) = s_{in}(t) - s_{in}(t - T), \tag{42}$$

where $s_{in}(t)$ is the input signal. When $s_{in}(t)$ corresponds to the signal vector of Eq. (41), the vector for the output signal is

$$\bar{y}_i = \bar{s}_i - \bar{s}_{i-1}, \qquad i = 1, 2, \cdots, N. \tag{43}$$

The optimum weight vector for the cascaded filters is

$$\mathbf{W}_{\text{opt}} = \mathbf{M}_{CN}^{-1} \overline{\mathbf{Y}}^* \tag{44}$$

where $\mathbf{M}_{CN}^{-1}$ is the inverse of the clutter-plus-noise covariance matrix of Eq. (26) and $\overline{\mathbf{Y}}^*$ is the complex conjugate of the signal vector defined by Eqs. (43) and (41).

Optimal Filter Performance

The difference between the optimal filter bank of this report and the optimal filters of Ref. 4 is that in this report the filter bank is preceded by an MTI canceler, to take advantage of the platform motion compensation techniques. As has been stated, this motion-compensated canceler achieves essentially the same improvement factor as an MTI canceler without platform motion. In this way, only the succeeding doppler filters must contend with the platform motion spectrum.

The effects of the MTI canceler on total system performance are most conveniently evaluated in terms of z-transforms [10]. In the z-plane, the filters considered in this report are defined in terms of the location of the zeros of their transfer functions. The number of zeros is $N - 1$, where N is the number of pulses processed. The optimization can be thought of as optimally locating these zeros.

When the radar is not on a moving platform, an MTI canceler corresponds to a zero on the unit circle at zero doppler shift. If this canceler is followed by an optimized doppler filter bank, the location of the zeros of the filter bank are then determined under the constraint that a zero already exists on the unit circle at zero doppler shift. This constraint leads to somewhat less performance than could be achieved without the constraint. Figure 13a shows the improvement factor of an optimized 16-pulse processor without this constraint. The covariance matrix for the Gaussian clutter spectrum is given by

$$m_{CN}(i, k) = (1 - E_w)e^{-2\pi^2 \sigma_c^2 T^2 (k-i)^2} + E_w \delta(k - i). \tag{45}$$

The signal covariance matrix is derived from Eq. (27). Figure 13b shows the improvement factor of an optimized 16-pulse processor with the zero at zero doppler shift constraint. Equations (26) and (29) are used to compute the clutter-plus-noise covariance matrix and the signal covariance matrix, respectively. The following parameters are used:

$$E_w = 10^{-6}$$

$$E_{SL} = 0$$

$$B = 1/NT, \qquad N = 16$$

$$f_c = (n - 1)/NT, \text{ for the nth filter}$$

$$0.005 \leqslant \sigma_c T \leqslant 0.5.$$

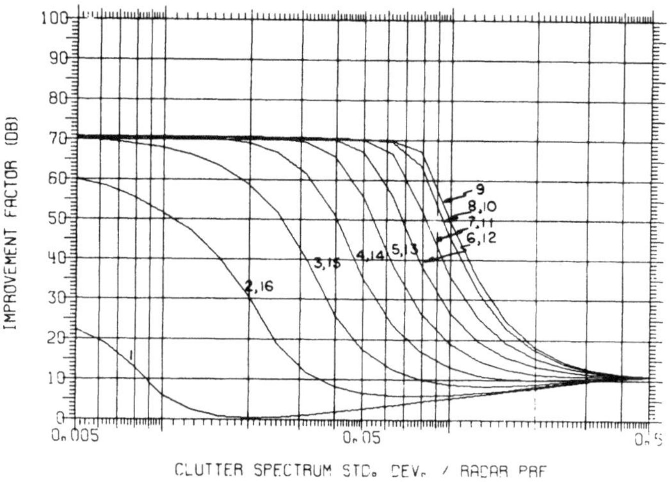

(*a*) Optimal doppler filter bank

(*b*) MTI canceler cascaded with optimal filter bank

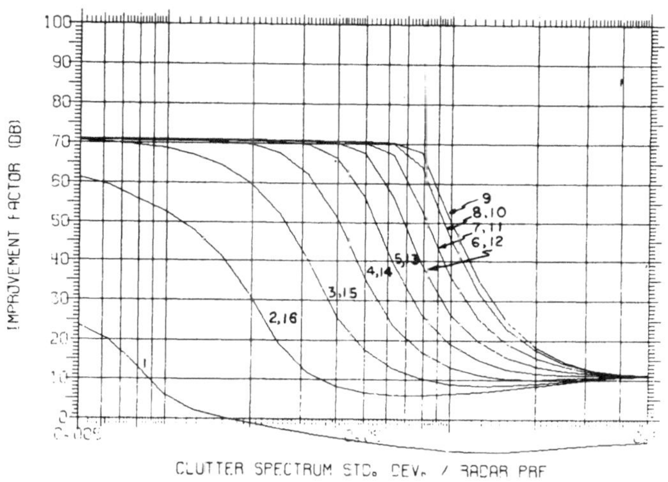

Fig. 13—Improvement factor of an optimally weighted filter bank for 16 pulses. The clutter-to-noise ratio is 60 dB.

Comparing Figs. 13a and 13b shows that the filters very near the clutter at zero doppler shift are most affected by the canceler. Filter 1 of Fig. 13b is most affected. Filters 2 and 16 are affected somewhat. The other filters are not noticeably affected. When the clutter spectrum is very broad, filter 1 in Fig. 13b actually has a negative improvement factor. The optimum filter in Fig. 13a when the clutter spectrum is very broad approaches a $(\sin\psi)/\psi$ shape [9]. However, the MTI null at zero doppler keeps the total system response of filter 1 in Fig. 13b from approaching the $(\sin\psi)/\psi$ shape for wide clutter spectra. This null has less effect on the other filters, and therefore the improvement factor of those filters is affected less. From Fig. 13 it can be seen that the MTI canceler would probably not be used except on a moving platform.

Figure 14 shows the improvement factor of a motion-compensated canceler followed by an N-pulse optimal doppler filter bank for $N = 16$, 32, and 64. The weights for the optimal filters are computed from Eq. (44), under the conditions of Eqs. (31) and (33). These are the same conditions used for Figs. 7 and 10. When the number of pulses is doubled in Fig. 14, the increase in improvement factor ranges from about 3 dB for the filters near the center of the ambiguous doppler region ($\sigma_c T = 0.5$), to nearly 10 dB for $\sigma_c T$ of about 0.17, and then to very little (0 dB) near the canceler notch ($\sigma_c T < 0.1$).

Comparing Fig. 14 with Fig. 10 indicates that the advantage of the optimum weighted filters over the FFT is in the performance of the filters in region of $0.15 < \sigma_c T < 0.85$, remembering that the other half of the doppler response ($0.5 < \sigma_c T < 1.0$) is simply the mirror image of the response in the region shown ($0 \leqslant \sigma_c T \leqslant 0.5$). The optimum filters achieve their maximum over a larger part of the doppler region, which results in a more constant overall response for targets of different velocities. Also, the maximum is larger for the optimum filters (about 2-5 dB), with a greater increase for $N = 16$ and a smaller one for $N = 64$.

In Fig. 15 the sidelobe clutter is increased 10 dB, and the loss in improvement factor is also about 10 dB, as compared with Fig. 14. This shows that the optimum filters are limited by sidelobe clutter even for the low sidelobe level of Fig. 14. The FFT filters of Fig. 10 are not limited by sidelobe clutter under these same conditions.

Comparing Fig. 15 with Fig. 11 shows that with the high sidelobe clutter there is only a small increase in improvement factor for the optimum weighted filters as compared to FFT filters. This means that the FFT filters achieve very near optimum performance for high sidelobe clutter.

In Fig. 16, as in Figs. 9 and 12, the following conditions are changed:

$$\theta_a = 0$$
$$E_{SL} = 0$$
$$E_W = 10^{-6}.$$

Again, the maximum improvement factor under these conditions is 72 dB for $N = 16$, 75 dB for $N = 32$, and 78 dB for $N = 64$. These values are achieved by all the filters except the ones very near the canceler notch, and therefore the improvement factor for these filters is limited by the clutter-to-noise ratio under these conditions. Comparing Fig. 16

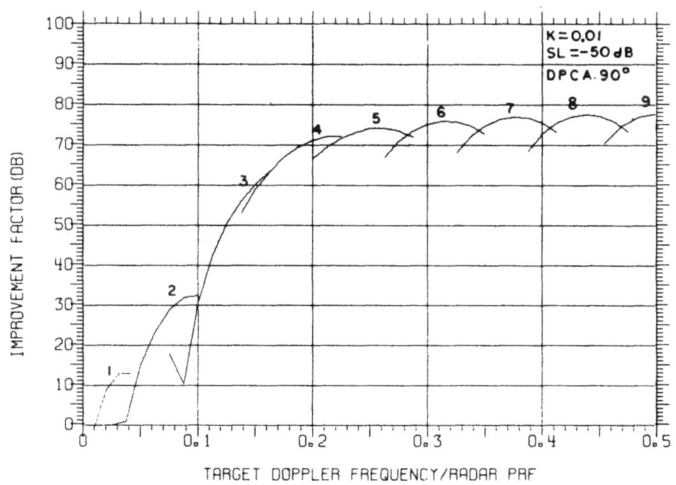

(*a*) MTI canceler and a 16-pulse optimal filter bank

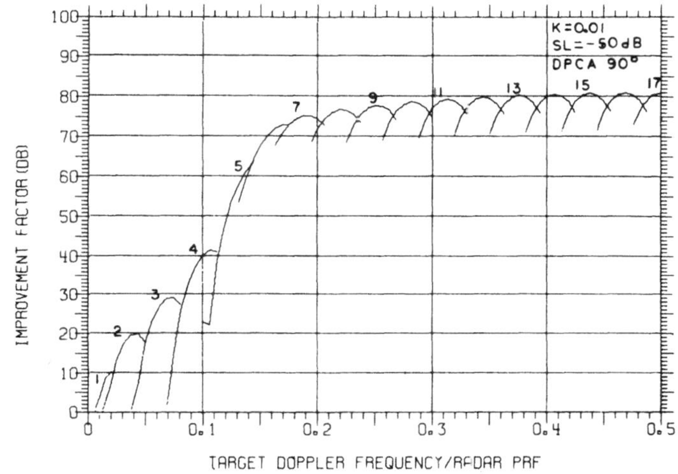

(*b*) MTI canceler and a 32-pulse optimal filter bank

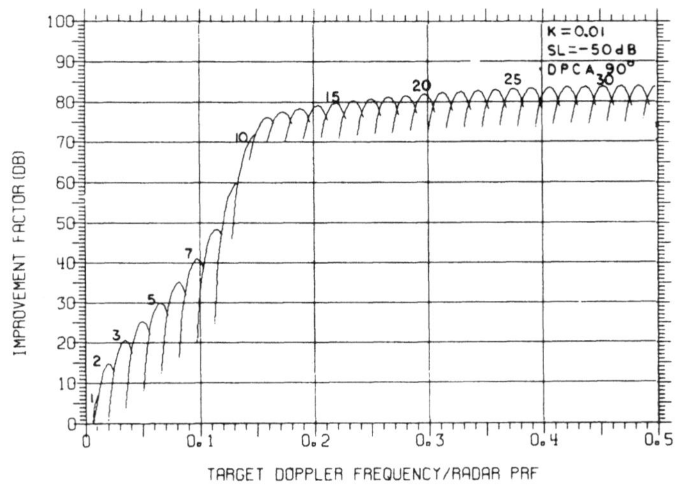

(*c*) MTI canceler and a 64-pulse optimal filter bank

Fig. 14—Improvement factor of a motion-compensated MTI canceler cascaded with an optimally weighted filter bank. Platform motion is as defined by Eq. (1), the antenna pointing angle is 90°, and the ratio K of clutter spectral width to radar PRF is 0.01. Sidelobe clutter is at the -50-dB level.

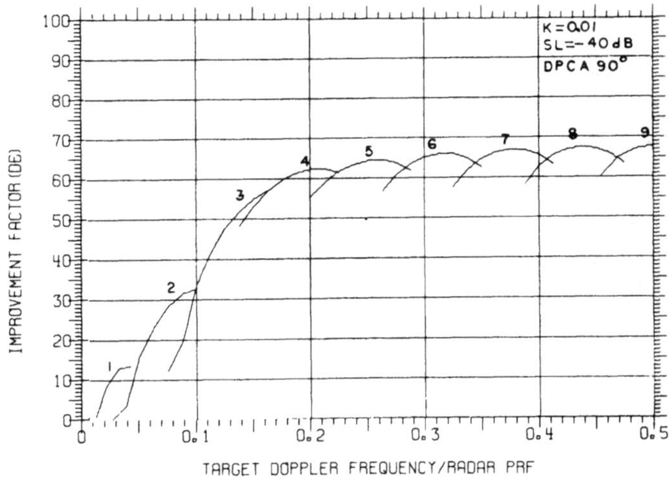

(*a*) MTI canceler and a 16-pulse optimal filter bank

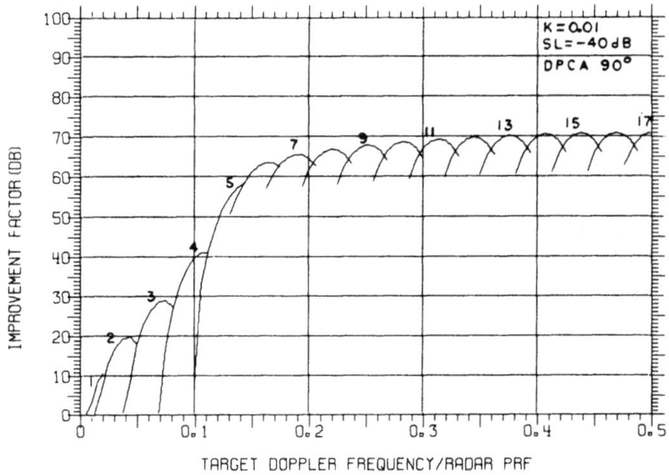

(*b*) MTI canceler and a 32-pulse optimal filter bank

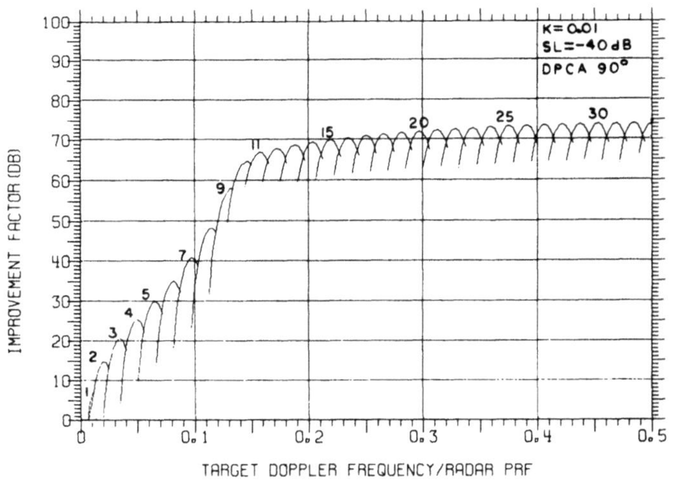

(*c*) MTI canceler and a 64-pulse optimal filter bank

Fig. 15—Improvement factor of a motion-compensated MTI canceler cascaded with an optimally weighted filter bank. Platform motion is as defined by Eq. (1), the antenna pointing angle is 90°, and the ratio K of clutter spectral width to radar PRF is 0.01. Sidelobe clutter is at the -40-dB level.

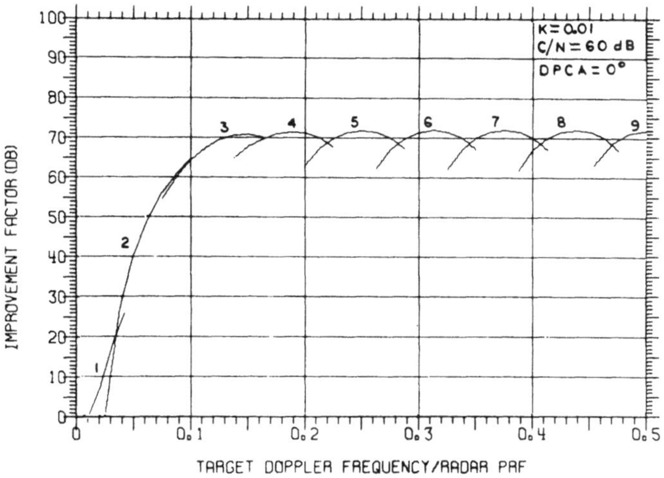

(*a*) MTI canceler and a 16-pulse optimal filter bank

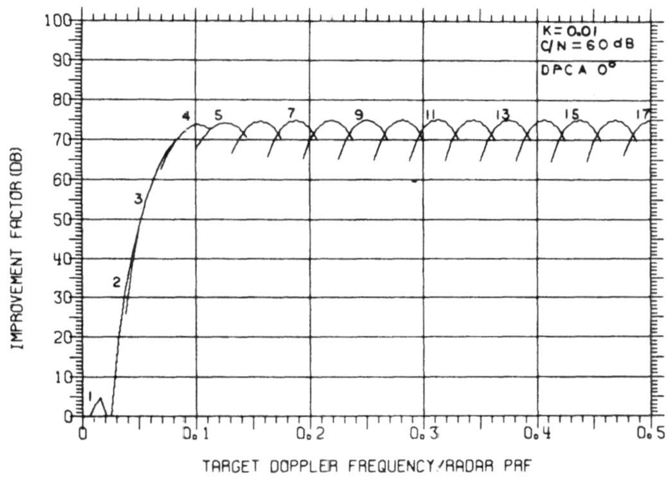

(*b*) MTI canceler and a 32-pulse optimal filter bank

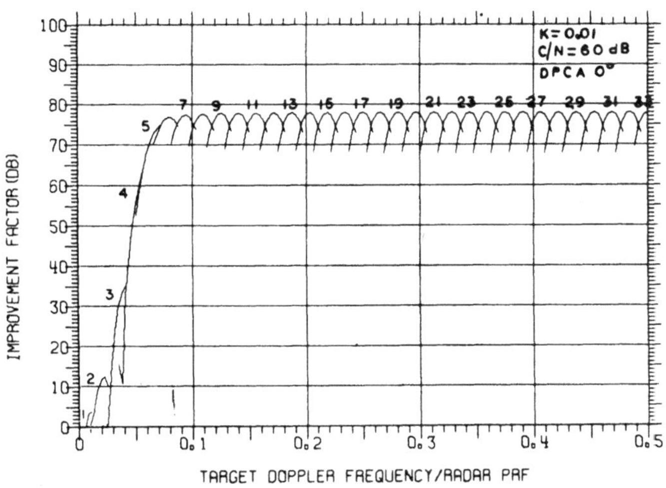

(*c*) MTI canceler and a 64-pulse optimal filter bank

Fig. 16—Improvement factor of an MTI canceler cascaded with an optimal filter bank. There is no platform motion, and the ratio *K* of clutter spectral width to radar PRF is 0.01. The clutter-to-noise ratio is 60 dB.

with Fig. 14 demonstrates that the width of the notches is much less for $\theta_a = 0$ due to the narrower clutter spectrum at this antenna pointing angle. Comparing Fig. 16 with Fig. 12 shows that the maximum improvement factor is achieved over a much wider range of doppler shifts for optimally weighted filters; the result is a more constant response for targets of different velocities.

CONCLUSIONS

This investigation has considered three configurations of doppler processing for airborne applications. Each includes a motion-compensated MTI canceler to take advantage of motion-compensation techniques that for most applications essentially eliminate the effects of platform motion on the cancellation of mainlobe clutter. It was shown that, although the improvement factor of the canceler against mainlobe clutter was not appreciably reduced by platform motion, the effects of platform motion are still evident in the clutter residue at the output of the canceler. Therefore, the succeeding doppler processing will be affected by platform motion.

The succeeding doppler filters for the three configurations consist of (a) FFT filters, (b) an additional canceler and FFT filters, and (c) optimally weighted doppler filters. It has been shown that the most important factors affecting the selection of a configuration are the sidelobe level of the antenna correction pattern and the clutter-to-noise ratio. It should be remembered that the sidelobe level of the correction pattern depends on the sidelobe level of the primary antenna pattern, according to the optimal design procedure of Ref. 5. Both types of interference are incoherent, so that if the mainlobe clutter is canceled enough, the residue is dominated by one of these factors. Then the additional improvement factor is just the coherent integration gain against this incoherent residue (see Eq. (37)). When this is the case, the smallest amount of doppler processing that achieves this limit is preferred.

Figs. 7, 10, and 14 illustrate the performance of the three configurations with very low sidelobes and no receiver noise. These figures show that only doppler filters with optimum weights achieve a large enough improvement factor to be limited by sidelobe clutter. Under these conditions the optimum filters would be preferred, since (depending on target velocity) an additional improvement factor of about 5-15 dB can be achieved with the same number of pulses.

Figures 8, 11, and 15 illustrate the performance of the three configurations under the same conditions, except that sidelobe clutter is increased by 10 dB. These levels might be considered moderate. Under these conditions the difference between the performance of the double-canceler/FFT configuration (Fig. 11) and the single-canceler/optimum-filter configuration (Fig. 15) is less pronounced and the preferable system is not so obvious. Although the maximum improvement factor is about the same for both, more of the optimum filters achieve this maximum. The optimum filters might be selected on this basis. On the other hand, the double canceler/FFT configuration might be selected for its simplicity.

In Fig. 17 sidelobe clutter is increased 10 dB above the levels of Figs. 8, 11, and 15. Under these conditions, a double-canceler/16-pulse FFT and a single-canceler/16-pulse

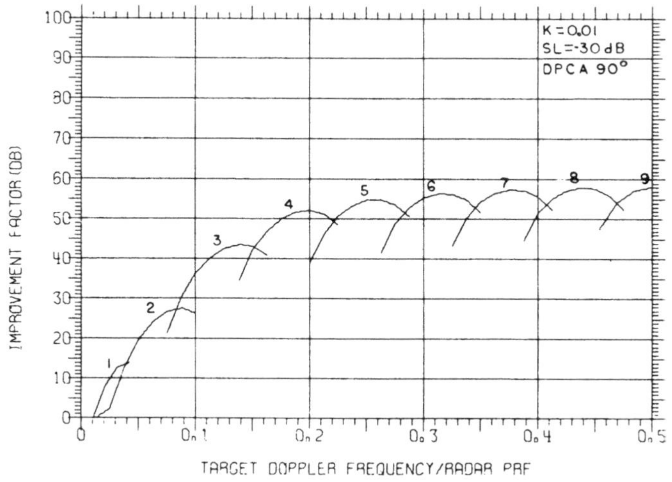

(*a*) Double canceler and a 16-pulse FFT

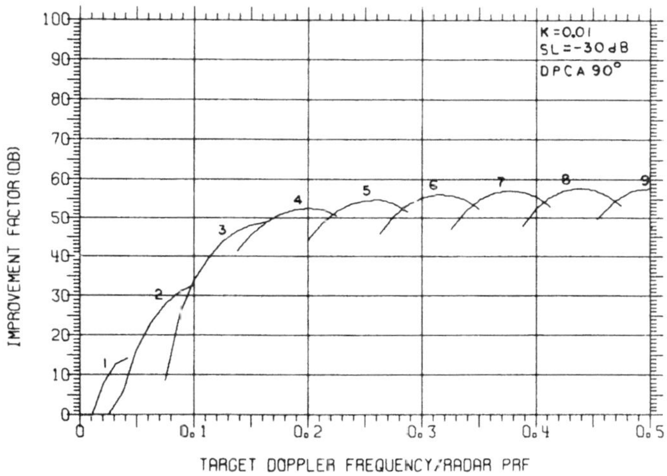

(*b*) Single canceler and a 16-pulse optimal filter bank

Fig. 17—Comparison of the improvement factor of a double canceler/16-pulse FFT and a single canceler/16-pulse optimal filter bank for a high sidelobe clutter level. Platform motion is as defined by Eq. (1), the antenna pointing angle is 90°, and the ratio K of clutter spectral width to radar PRF is 0.01. Sidelobe clutter level is at the -30-dB level.

optimum filter bank are compared. There is no appreciable difference in the performance of the two systems. Therefore, for relatively high antenna sidelobes, the double-canceler/FFT processor would be preferred for its simplicity.

These results are summarized with Figs. 18 and 19. Figure 18 illustrates the performance of the double-canceler/FFT processor. Curve 1 is Fig. 17a redrawn and shows the performance of a double-canceler/16-pulse FFT for a system with relatively high sidelobes. Curves 2, 3, and 4 are Fig. 10 redrawn and show the increased improvement factor that can be achieved by reducing the sidelobe clutter 20 dB (which corresponds to a 10-dB reduction in sidelobe level) and also increasing the number of pulses processed by the FFT. Figure 19 illustrates the advantage of optimally weighted filters for a low-sidelobe system. All three curves refer to a 16-pulse filter bank. Curves 1 and 2 are the same as curves 1 and 2 of Fig. 18 and are shown for reference. Curve 3 is Fig. 14a redrawn to show the additional improvement factor of the optimum filters.

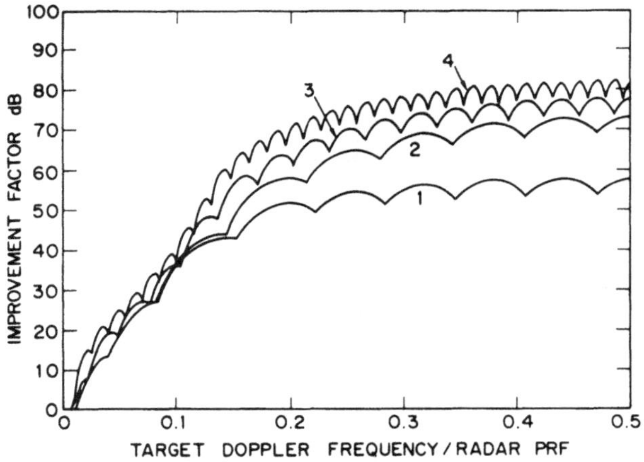

Fig. 18—Comparison of improvement factor achieved by decreasing the sidelobe clutter and by increasing the number of pulses of the FFT filter bank. Curve 1 is Fig. 17a redrawn (double canceler and a 16-pulse FFT with -30-dB sidelobe clutter). Curve 2 is Fig. 10a redrawn (sidelobe clutter is reduced to the -50-dB level). Curves 3 and 4 are Fig. 10b and 10c redrawn showing a 32-pulse FFT and a 64-pulse FFT.

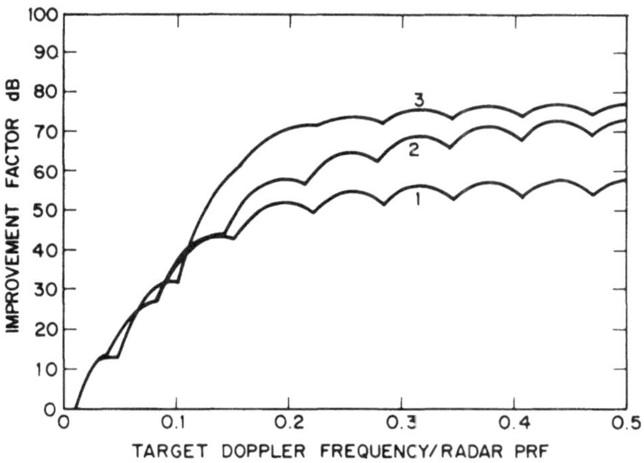

Fig. 19—Comparison of improvement factor achieved by double canceler/FFT and a single canceler/optimal filter bank for low sidelobe clutter. Curves 1 and 2 are curves 1 and 2 of Fig. 18 redrawn for reference. Curve 3 represents a single canceler/16 pulse optimal filter bank.

REFERENCES

1. G. A. Andrews, "Airborne Radar Motion Compensation Techniques: Evaluation of TACCAR," NRL Report 7407, Apr. 12, 1972.
2. G. A. Andrews, "Airborne Radar Motion Compensation Techniques: Evaluation of DPCA," NRL Report 7426, July 20, 1972.
3. G. A. Andrews, "Performance of Cascaded MTI and Coherent Integration Filters in a Clutter Environment," NRL Report 7533, Mar. 27, 1973.
4. G. A. Andrews, "A Detection Philosophy for AMTI Radar," IEEE 1975 International Radar Conference Proceedings, Institute of Electrical and Electronics Engineers, Inc., New York, Apr. 1975.
5. G. A. Andrews, "Airborne Radar Motion Compensation Techniques: Optimum Array Correction Patterns," NRL Report 7977, Mar. 16, 1976.
6. F. E. Nathanson, *Radar Design Principles*, McGraw-Hill Book Co., New York, 1969.
7. R. C. Emerson, "Some Pulsed Doppler MTI and AMTI Techniques," Report R-274, The Rand Corporation, Mar. 1954.
8. L. E. Brennan and I. S. Reed, "Theory of Adaptive Radar," *IEEE Trans. Aerosp. Electron. Syst.* AES-9(2), 237-252 Mar. 1973.
9. G. A. Andrews, "Optimal Radar Doppler Processors," NRL Report 7727, May 29, 1974.
10. Bernard Gold and C. M. Rader, *Digital Processing of Signals*, McGraw-Hill Book Co., New York, 1969.

ADAPTIVE MTI AND DOPPLER FILTER BANK CLUTTER PROCESSING

FRANK F. KRETSCHMER, JR.
BERNARD L. LEWIS
FENG-LING C. LIN

Naval Research Laboratory, Washington, D. C.

ABSTRACT

In many situations radar targets are embedded in a clutter background and clutter processing is required. If the clutter parameters are known but the target velocity is unknown a priori, a bank of doppler filters can be designed with each filter optimized in some sense. In general however, the clutter parameters, which vary with range as well as azimuth, are unknown or only partially known. In this situation adaptive processing is important.

An adaptive MTI is described which makes use of Gram-Schmidt adaptive open-loop processors. It is shown in a clutter limited environment and for a fixed total number of pulses that improved performance can be obtained by using an adaptive MTI prior to doppler filtering. It is also shown that the adaptive MTI provides better performance than a fixed binomial-weight MTI prior to doppler filtering.

INTRODUCTION

Clutter is a significant problem with many radars which, for satisfactory detection of targets, requires processing to provide a sufficient improvement in the target-to-clutter ratio T/C. Earlier radars generally attempted to achieve this by employing a fixed binomial-weighted MTI. More recently, the trend has been toward utilization of a doppler filter bank.

For special cases where the clutter parameters are known, each filter of the filter bank can be individually optimized using any of several criteria. In general however, the clutter varies with range as well as azimuth and the parameters are unknown. An alterative is then to use a doppler filter bank consisting of identical filters except for a frequency translation. These filters can be realized using an FFT which is usually weighted to equally reduce the sidelobes of each filter.

An N-point FFT results in N doppler filters and a signal-to-noise S/N improvement of N for a target having a doppler frequency which is at the peak of the filter response. If the filter sidelobes provide enough clutter rejection, then a suitable improvement in T/C is achieved.

In some situations however, the filter bank may not be capable of providing adequate clutter rejection and additional clutter processing is required. This can be achieved by using an MTI prior to doppler filtering. For a fixed number of transmitted pulses per beam position, a small loss in S/N results.

An adaptive MTI is described which has optimum properties that are based on digital open-loop processors developed at NRL. A comparison is made between using a doppler filter bank alone (DFB), an adaptive MTI prior to a doppler filter bank (AMTI/DFB), and a fixed-weight MTI prior to a doppler filter bank (MTI/DFB). This comparison is made in terms of the T/C improvement factor, I, for several clutter scenarios where the number of pulses is a constant. I is computed for a Gaussian clutter spectrum having different spectral widths, and for two simultaneous clutter spectra corresponding, for example, to ground and rain clutter.

ADAPTIVE MTI

While adaptive MTI may be performed using various techniques, we shall focus on the concept of an open-loop digital adaptive processor and the configuration of multiple processors for cancellation of clutter using multiple degrees of freedom (number of time delay taps).

The digital open-loop adaptive processor [1] develops an optimum weight based on the correlation between a primary or main complex signal M and an auxiliary complex signal A such that when the weighted auxiliary signal is subtracted from the main signal the average residue power is a minimum. The ith residue signal R_i may be expressed

$$R_i = M_i - W_i A_i$$

where

$$W_i = \sum_{k=0}^{n-1} M_{i-k} A^*_{i-k} \Big/ \sum_{k=0}^{n-1} |A_{i-k}|^2 \qquad (1)$$

n = number of samples averaged
* = complex conjugate.

For a single degree of freedom, or one PRI delay, the M and A signals are taken from the taps at either end of the delay (e.g., shift registers).

It should be noted that the complex signals are formed from the sampled I and Q video channels and one complex sample is obtained per range cell. It has been found that the digital open-loop adaptive processor converges to the correct weight very rapidly.

The optimality of the weights for a minimum residue power is readily demonstrated by making use of the orthogonality principle [2] which states that the optimum weight is achieved when the residue or error signal is orthogonal to the auxiliary signal.

That is, from (1)

$$\overline{RA^*} = \overline{MA^*} - W_{OPT}\,\overline{AA^*} = 0$$

$$\text{or} \quad W_{OPT} = \frac{\overline{MA^*}}{\overline{AA^*}} \quad . \tag{2}$$

The overbar denotes a statistical average which we have replaced by an n-sample average in the digital processor. The digital open-loop processor has been implemented at NRL and a block diagram illustrating a two sample average is shown in Fig. 1. This processor may be thought of as performing a Gram-Schmidt operation since M can be represented by the the sum of the two orthogonal signals R and WA.

The configuration of these adaptive digital processors (AP) for multiple degrees of freedom is based on an extension of the Gram-Schmidt concept and is shown in Fig. 2 for three degrees of freedom. The output residue power is a minimum which may be again shown by use of the orthogonality principle. In this configuration, each adaptive processor is identical and each computes an optimum weighting according to its main (vertical input) and auxiliary (horizontal input) signals. Figure 3 shows the transfer characteristic for a 3-pulse adaptive MTI, for a Gaussian input-clutter spectrum centered at f/PRF equal to 0.4, as the normalized standard deviation σ/PRF is varied.

PERFORMANCE CALCULATIONS

We assume that the total number of transmitted pulses available per range cell for both the AMTI and doppler-filter bank processing is a constant equal to k. Also we let the k complex input-data samples, corresponding to a given range cell sampled at the interpulse interval T be represented by the vector X. Next, we assume X to be a zero-mean complex Gaussian random variable having a normalized covariance matrix M_X defined by

$$M_X = \overline{XX^{*t}} \tag{3}$$

where t denotes a transpose operator. For the case of two simultaneous stationary clutter spectra having clutter powers C_1 and C_2, the $(m,n)^{th}$ element of the matrix M_X is written as

$$M_X(m,n) = [C_1\,\rho_1(m,n) + C_2\,\rho_2(m,n)]/(C_1+C_2) \tag{4}$$

where $\rho_i(m,n)$ denotes the normalized complex correlation function of the ith clutter source evaluated at time (m−n)T. For the single power spectrum case considered, C_2 is set equal to zero in (4). The clutter spectra in (4) are each assumed to be Gaussian having a standard deviation σ_i, a mean μ_i and given by

$$S_i(f) = \frac{1}{\sqrt{2\pi}\sigma_i}\; e^{-\frac{1}{2}\left(\frac{f-u_i}{\sigma_i}\right)^2} \quad . \tag{5}$$

The correlation function is then given by the Fourier transform of (5) as

$$\rho_i(\tau) = \exp\,(-2\pi^2\,\sigma_i^{\,2}\,\tau^2 - j2\pi\mu_i\tau). \tag{6}$$

From (6) the elements of the covariance matrix M_X can be determined.

For an ℓ-pulse AMTI we let the effective steady state delay-tap weightings spaced T sec. apart be designated by $g_1,\ g_2\ \dots\ g_\ell$. These are found from the submatrix consisting of the first ℓ rows and columns of M_X and the use of W_{OPT} in computing the weights of the adaptive processors used in a Gram-Schmidt configuration. From a knowledge of the processor weights, the effective tap weights may be determined. For the fixed weight case, the tap weights are the binomial weights. The output of an ℓ-pulse AMTI or MTI, denoted by Y^t is then

$$Y^t = [x_1\ x_2\ \dots\ x_k]\begin{bmatrix} g_1 & 0 & \dots & 0 \\ \vdots & g_1 & & \vdots \\ \vdots & \vdots & & \vdots \\ g_\ell & \vdots & & \\ 0 & g_\ell & & 0 \\ \vdots & 0 & & g_1 \\ \vdots & \vdots & & \vdots \\ 0 & 0 & & g_\ell \end{bmatrix}$$

$$= [y_1\ y_2\ \dots\ y_{k-\ell+1}] = X^tG \tag{7}$$

where G is a k x (k−ℓ+1) matrix.

Next the Y data is weighted with the amplitudes and phases of the weighted doppler-filter bank denoted by

$$D = [d_1\ d_2\ \dots\ d_{k-\ell+1}]$$

where d_i is a column vector of length (k−ℓ+1) containing the ith doppler-filter bank complex weights. The doppler-filter output values Z^t are then

$$Z^t = [z_1\ z_2\ \dots\ z_{k-\ell+1}] = X^tGD. \tag{8}$$

The overall weighting of the input X data by the AMTI or MTI and the ith doppler filter is the ith column of GD in (8) which we call u_i so that we may write

$$Z^t = X^tU \tag{9}$$

where $U = [u_1,\ u_2\ \dots\ u_{k-\ell+1}]$.

The output clutter power from the ith filter [3] may then be computed from

$$C_o = C_{in}\, u_i^t\, M_X\, u_i^* \quad (10)$$

where C_{in} is the total input clutter power. Likewise, the target output power may be determined from

$$T_o = T_{in}\, u_i^t\, M_S\, u_i^* \quad (11)$$

where M_S is the target covariance matrix. To determine M_S, we assume that the target is equally likely to be anywhere in the passband of the ith doppler filter bandwidth defined by the filter crossover points. The signal correlation function then takes the form

$$\rho_S(\tau) = \exp(j2\pi f_i \tau)\, \sin(\pi B\tau)/\pi B\tau \quad (12)$$

where $B = 1/(NT)$
N = number of doppler filters.
$f_i = (i-1)/NT$.

From (12) the elements of M_S can be found.

The improvement factor for the ith channel I_i is then given by

$$I_i = \frac{u_i^t\, M_S\, u_i^*}{u_i^t\, M_X\, u_i^*} \quad . \quad (13)$$

For illustrative purposes it is assumed in the paper that there are a total of 10 PRIs or pulses per range cell. We first consider a single Gaussian spectrum having a mean equal to zero. Figure 4 shows the improvement factor as a function of σ/PRF for each of the 9 doppler-filter outputs for the case where a total of 10 pulses are processed using a 2-pulse adaptive MTI followed by an unweighted doppler-filter bank. The first useful pulse for doppler filtering occurs after the second input pulse. The curve in Fig. 4 marked with squares represents the average improvement factor for all filters. In the remainder of the paper performances will be compared based on the average improvement factor. Figures 5(a) and 5(b) show a plot of the average improvement factor for a 2 and 3-pulse adaptive MTI (solid curves) preceding a doppler filter bank having several different weightings. This is compared with a 10-pulse doppler filter bank without an MTI (dotted curves). It is seen that for clutter spectral widths less than approximately 0.2, the adaptive MTI is significantly advantageous in terms of improving T/C. For this special case of a single Gaussian spectrum centered at zero frequency, a fixed binomial-weighted MTI would perform nearly as well.

Next we consider two simultaneous spectra which might be due for example, to ground clutter and rain. We present the results where the ground clutter power C_g, equals 1000 and the rain clutter power C_r equals 100. Actually, only the relative power is important since we are assumming a clutter dominated environment where the clutter is much greater than the noise. Also, the ground clutter normalized spectral width $\sigma_g T$ is 0.0025 and the rain spectral width $\sigma_r T$ is 0.025. The result of using a 3-pulse adaptive MTI prior to doppler filtering is compared with using the doppler filter bank alone in Fig. 6(a) where the ground clutter is centered at zero frequency and in Fig. 6(b) where the normalized ground clutter frequency is offset by 0.0625. The abscissa in these plots denotes the mean frequency of the rain spectrum which is varied from 0.05 to 0.5. The results are symmetrical about the point where $\sigma_r T$ equals 0.5. Here again it is seen that a substantial improvement is obtained by using some of the available pulses to perform an adaptive MTI prior to doppler filtering.

For the bimodal clutter distribution the adaptive MTI is superior to the fixed-weight MTI as shown in Fig. 7(a) for the ground clutter centered at zero frequency, which corresponds to the center of the binomial-weighted MTI cancellation notch. The adaptive MTI compares even more favorably in Fig. 7(b) where the ground clutter is offset in frequency to 0.0625.

SUMMARY

It was found in a clutter limited environment and for a fixed number of pulses processed that improved clutter rejection was achievable by using an adaptive MTI prior to doppler filtering. The adaptive MTI would also reduce the dynamic range requirements of the doppler-filter bank. A particular form of an adaptive MTI based on Gram-Schmidt orthgonalization was described. This adaptive MTI is capable of very fast convergence and was shown to have optimum properties. A problem area with adaptive MTI is that of inadvertently canceling desired targets. However, there are several techniques available to prevent this, such as offsetting the processor weight one range cell so that it is applied to a range cell not included in determining the weight.

Finally, it was shown for bimodal clutter spectra that an adaptive MTI is capable of much better performance than a fixed-weight binomial MTI.

REFERENCES

[1] F. F. Kretschmer, Jr., and B. L. Lewis, "A Digital Open-Loop Processor", IEEE Trans. Aerospace and Electronic Systems, pp 165-171, Jan., 1978.

[2] A. Papoulis, Probability, Random Variables and Stochastic Processes, McGraw-Hill, N.Y., 1965.

[3] D. C. Schleher, (editor), MTI Radar, Artech House, 1978.

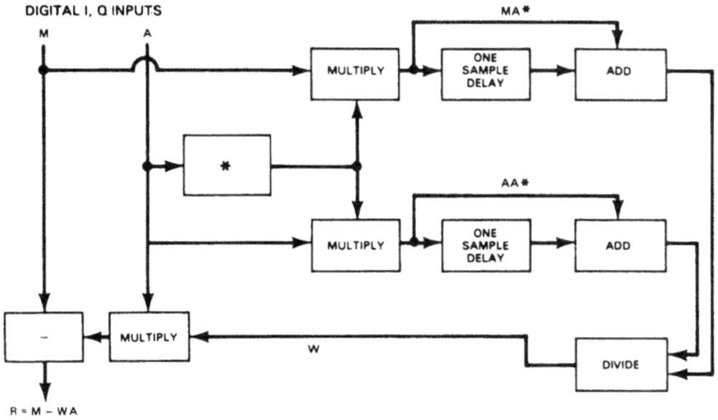

Fig. 1. Digital adaptive processor.

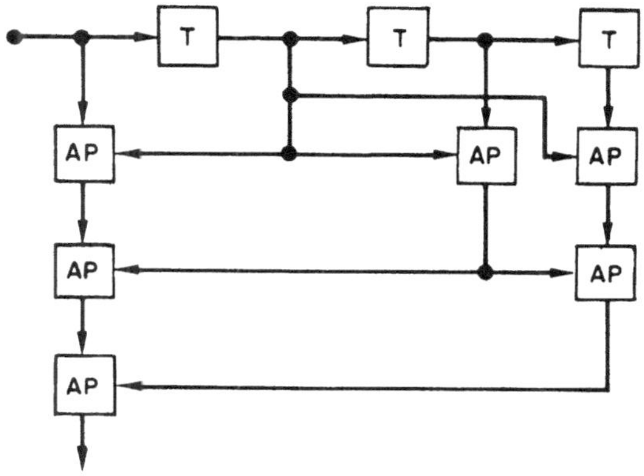

Fig. 2. Gram-Schmidt adaptive MTI.

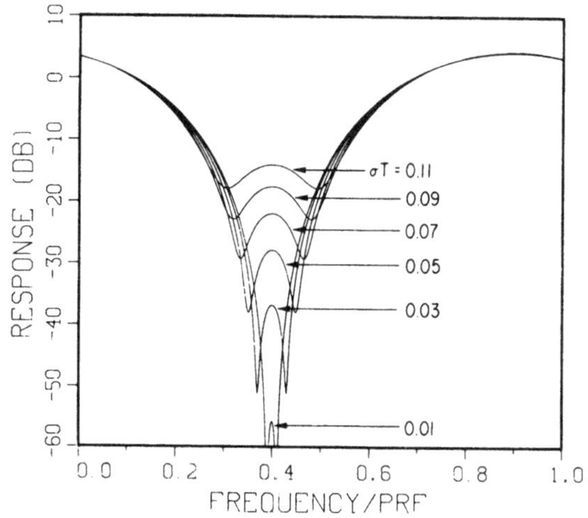

Fig. 3. Transfer characteristic for 3-pulse adaptive MTI.

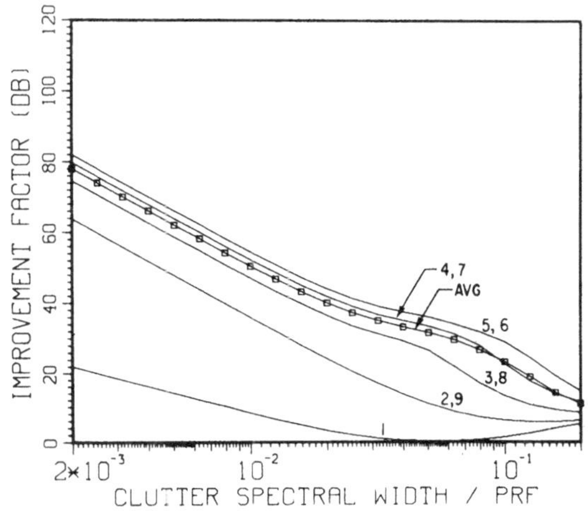

Fig. 4. 2-pulse adaptive MTI plus DFB (uniform weighting).

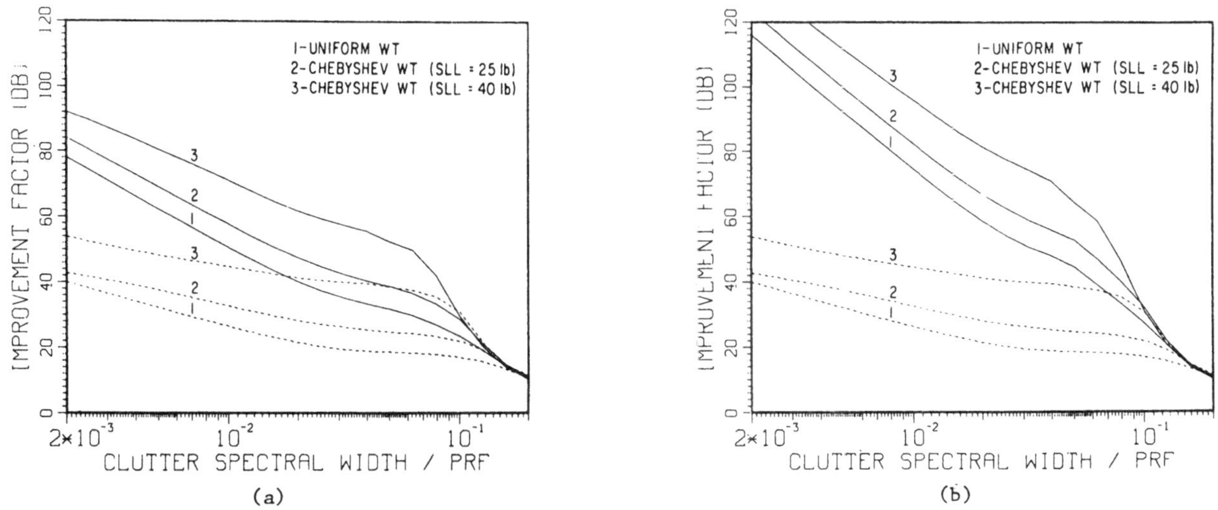

Fig. 5. Adaptive MTI plus DFB (solid curves) compared to DFB (dotted curves), (a) 2-pulse AMTI/DFB, (b) 3-pulse AMTI/DFB.

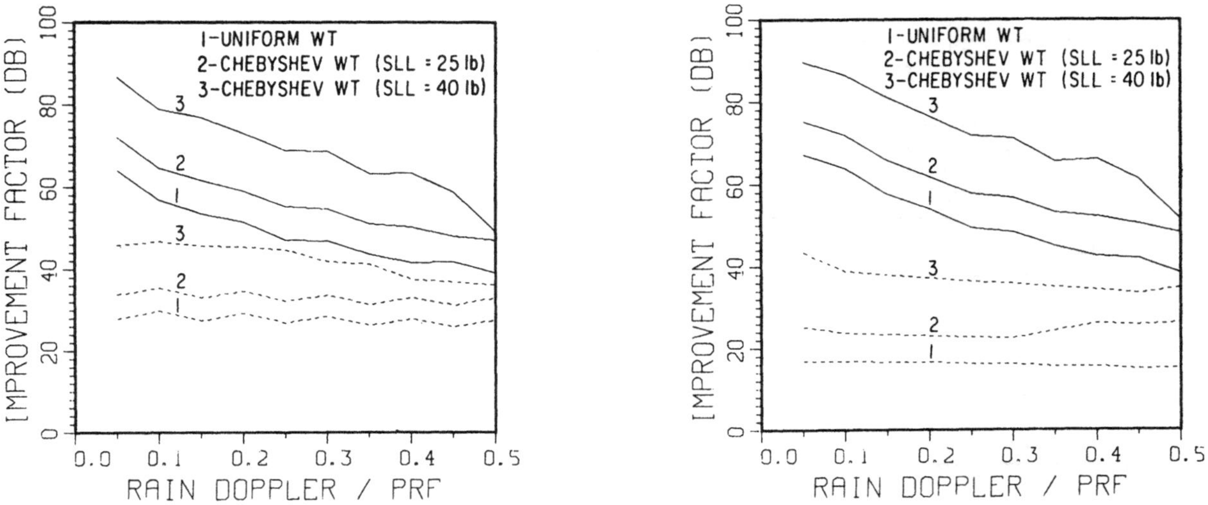

(a) (b)

Fig. 6. 3-pulse adaptive MTI plus DFB (solid curves) compared to DFB (dotted curves), (a) $\mu_g T = 0$, (b) $\mu_g T = 0.0625$.

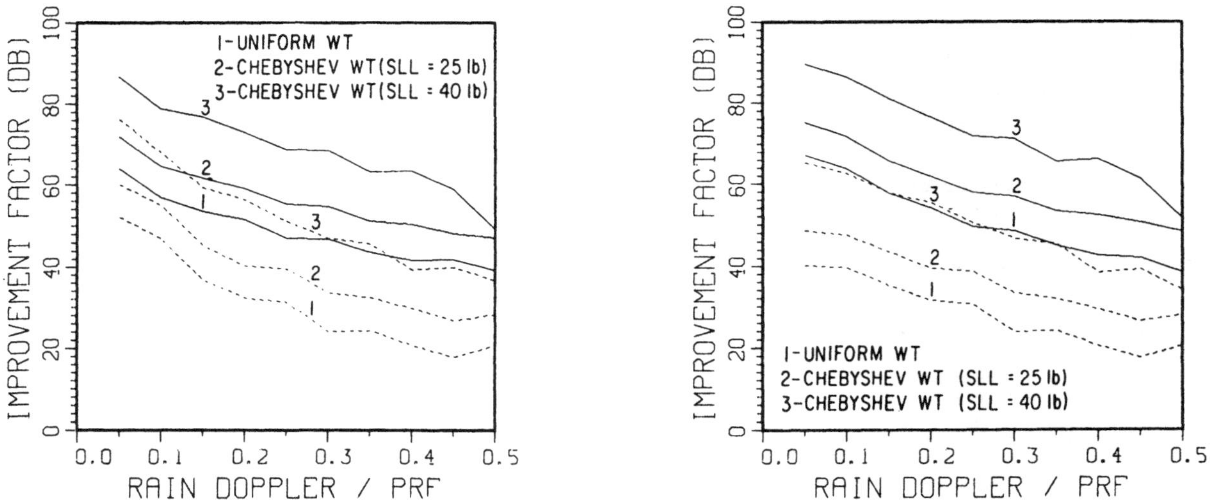

(a) (b)

Fig. 7. 3-pulse adaptive MTI plus DFB (solid curves) compared to 3-pulse fixed-weight MTI plus DFB (dotted curves), (a) $\mu_g T = 0$, (b) $\mu_g T = 0.0625$.

RANGE-DOPPLER COUPLED MOVING TARGET INDICATOR (MTI) ANALYSIS AND ASSESSMENT

INTRODUCTION

In conventional long range search radars, the pulse repetition frequency (PRF) is such that unambiguous range information is obtained. However, at typical search radar transmit frequencies, the PRF is insufficient to prevent aliasing of the target's doppler spectrum. The consequence of this is that the velocity of the target is ambiguous and, more seriously, the target's doppler spectrum may fold over onto the spectrum of the clutter thereby precluding proper operation of the MTI which attempts to notch-out the clutter return. This undesirable situation is usually handled by consecutively transmitting slightly different constant PRF pulse trains or by transmitting a pulse-to-pulse staggered PRF.

Although transmitting pulse trains at different PRFs is generally simpler in terms of the MTI design and implementation, it tends to be more wasteful of pulses than the staggered-pulse MTI. However, a limitation of the staggered-pulse MTI is that ambiguous range clutter will not properly cancel. On the other hand, the multiple-pulse train MTI can perform properly in the presence of ambiguous range clutter if additional filler pulses are used to stabilize the ambiguous range clutter. However, in this case, a magnetron transmitter cannot be used since coherence is required on a pulse-to-pulse basis.

NRL Report 8592 [1] presents a different technique which eliminates blind speeds in an MTI resulting from targets whose doppler is a multiple of the PRF. This technique makes use of the well-known range-doppler coupling (RDC) which occurs with a frequency-swept waveform such as the chirp or the stepchirp waveform.

This report (8789) also reviews the underlying concepts and include the results of additional detailed analysis of the RDC MTI. The use of polyphase codes is described, and the applications and limitations of the technique are discussed.

RANGE-DOPPLER COUPLED MTI CONCEPT

Figure 1 depicts a simplified ambiguity diagram for a chirp waveform. The ordinate corresponds to the doppler shift of the received signal, while the abscissa represents time referened to a value of zero for the nominal round-trip time delay of a point target. Figure 1 shows the trajectory of the peak response at the output of a matched filter for different doppler-shifted receive signals. For the upchirp signal denoted by the dashed line, a doppler shift of f_d Hz results in a compressed pulse peak which is delayed by τ from the true round-trip propagation time of the radar signal. For a compressed downchirp radar signal, represented by the solid line in Fig. 1, the same doppler shift causes the output signal to compress τ s ahead of the nominal time. This provides the basis for the RDC MTI since two consecutive pulses can be transmitted like a normal MTI except that one pulse is an upchirp signal and the other is a downchirp signal. By separately compressing each of the pulses (Fig. 2) and subtracting the signals noncoherently (or coherently), the resultant output signals will cancel completely only if the input signals are not doppler shifted. For doppler-shifted. For doppler-shifted returns, the compressed signals will not cancel completely since they are displaced from each other and will appear, as shown in Fig. 2, when the pulses are separated by more than a compressed pulse width.

Manuscript approved October 17, 1983.

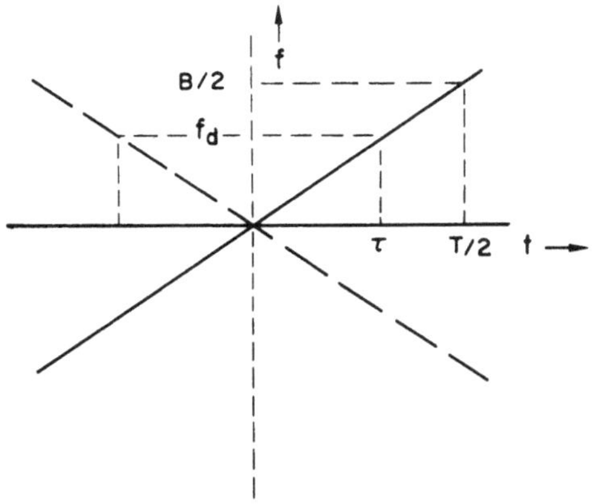

Fig. 1 — Range-doppler coupling

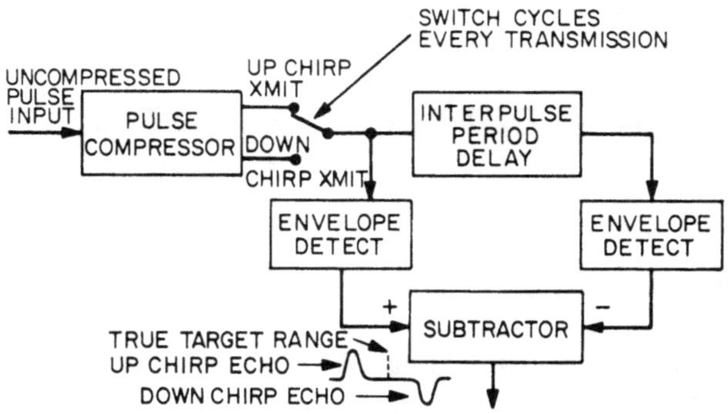

Fig. 2 — Two-pulse range-doppler coupled MTI (noncoherent) and response to incoming target

Figure 3 shows the amplitude of the resultant pulse, or pulses when they become separate, at the output of the subtracter in Fig. 2 for idealized bandlimited compressed pulses. For a velocity higher than V_m, the response is flat so that there are no blind velocity targets. Note that clutter signals having a small velocity will be attenuated by the RDC MTI filtering. The velocity V_m is next derived by referring to the upchirp response (solid line) in Fig. 1. The slope of this line is

$$k = B/T, \tag{1}$$

where B is the signal bandwidth and T is the uncompressed pulse length; thus,

$$f_d = k\tau \tag{2}$$

from which

$$\tau = f_d/k = (f_d/B)\,T. \tag{3}$$

The number of range cells ΔR corresponding to a delay of τ s is then given by

$$\Delta R = \tau B = (f_d/B)\rho = f_d T, \tag{4}$$

where ρ is the pulse compression ratio TB. To resolve the compressed pulses after subtraction, we assume that they are separated by a pulse width or that each pulse is shifted oppositely in time by $\tau = 1/2B$. Therefore,

$$\Delta R = \tau B = 1/2 = f_d T = 2(V_m/\lambda)\,T, \tag{5}$$

and

$$V_m = \lambda/(4T) = C/(4Tf_0). \tag{6}$$

From the relation

$$V_m = C/(4Tf_0),$$

V_m is decreased by increasing T and/or f_0. For example, for $V_m = 300$ m/s, we require that at X-band T = 25 μ and at S-band T = 75 μs. Note that T can also be increased without changing the bandwidth or duty cycle by leaving spaces in the transmitted waveform. This is more readily achievable with the polyphase coding which is described subsequently.

Figure 3 shows the slope of the response curve in the notch region is $1/V_m$. The amplitude response A may therefore be written as

$$A = V/V_m.$$

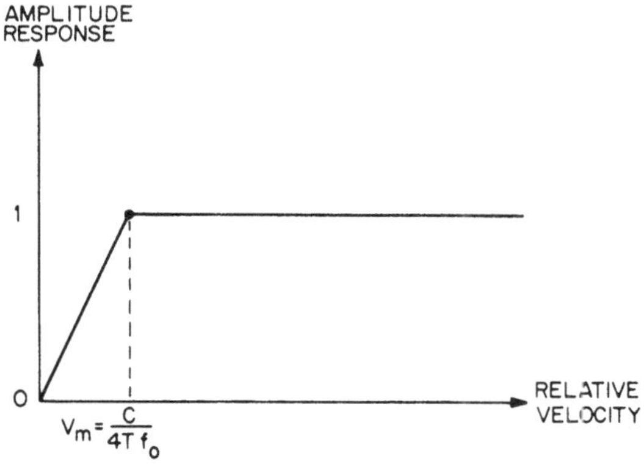

Fig. 3 — Response of noncoherent range-doppler coupled MTI

In this region, the ratio of amplitude response is the same as the ratio of the relative velocities. Thus a doubling of the relative velocity results in a 6-dB increase in peak signal power.

Figure 4 shows the response curve or the coherent RDC MTI using idealized pulses for an interpulse period T_I equal to 10 times the uncompressed pulse width T. In this case, the phase of the target returns from pulse to pulse must be taken into account. The first peak of the cyclic response in Fig. 4 occurs at a velocity V_1 when there is a π phase shift between successive pulses as is the case with the normal coherent MTI canceler. When there is a 2π phase shift between successive pulses, the normal MTI response goes to zero and this corresponds to the first blind velocity. For the RDC MTI, however, the response does not go to zero since the compressed pulses have moved apart due to range-doppler coupling before subtraction. A peak in the cyclic response occurs at odd multiples of π phase shift from pulse to pulse. When a π phase shift occurs within the uncompressed pulse, the compressed pulses are separated by one pulse width and the response is at the peak. This is evident from Eq. (5) where we had

$$f_d T = 1/2$$

so that total phase shift across the uncompressed pulse $\Delta\phi$ is

$$\Delta\phi = 2\pi f_d T = \pi.$$

The number of cyclic peaks occurring in a coherent RDC response curve as shown in Fig. 4 is then given by one-half of the ratio of T_I to the uncompressed pulse length. The lower part of the cyclic swing is the same as the noncoherent response shown in Fig. 3. Also the upper part of the cyclic swing is bounded by the upper dashed line in Fig. 4 which is the symmetrical counterpart of the lower bound. This indicates that the target response of the coherent RDC MTI is equal to or larger than for the noncoherent MTI when $V < V_m$, and for $V > V_m$ the two are the same.

The noncoherent RDC MTI allows determination of the velocity direction by noting the signs of the pulse pair after the subtractor in Fig. 2. Also, the noncoherent RDC MTI does not require a clutter background for proper operation as the normal noncoherent MTI does. The normal noncoherent MTI tends to cancel targets in the clear if strong clutter is not present. The RDC MTI case permits detection of higher velocity targets because of the time separation of the output pulses due to range-doppler coupling.

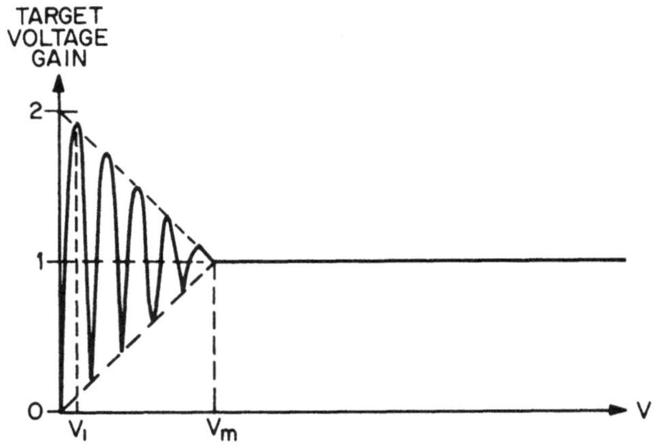

Fig. 4 — Response of coherent 2-pulse range-doppler coupled MTI with $T/T_I = 10$

PERFORMANCE ANALYSIS

Chirp and Polyphase Code Response Curves

Figure 5 shows the computer-simulated result of noncoherently subtracting two compressed Hanning-weighted chirp pulses having a total separation of approximately 0.1 and 0.6 of a pulse width. Figure 6 shows a plot of the peak signal after noncoherent subtraction of the Hanning-weighted chirp signals.

Since the polyphase codes recently investigated by the authors [2] also exhibit range-doppler coupling properties, they can also be used in an RDC MTI. Figure 7 shows a compressed pulse for a 100-element ($\rho=100$) P2 polyphase code [2] where the sample numbers correspond to range resolution cells. Figure 8 shows the results on an amplitude scale of noncoherently subtracting an upsweep and a downsweep 100-element P2 code after compression for two different target velocities. The peak amplitude of the compressed pulses for zero doppler is 100 prior to subtraction. The downsweep version of the P2 code or any other polyphase code is obtained by taking the conjugate (negative) phases of the upsweep or normal version of the polyphase codes. Figure 9 shows the RDC MTI response curve for the 100-element P2 code for the coherent and noncoherent cases. The cyclic minima for the coherent case do not equal the noncoherent case as they do for a chirp signal. Figure 10 shows an approximation to a chirp signal where we have used a polyphase code obtained by using 2ρ phase samples (rather than ρ) taken from the quadratic phase characteristic of a chirp signal at twice the Nyquist rate. The compressed waveform for this case has the same peak sidelobes and behaves approximately the same as the chirp waveform.

The polyphase codes differ in several other respects from the chirp signals. Figure 9 shows that the polyphase code response is full amplitude (100 units) for a normalized doppler shift of approximately 0.01 rather than 0.005 for the simulated chirp signal shown in Fig. 10. This is due to the peak response of the polyphase code splitting into two adjacent cells with a reduced amplitude for a doppler shift of 0.005. At a doppler shift of 0.01, there is one peak at full amplitude. This corresponds, in general, to a total range separation between the upsweep and downsweep versions of the polyphase code being two range cells rather than one range cell for the chirp signal. Also, for higher doppler velocities, Fig. 9 shows the amplitude of the polyphase codes cycles for doppler shifts higher than 0.01. This is a property of the polyphase codes [2] which does not occur with the chirp signal. The cycling shown in Fig. 10 beyond the normalized doppler frequency of 0.01 is due to using a polyphase code in approximating a chirp signal.

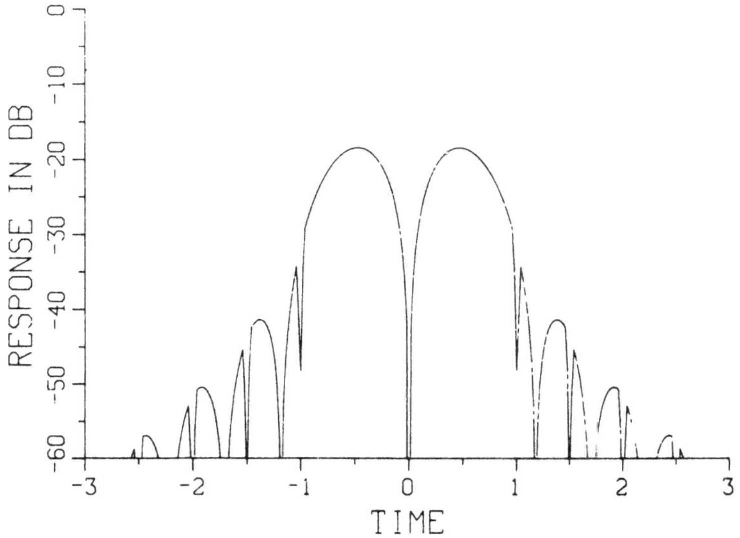

(a)

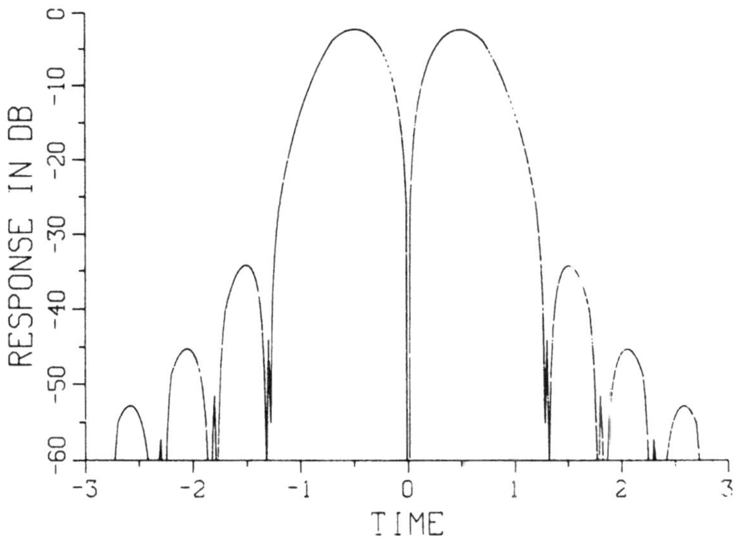

(b)

Fig. 5 — Noncoherent subtraction of two range-doppler coupled Hanning-weighted chirp pulses. Time scale is in units of a pulse width. (a) Separated by approximately 0.1 pulse width (b) Separated by 0.6 pulse width

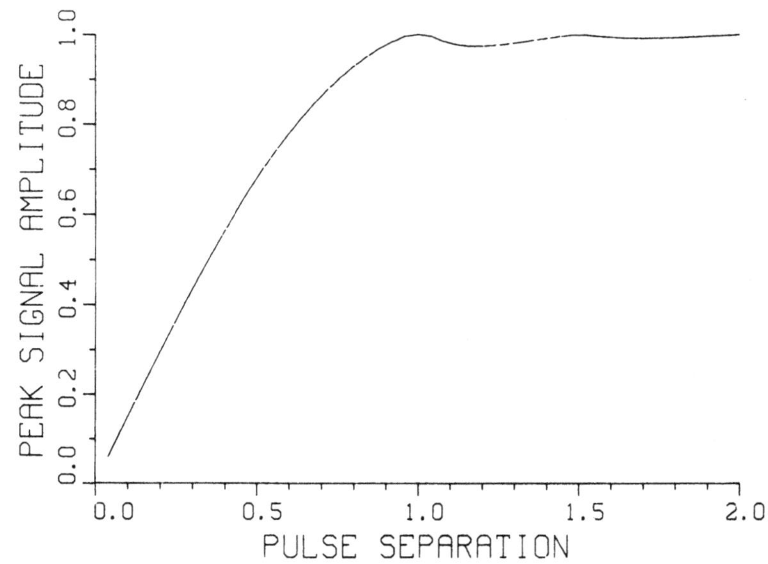

Fig. 6 — Peak response of noncoherent range-doppler coupled MTI using Hanning-weighted chirp signals. Pulse separation is in units of a pulse width

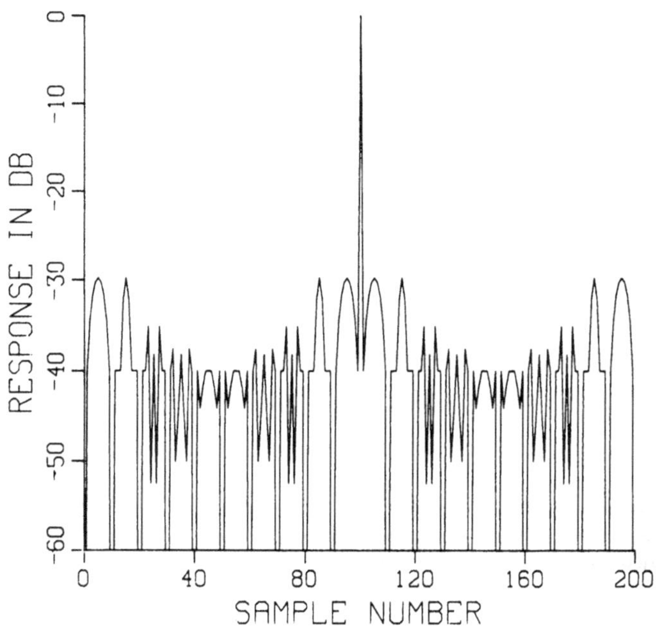

Fig. 7 — Compressed 100-element P2 polyphase code

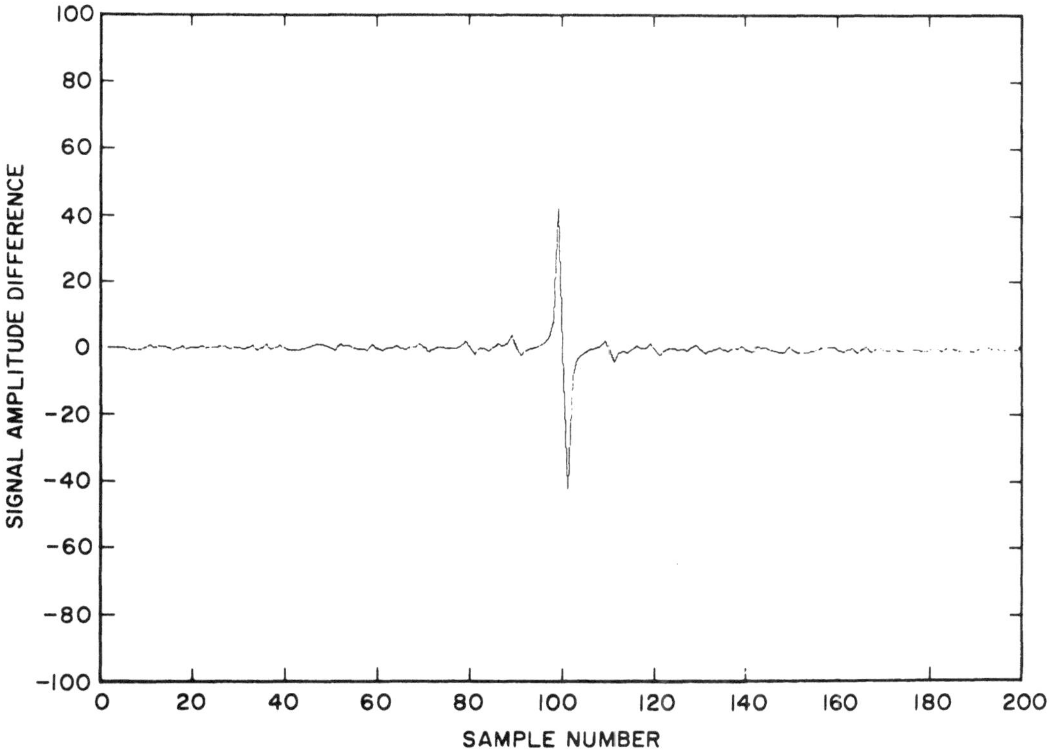

(a)

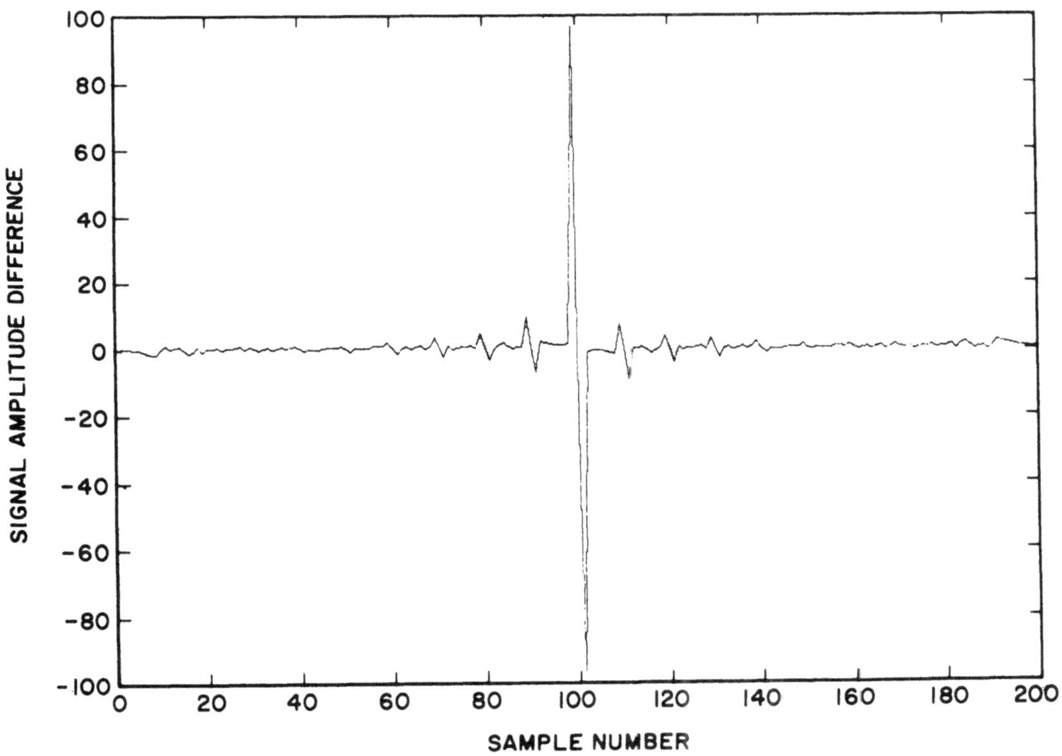

(b)

Fig. 8 — Target response for 2-pulse range-doppler coupled MTI using a 100-element P2 polyphase coded waveform (a) Normalized target doppler is 0.005 B (b) Normalized target doppler is 0.01 B

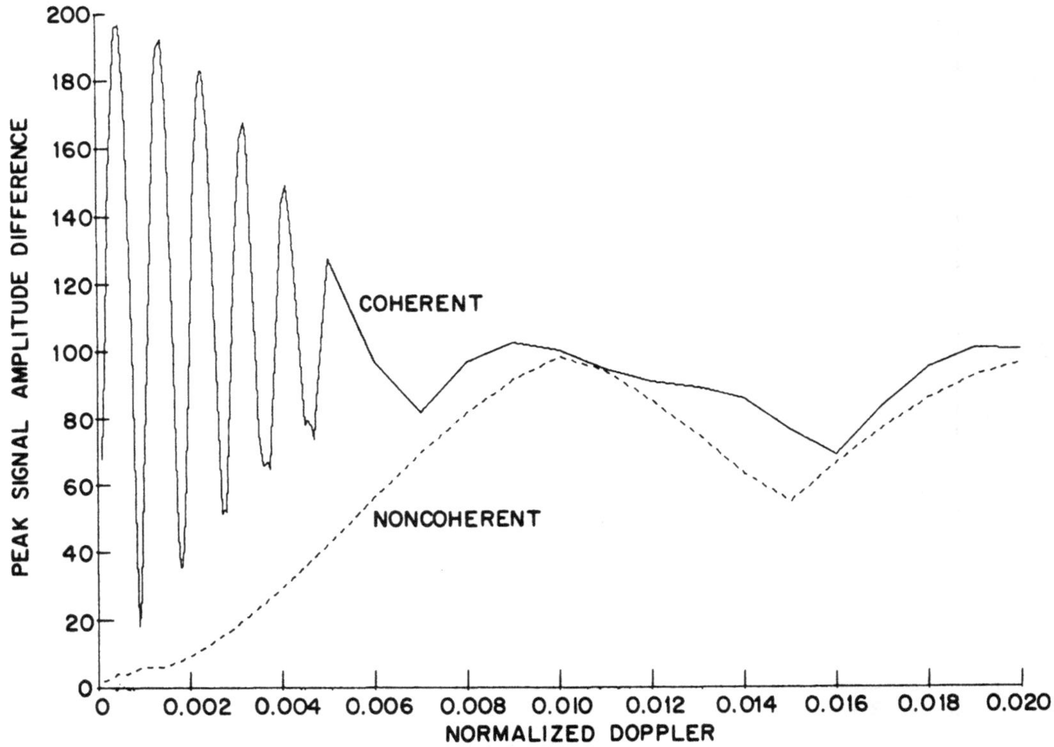

Fig. 9 — Response of a range-doppler coupled MTI for a doppler-shifted target using a 100-element P2 polyphase coded waveform

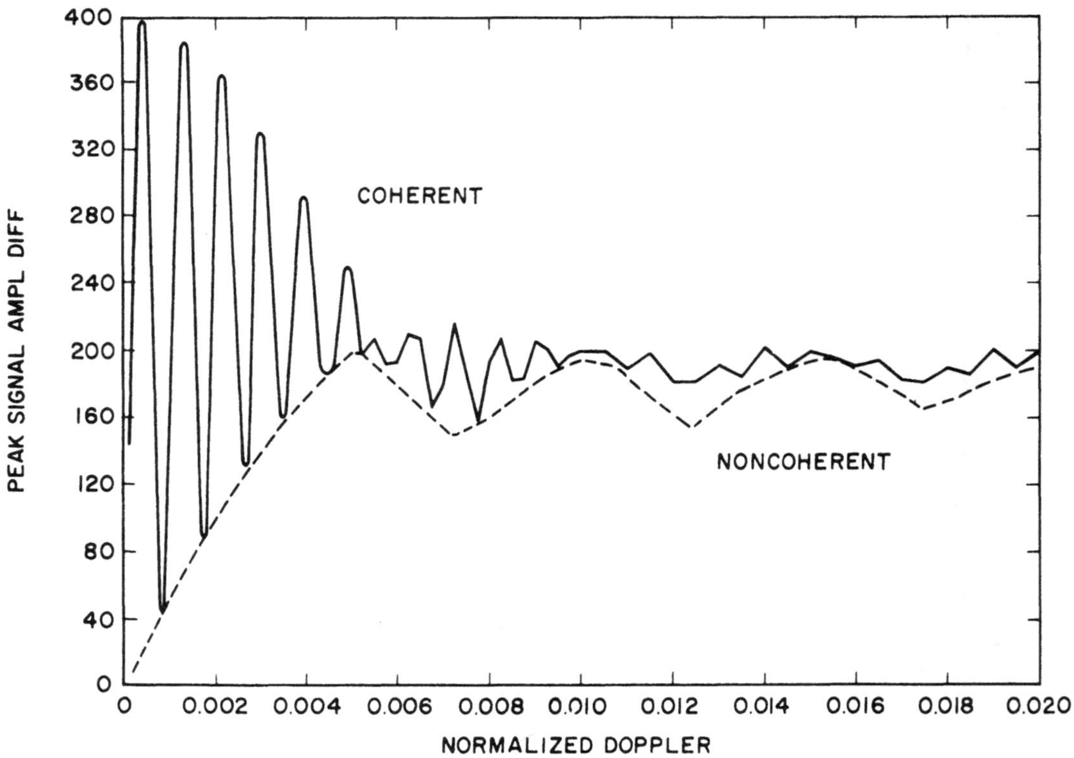

Fig. 10 — Response for a range-doppler coupled MTI for a chirp signal approximated by a 200-element polyphase code

Evaluation of the RDC MTI Improvement Factor

An analysis was performed to determine the effect on the MTI cancellation ratio (CR) of the range-doppler coupling effect. This is summarized in the appendix of this report. The range-doppler coupling of the clutter for typical clutter spectral widths (assuming a Gaussian power spectral density function) had a negligible effect on the cancellation ratio. This is attributed to the transfer function of the range-doppler coupled MTI being approximately the same as the normal coherent or noncoherent MTI counterparts in the small doppler region near zero that is occupied by the clutter.

The average improvement factor I for a coherent MTI is computed from the relation

$$I = CR \cdot TE,$$

where TE denotes the average target gain or enhancement over the region of interest. Referring to Fig. 4, the target enhancement evaluated from 0 doppler to a doppler $F > F_m$ $(F_m = 2\ V_m/\lambda)$ for the 2-pulse coherent RDC MTI using chirp signals can be determined from

$$TE = \frac{1}{F}\left[\int_0^{F_M} E^2(f)\,df + 1\right], \tag{7}$$

where

$$E(f) = [(1 - f/F_m) \sin \pi fT + 1].$$

Substituting for $E(f)$ in Eq. (7) results in

$$TE = \frac{1}{F}\Bigg[\int_0^{F_M} [(1 - 2f/F_m + f^2/F_m^2) \sin^2 \pi ft + 2(1 - f/F_m) \sin \pi fT + 1]\, df + 1\Bigg]. \tag{8}$$

This can be written as the sum of integrals which are in standard form. However, an approximate solution can be obtained by considering Fig. 4 and noting that by integrating over a large doppler region compared to F_m, TE approaches unity and represents a 3-dB loss in improvement factor relative to the normal MTI. For higher order cancelers, the loss is less. For polyphase codes, there is an additional loss due to the cycling of the resolved peak signals as previously discussed. However, in either case, the tradeoff is that the velocity response has been extended to prevent target blind speeds.

It was found in the simulations described in the appendix of this report that the noncoherent RDC MTI performed the same way relative to the coherent RDC MTI as it does for the normal MTI [3]. The improvement factors of the noncoherent and coherent RDC MTIs are the same for the two-pulse canceler but differ from the normal MTI as described in the preceding section. According to Ref. 3, the noncoherent MTI improvment factor degrades relative to the coherent MTI for binomial weighting as the number of pulses exceeds two.

Use of Polyphase Codes in Extended Clutter

A weighted P4 polyphase code was used in a 2-pulse RDC MTI simulation in which extended clutter having zero doppler was generated. It was assumed that the clutter power per range cell (or sample number) was unity and that the clutter had an independent Gaussian distribution from range cell to range cell.

Figure 11 shows the results of this simulation where the target doppler is such that it range-doppler couples one range cell and hence the target occupies two range cells at the output of the RDC MTI. However, as shown in Fig. 11, the clutter was only cancelled by approximately 20 dB; this is because the P4 code is not symmetrical about the center of the code. This in turn results in a compressed pulse which is complex, and the upchirp and downchirp compressed pulses are not identical. Whereas the P2 is a symmetrical code, the unsymmetrical P1, P3, and P4 codes can be made symmetrical by shifting the initial sampling point one-half of a subpulse width in deriving these codes from the underlying phase characteristic. The symmetrical version of these codes is denoted by the prefixed letter P (for palindromic) such as the PP4 code. Figure 12 illustrates the derivation of the P4 and PP4 code phases from the sampled quadratic-phase characteristic of a linear chirp signal.

By using symmetrical polyphase codes, the compressed pulse is real and symmetrical, which results in identical upchirp and downchirp versions. For this case, the nonmoving extended clutter cancelled completely, provided the extended clutter is unambiguous in range (Figs. 13 and 14).

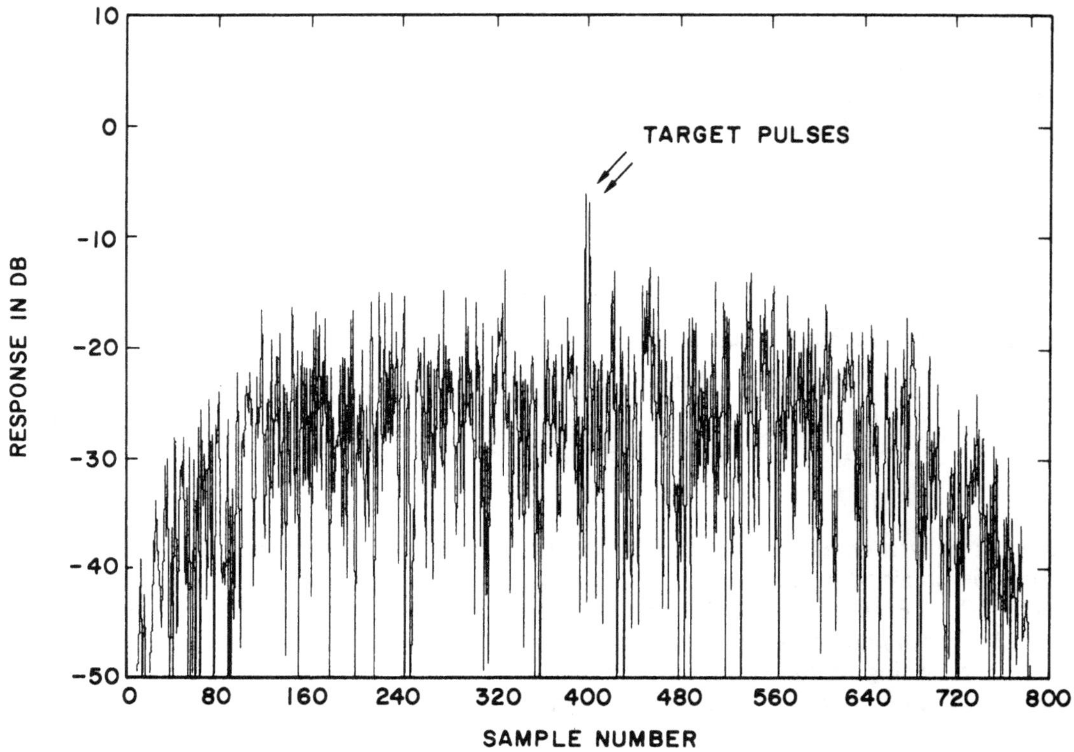

Fig. 11 — Simulated output of a 2-pulse range-doppler coupled MTI for a target embedded in zero velocity distributed clutter using a weighted P4 code. The target's normalized doppler is 0.01 B.

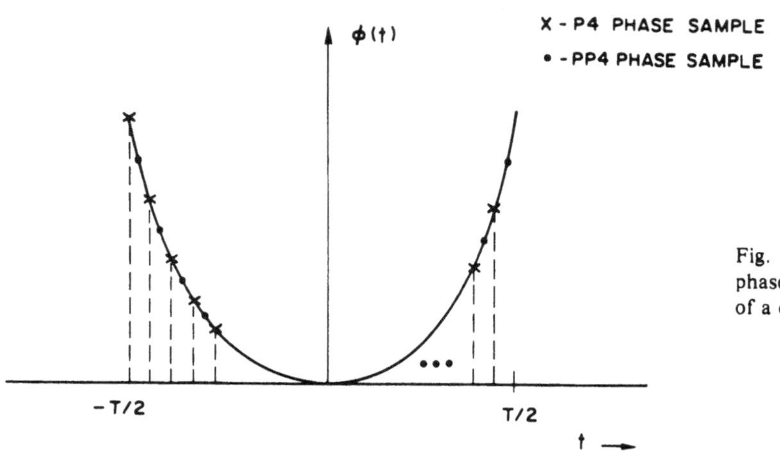

Fig. 12 — Derivation of P4 and PP4 polyphase code phases by sampling the quadratic phase characteristic of a chirp signal

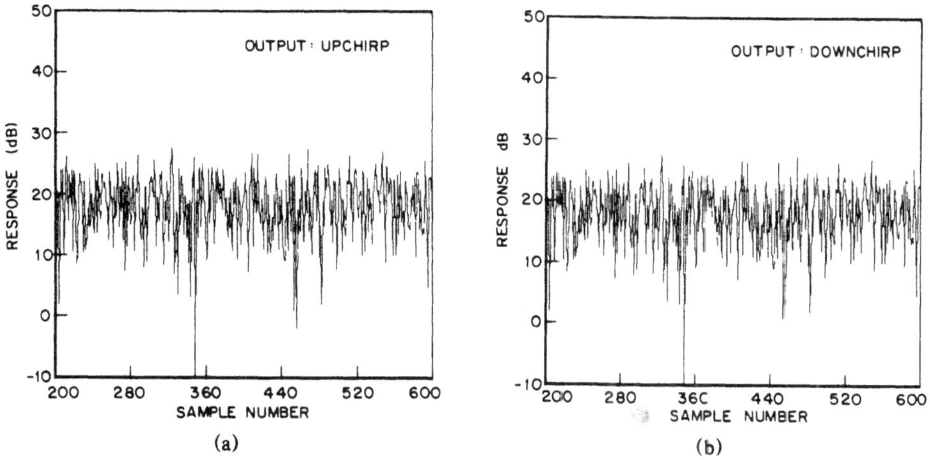

Fig. 13 — Output of the (a) upchirp and (b) downchirp pulse compressors for distributed clutter and a target using a 100 element PP4 code. Clutter-to-target ratio is 20 dB. Clutter doppler is zero and target doppler is 0.01 B.

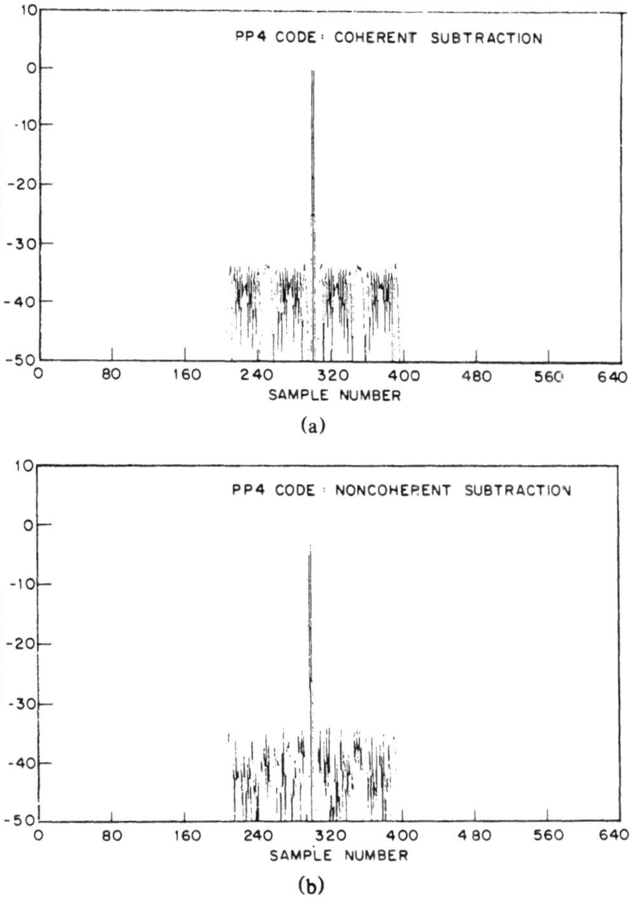

Fig. 14 — Output of range-doppler coupled MTI showing target alone (a) Coherent subtraction (b) Noncoherent subtraction

RANGE-AMBIGUOUS EXTENDED CLUTTER

Simulations were performed for ambiguous range-extended clutter. The upchirp and downchirp signals were assumed to be transmitted simultaneously for convenience as shown in Fig. 15. If the pulses were transmitted with a separation of τ s, the channel matched to the first transmitted pulse would be delayed by τ. The important result, however, is that range-ambiguous stationary clutter does not properly cancel. This is shown in Fig. 16 where 0 dB corresponds to the input clutter power level. This residue is attributed to the cross responses not cancelling, i.e., the response of the upchirp signal in the filter matched to the downchirp signal and vice versa. It is important to realize that for ambiguous range clutter, one cannot compress each upchirp or downchirp pulse separately as shown in Fig. 2 for unambiguous clutter. The cross responses have the same amplitudes, but unequal phases which are symmetrical about any constant phase angle of the input signal. Figure 17 shows the magnitude responses for the matched and mismatched channels For the special case of a zero-degree constant phase angle associated with the received waveform, the phases in the mismatched channels are conjugates of each other for a point target return. The fact that proper cancellation does not take place for ambiguous range returns places a performance limitation on the RDC MTI similar to that for a pulse-to-pulse staggered waveform used to remove blind speeds in a conventional MTI.

Although the RDC MTI does not work properly in range-ambiguous extended clutter, it may have applications in making velocity measurements on a target in the clear.

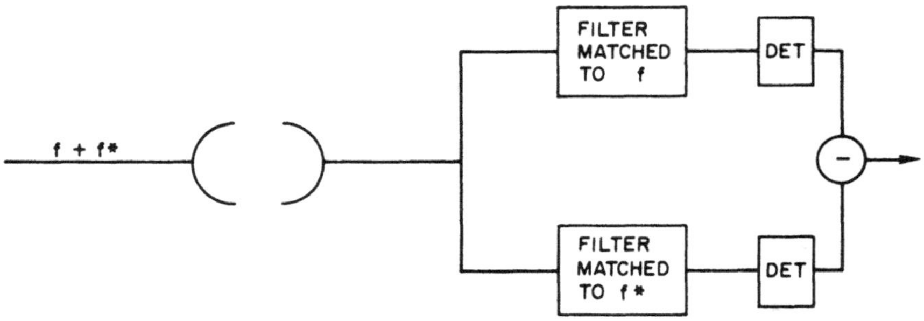

Fig. 15 — Range-doppler coupled MTI for ambiguous range case

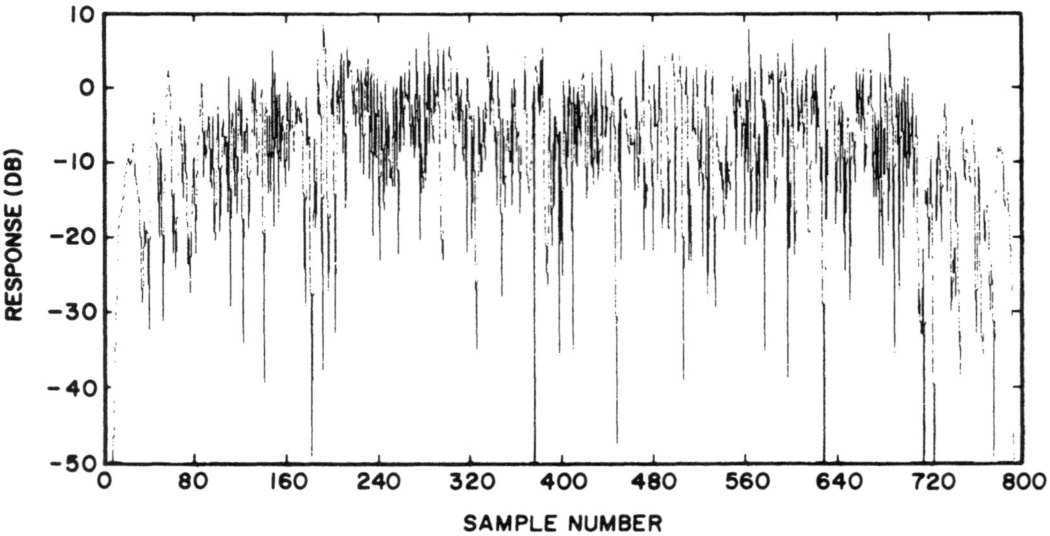

Fig. 16 — Range-doppler coupled MTI output for nonmoving distributed clutter

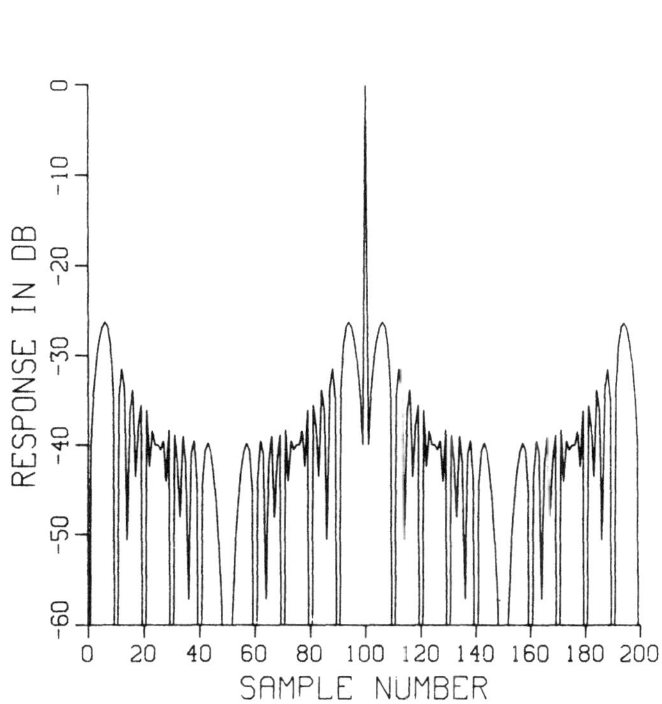

(a)

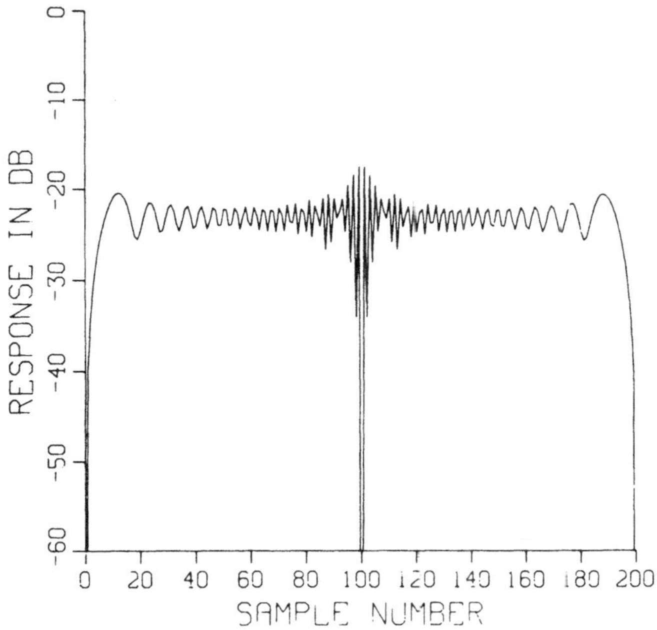

(b)

Fig. 17 — Outputs of range-doppler coupled MTI pulse compressors for 100-element PP4 code (a) Matched channel (b) Mismatched channel

SUMMARY AND CONCLUSIONS

The concept of a range doppler coupled MTI was presented where it was shown that target blind speeds could be prevented. It was shown that a relatively flat target response is obtained for targets having doppler velocities above V_m which corresponds to a range shift, due to range-doppler coupling, of 1/2 or 1 range cell depending on whether chirp signals or polyphase codes are used. V_m, in turn, was shown to be inversely proportional to the product of Tf_o, where T is the uncompressed pulse width and f_o is the transmit center frequency. For chirp signals, for example, with $V_m = 300$ m/s, we require T = 25 μs or 75 μs at X band and S band. At the lower frequencies, where the pulse length becomes larger, it is not necessary to transmit continuously for a time T. The overall transmission time T could consist of transmissions and spaces with the appropriate matched filtering. Note that an improvement in the target-to-clutter ratio is also achieved for target velocities less than V_m as long as the target velocity is larger than that of clutter. This means that the time duration T can be traded off.

The problem of multiple targets could cause some confusion in the coherent RDC MTI. However, there are several mitigating factors. In the previous example, a 600-m/s target would result in two pulses separated by only one range cell, and for Swerling I or Swerling III targets would have similar amplitudes which may be helpful in associating the pulses. For noncoherent RDC MTI either the positive or the negative pulses of the pulse pair from each target could be eliminated and then only one pulse would be associated with each target as in a regular MTI system using pulse compression.

It was shown that the range-doppler coupling effect of the clutter is negligible and that the RDC MTI has a cancellation ratio which is approximately the same as the normal MTI for both the coherent and noncoherent cases. However, the improvement factor, which consists of the cancellation ratio times the target enhancement factor, is less than the regular coherent or noncoherent MTI when averaging the target response over large doppler frequencies. This is mainly due to the RDC MTI target transfer function being approximately unity for a two-pulse canceler with resolved pulses at the RDC MTI output. Thus, for a two-pulse binomial weighting, the loss in target enhancement and also the resultant improvement factor is approximately 3 dB. However, one reduces the blind speed problem by using this technique. There also appears to be a loss associated with the staggered pulse MTI; however, we have not quantitatively analyzed this loss.

An advantage of the noncoherent RDC MTI over its normal noncoherent MTI counterpart is that it does not require the presence of clutter for proper operation. The normal noncoherent MTI tends to cancel targets in the clear unless strong clutter is present.

It was shown that, like the staggered pulse MTI, the RDC MTI does not work properly in range ambiguous extended clutter. It was also shown that for unambiguous extended clutter that the unsymmetrical polyphase codes could not be used since the compressed pulses for the unsymmetrical codes were complex and differed in phase between the upchirp and downchirp versions of the compressed codes. Symmetrical phase codes were derived which are suitable for use with the RDC MTI.

Although there is no reason to recommend using the RDC MTI in preference to the pulse-to-pulse staggered MTI for eliminating target blind speeds, the RDC MTI provides an alternate technique which may have other useful applications, such as providing velocity information on isolated targets in the clear.

REFERENCES

1. B. L. Lewis, F. F. Kretschmer, Jr., and F. C. Lin, "Range-Doppler-Coupled Moving Target Indicator," NRL Report 8592, June 1982.

2. F. F. Kretschmer, Jr., and B. L. Lewis, "Doppler Properties of Polyphase Coded Pulse Compression Waveforms," *IEEE Trans. Aerospace and Electronic Systems* **AES-19** (4), July 1983.

3. F. F. Kretschmer, Jr., F. C. Lin, and B. L. Lewis, " A Comparison of Noncoherent and Coherent MTI Improvement Factors," *IEEE Trans. Aerospace and Electronic Systems* **AES-19** (3) May 1983.

Appendix

Previous methods [A1] that were used to simulate coherent and noncoherent MTI performance were inappropriate for analyzing the RDC MTI. For this analysis, a different approach was taken. To assess the range-doppler coupling effects of the clutter, the narrowband Gaussian clutter spectrum shown in Fig. A1 was approximated by the discrete line spectra as shown in Fig. A2 [A2]. The voltage for this approximation is then given by

$$c(t) = \sum_{m=1}^{M} [2G(f_1 + m\Delta f)\Delta f]^{1/2} \cos [(\omega_1 + m\Delta\omega)\, t + \theta_m],$$

where θ_m is an independent and uniformly distributed phase term. We may also write $c(t)$ as

$$c(t) = x(t)\cos \omega_c t - y(t)\sin \omega_c t,$$

where

$$x(t) = \sum_{m=1}^{M} [2G(f_1 + m\Delta f)\Delta f]^{1/2} \cos [(\omega_m - \omega_c)t + \theta_m],$$

$$y(t) = \sum [2G(f_1 + m\Delta f)\Delta f]^{1/2} \sin [(\omega_m - \omega_c)t + \theta_m],$$

and

$\omega_m = \omega_1 + m\Delta\omega$, and ω_c = carrier frequency.

It may be shown that x(t) and y(t) have a Gaussian probability distribution [A2] as desired and that c(t) has the proper power density spectrum which in turn defines the underlying correlation function of the clutter. This approach allowed us to compute the effect of RDC by processing each spectral line separately.

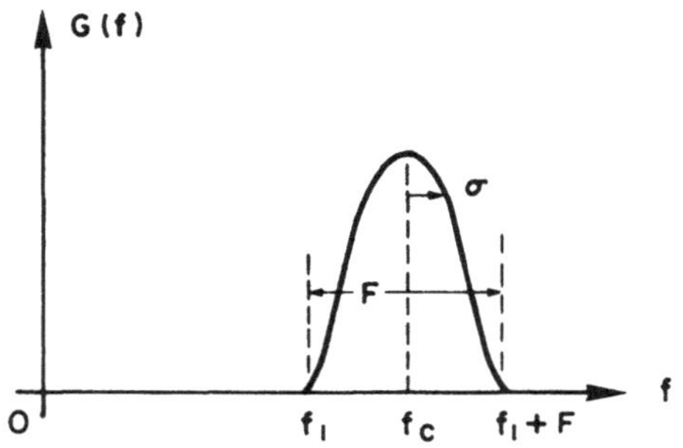

Fig. A1 — Narrowband Gaussian clutter spectrum

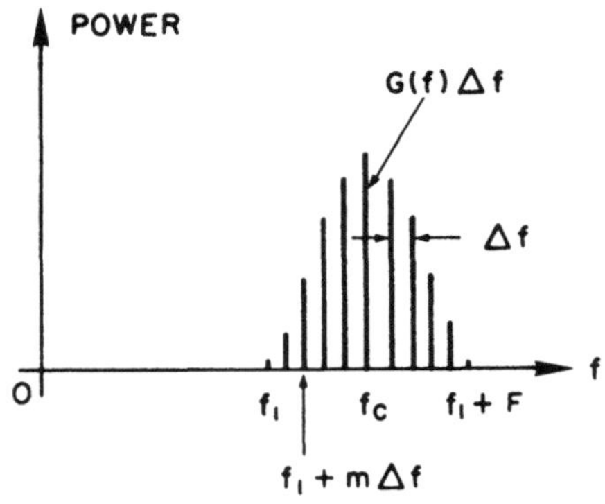

Fig. A2 — Discrete approximation to the narrowband spectrum

Simulations of the normal MTI were performed by using an analytical approach, a Monte Carlo simulation as described in Ref. A1, and the approach based on the approximated power spectrum as previously described; agreement was obtained. Approximately 11 spectral lines, and the average of approximately 500 sets of 11 random phases gave a good approximation.

Various Monte Carlo simulations were performed for the RDC MTI using the polyphase codes. In one simulation a target was embedded in clutter which appeared in four adjacent range cells and had a C/T of 20 dB. Improvement factors were computed over several doppler regions of the response curve, such as shown in Fig. A3, by averaging the results over the specified doppler regions. In this simulation the range shifted target, due to RDC, competed with clutter. The simulation took into account the superposition of the reflections of an uncompressed PP4 code from the randomized clutter in the different range cells. The clutter was assumed to have a complex Gaussian probability distribution in each range cell and to be independent from cell to cell. Each spectral line of the approximated clutter spectrum gave rise to a doppler-shifted uncompressed pulse which was compressed and then processed according to the RDC MTI as in Fig. A4. The residue on a given trial was obtained from the sum of the residues from the individual spectral lines. This was repeated 500 times, using different phase distributions for the spectral lines for each range cell. In this simulation, we could account for the doppler shift of the clutter within the pulse, and hence, the RDC corresponding to each spectral line, or we could ignore it to determine the difference. In either case, the phase shift of the clutter, due to doppler, between two pulses separated by an interpulse period was taken into account.

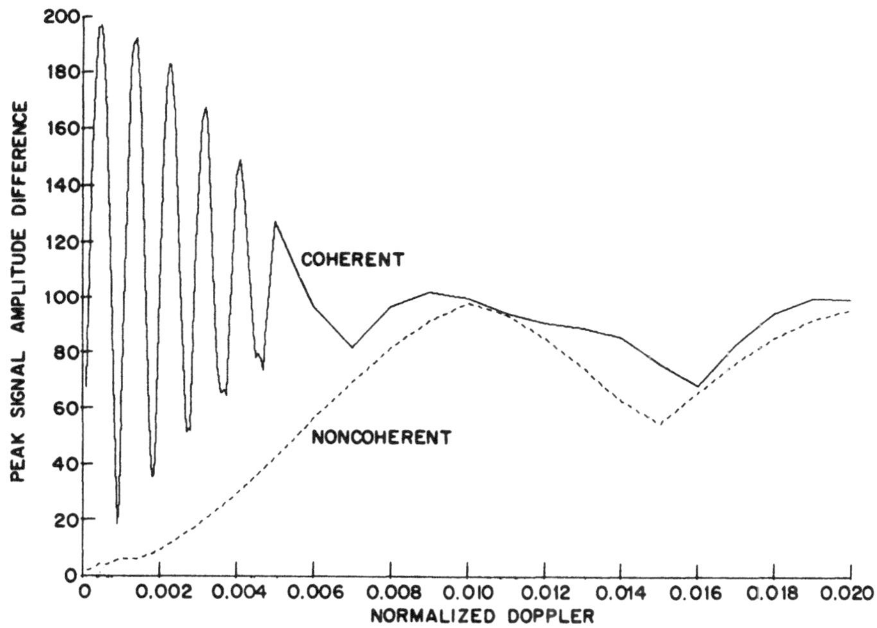

Fig. A3 — Response of a range-doppler coupled MTI for a doppler-shifted target using a 100-element P2 polyphase coded waveform

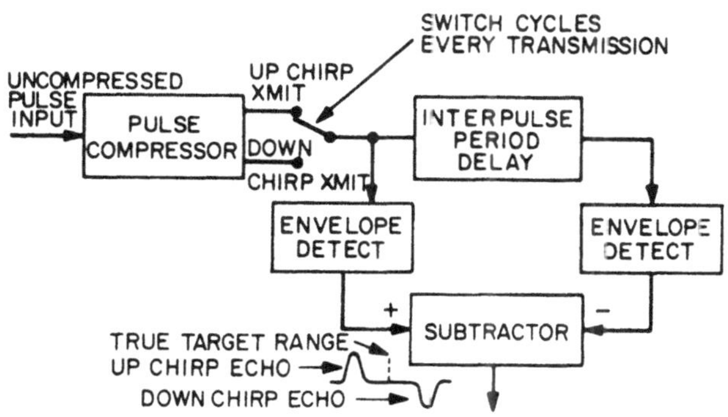

Fig. A4 — Two-pulse range-doppler coupled MTI (noncoherent) and response to incoming target

Table A1 summarizes these results for a two-pulse range-doppler coupled MTI using a 16-element PP4 code. Both the coherent and noncoherent cases were simulated. In this simulation, the ratio of the standard deviation of the Gaussian spectrum to the PRF was assumed to be 0.005. In Table A1, IPP = 1 means that the RDC effect on the clutter was included. IPP = 0 means that only the pulse-to-pulse phase shift of the clutter was taken into account. The improvement factor was computed for different regions of the target doppler. Table A1 shows the results for the improvement factor averaged over one PRF near 0 doppler and over the same interval near the corner frequency F_m corresponding to the velocity V_m.

Table A1 — Simulation of Clutter Attenuation (CA) and Improvement Factor (IMP) for RDC MTI using a 16-element PP4 Code at Different Target Doppler Frequencies

Target Doppler	IPP	Coherent		Noncoherent	
		CA	IMP	CA	IMP
Near 0.0	0	30.19	33.31	33.04	33.13
Near F_m	0	29.81	29.28	32.28	28.45
Near 0.0	1	30.35	33.45	33.03	33.01
Near F_m	1	29.93	29.25	32.71	28.98

For IPP = 0 the clutter attenuation and the improvement factor for the target averaged over one PRF interval of the MTI response near zero doppler is approximately the same as we obtained for the normal MTI for both the coherent and the noncoherent cases (see Table A1). It is seen that taking into account the RDC effect of the clutter (IPP = 1) had a negligible effect. This was also observed in other simulations that were performed.

Another result shown in Table A1 is that the clutter attenuation is the same, as it should be, for the target dopplers averaged over one PRF at zero frequency and at the corner frequency F_m. However, there is an approximate loss of 3 dB in the improvement factor for the coherent case at F_m due to the reduced average target response in this region. This is an expected result which can be deduced from the reponse curves, such as Fig. A5, and the fact that the clutter attenuation and the improve-

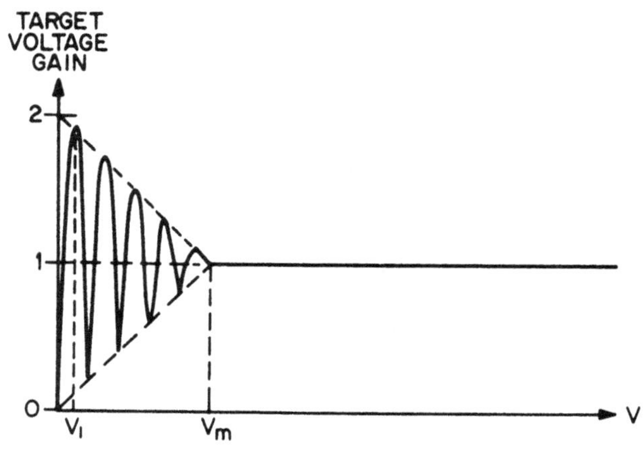

Fig. A5 — Response of coherent 2-pulse range-doppler coupled MTI with $T/T_l = 10$

ment factors were approximately the same as the normal coherent MTI (see Table A1). For the 2-pulse normal MTI, the target enhancement factor is theoretically 3 dB, and in Fig. A5 the response averaged over one PRF at $f = F_m$ is unity thereby representing a 3-dB reduction in the target enhancement and, hence, in the associated improvement factor.

REFERENCES

A1. F. F. Kretschmer, Jr., F. C. Lin, and B. L. Lewis, "A Comparison of Noncoherent and Coherent MTI Improvement Factors," *IEEE Trans., Aerospace and Electric Systems,* **AES-19**, No. 3, May 1983.

A2. M. Schwartz, "Information Transmission, Modulation and Noise," pp. 363-369, 2nd edition, McGraw-Hill, New York, 1970.

NONLINEAR SPECTRAL ANALYSIS AND ADAPTIVE ARRAY SUPERRESOLUTION TECHNIQUES

INTRODUCTION

Nonlinear spectral analysis techniques are currently of intense interest because of reported "superresolution" capabilities beyond the conventional periodogram or the Blackman-Tukey windowed Fourier transform [1]. Two methods, in particular, which have demonstrated a considerable increase in resolution are the maximum entropy spectral analysis (MESA) technique introduced by J. P. Burg [2,3], and the maximum likelihood method (MLM) demonstrated by J. Capon [4-6]. Since these techniques are most significant when processing short data sets, it is natural to consider their use for RF array antennas with a modest number of elements [7,8].

Adaptive processing techniques have been associated with these spectral estimation methods to some extent [9-11], but the literature indicates that cross-fertilization has been rather sparse. This situation is surprising, because both MESA and MLM bear a very close relationship to nonlinear adaptive array processing techniques. It is the purpose of this report to relate the MESA and MLM methods to their similar adaptive array antenna counterparts. The comparison analysis permits an examination of their principles of operation from the antenna array spatial pattern viewpoint, and helps to qualify their superresolution performance behavior. The real-time adaptive resolution of two incoherent sources located within a beamwidth has been simulated, and results are presented over an array output signal-to-noise ratio (SNR) range of 0 to 40 dB. The difficulties involved in resolving more than two closely spaced sources are also treated.

In addition to a discussion on the similarities between MESA, MLM, and adaptive array processing, some attention is given to the significant differences, which include the matter of two-dimensional data and the particular manner of averaging or estimating interelement signal correlations.

Alternate techniques for estimating spatial spectra have suggested themselves during the course of this study, and two of these are briefly described in the section on alternate adaptive processing (pp. 19, 20): phase center prediction utilizing a circular array, and a new adaptive "thermal noise" algorithm.

BURG MESA LINEAR PREDICTION FILTER

The Burg MESA method has been shown by van den Bos [12] to be equivalent to least mean square (LMS) error linear prediction. It runs a K-point linear prediction filter across a data sequence of N samples, where N should be at least twice the value of K. In the

Manuscript submitted July 24, 1979.

discrete filter diagram of Fig. 1, an optimum K-point prediction filter predicts the nth value of the sequence from K past values,

$$\hat{X}_n = \sum_{k=1}^{K} A_k X_{n-k}, \tag{1}$$

where $\hat{X}_n$ is the predicted sample, the A_k are optimum weighting coefficients, and the K past samples of X_{n-k} are presumed known. To define the difference between this predicted value and the current true value of X_n as the error ϵ_n, which is to be LMS minimized, we set

$$\epsilon_n = X_n - \hat{X}_n . \tag{2}$$

We minimize the total squared error E over the complete data sequence of N samples,

$$E = \sum_{n=K}^{N-1} \epsilon_n^2 \tag{3}$$

and

$$\frac{\partial E}{\partial A_i} = 0, \qquad 1 \leqslant i \leqslant K, \tag{4}$$

thus obtaining a set of K equations in K unknowns, i.e., the A_k filter weights,

$$\sum_{k=1}^{K} A_k \phi_{ki} = -\phi_{0i} \qquad \begin{matrix} 1 \leqslant k \leqslant K \\ 1 \leqslant i \leqslant K \end{matrix} \tag{5}$$

$$\phi_{ki} = \sum_{n=k}^{N-1} X_{n-k} X_{n-i} . \tag{6}$$

There are several different techniques, including Burg's, for manipulating this set of equations to solve for the optimum A_k filter weights, and Ref. 1 is recommended if the reader is interested in pursuing the details further. When this error has been minimized, its power spectrum will be equivalent to "white" noise. Thus, the uncertainty in ϵ_n has been maximized, hence we have a maximum entropy filter.

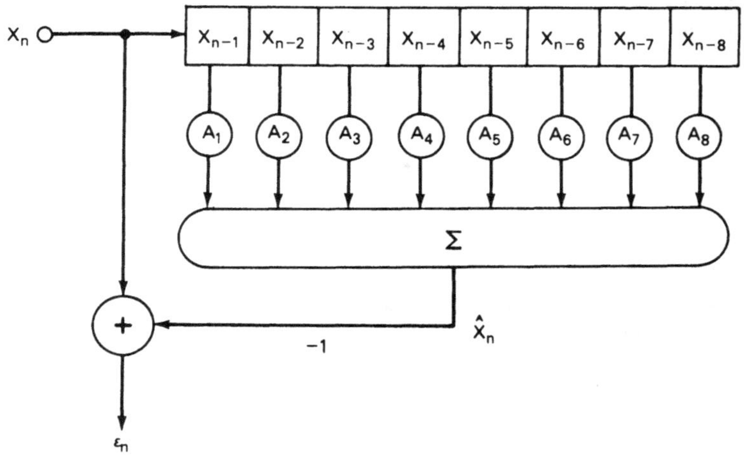

Fig. 1—Maximum-entropy filter with one-step linear prediction estimator

Upon substituting Eq. (1) into Eq. (2), we readily see that we have the form of a discrete convolution

$$\epsilon_n = -\sum_{k=0}^{K} A_k X_{n-k}, \tag{7}$$

where $A_o = -1$. The associated Z-transforms may be written,

$$\mathcal{E}(Z) = \left(1 - \sum_{k=1}^{K} A_k Z^{-k}\right) X(Z), \tag{8}$$

where the expression within the parentheses may be defined as the filter transform function H(Z),

$$H(Z) = \left[1 - \sum_{k=1}^{K} A_k Z^{-k}\right], \tag{9}$$

or

$$\mathcal{E}(Z) = H(Z)\ X(Z). \tag{10}$$

Note that H(Z) is a polynomial in Z which will have K roots or zero factors. Since "white" noise has a power spectrum known to be equal to a constant, then from Eq. (10) it is evident that we can solve for our unknown input power spectrum if the filter function is known, i.e.,

$$|X(\omega)|^2 = \frac{|\mathcal{E}(\omega)|^2}{|H(\omega)|^2} = \frac{(\text{constant})}{\left|\prod_{k=1}^{K} (1 - D_k e^{-j\omega})\right|^2}, \tag{11}$$

where the peaks (poles) of the unknown power spectrum will occur at the zeros of the filter function. This permits us to model the input sequence with the powerful, discrete, all-pole, linear prediction filter illustrated in Fig. 2, both in the frequency domain and in the time domain. The prediction filter is driven by white noise.

If one considers the action of such a filter upon sinusoids corrupted with a small amount of noise, it is evident that the filter can synchronize with even a short section of the sampled time waveform of the sinusoids and, once synchronized, can then proceed to "predict" many additional samples of the waveform with little error.

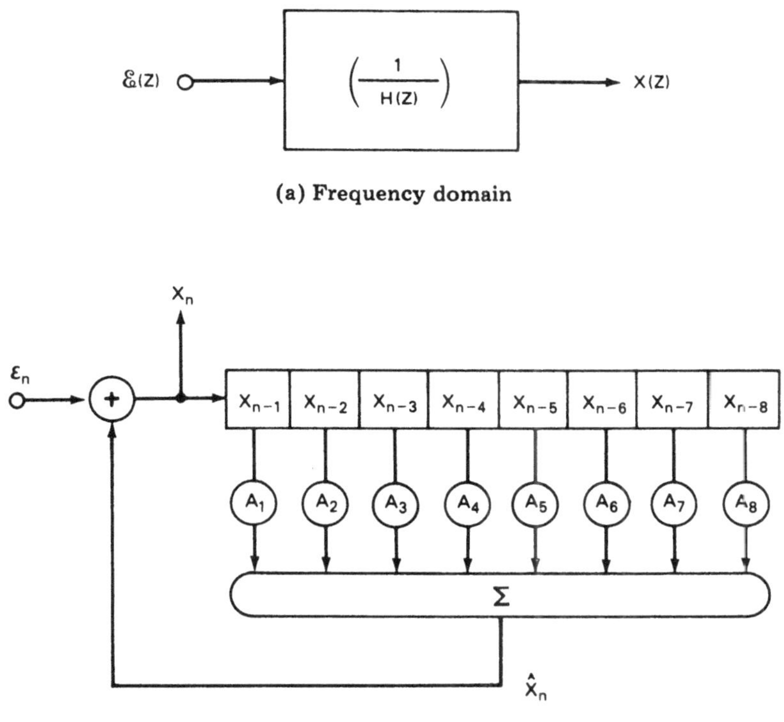

(b) Time domain

Fig. 2—Discrete all-pole linear prediction filter model

Another aspect of this filter is that it is a deconvolution filter, so-called because it estimates the unknown spectrum directly from the reciprocal of the filter transform function. Note in Eq. (10) that the error spectrum results from a simple multiplication of the unknown spectrum with the filter function, and that no convolution of the two occurs. In the conventional windowed Fourier transform method, on the other hand, the unknown spectrum is estimated by convolving the spectrum with the window filter transform function, and the convolution usually smears or destroys the fine detail of peaked spectra.

Reference 1 is recommended for those readers who want additional information on spectrum estimation filters.

LINEAR PREDICTION FILTERS AND ADAPTIVE SIDELOBE CANCELLERS

Conversion of the above MESA linear prediction filter to a weighted linear array of spatial sensors is straightforward, with the simplest configuration illustrated in Fig. 3. The element signal samples will be correlated in both space and time, giving rise to a two-dimensional data problem, but we convert this to spatial domain only by assuming that narrowband filtering precedes our spatial domain processing. For example, one could perform fast Fourier transform (FFT) on the element data prior to spatial processing. Also, we assume that our elements are equally spaced.

The nth "snapshot" signal sample at the kth element will consist of independent Gaussian receiver noise η_{kn} plus I incoherent source voltages,

$$E_{kn} = \eta_{kn} + \sum_{i=1}^{I} J_i e^{j(ku_i + \phi_{in})} \qquad 1 \leqslant k \leqslant K, \tag{12}$$

where

$u_i = 2\pi \left(\frac{d}{\lambda}\right) \sin \theta_i$

d = element spacing, assumed near $\lambda/2$

λ = wavelength

θ_i = spatial location angle of ith source

J_i = amplitude of ith source†

ϕ_{in} = random phase of ith source, nth sample

k = element index

n = snapshot sample index.

A "snapshot" is defined as one simultaneous sampling of the aperture signals at all array elements, and we assume that N snapshots of data are available.

†J_i has a constant rather than random amplitude because of a concurrent measurement program involving CW sources sampled at random times.

A brief examination of Fig. 3 from the standpoint of adaptive arrays leads to the conclusion that it is identical in configuration to a special subclass commonly referred to in the literature as a *sidelobe canceller* [13,14]. A typical sidelobe canceller configuration from Applebaum [14] is illustrated in Fig. 4. For the benefit of those who may not be familiar with them, it should be noted that the unweighted main-beam "element" is usually different and of much higher gain than the others, and the elements may or may not be equally spaced. They are designed to be operated on the basis of many successive snapshots (assuming digital operation) because their environment generally involves weak desired signals and an abundance of interference source data. They are a prediction filter in the sense that, after convergence, they are predicting the signal at the phase center of the main-beam element.

The adaptive sidelobe canceller is pertinent to our linear prediction filter because its spatial filter pattern analysis is well developed and can be applied directly to achieve a better understanding of the superresolution performance behavior. A further point is that real-time operation is readily achieved via most of the current adaptive algorithms, provided that the number of snapshots is enough to reach convergence in whitening ϵ. Convergence may require as few as two snapshots or as many as several thousand, depending on the particular algorithm and the parameters of the source distribution. Several examples will be discussed.

SPATIAL FILTER PATTERNS

The spatial filter function for the array of Fig. 3 is simply the adapted pattern after convergence, which is commonly referred to as the steady-state adapted pattern and may readily be computed from the inverse of the sample covariance matrix [14],

$$\mathbf{W}_o = \mu \mathbf{M}^{-1}\mathbf{S}^* \tag{13}$$

$$\mathbf{S}^{*t} = [0,0,0,0,0,0,0,1] \tag{14}$$

$$\mathbf{M} = \frac{1}{N}\sum_{n=1}^{N} \mathbf{M}_n \tag{15}$$

$$\mathbf{M}_n = \left[\mathbf{E}_n^* \cdot \mathbf{E}_n^t\right], \tag{16}$$

where $\mathbf{E}_n$ is the nth "snapshot" signal sample vector whose components are given by Eq. (12), $\mathbf{M}_n$ is the nth snapshot contribution to the covariance matrix, $\mathbf{M}$ is the sample covariance matrix averaged over N shapshots, $\mathbf{S}^*$ is the quiescent weight steering vector, μ is a scalar quantity, and $\mathbf{W}_o$ is the optimum weight vector. Note that the steering vector $\mathbf{S}^*$ injects zero weight on every element except for the end element, causing the quiescent pattern of the array to be that of the single end element.

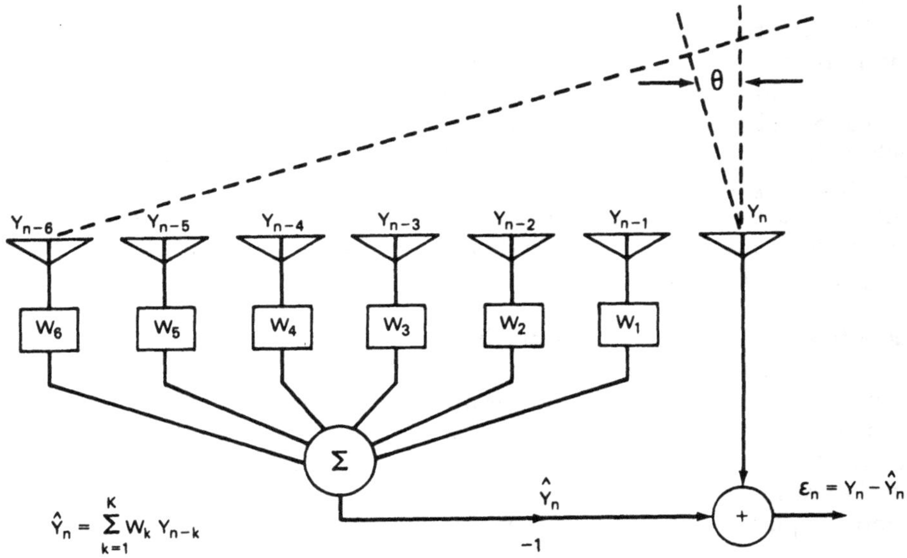

Fig. 3—Array aperture linear prediction spatial filter model

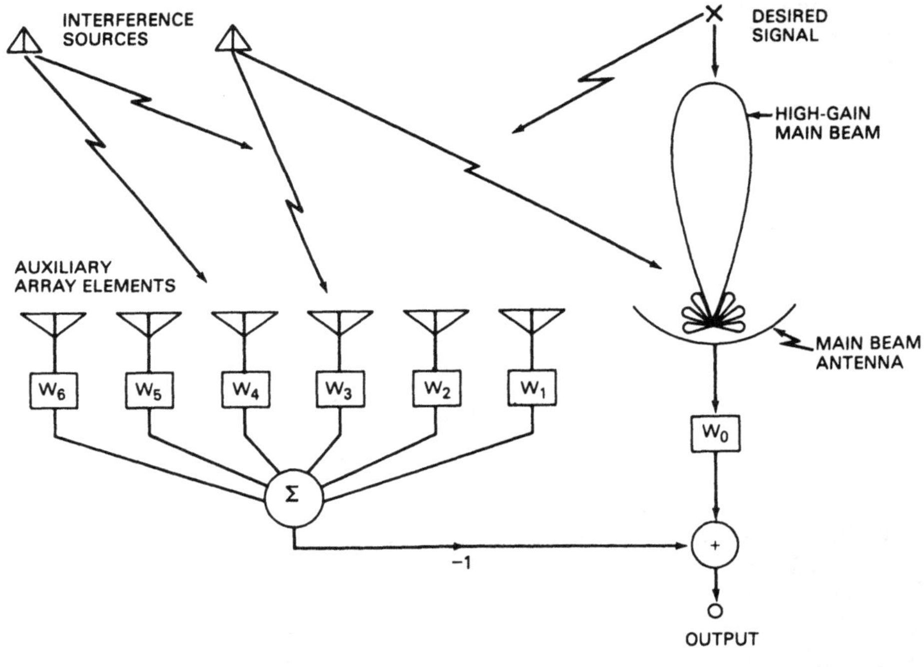

Fig. 4—Typical adaptive array sidelobe canceller configuration

Figure 5 shows a typical quiescent (single-element) pattern and an adapted pattern obtained from an 8-element linear array with two far-field, incoherent, 30-dB sources located at 18° and 22°. The adapted pattern weights were computed per Eq. (13) from the inverse of the covariance matrix averaged over 1024 simulated snapshots. Note that the two pattern nulls (zeros) align perfectly with the locations of the two sources. Of course, the array signals in this simulation were corrupted only by receiver noise (no element errors are included), and an average over 1024 snapshots is indeed steady state. Another important point to note is that nulls in such an adapted pattern may be located arbitrarily close together in terms of beamwidth, without violating any physical principle. Yet, because the nulls have served to locate two sources within a beamwidth, one may describe this as a "superresolution" pattern.

It is readily shown that this adapted pattern is obtained by subtracting the summed array output pattern from the element (main beam) pattern and, furthermore, that the summed array pattern consists of properly weighted "eigenvector beams" [15]. In terms of the eigenvector weights, we can express the optimum weights in the form,

$$\mathbf{W}_o = \mathbf{S}^* - \sum_{i=1}^{K} \left(\frac{B_i - B_o}{B_i + B_o} \right) \hat{\mathbf{W}}_{qi} \mathbf{e}_i \tag{17}$$

$$\hat{\mathbf{W}}_{qi} = \left(\mathbf{e}_i^{*t} \, \mathbf{S}^* \right)$$

where $\mathbf{e}_i$ is the ith eigenvector of the covariance matrix, B_i is the ith eigenvalue, and B_o is the smallest eigenvalue corresponding to receiver noise power. Note that only the significant eigenvectors corresponding to $B_i > B_o$ need be considered here. An adaptive array forms one such eigenvector beam for each degree of freedom consumed in nulling out the spatial source distribution. Figure 6 shows the two eigenvector beams required for this two-source example. It should be emphasized here that the true resolution and signal gain of the array are reflected in these eigenvector beams. They demonstrate the importance of having as wide an aperture as possible because the superresolution capability in the adapted pattern is a percentage of the true resolution of these beams. Also, since the superresolution nulls are formed by subtracting these beams of conventional width, it follows that the nulls will be rather delicate and very sensitive to system imperfections and signal fluctuations.

The desired "spatial spectrum pattern" is then obtained from Eq. (11) as simply the inverse of the adapted pattern. Figure 7 shows this inverse for the two-source example and compares it with the output of a conventional beam scanned through the two sources. Several comments are in order concerning such inverse patterns.

1. They are not true antenna patterns because there is no combination of the element weights that could produce such a peaked spatial pattern. They are simply a function computed from the reciprocal of a true antenna pattern.

2. Linear superposition does not hold in either the inverse or the original adapted pattern because of the nonlinear processing involved in the inverse of the covariance matrix (or the equivalent).

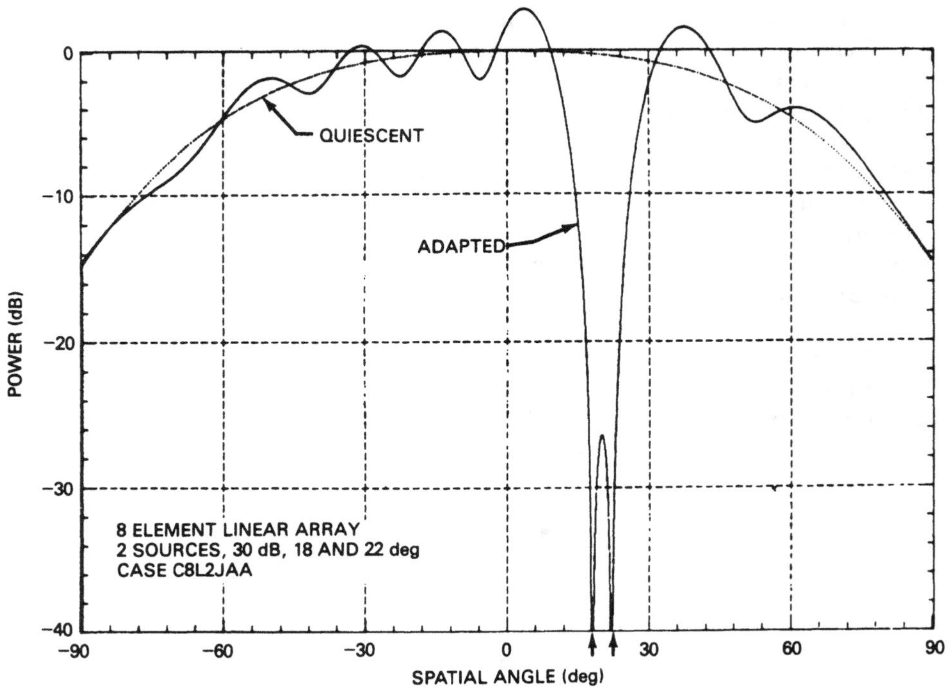

Fig. 5—Quiescent (single-element) and adapted patterns for two-source case, covariance matrix inverse algorithm, 1024 datasnaps

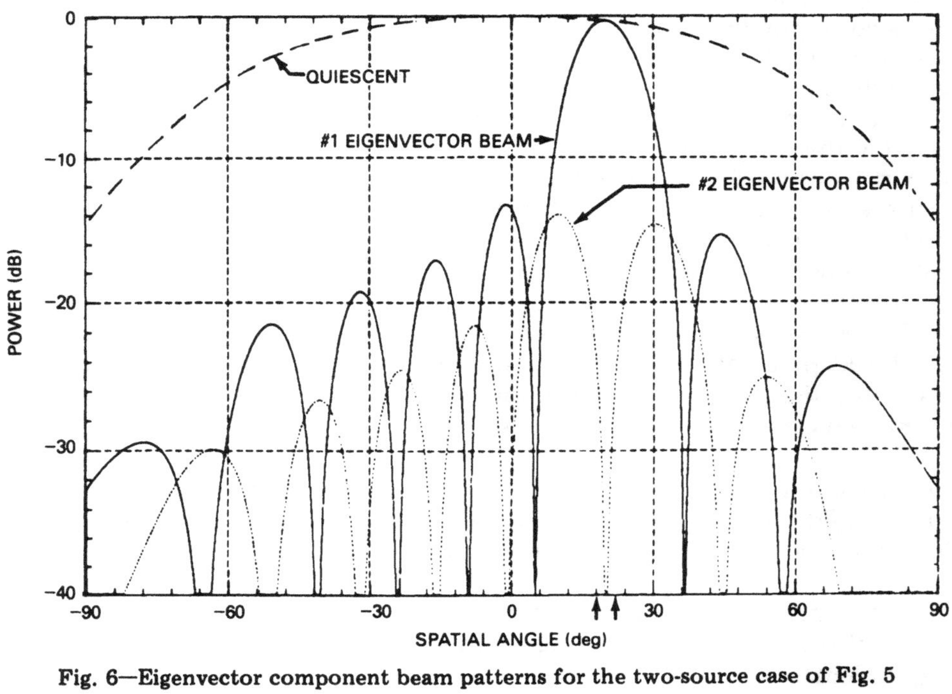

Fig. 6—Eigenvector component beam patterns for the two-source case of Fig. 5

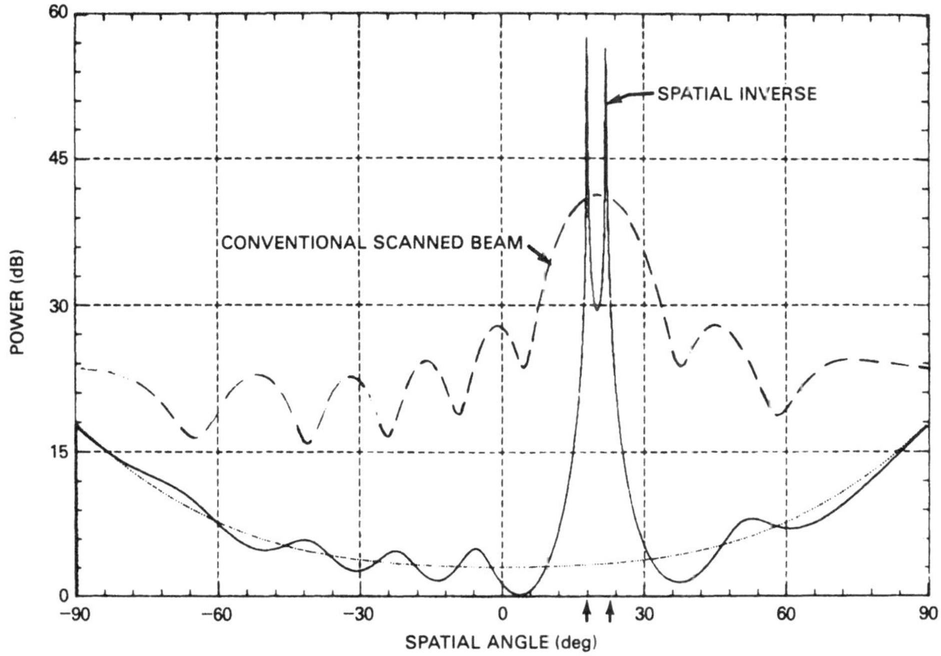

Fig. 7—Spatial spectrum inverse pattern for the two-source case of Fig. 5, and comparison with output of conventional scanned beam

3. The heights of the peaks do not correspond with the relative strengths of the sources because the depths of the adapted pattern nulls do not. In general, the adaptive null depth will be proportional to the square of the SNR of a source [15], but even this relationship fails when there are multiple sources closely spaced.

4. There is no real-signal output port associated with such a pattern because it is not a true antenna pattern. An output could be simulated, of course, by implementing the equivalent all-pole filter of Fig. 2 and driving it with white noise.

5. They do emphasize the locations of the zeros (nulls) of the adaptive array filter polynomial.

6. They are inherently capable of superresolution.

7. They achieve good "contrast" with the quiescent pattern background ripple because of the aforementioned proportionality to the square of source strengths.

8. Spatial information is gained beyond that obtained from a conventional array beam scanned through the sources because the array degrees of freedom are utilized in a more effective, data-adaptive manner.

REAL-TIME FILTER OPERATION

To get a feel for real-time operation performance with realistic weight update averaging, we ran simulations in which an eight-element array had its weights computed from the simple Howells—Applebaum algorithm in recursive digital form as diagrammed in Fig. 8. The associated recursive relationship for the kth weight may be written

$$(1 + \tau)\, W_k(n) = \tau W_k(n-1) + S_k^*(n) - \left(\frac{1}{B_o}\right) E_k^*(n)\, Y(n) \tag{18}$$

where

$$Y(n) = \sum_{k=1}^{K} E_k(n)\, W_k(n-1). \tag{19}$$

$Y(n)$ is the current array output, $E_k(n)$ is the current snapshot signal sample at the kth element (similar to Eq. (12)), $W_k(n\text{-}1)$ is the previous value of the kth weight, $S_k^*(n)$ is the injected kth beam-steering weight, and B_o is a constant equal to receiver noise power.

The digital integration loop shown in Fig. 8 is designed to simulate a simple low-pass RC filter with a time constant of τ, but we choose to make τ dynamic to get faster convergence for most situations. Thus, let τ become $\tau(n)$,

$$\tau(n) = \tau_o + TP_r(n) \tag{20}$$

where

T = high-power, fast time constant

τ_o = quiescent conditions, slow time constant

$P_r(n)$ = snapshot SNR (power ratio).

This formulation permits us to satisfy the 10% bandwidth criterion at high power levels to avoid noisy weights [15] by choosing the value of $T = 3.2$, and yet the quiescent condition time constant need be no worse than $\tau_0 = 200$. The larger value for τ_0 is necessary in order to have a relatively stable quiescent pattern. Actual weight update averaging is performed in accordance with the reciprocal of the closed-loop bandwidth α,

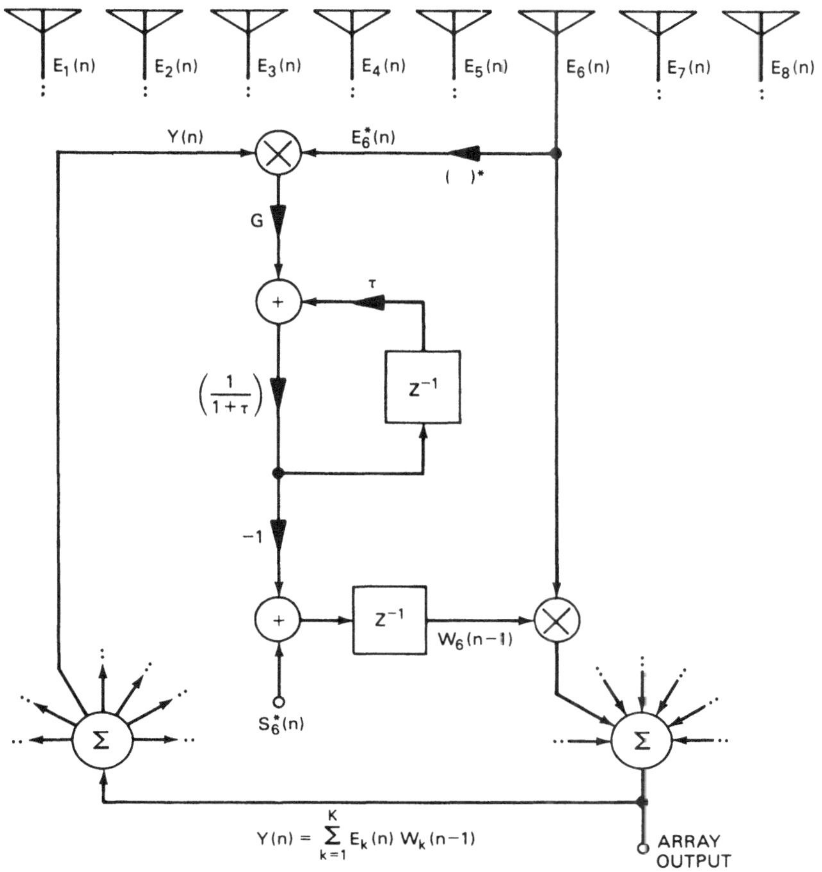

Fig. 8—The Howells-Applebaum algorithm in a recursive digital form

$$\frac{1}{\alpha} = \frac{\tau}{1 + P_r(n)} = \frac{\tau_o + TP_r(n)}{1 + P_r(n)} \quad , \tag{21}$$

where we approach the value $\tau_o/2$ under quiescent conditions when $P_r = 1$, and we approach the value of T when $P_r >> 1$.

A typical time plot of snapshot output power in decibels above receiver noise level is illustrated in Fig. 9 for the case of two 30-dB sources located at 18° and 22°. Note that the system converges and nulls out the two strong sources within about 60 snapshots after processing commences. For comparison, the rms orthonormal eigenvector solution is also shown plotted, but this solution used a fixed τ = 48,000 and does not converge as rapidly. The principal point demonstrated here is that convergence occurs reasonably fast in terms of snapshot counts, even for this simple LMS adaptive algorithm and, after convergence,

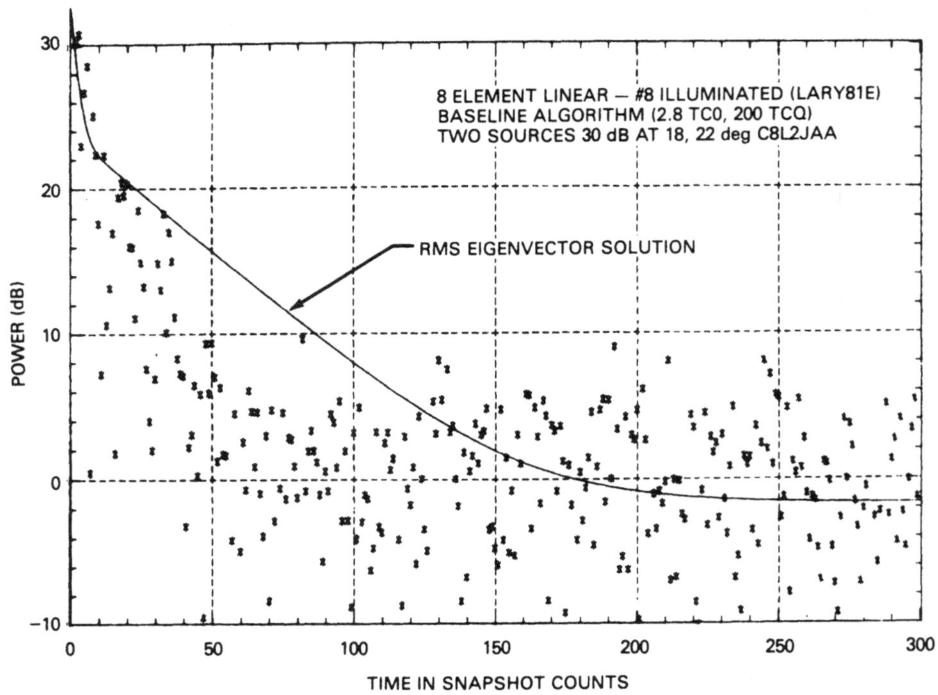

Fig. 9—Time-sequence snapshot output of array in decibels above receiver noise power; Howells-Applebaum recursive algorithm with dynamic time constant

any of the snapshot weight sets can be used to compute the spatial spectrum. Figure 10 shows the spatial spectrum plots associated with snapshots 100, 200, and 300. In comparing Fig. 10 against Fig. 7, note that the pattern has changed very little, in spite of the fact that the integration or averaging has been reduced by two orders of magnitude, i.e., from a value of 1024 snapshots in Fig. 7 to about 3 snapshots in Fig. 10 (albeit a decaying average inherent in the low-pass filter). The greatest effect of this reduced averaging is that the heights of the peaks are reduced and the peaks fluctuate from snapshot to snapshot because of the perturbation of the noise on each snapshot weight update.

In rounding out the example of these two incoherent sources spaced 4° apart (about 0.27 beamwidths apart), Fig. 11 illustrates what happens as we reduce the SNR strength of the sources. Figure 11a at 20 dB shows increasing peak fluctuations in magnitude, which merely reflect the null fluctuations in the adapted pattern, although the spatial locations of the sources are still accurate. Figure 11b at 10 dB shows even greater fluctuations in peak magnitude, but now the patterns are deteriorating in both shape and peak locations, indicating that the resolution capability is nearing its limit, i.e., if the source power levels are reduced further, then the adaptive array can no longer resolve them accurately at that particular spacing.

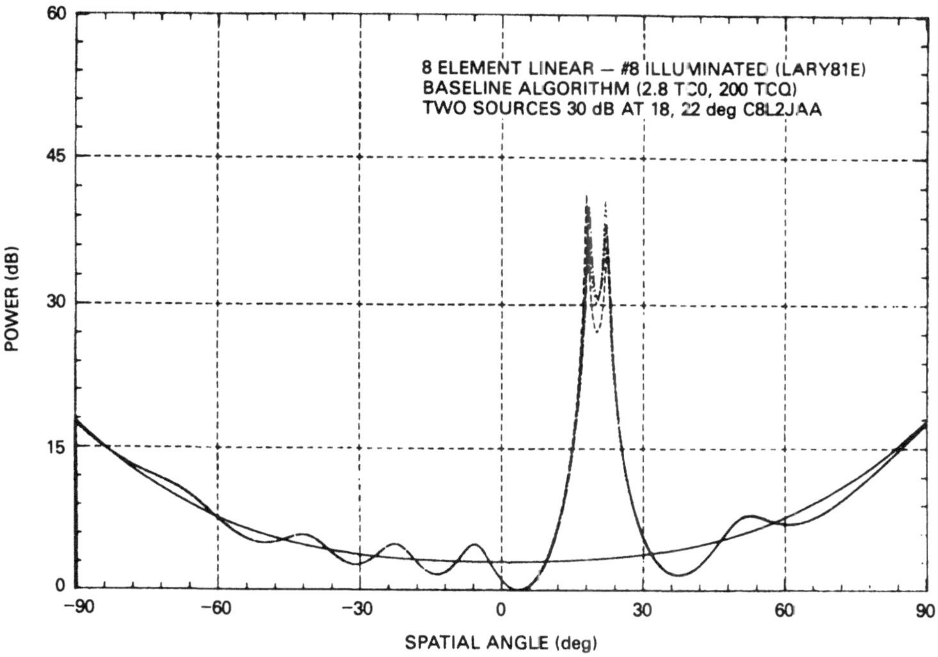

Fig. 10—Typical spatial spectrum snapshot plots after convergence, snapshot weight sets 100, 200, and 300. Two 30-dB sources were located at 18° and 22°

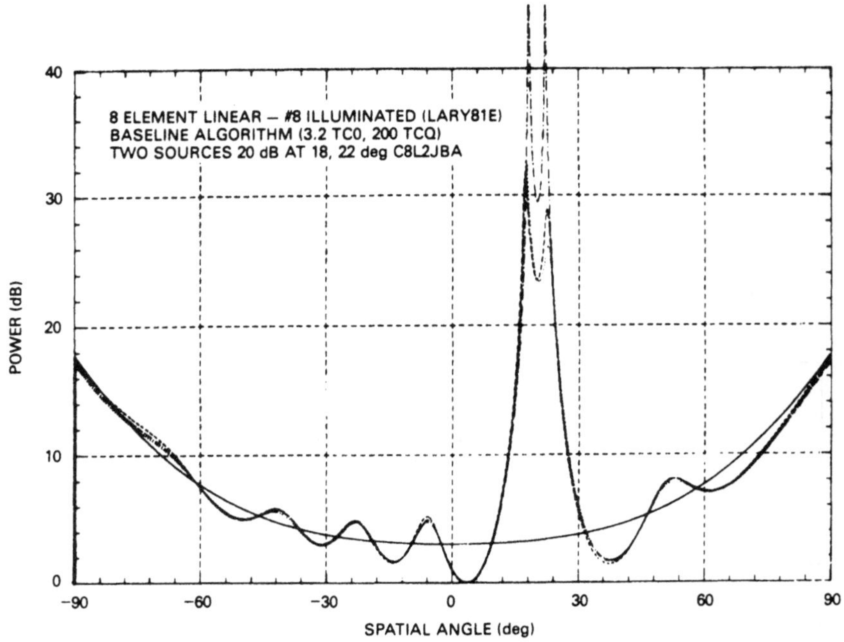

(a) 20-dB SNR sources

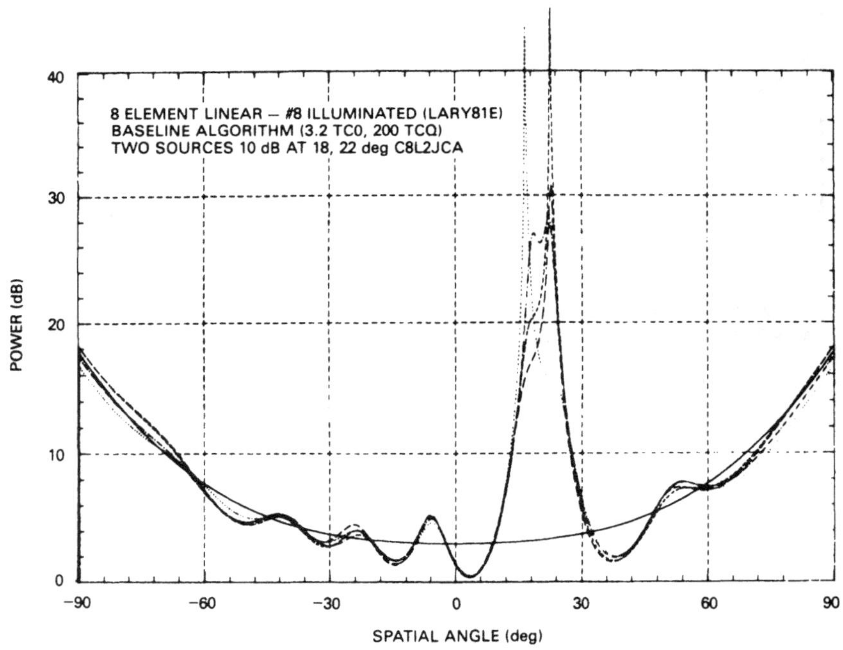

(b) 10-dB SNR sources

Fig. 11—Typical spatial spectrum snapshot plots after convergence. Two sources located at 18° and 22°

A summary of the approximate resolution capability limit for the adaptive array spatial filter operating against two incoherent sources is shown in Fig. 12. This performance curve is universal because the abscissa is source separation in beamwidths and the ordinate is source SNR measured at the array output, i.e., element SNR multiplied by the number of elements in the array. Thus, the curve can be utilized for any number of array elements in a linear array configuration. Note that at low ordinate SNR values, we actually have negative SNR at the elements. The curve tells us that we can separate two sources at arbitrarily small spacings, provided we have sufficient SNR and, also, provided that our element data samples are sufficiently accurate. Recall that the simulations involved here did not include any element errors.

If there are more than two sources within a beamwidth, then difficulties mount rapidly and the filter null points may not accurately represent the spatial locations of the sources. For example, the above simple LMS algorithm was tested against three sources of 30-dB strength spaced 4° apart, and it could not resolve all three even after 2048 snapshots because it never converged sufficiently (the eigenvalue spread was too great). To separate the three sources in a reasonable number of snapshots requires use of an adaptive algorithm of faster convergence [16], such as the matrix inverse of Eq. (9), wherein we use a "sliding window" averaging of the sample covariance matrix, or the Gram—Schmidt algorithm

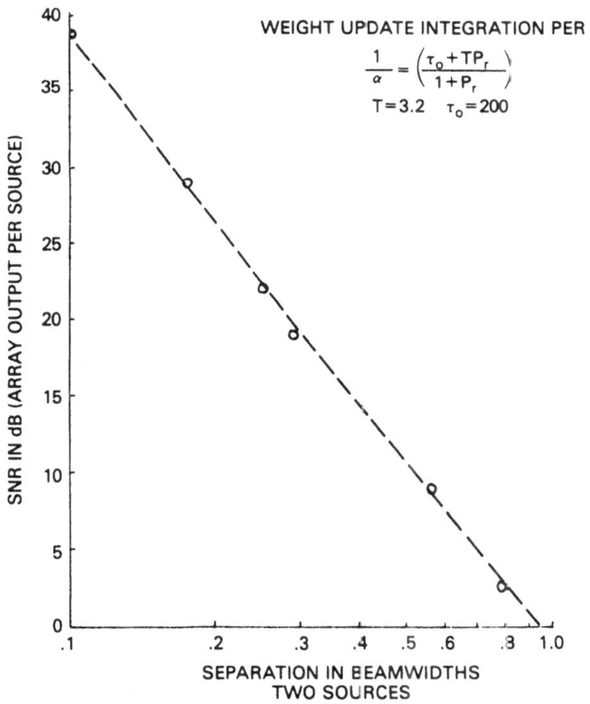

Fig. 12—Universal approximate resolution limit for two incoherent sources. Simulation conditions: narrowband, no array errors, λ/2 element spacing, linear array, Gaussian receiver noise.

described by Alam [11,17]. Figure 13 shows the resolution of the three sources via the Gram—Schmidt algorithm, but it should be noted that, even with this fast algorithm, resolution was not achieved until after 200 snapshots and the locations of the peaks fluctuates considerably. This was a good illustration of the "delicacy" of null formation for the case of closely spaced multiple sources.

MLM AND ADAPTIVE DIRECTIONAL CONSTRAINTS

The maximum likelihood spectral estimate is defined as a filter designed to pass the power in a narrow band about the signal frequency of interest and minimize or reject all other frequency components in an optimal manner [4,5]. This is identical to the use of a zero-order main-beam directional gain constraint in adaptive arrays [18,19], where the "spatial spectrum" would be estimated by the output residual power P_o from the optimized adapted array weights,

$$P_o = \mathbf{W}_o^{*t} \mathbf{M} \mathbf{W}_o \tag{22}$$

where

$\mathbf{W}_o$	=	$\mu \mathbf{M}^{-1} \mathbf{S}^*$(optimized weights)
$\mathbf{M}$	=	covariance matrix estimate
$\mathbf{S}^*$	=	main-beam direction-steering vector
μ	=	scalar quantity.

Under the zero-order gain constraint, we require $\mathbf{S}^t \mathbf{W}_o = 1$, whereupon μ becomes

$$\mu = (\mathbf{S}^t \mathbf{M}^{-1} \mathbf{S}^*)^{-1}. \tag{23}$$

Substituting μ and $\mathbf{W}_o$ into Eq. (22), then, results in

$$P_o = \frac{1}{\mathbf{S}^t \mathbf{M}^{-1} \mathbf{S}^*} . \tag{24}$$

Upon sweeping the steering vector $\mathbf{S}^*$ for a given covariance matrix inverse, P_o will estimate the spatial spectrum. Interestingly, this result is identical (within a constant) to the spectrum obtained from the *inverse* of the output residual power from an unconstrained optimized adapted array.

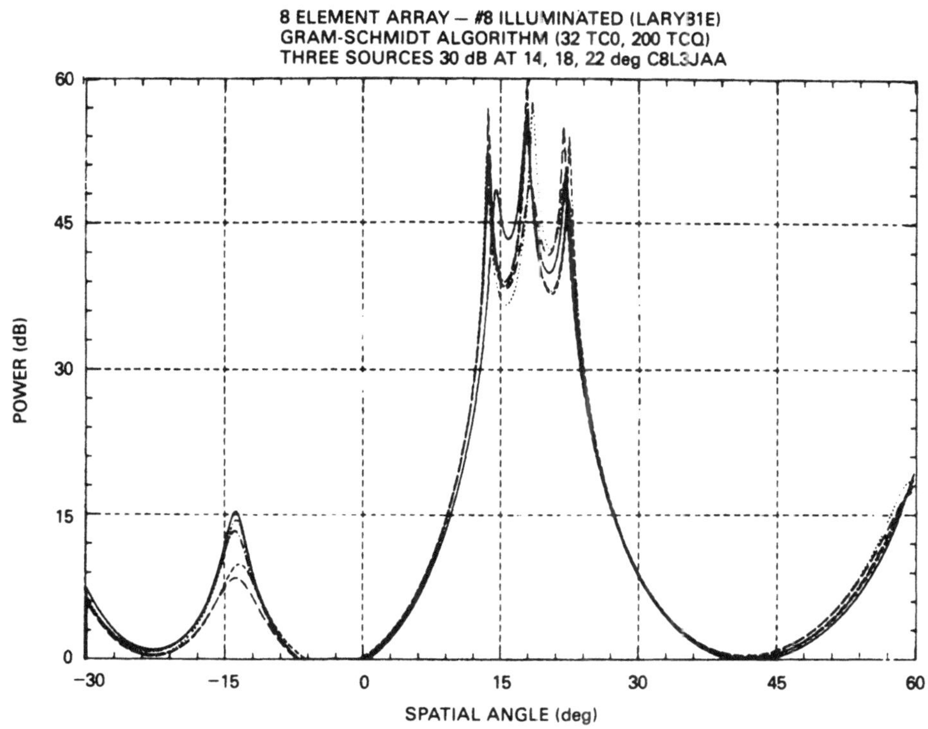

Fig. 13—Typical spatial spectrum snapshot plots after convergence; Gram-Schmidt algorithm, three 30-dB sources located at 14°, 18°, and 22°.

Referring back to the fourth section and the eigenvector beams shown in Fig. 6, one may describe the principle of operation in terms of subtracting the eigenvector beams from the quiescent uniform-illumination steering vector "main beam" for each position of the main beam. Thus, you have a continuously changing adapted pattern as the main beam is scanned, subject to the above gain constraint in the steering direction.

Figure 14 illustrates the output spectrum plotted from P_o for the two-source case utilized for Figs. 5, 6, and 7. Note that in comparison with Fig. 7, this MLM spectrum has peaks that are about 18 dB lower and are thus of lower resolution capability. However, the two peaks have located the sources correctly and, in addition, the peak values reflect the true power levels of the sources. This is in agreement with the observations of Lacoss [5] and others. Although this technique has poorer resolution than the previous one and requires more computation in plotting the output spectrum, it does offer several rather significant advantages.

1. The output power is directly referenced to receiver noise power, thus permitting calibration and measurement of relative source strength.

2. If the sources can be resolved, then a psuedolinear superposition holds at the peaks, and they should reflect the true relative source strengths.

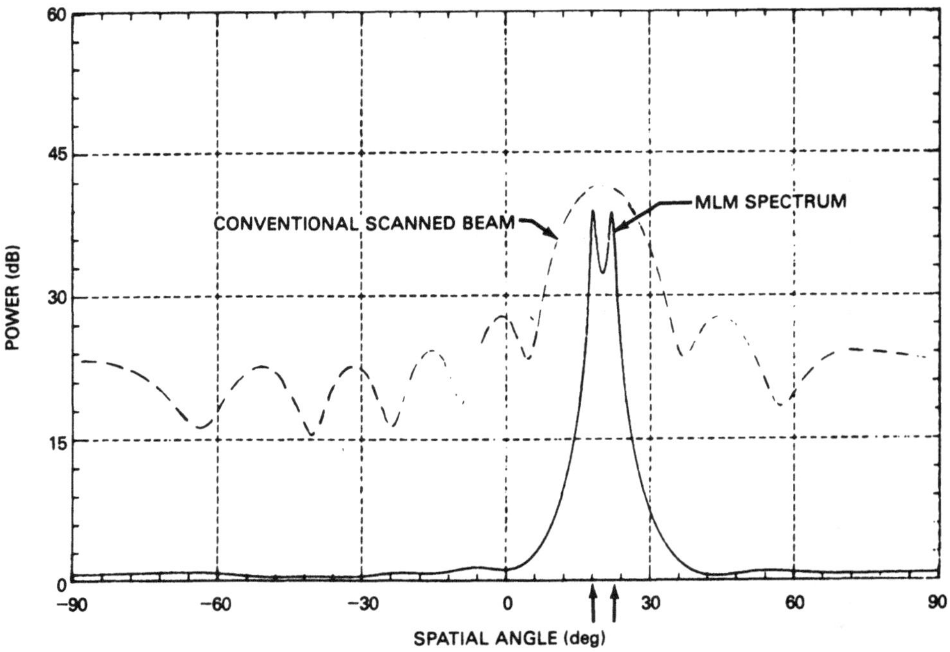

Fig. 14—MLM spatial spectrum plotted from residual power of adaptive zero-order main-beam constraint for the two-source case of Fig. 5

3. The output of this filter is a real signal, and if the filter passband is steered to a particular source, one can monitor that source at full array gain while rejecting outputs of all other sources.

4. The residual background spatial ripple is very low and well behaved.

5. It is not necessary to have the elements equally spaced. Thus, one should take advantage of this property to spread them out for a wider aperture and substantially increase the resolution for a given number of elements. (This is done in the field of geophysics [4].) If this is done, it is very likely that this method could equal or surpass the resolution of the previous technique.

ALTERNATE ADAPTIVE PROCESSING FOR SPATIAL SPECTRA

Phase Center Prediction

The adaptive array processing described in the fourth section did not use the configuration of Fig. 3 in the true sense of a K-point linear prediction filter that runs across a larger aperture of data samples. The K elements involved were the total aperture, and a series of N snapshots of data were used to estimate or predict the signal at the phase center of the unweighted main-beam element. Phase-center prediction of this type is very flexible in that the location of the phase center of the main beam is rather arbitrary, and it is not necessary for the elements to be equally spaced. For example, the main beam may be a weighted summation of some or all of the elements in the array.

If phase-center prediction is carried to its logical conclusion from the standpoint of estimating spatial spectra, it would appear that the ideal element configuration is a circular aperture array with an omni-mode main beam, i.e., the array would be predicting to the exact center of the circle. An example of such an array is contained in Ref. 15, and it could readily utilize the processing described in the previous two sections.

The "Thermal Noise" Algorithm

In discussing the MLM spectral estimate in the preceding section and its identical relationship to a zero-order, main-beam, directional constraint, it was mentioned that Eq. (24) was equal (within a constant) to the spectrum obtained from the inverse of the output residual power from an unconstrained optimized adapted array. If we define this output residual power as R_o, then

$$R_o = \left(\mathbf{S}^t \mathbf{W}_o \right) = \mu \left(\mathbf{S}^t \mathbf{M}^{-1} \mathbf{S}^* \right) , \tag{25}$$

and we have the reciprocal relationship to Eq. (24) except for μ.

An interesting feature about R_o is that it approaches zero whenever the steering vector sweeps through the position of a source, and the reason it approaches zero is that the optimum weight vector W_o approaches zero. Therefore, it appears prudent to formulate the dot product of W_o with its own conjugate and define this product as the adapted thermal noise power output N_o,

$$N_o = \mathbf{W}_o^{*t} \mathbf{W}_o . \tag{26}$$

The reciprocal of N_o then estimates the spatial spectrum, and we may refer to this as the "thermal noise" algorithm for spectrum estimation. Figure 15 illustrates the application of this algorithm to the same two-source case as was used for Figs. 5, 6, 7, and 14. Contrasting it with the equivalent MESA and MLM techniques, note that it exhibits resolution peaks fully equal to MESA, and yet retains the very low residual background ripple of MLM referenced to quiescent receiver noise power level. Like MESA, it cannot measure relative source strength, but it should be proportional to the square of source strengths whenever it can resolve the sources. Thus, the "thermal noise" algorithm appears to possess an interesting combination of the characteristics of both MESA and MLM.

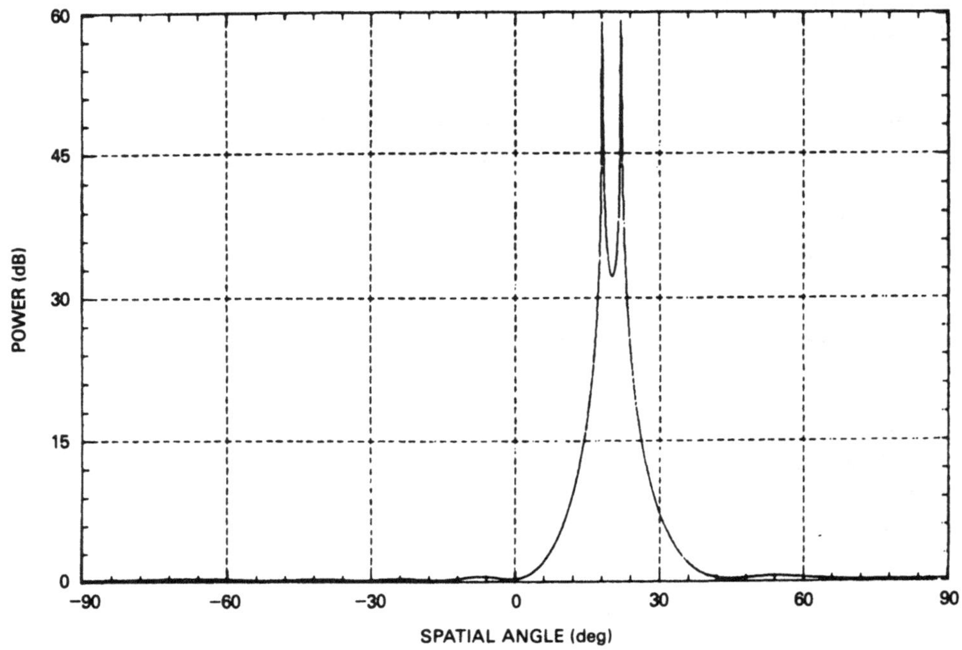

Fig. 15—"Thermal noise" algorithm spatial spectrum plotted for the two-source case of Fig. 5

SIGNIFICANT PROCESSING DIFFERENCES

Although the similarities are extensive enough to create a favorable climate for technique interchanges, there are also some significant differences which arise from the very nature of the applications and their data. For example, assume that we utilize the Burg MESA technique to run a K-point prediction filter across an RF aperture of M elements, where K must be smaller than M by at least 50% in order to obtain a reasonable averaged estimate for determination of the K filter weights. This type of processing has a great advantage in being able to operate with a single snapshot of element data, but it is "unthinkable" from an RF array point of view because it is wasting expensive aperture elements. It is far more preferable to operate on the basis of many snapshots of data from the smallest number of elements possible. In fact, a recent study of the Burg MESA technique by King [8], as applied to an RF spatial array, found that single snapshot results were seldom satisfactory, and that it was usually necessary to average the results from 10 or 20 snapshots in order to achieve a stable spectrum estimate. This comment is not meant to imply that the Burg technique is not applicable, but only that single-snapshot operation is not very practical for RF arrays.

A related difference is simply the fact that RF array element signal samples are correlated in both space and time, thus giving rise to a two-dimensional data problem [7] that does not exist in spectrum analysis. To overcome this problem usually requires filtering in both domains. For example, one may handle the time domain via tapped delay lines, an FFT operation, or actual narrowband filters at each array element. Note that, in each case, we imply many aperture snapshots in order to handle this two-dimensional data problem. Fortunately, the applications of RF array systems are such as to produce an abundance of data snapshots, so that processing is naturally designed to operate on this premise.

Another area of significant difference is the manner in which correlation matrix coefficients are estimated. In spectral analysis, one generally deals with a single data sequence in X_n of M samples, from which is computed the autocorrelation matrix coefficients of the form

$$\phi_\tau = \frac{1}{M} \sum_{n=0}^{N-\tau-1} X_n X^*_{n+\tau}, \qquad 0 \leqslant \tau \leqslant L \tag{27}$$

where the lag τ is restricted to some maximum value L, which is a fraction of M. In adaptive array processing, on the other hand, one generally deals with N snapshots of data sampled from K elements, from which is computed covariance matrix coefficients of the form

$$\phi_{kl} = \frac{1}{N} \sum_{n=1}^{N} X^*_{kn} X_{ln} \qquad \begin{matrix} 1 \leqslant k \leqslant K \\ 1 \leqslant l \leqslant K. \end{matrix} \tag{28}$$

Equations (27) and (28) represent two different types of interelement signal correlation averages, and they can lead to different spectrum estimates, particularly if coherence exists among the spatial sources.

Therefore, some care must be exercised when borrowing a given technique from one field for use in the other.

CONCLUSIONS

The Burg MESA and MLM nonlinear spectral analysis techniques have been related to their similar nonlinear adaptive array antenna counterparts, which consist of the sidelobe canceller and directional gain constraint techniques. The comparison study was conducted in the interest of achieving some cross-fertilization by examining their principles of operation from the antenna array spatial pattern viewpoint, and the analysis helps to qualify their superresolution performance behavior. It was shown that the superresolution derives from the subtraction of eigenvector beams that embody the true conventional resolution and signal gain of the array, such that physical principles are not really violated. Resolution is still proportional to the width of the array aperture. The spatial spectrum patterns are not true antenna patterns but are simply the inverse of the adaptive filter patterns. They are capable of superresolution, and spatial information is gained beyond that obtained from a conventional array beam, which is scanned through the sources.

The adaptive array counterpart is naturally suited to real-time spectral estimation via most of the current adaptive algorithms, and the case of two incoherent sources located within a beamwidth was simulated over a SNR range of 0 to 40 dB. A universal super-resolution performance curve, Fig. 12, was developed for this particular case, which can be utilized for linear arrays of any number of elements. If there are more than two sources within a beamwidth, difficulties mount rapidly and the filter null points may not accurately represent source locations.

In addition to the direct adaptive counterparts, two alternate adaptive spatial spectrum estimators were suggested. One is a circular array aperture arrangement which predicts to the center of the circle, and the other is a new adaptive "thermal noise" algorithm which appears to possess an interesting combination of both MESA and MLM characteristics.

There are some significant differences between spectral analysis techniques and adaptive array techniques that relate to the nature of their applications and the two-dimensional data problem. However, it appears that there is much to be gained through careful analysis of the other's techniques. For example, in addition to the obvious applications in target detection, DF (direction finding) systems, and source classification, spectral analysis techniques should be of benefit in data extension and coherence effects investigations.

REFERENCES

1. *Modern Spectrum Analysis*, D. G. Childers, ed., IEEE Press, New York, 1978. (Note: This book contains complete copies of references 2—7, 9, 11, 12.)

2. J. P. Burg, "Maximum Entropy Spectral Analysis," in *Proc. 37th Meeting of the Society of Exploration Geophysicists*, 1967, Oklahoma City.

3. J. P. Burg, "A New Analysis Technique for Time Series Data," Paper presented at the NATO Advanced Study Institute on Signal Processing with Emphasis on Underwater Acoustics, Enschede, Netherlands, 1968.

4. J. Capon, "High-Resolution Frequency-Wavenumber Spectrum Analysis," *Proc. IEEE* 57, 1408—1418 (Aug. 1969).

5. R. T. Lacoss, "Data Adaptive Spectral Analysis Methods," *Geophysics* 36, 661—675 (Aug. 1971).

6. J. P. Burg, "The Relationship between Maximum Entropy Spectra and Maximum Likelihood Spectra," *Geophysics* 37, 375—376 (Apr. 1972).

7. R. N. McDonough, "Maximum-Entropy Spatial Processing of Array Data," *Geophysics* 39, 843—851 (Dec. 1974).

8. W. R. King, "Maximum Entropy Wavenumber Analysis," NRL Report 8298, Mar. 1979.

9. L. J. Griffiths, "Rapid Measurement of Digital Instantaneous Frequency," *IEEE Trans.* **ASSP-23,** 207–222 (Apr. 1975).

10. D. R. Morgan and S. E. Craig, "Real-Time Adaptive Linear Prediction Using the Least Mean Square Gradient Algorithm," *IEEE Trans.* **ASSP-24,** 494–507 (Dec. 1976).

11. M. A. Alam, "Adaptive Spectral Estimation," in *Proc. 1977 Joint Automatic Control Conf.*, June 1977, San Francisco, Calif.

12. A. van den Bos, "Alternative Interpretation of Maximum Entropy Spectral Analysis," *IEEE Trans.* **IT-17,** 493–494 (July 1971).

13. P. W. Howells, "Explorations in Fixed and Adaptive Resolution at GE and SURC," *IEEE Trans.* **AP-24,** 575–584 (Sept. 1976).

14. S. P. Applebaum, "Adaptive Arrays," *IEEE Trans.* **AP-24,** 585–598 (Sept. 1976).

15. W. F. Gabriel, "Adaptive Arrays—An Introduction," *Proc. IEEE* **64,** 239–272 (Feb. 1976).

16. I. S. Reed, J. D. Mallett, and L. E. Brennan, "Rapid Convergence Rate in Adaptive Arrays," *IEEE Trans.* **AES-10,** 853–863 (Nov. 1974).

17. M. A. Alam, "Orthonormal Lattice Filter—A Multistage Multichannel Estimation Technique," *Geophysics* **43,** 1368–1383 (Dec. 1978).

18. O. L. Frost, III, "An Algorithm for Linearly Constrained Adaptive Array Processing," *Proc. IEEE.* **60,** 926–935 (Aug. 1972).

19. S. P. Applebaum and D. J. Chapman, "Adaptive Arrays with Main Beam Constraints," *IEEE Trans.* **AP-24,** 650–662 (Sept. 1976).

EFFECTS OF THE MAIN TAP POSITION IN ADAPTIVE CLUTTER PROCESSING

Frank F. Kretschmer, Jr. and Feng-ling C. Lin

Naval Research Laboratory, Washington, D.C.

ABSTRACT

Adaptive MTI is superior to a fixed-weight MTI in the general situation where the unknown clutter parameters can vary in both range and azimuth. Adaptive MTI processing is usually performed using a linear predictor whereby the signals appearing at different time-delay taps are weighted and subtracted from the signal appearing at the reference or main tap to achieve clutter cancellation. Usually, the main tap position is taken to be the first tap which contains the undelayed return signal. In this paper we investigate the effects of varying the main tap position and show that a significant improvement in clutter cancellation is obtained by using the center tap.

Optimum weights are derived taking into account the presence of noise, and the weights are evaluated by using standard matrix-inverse methods and by using an orthogonalization technique developed at the Naval Research Laboratory (NRL). An adaptive MTI system based on the orthogonalization technique is described. Quantitative results are presented for 3- and 5-pulse cancelers in terms of the standard MTI improvement factor I_c and in terms of a target-to-interference improvement factor I_{c+n}, where the interference consists of clutter plus noise.

INTRODUCTION

In digital adaptive clutter processing work, a linear predictor is frequently employed using a digital memory having uniform tap spacings which are separated by the pulse repetition interval. By assuming a "main tap" position, the function of the linear predictor is to linearly weight the clutter appearing in the remaining taps so as to provide the best estimate of the main tap clutter in the least squares sense. This is illustrated in Fig. 1 where it is seen that for the usual case where the main tap is located at the first tap position, the present input is estimated by a linear combination of the previous inputs to provide a one-step prediction. However, we are at liberty to choose other positions for the main tap position.

For these reasons, it is desirable to determine and quantify the effect of the main tap location on adaptive moving target indicator (MTI) processing. This problem was also addressed by Ueno (Ref. 1); however, our calculations, performed two independent ways, disagree with Ref. 1. In this paper we present our calculations and extend the results of Ref. 1, which is in terms of the target-to-clutter improvement factor I_c, to discuss the significance of I_c and to determine the target-to-interference improvement factor I_{c+n}, where the interference consists of clutter plus noise. Also we present the theory and implementation of an adaptive MTI which uses digital open-loop adaptive processors.

OPTIMUM STEADY STATE WEIGHTS

The determination of the optimum weighting to linearly predict the value of the main tap weighting may be derived as follows. Without loss in generality, we assume in Fig. 1 that the main tap data is $x_2(t)$ rather than $x_1(t)$. Then the prediction error is

$$e(t) = x_2(t) - W_1 x_1(t) - W_3 x_3(t) - W_4 x_4(t) \tag{1}$$

and we wish to minimize the mean squared error $\overline{|e(t)|^2}$.

It is well known (Ref. 2) that the error $\overline{|e(t)|^2}$ is minimized when e(t) is orthogonal to each of the input processes $x_i(t)$, excluding the main tap process. We, therefore, require*

$$\begin{aligned}\overline{ex_1^*} &= \overline{x_2x_1^*} - W_1\overline{x_1^*x_1} - W_3\overline{x_1^*x_3} - W_4\overline{x_1^*x_4} = 0\\ \overline{ex_3^*} &= \overline{x_2x_3^*} - W_1\overline{x_3^*x_1} - W_3\overline{x_3^*x_3} - W_4\overline{x_3^*x_4} = 0\\ \overline{ex_4^*} &= \overline{x_2x_4^*} - W_1\overline{x_4^*x_1} - W_3\overline{x_4^*x_3} - W_4\overline{x_4^*x_4} = 0\ .\end{aligned} \tag{2}$$

These equations may be written in matrix form as

$$\begin{bmatrix}\overline{x_1^*x_1} & \overline{x_1^*x_3} & \overline{x_1^*x_4}\\ \overline{x_3^*x_1} & \overline{x_3^*x_3} & \overline{x_3^*x_4}\\ \overline{x_4^*x_1} & \overline{x_4^*x_3} & \overline{x_4^*x_4}\end{bmatrix}\begin{bmatrix}W_1\\ W_3\\ W_4\end{bmatrix} = \begin{bmatrix}\overline{x_2x_1^*}\\ \overline{x_2x_3^*}\\ \overline{x_2x_4^*}\end{bmatrix} \tag{3}$$

which may be succinctly written, for the case where x consists of clutter alone, as

$$M_c\ W = \Phi \tag{4}$$

where M_c denotes the clutter covariance matrix. The solution to this equation is then

$$W = M_c^{-1}\ \Phi\ . \tag{5}$$

The above equation can be readily modified to account for additive thermal noise which is assumed to be uncorrelated between different channels but to have the same power N in each channel. This is achieved by adding the noise-to-input clutter power

*For notational simplicity, the time variable is omitted in the subsequent equations.

ratio N/C to the diagonal elements of M_c to obtain the clutter-plus-noise or interference covariance matrix M_{c+n}. It is also obvious how the matrices are modified to allow for any other main tap position.

There are many ways of implementing Eq. 5 that are described in the literature. These include the direct matrix inverse, which actually inverts the M_c matrix, and various other algorithms (Ref. 3). To evaluate Eq. 5 a double precision matrix-inverse routine was used on the ASC digital computer at NRL. An independent method originally developed at NRL by B. L. Lewis and F. F. Kretschmer, commonly referred to as the Gram-Schmidt method was also used. This technique, which is described next, is less susceptible to inaccuracies which can occur with the inverse-matrix method when the matrix becomes poorly conditioned. Moreover, an MTI based on the Gram-Schmidt technique may be easily implemented with digital open-loop adaptive processors (Refs. 4,5).

A two-tap or two-pulse adaptive MTI making use of the Gram-Schmidt technique is shown in Fig. 2. The adaptive processor AP computes the error or residual signal

$$e = x_1 - W x_2.$$

In order for W to provide the minimum error, it must satisfy the orthogonalization principle

$$\overline{e\, x_2^*} = \overline{x_1\, x_2^*} - W\, \overline{x_2\, x_2^*} = 0 \tag{6}$$

$$W = \overline{x_1\, x_2^*} \;/\; \overline{|x_2|^2}. \tag{7}$$

A representation of this process is shown in Fig. 3 for the two dimensional signals s_1 and s_2. The signal s_2 is weighted by the real weight W to minimize the error e which is seen to be orthogonal or uncorrelated with the signal s_2. For perfectly correlated signals, e is equal to zero. e may be nonzero due to uncorrelated thermal noise or the presence of any other uncorrelated signals which comprise s_1 and s_2. It is observed that the optimization procedure described above is the same as the Gram-Schmidt orthogonalization technique. For multiple signal channels, the channels exclusive of the main tap channel may be mutually orthogonalized by extending the Gram-Schmidt procedure described above. This processing configuration is shown in Fig. 4 for a four-tap system which incorporates multiple adaptive processors that are identical in operation to the one shown in Fig. 2. In Fig. 4, the main tap signal is x_1. It may be readily shown that the output residue or error signal is orthogonal to each of the input signals and hence the error signal is a minimum.

PERFORMANCE COMPUTATIONS

Computations were performed on the NRL ASC digital computer to evaluate the MTI improvement factor for three- and five-pulse MTIs using different main tap positions. In these computations, we assumed a Gaussian clutter spectrum, with zero mean doppler frequency, having a specified standard deviation σ which was normalized to the radar prf. The MTI clutter improvement factor I_c is defined as the ratio of the target-to-clutter ratio out of the processor to the target-to-clutter ratio at the input to the processor at each of the tap positions. The target doppler is assumed to be uniformly distributed and I_c is obtained by averaging over all doppler frequencies. In mathematical terms, we have

$$I_c = \overline{(T_o/C_o)/(T_{in}/C_{in})} = \overline{(T_o/T_{in})}(C_{in}/C_o) = TE \cdot CR_c \tag{8}$$

where TE is the target enhancement factor and CR_c is the clutter cancellation ratio. TE is equivalent to the ratio of the output noise to the input noise power which is given by $W^T W^*$. The output clutter power C_o is computed from

$$C_o = C_{in}\, \bar{W}\, M_c\, W^*,$$

so that I_c may be expressed as

$$I_c = W^T W^* / W^T M_c W^*. \tag{9}$$

Computations were performed by two independent methods using a double precision matrix-inverse routine and the Gram-Schmidt procedure for the three- and five-pulse canceler cases for various clutter-to-noise ratios C/N. Excellent agreement was obtained using the two methods and the results are shown in Figs. 5 and 6 for the 3- and 5-pulse cases respectively, where we compare I_c for the first and center-tap positions. It is seen that appreciable improvement in I_c is achieved by using the center-tap position. The improvement factors shown in Figs. 5 and 6 are generally better than those computed in Ref. 1. It appears that in Ref. 1 the thermal noise contributions may have been included in the calculation of the clutter covariance matrix M_c in Eq. 9.

It is claimed in Ref. 1 that better transient performance is also achieved by using the center-tap position and the LMS algorithm. However, the direct matrix-inverse and the Gram-Schmidt techniques do not suffer from the transient response problem as the LMS algorithm and it is not of concern in this paper.

We note from Figs. 5 and 6 and the data that the cancellation ratio, which is within 10 dB of the improvement factor for σ/prf less than 10^{-2}, is greater than C/N so that the clutter is canceled below the noise level. For example, for the 3-pulse case in Fig. 5a, having C/N equal to 30 dB and for σ/prf = 0.005, the cancellation ratio was computed to be 30 dB for the main tap taken to be the first tap and 62 dB for the main tap taken to be the middle tap position. This means that the clutter is canceled 32 dB below the noise when the center tap is used as the main tap. In typical systems where multiple pulses are integrated for target detection, it is desirable that the clutter residue be sufficiently below the noise so that it does not integrate back up to the noise level or above. This can be a potential problem with highly correlated clutter residues which might occur with strong nonmoving clutter, but is less likely to be an important consideration for rain or chaff clutter having a wide spectral spread that might result from large wind shear conditions.

For the situation where little or no pulse integration takes place, it is of interest to determine the effect of the main tap position on the target-to-interference improvement factor I_{c+n}. This may be stated as

$$I_{c+n} = \overline{(T_o/IN_o)/(T_{in}/IN_{in})} = \overline{(T_o/T_{in})}(IN_{in}/IN_o)$$

$$= TE \cdot CR_{c+n} = W^T W^*/W^T M_{c+n} W^* \tag{10}$$

where IN_o and IN_{in} are the output and input interference, CR_{c+n} is the interference cancellation ratio, and M_{c+n} is the interference covariance matrix.

I_{c+n} has been evaluated and the results for the 3- and 5-pulse cancelers are shown in Figs. 7 and 8, respectively. We observe that in terms of the target-to-interference ratio, the results are not very dependent on the main tap position. For large C/N, the ratio of IN_{in}/IN_o depends primarily on the ability to cancel the clutter having a stated spectral width. However, for smaller values of C/N the ratio of IN_{in}/IN_o is limited by C/N and properly takes into account that the output residue is noise limited.

SUMMARY AND CONCLUSIONS

We have assessed the effect of the main tap position in adaptive MTI processing and presented the results for the three- and five-pulse cases. It was shown that since the adaptive processor finds the proper weighting to minimize the residual clutter, it is equivalent to a linear prediction. Using the center tap as the main tap position corresponds to a smoothing of the underlying clutter process. This results in better cancellation of clutter than using the first tap which corresponds to a prediction filter. The effect of tap position was independently collaborated by two independent techniques: the direct inverse-matrix method and the Gram-Schmidt method, and presented in terms of the standard MTI improvement factor I_c. This parameter specifies the improvement of the target-to-clutter ratio achieved by the adaptive processing. It was shown that I_c can be significantly improved by using the center-tap position rather than the end tap.

It was shown that in terms of the target-to-interference improvement parameter I_{c+n}, where the interference consists of clutter plus noise, the performance was not very sensitive to the tap position. The parameter I_{c+n} correctly indicates that even though improved clutter attenuation is achieved by using the center tap, the performance is noise limited when the clutter is reduced below the noise level.

Using the center tap (or the tap closest to the center when there are an even number of taps) is a recommended procedure, particularly if the MTI is followed by integration which could increase the residual clutter level. There is no additional processing or complexity required in using the center tap.

REFERENCES

1. M. Ueno, "Clutter Covariance Matrix Form Effect on Adaptive MTI Performance", IEEE Trans. Aerospace and Electronic Systems, pp 348-353, May 1983.

2. A. Papoulis, Probability, Random Variables and Stochastic Processes, McGraw-Hill, 1965.

3. R. A. Monzingo and T. W. Miller, Introduction to Adaptive Arrays, John Wiley and Sons, 1980.

4. F. F. Kretschmer, Jr., AND B. L. Lewis, "A Digital Open-Loop Processor", IEEE Trans. Aerospace and Electronic Systems, pp 165-171, Jan 1978.

5. F. F. Kretschmer, Jr., B. L. Lewis and F. C. Lin, "Adaptive MTI and Doppler Filter Bank Clutter Processing", Proceedings of the 1984 IEEE National Radar Conference, pp 69-73, Mar 1984.

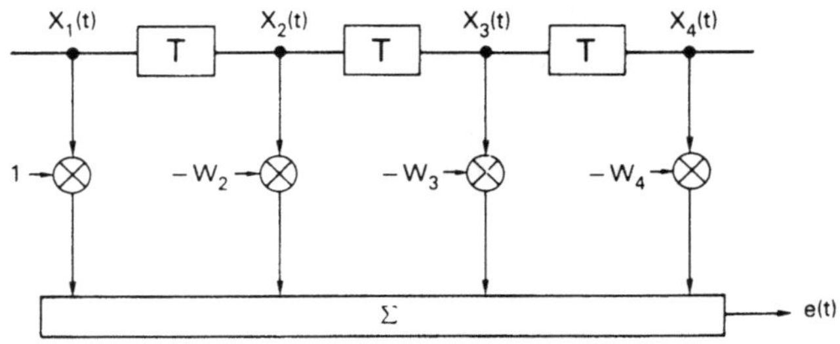

Fig. 1. 4-pulse linear predictor

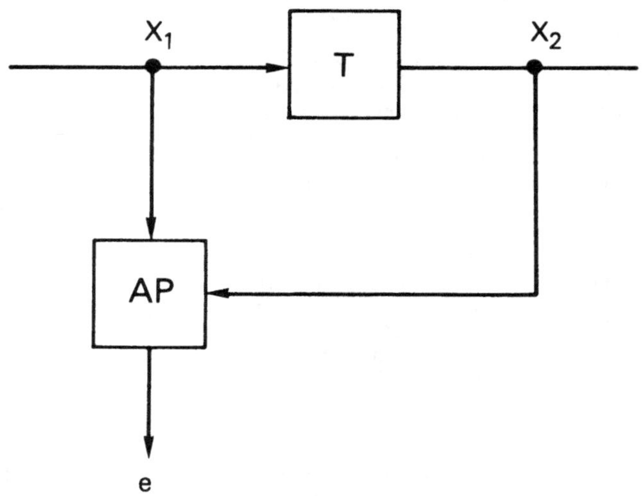

Fig. 2. 2-pulse adaptive MTI

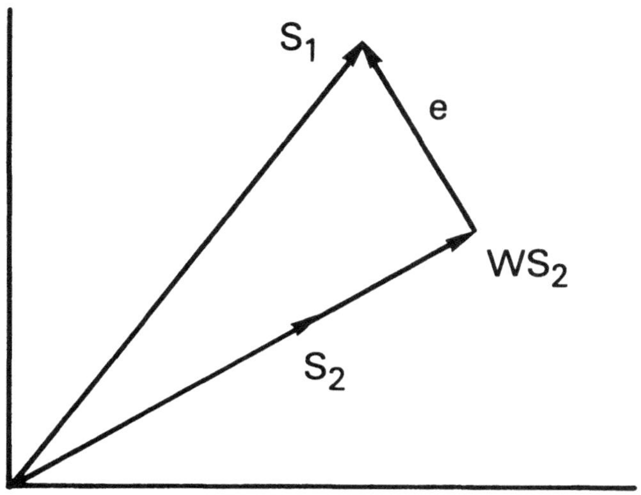

Fig. 3. Error signal minimization

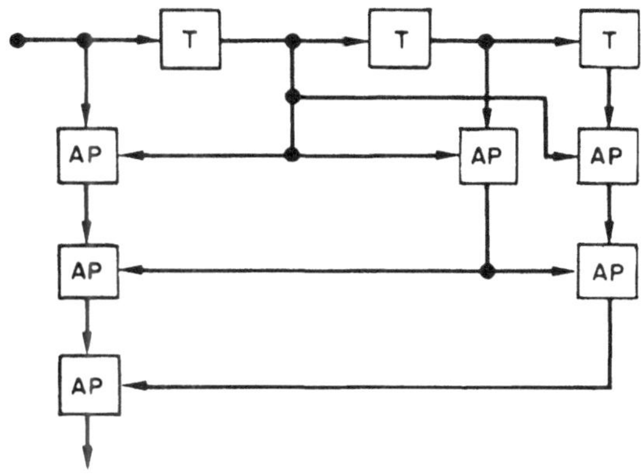

Fig. 4. 4-pulse adaptive MTI

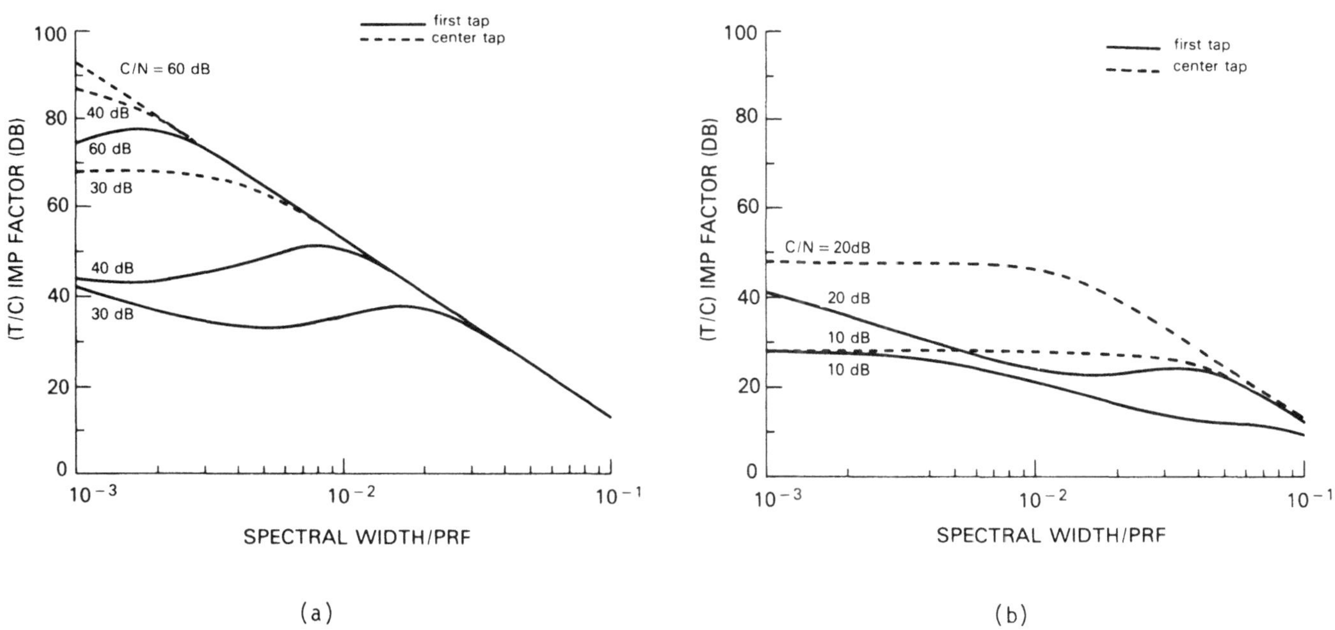

Fig. 5. I_c for 3-pulse adaptive MTI

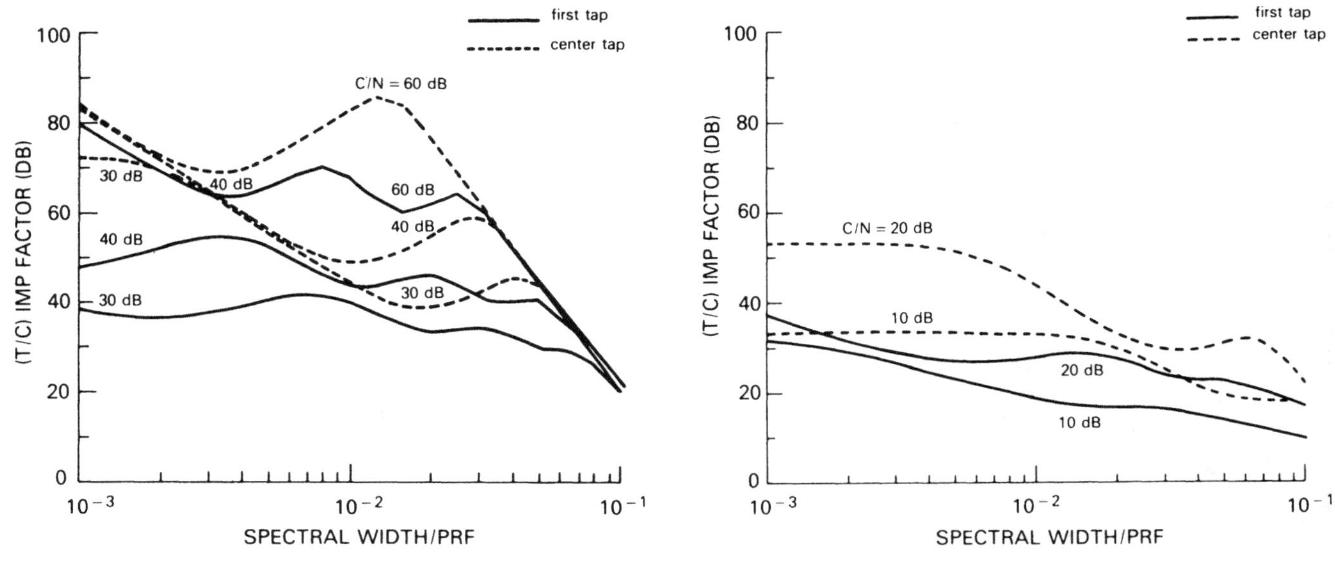

(a) (b)

Fig. 6. I_c for 5-pulse adaptive MTI

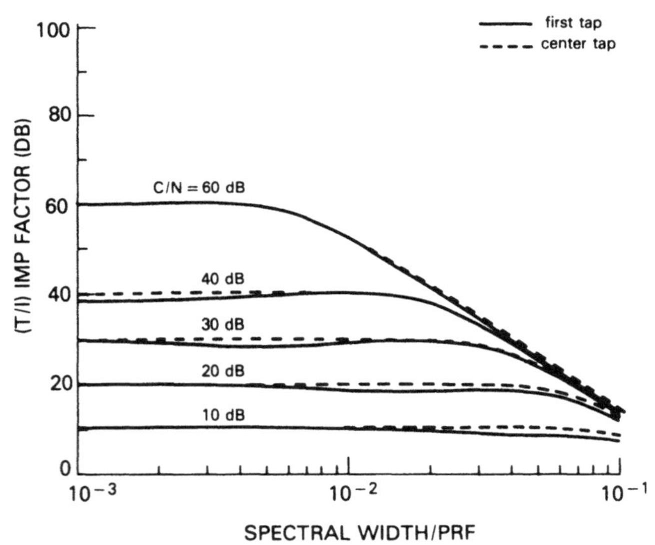

Fig. 7. I_{c+n} for 3-pulse adaptive MTI

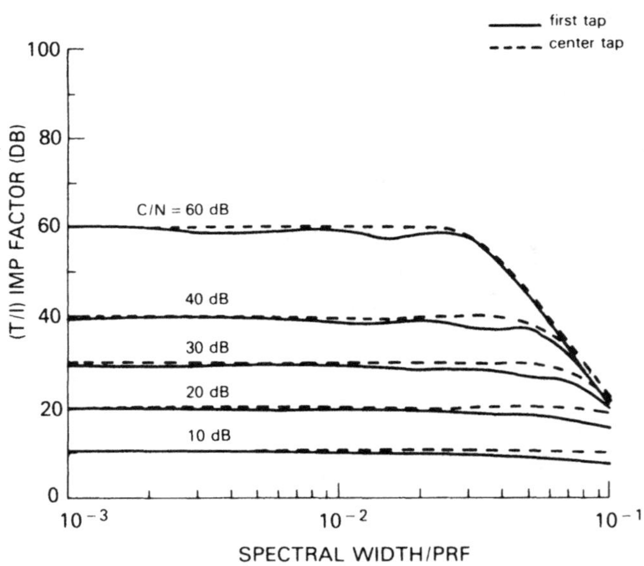

Fig. 8. I_{c+n} for 5-pulse adaptive MTI

Index

www.ingramcontent.com/pod-product-compliance
Lightning Source LLC
LaVergne TN
LVHW082016200926
848481LV00010B/2867